Functional Fractional Calculus

Shantanu Das

Functional Fractional Calculus

Springer

Author

Shantanu Das
Scientist Reactor Control Division BARC
Mumbai 400085
India
E-mail: shantanu@barc.gov.in

ISBN 978-3-642-20544-6 e-ISBN 978-3-642-20545-3

DOI 10.1007/978-3-642-20545-3

Library of Congress Control Number: 2011926534

Typeset & Cover Design: Scientific Publishing Services Pvt. Ltd., Chennai, India.

Printed on acid-free paper

9 8 7 6 5 4 3 2 1

springer.com

This work is dedicated to my blind father (Late) Sri Soumendra Kumar Das, my mother Purabi, my wife Nita, my son Sankalan, my sister Shantasree, brother-in-law Hemant, and to my little niece Ishita

Preface

This book is aimed at to entice multidisciplinary pure and applied science researchers to think differently and to discover and describe nature and natural phenomena (sociological, economical, biological, physical or chemical processes) with wonderful tool of mathematics called Fractional Calculus, which is in fact a generalization of the classical Newtonian Calculus, what we all are used to. The greatest discovery by Newton was to tell the world is 'that nature follows mathematics'; well the question is "which mathematics"? This question of finding "which mathematics" nature should follow is never ending quest for the pure and applied science researchers and this book will entice them. This work is aimed at, to make this subject popular and acceptable to engineering and science community to appreciate the universe of wonderful mathematics, which is in between classical integer order differentiation and integration, which till now is not much acknowledged, and is hidden from scientists and engineers. This book will generate 'Physical and Engineering Essence of Fractional Calculus'.

This work is inspired by thought to have overall fuel-efficient nuclear plant control system. I picked up the topic more than a decade ago while deriving reactor control laws, which aimed at fuel efficiency. Controlling the nuclear reactor close to its natural behavior by concept of exponent shape governor, ratio control and use of logarithmic logic, aims at the fuel efficiency. The power maneuvering trajectory is obtained by normalized shaped reactor period function, and this defines the road map on which reactor should be governed. The experience of this concept governing Atomic Power Plant of Tarapur Atomic Power Station gives lesser overall gains compared to older plants where conventional proportional integral and derivative type (PID) scheme is employed. Therefore this motivation led me to design scheme for control other than conventional schemes to aim at overall plant efficiency. Thus I felt need to look beyond PID and obtained the answer in fractional order control system, requiring fractional calculus (a three hundred year old subject). This work is taken from large number of studies on fractional calculus and here aimed at giving application-oriented treatment, to understand this beautiful old new subject. My contribution, in having fractional divergence concept to describe neutron flux profile in nuclear reactors and to make efficient controllers based on fractional calculus, to have physical sense to solve fractional differential equation, to interpret laws of nature in simple way to make fractional calculus subject attractived and to give picturesque sense to sometime complex looking mathematics, are some minor contribution in this vast (hidden) area of science.

This work is due to blessings of Prof Michelle Caputo, the founder of Caputo's fractional derivative. His blessings for the first edition encouraged me to carry out further research and development on the scientific and engineering aspects of Fractional Calculus further.

This second edition carries most of the advances which I could carry out with my colleagues, publish papers, make circuits and carry out experiments and filing patents. The photograph of one such system is shown below; the system is now under industrialization. This edition is improvement on the contents of first edition having advance work on this subject. Some mistakes that appeared in the first edition have been corrected too. In several places I have added extra explanations with derivations as enquired by various users of the first edition.

The extra portions of this edition is from, advanced work done during past years by the students and faculty of various Universities and Institutes who have taken up research work on this subject with me, and also from various invited talks and lecture series that I have taught at post graduate level at Universities and deliberated in past, as a 'Resource Faculty' of Ministry of Human Resource Development Government of India, or deliberated as Invited Speaker at various scientific forums; on this subject.

This photograph is of DC motor position servo with analog circuits of fractional order PID, under industrialization process. Well, now it is a real reality, once opined as a paradox three hundred years ago by Leibniz.

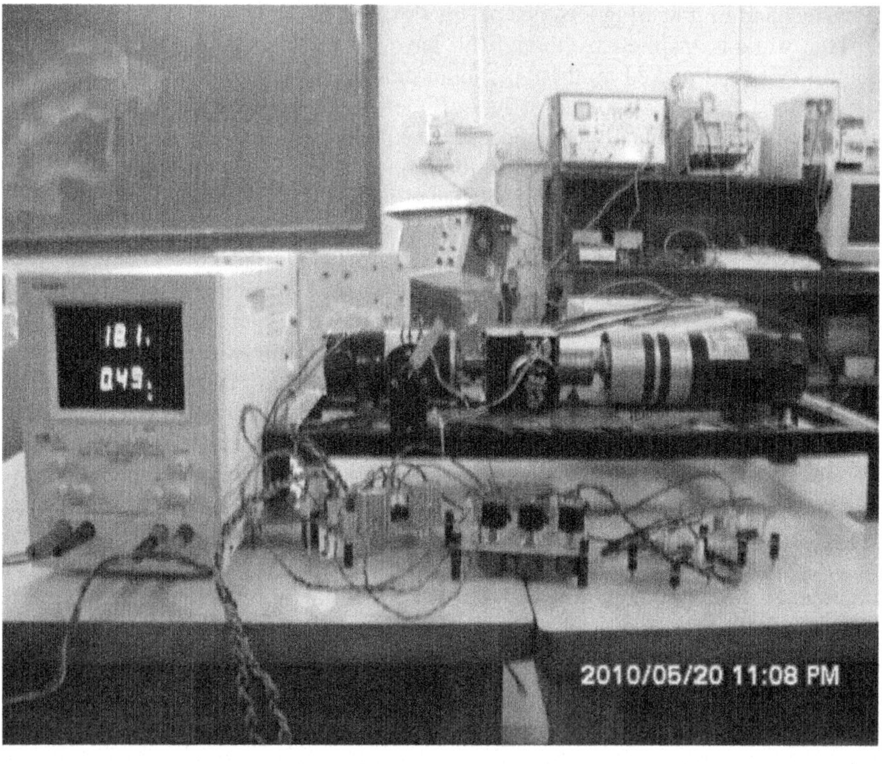

The treatment here in this book is very simple, with physical reasoning and neat derivations, for almost all the concepts; a final year student can take this for his research.

Enjoy the wonderful world of Fractional Calculus.

January-2011 Shantanu Das

Acknowledgements

I am inspired by encouragements from Dr. Srikumar Banerjee, Chairman Atomic Energy Commission, (AEC) and Dr. Ratan Kumar Sinha, Director, Bhabha Atomic Research Centre (BARC) for guiding in doing curiosity driven research and applying them for technological achievements. I acknowledge the encouragement received from, Prof. P.N. Ghosh, Vice Chancellor of University of Jadavpur and Prof Sujata Tarafdar, Department of Physics of University of Jadavpur Kolkata; for instituting scholarship for M.Sc. pure science students, to take this subject as special course and do a semester project, at University. I acknowledge Prof. Manoj Kumar Mitra Dean Faculty of Engineering and Prof. Amitava Gupta (Power Engineering Department) of Jadavpur University, for accepting the concept of fractional calculus for control engineering research. I wish to show my gratitude to Prof. Mohan Aware and Prof. Ashwin Dhabale Department of Electrical Engineering of Visvesvarya National Institute of Technology (VNIT)-Nagpur to have taken challenge in developing analog circuits with me for Fractional Order Controls which are under patenting process; the photographs of these systems are given in the book. I thank Prof. Susmita Sarkar and Prof. Uma Basu of Department of Applied Mathematics of Calcutta University, for planning to formally induct this book for academic curriculum for PhD in applied mathematics, and also to give me opportunity to teach the subject at detailed class room sessions for post graduate PhD students at University of Calcutta and University of Jadavpur. I acknowledge the encouragement received from Prof. Dr. Siddharta Sen and Dr. Karabi Biswas, (Electrical Engineering Department), Indian Institute of Technology (IIT) Kharagpur to make this book for ME and PhD students who will carry this knowledge for research in instrumentation and control science. I thank students Sri Suman Saha (PhD-scholar) and Sri Saptarishi Das (ME & PhD-scholar), of University of Jadavpur to develop the control system for nuclear industry, aimed at increasing efficiency and robustness of total plant.

From Department of Atomic Energy I wish to acknowledge the encouragement received from Dr. M.S. Bhatia (LPTD BARC and Professor at HBNI, Bombay University PhD guide) Dr. Abhijit Bhattacharya (ATSS-BARC), Supratim Majumder (ATSS-BARC), Dr. S.K.H Aulluck (Head-ATSS-BARC), Dr. M.K. Srivastava (Theoretical Physics Division BARC), Dr. Sudhir Jain (NPD BARC), Dr. Zafer Ahmed (NPD BARC), and Dr. S.K. Ghosh (Head Theoretical Physics Division BARC), for recognizing the richness and potential for research and development in physical science and control systems. I am obliged to Sri A.K. Chandra Sr. Executive Director NPCIL for recognizing the potential of this topic to have invited me to present the control concepts at NPCIL R&D and to have this new control scheme developed for NPCIL plants. I am obliged to Sri G.P. Srivastava (Director EIG-BARC),

Sri Shiben Bhattacharya (AD-T, EIG BARC) and Sri B.B. Biswas (Head of RCnD-BARC) for their guidance and encouragement in expanding the scope of this R&D with various universities and colleges, and to write this book. I thank Dr. D. Datta (HPD-BARC), and Sri Subroto Dutta (BRNS) my batch-mate of 1984 BARC Training School to have first appreciated logic in this concept of fractional calculus and encouraged me with interesting discussions on this subject.

Without acknowledging the work of several scientists dealing to renew and enrich this particular subject all over the globe the work will remain incomplete. Especially, Dr Michelle Caputo who read my first edition with interest and gave directions and encouragement for engineering & scientific research on this subject. My sincere thanks to Prof. M. Caputo to have gifted me with reprints of his pioneering work; which has been source of inspiration. I especially thank Dr. Ivo Petras Department of Informatics and process control BERG facility Technical University Kosice Slovak Republic, to have helped me to deal with doubts in digitized controller in fractional domain. I took inspiration and learnt the subject from several presentations and works of Prof. Michelle Caputo Instituto de Fisica, Universita degli Studi, Bologna, Prof. Dr. V. Balakrishnan Dept. of Physics I.I.T. Madras, Dr. Alain Oustaloup CNRS-University Bordeaux, Dr. Francesco Mainardi, University Bologna Italy, Dr. Stefan G. Samko University do Algarve Portugal, Dr. Katsuyuki Nishimoto Institute of Applied Mathematics Japan, Dr. Igor Podlubny Kosice Slovak Republic, Dr. Kiran M. Kolwankar and Dr. Anil D. Gangal Department of Physics University of Pune India. The effort of Dr. Carl F. Lorenzo Glen Research Center Cleveland Ohio, Dr. Tom T. Hartley University of Akron Ohio that has popularized this old (new) subject of fractional calculus is worth acknowledging. I acknowledge applied work on anomalous diffusion by Prof. R.K. Saxena (Jai Narain Vyas University Rajasthan) Dr. Santanu Saha Ray (of NIT Rourkela) and Prof Dr. Rasajit Kumar Bera (Heritage Institute of Technology Kolkata), who are source of inspiration. Especially I have learned the solution of differential equations by decomposition techniques from pioneering works of Prof. Rasajit Kumar Bera and Dr. S. Saha Ray. All their pioneering work is listed in detailed bibliography. I consider these scientists as fathers of modern fractional calculus of the XXI century and salute them.

After the first edition was published on 16.10.2007, several researchers have used this book for control system research and pure and applied science research, I acknowledge their contribution. In this part of the globe, several students took up this subject for higher studies and are doing projects, I acknowledge all of them. I acknowledge contribution of the students, Ms. Rituja Dive of VNIT-Nagpur, Sri Tridip Sardar of Heritage Institute of Technology Calcutta, Sri Jitesh Khanna and Sri Vamsi of IIT Kharagpur, Sri Indranil Pan, Sri Basudeb Mazumder and Sri Sumit Mukherjee of Department of Power Engineering University of Jadavpur; Sri Subrata Chandra, Ms. Moutushi Dutta Choudhury, Ms. Soma Nag of Department of Physics University of Jadavpur, who have contributed for development of products based on Fractional Calculus.

January-2011

Shantanu Das
Scientist Reactor Control Division BARC

About Contents of This Book

The book is organized as eleven chapters. The book aims at giving a feel of this beautiful subject of fractional calculus to scientists and engineers and should be taken as start point for research in application of fractional calculus. The book is aimed for appreciation of this fractional calculus and thus is made as application oriented, from various science and engineering fields. Therefore, the topics restrict the use of too formal mathematical symbolism and mathematical formal theorem stating language. The chapter three and four gives overview of application of fractional calculus before dealing in details about fractional differintegrations and initialization issues. These two chapters deal with all types of differential operations including fractional divergence application and usage of fractional curl.

Chapter one is basic introduction, dealing with development of the fractional calculus. Several definitions of fractional differintegrations and the most popular ones, are introduced here. A table in this, chapter gives feel of fractional differentiation of some functions i.e. how they look. To aid the understanding diagrams are given. In this chapter the solution of Fractional Differential Equation is given to draw similarity between uses of indicial polynomials used to solve integer order differential equations, with classical example of 'tautochrone' problem, solved by Abel. Also in this chapter classical physical law of transport where conservation of probability is generalized is introduced deriving the Fractional Differential equation. The laws of irreversibility and its relation to non local character of fractional calculus are highlighted. This introductory chapter also gives idea of scale invariance, generalization of normal statistical laws, useful in fractional calculus context. Here simple derivation is done to show that normal integer order diffusion equation has inbuilt 'half' derivative thus paving way by indexing the order of fractional half derivative, to adapt a concept of 'fractional Brownian motion' a motion with memory.

Chapter two deals with the important functions relevant to fractional calculus basis. Few examples of these functions are highlighted as to aid understanding in utilizing these for solving Fractional Order Differential equations. Laplace transformation is given for each function, which are important in analytical solution. In this chapter interesting discussions are put as for memory integrals, which are essence of various physical systems relaxing through non-exponential power laws. Also brief introduction is made regarding regularizing of irregular and rough functions with 'fractal' non-integer order dimension. Definitions of various dimensions are also placed here.

Chapter three gives observation of fractional calculus in physical systems (like electrical mechanical, thermal, control system etc.) description. This chapter is made so that readers get feel of reality. In this chapter the anomalous diffusion laws are introduced, along with synthesis of fractional order element as approximate to continued fraction expansion. The derivation is done in simple manner for dynamics of chain network, and its average displacement for infinite case is related to half differintegral with external force. The same is also derived for 'charged' chain in an Electric Field, indicating different fractional order differintegral (other than half) relating average displacement with the external uniform electric field.

Chapter four is extension of chapter three where concept of fractional divergence and curl operator is elucidated with application in nuclear reactor and electromagnetism. With this the reader gets a broad feeling about the subject's wide applicability in field of science and engineering. In this chapter Fractional Derivative and its spectral derivation is made through Fractional Discrete Difference concept, in random walk context; with extension giving super diffusion and sub diffusion. The concept of persistent and anti-persistent diffusion with memory is derived. The fractional reactor kinetics equations are derived and solved and possibility of fractional criticality is elucidated.

Chapter five is dedicated to insight of fractional integration fractional differentiation and fractional differintegral with physical and geometric meaning for these processes. Explanations are made to elaborate concept of fractional differintegration by computing area under shape changing curve, and concept of imaging (projection), and non-uniform flow of time. The concept of Fractional Order Signal Processing and Fractional Noise is introduced to identify the Fractional Stochastic Processes. The Local Fractional Derivative and its relation to fractal dimension of graph along its utility to enlarge critical points study of physical systems is elaborated. In this chapter example of numerical calculation of simple fractional order differential equation is presented, along with method to extract exponent describing local singularity at a point is derived. Along with this basics are developed to do integration on fractal distribution, thereby giving concepts of fractional line surface and volume integrations, with respect to fractal dimensions. Naturally the generalization of Stroke's law and Gauss's law is derived for systems with 'fractal distribution'.

Chapter six tries to generalize the concept of initialization function, which actually embeds hereditary and history of the function. Here attempt is made to give some light into decomposition properties of the fractional differintegration. Generalization is called as this fractional calculus theory with the initialization function becomes general theory and does cover the integer order classical calculus. The periodic signals are fractionally differintigrated with different initial start points. Here conflicting issues of initialization of Riemann-Liouvelli vis-à-vis Caputo formulation is elaborated, giving physical interpretability. With the assumed fractional initial conditions demonstration to solve Fractional Advection Dispersion Equation, and concept of modeling fractional stochastic process as example dynamic delays of computer network system and its physics are derived.

Chapter seven gives the Laplace transform theory a general treatment to cover initialization aspects. In this chapter also described is concept of w-plane on which fractional control system properties are studied. In chapter seven elaborate dealing is carried on for scalar initialization and vector initialization problems. Elaborate block diagrams are given for aid to understand these concepts in chapter six and seven. In this chapter fundamental fractional differential equation is taken and impulse response to that is obtained, along with derivation of interlacing of poles and zeros to get arbitrary fractional order operator realized practically is elaborated.

Chapter eight gives application of fractional calculus in electrical circuits and electronic circuits and electromagnetism. In this chapter the concept of Generating-function is presented which gives transfer function realization for digital realization in real time application of controls Realization of transfer function for implementation of Fractional Order PID is given here. Interesting elaborations of laws of Electromagnetism is included where electric charges and magnetic flux is distributed in fractal media.

Chapter nine deals with application of fractional calculus in other fields of science and engineering for system modeling and control. In this chapter modern aspects of multivariate controls are touched to show the applicability in fractional feed back controllers and state observer issues. Also the theory of iso-damping is elaborated by practical examples of design of fractional phase shaper by Bode's integral. In this chapter the derivations of fractional laws of viscoelastic behavior of fractal chain network, and anomalous electrochemical behavior are discussed, along with memory based relaxation of non-Newtonian fluid.

Chapter ten gives detailed treatment of order of a system and its identification approach, with concepts of fractional resonance ultra-damped and hyper-damped systems. Also a brief is presented on future formalization of research and development for variable order differintegrations and continuous order controller that generalizes conventional control system. This chapter also deals with complex non-linear relaxation of condense matter and methods to identify these behaviors. The introduction of complex order calculus in system identification is given, in this chapter. The identification technique of important parameters for irregular stochastic process is developed here, in order to fit Levy statistic with long range dependency with lingering tail, and with memory effect.

The chapter eleven deals with the solution of Generalized Differential Equation System by decomposition technique by physical principles of action reaction series. The solutions thus obtained are analytical approximates for the system. Here the method highlights uniformity in Riemann-Liouvelli and Caputo formulations where only integer order initial states are required and the solution is close to physics of action reaction.

References contains list of work in fractional calculus and applications a few of them. Several other authors draw some of the references indirectly on their research papers. Undoubtedly it is not possible to include all references and work of past three hundred years.

Undoubtedly this is an emerging area or research (not so popular at present in India), but in next decade perhaps will see plethora of applications based on this field. May be XXI century will speak language of nature that is fractional calculus.

January-2011 Shantanu Das

Contents

Chapter 1
Introduction to Fractional Calculus

1.1 Introduction

Fractional calculus is three centuries old as the conventional calculus, but not very popular amongst science and or engineering community. The beauty of this subject is that fractional derivatives (and integrals) are not a local (or point) property (or quantity). Thereby this considers the history and non-local distributed effects. In other words perhaps this subject translates the reality of nature better! Therefore to make this subject available as popular subject to science and engineering community, adds another dimension to understand or describe basic nature in a better way. Perhaps fractional calculus is what nature understands and to talk with nature in this language is therefore efficient. For past three centuries this subject was with mathematicians and only in last few years, this is pulled to several (applied) fields of engineering and science and economics. However recent attempt is on to have definition of fractional derivative as local operator specifically to fractal science theory. Next decade will see several applications based on this three hundred years (old) new subject, which can be thought of as superset of fractional differintegral calculus, the conventional integer order calculus being a part of it. Differintegration is operator doing differentiation and sometimes integrations in a general sense. Also the applications and discussions are limited to fixed fractional order differintegrals and the variable order of differintegration is kept as future research subject. Perhaps the Fractional Calculus will be the calculus of XXI century. In this book attempt is made to make this topic application oriented for regular science and engineering applications. Therefore rigorous mathematics is kept minimal. In this introductory chapter list in tabular form is provided to readers to have feel of fractional derivatives of some commonly occurring functions. Here various definitions of fractional differ-integrations are introduced, and some applications of Fractional Differential Equations are deliberated. Also, discussed are the law of irreversibility, non-locality and conservation of probability, stable random variables, scale invariance and generalization of normal probability density function to power law distributions. The simple derivation is shown here to elucidate the fact that integer order diffusion equation Fick's law has half order differentiation embedded into it, and its extension to have concept of anomalous transport, and Fractional Brownian Motion. This treatment will give the readers the feel of fractional calculus to explain natural laws, stochastic processes of several manifestations and the subject's relevance to basic physical laws of nature. To start with let us think for a moment, about the normal derivative d/dt is representing the rate of accumulation or loss; that is, gain rate minus loss rate, at infinitesimal bounded space. Well if that infinitesimal space is having traps (of various sizes) where our particle money or any variable what we are studying is temporarily parked then will the d/dt replicate real picture of accumulation or loss? Similarly these trap pictures could be islands or forbidden zones in the infinitesimal space where the variable (particle, money, flux etc.) cannot reside; then also the rate of accumulation or loss will be different than d/dt. Well then the fractional differentiation d^{α}/dt^{α}, where $\alpha \in \Re e$ may give the sub- or super-rate of accumulation or loss with index α representing the heterogeneity distribution of the infinitesimal space (traps or islands)!

1.2 Birth of Fractional Calculus

In a letter dated 30th September 1695, L'Hopital wrote to Leibniz asking him particular notation he has used in his publication for the n-th derivative of a function

$$\frac{D^n f(x)}{Dx^n}$$

i.e. what would the result be if $n = 1/2$. Leibniz's response "an apparent paradox from which one day useful consequences will be drawn." In these words fractional calculus was born. Studies over the intervening three hundred years have proven at least half right. It is clear, that with in the XX century, especially numerous applications have been found. However, these applications and mathematical background surrounding fractional calculus are far from paradoxical. While the physical meaning is difficult to grasp, the definitions are no more rigorous than integer order counterpart.

Later the question became: Can n, be any number: fractional, irrational, or complex? Because the latter question was answered affirmatively, the name 'fractional calculus' has become a misnomer and might better be called 'integration and differentiation of arbitrary order' or 'arbitrary ordered differ-integrations'. Anyway mathematics is art of giving misleading names!

In 1812, P.S. Laplace defined a fractional derivative of arbitrary order appeared in Lacroix's (1819) writings. He developed a mere mathematical exercise generalizing from a case of integer order. Starting with $y = x^m$, where m a positive integer, Lacroix easily develops n th derivative:

$$\frac{d^n y}{dx^n} = \frac{m!}{(m-n)!} x^{m-n}, m \geq n .$$

Using Legengdre's symbol for the generalized factorial (the complete Gamma function), Lacroix gets:

$$\frac{d^n y}{dx^n} = \frac{\Gamma(m+1)}{\Gamma(m-n+1)} x^{m-n}.$$

He then gives example for $y = x$ and $n = 1/2$, and obtains:

$$\frac{d^{1/2} y}{dx^{1/2}} = \frac{2\sqrt{x}}{\sqrt{\pi}}.$$

It is interesting to note that the result of Lacroix in the manner typical of the classical formalists of the periods is same as that yielded by the formalists Riemann-Liouvelli definition of fractional derivative. This expression of Lacroix is also referred to as Euler's formula (1730). Let us try and use this to evaluate fractional derivative of $f(t) = \exp(t)$.

The exponential function is represented as series

$$f(t) = e^t = \sum_{k=0}^{\infty} \frac{t^k}{k!},$$

applying this term to the Euler expression (as above) we get,

$$\frac{d^v}{dt^v} e^t = \sum_{k=0}^{\infty} \frac{t^{k-v}}{\Gamma(k-v+1)},$$

where $v > 0$ and $v \in \mathbb{R}$ (real number). The fractional derivative of exponential function does not return exponential function. The expression of the series function is one of the Higher Transcendental Function, and in this case Miller-Ross function is represented as $E_t(-v,1)$.

1.3 Fractional Calculus a Generalization of Integer Order Calculus

Let us consider n an integer and when we say x^n we quickly visualize x multiply n times will give the result. Now we still get a result if n is not an integer but fail to visualize how. Like to visualize 2^{π} is hard to visualize, but it exists. Similarly the fractional derivative we may say now as

$$\frac{d^{\pi}}{dx^{\pi}} f(x)$$

though hard to visualize (presently), does exist. As real numbers exists between the integers so does fractional differintegrals do exist between conventional integer order derivatives and n fold integrations. We see the following generalization from integer to real number on number line as

$$x^n = \underbrace{x.x.x.x.........x}_{n} \qquad n \text{ is integer}$$

$$x^n = e^{n \ln x} \qquad n \text{ is real number}$$

$$n! = 1.2.3....(n-1)n \qquad n \text{ is integer}$$

$$n! = \Gamma(n+1) \qquad n \text{ is real}$$

and Gamma Functional is $\Gamma(x) = \int_0^\infty e^{-t} t^{x-1} dt$

Therefore the above generalization from integer to non-integer is what is making number line general (i.e. not restricting to only integers). Figure 1.1 demonstrates the number line and the extension of this to map any fractional differintegrals. The negative side extends to say integration and positive side to differentiation.

$$f, \frac{df}{dt}, \frac{d^2 f}{dt^2}, \frac{d^3 f}{dt^3}, \ldots \rightarrow$$

$$\leftarrow \ldots, \int dt \int dt \int f dt, \int dt \int f dt, \int f dt, f$$

Writing the same in differintegral notation as represented in number line we have

$$\leftarrow \ldots \frac{d^{-3} f}{dt^{-3}}, \frac{d^{-2} f}{dt^{-2}}, \frac{d^{-1} f}{dt^{-1}}, f, \frac{df}{dt}, \frac{d^2 f}{dt^2}, \frac{d^3 f}{dt^3}, \ldots \rightarrow$$

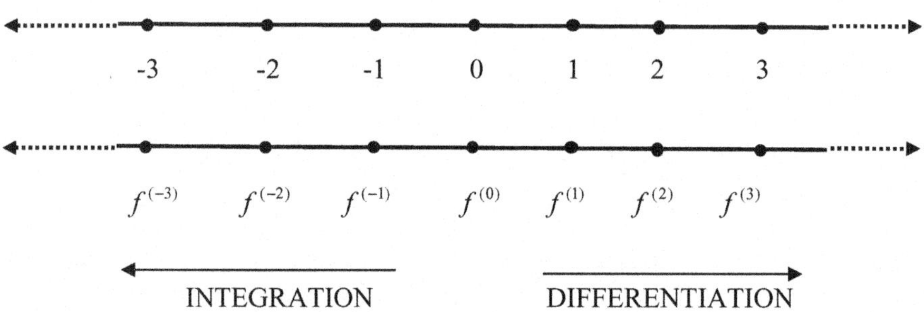

Fig. 1.1 Number line & Interpolation of the same to differintegrals of fractional calculus

Heaviside (1871) states that, there is universe of mathematics lying between the complete differentiation and integration, and fractional operators push themselves forward sometimes and are just as real as others.

Mathematics is art of giving things misleading names. The beautiful-and at first glance mysterious-name, the fractional calculus is just one of those misnomers, which are essence of mathematics. We know such names as natural number and real numbers. We use them very often; let us think for a moment about these names. The notion of natural number is a natural abstraction, but is the number natural itself a natural? The notion of a real number is generalization of the notion of a natural number. The real emphasizes that we pretend that they reflect real quantities, but cannot change the fact that they do not exist. If one wants to compute something he immediately discovers for himself that there is no place for real numbers in this real world. On a computer he can work with finite set of finite fractions, which serves as approximations to unreal real number.

Fractional calculus does not mean the calculus of fractions, nor does it mean a fraction of any calculus differentiation, integration or calculus of variations. The fractional calculus is a name of theory of integrations and derivatives of arbitrary order, which unify and generalize the notion of integer order differentiation and n-fold integration. So we call generalized differintegrals.

1.4 Historical Development of Fractional Calculus

Fractional order systems, or systems containing fractional derivatives and integrals, have been studied by many in engineering and science area. Heaviside 1922, Bush 1929, Goldman 1949, Holbrook 1966, Starkey 1954, Carslaw and Jeager 1948, Scott 1955 and Mikuniski 1959. Oldham and Spanier 1974, Miller and Ross 1993 present additionally very reliable discussions devoted specifically to the subject. It should be noted that there are growing number of physical systems whose behavior can be compactly described using fractional calculus system theory. Of specific interest to electrical engineers are long electrical lines (Heaviside 1922), electrochemical process (Ichise, Nagayanagi and Kojima 1971; Sun, Onaral, and Tsao 1984), dielectric polarization (Sun, Abdelwahab and Onaral 1984), colored noise (Manderbolt 1967), viscoelestic materials (Bagley and Calico 1991; Koeller 1986; Skaar, Michel, and Miller 1988) and chaos (Hartley, Lorenzo and Qammar 1995), electromagnetism fractional poles (Engheta 1999). During the development of the fractional calculus applied theory for past three hundred years the contributions from N, Ya. Sonnin (1869), A.V. Letnikov (1872), H. Laurent (1884), N. Nekrasov (1888), K. Nishimoto (1987), H.M. Srivastava, R P Agarwal (1953), S.C Dutta Roy (1967), Miller and Ross (1993), Kolwankar and Gangal (1994), Oustaloup (1994), L. Debnath (1992), Igor Podlubny (2003), Carl Lorenzo (1998) Tom Hartley (1998), R.K. Saxena (2002), Mariandi (1991), R.K. Bera and S Saha Ray (2005), and several others are notable. Author has tried to apply the fractional calculus concepts to describe the Nuclear Reactor constitutive laws and apply the theory for obtaining efficient automatic control for nuclear power plants. Following are some of the notations and formalization efforts since late seventeenth century by several mathematicians:

Since 1695 after L'Hopital's question regarding the order of the differentiation Leibniz was the first to start on this direction. Leibniz (1695-1697) mentioned possible approach to fractional order differentiation, in that sense that for non-integer (n) the definition could be following. This letter he wrote to J. Wallis and J. Bernulli.

$$\frac{d^n e^{mx}}{dx^n} = m^n e^{mx}$$

L. Euler (1730) suggested using relation ship for negative or non-integer (rational) values, taking $m = 1$ and $n = 1/2$ he obtained the following:

$$\frac{d^n x^m}{dx^n} = m(m-1)(m-2)...(m-n+1)x^{m-n}$$
$$\Gamma(m+1) = m(m-1)...(m-n+1)\Gamma(m-n+1)$$
$$\frac{d^n x^m}{dx^n} = \frac{\Gamma(m+1)}{\Gamma(m-n+1)} x^{m-n}$$
$$\frac{d^{1/2} x}{dx^{1/2}} = \sqrt{\frac{4x}{\pi}} = \frac{2}{\sqrt{\pi}} x^{1/2}$$

First step in generalization of notation for differentiation of arbitrary function was conceived by J.B.J. Fourier (1820-1822), after introduction of:

$$f(x) = \frac{1}{2\pi} \int_{-\infty}^{+\infty} f(z)dz \int_{-\infty}^{+\infty} \cos(px - pz)dp \; .$$

His definition of fractional operations was obtained from this integral representation of $f(x)$. Now, $\frac{d^n}{dx^n} \cos p(x-z) = p^n \cos[p(x-z) + \frac{1}{2} n\pi]$ for n, an integer.

He made a remark as:

$$\frac{d^n f(x)}{dx^n} = \frac{1}{2\pi} \int_{-\infty}^{+\infty} f(z)dz \int_{-\infty}^{+\infty} p^n \cos[p(x-z) + n\frac{\pi}{2}]dp$$

and this relationship could serve as a definition of n-th order derivative for non-integer order n. Fourier states that "The number n that appears in above will be regarded as any quantity whatsoever, positive or negative".

N.H. Abel (1823-1826) introduced integral as

$$\int_0^x \frac{S'(\eta)d\eta}{(x-\eta)^\alpha} = \psi(x)$$

He in fact solved the integral for an arbitrary α and not just for ½ he obtained

$$S(x) = \frac{\sin(\pi\alpha)}{\pi} x^\alpha \int_0^1 \frac{\psi(xt)}{(1-t)^{1-\alpha}} dt \; .$$

After that Abel expressed the obtained solution with help of an integral of order of α.

$$S(x) = \frac{1}{\Gamma(1-\alpha)} \frac{d^{-\alpha} \psi(x)}{dx^{-\alpha}} .$$

Abel applied the fractional calculus in the solution of an integral equation that arises during the formulation of tautochrone problem, the problem of finding a shape of curve such that the time of descent of a frictionless point ball mass, sliding down the curve under action of gravity is independent of the start point. If the time of slide is a known constant (T), then Abel's equation is

$$k = \int_0^x (x-t)^{-\frac{1}{2}} f(t) dt .$$

This equation, except the multiplicative factor $1/\Gamma(1/2)$, is a particular case of definite integral that defines fractional integral of order ½. Abel wrote the RHS of above integral as $\sqrt{\pi}[d^{-1/2} / dx^{-1/2}] f(x)$. Then he operated both sides with $d^{1/2} / dx^{1/2}$, to get:

$$\frac{d^{1/2}}{dx^{1/2}} k = \sqrt{\pi} f(x) .$$

Thus when fractional derivative of order ½ is computed (in above) $f(x)$ is determined. A remarkable contribution by Abel is that fractional derivative of constant need not be zero.

J. Liouvilli (1832-1855) gave three approaches. The first one in following Leibniz's formulation as follows.

$$\frac{d^m e^{ax}}{dx^n} = a^m e^{ax}$$

$$f(x) = \sum_{n=0}^{\infty} c_n e^{a_n x} ; \operatorname{Re} a_n > 0$$

$$\frac{d^\gamma f(x)}{dx^\gamma} = \sum_{n=0}^{\infty} c_n a_n^\gamma e^{a_n x}$$

Here the function is decomposed by infinite set of exponential functions. J. Liouville introduced the integral of non-integer order as the second approach is noted below:

$$\int^{\mu}\varphi(x)dx^{\mu} = \frac{1}{(-1)^{\mu}\Gamma(\mu)}\int_{0}^{\infty}\varphi(x+\alpha)\alpha^{\mu-1}d\alpha$$

$$\int^{\mu}\varphi(x)dx^{\mu} = \frac{1}{\Gamma(\mu)}\int_{0}^{\infty}\varphi(x-\alpha)\alpha^{\mu-1}d\alpha$$

$$\tau = x+\alpha, \& \tau = x-\alpha$$

$$\int^{\mu}\varphi(x)dx^{\mu} = \frac{1}{(-1)^{\mu}\Gamma(\mu)}\int_{x}^{\infty}(\tau-x)^{\mu-1}\varphi(\tau)d\tau$$

$$\int^{\mu}\varphi(x)dx^{\mu} = \frac{1}{\Gamma(\mu)}\int_{-\infty}^{x}(x-\tau)^{\mu-1}\varphi(\tau)d\tau$$

The third approach given by Liouville is the definitions of derivatives of non-integer order as:

$$\frac{d^{\mu}F(x)}{dx^{\mu}} = \frac{(-1)^{\mu}}{h^{\mu}}\left(F(x)-\frac{\mu}{1}F(x+h)+\frac{\mu(\mu-1)}{1.2}F(x+2h)-...\right)$$

$$\frac{d^{\mu}F(x)}{dx^{\mu}} = \frac{1}{h^{\mu}}\left(F(x)-\frac{\mu}{1}F(x-h)+\frac{\mu(\mu-1)}{1.2}F(x-2h)-...\right)$$

$$\lim.h \to 0$$

Liouville was first to point the existence of right sided and left sided differential and integrals.

G.F.B. Riemann (1847) used a generalization of Taylor series for obtaining a formula for fractional order integration. Riemann introduced an arbitrary "complimentary" function $\psi(x)$ because he did not fix the lower bound of integration. This disadvantage he could not solve. From here the initialized fractional calculus was born lately in the later half of the twentieth century, Riemann's notation is as follows with the complimentary function.

$${}_{c}D_{x}^{-\nu}f(x) = {}_{c}I_{x}^{\nu}f(x) = \frac{1}{\Gamma(\nu)}\int_{c}^{x}(x-t)^{\nu-1}f(t)dt + \psi(x);[\Re e\nu > 0]$$

Cauchy formula for n-th derivative in complex variables is:

$$D^{n}f(z) = f^{n}(z) = \frac{n!}{\mathrm{j}2\pi}\oint\frac{f(t)}{(t-z)^{n+1}}dt\,,$$

for non-integer $n = \nu$ a branch point of the function $(t-z)^{-\nu-1}$, appears instead of pole:

$$D^{\nu} f(z) = \frac{\Gamma(\nu+1)}{\mathrm{j}2\pi} \int_{C}^{x+} \frac{f(t)}{(t-z)^{\nu+1}} dt$$

Generally to understand the dynamics of any particular system, we often consider the nature of the complex domain singularities (poles). Consider a complex function $G(z) = (z^q + a)^{-1}$, where $q > 0$ and is fractional number. This particular function of the complex variable does not have any poles on the primary Riemann sheet of the complex plane $z = r\exp(\mathrm{j}\theta)$, i.e. with in $|\theta| < \pi$. It is impossible to force the denominator $z^q + a$ to zero anywhere in complex plane $|\theta| < \pi$. Consider for $q = 0.5$, the denominator $z^{0.5} + 1$ does not go anywhere to zero in primary Riemann sheet, $|\theta| < \pi$. It becomes zero on secondary Riemann sheet at $z = \exp(\pm \mathrm{j}2\pi) = 1 + \mathrm{j}0$.

Normally to get to secondary Riemann sheet it is necessary to go through a 'branch cut', on the primary Riemann sheet. This is accomplished by increasing the angle in complex plane z. Increasing the angle to $\theta = +\pi$ gets us to the 'branch-cut' on the $z-$ complex plane. This can also be accomplished by decreasing the angle until $\theta = -\pi$, which also gets us to the 'branch-cut'. This 'branch-cut' lies at $z = r\exp(\pm \mathrm{j}\pi)$ for all positive r. Further increasing the angle eventually gets to $\theta = \pm \mathrm{j}2\pi$. Further increasing the angle $\theta > \pi$ makes us go 'underneath', the primary Riemann sheet, inside the negative real axis of $z-$ complex plane.

The behavior of the function $(z^{0.5} + 1)^{-1}$ is described by thus two Riemann sheets. Return to the first Riemann sheet on the $z-$ complex plane, the branch cut begins at $z = 0$, the origin and extends out to the negative real axis to infinity. The end of the branch cuts are called 'branch-points', which are then at the origin and at minus infinity in the $z-$ plane.

The 'branch-points', can be considered as singularities on the primary Riemann sheet of the $z-$plane as well, but the function $(z^{0.5} + 1)^{-1}$ does not go to infinity then. Therefore to obtain the plot of the pole one has to wrap around these branch points and go to secondary-Riemann sheet (in this case at $1 + \mathrm{j}0$ at $\theta = \pm 2\pi$).

An appropriate contour would then require a branch cut. An open circuit contour on a Riemann surface with contour integration produces Riemann fractional integral definition. This work of complex variable approach, with contour integration, was from H. Laurent (1884), extending Letnikov' and Sonin's work on Cauchy's formula to generalize derivative to arbitrary order.

Marchaud (1927) was trying to define fractional derivative from Riemann-Liouvelli fractional integral definition that is, by taking ν as ν and wrote:

$${}_0D_{\infty}^{\nu} f(x) \equiv {}_0I_{\infty}^{-\nu} f(x) = \frac{1}{\Gamma(-\nu)} \int_0^{\infty} u^{-\nu-1} f(x-u) du \text{, for } \nu > 0 .$$

No matter how smooth $f(x)$ might be, the above representation (integral) diverges due to $u^{-\nu-1}$ at the origin. Later this definition was modified only for $0<\nu<1$, by considering finite part of the above diverging integral in the sense that one should subtract the part which makes the integral diverge; namely: $\int_{\varepsilon}^{\infty} u^{-\nu-1} f(x)du = \frac{f(x)}{\nu\varepsilon}$ with $\varepsilon \to 0^+$ .Thus,

$$_0D_x^\nu f(x) = \lim_{\varepsilon\to 0^+} \frac{1}{\Gamma(-\nu)} \int_{\varepsilon}^{\infty} u^{-\nu-1}[f(x-u)-f(x)]du \text{ , using } \Gamma(1-\alpha) = -\alpha\Gamma(-\alpha).$$

The Marchaud definition is:

$$_{0^+}D_x^\nu f(x) = \lim_{\varepsilon\to 0^+} \frac{1}{\Gamma(1-\nu)} \int_{\varepsilon}^{\infty} u^{-\nu} f'(x-u)du = {}_{-\infty}I_x^{1-\nu} f'(x)$$

Riesz (1949) formulated potential expression for arbitrary order with fractional integration as:

$$I_\alpha f(x) \overset{\text{def}}{=} \frac{\Gamma((n-\alpha)/2)}{2^\alpha \pi^{n/2}\Gamma(\alpha/2)} \int_{-\infty}^{\infty} \frac{f(u)}{|x-u|^{n-\alpha}} du \text{ , with } n=1 \text{ the definition is as follows:}$$

$$R^\alpha f(x) \overset{\text{def}}{=} \frac{1}{2\Gamma(\alpha)\cos[\alpha\pi/2]} \int_{-\infty}^{\infty} \frac{f(u)}{|x-u|^{1-\alpha}} du .$$

This is also called Riesz's potential, which generalizes to several variables the Rieman-Liouvelli integral. These singular integrals are well defined provided the f decays rapidly at infinity.

1.4.1 The Popular Definitions of Fractional Derivatives/Integrals in Fractional Calculus

1.4.1.1 Riemann-Liouville

$$_aD_t^\alpha f(t) = \frac{1}{\Gamma(n-\alpha)}\left(\frac{d}{dt}\right)^n \int_a^t \frac{f(\tau)}{(t-\tau)^{\alpha-n+1}} d\tau.$$

$$(n-1) \le \alpha < n$$

Where n is integer and α is real number. This is a forward derivative.

1.4.1.2 Grunwald-Letnikov: (Differintegrals)

$$ {}_aD_t^{\alpha} f(t) = \lim_{h\to 0} \frac{1}{h^{\alpha}} \sum_{j=0}^{\left[\frac{t-a}{h}\right]} (-1)^j \binom{\alpha}{j} f(t-jh) $$

$$ \left[\frac{t-a}{h}\right] \to \text{INTEGER} $$

1.4.1.3 M. Caputo (1967)

$$ {}_a^C D_t^{\alpha} f(t) = \frac{1}{\Gamma(n-\alpha)} \int_a^t \frac{f^{(n)}(\tau)}{(t-\tau)^{\alpha+1-n}} d\tau, (n-1) \le \alpha < n, $$

Where n is integer and α is real number.

1.4.1.4 Oldham and Spanier (1974)

Fractional derivatives scaling property is:

$$ \frac{d^q f(\beta x)}{dx^q} = \beta^q \frac{d^q f(\beta x)}{d(\beta x)^q} $$

This makes it suitable for the study of scaling and scale invariance. There is connection between local-scaling, box-dimension of an irregular function and order of Local Fractional Derivative.

1.4.1.5 K.S. Miller and B. Ross (1993)

$$ D^{\alpha} f(t) = D^{\alpha_1} D^{\alpha_2} ... D^{\alpha_n} f(t) $$
$$ \alpha = \alpha_1 + \alpha_2 + ... + \alpha_n $$
$$ \alpha_i < 1 $$

This definition of sequential composition is very useful concept for obtaining fractional derivative of ant arbitrary order. The derivative operator can be any definition RL or Caputo.

1.4.1.6 Kolwankar and Gangal (1994)

Local fractional derivative is defined by Kolwankar and Gangal (KG) as to explain the behavior of 'continuous but nowhere differentiable' function. The other definition of fractional derivative in classical sense is integral-derivatives and is non-local property.

For $0 < q < 1$, the local fractional derivative at point $x = y$, for $f:[0,1] \to \mathbb{R}$ is:

$$\mathbf{D}^q f(y) = \lim_{x \to y} \frac{d^q \left(f(x) - f(y) \right)}{d(x-y)^q}$$

There is a logical extension of this definition to subject of Calculus on Fractals.

1.5 About Fractional Integration Derivatives and Differintegration

All the efforts to realize fractional differintegration is "interpolating" the operations between the two integer order operations. In the limit when the order of the operator approaches the nearest integer the "generalized" differintegrals tends to normal integer order operations.

1.5.1 Fractional Integration Riemann-Liouville (RL)

The repeated n-fold integration is generalized by Gamma function for the factorial expression, when the integer n is real number α.

$$D^{-n} f(t) = I^n f(t) = f_n(t) = \frac{1}{(n-1)!} \int_0^t (t-\tau)^{n-1} f(\tau) d\tau$$

$$D^{-\alpha} f(t) = I^\alpha f(t) = f_\alpha(t) = \frac{1}{\Gamma(\alpha)} \int_0^t (t-\tau)^{\alpha-1} f(\tau) d\tau$$

Defining power function as

$$\Phi_\alpha(t) = \frac{t^{\alpha-1}}{\Gamma(\alpha)}$$

and using definition of convolution integral, the expression for the fractional integration we can therefore write as the convolution of the function and the power function.

$$D^{-\alpha} f(t) = \Phi_\alpha(t) * f(t) = \int_0^t \Phi_\alpha(t) f(t-\tau) d\tau .$$

This process is depicted in Figure 1.2 where $\boldsymbol{L}$ is Laplace operator and $\boldsymbol{L}^{-1}$ is inverse Laplace operator.

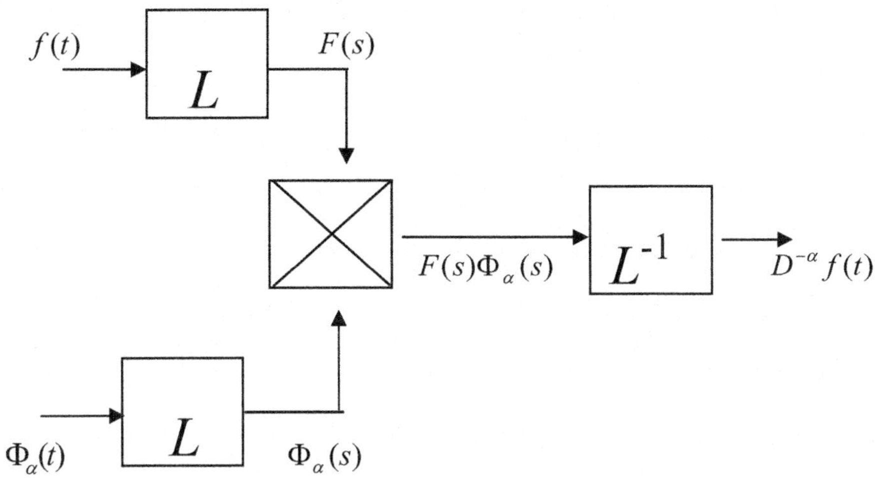

Fig. 1.2 Block diagram representation of fractional integration process by convolution

Convolution means folded-up or rolled-up condition. Non exponential relaxation processes, imply memory that the underlying fundamental relaxation processes are of Non-Markovian nature. Natural way to incorporate such memory effect is by 'fractional calculus', By the involved convolution integration in time, the present state is being influenced by all states and the system has been running through at times $\tau = 0,1,2......t$. The power-law kernel defining the fractional expression represents a particular long memory.

The above is Cauchy's generalization of *n*-fold multiple integral. The fractional integral represents (as in Fig: 1.2) a convolution of power-law with a function $f(t)$ defined as:

$$ {}_aD_t^{-\alpha} f(t) = \left(\frac{t^{\alpha-1}}{\Gamma(\alpha)}\right) * f(t) \stackrel{\text{def}}{=} \frac{1}{\Gamma(\alpha)} \int_a^t \frac{f(\tau)}{(t-\tau)^{1-\alpha}} d\tau $$

The lower limit of integration $a = 0$ implies 'Riemann' system, (without complementary function $\psi(x)$) and corresponding convolution definition is 'Laplace Convolution'. The lower limit of integration when $a = -\infty$ represents 'Liouvelli' fractional integral definition with convolution, as Fourier's convolution. Convolution algebra plays an important role in solution of Fractional Differential equation. Convolution can be interpreted as Memory Integrals, i.e., all instances from $\tau = 0$ to $\tau = t$, contribute to situation at present instance $\tau = t$. Any dynamic time evolution process can be generalized via following convolution integral with memory kernel $K(t)$

$$\frac{d\phi}{dt} = -\int_0^t K(t-\tau)\phi(\tau)d\tau$$

The memory Kernel $K(t)$ in the Memory-Integral of the definition of fractional integral is power law $t^{\alpha-1}$.

The definition of integration has two types, based on limits of integration, the Riemann-Liouvelli definition is :(for $\alpha > 0$)

$${}_aI_x^\alpha f(x) = {}_aD_x^{-\alpha} f(x) = \frac{1}{\Gamma(\alpha)}\int_a^x (x-u)^{\alpha-1} f(u)du$$ is 'forward integration'

$${}_xI_b^\alpha f(x) = {}_xD_b^{-\alpha} f(x) = \frac{1}{\Gamma(\alpha)}\int_x^b (u-x)^{\alpha-1} f(u)du$$ is backward integration

Note that the Kernel in backward integration is $(u-x)^{\alpha-1}$, and for the forward integration the Kernel is $(x-u)^{\alpha-1}$. More generally ${}_aI_x^\alpha f(x)$ is called Riemann-Liouvelli definition.

1.5.2 Fractional Integration Weyl's (W)

The Weyl's definition for right hand and left hand integration is:

$${}_xW_\infty^\alpha f(x) = {}_xI_\infty^\alpha f(x) = \frac{1}{\Gamma(\alpha)}\int_x^\infty (u-x)^{\alpha-1} f(u)du$$ is 'backward integration'

$${}_{-\infty}W_x^\alpha f(x) = {}_{-\infty}I_x^\alpha f(x) = \frac{1}{\Gamma(\alpha)}\int_{-\infty}^x (x-u)^{\alpha-1} f(u)du$$ is 'forward integration'.

The forward and backward integration i.e., ${}_aI_x^\alpha f(x)$ and ${}_xI_b^\alpha f(x)$ are related by Parseval equality (fractional integration by parts). For $a = 0$ and $b = \infty$ the relation is:

$$\int_0^\infty f(x)\left({}_0I_x^\alpha g\right)(x)dx = \int_0^\infty \left({}_xW_\infty^\alpha f\right)(x)g(x)dx$$

1.5.3 Nature of Kernel for Fractional Integration

Refer Figure 1.3 where the kernel, the power function

$$\Phi_\alpha(t) = \frac{t^{\alpha-1}}{\Gamma(\alpha)}$$

is plotted. It is worth noting that when the fractional order tends to unity, the kernel of the power function tends to Heaviside's unit step function and in case the fractional order tends to zero the power function or the kernel tends to become delta function. That is

$\lim_{\alpha\to 1}\Phi_\alpha(t) = H(t)$, where $H(t) = 1.0$ for $t \geq 0$ and $H(t) = 0$ for $t < 0$, which also implies the identity as $\lim_{\alpha\to 1} {}_0I_t^\alpha f(t) = \lim_{\alpha\to 1} {}_0D_t^{-\alpha} f(t) = \int_0^t f(\tau)d\tau$ is normal integer order integration of order one.

$\lim_{\alpha\to 0}\Phi_\alpha(t) = \delta(t)$, gives identity as $\lim_{\alpha\to 0} {}_0I_t^\alpha f(t) = \lim_{\alpha\to 0} {}_0D_t^{-\alpha} f(t) = f(t)$. Here we have used the property that convolution of delta function with any function results in function itself.

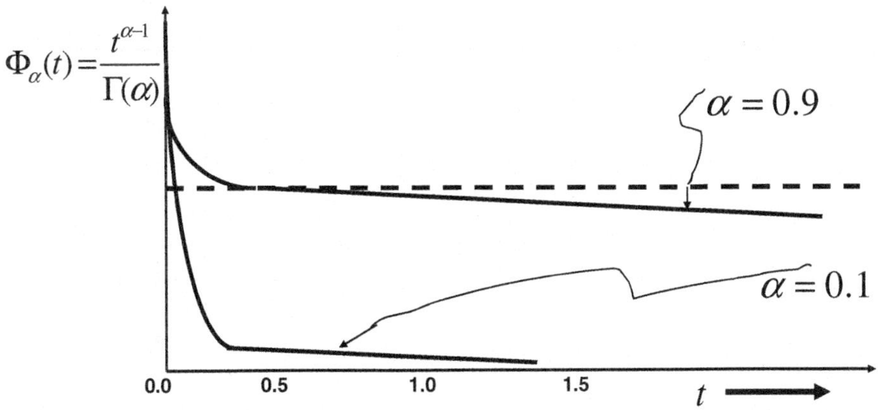

Fig. 1.3 Nature of Kernel (Power-function) for fractional integration

1.5.4 Fractional Derivatives Riemann-Liouville (RL) Left Hand Definition (LHD)

The formulation of this definition is:

Select an integer m greater than fractional number α

(i) Integrate the function $(m-\alpha)$ folds by RL integration method.
(ii) Differentiate the above result by m.

Expression is given as:

$$D^{\alpha} f(t) = \frac{d^m}{dt^m}\left[\frac{1}{\Gamma(m-\alpha)}\int_0^t \frac{f(\tau)}{(t-\tau)^{\alpha+1-m}}\, d\tau\right]$$

The Figure 1.4 gives the process block diagram & Figure 1.5 gives the process of differentiation 2.3 times for a function.

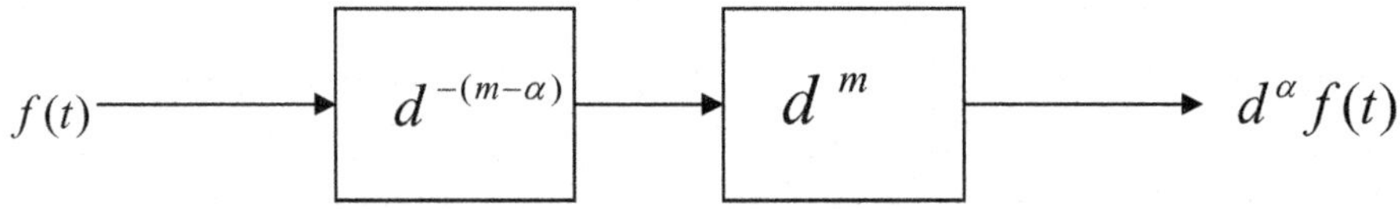

Fig. 1.4 Fractional differentiation Left Hand Definition (LHD) block diagram

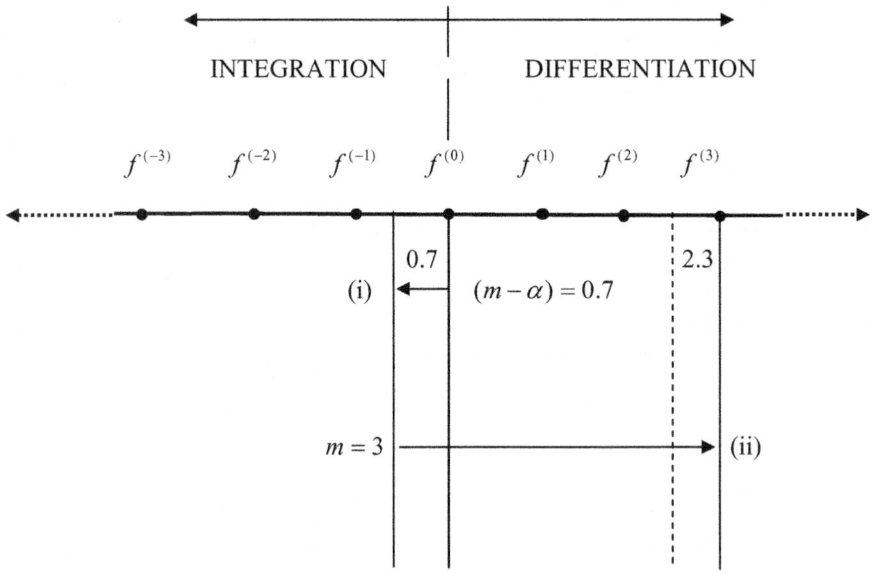

Fig. 1.5 Fractional differentiation of 2.3 times in LHD

In this LHD the limit of integration is from 0 to t. We thus denote the derivative by notation ${}_0D_t^{\alpha} f(t)$. In fractional calculus we find limit of derivative i.e., derivatives are taken in interval. We call this as 'forward derivative'. Now if the limits of integration are changed to (t to 0) the derivative is denoted as ${}_tD_0^{\alpha} f(t)$ the 'backward derivative'. The backward derivative is related to forward derivative by

$$ {}_{t}D_{0}^{\alpha} f(t) = (-1)^{m} \frac{d^{m}}{dt^{m}} {}_{t}I_{0}^{m-\alpha} f(t) $$

Therefore, in order to obtain fractional derivative of a function at a point (say 0) we should have the values of these two derivatives the same: forward derivative should equal the backward derivative. This implies that not only one should know the function from past to the point of interest (say 0) but also the function should be known into the future in order to have point fractional derivative at a point.

Fractional derivative of purely imaginary order i.e., $\alpha = \mathrm{i}\theta$, ($\theta \neq 0$) with real part as 'zero' is expressed in Riemann-Liouvelli notation (with $m = 1$):

$$ {}_{a}D_{x}^{\mathrm{i}\theta} f(x) = \frac{1}{\Gamma(1-\mathrm{i}\theta)} \frac{d}{dx} \int_{a}^{x} \frac{f(u)}{(x-u)^{\mathrm{i}\theta}} du $$

and associated integral of purely imaginary order in Riemann-Liouvelli definition is:

$$ {}_{a}D_{x}^{-\mathrm{i}\theta} f(x) = {}_{a}I_{x}^{\mathrm{i}\theta} f(x) = \frac{d}{dx} {}_{a}I_{x}^{1+\mathrm{i}\theta} f(x) = \frac{1}{\Gamma(1+\mathrm{i}\theta)} \frac{d}{dx} \int_{a}^{x} (x-u)^{\mathrm{i}\theta} f(u) du $$

1.5.5 Fractional Derivatives Caputo Right Hand Definition (RHD)

The formulation is exactly opposite to LHD.

Select an integer m greater than fractional number

(i) Differentiate the function m times.

(ii) Integrate the above result $(m-\alpha)$ fold by RL integration method.

In LHD and RHD the integer selection is made such that $(m-1) < \alpha < m$. For example differentiation of the function by order π will select $m = 4$. The formulation of RHD Caputo is as follows:

$$ {}^{C}D^{\alpha} f(t) = \frac{1}{\Gamma(m-\alpha)} \int_{0}^{t} \frac{\frac{d^{m} f(\tau)}{d\tau^{m}}}{(t-\tau)^{\alpha+1-m}} d\tau = \frac{1}{\Gamma(m-\alpha)} \int_{0}^{t} \frac{f^{(m)}(\tau)}{(t-\tau)^{\alpha+1-m}} d\tau $$

Figure 1.6 gives the block diagram representation of the RHD process and Figure 1.7 represents graphically RHD used for fractionally differentiating function 2.3 times.

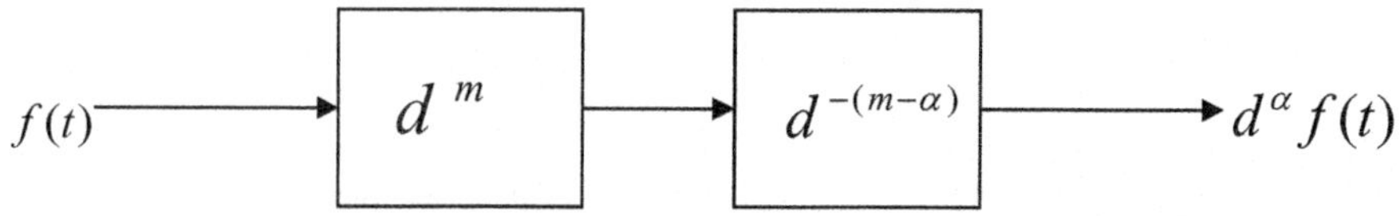

Fig. 1.6 Block diagram representation of RHD Caputo

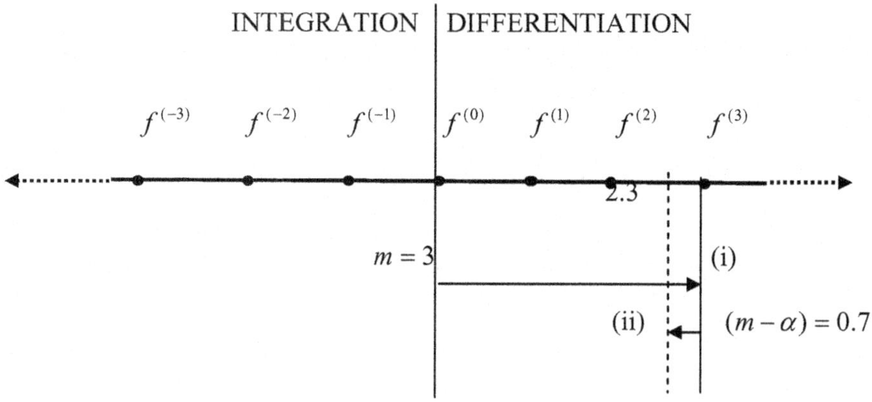

Fig. 1.7 Differentiation of 2.3 times by RHD

The definitions of Riemann-Liouville of fractional differentiation played an important role in development of fractional calculus. However the demands of modern science and engineering require a certain revision of the well established pure mathematical approaches. Applied problems require definitions of fractional derivatives allowing the utilization of physically interpretable "initial conditions" which contain $f(a), f^{(1)}(a), f^{(2)}(a)$ and not fractional quantities (presently unthinkable!). The RL definitions require

$$\lim_{t\to a} {}_aD_t^{\alpha-1} f(t) = b_1$$

$$\lim_{t\to a} {}_aD_t^{\alpha-2} f(t) = b_2$$

In spite of the fact that initial value problems with such initial conditions can be successfully solved mathematically, their solutions are practically useless, because at present there is no known physical interpretation for such initial conditions, presently. It is hard to interpret.

RHD is more restrictive than LHD. For RL $f(t)$ is causal. For LHD as long as initial function of t satisfies $f(0) = 0$. For RHD because $f(t)$ is first made to m-th derivative i.e. $f^{(m)}(t)$, the condition $f(0) = 0$ *and* $f^1 = f^2 = ...f^m = 0$ is required. In mathematical world this is vulnerable for RHD may be deliberating. For LHD

$$D^{\alpha}C \neq 0 = \frac{Ct^{-\alpha}}{\Gamma(1-\alpha)},$$

the derivative of constant C is not zero. This fact led to using the RL or LHD approach with lower limit of differentiation $a \to -\infty$ in physical world this posses problem. The physical meaning of this lower limit extending towards minus infinity is starting of physical process at time immemorial! In such cases transient effects cannot be then studied. However making $a \to -\infty$ is necessary abstraction for consideration of steady state process, for example for study of sinusoidal analysis for steady state fractional order system.

Note that non-local character of Fractional Derivative definitions and the fact that in LHD the Riemann-Liouvelli definitions, the fractional derivative of constant being non-zero makes scaling information somewhat difficult. Sometimes it is desirable to have 'local-character' of a function i.e., to have local fractional derivatives defined-in wide range of applications ranging from structure of differentiable manifolds to various physical models. Therefore extensions to this LHD are made in KG definition to have Local Fractional Derivative. This KG definition too takes care of this LHD i.e., Fractional Derivative of constant being non-zero gives a consequence to change magnitude of Fractional Derivative of a function at a point when different constants are added. The Local Fractional Derivative of KG definition with LHD formulation addresses these issues.

While today we are familiar with interpretation of the physical world with integer order differential equations; we do not (currently) have practical understanding of the world with fractional order differential equations. Our mathematical tools go beyond practical limitation of our understanding.

Therefore still process is on to 'generalize' the concepts for use in practical world.

1.5.6 Fractional Derivatives of Same Order but Different Types RL-Caputo

We have seen two types of fractional derivative and discussed them in above section. The following formula gives fractional derivative definition with in RL-Caputo type, with β as type defining parameter as $0 \le \beta \le 1$. When $\beta = 1$ the definition is Caputo definition with symbol, ${}^{C}D_t^{\alpha} f(t)$ while $\beta = 0$ gives RL fractional derivative with symbol $D_t^{\alpha} f(t)$. The generalized definition is as following:

$${}_0^{\beta}D_t^{\alpha} f(t) = {}_0I_t^{\beta(1-\alpha)} \frac{d}{dt}\left[{}_0I_t^{(1-\beta)(1-\alpha)} f(t)\right]$$

The α is the order of derivative, and here in this definition is $0 < \alpha < 1$ and β is the type of derivative with $0 \le \beta \le 1$. Therefore the nearest integer is $m = 1$ in the above generalization.

1.5.7 Fractional Differintegrals Grunwald Letnikov (GL)

Differintegration process as described below is differentiation for positive index and integration for negative index for the differintegral (generalized) operator.

$$f^1(x)=\lim_{h\to 0}\frac{f(x+h)-f(x)}{h}$$

$$f^2(x)=\lim_{h\to 0}\frac{f^1(x+h)-f^1(x)}{h}=\lim_{h_1\to 0}\frac{\lim\limits_{h_2\to 0}\frac{f(x+h_1+h_2)-f(x+h_1)}{h_2}-\lim\limits_{h_2\to 0}\frac{f(x+h_2)-f(x)}{h_2}}{h_1}$$

$$h_1=h_2=h$$

$$f^2(x)=\lim_{h\to 0}\frac{f(x+2h)-2f(x+h)+f(x)}{h^2}$$, Continuing for n times we have:

$$f^n(x)=D^n f(x)=\lim_{h\to 0}\frac{1}{h^n}\sum_{m=0}^{n}(-1)^m\binom{n}{m}f(x-mh)$$

$$\binom{n}{m}=\frac{n!}{m!(n-m)!}$$

This can be replaced by Gamma functions as $\dfrac{\Gamma(\alpha+1)}{m!\Gamma(\alpha-m+1)}$ for non- integer n i.e. α. Therefore differentiation in fractional order is:

$$_aD^\alpha f(x)=\lim_{h\to 0}\frac{1}{h^\alpha}\sum_{m=0}^{\left[\frac{x-a}{h}\right]}(-1)^m\frac{\Gamma(\alpha+1)}{m!\Gamma(\alpha-m+1)}f(x-mh)$$

For negative α the process will be integration.

$$\binom{-n}{m}=\frac{-n(-n-1)(-n-2)\ldots(-n-m+1)}{m!}=(-1)^m\frac{n(n+1)(n+2)(n+3)\ldots(n+m-1)}{m!}$$

$$=(-1)^m\frac{(n+m-1)!}{m!(n-1)!}\to(-1)^m\frac{\Gamma(\alpha+m)}{m!\Gamma(\alpha)}$$

Therefore for integration we write:

$$_aD^{-\alpha} f(x)=\lim_{h\to 0}h^\alpha\sum_{m=0}^{\left[\frac{x-a}{h}\right]}\frac{\Gamma(\alpha+m)}{m!\Gamma(\alpha)}f(x-mh).$$

The part $\left[\dfrac{x-a}{h}\right]$ is integer part (floor-function). That is the upper limit of the summation is the integer part of the fraction.

The RL GL RHD (Caputo) LHD are they equivalent? The answer to this is yes.

1.5.8 *Fractional Derivative Weyl's*

Like Riemann-Liouvelli definition the Weyl's derivative of function is defined, for an integer just greater than the fractional index of derivative. For $m-1 \le \alpha < m$, where $m \in \mathbb{N}$, the Weyl's definition is:

$$ {}_xD_\infty^\alpha f(x) = (-1)^m \frac{d^m}{dx^m}\, {}_xW_\infty^{m-\alpha} f(x) $$

The ${}_xW_\infty^{m-\alpha} f(x)$ is Right Hand definition of Weyl's fractional integration of order $(m-\alpha)$.

Let $f(x)=e^{-px}$, with $p>0$, and under the substitution $u-x=(y/p)$, and with integral representation of Gamma function as defined as

$$ \Gamma(x) = \int_0^\infty e^{-t} t^{x-1} dt, $$

we get the following

$$ {}_xW_\infty^\alpha e^{-px} = \frac{1}{\Gamma(\alpha)} \int_x^\infty (u-x)^{\alpha-1} e^{-pu} du = \frac{e^{-px}}{p^\alpha \Gamma(\alpha)} \int_0^\infty y^{\alpha-1} e^{-y} dy = \frac{e^{-px}}{p^\alpha}. $$

This yields:

$$ {}_xD_\infty^\alpha e^{-px} = (-1)^m \frac{d^m}{dx^m} \left\{ \frac{e^{-px}}{p^{(m-\alpha)}} \right\} = (-1)^m \left[\frac{(-p)^m e^{-px}}{p^m p^{-\alpha}} \right] = p^\alpha e^{-px} $$

Earlier in this chapter we have seen the fractional derivative of exponential function e^x returning Miller-Ross function.

$$ {}_0D_x^v e^x = \sum_{k=0}^\infty \frac{x^{k-v}}{\Gamma(k-v+1)} = E_x(-v,1), $$

for $v>0$ and $v \in \mathbb{R}$, whereas, ${}_{-\infty}D_x^v e^x = e^x$. Thus it is observed that the lower limit of fractional differentiation is very important.

1.5.9 Scale Invariance and Power Law

By scale change of the function $f(x)$ w.r.t. a lower limit with a we mean replacement by $f(\beta x-\beta a+a)$, where β is a constant termed as the 'scaling factor'. To clarify this consider, $a=0$. Then scale change converts $f(x)$ to $f(\beta x)$. This is in contrast to homogeneity property where if $f(x)$ is changed to $Cf(x)$ then $D^q\left[Cf(x)\right]=CD^q f(x)$.

We seek a procedure by which the effect of generalized differ-integration operation ${}_aD_x^q$ operation be found if ${}_aD_x^q f(x)$ is known. Let $X\equiv x+[a-a\beta]/\beta$. Then by Riemann-Liouvelli definition we write:

$$
\begin{aligned}
{}_aD_x^q f(\beta X) = {}_aD_x^q f(\beta x-\beta a+a) &= \frac{1}{\Gamma(-q)}\int_a^x \frac{f(\beta y-\beta a+a)dy}{[x-y]^{q+1}} \\
&= \frac{1}{\Gamma(-q)}\int_a^{\beta X} \frac{f(Y)\{dY/\beta\}}{\{[\beta X-Y]/\beta\}^{q+1}} \\
&= \frac{\beta^q}{\Gamma(-q)}\int_a^{\beta X} \frac{f(Y)dY}{[\beta X-Y]^{q+1}} \\
&= \beta^q\, {}_aD_{\beta X}^q f(\beta X)
\end{aligned}
$$

When $a=0$, $X=x$ and the scale change is simply multiplication of independent variable by a constant, i.e. ${}_0D_x^q f(\beta x)=\beta^q\,{}_0D_{\beta x}^q f(\beta x)$.

When $a\neq 0$, the effect requires additional process of translation, i.e., $f(x)$ to $f(x+A)$ which is a difficult process to generalize.

$$
{}_aD_x^q f(x+A)=\frac{1}{\Gamma(-q)}\int_a^x\frac{f(y+A)dy}{[x-y]^{q+1}}=\frac{1}{\Gamma(-q)}\int_{a+A}^{x+A}\frac{f(Y)dY}{[x+A-Y]^{q+1}},\quad Y=y+A
$$

That is shift the upper limit to $x+A$ and lower limit to $a+A$.

This implies the study of self-similar processes, objects and distributions too. Let $y(t)=t^\beta$, then $y(\lambda t)=(\lambda t)^\beta=\lambda^\beta y(t)$. A scaling in time axis results in simple scaling of ordinate y. The law $y(t)=t^\beta$ is not altered. In log-log plot this means a shift $\log\lambda$ on the $\log t$ axis and shift $\beta\log\lambda$ in the ordinate. This 'scale invariance', however, is the cause of stability of systems following power laws. The kernel of fractional integral, defined above

$$\Phi_\alpha(t) = \frac{t^{\alpha-1}}{\Gamma(\alpha)}$$

is also a power-law-and is scale-invariant.

Power law relaxation is observed in various physical processes. Thus often one encounters algebraic relaxation function as $\phi(t) \propto t^{-\alpha}$, with $0 < \alpha < 1$. The power-law is any polynomial that exhibit 'scale-invariance'. If $f(x) \approx ax^k$, with k as scaling exponent, examples being inverse square law (gravity, electrostatic), area of circle. Power law distributions in most general sense have a probability distribution of form: $p(x) \propto L(x)x^{-\alpha}$, with $\alpha > 1$ and $L(x)$ as a slowly varying function. Implying $p(x)$ be asymptotically 'scale-invariant' thus $L(x)$ controls the shape and finite extent of lower-tail. These are also called long-tailed distributions with no defined average or standard deviation, as compared to normal-distribution. It is convenient to assume sometimes lower-bounds $x_{\min}$ from which the power law holds as:

$$p(x) = \frac{\alpha-1}{x_{\min}}\left(\frac{x}{x_{\min}}\right)^{-\alpha}, \text{ with } \alpha > 1$$

Moments of power law distributions are

$$\left\langle x^m \right\rangle = \int_{x_{\min}}^{\infty} x^m p(x) dx = \frac{\alpha-1}{\alpha-1-m}(x_{\min})^m,$$

which is only well defined (finite) for $m < \alpha - 1$. That means all moments for $m \geq \alpha - 1$ diverge. Say when $\alpha < 2$, the average (mean) and all other moments diverge (not-finite). When $2 < \alpha < 3$, the mean exists but variance $\left\langle x^2 \right\rangle$ and higher order moments are infinite. These power-law distributions with diverging (integer order) moments are characteristics of fractional order differential equations defining relaxation in disordered system, thus paving way to have 'Fractional Lower Order Statistics'.

The most discussed power-law distribution is 'Levy stable distribution'; which has asymptotic form as $f_{\alpha,\beta}(x) \sim A_{\alpha,\beta} / |x|^{1+\alpha}$, where $\alpha < 2$ is called the Levy index or the characteristic exponent. The limiting case with $\alpha = 2$ corresponds to Gaussian normal distribution governed by central limit theorem. For all Levy stable laws with $0 < \alpha < 2$, the variance diverges $\left\langle x^2 \right\rangle \to \infty$. Conversely the fractional moments of the absolute value of x exists. Meaning $\left\langle |x|^\delta \right\rangle < \infty$ for $0 \leq \delta < \alpha \leq 2$.

The processes are like:

(i) Transport of charge carriers in amorphous semiconductors.
(ii) Behavior of electrical current at rough blocking electrodes (Warburg electrochemical impedance in battery).
(iii) Di-electric relaxation of liquids and solids.
(iv) Microscopic and Macroscopic dynamic behavior of macromolecular systems (such as linear polymers or gel).
(v) Attenuation of seismic waves.
(vi) Transient Electromagnetic Method for geophysical exploration, the late time decay of receiver voltage profile from conductive half space, and several other decay processes.
(vii) Scale invariance in 'fractional' stochastic process of financial returns or fluctuating dynamics of delay generating systems in computer network.

Fourier transform of power-law is also a power law, the Magnitude and (phase) angle of a Fourier transformation is depicted in an example as:

$$\left|\Im\left\{t^{\alpha}H(t)\right\}\right| = \Gamma(1+\alpha)\omega^{-(\alpha+1)} \text{ and } \angle\Im\left\{t^{\alpha}H(t)\right\} = -\frac{\pi}{2}(\alpha+1),$$

Here $H(t)$ is Heaviside's step function that is $H(t)=1$ for $t \geq 0$ and $H(t)=0$ for $t<0$.

Therefore often one encounters functional relations in the frequency domain of the form $\omega^{-\alpha}F(\omega)$, for example $\omega^{-\alpha}\exp(-\omega\tau)$. Such relations back-transformed by generalized differentiation theorem gives 'fractional integral' expression as:

$$\omega^{-\alpha}F(\omega) \leftrightarrow {}_0D^{-\alpha}f(t)$$

Fractional expressions can arise quite naturally from functional relations observed in 'spectral-domain' analysis and experiments. In time domain a typical example of asymptotic fractal is of the form $G(t) \sim \left(\frac{t}{\tau}\right)^{-\beta} \gamma^*(\beta, t/\tau)$, with $\beta > 0$ and γ^* as incomplete Gamma function, or any other higher transcendental function. This form has late times a fractional power decay law. Power law is also referred as Long-Tailed distribution and Long Range Dependence (LRD) as its tail lingers as if having memory. This is in comparison with exponential decay curves associated with Integer Order Linear Differential Equations, which decays very fast as though having no-memory. Power-law decays are associated with Fractional Order Decay Equations and those are having long lingering memory kernels.

Why the night sky is irregularly illuminated (Olber's paradox), why income is non-uniformly distributed (Pareto's law), that is why rich get richer and poor get poorer, is all follows self-similar power-law distribution. Mandelbrot coined the term fractal, to take cognizance of the fact that a large class of natural and social phenomena that traditional statistical physics is not equipped to describe. The fractals and fractional calculus are related.

1.5.10 Fourier Transform of Fractional Derivative

The lower limit of fractional differ-integration is commonly set to zero or minus infinity. The zero bound is most commonly used for time dependent functions that are causal, and nonzero only for $t \geq 0$. For spatial functions an infinite bound is typically used and for simplicity the limit is eliminated from the subscript and only the direction of fractional differ-integration is noted. For infinity bound, fractional differ-integration Fourier transformation is the technique many times used to solve fractional differential equations. The Fourier transform is expressed as

$$\Im\{f(x)\} = \hat{f}(k) = \int_{-\infty}^{+\infty} e^{\mathrm{i}kx} f(x)dx\,,\ \mathrm{i} = \sqrt{-1}$$

and the Fourier transform of integer (n) order derivatives is:

$$\Im\{D_{+}^{n} f(x)\} = \Im\left\{\frac{d^{n} f(x)}{dx^{n}}\right\} = (\mathrm{i}k)^{n} \hat{f}(k)$$

$$\Im\{D_{-}^{n} f(x)\} = \Im\left\{\frac{d^{n} f(x)}{d(-x)^{n}}\right\} = (-\mathrm{i}k)^{n} \hat{f}(k)$$

Staying with the definition of the positive direction fractional derivative that is:

$$D_{+}^{q} f(x) = \frac{d^{q}}{dx^{q}} f(x) = \frac{1}{\Gamma(n-q)} \frac{d^{n}}{dx^{n}} \int_{-\infty}^{x} (x-\zeta)^{n-q-1} f(\zeta) d\zeta\,,$$

substituting $y = x - \zeta$, for $-1 < p < 0$, we have

$$D_{+}^{p} f(x) = \frac{1}{\Gamma(-p)} \int_{\infty}^{0} y^{-p-1} f(x-y)(-dy)\,.$$

We re-write this expression as $D_{+}^{p} f(x) = \frac{1}{\Gamma(-p)} \int_{-\infty}^{+\infty} y^{-p-1} H(y) f(x-y) dy$, where $H(y)$ is Heaviside's step function with value 1 for $y \geq 0$ and 0 for $y < 0$. The Heaviside's step function multiplied with the function in the integrand changes the limits of integration as $-\infty$ to $+\infty$, required to perform Fourier transformation. This derived integral is convolution of product of Fourier transforms of two functions namely $x^{-1-p} H(x)/\Gamma(-p)$ and $f(x)$. From the standard tables of Fourier transform we have

$$\Im\left\{x^{-p-1}H(x)/\Gamma(-p)\right\}=|k|^{p}\exp\left((\mathrm{i}\pi p/2)\operatorname{sgn}(k)\right)$$

Since $\exp\left[(\mathrm{i}\pi p/2)\operatorname{sgn}(k)\right]=\mathrm{i}^{p}$, when $k>0$ and is equal to $(-\mathrm{i})^{p}$ when $k<0$, we have:

$$\Im\left\{D_{+}^{p}f(x)\right\}=(\mathrm{i}k)^{p}\hat{f}(k)\,.$$

Similarly one can draw $\Im\left\{D_{-}^{p}f(x)\right\}=(-\mathrm{i}k)^{p}\hat{f}(k)$

For other fractional values we can construct the Fourier identities as:

$$\Im\left\{D_{\pm}^{q}f(x)\right\}=\Im\left\{D_{\pm}^{n}\left[D_{\pm}^{p}f(x)\right]\right\}=(\pm\mathrm{i}k)^{n}(\pm\mathrm{i}k)^{p}\hat{f}(k)=(\pm\mathrm{i}k)^{n+p}\hat{f}(k)=(\pm\mathrm{i}k)^{q}\hat{f}(k)$$

The Fourier transform of Fractional derivative of Dirac delta is thus:

$$\Im\left\{D_{\pm}^{q}\delta(x)\right\}=(\pm\mathrm{i}k)^{q}$$

The convolution ($\otimes$) of function f and g with respect to Fourier transform is given by

$$f(x)\otimes g(x)=\int_{-\infty}^{+\infty}g(x-\zeta)f(\zeta)d\zeta\,.$$

One important convolution relation with Dirac delta function is,

$$\left[D_{\pm}^{q}\delta(x)\right]\otimes f(x)=\delta(x)\otimes\left[D_{\pm}^{q}f(x)\right]=D_{\pm}^{q}f(x)\,.$$

1.5.11 Composition and Property

In this book the symbols for fractional differintegration have been standardized as follows.

$${}_{c}D_{t}^{q}f(t)$$

Represents initialized q-th order differintegration of $f(t)$ from start point c to t.

$${}_{c}d_{t}^{q}f(t)$$

Represents un-initialized generalized (or fractional) *q-th* order differintegral. This is also same as

$$\frac{d^q f(t)}{[d(t-c)]^q} \equiv {}_c d_t^q f(t),$$

shifting the origin of function at start of the point from where differintegration starts. This un-initialized operator can also be short formed as $d^q f(t)$.

The index $q > 0$ is differentiation and index $q < 0$ is integration process. For q as integer, the process is integer order classical differentiation and integration.

Miller and Ross (1993) with sequential fractional derivatives tried to give formal properties and to have composition methods for generalized differintegrals. Decomposition ${}_aD_t^{\alpha} y(t) = {}_aD_t^{m}\, {}_aD_t^{\alpha-m} y(t)$ also to some extent the index commutation (under certain conditions) $D^{-\alpha}D^{-\beta} = D^{-(\alpha+\beta)} = D^{-\beta}D^{-\alpha}$ well true for fractional integration. But fractional derivatives do not commute always i.e. $D^{\alpha}D^{\beta} \neq D^{\beta+\alpha} \neq D^{\alpha+\beta}$, (except at zero initial conditions). Integer operator (m) commutes with fractional operator (α) i.e. $D^m D^{\alpha} = D^{m+\alpha}$, are some of the basic composition properties. The desirable properties of fractional derivatives and integrals are the following:

a. If $f(z)$ is an analytical function of z, then its fractional derivatives ${}_0D_z^{\alpha} f(z)$ is an analytical function of z and α.
b. For $\alpha = n$, where n is integer, the operation ${}_0D_z^{\alpha} f(z)$ gives the same result as classical differentiation, or integration of integer order n.
c. For $\alpha = 0$ the operation ${}_0D_z^{\alpha} f(z)$ is identity operator i.e. ${}_0D_z^{0} f(z) = f(z)$
d. Fractional differentiation and fractional integration are linear operations: ${}_0D_z^{\alpha} af(z) + {}_0D_z^{\alpha} bg(z) = a\,{}_0D_z^{\alpha} f(z) + b\,{}_0D_z^{\alpha} g(z)$
e. The additive index law: ${}_0D_z^{\alpha}\, {}_0D_z^{\beta} f(z) = {}_0D_z^{\beta}\, {}_0D_z^{\alpha} f(z) = {}_0D_z^{\alpha+\beta} f(z)$ holds under some reasonable constraints on the function $f(z)$.

The above desirable properties are valid under causality that is the function is differintegrated at the start point of function itself (with initialization function being zero).

For a function f differentiable in open-interval (a,b) by fundamental theorem of calculus we can express the function as:

$$f(x) = \left(\int_a^x f'(u)du \right) + f(a) = {}_aI_x^1 f'(x) + f(a)$$

On the both sides carrying out fractional integral operation ${}_aI_x^{1-\alpha}$ where $0 < \alpha < 1$ and making use of commutation of fractional integration i.e., $I^{1-\alpha}$ commuting with I^1 we write the above expression as:

$$ {}_aI_x^{1-\alpha} f(x) = {}_aI_x^1 \, {}_aI_x^{1-\alpha} f'(x) + \frac{(x-a)^{1-\alpha}}{\Gamma(2-\alpha)} f(a) $$

Differentiating the above by one integer order, and applying Riemann-Liouvelli definition of fractional differentiation on LHS of the above we obtain:

$$ {}_aD_x^{\alpha} f(x) = \frac{d}{dx} {}_aI_x^{1-\alpha} f(x) = {}_aI_x^{1-\alpha} f'(x) + f(a)\frac{(x-a)^{-\alpha}}{\Gamma(1-\alpha)} = {}_aI_x^{1-\alpha} \frac{d}{dx} f(x) + f(a)\frac{(x-a)^{-\alpha}}{\Gamma(1-\alpha)} $$

This shows that ${}_aI_x^{1-\alpha}$ and ${}_aD_x^1 \equiv d/dx$ do not commute, except when initial condition is zero. They are related by initial condition. The above expression also shows the relation between Riemann-Liouvelli and Caputo derivatives (for $0 < \alpha < 1$), that is:

$$ {}_aD_x^{\alpha} f(x) = {}_a^CD_x^{\alpha} f(x) + f(a)\frac{(x-a)^{-\alpha}}{\Gamma(1-\alpha)} $$

1.5.12 Fractional Derivative for Some Standard Function

The table lists Riemann-Liouvelle fractional derivatives of some functions, which are used very often. In most cases the order of differentiation α may be real number, so replacing it with $-\alpha$ gives Riemann-Liouville fractional integral. The table can be used to find Grunwald-Letnikov, fractional derivatives, Caputo fractional derivatives and Miller-Ross sequential fractional derivatives. In such cases α should be taken between 0 and 1, and Riemann-Liouvelli fractional derivatives should be properly combined (composed) with integer order derivatives with considered definition (of composition). Table 1.1 gives RL derivative with lower terminal at 0, and Table 1.2 gives fractional RL derivatives with lower terminal at $-\infty$. In the list H (t) is unit step Heaviside function. E is Mittag-Leffler function, ψ-is 'psi' function. These tables' gives a feel of how fractional differintegration will look like in analytical expressions.

We demonstrate the derivative calculation of Dirac's delta function $\delta(x-c)$ with $b \le c \le d$, defined by

$$ \int_b^d \delta(x-c) f(x) dx = f(c) $$

Using

$$ D_{x+}^q \left\{ x^u \right\} = \frac{\Gamma(u+1)}{\Gamma(u-q+1)} x^{u-q} . $$

The fractional derivative with order $0<q$ of $\delta(x-c)$ is directly obtained as follows.

$$D_{+}^{q}\delta(x-c)=\frac{1}{\Gamma(n-q)}\frac{d^{n}}{dx^{n}}\int_{-\infty}^{x}\delta(\zeta-c)(x-\zeta)^{n-q-1}d\zeta=\frac{1}{\Gamma(-q)}\begin{cases}0 & x<c\\(x-c)^{-q-1} & x\geq c\end{cases}$$

$$D_{-}^{q}\delta(x-c)=\frac{(-1)^{n}}{\Gamma(n-q)}\frac{d^{n}}{dx^{n}}\int_{-\infty}^{x}\delta(\zeta-c)(\zeta-x)^{n-q-1}d\zeta=\frac{1}{\Gamma(-q)}\begin{cases}(c-x)^{-q-1} & x\leq c\\0 & x>c\end{cases}$$

The higher transcendental functions play important role in fractional calculus, as appearing in the tables.

Table 1.1 RL Derivative with lower terminal 0 i.e. ${}_0D_t^{\alpha}f(t)$ for $t>0$

Function f (t)	${}_0D_t^{\alpha}f(t)$. Fractional derivative
$H(t)$	$\dfrac{t^{-\alpha}}{\Gamma(1-\alpha)}$
$H(t-a)$	$\begin{cases}\dfrac{(t-a)^{-\alpha}}{\Gamma(1-\alpha)},(t>a)\\0,(0\leq t\leq a)\end{cases}$
$H(t-a)f(t)$	$\begin{cases}{}_aD_t^{\alpha}f(t),(t>a)\\0,(0\leq t\leq a)\end{cases}$
$\delta(t)$	$\dfrac{t^{-\alpha-1}}{\Gamma(-\alpha)}$
$\delta^{(n)}(t)$	$\dfrac{t^{-\alpha-n-1}}{\Gamma(-\alpha-n)}$
$\delta^{(n)}(t-a)$	$\begin{cases}\dfrac{(t-a)^{-\alpha-n-1}}{\Gamma(-n-\alpha)},(t>a)\\0,(0\leq t\leq a)\end{cases}$
t^{v}	$\dfrac{\Gamma(v+1)}{\Gamma(v+1-\alpha)}t^{v+\alpha}$ $\quad v>-1$
$e^{\lambda t}$	$t^{-\alpha}E_{1,1-\alpha}(\lambda t)=E_t(-\alpha,\lambda)$
$\cosh(\sqrt{\lambda t})$	$t^{-\alpha}E_{2,1-\alpha}(\lambda t^2)$
$\dfrac{\sinh(\sqrt{\lambda t})}{\sqrt{\lambda t}}$	$t^{1-\alpha}E_{2,2-\alpha}(\lambda t^2)$
$\ln(t)$	$\dfrac{t^{-\alpha}}{\Gamma(1-\alpha)}\left(\ln(t)+\psi(1)-\psi(1-\alpha)\right)$

Table 1.1 (*continued*)

$t^{\beta-1}\ln(t)$	$\frac{\Gamma(\beta)t^{\beta-\alpha-1}}{\Gamma(\beta-\alpha)}\left(\ln(t)+\psi(\beta)-\psi(\beta-\alpha)\right)$
$t^{\beta-1}E_{\mu,\beta}(\lambda t^{\mu})$	$t^{\beta-\alpha-1}E_{\mu,\beta-\alpha}(\lambda t^{\mu})$

The $E_{1,1-\alpha}(\lambda t), E_t(-\alpha,\lambda), \psi$ are higher transcendental functions, used in fractional calculus, namely Mittag-Leffler, Miller-Ross, and 'psi' functions.

Table 1.2 RL derivative with lower terminal at $-\infty$ i.e. ${}_{-\infty}D_t^{\alpha} f(t)$,

Function $f(t)$	${}_{-\infty}D_t^{\alpha} f(t)$ Derivative
$H(t-a)$	$\begin{cases} \frac{(t-a)^{-\alpha}}{\Gamma(1-\alpha)}, (t>a) \\ 0, (t \le a) \end{cases}$
$H(t-a)f(t)$	$\begin{cases} {}_aD_t^{\alpha} f(t), (t>a) \\ 0, (t \le a) \end{cases}$
$e^{\lambda t}$	$\lambda^{\alpha} e^{\lambda t}$
$e^{\lambda t+\mu}$	$\lambda^{\alpha} e^{\lambda t+\mu}$
$\sin \lambda t$	$\lambda^{\alpha}\sin\left(\lambda t+\frac{\pi\alpha}{2}\right)$
$\cos \lambda t$	$\lambda^{\alpha}\cos\left(\lambda t+\frac{\pi\alpha}{2}\right)$
$e^{\lambda t}\sin \mu t$	$r^{\alpha} e^{\lambda t}\sin(\mu t+\alpha\phi)$ $r=\sqrt{\lambda^2+\mu^2}$ $\tan\phi=\frac{\mu}{\lambda}$ $(\lambda,\mu>0)$
$e^{\lambda t}\cos \mu t$	$r^{\alpha} e^{\lambda t}\cos(\mu t+\alpha\phi)$ $r=\sqrt{\lambda^2+\mu^2}$ $\tan\phi=\frac{\mu}{\lambda}$ $(\lambda,\mu>0)$

1.6 Solution of Fractional Differential Equations

Fractional differential equations appear in several physical systems. Solution to these is no more rigorous than its integer order counterpart. The Laplace transformation technique is very popular though yet several analytical approaches do exist. Numerical evaluation with "short memory principle" is one amongst them popular for computer programming and numerical regression. Mellin transform, power series expansion method approach using fractional Green's function, Babenko's symbolic method, Orthogonal polynomial method, Complementary polynomial (changing fractional differential equation to integer order differential equations) Reisz fractional potential method, method with Wright's function and finite-part integral method are some of the mathematician's tool for obtaining the fractional differintegrals and solution of fractional differential equation. However in this book only Laplace is considered as most easily understood and being popular amongst engineers and scientists. A very recent approach to solve General Dynamic Systems with action reaction process is described in the Chapter-11, which gives approximate analytic solution close to physics; the mathematics of Adomian Decomposition Method is applied to describe the physical reaction chain. A simple examples in this section elaborates the concept of Fractional Differential equation.

1.6.1 Abel's Fractional Integral Equation of Tautochrone

The Abel's problem is to find a curve where the time of decent is same irrespective of the position of release of ball in frictionless system. S is the arc length measured along curve C from point O (the origin) to an arbitrary point Q on C (Figure 1.8). The gain in Kinetic Energy while the ball decends is loss in the potential energy and is given by:

$$\frac{1}{2}m\left(\frac{ds}{dt}\right)^2 = mg[y-\eta]$$

$$ds = -\sqrt{2g(y-\eta)}dt$$

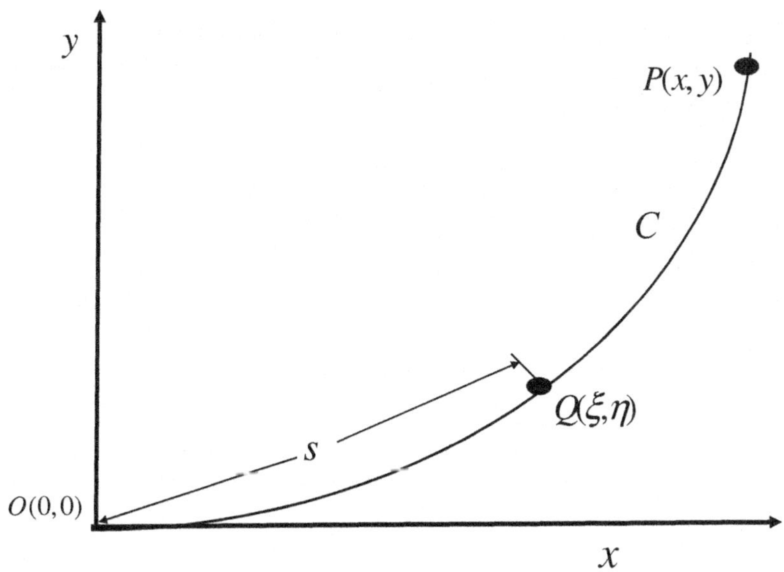

Fig. 1.8 Abel's tautochrone curve

The negative square root indicates that the distance (arc-length) decreases as the time increases.

The equation thus to be solved is

$$dt = -\frac{1}{\sqrt{2g(y-\eta)}}ds$$

The time of decent from 'P' to 'O' is T , which is constant and is

$$T = -\frac{1}{\sqrt{2g}}\int_{P}^{O}\frac{1}{\sqrt{y-\eta}}ds$$

Now taking, the arc length as a function $s = h(\eta)$, where h depends on the shape of 'C'. We can write in differential form the curve as $ds = h^{/}(\eta)d\eta$, so this substitution gives:

$$T = -\frac{1}{\sqrt{2g}}\int_{y}^{0}(y-\eta)^{-1/2}[h^{/}(\eta)d\eta]$$

Re-writing above expression

$$(\sqrt{2g})T = \int_{0}^{y}(y-\eta)^{-1/2}h^{/}(\eta)d\eta\,.$$

Meaning that the integral of RHS of above, when a constant will be solution to get constant time of decent (T). In the above expression writing, with k as constant the integral of RHS is:

$$k \equiv \int_{0}^{x}(x-t)^{-1/2}f(t)dt\,.$$

This is Abel's integral equation.

With

$$\frac{d^{1/2}}{dx^{1/2}}k = \sqrt{\pi}f(x)\,.$$

Thus when half derivative of a constant is computed the function $f(x)$, is obtained.

Solution to

$$(\sqrt{2g})T = \int_{0}^{y}(y-\eta)^{-1/2}h^{/}(\eta)d\eta\,,$$

can be obtained when $h'(\eta) = {}^{ds}\!/_{d\eta}$, and letting $f(y) \equiv h'(y)$, then dividing both sides of expression by Gamma half and applying Riemann-Liouvelli definition of fractional integration, that is; as follows:

$$D^{-\frac{1}{2}} f(x) = \frac{1}{\Gamma(\frac{1}{2})} \int_0^x (x-t)^{-\frac{1}{2}} f(t) dt$$

$$\sqrt{\pi} D^{-\frac{1}{2}} f(x) = \int_0^x (x-t)^{-\frac{1}{2}} f(t) dt$$

and thereby inverting the operator we get:

$$\frac{\sqrt{2g}}{\Gamma\left(\frac{1}{2}\right)} T = D^{-\frac{1}{2}} f(y)$$

$$D^{\frac{1}{2}} \sqrt{\frac{2g}{\pi}} T = f(y)$$

Put half derivative of constant as:

$$D^{\frac{1}{2}} T = {}^{T}\!/_{\sqrt{\pi y}},$$

to get:

$$f(y) = \frac{\sqrt{2g}}{\pi} \frac{T}{\sqrt{y}}.$$

Doing algebraic manipulation the following is obtained.

$$f(y) \equiv h'(y) = \frac{ds}{dy} = \frac{\sqrt{dx^2 + dy^2}}{dy} = \sqrt{1 + \left(\frac{dx}{dy}\right)^2}$$

$$\frac{dx}{dy} = \sqrt{[f(y)]^2 - 1}$$

$$x = \int_0^y \sqrt{\frac{2gT^2}{\pi^2 \eta} - 1} d\eta + c$$

Solution continuation with

$$c = 0, x = 0 = y$$

$$x = \int_0^y \sqrt{\frac{2gT^2}{\pi^2 \eta} - 1} d\eta$$

Letting $a = {gT^2}/{\pi^2}$, and change of variables $\eta = 2a\sin^2\xi$, gives $x = 4a\int_0^{\beta}\cos^2\xi d\xi$, with $\beta = \sin^{-1}\left(\sqrt{\frac{y}{2a}}\right)$. The parametric equation of the cycloid is thus obtained as:

$$x = 2a(\beta + \frac{1}{2}\sin 2\beta)$$
$$y = 2a\sin^2\beta$$
$$\theta = 2\beta, a = \frac{gT^2}{\pi^2}$$

$$x = a(\theta + \sin\theta)$$
$$y = a(1 - \cos\theta)$$

Cycloid is the tautochrone.

A point be mentioned about the cycloid curve shape C is that this is too a curve for brachistochrone problem solved by Bernoulli. That is determination of shape of the curve giving 'minimum' time of descent. Therefore, the constant time T is also minimum time of descent in a cycloid.

1.6.2 Fractional Damped Motion

Consider a ball falling freely under gravity in viscous fluid having constituent equation as:

$$\frac{d}{dt}v(t) + \frac{d^{\alpha}}{dt^{\alpha}}v(t) + v(t) = 1$$

With initial condition as: $v(0) = 0$ and with generalized Laplace Transformation (let us believe $\mathcal{L}\left\{d^{\pm\alpha}f(t)\right\} \equiv s^{\pm\alpha}F(s)$, with zero initial condition) and some algebraic manipulation the following is obtained:

$$V(s) = \frac{1}{s(1+s+s^{\alpha})} = \frac{[1-(-s^{-1}-s^{\alpha-1})]^{-1}}{s^2}$$

Expanding numerator as binomial series, where $[(s^{-1}+s^{\alpha-1})] < 1$ and $\alpha < 1$, for large s; the expansion is:

$$V(s) = \sum_{n=0}^{\infty}(-1)^n\sum_{r=0}^{\infty}\binom{n}{r}\frac{1}{s^{n+2-r\alpha}}$$

Using Laplace Inverse term and by consolidating term one obtains temporal response:

$$v(t) = \sum_{n=0}^{\infty}(-1)^n \sum_{r=0}^{\infty}\binom{n}{r}\frac{t^{n+1-r\alpha}}{\Gamma(n+2-r\alpha)}$$

Consider another fractional damping system where the inertia plays negligible role given by

$D^{\frac{1}{3}}x(t)+x(t)=f(t)$, with $x(0)=0$ and $f(t)=H(t)$ Heaviside's step input.

The Laplace transforms gives

$$X(s)=\frac{1}{s(1+s^{1/3})}=\frac{[1-(-s^{-1/3})]^{-1}}{s^{4/3}}$$

With $|s|<<1$, the above is expanded as series, giving:

$X(s)=\sum_{n=4}^{\infty}\frac{(-1)^n}{s^{n/3}}$, and using the Laplace pair $t^n \leftrightarrow \frac{n!}{s^{n+1}}$, the temporal expression for above fractional dynamics equation's solution is:

$$x(t)=\sum_{n=4}^{\infty}(-1)^n\frac{t^{(n/3)-1}}{\Gamma(n/3)}$$

1.6.3 Formal Definition of Fractional Differential and Fractional Integral Equation

Consider Ordinary linear differential equation (ODE), homogeneous one, with constant coefficients $D^2y(t)+aDy(t)+by(t)=0$. Then if α and β are distinct zero's (roots) of indicial polynomial $P(x)=x^2+ax+b$, then we know that $\exp(\alpha t)$ and $\exp(\beta t)$ are linearly independent solutions of the ODE and if $\alpha=\beta$ then $\exp(\alpha t)$ and $t\exp(\alpha t)$ are the linearly independent solutions of ODE.

As an attempt to define a fractional differential equation (FDE), let $r_m, r_{m-1},....r_1, r_0$ be strictly, decreasing sequence of non-negative numbers. Then for $a_1, a_2,....a_m$ constants

$\left[D^{r_m}+a_1D^{r_{m-1}}+....+a_mD^{r_0}\right]y(t)=x(t)$ is candidate to represent general fractional differential equations. This equation too is complex looking, so we try to simplify, by addition of imposed condition that r_j be rational numbers. Thus if q is LCM of the denominators of the nonzero r_j's, then we define formally the fractional differential equation as:

$$\left[D^{nv}+a_1D^{(n-1)v}+....+a_nD^0\right]y(t)=x(t)\,,\ t\geq 0 \text{ where } v=(q)^{-1}$$

The order of this fractional differential equation is of (n,q), and is of constant coefficients. If $q=1$, then $v=1$ and the equation is simply ordinary differential equation of order n with constant coefficients. The above definition is for Non-Homogeneous, fractional differential equation with constant coefficients. With $x(t)=0$, vanishing source term the equation is Homogeneous Fractional Differential Equation with constant coefficients, first this has to be solved for obtaining Green's function, and then for source term standard convolution of this Green's function, with source function will yield solution.

For convenience introduce the indicial polynomial

$$P(x)=x^n+a_1x^{n-1}+...+a_n\,,$$

then $P\left(D^v\right)$ is fractional differential operator as

$$P\left(D^v\right)\equiv D^{nv}+a_1D^{(n-1)v}+...+a_nD^0\,.$$

With this, the FDE is compactly described as:

$$P\left(D^v\right)y(t)=x(t)\,.$$

Consider a simple FDE of order $(4,3)$ as $D^{4/3}y(t)=0$. Then if C_1 and C_2 are arbitrary constants $y(t)=C_1t^{1/3}+C_2t^{-2/3}$ is a solution. The first solution is $y_1(t)=C_1t^{1/3}$. Using this in the FDE the following is obtained:

$$D^{4/3}y_1(t)=C_1\frac{\Gamma\left(\frac{1}{3}+1\right)t^{\frac{1}{3}-\frac{4}{3}}}{\Gamma\left(\frac{1}{3}-\frac{4}{3}+1\right)}=C_1\frac{\Gamma\left(\frac{1}{3}\right)}{\Gamma(0)}t^{-1}=0$$

This satisfies as one solution.

Derivative of first solution shall be second that is following from ODE.

$$Dy_1(t)=Dt^{\frac{1}{3}}=C_2t^{-\frac{2}{3}}\,,$$

is the solution. The combination of these two gives solution for FDE as mentioned above.

One may draw parallel to ODE. Consider $D^1 y(t) = 0$ it has one solution, namely $y(t) = C$, a constant. Take $D^2 y(t) = 0$ it has two solutions, namely $y_1(t) = C_1 t$ and $y_2(t) = C_2$. Here $Dy_2(t) = y_1(t)$ and the composite solution is $y(t) = C_1 t + C_2$. Similarly, $D^3 y(t) = 0$ has three solutions, and the combined solution is $y(t) = C_1 t^2 + C_2 t + C_3$. General n order ODE has n linearly independent solutions. The FED $D^{1.333} y(t) = 0$ has two solutions and not 1.333.. solutions. Here $nv = 4/3 = 1.333..$ must logically have integer number of solutions $N \geq nv$, in this FDE is two.

Let m and q be positive integers and let $v = (q)^{-1}$. Then the Fractional Integral Equation is defined as:

$$\left[D^0 + a_1 D^{-v} + \dots + a_m D^{mv}\right] y(t) = x(t)$$

1.7 Fractional Calculus and Law of Irreversibility Non-locality

A large number of problems of physics and engineering, including Schrödinger's, Maxwell's, Newton's, Fourier's Heat Flow, Brownian Motion, Bio-Physical Systems etc. can be formulated as initial value problem for dynamical evolution equation as:

$$\frac{d}{dt} f(t) = \mathrm{B} f(t)$$

where $t \in \mathbb{R}$ denotes time and B is an operator on Banach space. This classical equation of evolution represents basic symmetry, principle of locality and irreversibility. This state equation need not be of first order in time, and can be of wave equation of $g(x,t)$ as:

$$\frac{\partial^2 g}{\partial t^2} = c^2 \frac{\partial^2 g}{\partial x^2}$$

It can be casted into the first order by taking

$$f = \begin{bmatrix} g \\ h \end{bmatrix}; \mathrm{B} = \begin{bmatrix} 0 & 1 \\ 1 & 0 \end{bmatrix} c^2 \frac{\partial}{\partial x}$$

For example consider distributed loss less transmission line with repeated per unit series inductance and shunt capacitances. Then $g(x,t) = v(x,t)$ is per unit cell voltage difference and $h(x,t) = i(x,t)$ is the current, and $c^2 = (LC)^{-1}$, gives, wave equation $\partial_{tt}^2 v = (LC)^{-1} \partial_{xx}^2 v$. Depending on the initial data $f(0) = f_0$

describing the state or observables of the system at initial start time, the problem is to find the state or observables $f(t)$ of the system at later times. The generalization of the above evolution equation to fractional differential equation is:

$$\frac{d^{\alpha}}{dt^{\alpha}} f(t) = \mathrm{B} f(t)$$

where the order of derivative is of fractional order or complex order raises several fundamental questions regarding non-local behavior and symmetry and conservations. Let us postulate that all time evolutions of physical systems are irreversible. Obviously law of irreversibility must be considered to be an empirical law of nature equal in rank with the law of energy conservation. Reversible behavior is idealization; its validity or applicability in physical systems depends on the degree to which the system can be isolated from past history and its environment. The order of the fractional derivative $0 < \alpha < 1$ classifies and quantifies the influence of past history. Note that Fractional Derivative definitions have inbuilt integration, meaning that the operation is non-local, requires information about past states. For unity value of the fractional order (or integer order of one) the influence of past history is minimal, in the sense that, then the process (evolution) enters only through the present state. Small value of α corresponds to a strong influence of past history. This is too embedded in definitions of fractional differ-integrals by Grunwald-Letnikov (GL), the lesser the order the more are the weights as compared to unity order of differentiation effect of past history and heredity comes in while fractionally differentiating the function!

1.8 Stable Random Variables and Generalization of Normal Probability Density Function

De Movie's central limit theorem (CLT) for finite variance, independent identically distributed (iid) random variables X_i, with μ, σ as constants, states that

$$\frac{X_1 + X_2 + \ldots + X_n - n\mu}{\sigma n^{1/2}}$$

converges to Z, as standard normal random variable, can be generalized by removing the assumption of finite variance

$$\frac{X_1 + X_2 + \ldots + X_n - n\mu}{\sigma n^{1/\alpha}},$$

converges to Y as a standard α-stable variable that has infinite variance. The CLT was generalized by Levy (1937). Power-law distributions are observed in various fields of science, engineering, sociology, and economics and these are connected to fractional calculus. Densities of α stable variables are generalization of the normal probability density function, the 'characteristic function' of which is defined as,

$$\psi(k) = \exp\left\{-|k|^{\alpha}\sigma^{\alpha}\left[1 - \mathrm{i}\beta\,\mathrm{sgn}(k).\tan(\pi\alpha/2)\right] + \mathrm{i}k\mu\right\}, 1 < \alpha \le 2.$$

The parameters σ, β and μ describe the spread, the skewness and the location of the density, respectively. The skewness can be viewed from picture of random walks, as this factor denotes the probability of particles moving either ahead or behind the mean. If the forward moving probability and the backward moving probabilities are same then there is no skewness. This expression of $\psi(k)$ is characteristic function (that is the Fourier transform with k the Fourier (normal) coordinates changed to $-k$), for the stable density. The $\mathrm{sgn}(k)$ function is -1 for $k < 0$ and $+1$ for $k > 0$. The characteristic function for $\alpha = 1$ the Cauchy distribution is slightly different from the above expression.

When the density is symmetric, the skewness parameter $\beta = 0$, the symmetric characteristic function is $\psi(k) = \exp\left(-\sigma^{\alpha}|k|^{\alpha} + \mathrm{i}k\mu\right)$. A standard α-stable density function has unit "spread" and is centered on the origin, so $\sigma = 1$ and $\mu = 0$ is $s_{\alpha}(x) = \exp(-|k|^{\alpha})$. A nonstandard α-stable density $f_{\alpha}(x)$ is related to standard $s_{\alpha}(x)$ by $f_{\alpha}(x) = \sigma^{-1}s_{\alpha}\left[(x-\mu)/\sigma\right]$. A standard symmetric α stable distribution is characterized by $\psi(k) = \exp(-|k|^{\alpha})$. In this form it is easy to see that the Gaussian (normal) density is α-stable with $\alpha = 2$. Note, however, that when the scale factor of the stable law, $\sigma = 1$ the standard deviation of the normal ($\alpha = 2$) distribution ($\mathcal{N}$) is $\sqrt{2}$.

$$\mathcal{N}(k) = \exp(-2\sigma^2 k^2 + \mathrm{i}k\mu) \text{ is normal distribution.}$$

The most important feature of α-stable distribution is the characteristic exponent, also called index of stability α. The value of α determines how "non-Gaussian" a particular density becomes as the value of this index of stability decreases from a maximum of 2, more probability density shifts towards tail; that is the density becomes a long-tailed distribution or LRD (long range dependence).

A simple test of infinite (diverging) variance is to plot running sample variance estimate S_n with respect to number of points n, where

$$S_n^2 = \sum_{k=1}^{n}(x_k - \bar{x}_n)^2/(n-1)$$

and the average is

$$\overline{x}_n = \sum_{k=1}^{n} x_k / n .$$

If for large n, the value S_n does not converge (rather diverges) then the process is with infinite variance and a non-Gaussian one. In fact in non-Gaussian stable distribution with characteristic exponent α, only moments of order less than α are finite. Therefore, variance can no longer be used as a measure of dispersion in many standard signal processing techniques such as spectral analysis with least square based methods may give misleading results!

1.9 Conservation of Probability

The statement about the conservation of probability is that a particle will occupy a specific location that is at any particular time the sum of probabilities at all locations must be equal to unity. So if the probability changes in one location from one time moment to the next time moment, the probability must also change in vicinity to conserve the probability. An ensemble of particles that is very large numbers can fulfill the probabilities and this statement becomes conservation of mass. Fokker-Plank-Equation (FPE) is the outcome of this statement of conservation of probabilities. Derivation of FPE starts with simple mathematical statement of how a random measure changes state from one moment to the next moment after some event has occurred. In this case we are interested in the (probability) that a particle has moved from location x_0 to x_2 in time t_0 to t_2 or $p(x_2 - x_0; t_2 - t_0)$. This probability will be referred to as transitional density, which is conditional on the initial position x_0 and the time interval. The particle must move through an intermediate location x_1; so the probability can be found by summing the over all possible intermediate points x_1. If the process is defined by Markovian process, in which the movement of a particle is independent of past movements; then the probability of making both transitions (x_0 to x_1 to x_2) is the product of the single transition probabilities, giving the Chapman-Kolmogorov Equation.

$$p(x_2 - x_0; t_2 - t_0) = \int p(x_2 - x_1; t_2 - t_1) p(x_1 - x_0; t_1 - t_0) dx_1$$

The relationship between transition density and the particle position density (also called propagator or plume) is that the particle density has moved from (and must incorporate) the initial condition (position) and is defined as x_0 at time $t_0 = 0$. Placing this in above expression we have

$$p(x - x_0; t) = \int p(x - x_1; t - t_1) p(x_1 - x_0; t_1) dx_1 .$$

Similar to equation for conservation of mass, the FPE is a statement of conservation of probability of single particle's location over a brief period of time. Using the short hand notation for propagator

$$p(x - x_0;t) \equiv P(x,t)\,,$$

that is transition from initial condition to the present time and replacing x_1 with ζ and signifying the interval $(t - t_1)$ by Δt gives a compact form of above obtained expression as,

$$P(x,t+\Delta t) = \int p(x-\zeta;\Delta t)P(\zeta,t)d\zeta\,.$$

To make the next step of derivation, a requirement on the relative size of total time t versus the transition time Δt is needed. For a general transition density, Δt must be much smaller than the total time, and Δt is called a 'mixing' time. Then an appropriate limit theorem such as central limit theorem (CLT) can be invoked, implying that a large number of transitions are integrated and the limit distribution is approached. The assumption of Markovian process requires that all time dependence is contained in this mixing time so that the convolution still holds. If the transitions are independent identically distributed (iid) stable variables, the standard normal for example; then no restrictions are placed on the transition time and the convolution is always satisfied. For example, iid Gaussian transitions results immediately in a Gaussian propagator.

By taking $\Delta t \approx 0$, we will know change in $P(x,t)$ over a very short time period, resulting in a differential equation. The time derivative of p is defined as:

$$\frac{\partial p(x-x_0;t)}{\partial t} = \lim_{\Delta t \to 0} \frac{1}{\Delta t}\left[p(x-x_0;t+\Delta t) - p(x-x_0;t)\right]$$

The limits and moments of the particle transition probability must be correctly identified. One should expect that a particle that travels along a fractal path and/or requires power law times to complete individual walks will have different limiting behavior than a typical Gaussian or finite variance process.

The density $p(x-x_0;t+\Delta t)$ can be replaced by Markov relation, that is

$$p(x-x_0;t+\Delta t) = \int p(x-\zeta,\Delta t)p(\zeta - x_0;t-t_0)d\zeta\,.$$

Using this in the above expression and recalling that $p(x-x_0;t) \equiv P(x,t)$, gives the differential probability change:

$$\frac{\partial P(x,t)}{\partial t} = \lim_{\Delta t \to 0} \frac{1}{\Delta t}\left(\int p(x-\zeta;\Delta t)P(\zeta,t)d\zeta - P(x,t)\right)$$

The instantaneous transition density has the limit

$$\lim_{\Delta t \to 0} p(x-\zeta;\Delta t) = \delta(x-\zeta)\,,$$

where $\delta(x-\zeta)$ is the Dirac delta function, which means that as the transition time tends to zero, the probability that the particle does not move goes to unity. Equivalently, the Fourier transform of the density has a limit of unity, that is

$$\lim_{\Delta t\to 0} \Im\{p(x-\zeta;\Delta t)\} = \lim_{\Delta t\to 0} \hat{p}(k;\Delta t) = 1 .$$

This density has all the positive moments equal to zero. When $\Delta t \neq 0$, the density has higher order moments. The first moment of this instantaneous transition density is defined by the expected value of the particle's new position minus the initial position; let it be called as $A(\Delta t)$ a function of Δt, that is

$$A(\Delta t) = \int (x-\zeta) p(x-\zeta;\Delta t) dx .$$

The second moment of many power law functions and most of the Levy's α-stable instantaneous transition densities is infinite (diverging); therefore we choose a coefficient(again a function of Δt) as $B(\Delta t)$ that is measure of the spread of density similar to the second moment of a Gaussian distribution. A very general transition density with finite or infinite variance has Fourier transform,

$$\hat{p}(k;t) = 1 - A(\Delta t)(ik) + \frac{1}{2} B(\Delta t)\left[(1+\beta)(ik)^{\alpha} + (1-\beta)(-ik)^{\alpha}\right] + O(\Delta t)$$

where $-1 \le \beta \le 1$ indicates the relative weight of forward versus backward transition probability and $1 < \alpha \le 2$ is the scaling exponent in 1-Dimensional space. The final term $O(\Delta t)$ indicates higher-order terms that diminish to zero faster than Δt. Since $\exp(-x) = 1 - x + x^2/2 + \ldots$, this density contains the expansion for the α-stables as subset. A finite variance density requires that $\alpha = 2$, so the above Fourier transformed density reduces to the transition density used to derive classical FPE, and that is:

$$\hat{p}(k;\Delta t) = 1 - A(\Delta t)(ik) + B(\Delta t)(ik)^2 + O(\Delta t) ,$$

where $2B(\Delta t)$ is equal to the second moment of particle excursion distance when $\alpha = 2$, related to mean squared displacement (MSD). This special case instantaneous transition density has an inverse transform that contains derivatives of the Dirac delta function:

$$p(x-\zeta;\Delta t) = \delta(x-\zeta) - A(\Delta t)\delta'(x-\zeta) + B(\Delta t)\delta''(x-\zeta) + O(\Delta t)$$

This transition density when substituted into

$$\frac{\partial P(x,t)}{\partial t} = \lim_{\Delta t\to 0} \frac{1}{\Delta t}\left(\int p(x-\zeta;\Delta t) P(\zeta,t) d\zeta - P(x,t) \right),$$

leads to classical FPE for Brownian motion with advective drift. The derivation is similar to that done for the generalized case below. Power law and α stable transitions with $\alpha < 2$ have a real space density for small time (that is inverse of General Fourier expansion), as:

$$p(x-\zeta;\Delta t) = \delta(x-\zeta) - A(\Delta t)\delta'(x-\zeta) + \frac{1}{2}(1+\beta)B(\Delta t)D_{+}^{\alpha}\delta(x-\zeta) \\ +\frac{1}{2}(1-\beta)B(\Delta t)D_{-}^{\alpha}\delta(x-\zeta) + \mathrm{O}(\Delta t)$$

where $D_{\pm}^{\alpha}$ are the α th order fractional derivative. The terms of these fractional derivatives can be evaluated for the delta function as proportional to $(|x-\zeta|)^{-1-\alpha}$, which shows the power function density of the instantaneous transition approximations. For symmetric jumps $\beta = 0$. The coefficients $A(\Delta t)$ and $B(\Delta t)$ vanish as Δt goes to zero, as required by

$$\lim_{\Delta t \to 0} p(x-\zeta;\Delta t) = \delta(x-\zeta).$$

We assume that the rate of change of the drift and scale coefficients for small time are constants, so $A(\Delta t)$ and $B(\Delta t)$ are linear with Δt for small transition times, leading to

$$v \equiv \lim_{\Delta t \to 0}\left(A(\Delta t)/\Delta t\right) \text{ and } \mathbb{D} \equiv \lim_{\Delta t \to 0}\left(B(\Delta t)/\Delta t\right).$$

Nonlinear power function scaling of $A(\Delta t)$ and $B(\Delta t)$ with Δt can be used to model non-Markovian, long-term temporal correlations (LRD) yielding fractional derivative of time; this will be elaborated in Chapter-4. For simplicity we assume that the time correlation is thin tailed and relative to spatial correlation, so that v and $\mathbb{D}$ as represented above are good approximates. Placing the expansion of $p(x-\zeta;\Delta t)$, expressed with fractional derivative of delta function in the expression of $\partial P(x,t)/\partial t$ we get:

$$\frac{\partial P(x,t)}{\partial t} = \lim_{\Delta t \to 0}\frac{1}{\Delta t}\left(\int p(x-\zeta;\Delta t)P(\zeta,t)d\zeta - P(x,t)\right)$$

$$= \lim_{\Delta t \to 0}\frac{1}{\Delta t}\left(\begin{array}{l}\displaystyle\int\left\{\begin{array}{l}\delta(x-\zeta) - A(\Delta t)\delta'(x-\zeta) + \frac{1}{2}(1+\beta)B(\Delta t)D_{+}^{\alpha}\delta(x-\zeta)\\ +\frac{1}{2}(1-\beta)B(\Delta t)D_{-}^{\alpha}\delta(x-\zeta)\end{array}\right\}P(\zeta,t)d\zeta\\ -P(x,t)\end{array}\right)$$

Recognizing that

$$\int \delta(x-\zeta)P(\zeta,t)d\zeta$$

is convolution of delta function with $P(x,t)$ giving $P(x,t)$. Then, using, Fourier Identity

$$\left[D_{\pm}^{q}\delta(x)\right]\otimes f(x)=\delta(x)\otimes\left[D_{\pm}^{q}f(x)\right]=D_{\pm}^{q}f(x),$$

and invoking the limits we get:

$$\lim_{\Delta t\to 0}\frac{1}{\Delta t}\int A(\Delta t)\delta'(x-\zeta)P(\zeta,t)d\zeta=\lim_{\Delta t\to 0}\frac{\partial}{\partial x}\big(A(\Delta t)P(x,t)\big)=\frac{\partial}{\partial x}vP(x,t)$$

and

$$\lim_{\Delta t\to 0}\frac{1}{\Delta t}\int B(\Delta t)D_{\pm}^{\alpha}\delta(x-\zeta)P(\zeta,t)d\zeta=\lim_{\Delta t\to 0}\frac{1}{\Delta t}\frac{\partial^{\alpha}}{\partial(\pm x)^{\alpha}}\left[B(\Delta t)P(x,t)\right]=\frac{\partial^{\alpha}\mathbb{D}P(x,t)}{\partial(\pm x)^{\alpha}},$$

substituting these we obtain the Fractional Fokker Plank Equation, a generalized conservation of probability.

$$\frac{\partial P}{\partial t}=-\frac{\partial}{\partial x}vP+\frac{1}{2}(1+\beta)\frac{\partial^{\alpha}}{\partial x^{\alpha}}\mathbb{D}P+\frac{1}{2}(1-\beta)\frac{\partial^{\alpha}}{\partial(-x)^{\alpha}}\mathbb{D}P$$

The function v is the drift of the process, i.e. the mean advective velocity. If the particle transition is modeled by an infinite variance probability then the nonlinear growth of the particle propagator is incorporated by fractional derivative, rather than the leading parameter $\mathbb{D}$.

1.10 Half Order Fractional Differentiation Embedded in Standard Fick's Law and Its Extension to Describe Anamolous Diffusion

In the previous section, we have discussed in generalized Fokker Plank Equation; which in force free or field free situation reduces to Fick's diffusion equation. In the above assume drift component is zero, then we get the classical and generalized diffusion in field free environment. Let us assume that $P(r,t)$ represents density function per unit space r, then the mass is integral of $P(r,t)$ represented as

$$M(r,t) = \int_0^r dr P(r,t)\,.$$

The continuity equation is rate of change of mass and is related to mass current density as,

$$\frac{d}{dt}M(r,t) = -j(r,t)$$

where constitutive equation is

$$j(r,t) = -\mathbb{D}_0 \frac{\partial P(r,t)}{\partial r}\,.$$

From these two equations we obtain the Fick's law of diffusing species as

$$\frac{\partial}{\partial t}P(r,t) = \mathbb{D}_0 \frac{\partial^2}{\partial r^2}P(r,t)$$

This has got standard solution as Gaussian plume at any other r and any other time t , when at initial instant at $t = 0$ a delta function is placed at origin, $P(0,0) = \delta(r,0)$; that is

$$P(r,t) = \frac{1}{\sqrt{\pi \mathbb{D}_0 t}} e^{\left(-\frac{r^2}{4\mathbb{D}_0 t}\right)}$$

Taking Laplace of this Gaussian plume (refer Laplace table of Chapter-2), we get

$$P(r,s) = \mathcal{L}\{P(r,t)\} = \mathcal{L}\left\{\frac{1}{\sqrt{\pi \mathbb{D}_0 t}} e^{\left(-\frac{r^2}{4\mathbb{D}_0 t}\right)}\right\} = \left(\frac{1}{\sqrt{\mathbb{D}_0 s}}\right) e^{\left(-r\sqrt{\frac{s}{\mathbb{D}_0}}\right)}$$

Differentiating the above we obtain

$$\frac{d}{dr}P(r,s) = \frac{-1}{\sqrt{\mathbb{D}_0 s}}\sqrt{\frac{s}{\mathbb{D}_0}}e^{-r\sqrt{s/\mathbb{D}_0}} = -\frac{1}{\mathbb{D}_0}e^{-r\sqrt{s/\mathbb{D}_0}}$$

Using Laplace pair

$$\sqrt{s} \leftrightarrow \mathcal{L}\left\{d^{1/2} / dt^{1/2}\right\},$$

taking Laplace of constitutive equation, that is $j(r,s) = -\mathbb{D}_0 \partial P(r,s) / \partial r$, and substituting above obtained derivative that is $dP(r,s) / dr$, we obtain (after simplification and simple algebra) $j(r,s)=\sqrt{s\mathbb{D}_0}P(r,s)$. The inverse Laplace transform gives, a constitutive equation expressed with half (fractional) derivative as:

$$j(r,t)=\sqrt{\mathbb{D}_0}\frac{\partial^{1/2}P(r,t)}{\partial t^{1/2}}$$

Using this fractional derivative equation for constitutive equation, we can reconstruct easily the fractional derivative counterpart of Fick's law as:

$$\frac{\partial^{1/2}P(r,t)}{\partial t^{1/2}}=-\sqrt{\mathbb{D}_0}\frac{\partial P(r,t)}{\partial r}$$

The half (fractional) derivative is imbibed in this isotropic uniform normal diffusion equation of Fick's law, where the continuous time random walk (CTRW) is Brownian Motion (BM).

1.11 Fractional Brownian Motion

Let us re-write the Gaussian plume (as expressed in previous section) with parameter d_w, as following with FBM as full form of 'Fractional Brownian Motion'.

$$P_{\mathrm{FBM}}(x,t)=\frac{1}{\sqrt{4\pi\mathbb{D}_0t^{2/d_w}}}e^{\left(-\frac{x^2}{4\mathbb{D}_0t^{2/d_w}}\right)}$$

Then the corresponding scaling of mean square displacement (MSD) is

$$\langle x^2(t)\rangle\equiv 2\mathbb{D}_0t^{2/d_w},$$

With $2\le d_w<\infty$. When $d_w=2$, we get the Gaussian plume back with MSD scaling linearly with time, and we return to normal Fick's law corresponding to normal brownian motion (BM) as obtained in previous section, that is:

$$\frac{\partial^{1/2}P_{\mathrm{BM}}(x,t)}{\partial t^{1/2}}=-\sqrt{\mathbb{D}_0}\frac{\partial P_{\mathrm{BM}}(x,t)}{\partial|x|}$$

For any type of anomalous transport, (diffusion) with exponent other than two we may write fractional diffusion equation from our above explanation as:

$$\frac{\partial^{1/d_w} P_{\mathrm{FBM}}(x,t)}{\partial t^{1/d_w}} = -\mathrm{A}\frac{\partial P_{\mathrm{FBM}}(x,t)}{\partial |x|}$$

We may thus tempt to relate the brownian motion (BM) and fractional brownian motion (FBM), with integral transform as,

$$x_{\mathrm{FBM}}(t) = \int_{-\infty}^{t} K_M(t-\tau)dx_{\mathrm{BM}}(\tau),$$

with kernel defined as follows:

$$K_M(t-\tau) = \begin{cases} (t-\tau)^{(1/d_w)-(1/2)} - (-\tau)^{(1/d_w)-(1/2)} & ;\tau < 0 \\ (t-\tau)^{(1/d_w)-(1/2)} & ;0 < \tau < t \end{cases}$$

This is Mandelbrot's kernel, relating BM and FBM, similar to singular kernel of Riemann-Liouvelli, fractional integral that is

$$I_t^{\phi} f(t) = d_t^{-\phi} f(t) = \frac{1}{\Gamma(\phi)}\int_0^t (t-\tau)^{\phi-1} f(\tau)d\tau.$$

When the anomalous index $d_w = 2$, then, memory kernel is $K_M(t-\tau) = 1$ and $x_{\mathrm{FBM}}(t) = x_{\mathrm{BM}}(t)$, that is walk without memory. However, this Mandelbrot's kernel is different from memory kernel that we had formalized for fractional calculus previously, the concept of memory however remains the same.

Thus a walker undergoing FBM remembers his past and its motion having memory kernel, while a walker undergoing BM does not remember its past. Well a walker can remember its past and have preferences in same direction giving persistence walk, or a walker remembering its past can change its direction giving anti-persistence walks, which are the cases of anomalous transport. This gives concept of sub- or super-diffusion, in FBM context. The variable x_{FBM} could be computer network delay; could be price-index of stock market; could be random returns of insurance system; could be infected population by swine flu or could be stochastic noise. In general this could be variable for physical systems represented by random stochastic process.

Dynamics of financial assets demonstrate stochastic behavior; the first attempt to describe stochastically financial dynamics was made by L. Bachelier in 1900. He proposed the Brownian motion to model the stochastic process of returns x_{BM} as $G(t) \equiv G_{\Delta t}(t)$ over time scales Δt as the forward change in the logarithmic of price or market index $S(t)$, as $G_{\Delta t}(t) = \ln S(t+\Delta t) - \ln S(t)$. Thus $G_{\Delta t}(t) \equiv x_{\mathrm{BM}}$ as Bachelier's approach is natural if one considers the return over time scale Δt to be result of many independent shocks which then lead by central limit theorem

(CLT) to a Gaussian distribution of returns. Well, with spiky nature of the returns where the fluctuations are too spiky will be different than Gaussian, with pronounced tail and long range dependent (LRD), and in that case $G_{\Delta t}(t) \equiv x_{\mathrm{FBM}}$ will be a Fractional Brownian Process. These cases will be taken up subsequently.

The MSD, in case of Brownian motion scales as proportional to time, that is $\left\langle {x_{\mathrm{BM}}}^2 \right\rangle \cong t$. The fractional brownian case was stated as anomalous case, when MSD is non linear with time; can be expressed with Hurst exponent, H as

$$\left\langle {x_{\mathrm{FBM}}}^2 \right\rangle \cong t^{2H} = t^{(2/d_w)}.$$

With $H = 0.5$, giving Brownian case, with $0 < H < 0.5$ indicating anti-persistent walk (slower diffusion than Brownian case), $0.5 < H < 1$ indicating persistent walk (faster diffusion than Brownian case). Persistent walk is case when walker remembers the step, and next step being in the same direction is more probable, in case of anti-persistent case the step is remembered and next step being in reverse direction is more likely. The Brownian case the walk is 'without memory'. These concepts will be used in subsequent chapters and will be elaborated.

1.12 A Thought Experiment

From aircraft, we see the city roads and observe the vehicular traffic movement. The vehicle seems to move in one straight line. Therefore as an observer we draw the velocity curve by simple one order integer derivative of displacement and find that maps straight lines. In Figure 1.9 the pair of straight lines gives the velocity trajectory of the up stream vehicle and down stream vehicle, as observed in macroscopic scale.

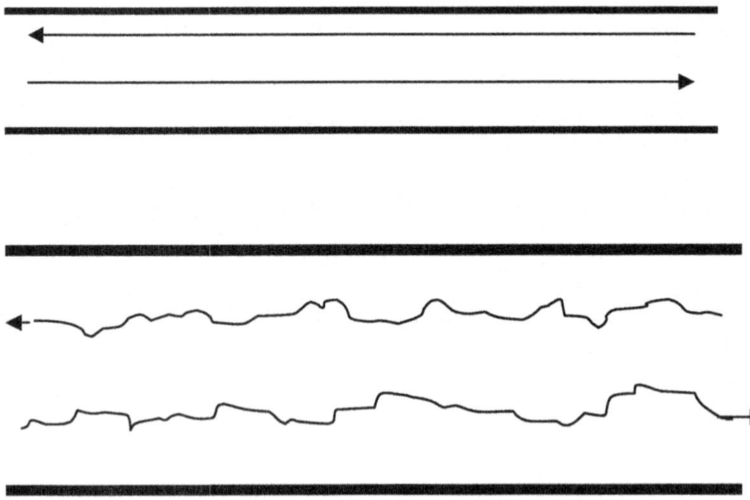

Fig. 1.9 Macroscopic and Microscopic view of moving vehicles on road

The same vehicle when looked with enlarged view tells us its continuous movement but to avoid road heterogeneity travels in zigzag fashion. The curve in lower frame of the Figure 1.9 maps this picture. Here scale is enlarged. The velocities for up stream and down stream vehicles are not pair of straight lines, but follow a continuous but nowhere differentiable curve. So will the dx/dt give the true picture of velocity or will it be ($d^{1+\alpha}x/dt^{1+\alpha}$), where $0<\alpha<1$, give the representation of the actual zigzag pattern, is the thought experiment. Now the question about the dimensions of velocity, in the thought experiment when defined as fractional derivative of displacement, is matter of another thought. In the present understanding as per uniform time scales the quantity dx/dt is velocity and d^2x/dt^2 is the acceleration, however the quantification of $d^{1.23}x/dt^{1.23}$ is hard to visualize. This fractional differentiation is in between velocity and acceleration perhaps a velocity in some transformed time scale, which is non-uniform-enriching thought for physical understanding of fractional quantities. The nature of zigzag pattern shown is somewhat called fractal curve, actually a continuous and nowhere differentiable function. Relation of fractal dimensions and the fractional calculus is evolving field of science at present. The macroscopic view presented above gives a thought of explanation of discontinuity and singularity formations in nature, in classical integer order calculus. Can fractional calculus be an aid for explanation of discontinuity formation and singularity formation- is enriching thought experiment.

In this chapter definition of fractional derivative and fractional integration is given as brief outline. Note may be taken as fractional differentiation embeds the fractional integration (in both the cases of RL as well as Caputo definitions). Looking at the formulation of fractional integration one sees it as convolution of power function $\Phi_\alpha(t)$ with the function $f(t)$ giving fractional integration ${}_0I_t^{\alpha}f(t)$. The Figure 1.3 here demonstrates the nature of power function $\Phi_\alpha(t)$. Consider for a thought experiment that $f(t)$ is velocity of moving vehicle and $\Phi_\alpha(t)$ gives 'time-interval' flowing with time. Now in the limit $\alpha \to 1$, the Figure 1.3 shows that $\Phi_\alpha(t)=H(t)$, meaning that with the flow of time the 'time-interval' is always unity. This condition may also be viewed as 'time-interval' flow is uniform and the 'speed of time-interval' is zero! That is at every instant the 'time-interval is unchanged'. In this scenario what we measure the distance as

$$ {}_0I_t^1 f(t) = {}_0D_t^{-1} f(t) = \int_0^t f(t)dt $$

which is normal way of estimating distance. This system has 'homogeneous flow of time'. The above argument gives us a thought that there can be condition when $\alpha<1$ giving rise to a convolution of velocity with power function as fractional integration of velocity as displacement! Well, in this particular case, then, the 'time-interval' is not unity always and well, 'time-interval' has a variable velocity or we can say 'time-flow' is non-uniform! Here the distance traveled will be

${}_0I_t^{\alpha} f(t) = {}_0D_t^{-\alpha} f(t)$, in terms of fractional integral or we calculate velocity by fractional differentiation of distance traveled. Can 'time-interval' have dynamic speed? Perhaps in system where heterogeneity exists the movement of particles appears to have 'non-homogeneous' time flow!

1.13 Quotable Quotes about Fractional Calculus

Expressed differently we may say that nature works with fractional derivatives.

We may express our concepts in Newtonian terms if we find convenient but if we do so, we must realize that we have made translation into a language, which is foreign to the system we are studying.

All systems need a fractional time derivative in the equation describing them. System have memory of all earlier events is thus necessary to include this records of earlier events to predict the future. Conclusion is obvious and unavoidable: 'Dead matter has memory'.

Fractional calculus is the calculus of XXI century.

1.14 Concluding Comments

This field of science is evolving and particularly as per author's intuition, this calculus will be the language of XXI century for physical system description and controls. In this chapter observation points towards evolving nature of the science of fractional calculus definitions, though born three hundred years ago. The 'ifs and buts' related to fractional calculus as on today, is due to our own limitation of understanding. This will have clearer picture tomorrow when products based on this subject will be used at industry. This introduction chapter has given the thought that there is wonderful universe of mathematics staying within the boundary of 'one complete differentiation' and 'one complete integration'. The science maturity will absorb the richness in this fractional calculus may be in coming years of XXI century. The chapter gives idea that this fractional calculus is as rigorous as its counterpart classical integer order differentiation and integration, with subject's richness for scientific research for future. Also highlights the general laws of nature where integer order differential equation descriptions are idealism, and in real cases anomalous nature of physical behavior are a generalization with fractional differentiation or fractional integration.

Chapter 2
Functions Used in Fractional Calculus

2.1 Introduction

This chapter presents a number of functions that have been found useful in the solution of the problems of fractional calculus. The base function is the Gamma function, which generalizes the factorial expression, used in multiple differentiation and repeated integrations, in integer order calculus. The Mittag-Leffler function is the basis function of fractional calculus, as exponential function is to integer order calculus. Several modifications of the Mittag-Leffler functions, along with other variants are introduced which are developed since 1903, for study of the fractional calculus. These functions are called Higher Transcendental Functions and its use in solving Fractional Differential Equations is as similar to use of transcendental functions for solving Integer Order Differential Equations. Use of these functions is demonstrated for solving Fractional differential equations with Laplace Transform Technique. Here, some interesting physical interpretation is given, for memory integrals for relaxation laws for generalized system dynamics (with memory); along with basic definition and physical interpretation of rough functions, and its fractal dimension. Several examples are solved to get fractional integration and fractional differentiation of standard function and use of introduced higher transcendental functions is demonstrated especially for solving Fractional Differential Equations.

2.2 Functions for the Fractional Calculus

2.2.1 Gamma Function

One of the basic functions of the fractional calculus is Euler's gamma function. This function generalizes the factorial $n!$, and allows n, to take non-integer values.

2.2.1.1 Representation of Gamma Function

Definition of the Gamma function in integral form is: $\Gamma(z)=\int_0^{\infty} e^{-t}t^{z-1}dt$, Which converges in the right half of the complex plane $\Re e(z)>0$ considering z to be real number the above statement implies that gamma function is defined continuously for positive real values of z.

For z, small, between 0 and 1, $(\Gamma(z))^{-1}\approx z$.

2.2.1.2 Basic Properties of Gamma Function

$$\Gamma(z+1)=z\Gamma(z)$$

$$\Gamma(z+1)=\int_0^{\infty} e^{-t}t^{(z+1)-1}dt=\int_0^{\infty} e^{-t}t^{z}dt=\left[-e^{-t}t^{z}\right]_{t=0}^{t=\infty}+z\int_0^{\infty} e^{-t}t^{z-1}dt=z\Gamma(z)$$

The above is obtained by integration by parts. Obviously $\Gamma(1)=1$, and using the above property we obtain values for $z=1,2,3,\ldots$

$$\begin{aligned}&\Gamma(2)=1.\Gamma(1)=1!\\&\Gamma(3)=2.\Gamma(2)=2!\\&\Gamma(4)=3.\Gamma(3)=3!\\&\ldots\ldots\ldots\ldots\ldots\ldots\\&\Gamma(n+1)=n.\Gamma(n)=n.(n-1)!=n!\end{aligned}$$

The above property valid for positive values of z. Another important property of the gamma function is that it has simple poles at $z=0,-1,-2,-3,\ldots$

The proof is explained as splitting the function in two intervals as indicated below:

$$\Gamma(z)=\int_0^{1} e^{-t}t^{z-1}dt+\int_1^{\infty} e^{-t}t^{z-1}dt$$

The first integral can be evaluated by using series expansion for the exponential function. If $\Re e(z)=x>0$, then $\Re e(z+k)=x+n>0$ and thus $t^{z+k}\big|_{t=0}=0$. Therefore,

$$\int_0^{1} e^{-t}t^{z-1}dt=\int_0^{1}\sum_{k=0}^{\infty}\frac{(-t)^k}{k!}t^{z-1}dt=\sum_{k=0}^{\infty}\frac{(-1)^k}{k!}\int_0^{1}t^{k+z-1}dt=\sum_{k=0}^{\infty}\frac{(-1)^k}{k!(k+z)}$$

The second integral may be represented as an "entire-function"

$$\phi(z)=\int_1^\infty e^{-t}t^{z-1}dt$$

$$\Gamma(z)=\sum_{k=0}^{\infty}\frac{(-1)^k}{k!}\frac{1}{k+z}+\phi(z)=\phi(z)+\frac{(-1)^0}{0!}\frac{1}{0+z}+\frac{(-1)^1}{1!}\frac{1}{1+z}+\frac{(-1)^2}{2!}\frac{1}{2+z}+.....$$

Clearly indicating simple poles at 0, -1, -2, -3…this means that at negative integer points the gamma function asymptotically approaches infinity and is discontinuous at those negative integer values.

Besides the obvious relation $\Gamma(z+1)=z\Gamma(z)$, we observe that if n is an integer, then $\Gamma(z+n)\Gamma(-z-n+1)=(-1)^n\Gamma(z)\Gamma(1-z)$, for all n. Using $\alpha!=\Gamma(\alpha+1)$, when α is not a positive integer, then we may generalize the binomial coefficients as:

$$\binom{-z}{u}=\frac{\Gamma(1-z)}{\Gamma(u+1)\Gamma(1-z-u)}$$

In particular if u is nonnegative integer, say n then using the above identity, we have:

$$\binom{-z}{n}=\frac{\Gamma(1-z)}{n!\Gamma(1-z-n)}=(-1)^n\frac{\Gamma(z+n)}{n!\Gamma(z)}=(-1)^n\binom{z+n-1}{n}$$

The other properties of gamma function are:

Duplication formula:

$$\Gamma(2z)=\pi^{-1/2}2^{2z-1}\Gamma(z)\Gamma\left(z+\frac{1}{2}\right)$$

Reflection formula:

$$\Gamma(z)\Gamma(1-z)=\frac{\pi}{\sin\pi z}$$

For $x>0$, the asymptotic representation of gamma function is

$$\Gamma(x+1)\sim x^x e^{-x}\sqrt{2\pi x}$$

The reciprocal gamma function is defined as

$$\frac{1}{\Gamma(z)}=ze^{\gamma z}\prod_{k=1}^{\infty}\left(1+\frac{z}{k}\right)^{-z/k},$$

where $\gamma=0.577215666....$ is Euler's constant.

This reciprocal gamma function is continuous everywhere for real values of z. The value of reciprocal Gamma function is zero at negative integer points and at zero, an important property. Whereas, the gamma function has discontinuities at all negative integer points at zero.

The asymptotic representation is

$$\frac{1}{\Gamma(x)} \sim \frac{x^{\frac{1}{2}-x}}{\sqrt{2\pi}}$$

The gamma function can also be represented as limit as follows:

$$\Gamma(z) = \lim_{n\to\infty} \frac{n!.n^z}{z(z+1)...(z+n)},$$

Here we initially assume right half plane $\Re e(z) > 0$, or in case of real number positive values.

Let us introduce an auxiliary function as below to prove this part,

$$f_n = \int_0^n \left(1-\frac{t}{n}\right)^n t^{z-1} dt .$$

Substitute $\tau = t/n$, and then performing integration by parts we get the following:

$$\begin{aligned} f_n(z) &= n^z \int_0^1 (1-\tau)^n \tau^{z-1} d\tau \\ &= \frac{n^z}{z} n \int_0^1 (1-\tau)^{n-1} \tau^z d\tau \\ &= \frac{n^z n!}{z(z+1)...(z+n-1)} \int_0^1 \tau^{z+n-1} d\tau \\ &= \frac{n^z n!}{z(z+1)...(z+n-1)(z+n)} \end{aligned}$$

Taking into account the well known $\lim_{n\to\infty}\left(1-\frac{t}{n}\right)^n = e^{-t}$, we expect the following:

$$\lim_{n\to\infty} f_n(z) = \lim_{n\to\infty} \int_0^n \left(1-\frac{t}{n}\right)^n t^{z-1} dt = \int_0^\infty e^{-t} t^{z-1} dt = \Gamma(z)$$

The Stirling's approximation for gamma function ratio is given by Stirling's number as:

$$\frac{\Gamma(j-q)}{\Gamma(-q)\Gamma(j+1)} = \frac{(-1)^j}{j!} \sum_{m=0}^{j} S_j^{(m)} q^m \,,\; S_j^{(m)}$$

is Stirling's number of first kind with recurring relation as: $S_{j+1}^{(m)} \equiv S_j^{(m-1)} - jS_j^{(m)}$, $S_0^{(m)} = S_j^{(0)} = 1$, except $S_0^{(0)} = 1$. The approximation, of ratio of gamma function as polynomial in q establishes ratio of gamma function as finite and single value function for all q and j. Two examples of this Stirling's approximations are:

$\frac{\Gamma(x+a)}{\Gamma(x+b)} = x^{a-b}\left[1+O(x^{-1})\right]$ and $\frac{\Gamma(x-q)}{\Gamma(x+1)} = x^{-(1+q)}\left[1+\frac{q(q+1)}{2x}+O(x^{-2})\right]$, for $x \to \infty$.

The incomplete gamma function is defined, in series form as:

$$\gamma^*(v,z) = e^{-z} \sum_{k=0}^{\infty} \frac{z^k}{\Gamma(v+k+1)} .$$

In particular

$$\gamma^*(v,0) = \frac{1}{\Gamma(v+1)}$$

If p is nonnegative integer, the following properties are deduced from series definition as:

$$\gamma^*(p,at) = e^{-at} \sum_{k=p}^{\infty} \frac{(at)^{k-p}}{k!} \quad \text{and} \quad \gamma^*(-p,at) = (at)^p .$$

The special values of this incomplete gamma functions are:

$$\gamma^*(1,at) = \frac{1-e^{-at}}{at} ,\; \gamma^*(0,at) = 1 \text{, and } \gamma^*(-1,at) = at$$

Incomplete gamma function is an entire function of both z and v and its integral representation is:

$$\text{For } \Re e v > 0 ,\; \gamma^*(v,t) \stackrel{\text{def}}{=} \frac{1}{\Gamma(v)t^v} \int_0^t u^{v-1} e^{-u} du \text{ and } \lim_{t\to\infty} t^v \gamma^* = 1 .$$

Some elementary relations that exist among the incomplete gamma functions are:

$$\gamma^*(v-1,t)-t\gamma^*(v,t)=e^{-t}\sum_{k=0}^{\infty}\frac{t^k}{\Gamma(v+k)}-e^{-t}\sum_{k=0}^{\infty}\frac{t^{k+1}}{\Gamma(v+k+1)}=\frac{e^{-t}}{\Gamma(v)},$$

Replacing t, by at, we get

$$\gamma^*(v-1,at)-(at)\gamma^*(v,at)=\frac{e^{-at}}{\Gamma(v)};$$

iterating this $(p-1)$ times we arrive at:

$$\gamma^*(v,at)=(at)^p\gamma^*(v+p,at)+e^{-at}\sum_{k=0}^{\infty}\frac{(at)^k}{\Gamma(v+k+1)}.$$

Using the series definition of incomplete gamma function on the above expression we get:

$$(at)^v e^{at}\gamma^*(v,at)=\sum_{k=0}^{\infty}\frac{(at)^{k+v}}{\Gamma(v+k+1)}$$

and taking p-th derivative, leads to:

$$\frac{d^p}{dt^p}\left[t^v e^{at}\gamma^*(v,at)\right]=t^{v-p}\sum_{k=0}^{\infty}\frac{(at)^k}{\Gamma(v+k+1-p)}=t^{v-p}e^{at}\gamma^*(v-p,at).$$

From this and earlier identity we write:

$$\frac{d^p}{dt^p}\left[t^v e^{at}\gamma^*(v,at)\right]=a^p\left[t^v e^{at}\gamma^*(v,at)\right]+t^{v-p}\sum_{k=0}^{p-1}\frac{(at)^k}{\Gamma(v+k+1-p)}.$$

In this form we see that the incomplete gamma function on the both sides of equation has the same arguments. From this with $p=1$, we find the integral form as:

$$\int_0^t u^v e^{au}\gamma^*(v,au)du=t^{v+1}e^{at}\gamma^*(v+1,at),\ \Re e(v)>-1$$

The incomplete Gamma function is used in obtaining fractional differentiation and fractional integration of periodic functions, used as sinusoidal response studies of fractional operators.

Let us try to find fractional integral of order $v>0$ of exponential function $f(t)=e^{at}$

$$ {}_0D_t^{-v}e^{at} = \frac{1}{\Gamma(v)}\int_0^t (t-u)^{v-1}e^{au}du . $$

Make substitution as $x = t-u$, then integral expression is:

$$ {}_0D_t^{-v}e^{at} = \frac{e^{at}}{\Gamma(v)}\int_0^t x^{v-1}e^{-ax}dx . $$

Clearly this integral is not elementary function, but is closely related to incomplete gamma function and thus fractional integral of exponential function is:

$$ {}_0D_t^{-v}e^{at} = t^v e^{at}\gamma^*(v,at) $$

The derived functions from complete gamma function, is called Beta function defined as

$$ \mathrm{B}(p,q) = \int_0^1 u^{p-1}(1-u)^{q-1}du = \frac{\Gamma(p)\Gamma(q)}{\Gamma(p+q)} , $$

Also there is incomplete Beta function defined as:

$$ \mathrm{B}_\tau(x,y) \overset{\text{def}}{=} \int_0^\tau t^{x-1}(1-t)^{y-1}dt , \; 0<\tau<1 $$

If $\Re e(v) > 0$ and k is nonnegative integer then

$$ \int_0^t u^{v-1}(t-u)^k du = \mathrm{B}(k+1,v)t^{k+v} = \frac{\Gamma(v)k!t^{k+v}}{\Gamma(v+k+1)} . $$

Hence we may write the same as:

$$ \frac{1}{\Gamma(v)k!t^v}\int_0^t u^{v-1}(t-u)^k du = \frac{t^k}{\Gamma(v+k+1)} $$

Using the series definition of incomplete gamma function with integrating by parts, we obtain:

$$ e^t\gamma^*(v,t) = \frac{1}{\Gamma(v)t^v}\left\{\int_0^t u^{v-1}du + \frac{1}{v}\int_0^t u^v\left[\sum_{j=0}^{\infty}\frac{(t-u)^j}{j!}\right]du\right\} = \frac{1}{\Gamma(v)t^v}\int_0^t u^{v-1}e^{(t-u)}du $$

Thus we see that integral representation of incomplete gamma function as described before.

The derivative of complete gamma function a 'psi' function is defined as:

$$\psi(x)=\frac{1}{\Gamma(x)}\frac{d\Gamma(x)}{dx} \text{ or } \psi(x)=D\ln\Gamma(x)=\frac{D\Gamma(x)}{\Gamma(x)}$$

The utility of these two functions is demonstrated for evaluating fractional derivatives of power function $f(x)=x^b$, with $b>-1$ and $f(x)=\log x$.

From Riemann-Liouvelli definition for fractional derivative, with $m-1\le\alpha<m$, where $m\in\mathbb{N}$, we write:

$${}_0D_x^\alpha x^b=\frac{d^m}{dx^m}\left\{\frac{1}{\Gamma(m-\alpha)}\int_0^x(x-u)^{m-\alpha-1}u^b du\right\}=\frac{1}{\Gamma(m-\alpha)}\frac{d^m}{dx^m}\left\{x^{m-\alpha+b}\int_0^1(1-v)^{m-\alpha-1}v^b dv\right\}$$

Using

$$\mathrm{B}(p,q)=\int_0^1 u^{p-1}(1-u)^{q-1}du=\frac{\Gamma(p)\Gamma(q)}{\Gamma(p+q)} \text{ we get:}$$

$$\begin{aligned}{}_0D_x^\alpha x^b&=\frac{1}{\Gamma(m-\alpha)}\frac{\Gamma(b+1)\Gamma(m-\alpha)}{\Gamma(b+1+m-\alpha)}\left(\frac{d^m}{dx^m}\right)x^{m-\alpha+b}\\&=\frac{\Gamma(b+1)}{\Gamma(b+1+m-\alpha)}x^{b-\alpha}\frac{\Gamma(m-\alpha+b+1)}{\Gamma(m-\alpha+b-m+1)}\\&=\frac{\Gamma(b+1)}{\Gamma(b-\alpha+1)}x^{b-\alpha}\end{aligned}$$

For $b=0$, we obtain ${}_0D_x^\alpha\{1\}=[\Gamma(1-\alpha)]^{-1}(x)^{-\alpha}$, for $\alpha>0$ thus fractional differentiation of a constant is 'zero'; only for positive integer values of $\alpha=n\in\mathbb{Z}^+$ as reciprocal of complete gamma function at zero point and negative integer points are zero $[\Gamma(1-n)]^{-1}=0$. When $\alpha\notin\mathbb{Z}^+$, ${}_0D_x^\alpha\{1\}=[\Gamma(1-\alpha)]^{-1}(x)^{-\alpha}\neq0$.

On the other hand for any arbitrary positive real number $\alpha>0$ the fractional derivative, of a function, is zero, when difference between the order of fractional differentiation and order of power function is integers 1, 2, 3....

That is, ${}_0D_x^\alpha\{x^b\}=0$ if $\alpha-b\in\mathbb{Z}^+$; for example $f(x)=x^b=x^{\alpha-k}$ and $k=1,2,...(1+\alpha)$.

Meaning that the difference between the order of fractional differentiation and the order of power function i.e. $\alpha-b$ is integers 1, 2, 3..., then the fractional derivative of power function goes to zero. This also demonstrates the role of reciprocal gamma function, whose values at negative integer points and zero point are zero.

Let $f(x)=\log x$, we find first the Riemann-Liouvelli fractional integration with change of variable $u=x(1-v)$:

$$
\begin{aligned}
{}_0D_x^{-\alpha} f(x) = {}_0I_x^{\alpha} &= \frac{1}{\Gamma(\alpha)}\int_0^x (x-u)^{\alpha-1}\log(u)du \\
&= \frac{(\log x)x^{\alpha}}{\Gamma(\alpha)}\int_0^1 v^{\alpha-1}dv + \frac{x^{\alpha}}{\Gamma(\alpha)}\int_0^1 v^{\alpha-1}\log(1-v)dv \\
&= \frac{x^{\alpha}}{\Gamma(\alpha+1)}\log x - \frac{x^{\alpha}}{\alpha\Gamma(\alpha)}\int_0^1 \log(10v)d(1-v^{\alpha}) \\
&= \frac{x^{\alpha}}{\Gamma(\alpha+1)}\left\{\log x - \left[(1-v^{\alpha})\log(1-v)\right]_0^1 - \int_0^1 \frac{1-v^{\alpha}}{1-v}dv\right\}
\end{aligned}
$$

By doing integration by parts and noting

$\int_0^1 \frac{v^x - v^y}{1-v}dv = \psi(y+1) - \psi(x+1)$, with $\Re e(x), \Re e(y) > -1$, with 'psi' function defined as above, obeys recursion.

$\psi(x+1) - \psi(x) = (x)^{-1}$ and $-\psi(1) = \gamma = 0.5772156666....$ is Euler's constant.
We write

$$
{}_0D_x^{-\alpha}\log x = {}_0I_x^{\alpha}\log x = \frac{x^{\alpha}}{\Gamma(\alpha+1)}\{\log x - \psi(\alpha+1) + \psi(1)\}
$$

Thus for $m-1 \le \Re e(\alpha) \le m$

$$
\begin{aligned}
{}_0D_x^{\alpha}\log x &= \frac{d^m}{dx^m}\, {}_0D_x^{-(m-\alpha)}\log x \\
&= \frac{d^m}{dx^m}\left\{\frac{x^{m-\alpha}}{\Gamma(m-\alpha+1)}\left[\log x - \psi(m-\alpha+1) + \psi(1)\right]\right\}
\end{aligned}
$$

Combining the result with ${}_0I_x^{\alpha}\log x = \frac{x^{\alpha}}{\Gamma(\alpha+1)}\left[\log x - \psi(\alpha+1) + \psi(1)\right]$ following is obtained:

$$
{}_0D_x^{\alpha}\log x = \frac{x^{-\alpha}}{\Gamma(1-\alpha)}\{\log x - \psi(1-\alpha) - \gamma\}
$$

When $\alpha = n \in \mathrm{N}$ the fractional derivative for RHS of above expression should be interpreted as the limit for $\alpha \to n$. In fact

$$
\lim_{\alpha\to n}\frac{\psi(1-\alpha)}{\Gamma(1-\alpha)} = (-1)^{-n}\Gamma(n),
$$

the rule $\frac{d^n}{dx^n}\log x = -[(n-1)!](-x)^{-n}$ is fulfilled.

However, for $\alpha = -n; n \in \mathbb{N}$ there holds the classical result.

$$ {}_0D_x^{-n}\log x = {}_0I_x^n \log x = \frac{x^n}{n!}\left\{\log x - \sum_{j=1}^{n}\frac{1}{j}\right\}, $$

also observing that $\psi(n+1)-\psi(1) = \sum_{j=1}^{n}\frac{1}{j}$ and $\psi(x+1)-\psi(x) = \frac{1}{x}$ are the properties of 'psi' function. The 'psi' function is also termed as 'digamma' function. The other interesting properties are $\psi(0.5) = \psi(1) - \ln 4 = -\gamma - \ln 4$. If z is not negative integer then

$$ \psi(z+1) = -\gamma + \sum_{k=1}^{\infty}\frac{z}{k(z+k)} $$

2.2.2 *Hypergeometric Functions*

The hyper-geometric function and its generalization encompass an extensive class of analytical functions. The generalized hypergeometric series is defined as:

$$ {}_pF_q\left(a_1,\ldots,a_p,b_1,\ldots,b_q;z\right) = \frac{\Gamma(b_1)\ldots\Gamma(b_q)}{\Gamma(a_1)\ldots\Gamma(a_p)}\sum_{k=0}^{\infty}\frac{\Gamma(a_1+k)\ldots\Gamma(a_p+k)}{\Gamma(b_1+k)\ldots\Gamma(b_q+k)}\frac{z^k}{k!} $$

Provided that the b_i are not non positive integers. The series converges for all z if $p \le q$, converges for $|z|<1$ if $p = q+1$ and diverges for all nonzero z if $p > q+1$. If $p = 2$ and $q = 1$ then:

$$ {}_2F_1\left(a,b,c;z\right) = \frac{\Gamma(c)}{\Gamma(a)\Gamma(b)}\sum_{k=0}^{\infty}\frac{\Gamma(a+k)\Gamma(b+k)}{\Gamma(c+k)}\frac{z^k}{k!} $$

And all its analytical continuations are called the hypergeometric functions. The series converges for all z with $|z|<1$. If $\Re e(c) > \Re e(a) > 0$ the integral representation is:

$$ {}_2F_1\left(a,b,c;z\right) = \frac{\Gamma(c)}{\Gamma(a)\Gamma(c-a)}\int_0^1 t^{a-1}(1-t)^{c-a-1}(1-tz)^{-b}\,dt $$

In particular, if $\Re e(c) > \Re e(a+b)$ and c is not a nonpositive integer, then

$$ {}_2F_1(a,b,c;z) = \frac{\Gamma(c)\Gamma(c-a-b)}{\Gamma(c-a)\Gamma(c-b)} $$

The following are useful identities:

$$ {}_2F_1(a,b,c;z) = (1-z)^{c-a-b}\, {}_2F_1(c-a,c-b,c;z) \text{ and} $$

$$ {}_2F_1(a,b,c;z) = (1-z)^{-a}\, {}_2F_1\left(a,c-b,c;\frac{z}{z-1}\right). $$

In particular if $p = 1 = q$, then

$$ {}_1F_1(a,c;z) = \frac{\Gamma(c)}{\Gamma(a)} \sum_{k=0}^{\infty} \frac{\Gamma(a+k)}{\Gamma(c+k)} \frac{z^k}{k!} $$

is confluent hypergeometric function. It converges for all z provided that $c \neq 0,-1,-2,\ldots$. The name comes from the fact that it may be defined by limit as ${}_1F_1(a,c;z) = \lim_{b\to\infty} {}_2F_1\left(a,b,c;\frac{z}{b}\right)$.

If $\Re e(c) > \Re e(a) > 0$, the integral representation is

$$ {}_1F_1(a,c;z) = \frac{\Gamma(c)}{\Gamma(a)\Gamma(c-a)} \int_0^1 t^{a-1}(1-t)^{c-a-1} e^{zt}\, dt $$

Following are interesting identities:

$$ {}_1F_1(a,c;z) = e^z\, {}_1F_1(c-a,c;z),\quad D\, {}_1F_1(a,c;z) = \frac{a}{c}\, {}_1F_1(a+1,c+1;z) $$

Relation of elementary functions to hypergeometric functions are for example:

$$ (1-z)^{-a} = {}_1F_0(a;z),\ |z| < 1 $$

$$ \ln\frac{1+x}{1-x} = 2x\, {}_2F_1\left(\frac{1}{2},1,\frac{3}{2};x^2\right),\ 0 \le x < 1 $$

$$ \gamma^*(v,z) = \frac{1}{\Gamma(v+1)}\, {}_1F_1(v,v+1;z) = \frac{1}{\Gamma(v+1)} e^{-z}\, {}_1F_1(1,v+1;z) $$

$$ \mathrm{B}_\tau(x,y) = x^{-1}\tau^x\, {}_2F_1(x,1-y,x+1;\tau) = x^{-1}\tau^x(1-\tau)^y\, {}_2F_1(x+y,1,x+1;\tau) $$

2.2.3 Mittag-Leffler Function

In the integer order calculus equations the exponential function $\exp(z)$ plays important role. Similarly in the fractional order calculus the Mittag-Leffler function plays the important part.

For this new function $E_q[az], q>0$ Mittag-Leffler considered, the parameter a to be a complex number, as $a=|a|\exp(j\phi)$. As he studied this function it became apparent that this function is either stable (decays to zero), or unstable (goes to infinity) as z increases depending upon how the parameter a and q are chosen. The result was that the function remained bounded for increasing z if $|\varphi|\geq q\frac{\pi}{2}$.

2.2.3.1 One-Parameter Mittag-Leffler Function

Defined as

$$E_\alpha(z)=\sum_{k=0}^{\infty}\frac{z^k}{\Gamma(\alpha k+1)}$$

The expanded form is infinite series looks as follows:

$$E_\alpha(z)=1+\frac{z}{\Gamma(\alpha+1)}+\frac{z^2}{\Gamma(2\alpha+1)}+\frac{z^3}{\Gamma(3\alpha+1)}+\ldots$$

This function was introduced by Mittag-Leffler in 1903.

Mittag-Leffler function plays an important role in the study of fractional order systems and used to express several physical processes. This function incorporates between initially 'stretched exponentials' for small z that is, $E_\alpha(-z^\alpha)\sim\exp\left(-\frac{z^\alpha}{\Gamma(1+\alpha)}\right)$ and for large z the asymptotic approximation is 'inverse-power-law' that is, $E_\alpha(-z^\alpha)\sim\frac{z^{-\alpha}}{\Gamma(1-\alpha)}$. This property is widely used to study relaxation in 'condense matter' physics, fractional Brownian motion and ordering the disordered system. The Mittag-Leffler function is a possible model for a long (fat) tailed survival function. For $\alpha=1$, the function coincides with ordinary exponential function. The following are some asymptotic expressions of Mittag-Leffler function useful in system identification:

$$\Psi(\tau)=E_\beta(-\tau^\beta)\cong 1-\frac{\tau^\beta}{\Gamma(\beta+1)}\cong\exp\{-\tau^\beta/\Gamma(\beta+1)\}\text{, for small }\tau\text{, }0\leq\tau\ll 1$$

$$\Psi(\tau)=E_\beta(-\tau^\beta)\cong\frac{\sin(\beta\pi)}{\pi}\frac{\Gamma(\beta)}{\tau^\beta}, 0<\beta<1\quad\text{for large}\quad\tau\to\infty;\quad\text{power-law}$$

asymptote.

$$\psi(\tau)=-\frac{d}{d\tau}E_\beta(-\tau^\beta)\cong\frac{\tau^{-(1-\beta)}}{\Gamma(\beta)}\text{, for small }\tau\text{, }0\leq\tau\ll 1$$

$$\psi(\tau) = -\frac{d}{d\tau}E_\beta(-\tau^\beta) \cong \frac{\sin(\beta\pi)}{\pi}\frac{\Gamma(\beta+1)}{\tau^{\beta+1}}, 0<\beta<1 \text{ ,for large } \tau \to \infty .$$

2.2.3.2 Two Parameter Mittag-Leffler Functions

Two parameter Mittag-Leffler functions plays very important role in fractional calculus. This, function type was introduced by R P Agarwal and Erdelyi, in 1953-54 The two-parameter function is defined as:

$$E_{\alpha,\beta}(z) = \sum_{k=0}^{\infty}\frac{z^k}{\Gamma(\alpha k+\beta)} \qquad (\alpha>0, \beta>0)$$

$$E_{\alpha,1}(z) = \sum_{k=0}^{\infty}\frac{z^k}{\Gamma(\alpha k+1)} \equiv E_\alpha(z)$$, is one parameter Mittag-Leffler function.

Following identities follows from the definition:

$$E_{1,1}(z) = \sum_{k=0}^{\infty}\frac{z^k}{\Gamma(k+1)} = \sum_{k=0}^{\infty}\frac{z^k}{k!} = e^z$$

$$E_{1,2}(z) = \sum_{k=0}^{\infty}\frac{z^k}{\Gamma(k+2)} = \sum_{k=0}^{\infty}\frac{z^k}{(k+1)!} = \frac{1}{z}\sum_{k=0}^{\infty}\frac{z^{k+1}}{(k+1)!} = \frac{e^z-1}{z}$$

$$E_{1,3}(z) = \sum_{k=0}^{\infty}\frac{z^k}{\Gamma(k+3)} = \sum_{k=0}^{\infty}\frac{z^k}{(k+2)!} = \frac{1}{z^2}\sum_{k=0}^{\infty}\frac{z^{k+2}}{(k+2)!} = \frac{e^z-1-z}{z^2}$$

The above has general form as follows:

$$E_{1,m}(z) = \frac{1}{z^{m-1}}\left(e^z - \sum_{k=0}^{m-2}\frac{z^k}{k!}\right)$$

The trigonometric and hyperbolic are also manifestations of the two-parameter Mittag-Leffler function, indicated below:

$$E_{2,1}(z^2) = \sum_{k=0}^{\infty}\frac{z^{2k}}{\Gamma(2k+1)} = \sum_{k=0}^{\infty}\frac{z^{2k}}{(2k)!} = \cosh(z),$$

$$E_{2,2}(z^2) = \sum_{k=0}^{\infty}\frac{z^{2k}}{\Gamma(2k+2)} = \frac{1}{z}\sum_{k=0}^{\infty}\frac{z^{2k+1}}{(2k+1)!} = \frac{\sinh(z)}{z}$$

Generalized hyperbolic function of order n, are represented below:

$$h_r(z,n) = \sum_{k=0}^{\infty}\frac{z^{nk+r-1}}{(nk+r-1)!} = z^{r-1}E_{n,r}(z^n), \ldots (r = 1,2,3,\ldots,n)$$

Generalized trigonometric function of order n, are represented below:

$$k_r(z,n)=\sum_{m=0}^{\infty}\frac{(-1)^m z^{nm+r-1}}{(nm+r-1)!}=z^{r-1}E_{n,r}(-z^n),\ldots(r=1,2,3,\ldots,n)$$

Mathematical handbooks describe $\mathrm{erfc}(z)$ as following:

Error function defined as:

$$\mathrm{erf}(z)=\frac{2}{\sqrt{\pi}}\int_0^z e^{-t^2}dt$$

Note that $\mathrm{erf}(\infty)=1$.

And is represented by series as:

$$\mathrm{erf}(z)=\frac{2}{\sqrt{\pi}}\sum_{n=0}^{\infty}\frac{(-1)^n z^{2n+1}}{(2n+1)n!}=\frac{2}{\sqrt{\pi}}\left(z-\frac{z^3}{3}+\frac{z^5}{10}-\frac{z^7}{42}+\frac{z^9}{216}+\ldots\right)$$

Complimentary error function is defined as:

$$\mathrm{erfc}(z)=1-\mathrm{erf}(z)=1-\frac{2}{\sqrt{\pi}}\int_0^z e^{-t^2}dt=\frac{2}{\sqrt{\pi}}\int_z^{\infty} e^{-t^2}dt\,.$$

$$E_{(1/2),1}(z)=\sum_{k=0}^{\infty}(z^k)/\Gamma[(k/2)+1]=e^{z^2}\mathrm{erfc}(-z)$$

The series asymptotic expansion of complimentary error function is:

$$\mathrm{erfc}(z)=\frac{e^{-z^2}}{z\sqrt{\pi}}\left[1+\sum_{n=1}^{\infty}(-1)^n\frac{1.3.5\ldots(2n-1)}{(2z^2)^n}\right]=\frac{e^{-z^2}}{z\sqrt{\pi}}\left[1+\sum_{n=1}^{\infty}(-1)^n\frac{(2n)!}{n!(2z)^{2n}}\right]$$

Error function in terms of hypergeometric function is:

$$\mathrm{erf}(x)=2\pi^{-1/2}xe^{-x^2}{}_1F_1\left(1,\frac{3}{2};x^2\right)=2\pi^{-1/2}x\,{}_1F_1\left(\frac{1}{2},\frac{3}{2};-x^2\right)$$

Error function and incomplete gamma function are related as:

$$\gamma^*\left(\frac{1}{2},at\right)=(at)^{-1/2}\mathrm{erf}(\sqrt{at})$$

$$\mathrm{erf}(z)=z\gamma^*\left(\frac{1}{2},z^2\right)$$

This error function integral has no algebraic formula, so it is to be numerically evaluated and tabulated. This error function is a solution of classical Advection Dispersion Equation (ADE) $\frac{\partial C}{\partial t} = \nabla \bullet \left(-vC + \mathbb{D}\nabla C\right)$, for a continuous source initial condition. Say the step function initial condition is given as C_0, at $t = 0$, then the solution in terms of error function is

$$C = \frac{C_0}{2}\left[1 - \operatorname{erf}\left(\frac{x - vt}{2\sqrt{\mathbb{D}t}}\right)\right].$$

For continuity with this widely used formula in ADE, a similar solution can be framed for 'Fractional ADE' that is

$$\frac{\partial C}{\partial t} = \left(-v \bullet \nabla C + \mathbb{D}_\alpha \nabla^\alpha C\right),$$

for a continuous source initial condition same as for ADE. That is

$$C = \frac{C_0}{2}\left[1 - \operatorname{serf}_\alpha\left(\frac{x - vt}{(\mathbb{D}_\alpha t)^{1/\alpha}}\right)\right],$$

where we define α-stable error function $\operatorname{serf}_\alpha$ similar to the error function, call it generalized error function; that is twice the integral of α stable density for argument 0 to z as

$$\operatorname{serf}_\alpha(z) = 2\int_0^z s_\alpha(x)dx$$

where $s_\alpha(x) = \exp(-|x|^\alpha)$ is standard symmetric α stable density. The error function is thus special case when the probability density function is normal (Gaussian) with $\alpha = 2.0$, that is $\operatorname{erf}(z) = \operatorname{serf}_{2.0}(2z)$.

2.2.3.3 Variants of Mittag-Leffler Function

$$\xi_t(v,a) = t^v \sum_{k=0}^{\infty} \frac{(at)^k}{\Gamma(v+k+1)} = t^v E_{1,v+1}(at)$$

This function is important for solving fractional differential equations.

$$\ni_\alpha(\beta,t) = t^\alpha \sum_{k=0}^{\infty} \frac{\beta^k t^{k(\alpha+1)}}{\Gamma(\{k+1\}\{\alpha+1\})} = t^\alpha E_{\alpha+1,\alpha+1}(\beta t^{\alpha+1})$$

This function is called Rabotnov function and one special variant too.

$$Sc_{\alpha}(z)=\sum_{n=0}^{\infty}\frac{(-1)^{n}z^{(2-\alpha)n+1}}{\Gamma(\{2-\alpha\}n+2)}=zE_{2-\alpha,2}(-z^{2-\alpha})$$

The fractional sine function form-I

$$Cs_{\alpha}(z)=\sum_{n=0}^{\infty}\frac{(-1)^{n}z^{(2-\alpha)n}}{\Gamma(\{2-\alpha\}n+1)}=E_{2-\alpha,1}(-z^{2-\alpha})$$

The fractional cosine function form-I

$$\sin_{\lambda,\mu}(z)=\sum_{k=0}^{\infty}\frac{(-1)^{k}z^{2k+1}}{\Gamma(2\mu k+2\mu-\lambda+1)}=zE_{2\mu,2\mu-\lambda+1}(-z^{2})$$

The fractional sine function form-II

$$\cos_{\lambda,\mu}(z)=\sum_{k=0}^{\infty}\frac{(-1)^{k}z^{2k}}{\Gamma(2\mu k+\mu-\lambda+1)}=E_{2\mu,\mu-\lambda+1}(-z^{2})$$

The fractional cosine function form-II.

Generalization of the Mittag-Leffler function to two variables was suggested and was further extended by Srivastava to the following type of symmetric form.

$$\xi^{\nu,\sigma}_{\alpha,\beta,\lambda,\mu}=\sum_{m=0}^{\infty}\sum_{n=0}^{\infty}\frac{x^{m+\frac{\beta(\nu n+1)-1}{\alpha}}\,y^{n+\frac{\mu(\sigma m+1)-1}{\lambda}}}{\Gamma(m\alpha+(\nu n+1)\beta)\Gamma(n\lambda+(\sigma m+1)\mu)}$$

Several manifestations including several variables representation of Mittag-Leffler have been made for multi-dimensional studies on fractional calculus.

2.2.3.4 Laplace Transforms of Mittag-Leffler Function

Following expressions give some identities for Laplace transforms pairs of Mittag-Leffler functions

$$\mathcal{L}\left\{t^{\alpha k+\beta-1}E^{(k)}_{\alpha,\beta}(at^{\alpha})\right\}=\frac{s^{\alpha-\beta}k!}{(s^{\alpha}-a)^{k+1}};\quad \mathrm{Re}(s)>|a|^{(1/\alpha)}$$

$$\mathcal{L}\left\{t^{0.5k-0.5}E^{(k)}_{(0.5),(0.5)}(a\sqrt{t})\right\}=\frac{k!}{(\sqrt{s}-a)^{k+1}}$$

Here $E^{(k)}_{\alpha,\beta}=\dfrac{d^{(k)}}{dt^{(k)}}E_{\alpha,\beta}$.

For $k>0$ the operation is differentiation of Mittag-Leffler function, and for $k<0$ the operation is integration of Mittag-Leffler function.

$$\beta = 1, k = 0, \qquad E^{(k)}_{\alpha,\beta} \to E_{\alpha,1}(at^{\alpha});$$

$$\mathcal{L}\left\{E_{\alpha}(at^{\alpha})\right\} = \frac{s^{\alpha-1}}{s^{\alpha} - a}$$

$$\mathcal{L}\left\{E_{\alpha}(-\lambda t^{\alpha})\right\} = \frac{s^{\alpha-1}}{s^{\alpha} + \lambda} = \frac{s^{\alpha}}{s(s^{\alpha} + \lambda)}; \quad \alpha > 0$$

$$\mathcal{L}\left\{E_{\alpha}(-t^{\alpha})\right\} = \frac{s^{\alpha-1}}{s^{\alpha} + 1} = \frac{s^{\alpha}}{s(s^{\alpha} + 1)}$$

$$\mathcal{L}\left\{E_{\alpha}(\lambda t^{\alpha})\right\} = \frac{s^{\alpha-1}}{s^{\alpha} - \lambda} = \frac{s^{\alpha}}{s(s^{\alpha} - \lambda)}$$

$$\mathcal{L}\left\{\frac{d}{dt}E_{\alpha}(-\lambda t^{\alpha})\right\} = s\mathcal{L}\left\{E_{\alpha}(-\lambda t^{\alpha})\right\} = \frac{s^{\alpha}}{s^{\alpha} + \lambda}$$

$$\mathcal{L}\left\{\frac{d^k}{dt^k}E_{\alpha}(-t^{\alpha})\right\} = s^k\mathcal{L}\left\{E_{\alpha}(-t^{\alpha})\right\} = \frac{s^{\alpha+k-1}}{s^{\alpha} + 1}$$

$$\mathcal{L}\left\{\frac{d^{-1}}{dt^{-1}}E_{\alpha}(-t^{\alpha})\right\} = s^{-1}\mathcal{L}\left\{E_{\alpha}(-t^{\alpha})\right\} = \frac{1}{s}\frac{s^{\alpha-1}}{s^{\alpha} + 1} = \frac{s^{\alpha-2}}{s^{\alpha} + 1}$$

$$\mathcal{L}\left\{\frac{d^{-k}}{dt^{-k}}E_{\alpha}(-t^{\alpha})\right\} = s^{-k}\mathcal{L}\left\{E_{\alpha}(-t^{\alpha})\right\} = \frac{1}{s^k}\frac{s^{\alpha-1}}{s^{\alpha} + 1} = \frac{s^{\alpha-k-1}}{s^{\alpha} + 1}$$

$$\mathcal{L}\left\{\delta(t) - \frac{d}{dt}E_{\alpha}(-\lambda t^{\alpha})\right\} = 1 - \frac{s^{\alpha}}{s^{\alpha} + \lambda} = \frac{\lambda}{s^{\alpha} + \lambda}$$

2.2.4 Agarwal Function

The Mittag-Leffler function is generalized by Agarwal in (1953). This function is particularly interesting to the fractional order system theory due to its Laplace transform given by Agarwal. The function is defined as follows:

$$E_{\alpha,\beta}(t^{\alpha}) = \sum_{m=0}^{\infty} \frac{t^{\left(m+\frac{\beta-1}{\alpha}\right)}}{\Gamma(\alpha m + \beta)}; \quad \mathcal{L}\{E_{\alpha,\beta}(t^{\alpha})\} = \frac{s^{\alpha-\beta}}{s^{\alpha} - 1}$$

2.2.5 Erdelyi's Function

Erdelyi (1954) has studied the generalization of Mittag-Leffler function as:

$$E_{\alpha,\beta}(t) = \sum_{m=0}^{\infty} \frac{t^m}{\Gamma(\alpha m + \beta)}; \quad \alpha, \beta > 0 \quad \mathcal{L}\left\{E_{\alpha,\beta}(t)\right\} = \sum_{m=0}^{\infty} \frac{\Gamma(m+1)}{s^{m+1}\Gamma(\alpha m + \beta)}$$

Where the powers of t are integers.

2.2.6 Robotnov-Hartley Function

To effect the direct solution of the fundamental linear fractional order differential equations the following function was introduced by Robotnov and Hartley (1998)

$$F_q(-a,t) = t^{q-1}\sum_{n=0}^{\infty}\frac{(-a)^n t^{nq}}{\Gamma(nq+q)};\quad q>0\quad \mathcal{L}\{F_q(a,t)\} = \frac{1}{s^q - a}$$

This function is the 'impulse-response' of the fundamental fractional differential equation, and is used by control system analysis to obtain the forced or the initialized system reaction.

2.2.7 Miller Ross Function

Miller and Ross in 1993 introduced a function as the basis of the solution of fractional order initial value problem. It is defined as the v-th integral of the exponential function, that is:

$$E_t(v,a) = \frac{d^{-v}}{dt^{-v}}e^{at} = t^v\sum_{k=0}^{\infty}\frac{(at)^k}{\Gamma(v+k+1)}\qquad \mathcal{L}\{E_t(v,a)\} = \frac{s^{-v}}{s-a}\quad \mathrm{Re}(v)>1$$

Miller-Ross function in terms of hypergeometric function is:

$$E_t(v,a) = \frac{t^v}{\Gamma(v+1)}\,{}_1F_1(1,v+1;at)$$

The one-parameter Mittag-Leffler function is

$$E_w(t) = \sum_{n=0}^{\infty}\frac{t^n}{\Gamma(1+nw)},\ \text{for } w \geq 0 .$$

Expanding this series as following grouping we get:

$$E_w(ct^w) = \sum_{n=0}^{\infty}\frac{\left(ct^w\right)^n}{\Gamma(1+nw)}$$

$$= \sum_{n=0,q,2q,\ldots\ldots}^{\infty}\frac{\left(ct^w\right)^n}{\Gamma(1+nw)} + \sum_{n=1,q+1,2q+1,\ldots}^{\infty}\frac{\left(ct^w\right)^n}{\Gamma(1+nw)} + \ldots\ldots + \sum_{n=q-1,2q-1,3q-1,\ldots}^{\infty}\frac{\left(ct^w\right)^n}{\Gamma(1+nw)}$$

$$= E_t(0,c^q) + cE_t(w,c^q) + c^2E_t(2w,c^q) + \ldots + c^{q-1}E_t([q-1]w,c^q)$$

Therefore,

$$E_w(ct^w) = \sum_{k-0}^{q-1}c^k E_t(kw,c^q)$$

is relation between Mittag-Leffler and Miller-Ross functions.

In terms of incomplete gamma function we define the Miller-Ross function as:

$$E_z(v,a) = z^v e^{az} \gamma^*(v,az)$$

and $C_z(v,a) = \frac{1}{2}[E_z(v,\mathrm{i}a) + E_z(v,-\mathrm{i}a)]$, $S_z(v,a) = \frac{1}{2\mathrm{i}}[E_z(v,\mathrm{i}a) - E_z(v,-\mathrm{i}a)]$.

Where $C_z(v,a)$ and $S_z(v,a)$ are Generalized cosine and sine functions respectively, elaborated below. These identities of Miller-Ross expressions are similar to exponential function and its polar representation to get cosine and sine trigonometric functions. These functions plays crucial role in Fractional Differential Equation solutions, and are similar to the normal transcendental and trigonometric functions. In the above definition of Miller-Ross function let a be replaced by purely imaginary number $\mathrm{i}a$, where $\mathrm{i} = \sqrt{-1}$, Electrical Engineers call this imaginary unit as $\mathrm{j} = \sqrt{-1}$. In the series form we can re-write organizing even and odd terms together as:

$$E_t(v,\mathrm{i}a) = t^v \left[\sum_{k=even}^{\infty} \frac{(-1)^{k/2} (at)^k}{\Gamma(v+k+1)} + \mathrm{i} \sum_{k=odd}^{\infty} \frac{(-1)^{\frac{(k-1)}{2}} (at)^k}{\Gamma(v+k+1)} \right]$$

$$E_t(v,\mathrm{i}a) = C_t(v,a) + \mathrm{i}S_t(v,a)$$

$$E_t(0,\mathrm{i}t) = e^{\mathrm{i}t}$$

$$e^{\mathrm{i}t} = \cos t + \mathrm{i}\sin t$$

The above identities indicate similarity of Miller-Ross function with exponential function and trigonometric functions.

Integration of Miller-Ross function is:

$$\int_0^t E_u(v,a)du = E_t(v+1,a), \quad \Re ev > -1$$

$$\int_0^t u^w E_{t-u}(v,a)du = \Gamma(w+1)E_t(v+w+1,a) \quad \Re e(w) > -1, \Re e(v) > -1$$

$$D^{-p} E_t(v,a) = E_t(v+p,a), \; p = 1,2,3,\ldots \; \Re e(v) > -1$$

Differentiation of Miller Ross Function is:

$$DE_t(v,a) = E_t(v-1,a)$$

$$D^p E_t(v,a) = E_t(v-p,a) = a^p E_t(v,a) + \sum_{k=0}^{p-1} \frac{a^k t^{v+k-p}}{\Gamma(v+k+1-p)}, \quad p = 0,1,2,3...$$

$$D\left[tE_t(v,a)\right] = tE_t(v-1,a) + E_t(v,a)$$

$$D\left[t^\mu E_t(v,a)\right] = t^\mu E_t(v-1,a) + \mu t^{\mu-1} E_t(v,a)$$

$$D^p\left[t^\mu E_t(v,a)\right] = \sum_{k=0}^{p} \binom{p}{k} \frac{\Gamma(\mu+1)}{\Gamma(\mu-k+1)} t^{\mu-k} E_t(v+k-p,a), \quad p = 0,1,2,...$$

Special values of Miller-Ross function are:

$$E_t(0,a) = e^{at}$$

$$E_0(v,a) = 0, \quad \Re e(v) = 0$$

$$E_t(-1,a) = aE_t(0,a)$$

$$E_t(-p,a) = a^p E_t(0,a), \quad p = 0,1,2,3,...$$

$$E_t(1,a) = \frac{E_t(0,a) - 1}{a}$$

$$E_t\left(\frac{1}{2},a\right) = a^{-1/2} e^{at} \left\{\operatorname{erf}\sqrt{at}\right\}$$

$$E_t\left(-\frac{1}{2},a\right) = aE_t\left(\frac{1}{2},a\right) + \frac{1}{\sqrt{\pi t}}$$

$$E_t(v,0) = \frac{t^v}{\Gamma(v+1)}$$

Recursive relations of Miller-Ross functions are listed as follows:

$$E_t(v,a) = aE_t(v+1,a) + \frac{t^v}{\Gamma(v+1)}$$

$$E_t(v,a) = a^p E_t(v+p,a) + \sum_{k=0}^{p-1} \frac{t^{v+k}}{\Gamma(v+k+1)}, \quad p = 0,1,2,3,...$$

$$E_t(v,a) - E_t(v,b) = aE_t(v+1,a) - bE_t(v+1,b)$$

$$E_t(v,a) - E_t(v,b) = a^p E_t(v+p,a) - b^p E_t(v+p,b) + \sum_{k=1}^{p-1} \frac{(a^k - b^k)t^{v+k}}{\Gamma(v+k+1)}, \quad p = 0,1,2...$$

Integral representation of Miller-Ross function is

$$E_t(v,a) = \frac{1}{\Gamma(v)} \int_0^t u^{v-1} e^{a(t-u)} du, \quad \Re e(v) > 0$$

This Miller-Ross function $E_t(v,a)$ is solution of the Ordinary Differential Equation:

$$\frac{dy}{dt} - ay = \frac{t^{v-1}}{\Gamma(v)}, \quad v > 0$$

2.2.8 *Generalized Cosine and Sine Function*

In the Miller-Ross modification above the Generalized Cosine function is defined as:

$$C_t(v,a) = t^v \sum_{k=even}^{\infty} \frac{(-1)^{k/2}(at)^k}{\Gamma(v+k+1)}$$
$$C_t(0,a) = \cos at$$

Laplace of Generalized cosine function is

$$\mathcal{L}\{C_t(v,a)\} = \frac{s}{s^v(s^2+a^2)}, \quad \Re e(v) > -1$$

In the Miller-Ross modification above the Generalized Sine function is defined as

$$S_t(v,a) = t^v \sum_{k=odd}^{\infty} \frac{(-1)^{\frac{(k-1)}{2}}(at)^k}{\Gamma(v+k+1)}$$
$$S_t(0,a) = \sin at$$

Laplace of Generalized sine function is

$$\mathcal{L}\{S_t(v,a)\} = \frac{a}{s^v(s^2+a^2)}, \quad \Re e(v) > -2$$

Integral representation of generalized cosine and sine functions are:

$$C_t(v,a) = \frac{1}{\Gamma(v)}\int_0^t u^{v-1}\cos a(t-u)du \text{ and } S_t(v,a) = \frac{1}{\Gamma(v)}\int_0^t u^{v-1}\sin a(t-u)du,$$
$$\Re e(v) > 0$$

Special values of generalized cosine and sine functions are

$$C_t(0,a) = \cos at \text{ and } S_t(0,a) = \sin at$$
$$C_0(v,a) = 0, \ \Re e(v) > 0 \text{ and } S_0(v,a) = 0, \ \Re e(v) > -1$$
$$C_t(-1,a) = -a\sin at \text{ and } S_t(-1,a) = a\cos at$$

$$C_t(-p,a)=(-1)^{p/2}a^p\cos at \text{ and } S_t(-p,a)=(-1)^{p/2}a^p\sin at\ ,\ p=0,2,4,\ldots$$

$$C_t=(-p,a)=(-1)^{(p+1)/2}a^p\sin at \text{ and } S_t(-p,a)=(-1)^{(p-1)/2}a^p\cos at\ ,\ p=1,3,5,\ldots$$

$$C_t(1,a)=\frac{1}{a}\sin at \text{ and } S_t(1,a)=\frac{2}{a}\sin^2\frac{1}{2}at$$

$$C_t\left(-\frac{1}{2},a\right)=\frac{1}{\sqrt{\pi t}}-aS_t\left(\frac{1}{2},a\right) \text{ and } S_t\left(-\frac{1}{2},a\right)=aC_t\left(\frac{1}{2},a\right)$$

$$C_t(v,0)=\frac{t^v}{\Gamma(v+1)} \text{ and } S_t(v,0)=0$$

$$C_t\left(\frac{1}{2},a\right)=\sqrt{\frac{2}{a}}\left[(\cos at)C(z)+(\sin at)S(z)\right]$$

$$S_t\left(\frac{1}{2},a\right)=\sqrt{\frac{2}{a}}\left[(\sin at)C(z)-(\cos at)S(z)\right]$$

Where $z=\sqrt{\frac{2at}{\pi}}$ and $C(z)$, $S(z)$ are Fresnel integrals defined as

$$C(z)=\int_0^z\cos\frac{1}{2}\pi t^2dt \text{ and } S(z)=\int_0^z\sin\frac{1}{2}\pi t^2dt$$

Recursive relation of generalized cosine and sine functions are:

$$C_t(v-1,a)=-aS_t(v,a)+\frac{t^{v-1}}{\Gamma(v)} \text{ and } S_t(v-1,a)=aC_t(v,a)$$

$$C_t(v-1,a)+a^2C_t(v+1,a)=\frac{t^{v-1}}{\Gamma(v)} \text{ and } S_t(v-1,a)+a^2S_t(v+1,a)=\frac{at^v}{\Gamma(v+1)}$$

Integration of generalized cosine and sine functions is:

$$\int_0^t C_u(v,a)du=C_t(v+1,a),\ \Re e(v)>-1 \text{ and } \int_0^t S_u(v,a)du=S_t(v+1,a),\ \Re e(v)>-2$$

Differentiation of generalized cosine and sine is:

$$DC_t(v,a)=C_t(v-1,a) \text{ and } DS_t(v,a)=S_t(v-1,a)$$

$$D^mC_t(v,a)=C_t(v-m,a) \text{ and } D^mS_t(v,a)=S_t(v-m,a)\ ,\ m=0,1,2,\ldots$$

$$D\left[tC_t(v,a)\right]=tC_t(v-1,a)+C_t(v,a) \text{ and } D\left[tS_t(v,a)\right]=tS_t(v-1,a)+S_t(v,a)$$

$$D\left[t^{\mu}C_t(v,a)\right]=t^{\mu}C_t(v-1,a)+\mu t^{\mu-1}C_t(v,a)$$
$$D\left[t^{\mu}S_t(v,a)\right]=t^{\mu}S_t(v-1,a)+\mu t^{\mu-1}S_t(v,a)$$

This Generalized cosine and sine function $S_t(v,a)$ or $aC_t(v+1,a)$ is solution of the Ordinary Differential Equation:

$$\frac{d^2y}{dt^2}+a^2y=a\frac{t^{v-1}}{\Gamma(v)},\quad v>0$$

This detailed description of Miller-Ross E_t and Generalized cosine C_t and sine S_t functions is given in order to draw attention toward their similarity with exponential, cosine and sine functions. These similarities in property encourage us to select these functions as solution to fractional differential equations. Similarly one can construct generalized hyperbolic cosine and generalized hyperbolic sine functions with similar properties as described above for generalized cosine and sine functions.

Generalized Hyperbolic cosine and sine is constructed as follows:

$$HC_t(v,a)=\frac{1}{\Gamma(v)}\int_0^t u^{v-1}\cosh a(t-u)du$$

$$HS_t(v,a)=\frac{1}{\Gamma(v)}\int_0^t u^{v-1}\sinh a(t-u)du\ ,\ \Re e(v)>-1$$

2.2.9 Generalized R Function and G Function

It is of significant usefulness to develop a generalized function which when fractionally differentiated or integrated (differintegrated) by any order returns itself. Like exponential, trigonometric, hyperbolic functions of integer order calculus, the definitions of such generalized Mittag-Leffler functions are important in fractional calculus. In earlier section some variants of Mittag-Leffler is noted, here more generalized R function and G function is introduced.

$$R_{q,v}[a,c,t]=\sum_{n=0}^{\infty}\frac{(a)^n(t-c)^{(n+1)q-1-v}}{\Gamma\{(n+1)q-v\}}\equiv R_{q,v}[a,t-c]$$

Here t is independent variable c is the lower limit of fractional differintegration. Our interest in this function will be normally for the range $t>c$.

The Laplace Transforms of R function are:

$$\mathcal{L}\left\{R_{q,v}(a,0,t)\right\}=\frac{s^v}{s^q-a};\quad \mathrm{Re}(q-v)>0;\quad \mathrm{Re}(s)>0$$

$$\mathcal{L}\left\{R_{q,v}(a,c,t)\right\}=\frac{e^{-cs}s^v}{s^q-a};\quad c\ge 0\quad \mathrm{Re}(q-v)>0;\mathrm{Re}(s)>0$$

2.2.9.1 Relation to Elementary Functions

$$R_{1,0}(a,0,t)=e^{at}$$

$$aR_{2,0}(-a^2,0,t)=a\left\{t-\frac{a^2t^3}{3!}+\frac{a^4t^5}{5!}-\ldots\right\}=\sin(at)$$

$$R_{2,1}(-a^2,0,t)=\left\{1-\frac{a^2t^2}{2!}+\frac{a^4t^4}{4!}-\ldots\right\}=\cos(at)$$

$$aR_{2,0}(a^2,0,t)=\sinh(at)$$

$$R_{2,1}(a^2,0,t)=\cosh(at)$$

2.2.9.2 Relationship of *R* Function to Other Generalized Function

Mittag-Leffler Function:

$$\mathcal{L}\left\{E_q\left[-at^q\right]\right\}=\frac{1}{s}\left[\frac{s^q}{s^q+a}\right]=\frac{s^{q-1}}{s^q+a},q>0$$

$$\mathcal{L}\left\{E_q(-at^q)\right\}=\frac{s^{q-1}}{s^q+a}=\mathcal{L}\left\{R_{q,q-1}(-a,0,t)\right\};\quad {}_cd_t^{q-1}R_{q,0}(-a,c,t)=E_q[-a(t-c)^q]$$

Agarwal Function:

$$\mathcal{L}\left\{E_{q,p}(t^q)\right\}=\frac{s^{q-p}}{s^q-1}=\mathcal{L}\left\{R_{q,q-p}(1,0,t)\right\};\quad R_{q,q-p}(1,0,t)=E_{q,p}(t^q)$$

Erdelyi's Function:

$$t^{1-\beta}E_{q,\beta}(t^q)=R_{q,q-\beta}(1,0,t)=t^{1-\beta}\sum_{n=0}^{\infty}\frac{t^{nq}}{\Gamma(nq+1)}$$

Robotnov and Hartley Function:

$$\mathcal{L}\left\{F_q(-a,t)\right\}=\frac{1}{s^q+a}=\mathcal{L}\left\{R_{q,0}(-a,0,t)\right\};R_{q,0}(-a,0,t)=\sum_{n=0}^{\infty}\frac{(-a)^n t^{(n+1)q-1}}{\Gamma(\{n+1\}q)}=F_q(-a,t)$$

$$\mathcal{L}\left\{F_q\left[a,t\right]\right\}=\frac{1}{s^q-a},q>0$$

$$\mathcal{L}\left\{E_q\left[-at^q\right]\right\}=\frac{1}{s}\left[s^q\mathcal{L}\left\{F_q\left[-a,t\right]\right\}\right]$$

$$_0d_t^{q-1}F_q[a,t] = E_q\left[at^q\right]$$

$$\mathcal{L}^{-1}\left\{\frac{1}{s(s^q+a)}\right\} = \left\{\frac{1}{a}\left[H(t) - E_q(-at^q)\right]\right\} = \left\{_0d_t^{-q}E_q\left[at^q\right]\right\}$$

$$\mathcal{L}^{-1}\left\{\frac{s^q}{s^q+a}\right\} = {}_0d_t^qF_q[-a,t] = {}_0d_t^1E_q\left[-at^q\right] = \mathcal{L}^{-1}\left\{1 - \frac{a}{s^q+a}\right\} = \delta(t) - aF_q[-a,t]$$

Miller and Ross Function:

$$\mathcal{L}\{E_t(v,a)\} = \frac{s^{-v}}{s-a} = \mathcal{L}\{R_{1,-v}(a,0,t)\}; R_{1,-v}(a,0,t) = \sum_{n=0}^{\infty}\frac{(a)^n t^{n+v}}{\Gamma(n+v+1)} = E_t(v,a)$$

2.2.9.3 Further Generalized Function (*G* Function)

$$G_{q,v,r}(a,t) = \sum_{j=0}^{\infty}\frac{\{(-r)(-1-r)...(1-j-r)\}(-a)^j t^{(r+j)q-v-1}}{\Gamma(1+j)\Gamma(\{r+j\}q-v)}; \quad \mathcal{L}\{G_{q,v,r}(a,t)\} = \frac{s^v}{(s^q-a)}$$

$$\mathrm{Re}(qr-v) > 0 \quad \mathrm{Re}(s) > 0 \quad \left|a/s^q\right| > 0$$

2.2.10 *Bessel Function*

Perhaps among all higher transcendental functions the Bessel functions are the most ubiquitous; they appear very frequently in theoretical physics and engineering. The Bessel function, $J_v(z)$ of first kind and order v is defined as infinite series:

$$J_v(z) = \left(\frac{1}{2}z\right)^v \sum_{k=0}^{\infty}\frac{(-1)^k}{k!\Gamma(v+k+1)}\left(\frac{1}{2}z\right)^{2k}$$

The $J_v(z)$ is an entire function of v is solution of Bessel equation $z^2D^2w + zDw + (z^2 - v^2)w = 0$. The Bessel function $Y_v(z)$ of second kind and order v is also the solution of Bessel equation, linearly independent of $J_v(z)$. One of the integral representations of Bessel function is by Poisson's formula, as:

$$J_v(z) = \frac{2}{\Gamma\left(\frac{1}{2}\right)\Gamma\left(v+\frac{1}{2}\right)}\left(\frac{z}{2}\right)^v \int_0^1 (1-t^2)^{v-1/2}\cos zt\,dt\,,\ \Re e(v) > -\frac{1}{2}$$

Another representation is by Sonin's formula, as:

$$J_{\mu+v+1}(z) = \frac{z^{v+1}}{2^v\Gamma(v+1)}\int_0^{(1/2)\pi} J_\mu(z\sin\theta)\sin^{\mu+1}\theta\cos^{2v+1}\theta\,d\theta,\ \Re e(\mu), \Re e(v) > 1$$

Some special relations of elementary function to Bessel functions are:

$$J_{1/2}(z)=\sqrt{\frac{2}{\pi z}}\sin z \text{ and } J_{-1/2}(z)=\sqrt{\frac{2}{\pi z}}\cos z$$

The modified Bessel function $I_v(z)$ of the first kind and order v defined as infinite series is:

$$I_v(z)=\left(\frac{1}{2}z\right)^v\sum_{k=0}^{\infty}\frac{1}{k!\Gamma(v+k+1)}\left(\frac{1}{2}z\right)^{2k},$$

is solution of modified Bessel Equation, $z^2D^2w+zDw-(z^2+v^2)w=0$. The modified Bessel function $K_v(z)$ of the second kind is also solution of modified Bessel equation and is linearly independent of $I_v(z)$.

For $\Re e(v)>-\frac{1}{2}$ and $\Re e(z)>0$, the integral representation is:

$$K_v(z)=\frac{\pi^{1/2}\left(\frac{1}{2}z\right)^v}{\Gamma\left(v+\frac{1}{2}\right)}\int_0^{\infty}(\sinh\theta)^{2v}e^{-z\cosh\theta}d\theta=\int_0^{\infty}\cosh v\theta e^{-z\cosh\theta}d\theta$$

Particularly for nonnegative integer v we may write infinite series for $K_v(z)$. For example for $v=0$

$$K_0(z)=-\left(\ln\frac{1}{2}z\right)I_0(z)+\sum_{n=0}^{\infty}\frac{\psi(n+1)}{(n!)^2}\left(\frac{1}{2}z\right)^{2n}$$

Some relation special values in relation to the elementary functions are:

$$I_{1/2}(z)=\sqrt{\frac{2}{\pi z}}\sinh z,\ I_{-1/2}(z)=\sqrt{\frac{2}{\pi z}}\cosh z \text{ and } K_{1/2}(z)=K_{-1/2}(z)=\sqrt{\frac{\pi}{2z}}e^{-z}$$

Table 2.1 Lists of Higher Transcendental Functions with its Laplace Transforms

Function	**Time Expression** $f(t)$	**Laplace Transform** $F(s)$
Mittag-Leffler	$E_q(at^q)=\sum_{n=0}^{\infty}\frac{a^n t^{nq}}{\Gamma(nq+1)}$	$\frac{s^q}{s(s^q-a)}$
Agarwal	$E_{\alpha,\beta}(t^\alpha)=\sum_{m=0}^{\infty}\frac{t^{\left(m+\frac{\beta-1}{\alpha}\right)}}{\Gamma(\alpha m+\beta)}$	$\frac{s^{\alpha-\beta}}{s^\alpha-1}$
Erdelyi	$E_{\alpha,\beta}(t)=\sum_{m=0}^{\infty}\frac{t^m}{\Gamma(\alpha m+\beta)}$	$\sum_{m=0}^{\infty}\frac{\Gamma(m+1)}{\Gamma(\alpha m+\beta)s^{m+1}}$
Robotnov-Hartley	$F_q(a,t)=\sum_{n=0}^{\infty}\frac{a^n t^{(n+1)q-1}}{\Gamma(\{n+1\}q)}$	$\frac{1}{s^q-a}$
Miller-Ross	$E_t(v,a)=\sum_{k=0}^{\infty}\frac{a^k t^{k+v}}{\Gamma(v+k+1)}$	$\frac{s^{-v}}{s-a}$
Generalized Cosine	$C_t(v,a)=t^v\sum_{k=even}^{\infty}\frac{(-1)^{k/2}(at)^k}{\Gamma(v+k+1)}$	$\frac{s}{s^v(s^2+a^2)}$, $\Re e(v)>-1$
Generalized Sine	$S_t(v,a)=t^v\sum_{k=odd}^{\infty}\frac{(-1)^{\frac{(k-1)}{2}}(at)^k}{\Gamma(v+k+1)}$	$\frac{a}{s^v(s^2+a^2)}$ $\Re e(v)>-2$
Generalized R	$R_{q,v}(a,t)=\sum_{n=0}^{\infty}\frac{a^n t^{(n+1)q-1-v}}{\Gamma(\{n+1\}q-v)}$	$\frac{s^v}{s^q-a}$
Generalized G	$G_{q,v,r}(a,t)=\sum_{j=0}^{\infty}\frac{\{(-r)(-1-r)...(1-j-r)\}(-a)^j t^{(r+j)q-v-1}}{\Gamma(1+j)\Gamma(\{r+j\}q-v)}$	$\frac{s^v}{(s^q-a)^r}$

2.3 List of Laplace and Inverse Laplace Transforms Related to Fractional Calculus

Table 2.2 List of useful mathematical-physics functions with Laplace Transforms

Laplace Transform $F(s)$	**Time Expression** $f(t)$
$\frac{s^{\alpha-1}}{s^\alpha\mp\lambda},\ \Re(s)>\lvert\lambda\rvert^{1/\alpha}$	$E_{\alpha,1}(\pm\lambda t^\alpha)$
$\frac{k!s^{\alpha-\beta}}{(s^\alpha\mp\lambda)^{k+1}},\ \Re(s)>\lvert\lambda\rvert^{1/\alpha}$	$t^{\alpha k+\beta-1}E_{\alpha,\beta}{}^{(k)}(\pm\lambda,t^\alpha)$
$\frac{k!}{(\sqrt{s}\mp\lambda)^{k+1}},\ \Re(s)>\lambda^2$	$t^{\frac{k-1}{2}}E^{(k)}_{\frac{1}{2},\frac{1}{2}}(\pm\lambda\sqrt{t})$
$\frac{1}{s^\alpha}$	$\frac{t^{\alpha-1}}{\Gamma(\alpha)}$
$\arctan\frac{k}{s}$	$\frac{1}{t}\sin(kt)$

$\log\frac{s^2-a^2}{s^2}$	$\frac{2}{t}(1-\cosh at)$
$\log\frac{s^2+a^2}{s^2}$	$\frac{2}{t}(1-\cos at)$
$\log\frac{s-a}{s-b}$	$\frac{1}{t}(e^{bt}-e^{at})$
$\frac{e^{-k\sqrt{s}}}{\sqrt{s}(a+\sqrt{s})},(k\geq 0)$	$e^{ak}e^{a^2t}\operatorname{erfc}(a\sqrt{t}+\frac{k}{2\sqrt{t}})$
$\frac{ae^{-k\sqrt{s}}}{s(a+\sqrt{s})},(k\geq 0)$	$\operatorname{erfc}\left(\frac{k}{2\sqrt{t}}\right)-e^{ak}e^{a^2t}\operatorname{erfc}\left(a\sqrt{t}+\frac{k}{2\sqrt{t}}\right)$
$\frac{1}{s\sqrt{s}}e^{-k\sqrt{s}},(k\geq 0)$	$2\sqrt{\frac{t}{\pi}}e^{-\frac{k^2}{4t}}-(k)\operatorname{erfc}\left(\frac{k}{2\sqrt{t}}\right)$
$\frac{1}{\sqrt{s}}e^{-k\sqrt{s}},(k\geq 0)$	$\frac{1}{\sqrt{\pi.t}}e^{-\frac{k^2}{4t}}$
$\frac{1}{s}e^{-k\sqrt{s}},(k\geq 0)$	$\operatorname{erfc}\left(\frac{k}{2\sqrt{t}}\right)$
$e^{-k\sqrt{s}},(k\geq 0)$	$\frac{k}{2\sqrt{\pi.t^3}}e^{-\frac{k^2}{4t}}$
$\frac{1}{s^v}e^{k/s},(v>0)$	$\left(\frac{t}{k}\right)^{(v-1)/2}I_{v-1}(2\sqrt{kt})$
$\frac{1}{s^v}e^{-k/s},(v>0)$	$\left(\frac{t}{k}\right)^{(v-1)/2}J_{v-1}(2\sqrt{kt})$
$\frac{1}{s\sqrt{s}}e^{k/s}$	$\frac{1}{\sqrt{\pi.k}}\sinh 2\sqrt{kt}$
$\frac{1}{s\sqrt{s}}e^{-k/s}$	$\frac{1}{\sqrt{\pi.k}}\sin 2\sqrt{kt}$
$\frac{1}{\sqrt{s}}e^{k/s}$	$\frac{1}{\sqrt{\pi.t}}\cosh 2\sqrt{kt}$
$\frac{1}{\sqrt{s}}e^{-k/s}$	$\frac{1}{\sqrt{\pi.t}}\cos 2\sqrt{kt}$
$\frac{1}{s}e^{-k/s}$	$J_0(2\sqrt{kt})$
$\frac{k}{s^2+k^2}\coth\frac{\pi.s}{2k}$	$\lvert\sin kt\rvert$

$\frac{1}{\sqrt{s}}$	$\frac{1}{\sqrt{(\pi)(t)}}$
$\frac{1}{s\sqrt{s}}$	$2\sqrt{\frac{t}{\pi}}$
$\frac{1}{s^n\sqrt{s}}, (n=1,2,...)$	$\frac{2^n t^{n-(1/2)}}{1.3.5...(2n-1)\sqrt{\pi}}$
$\frac{s}{(s-a)^{3/2}}$	$\frac{1}{\sqrt{\pi.t}} e^{at}(1+2at)$
$\sqrt{s-a}-\sqrt{s-b}$	$\frac{1}{2\sqrt{\pi.t^3}}(e^{bt}-e^{at})$
$\frac{1}{\sqrt{s+a}}$	$\frac{1}{\sqrt{\pi.t}} - ae^{a^2t}\operatorname{erfc}(a\sqrt{t})$
$\frac{\sqrt{s}}{s-a^2}$	$\frac{1}{\sqrt{(\pi)(t)}} + ae^{a^2t}\operatorname{erf}(a\sqrt{t})$
$\frac{\sqrt{s}}{s+a^2}$	$\frac{1}{\sqrt{\pi.t}} - \frac{2a}{\sqrt{\pi}} e^{-a^2t} \int_0^{a\sqrt{t}} e^{\tau^2} d\tau$
$\frac{1}{\sqrt{s}(s-a^2)}$	$\frac{1}{a} e^{a^2t}\operatorname{erf}(a\sqrt{t})$
$\frac{1}{\sqrt{s}(s+a^2)}$	$\frac{2}{a\sqrt{\pi}} e^{-a^2t} \int_0^{a\sqrt{t}} e^{\tau^2} d\tau$
$\frac{b^2-a^2}{(s-a^2)(\sqrt{s}+b)}$	$e^{a^2t}\left[b-a\{\operatorname{erf}(a\sqrt{t})\}\right] - be^{b^2}\operatorname{erfc}(b\sqrt{t})$
$\frac{1}{\sqrt{s}(\sqrt{s}+a)}$	$e^{a^2t}\operatorname{erfc}(a\sqrt{t})$
$\frac{1}{\sqrt{s+b}(s+a)}$	$\frac{1}{\sqrt{b-a}} e^{-at}\operatorname{erf}(\sqrt{b-a}\sqrt{t})$
$\frac{b^2-a^2}{\sqrt{s}(s-a^2)(\sqrt{s}+b)}$	$e^{a^2t}\left[\frac{b}{a}\operatorname{erf}(a\sqrt{t})-1\right] + e^{b^2t}\operatorname{erfc}(b\sqrt{t})$
$\frac{(1-s)^n}{s^{n+(1/2)}}$	$\frac{n!}{(2n)!\sqrt{\pi t}} H_{2n}(\sqrt{t})$ $H_n(x) = e^{x^2}\frac{d^n}{dx^n}(e^{-x^2})$ Hermite Polynomial
$\frac{(1-s)^n}{s^{n+(3/2)}}$	$-\frac{n!}{(2n+1)!\sqrt{\pi}} H_{2n+1}(\sqrt{t})$

$\dfrac{\sqrt{s+2a}-\sqrt{s}}{\sqrt{s}}$	$ae^{-at}\left[I_1(at)+I_0(at)\right]$ $I_n(x)=j^{-n}J_n(jt)$ J_n Bessel function of first kind
$\dfrac{1}{\sqrt{s+a}\sqrt{s+b}}$	$e^{-\frac{1}{2}(a+b)t}I_0\left(\dfrac{a-b}{2}t\right)$
For $k>0$	$\sqrt{\pi}\left(\dfrac{t}{a-b}\right)^{k-(1/2)}e^{-\frac{1}{2}(a+b)t}I_{k-(1/2)}\left(\dfrac{a-b}{2}t\right)$
$\dfrac{1}{(s+a)^{1/2}(s+b)^{3/2}}$	$te^{-\frac{1}{2}(a+b)t}\left[I_0\left(\dfrac{a-b}{2}t\right)+I_1\left(\dfrac{a-b}{2}t\right)\right]$
$\dfrac{\sqrt{s+2a}-\sqrt{s}}{\sqrt{s+2a}+\sqrt{s}}$	$\dfrac{1}{t}e^{-at}I_1(at)$
$\dfrac{(a-b)^k}{(\sqrt{s+a}+\sqrt{s+b})^{2k}}$ For $k>0$	$\dfrac{k}{t}e^{-\frac{1}{2}(a+b)t}I_k\left(\dfrac{a-b}{2}t\right)$
$\dfrac{1}{\sqrt{s}\sqrt{s+a}(\sqrt{s+a}+\sqrt{s})^{2v}}$ For $k>0$	$\dfrac{1}{a^v}e^{-\frac{1}{2}at}I_v\left(\dfrac{a}{2}t\right)$
$\dfrac{1}{\sqrt{s^2+a^2}}$	$J_0(at)$
$\dfrac{1}{\sqrt{s^2-a^2}}$	$I_0(at)$, Modified Bessel function of the first kind zero order.
$\dfrac{(\sqrt{s^2+a^2}-s)^v}{\sqrt{s^2+a^2}}$ $(v>-1)$	$a^vJ_v(at)$
$\dfrac{1}{(\sqrt{s^2+a^2})^k}$ $k>0$	$\dfrac{\sqrt{\pi}}{\Gamma(k)}\left(\dfrac{t}{2a}\right)^{k-(1/2)}J_{k-(1/2)}(at)$
$(\sqrt{s^2+a^2}-s)^k$, $(k>0)$	$\dfrac{ka^k}{t}J_k(at)$
$\dfrac{(\sqrt{s^2-a^2}+s)^v}{\sqrt{s^2-a^2}}$, $(v>-1)$	$a^vI_v(at)$
$\dfrac{1}{(s^2-a^2)^k}$, $(k>0)$	$\dfrac{\sqrt{\pi}}{\Gamma(k)}\left(\dfrac{t}{2a}\right)^{k-(1/2)}I_{k-(1/2)}(at)$
$\dfrac{1}{s\sqrt{s+1}}$	$\mathrm{erf}(\sqrt{t})$ $\mathrm{erf}(x)=\dfrac{2}{\sqrt{\pi}}\int_0^x e^{-t^2}dt=1-\mathrm{erfc}(x)$

$\frac{1}{s+\sqrt{s^2+a^2}}$	$\frac{J_1(at)}{at}$
$\frac{1}{(s+\sqrt{s^2+a^2})^N}$	$\frac{NJ_N(at)}{a^N t}$
$\frac{1}{\sqrt{s^2+a^2}(s+\sqrt{s^2+a^2})}$	$\frac{J_1(at)}{a}$
$\frac{1}{\sqrt{s^2+a^2}(s+\sqrt{s^2+a^2})^N}$	$\frac{J_N(at)}{a^N}$

2.4 Paradoxial Conditions for Using Generalized Differentiation and Integration Expressions and Cautions

As with most mathematical expressions reasonable caution must be exercised. Let us try to see that if fractional derivative of order μ be obtained from fractional integral of order v, by replacing v with $-\mu$. This works with Euler's expression for x^b, with $b>-1$, but can this be generalized to any functions. That is ${}_0D_t^{\mu} f(t)=\left[{}_0D_t^{-v} f(t)\right]_{v=-\mu}$. For $v>0$, the fractional integration expression for $f(t)=t^{\lambda}, \lambda>-1$ is:

$${}_0D_t^{-v} t^{\lambda}=\frac{1}{\Gamma(v)}\int_0^t (t-u)^{v-1}u^{\lambda}du=\frac{\Gamma(\lambda+1)}{\Gamma(\lambda+v+1)}t^{\lambda+v}$$

By replacing v, with $-\mu$ ($\mu>0$) we get

$${}_0D_t^{\mu} t^{\lambda}=\frac{\Gamma(\lambda+1)}{\Gamma(\lambda-\mu+1)}t^{\lambda-\mu},$$

which is correct result.

But using this changed sign in integral expression, we get.

$${}_0D_t^{\mu} t^{\lambda}=\frac{1}{\Gamma(-\mu)}\int_0^t (t-u)^{-\mu-1}u^{\lambda}du,$$

which diverges (is absurd).

The conclusion is that Euler's expression is meaningful for positive or negative v (fractional differentiation and fractional integration); but the integral Riemann-Liouvelli representation for fractional integration is meaningful only for $v>0$.

Let us find the paradoxical condition, that is first we find integer n-th derivative of a function f, namely $D^n f(t)$, $n=1,2,\ldots$ and then from here we derive integral of f by replacing with negative one.

That is $D^{-1}f(t)=\int_0^t f(u)du$, this is incorrect.

Take for example $f(t)=\sin t$. Then $D^n f(t)=\sin\left(t+\frac{1}{2}n\pi\right)$, $n=1,2,3,\ldots$.

Replacing $n=-1$ yields $D^{-1}\sin t=\int_0^t \sin u du=\sin\left(t-\frac{1}{2}\pi\right)=-\cos t$.

But as per integer order calculus $\int_0^t \sin u du=1-\cos t$. This indicates paradox.

For another function, $f(t)=\ln t$, $D^n \ln t=(-1)^{n-1}\Gamma(n)t^{-n}$, for $n=1,2,3,\ldots$

Now letting $n=-1$, we get $D^{-1}\ln t=\int_0^t \ln u du=\Gamma(-1)t=\infty$, which is incorrect.

The paradox stems from the fact that we are computing ordinary derivatives of integer order, and then attempting to derive a result based on replacing the order of derivative by negative integer number. Integer order derivative is subset of generalized differ-integration, thus what we should try is to derive expression for fractional derivative $D^v f(t)$ which is true for all $v\in\mathbb{R}$, then deduce integer order expressions. In case of $f(t)=\sin t$, the $D^v\sin t=S_t(-v,1)$ (obtained from Higher Transcendental Miller-Ross expression).

Now if $v=-1$, we have $D^{-1}f(t)=\int_0^t \sin u du=S_t(1,1)=1-\cos t$, which is correct result. Putting $v=1$, we get $D\sin t=S_t(-1,1)=\cos t$, again correct result.

For the case $f(t)=\ln t$, $D^v\ln t=t^{-v}\left[\Gamma(1-v)\right]^{-1}\left[\ln t-\gamma-\psi(1-v)\right]$.

If we put $v=-1$ then $D^{-1}\ln t=\int_0^t \ln u du=t\left[\Gamma(2)\right]^{-1}\left[\ln t-\gamma-\psi(2)\right]=t(\ln t-1)$ is correct expression. Here the property of 'psi' function is used that is $\psi(z+1)-\psi(z)=z^{-1}$.

Now letting $v=1$ in the expression obtained above for $D^v\ln t$, the RHS of this becomes indeterminate. However, by using the property of 'psi' function and writing $z=1-v$, and with, $\Gamma(1-v)=(1-v)^{-1}\Gamma(v)$ we obtain $D\ln t=t^{-1}$ again getting correct result.

Consider the function $f(t)=e^t$, then

$$D^v e^t=D^v\sum_{k=0}^{\infty}\frac{t^k}{k!}=\sum_{k=0}^{\infty}\frac{t^{k-v}}{\Gamma(k-v+1)}=E_t(-v,1)\text{, for arbitrary } v.$$

Now if v is nonnegative integer say $v=n$, then

$$D^n e^t = \sum_{k=0}^{\infty} \frac{t^{k-n}}{\Gamma(k-n+1)} = \sum_{j=0}^{\infty} \frac{t^j}{j!} = e^t ,$$

which is independent of n. Here if we replace n by -1, we get $D^{-1}e^t = e^t$, which is not correct.

Thus if we replace $v = -1$, we get $D^{-1}e^t = \sum_{k=0}^{\infty} \frac{t^{k+1}}{\Gamma(k+2)} = e^t - 1$,

and is correct result, namely $D^{-1}e^t = \int_0^t e^u du = e^t - 1$.

2.5 Non-exponential Relaxation Power Law and Memory Integrals

Maxwell Debye relaxation process is standard process of reaction that is formulated as

$$\tau \frac{d\phi(t)}{dt} = -\phi(t) ,$$

for $t > 0$ and with initial condition $\phi(0) = \phi_0$,

The solution of this relaxation equation is $\phi(t) = \phi_0 e^{-t/\tau}$. In Integral formulation (Integrating the above differential equation) we re-write the standard relaxation law as

$$\phi(t) - \phi_0 = -\frac{1}{\tau}\frac{d^{-1}\phi(t)}{dt^{-1}} = -\frac{1}{\tau}\int_0^t \phi(t')dt'$$

These equations from integer order representation can be generalized with following replacement

$$-\frac{1}{\tau}\frac{d^{-1}\phi(t)}{dt^{-1}} \rightarrow -\frac{1}{\tau^\beta}\frac{d^{-\beta}\phi(t)}{dt^{-\beta}} .$$

This leads to fractional integral equation as:

$$\phi(t) - \phi_0 = -\frac{1}{\tau^\beta} {}_0D_t^{-\beta}\phi(t) .$$

This is non-exponential relaxation process. The curve of which can be fitted by Kohlrash-Williams-Watts (KWW) stretched exponential law as:

$$\phi(t) = \phi_0 \exp[(-t/\tau)^\alpha] \text{ , with } 0 < \alpha < 1 ,$$

or via asymptotic power law (Nutting Power Law)

$$\phi(t) = \phi_0 \frac{1}{\left(1 + {t}/{\tau}\right)^n}, \text{ with } 0 < n < 1,$$

and also via Mittag-Leffler or higher transcendental function types. This type of generalized relaxation processes are Non-Debye or Non Exponential relaxation processes. For sufficiently large experimental window, it is generally possible to observe transition from KWW to the Nutting-behavior or vice-versa. Glockle and Nonnenmacher in 1991 for the first time interpolates relaxation data by Mittag-Leffler (ML) function,

$$\phi(t) = \phi_0 \left(\frac{t}{\tau}\right)^{-\beta} E_{\alpha-\beta,1-\beta}\left(-\left(\frac{t}{\tau}\right)^{\alpha-\beta}\right) \;;$$

encompassing both KWW and Nutting-behavior. This is one special example of asymptotic fractal. Often one encounters functional relation in frequency domains of the form $\omega^{-\beta}F(\omega)$. Such relations back-transformed to time (space) by generalized differential theorem that gives fractional integral equation ${}_0D_t^{-\beta} f(t)$ and the temporal (special) relaxation of which can be put as asymptotic fractal expression of the form

$$\phi(t) \sim \left(\frac{t}{\tau}\right)^{-\beta} \gamma\left(\beta, \left({t}/{\tau}\right)\right),$$

where γ is incomplete Gamma function.

Non exponential relaxation implies memory, i.e. the underlying fundamental relaxation process is Non-Markovian. Natural way to incorporate memory effect is via fractional calculus, the involved convolution integral in time (space), where the present state of the system has been running through at times $t' = 0, 1, 2 \ldots t$. The memory integrals defined as:

$$\frac{d\phi(t)}{dt} = -\int_0^t K(t-t')\phi(t')dt' ,$$

encompassing all instances from $t' = 0$ to $t' = t$, contributing to the situation at $t' = t$. Following are interesting cases for memory kernel.

(i) Markovian Case i.e. memory breaking down (system without memory): The kernel in this case is $K(t-t') = K_0\delta(t-t')$, putting this kernel in the memory integral one gets

$$\frac{d\phi(t)}{dt} = -\int_0^t K_0\delta(t-t')\phi(t')dt' = -K_0\phi(t),$$

and the relaxation expression is exponential law of Maxwell-Debye type $\phi(t) = \phi_0 \exp(-K_0 t)$, with $\phi(0) = \phi_0$ for the corresponding homogeneous differential equation that is:

$$\frac{d\phi(t)}{dt} + K_0\phi(t) = 0$$

This relaxation process is like standard radioactive decay having one decay constant, and decay process is having no memory. Well if say the radioactive decay comprises of intermittent discharges with no average decay constant, then the system identification is of different kind, and there Fractional Differential Equation will play role to indicate degree of intermittency in relaxation process. Similarly this simple law will change if there are variety of paths for the relaxation (or decay).

(ii) Constant Memory Case, where kernel is $K(t) = K_0$, independent of time, leads to oscillatory solution as $\phi(t) = \phi_0 \cos(\sqrt{K_0} t)$; with $\phi(0) = \phi_0$, for the corresponding homogeneous differential equation that is:

$$\frac{d^2\phi(t)}{dt^2} + K_0\phi(t) = 0$$

Here the system remembers or holds its state (say energy); examples include one perfect $L\ C$ circuit holding forever the electromagnetic energy, oscillating between pure electrical energy and pure magnetic energy and a lossless ideal pendulum remembering its initial excitation force, that is $f(t) = \delta(t)$.

(iii) Slowly varying kernel as $K(t) = K_0 t^{-\gamma}\delta(t)$, gives KWW or stretched exponential law as:

$$\phi(t) = \exp\left(-K_0 t^{1-\gamma} / 1-\gamma\right)$$

If $\gamma = 0$, we have exponential relaxation, and memory kernel as delta function (as in (i)). For above law taking the Kernel as $K(t-t') = K_0 (t')^{-\gamma}\delta(t-t')$, gives non-linear homogeneous differential equation as:

$$\frac{d}{dt}\phi(t) + K_0 t^{-\gamma}\phi(t) = 0$$

(iv) Suppose the kernel is of type $K(t) = K_0 t^{q-2}$, with $0 < q \le 2$, then for $t > 0$, the Memory Integral is with RL fractional integral equation that can be casted as:

$$\frac{d\phi(t)}{dt} = -\tau^{-q} {}_0D_t^{1-q}\phi(t),$$

with $\tau^{-q} = K_0\Gamma(q-1)$. Now in the above equation we apply ${}_0D_t^{-1}$, i.e. integrating both sides, we obtain $\phi(t) - \phi_0(t) = -\tau^{-q} {}_0D_t^{-q}\phi(t)$; that is fractional integral equation. Now we apply fractional derivative operator ${}_0D_t^{q}$, to the both sides of fractional integral equation, with formula of fractional derivative of a constant ϕ_0 as $\phi_0 t^{-q}[\Gamma(1-q)]^{-1}$, and obtain fractional differential equation as:

$${}_0D_t^{q}\phi(t) - \frac{\phi_0 t^{-q}}{\Gamma(1-q)} = -\tau^{-q}\phi(t)$$

Thus we have derived the above fractional integral and fractional differential equations for a power-law memory kernel. All these concepts will be used in system identification with disorder, in Chapter 10.

2.6 Boltzmann's Superposition Principle

Another physical basis to come up with memory integrals is to start from Boltzmann's superposition principle, which formally incorporates memory via causal convolution. Power law relaxation observation $\phi(t) \sim t^{-\alpha}$ is wide spread in natural systems, and the asymptotic decays can be expressed as several forms of higher transcendental functions of Mittag-Leffler or related types. Consider an arbitrary history of external perturbation $\psi(t)$, and let us denote $G(t) = \phi_S(t)$ the response of the system to step like external perturbation $\psi(t) = H(t)$;(Heaviside step function). Then causal convolution system response is:

$$\phi(t) = \int_{-\infty}^{t} dt' G(t-t') \frac{d\psi(t')}{dt'}$$

$\phi(t)$ is through variable like current, stress and flow, $\psi(t)$ is across variable like voltage, strain and pressure. In the above causal convolution the $\psi(t)$ field starts at $-\infty$, at the distant past, and the above integral representation is of Weyl's notation. Whereas limiting $\psi(t) = \phi(t) = 0$ for $t \le 0$ leads to Riemann-Liouvelli formulation, in this case it is initial value problem. The above convolution expression is popular among control system engineers. The $G(t)$ is the 'impulse-response' for the control system block, where $\psi'(t)$ is the input, and $\phi(t)$ is the output. Obviously $\psi'(t)$ is impulse function for $\psi(t)$ as step input expressed above. In pure mathematics this $G(t)$ is Green's function; solution to

homogeneous system of differential equation system. In mathematics after evaluating Green's function, its convolution with source function gives particular solution to the system of non-homogeneous differential equation system.
Specifically let $G(t)$ obey a power law as:

$G(t) = \phi_S(t) = \dfrac{C}{\Gamma(1-\alpha)}\left(\dfrac{t}{\tau}\right)^{-\alpha}$, and comparing this with the causal convolution above, we find the response $\phi(t)$ to arbitrary history $\psi(t)$ as:

$\phi(t) = \dfrac{C\tau^{\alpha}}{\Gamma(1-\alpha)}\int_{-\infty}^{t} dt'(t-t')^{-\alpha}\dfrac{d\psi(t')}{dt'}$, which is fractional integral expression.

Thus the causal convolution for this specific power-law decay is.

$$\phi(t) = C\tau^{\alpha}\frac{d^{\alpha-1}}{dt^{\alpha-1}}\frac{d\psi(t)}{dt} = C\tau^{\alpha}\frac{d^{\alpha}}{dt^{\alpha}}\psi(t)$$

Thus the system response $\phi(t)$ follows by fractional differentiation of external perturbation $\psi(t)$. Such power law $G(t)$ as taken here can be found in practical systems of Electrical, Electronics, Mechanical Engineering and various other physical system dynamic behaviors. If one chooses the constituent elements (spring, dashpot, resistance, capacitance and inductances) to follow scaling law, one obtains the expression like chosen $G(t)$ -the basic idea leading to memory integral.

Let us exchange the roles of $\phi(t)$ and $\psi(t)$ and re-write the above obtained fractional differential equation, by applying $D_t^{-\alpha}$ to both the sides, to get:

$\psi(t) = C^{-1}\tau^{-\alpha}\dfrac{d^{-\alpha}}{dt^{-\alpha}}\phi(t)$, which can be further decomposed as:

$$\psi(t) = C^{-1}\tau^{-\alpha}\frac{d^{\alpha-1}}{dt^{\alpha-1}}\frac{d\phi(t)}{dt} = \frac{C^{-1}\tau^{-\alpha}}{\Gamma(1+\alpha)}\int_{-\infty}^{t} dt'(t-t')^{\alpha}\frac{d\phi(t')}{dt'},$$

we can infer immediately from this expression that 'impulse response' $G(t)$ for a step like perturbation of

$$\phi(t) = H(t) \text{ is } G(t) = \psi_S(t) = \frac{C^{-1}}{\Gamma(1+\alpha)}\left(\frac{t}{\tau}\right)^{\alpha}.$$

2.7 Motivation to Use Higher Transcendental Functions to Solve Fractional Differential Equations

If we have ordinary differential equation (consider homogeneous ODE), of order n, as $\left[D^n + a_1D^{n-1} + \ldots + a_nD^0\right]y(t) = P(D)y(t) = 0$, where $P(D) \equiv D^n + a_1 D^{n-1} + ... + a_n D^0$. We try, a solution $y(t) = \exp(ct)$, and by doing so we find that $P(D)[\exp ct] = P(c)\exp ct$, where $P(x) = x^n + a_1 x^{n-1} + ... + a_n$, is the indicial polynomial of order n. In other words if c is a root of indicial polynomial $P(x) = 0$, then $\exp ct$ is solution of the ODE. This is so because differentiations of exponential function return itself. If we apply fractional differentiation to exponential function i.e. $D^u\left[e^{ct}\right] = E_t(-u, c)$, we observe that the exponential function changes to Higher Transcendental function, does not retain itself on fractional differentiation. Therefore, $\exp ct$ will not be very useful candidate to try to solve fractional differential equation (FDE). However, if we try to fractionally differentiate Higher Transcendental Function (Miller-Ross) we observe that $D^u E_t(w, c) = E_t(w-u, c)$, retaining its form. Perhaps, this could be a candidate for solving FDE. Also from earlier section we take one more differentiating property of Miller-Ross function as $D^u tE_t(w, c) = tE_t(w-u, c) + uE_t(w-u+1, c)$, $w > -2$. These two expressions are similar to $De^{ct} = ce^{ct}$ and $Dte^{ct} = cte^{ct} + e^{ct}$. We also bring the special value of Miller-Ross function as $E_t(0, c) = e^{ct}$ so that $D^u E_t(0, c) = E_t(-u, c)$. Thus we are tempted to try a function $E_t(kv, c)$ where k is integer for FDE, (like e^{ct}, c an integer for ODE).

Consider a sample FDE $[D + aD^{0.5} + bD^0]y(t) = 0$, which has indicial polynomial as $P(x) = x^2 + ax + b$. This is FDE of order (2, 2) as described in Chapter 1, with $n = 2$ and $q = 2$, and operator as $P(D^{0.5}) \equiv D^1 + aD^{0.5} + bD^0$. By analogy of ODE let us try linear combination of $E_t(0, c)$, $E_t(-0.5, c)$.

That is $\psi_1(t) = AE_t(0, c) + E_t(-0.5, c)$ as possible candidate solution. Simple arithmetic by using the above mentioned properties of Miller-Ross function yields

$$P(D^{0.5})\psi_1(t) = (cA + ac + bA)E_t(0.c) + (c + aA + b)E_t(-0.5, c) + \frac{t^{-3/2}}{\Gamma(-0.5)}$$

Now letting $A = \lambda$ and $c = \lambda^2$, where λ is arbitrary (may be complex number) gives

$$P(D^{0.5})\psi_1(t) = \lambda P(\lambda)E_t(0, \lambda^2) + P(\lambda)E_t(-0.5, \lambda^2) + \frac{t^{-3/2}}{\Gamma(-0.5)}.$$

In particular if λ is root of $P(x)$ then $P(\lambda) = 0$ and the above expression takes the form:

$$P(D^{0.5})\psi_1(t) = \frac{t^{-3/2}}{\Gamma(-0.5)} \neq 0.$$

Thus $\psi_1(t)$ is still not the solution of this FDE (but perhaps a close one). Suppose that α and β are two roots of $P(x)$. Then with $\lambda=\alpha$, we write: $\psi_1(t)=\alpha E_t(0,\alpha^2)+E_t(-0.5,\alpha^2)$.

If we define $\psi_2(t)=\beta E_t(0,\beta^2)+E_t(-0.5,\beta^2)$, then proceeding as above we have:

$$P(D^{0.5})\psi_2(t)=\frac{t^{-3/2}}{\Gamma(-0.5)}.$$

Letting $\psi(t)=\psi_1(t)-\psi_2(t)$

$$=\alpha E_t(0,\alpha^2)-\beta E_t(0,\beta^2)+E_t(-0.5,\alpha^2)-E_t(-0.5,\beta^2)$$

Gives $P(D^{0.5})\psi(t)=\left[D^1+aD^{0.5}+bD^0\right]\psi(t)\equiv 0$. Thus if $\alpha\neq\beta$ the $\psi(t)$ is solution of FDE. With described properties of Miller-Ross function, the solution may also be expressed as $\psi(t)=\alpha e^{\alpha^2 t}\operatorname{erfc}(-\alpha\sqrt{t})-\beta e^{\beta^2 t}\operatorname{erfc}(-\beta\sqrt{t})$. From the solution we observe that $\psi(0)=\alpha-\beta, D^{-0.5}\psi(0)=0$, $D^{0.5}\psi(0)=\infty$ and $D\psi(0)=\infty$.

For equal roots $\alpha=\beta$ similar analogy as in ODE that solution is combination of $e^{\alpha t}$ and $te^{\alpha t}$, makes us to take linear combination of $E_t(0,\alpha^2)$, $E_t(-0.5,\alpha^2)$, $tE_t(0,\alpha^2)$, $tE_t(-0.5,\alpha^2)$, $E_t(0.5,\alpha^2)$ and $tE_t(0.5,\alpha^2)$, as possible logical candidate for FDE when roots are equal. Doing elaborate calculations, the solution to FDE when roots of the indicial polynomial are equal is obtained as:

$$\psi(t)=(1+2\alpha^2 t)E_t(0,\alpha^2)+\alpha E_t(0.5,\alpha^2)+2\alpha(-0.5,\alpha^2)$$
$$=(1+2\alpha^2 t)e^{\alpha^2 t}\operatorname{erfc}(-\alpha\sqrt{t})+\frac{2\alpha\sqrt{t}}{\Gamma(0.5)}.$$

2.8 Fractional Derivatives and Integrals of Important Functions with Use of Higher Transcendental Functions

Here in this section number of examples of fractional integrals and derivatives are tabulated. These are useful functions for physics and engineering. For simplicity assume that all variables are real and that x is positive and $x>0$. The fractional order v is assumed to be arbitrary (positive, negative or zero) unless stated explicitly. The constants a, c, λ and μ are assumed to be unrestricted unless otherwise indicated. The section demonstrates the use of Higher Transcendental functions $\Gamma(x)$, B_τ, ${}_2F_1$, E_t, $\psi(x)$, and others described in this chapter. The derivative in the table is Riemann-Liouvelli type.

Function $f(x)$	**Fractional Derivative /Fractional Integral expression.**
$(x-a)^{\lambda}$, $x>c>a$ and $c\ge 0$	${}_cD_x^{v}(x-a)^{\lambda}=\frac{(c-a)^{\lambda}}{\Gamma(1-v)}(x-c)^{-v}\times{}_2F_1\left(-\lambda,1,1-v;-\frac{x-c}{c-a}\right)$
$(x-a)^{\lambda}$, $x>c>a, c\ge 0$ $v>0$	${}_cD_x^{-v}(x-a)^{\lambda}=\frac{(x-a)^{\lambda+v}}{\Gamma(v)}\mathrm{B}_{\chi}(v,\lambda+1)$, $\chi=\frac{x-c}{x-a}$
$(x-a)^{\lambda}$, $x>c=a\ge 0$, $\lambda>-1$	${}_cD_x^{v}(x-a)^{\lambda}=\frac{\Gamma(\lambda+1)}{\Gamma(\lambda-v+1)}(x-a)^{\lambda-v}$
x^{λ}, $x>c=a=0$, $\lambda>-1$	${}_0D_x^{v}x^{\lambda}=\frac{\Gamma(\lambda+1)}{\Gamma(\lambda-v+1)}x^{\lambda-v}$
$f(x)=1, x>c\ge 0$	${}_cD_x^{v}(1)=\frac{(x-c)^{-v}}{\Gamma(1-v)}$
$f(x)=1$, $x>c=0$	${}_0D_x^{v}(1)=\frac{x^{-v}}{\Gamma(1-v)}$
$(a-x)^{\lambda}$, $a>x>c\ge 0$	${}_cD_x^{v}(a-x)^{\lambda}=\frac{(a-c)^{\lambda}}{\Gamma(1-v)}(x-c)^{-v}\times{}_2F_1\left(-\lambda,1,1-v;\frac{x-c}{a-c}\right)$
$(a-x)^{\lambda}, a>x>c\ge 0$, $v>0$	${}_cD_x^{-v}(a-x)^{\lambda}=\frac{(a-x)^{\lambda+v}}{\Gamma(v)}\mathrm{B}_{\chi}(v,-[\lambda+v])$, $\chi=\frac{x-c}{a-c}$
$(a-x)^{-1/2}$, $a>x$	${}_0D_x^{-1/2}(a-x)^{-1/2}=\frac{1}{\sqrt{\pi}}\ln\frac{\sqrt{a}+\sqrt{x}}{\sqrt{a}-\sqrt{x}}$
$(a-x)^{-1/2}$, $a>x$	${}_0D_x^{1/2}(a-x)^{-1/2}=\sqrt{\frac{a}{\pi t}}\frac{1}{a-x}$
$\exp ax$	${}_0D_x^{v}e^{ax}=E_x(-v,a)$
$\exp ax$, $a>0$	${}_0D_x^{-1/2}e^{ax}=a^{-1/2}e^{ax}\mathrm{erf}(\sqrt{ax})$
$\exp ax$, $\lambda>-1$	${}_0D_x^{v}t^{\lambda}e^{ax}=\frac{\Gamma(\lambda+1)}{\Gamma(\lambda-v+1)}x^{\lambda-v}{}_1F_1(\lambda+1,\lambda-v+1;ax)$
$x\exp ax$	${}_0D_x^{v}xe^{ax}=xE_x(-v,a)+vE_x(1-v,a)$
$E_x(\mu,a)$, $\mu>-1$	${}_0D_x^{v}E_x(\mu,a)=E_x(\mu-v,a)$
$x^{\lambda}E_x(\mu,a)$, $\lambda+\mu>-1$	${}_0D_x^{v}\left[x^{\lambda}E_x(\mu,a)\right]=\frac{\Gamma(\lambda+\mu+1)x^{\lambda+\mu-v}}{\Gamma(\mu+1)\Gamma(\lambda+\mu-v+1)}$ $\times{}_2F_2(\lambda+\mu+1,1,\mu+1,\lambda+\mu-v+1;ax)$
$xE_x(\mu,a)$, $\mu>-2$	${}_0D_x^{v}\left[xE_x(\mu,a)\right]=xE_x(\mu-v,a)+vE_x(\mu-v+1,a)$
$\cos ax$	${}_0D_x^{v}\cos ax=C_x(-v,a)$
$\cos ax$, $a>0$	${}_0D_x^{-1/2}\cos ax=\sqrt{\frac{2}{a}}\left[C(z)\cos ax+S(z)\sin ax\right]$, $z=\sqrt{\frac{2ax}{\pi}}$

$\cos^2 ax$	${}_0D_x^\nu \cos^2 ax = \frac{x^{-\nu}}{2\Gamma(1-\nu)} + \frac{1}{2}C_x(-\nu, 2a)$
$x\cos ax$	${}_0D_x^\nu [x\cos ax] = xC_x(-\nu, a) + \nu C_x(1-\nu, a)$
$C_x(\mu, a)$, $\mu > -1$	${}_0D_x^\nu C_x(\mu, a) = C_x(\mu-\nu, a)$
$xC_x(\mu, a)$, $\mu > -2$	${}_0D_x^\nu [xC_x(\mu, a)] = xC_x(\mu-\nu, a) + \nu C_x(\mu-\nu+1, a)$
$x^{-1/2}\cos\sqrt{x}$	${}_0D_x^\nu \left[x^{-1/2}\cos\sqrt{x}\right] = \sqrt{\pi}\left(2\sqrt{x}\right)^{-1/2-\nu} J_{-1/2-\nu}(\sqrt{x})$
$x^{-1/2}\cosh\sqrt{x}$	${}_0D_x^\nu \left[x^{-1/2}\cosh\sqrt{x}\right] = \sqrt{\pi}\left(2x^{1/2}\right)^{-1/2-\nu} I_{-1/2-\nu}(\sqrt{x})$
$\sin ax$	${}_0D_x^\nu \sin ax = S_x(-\nu, a)$
$\sin ax$, $a > 0$	${}_0D_x^{-1/2} \sin ax = \sqrt{\frac{2}{a}}[C(z)\sin ax - S(z)\cos ax]$, $z = \sqrt{\frac{2ax}{\pi}}$
$\sin^2 ax$	${}_0D_x^\nu \sin^2 ax = aS_x(1-\nu, 2a)$
$x\sin ax$	${}_0D_x^\nu [x\sin ax] = xS_x(-\nu, a) + \nu S_x(1-\nu, a)$
$S_x(\mu, a)$, $\mu > -2$	${}_0D_x^\nu S_x(\mu, a) = S_x(\mu-\nu, a)$
$xS_x(\mu, a)$, $\mu > -3$	${}_0D_x^\nu [xS_x(\mu, a)] = xS_x(\mu-\nu, a) + \nu S_x(\mu-\nu+1, a)$
$\sin\sqrt{x}$	${}_0D_x^\nu \left[\sin\sqrt{x}\right] = \frac{1}{2}\sqrt{\pi}\left(2x^{1/2}\right)^{1/2-\nu} J_{1/2-\nu}(\sqrt{x})$
$\sinh\sqrt{x}$	${}_0D_x^\nu \left[\sinh\sqrt{x}\right] = \frac{1}{2}\sqrt{\pi}\left(2x^{1/2}\right)^{1/2-\nu} I_{1/2-\nu}(\sqrt{x})$
$x^\lambda \ln x$, $\lambda > -1$	${}_0D_x^\nu \left[x^\lambda \ln x\right] = \frac{\Gamma(\lambda+1)}{\Gamma(\lambda-\nu+1)} x^{\lambda-\nu} [\ln x + \psi(\lambda+1) - \psi(\lambda-\nu+1$
$\ln x$	${}_0D_x^\nu \ln x = \frac{x^{-\nu}}{\Gamma(1-\nu)}[\ln x - \gamma - \psi(1-\nu)]$
$x^{-1/2}\ln x$	${}_0D_x^{-1/2}\left[x^{-1/2}\ln x\right] = \sqrt{\pi}\ln\frac{1}{4}x$
$\ln x$	${}_0D_x^{-1/2}\ln x = 2\sqrt{\frac{x}{\pi}}(\ln 4x - 2)$
$x^\lambda(\ln x)^2$, $\lambda > -1$	${}_0D_x^\nu\left[x^\lambda(\ln x)^2\right] = \frac{\Gamma(\lambda+1)}{\Gamma(\lambda-\nu+1)} x^{\lambda-\nu} \left\{\begin{array}{l} [\ln x + \psi(\lambda+1) - \psi(\lambda- \\ +D\psi(\lambda+1) - D\psi(\lambda-\nu \end{array}\right.$
$x^{\lambda/2}J_\lambda(\sqrt{x})$, $\lambda > -1$	${}_0D_x^\nu\left[x^{\lambda/2}J_\lambda(\sqrt{x})\right] = 2^{-\nu}x^{(\lambda-\nu)/2}J_{\lambda-\nu}(\sqrt{x})$
$x^{\lambda/2}I_\lambda(\sqrt{x})$, $\lambda > -1$	${}_0D_x^\nu\left[x^{\lambda/2}I_\lambda(\sqrt{x})\right] = 2^{-\nu}x^{(\lambda-\nu)/2}I_{\lambda-\nu}(\sqrt{x})$

2.9 Irregular Functions and Measure of Irregularity (Roughness) with Box Dimension, Holder and Hurst's Exponents

In fractal geometry, (irregular geometry) the fractal dimension is a statistical quantity that gives an indication of how completely a fractal appears to fill space, as one zooms down to finer and finer scales. The term fractal was coined in 1975 by Mandelbrot, from the Latin 'fractus', meaning "broken" or "fractured". A fractal is a geometric shape which is self-similar and has fractional (fractal) dimension. Fractals can be classified according to their self-similarity properties.

There is precise nature of connection between the dimensions of the curve (graph) of fractal (uni-fractal or multi-fractal) curve and fractional derivative. Here concept of local fractional derivative (LFD) at point will signify the nature of irregularity. The LFD gives notion to study fractional differentiability properties of irregular functions.

Numerous experimental and theoretical results suggest the response function diverge near 'critical point'. For instance in phase transition points of thermodynamic studies of material or at Curie point of studies for magnetism. Equivalently the response function $f(x)$ is found to vary as power law near critical point x_0. More precisely for almost all directions of approach to the critical point at x_0 one can find that at $\lim(x \to x_0)$ the function is $f(x) - f(x_0) \approx |x - x_0|^{1+\frac{1}{\delta}}$ with $\delta \geq 1$ as 'state exponent' in thermodynamics. The irregularity of this graph at the critical point is the idea of study and there exists relation to the measure of irregularity (or roughness) of graph at a point to the Fractional Derivative of that function at that point.

There are many self-similar structures in nature and several ways to describe them. A random walk in free space, or on a periodic lattice, a linear or branched polymer, is just few examples. When considering the vast category of these self-similar geometrical spaces it is tempting to regard them as basic spaces. Standard Euclidian space has translational symmetry while self similar space has dilation symmetry. As a consequence whereas Euclidian space is well characterized by one dimension d, integer number, self-similar space require the definition of (at least) three dimensions: d, the dimension of the embedding Euclidian space, d_f, the Hausdroff's dimension (of the irregular geometry and non-integer number), and the d_s, the Spectral dimensions giving the density of states of Spectral Graph (network) where the time series (space series) asymptotically gives power law relaxation to an excitation. The decay exponent for late times ($t^{-\alpha}$) for Green's function, or relaxation modulus function for solution of the system, gives indication of this spectral dimension. For instance a viscoelastic system described by $\sigma(t) = cT\left(\zeta / K\right)^{\alpha} D_t^{\alpha}\varepsilon(t)$, has asymptotic relaxation modulus as $G(t) \cong \left(cT / t^{\alpha}\right)\left(\zeta / K\right)^{\alpha}$, has spectral dimension as $d_s = 2\alpha$. The phenomena occurring in fractal lattice has this kind of constitutive equations, with fractional derivatives.

When the phenomena occur in Euclidian lattice then all the dimensions are same $d = d_f = d_s$. Wide range of natural phenomena occur in space of non-integer fractal Hausdroff's dimension. However, it is of greater importance to seek to discover how fundamental laws of nature are modified for fractal object. A fractal is described by various metric properties, like Hausdroff's dimension, which measures the density of (sites of) the fractal embedded in the Euclidian space. The physical phenomena occurring in fractal thus have length scale of non Euclidian non integer value, and the governing equations do get modified as fractional differential equations. Long-range dependent (LRD) time series (signal) say $\{Y(k)\} = \{Y(0), Y(1), Y(2)........Y(k)....\}$ that has slowly decaying auto-correlation function

$$R_{YY}(t) = \lim_{T\to\infty} \frac{1}{T} \int_{-T/2}^{+T/2} Y(t)Y(t+t')dt'$$

can also be described by a fractal dimension d_f which is related to the Hurst parameter H through $d_f = 2 - H$ .Here, the fractal dimension d_f can be interpreted as the number of dimensions the signal fills up. This Hurst index or (parameter H) signifies the degree of long range dependence (LRD), and pattern persistence for the time series $\{Y(t)\}$, and has relevance in subject Fractional Order Signal Processing. This will be elaborated in Chapter 4 and 5 subsequently, and its estimate in Chapter 10.

2.9.1 Measure of Roughness of Graph

Holder exponent or Box Dimensions of curve is related to maximum order of fractional derivative (LFD) which exists at a point. The LFD is explained in Chapter 5. An irregular function, of one variable at a particular point is best characterized by Holder exponent. For a function $f(x)$ if there exists a polynomial P_n of degree $n < h$ and a constant C such that $|f(x) - P_n(x - x_0)| \le C|x - x_0|^h$, the supremum of all exponents $h(x_0)$ such that $n < h(x_0) < n+1$ describes the 'local' regularity of $f(x)$ at x_0. For $h = 1$, the function satisfies Lipschiltz's condition; and for $h = 0$, the function is simply bounded. The rapid changes in the time series are called singularities of the signal and the singularity strength is measured by the local Holder exponent $h(x_0)$. Higher the value of Holder exponent h more regular is the local behavior. The irregular functions are characterized by Holder exponent. Some curves (graphs) have a range of Holder exponents, and such graphs are called multi-fractal graphs. While a curve having same Holder exponent is uni-fractal (fractal) graph. If a function has same Holder exponent ($\overline{h}$) at every point then the 'box-dimension' of the graph is $d_B = 1 - \overline{h}$. This box-dimension expression is derived in following sections. The Holder exponent is related to Hurst exponent as $\overline{h} = H - 1$.

Say, we generate a one-dimensional Brownian motion or 'white noise' with square root scaling law $|\Delta y_i| = (\Delta t_i)^{1/2}$, then the Hurst exponent is

$$H = \frac{\log|\Delta y_i|}{\log(\Delta t_i)} = \frac{1}{2},$$

with 'box-dimension' as, $d_B = 2 - H = 3/2$.

The irregular functions are continuous everywhere but no-where differentiable and classical example is sine or cosine Weierstrass Function defined as:

$$W_\lambda(x) = \sum_{k=1}^{\infty} \lambda^{(s-2)k} \sin \lambda^k x,$$

with $x \in \mathbb{R}$ (real number) and $\lambda > 1$. Here if $1 < s < 2$, this curve is no-where differentiable but continuous everywhere. This will be described in Chapter 5.

This function has box-dimensions s. The cosine Weierstrass Function is

$$f(x) = \sum_{n=0}^{\infty} a^n \cos(b^n \pi n x),$$

With $0 < a < 1$, $b > 0$ and $ab > 1 + \frac{3}{2}\pi$, with $d_B = \frac{\log a}{\log b + 2}$.

2.9.2 Generation of Irregular Graph

Now we generate a uni-fractal function a 'fractal noise' and non-Brownian motion with Hurst exponent H. The scaling law is $|\Delta y_i| = |\Delta t_i|^H$, with generator as $(\Delta t_1)^H - (\Delta t_2)^H + (1 - \Delta t_1 - \Delta t_2)^H = 1$. Generator starts at $(0,0)$ and ends at $(1,1)$. The Y-axis denotes the distance traveled in this motion at i-th jump with X-Axis denoting the time. The satisfying conditions for Δt_i and Δy_i are: $\Delta t_1 + \Delta t_2 + \Delta t_3 = 1$ and $\Delta y_1 + \Delta y_2 + \Delta y_3 = 1$. Here we impose uni-fractal scaling $|(\Delta y_i)| = (\Delta t_i)^H$ and take into account that generator segment (motion displacement) should alternatively go up, down, up and down so on, then the relation for Y-axis becomes $(\Delta t_1)^H - (\Delta t_2)^H + (\Delta t_3)^H = 1$.

From this relation we get $(\Delta t_1)^H - (\Delta t_2)^H + (1 - \Delta t_1 - \Delta t_2)^H = 1$. Once we specify Hurst exponent H and interval Δt_1 only one value of Δt_2 satisfies the scaling condition. Moreover scaling requires $0 < H < 1$.

Take $H = 0.25$ and $\Delta t_1 = 0.3$ solve for Δt_2 from

$$(0.3)^{0.25} - (\Delta t_2)^{0.25} + (1 - 0.3 - \Delta t_2)^{0.25} = 1,$$

to get $\Delta t_2 = 0.135608$.

From the equation $\Delta t_1 + \Delta t_2 + \Delta t_3 = 1$, we get

$$\Delta t_3 = 1 - \Delta t_1 - \Delta t_2 = 1 - 0.3 - 0.135608 = 0.564392.$$

Now we calculate the Δy_i from the fractal scaling $|\Delta y_i| = (\Delta t_i)^H$, with up, down, up, down rule and write.

$\Delta y_1 = 0.3^{0.25} = 0.740083$, $\Delta y_2 = -(0.135608)^{0.25} = -0.606836$ and
$\Delta y_3 = (0.564392)^{0.25} = 0.86675$.

The turning points are:

1. $(0,0)$
2. $(a,b) = (\Delta t_1, \Delta y_1) = (0.3, 0.740083)$
3. $(c,d) = (\Delta t_1 + \Delta t_2, \Delta y_1 + \Delta y_2) = (0.435608, 0.133247)$
4. Now Δt_3 has become new Δt_1 with $(0,0)$ shifted at (c,d). Proceed to get further points as repeat of above steps, and continue.

The plot of the above process is zigzag curve going regularly up and down sequentially, and when large number of points are plotted and then the curve is scaled down the graph looks like a plot of a highly irregular function. If a higher Hurst exponent is taken say $H = 0.75$ and the same steps are repeated to get same number of points, the graph will look smoother than with the lower Hurst exponent.

Note though Holder exponent h and Hurst parameter H are related to fractal dimensions but they are different. The h gives measure of local irregularity (character of local singularity) and scaling whereas H relates to spectral density and then fractal time series and its auto-correlation properties, relating to Long or Short Range Dependency. These will be elaborated in Chapter 4 and 5.

2.9.3 Determination of Box-Dimension of an Irregular Graph

Given an irregular shape curve or graph, what we do is cover with boxes and find how the number of boxes $N(r)$ changes with the size of the box r. Say for simple 1-dimmensional line-segment we expect $N(r) = 1/r$. It is $1/r$ instead of r because as the square boxes get smaller more will be needed to cover it. For a 2-dimmensional unit square we expect $N(r) = (1/r)^2$, and for radius the number of boxes will lead to a power law $N(r) = K\left(1/r\right)^d$. The box dimension (with this method) is defined as:

$$d_B = \frac{\log N(r)}{\log(1/r)}, \text{ with } N(r) \to \infty \text{ and } r \to 0$$

Let us write a procedure; of determining the box-dimension of a graph irregular graph (as generated in earlier section with Hurst exponent H is given). First take a section of graph and scale (map) the X-axis onto 0 to 1.00. Call this as function $y = f(x)$ (refer Figure 2.1). Then divide the X-axis (mapped onto 0-1.00) into r equal intervals. Above each interval make a column of width r . In this situation of scaling condition $\Delta y_i = (\Delta x_i)^H$, means in each of these columns of width r (as constructed) the graph (or normalized graph $y = f(x)$) passes through a height of about r^H . So the number of boxes (square boxes of each edges of size r) needed to cover the part of graph in that particular column is about (Height of graph) ÷ (Height of box)= $r^H / r = r^{H-1}$.

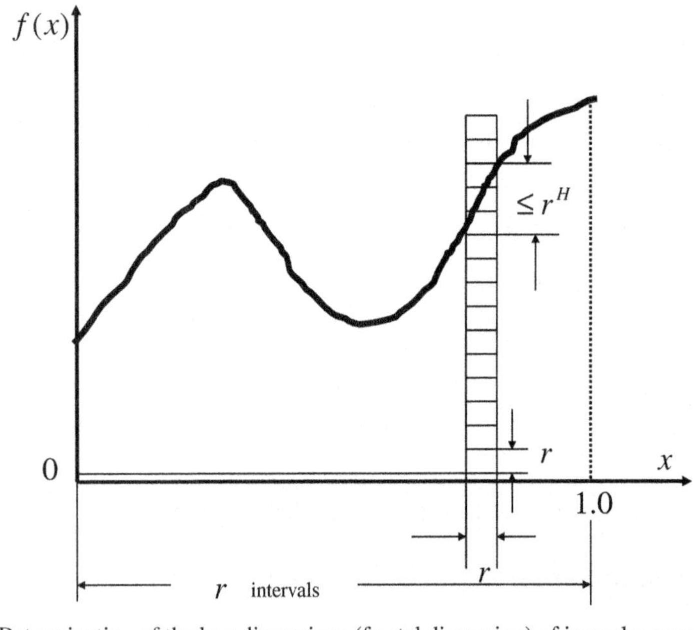

Fig. 2.1 Determination of the box dimensions (fractal dimension) of irregular curve

The height and width of the box is r . The number of columns in the X-axis of unit length is $1/r$. Therefore the number of boxes required of size r needed to cover the entire graph $y = f(x)$ is $N(r) = \left(r^{H-1}\right) \times \left(1/r\right) = r^{H-2}$. Thus the box-dimension is:

$$d_B = \frac{\log r^{H-2}}{\log(1/r)} = 2 - H$$

Actually a note may be taken that box-dimension is actually box counting dimension; since we count the boxes. For graph or curve to be of fractal self similar nature this figure should be greater than one ($d_B > 1$).

Geometrically one can arrive at same fractal dimension d_B, by 'Mandelbrot-Richardson' method, or 'Divider Step Method'. In this method the length of graph is measured by a scale, moved along the graph. When scale is reduced the more new features of the graph (irregularity) are observed and again length is determined. The length of 'divider' kept on decreasing and length is determined. The slope of log-log plot of total-length and length of divider step gives the idea of fractal dimension. The box counting method or this divider step method gives the same fractal dimension d_B. This d_B is Hausdroff's dimension of fractal graph d_f. The fractal dimension is physically described as how many new segments of an object are found as object is observed at finer resolution. For instance if N new pieces are revealed when the scale is reduced by factor F then $d_f = \log N / \log F$. An example of self-similar fractal is say take a line segment to start with; then you divide the segment into three equal parts and take out the middle portion and call it second stage. From this second stage you perform the above operation the two segments to get four segments, call it the third stage. Keep repeating this to get next stages one observes that each stage can be scaled (dilated) and can be overlapped to previous stage. This way we get self-similar pattern. Above divided line segment has fractal dimension as $d_B = -\log N(r)/\log(1/r)$, with $N(r) = 2^n$ a nd $(1/r) = (1/3)^n$, for n stages. Thus $d_B = -\log 2/\log(1/3) = 0.631 < 1$, which is less than Euclidian dimension in the case of line segment is 1. We talked of self-similar fractals, where the dilation symmetry is uniform. The generalization to self-similar fractal is self-affine fractal, where scaling by different amounts in X and Y directions is required. A structure is called self similar if the structure is invariant under the isotropic scale transformation that is, enlarging uniformly in every direction. The self–affine fractal is invariant under anisotropic transformation that is, rescaling is required with different scaling in different directions.

2.9.4 Difference in Persistent Anti Persistent Noise and Motion from Power law of Power Spectral Density

Fractional Gaussian Noise is defined by 'power-spectral' density of the form $S(f) \propto f^{-\beta}$, for a signal or random variable say position which scales as $X(t) = c^{-H} X(ct)$. H is called Hurst exponent and $\beta = 2H - 1$, valid for uni-fractal series. With $H = 1$ the time series $\{X(t)\}$ is 'pink' noise with power spectral density as $S(f) \cong (1/f)$. This type of $1/f$ noise is important since it is kind of threshold between persistent stable noise where $0.5 < H < 1$, with $0 < \beta < 1$ and non stationary noise $\beta > 1$.

A white noise (random noise) is characterized by $\beta = 0$ and $H = 0.5$. The noise characterized by $0 < H < 0.5$ is called anti-persistent, and characterized by $0.5 < H < 1$ is called persistent, whereas a random walk or Brownian motion (BM) is characterized by power spectral density $S(f) \cong f^{-2}$ that is, $\beta = 2$. Its Fourier Inverse is $X^2(t) \cong t$ giving indication of Mean Squared Displacement scales proportional to time. In the similar way as for persistent and anti-persistent noise the walks may be classified as a persistent walk if the spectral density index or the fluctuation is given by $1 < \beta < 2$ and anti-persistent walk given by $2 < \beta < 3$. We call these walks for fluctuating characteristic exponent $1 < \beta < 3$ because these walks may be obtained by 'integrating' the fractal noise defined by $0 < H < 1$.

Well the random walk in a continuous time can be expressed as $[dX(t)/dt] = \xi(t)$; where $\xi(t)$ is a stochastic quantity. If $\xi(t)$ is a white-noise, as we generated the same in previous section, with $H = 0.5$, then the walk is $X(t) = \int_0^t \xi(\tau)d\tau$. The generation of Brownian motion is thus as described now. Take a initial position $X(0)$ at t_1 add the stochastic quantity $\xi(t_1)$ (a white noise in time is $a_1\delta(t-t_1)$ to $X(0)$). This will give $X(t_1)$. Continuing it with several time intervals, this gives a time series as a 'white-noise' train. Integrate this to get Brownian motion. Now if the noise is $1/f$ or a fractal noise the same steps will generate train of pulses of fractal noise and integration of that curve will give Fractional Brownian Motion.

2.10 Concluding Comments

In this chapter the basis functions that are important in the study of the fractional order systems is introduced. Mostly the fundamental form is the Mittag-Leffler function can be stated as generalized exponential function. As the exponential function play basis role in integer order calculus, so does Mittag-Leffler function has role in the fractional calculus. Other compacted forms of the, variants of Mittag-Leffler variety is also listed, which find several applications of solution of fractional differential equations. All of these functions are of power-series expansions and fits variety of power law following processes. In conclusion the readers will be put to think about following reality. In circuit theory experiment we have made a low pass filter, with lumped resistance and lumped capacitor. The step response to this should have a pure exponential reaction, and mostly the recorders will show the similar reaction. The question is are we observing a pure exponential curve uniquely determined by unique time constant the product of lumped resistance and lumped capacitor used. We tend to believe that the observation is pure exponential and ode the aberration to non-linearity, instrument error, leakages, distributed effects and various others like parametric drifts of components. The aberration to exponential curve, if fitted by power series function of Mittag-Leffler type then naturally we ask a question about the descriptor equation which classically is integer order differential equation. The Mittag-Leffler type function is solution of fractional differential equation, thus if the basic circuit descriptor were of fractional order differential equation then we explain the reality, closely. However no capacitor is pure capacitor, no resistance is pure resistance and no system can have lumped characteristic, and the distributed parametric spread is reality. The same thoughts can be extended to various other relaxation processes of the nature about diffusion, reactor kinetics, electrochemistry and several others. The variants of Mittag-Leffler functions introduced here are developed in last four decade; several others may be developed in future to explain the physical processes of nature. The irregularity can be quantified by an irregularity exponent and we have seen the fractal dimension. The relation between the fractional derivatives of local character where the irregular function is not conventionally differentiable plays a role in characterizing various physical processes especially at the phase change. A regular irregularity is normal Brownian motion with fractal dimension 1.5, well a motion with memory can be regarded as Fractional Brownian Motion then, where the 'random' jumps remembers past and makes next jump accordingly! The relaxation (to stimulus) of natural phenomena well are not happening 'strongly' but has memory associated with it, giving 'weak' relaxation with long tailed responses; memory lingers; unlike conventional relaxation without memory. Well the nature is having disorder and thus the basic equations of integer order differential equations get changed to fractional differential equations (FDE); well is the order of FDE represent the background disorder, where the physical process is taking place? Here also we have observed that FDE can be solved in similar manner with same difficulty as Integer Order Differential Equations, and there lies these 'special functions' which are generalizations of the functions of ordinary calculus.

Chapter 3
Observation of Fractional Calculus in Physical System Description

3.1 Introduction

Fractional calculus allows a more compact representation and problem solution for some spatially distributed systems. Spatially distributed system representation allows a better understanding of the fractional calculus. The idea of fractional integrals and derivatives has been known since the development of regular calculus. Although not well known to most engineers, prominent mathematicians as well as scientists of the operational calculus have considered the fractional calculus. Unfortunately many of the results in the fractional calculus are given in language of advanced analysis and are not readily accessible to the general engineering and science community. Many systems are known to display fractional order dynamics. Probably the first physical system to be widely recognized as one demonstrating fractional behavior is the semi-infinite lossy (*RC*) transmission line. The current into the line is equal to the half-derivative of the applied voltage. That is impedance is

$$V(s) = \frac{1}{\sqrt{s}} I(s)$$

many studied this system, Heaviside (1871) considered it extensively using the operational calculus. He states that, "there is universe of mathematics lying in between the complete differentiations and integrations, and that fractional operators push themselves forward sometimes, and are just as real as others." Another equivalent system is diffusion of heat into semi-infinite solid. Here temperature looking in from the boundary is equal to the half integral of the heat rate there. Other systems that are known to display fractional order dynamics are viscoelasticity, colored noise, electrode-electrolyte polarization, dielectric polarization, boundary layer effects in ducts and electromagnetic waves. Because many of these systems depend upon specific material and chemical properties, it is expected that wide range of fractional order behaviors are possible using different materials. The classical

Bode's integral is introduced and discussed, which is a stepping stone for realization of fractional order element, giving constant phase for a band of (selected) frequency. This Bode's integral is a useful technique in plotting the gain and phase plots for system identification and controls, in frequency domain analysis. Fick's law is discussed and its modification by considering relaxation times (Catteneo's law) is discussed here; with derivation of its memory kernel and generalizing the same to have fractional diffusion equation. Also, concept of non-integer numbers in the generalized Laplacian operator is considered to be a possibility of defining 'fractional geometrical dimensions'. The fractional order behavior for semi-infinite system and its truncated version are discussed and developed in this chapter, also synthesizing the half-order fractional element and its relation to 'continued fraction approximation' is developed, along with dynamics of chain network and its fractional order behavior.

3.2 Temperature Heat Flux Relationship for Heat Flowing in Semi-infinite Conductor

The thermocouple consists of two pair of dissimilar metals with a common junction point. Because the wires are long and insulated they will be treated as "semi-infinite" heat conductors. The Figure 3.1 represents one such wire of thermocouple pair. The thick line in Figure 3.1 represents the semi-infinite heat conductor, the thermocouple wire measuring the temperature at $x=0$ the furnace wall, called as $T_{surf}(t)$, which dynamically varies with the time. The initial temperature is denoted by T_0.

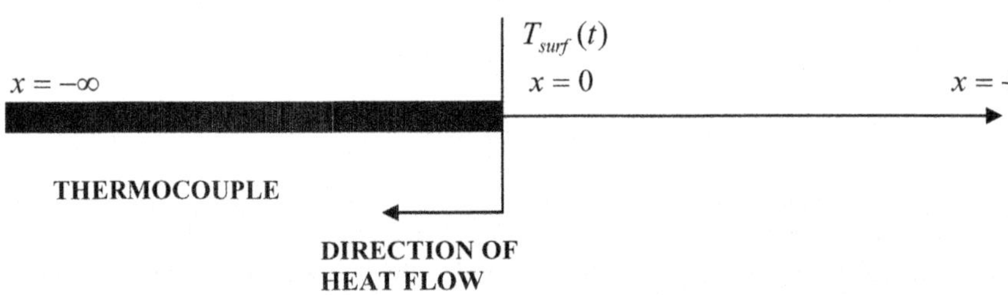

Fig. 3.1 Heat flow in semi-infinite wire thermocouple

The problem of heat conduction in the thermocouple wire is obviously one-dimensional. The following derivation shows how fractional calculus appears in the problem, of relating the conduction heat flux through semi-infinite thermocouple wire to the body temperature at the origin.

$$c\rho\frac{\partial T}{\partial t}=k\frac{\partial^2 T}{\partial x^2},$$
$$(t>0,\&-\infty<x<0)$$
$$T(0,x)=T_0$$
$$T(t,0)=T_{surf}(t)$$
$$\left|\lim_{x\to-\infty}T(t,x)\right|<\infty$$

Where:
t: is time[s], x is the spatial direction in the direction of heat flow [m], c is the specific heat or heat capacity [J kg^{-1}K^{-1}], ρ is density [kg m^{-3}], $T(t,x)$ is the temperature [K], and k is coefficient of heat conduction [W m^{-1}K^{-1}].

Let $u(t,x)=T(t,x)-T_0$. Substituting this, in above set we get equations as:

$$c\rho\frac{\partial u}{dt}=k\frac{\partial^2 u}{\partial x^2},$$
$$(t>0,\&-\infty<x<0)$$
$$u(0,x)=0$$
$$u(t,0)=T_{surf}(t)-T_0$$
$$\left|\lim_{x\to-\infty}u(t,x)\right|<\infty$$

Taking Laplace transforms for the above equation gives:

$$c\rho.sU(s,x)=k\frac{\partial^2 U(s,x)}{\partial x^2}$$
$$\frac{\partial^2 U(s,x)}{\partial x^2}-\frac{c\rho s}{k}U(s,x)=0$$

The bounded solution for x tends to $-\infty$ be

$$U(s,x)=U(s,0)\exp\left(x\sqrt{\frac{sc\rho}{k}}\right),$$

differentiating this we find

$$\frac{dU(s,x)}{dx}=U(s,0)\sqrt{\frac{sc\rho}{k}}\exp\left(x\sqrt{\frac{sc\rho}{k}}\right).$$

From these two expressions we get the following by putting $x=0$ and taking the inverse Laplace of $s^{-0.5}F(s)\to d^{-1/2}f(t)$, i.e. semi-integration, we obtain semi differential equation in time variable:

$$\frac{1}{\sqrt{s}}\frac{d}{dx}U(s,0)=\sqrt{\frac{c\rho}{k}}U(s,0)$$
$$\frac{d^{-1/2}}{dt^{-1/2}}\frac{\partial u(t,0)}{\partial x}=\sqrt{\frac{c\rho}{k}}u(t,0)$$
$$\frac{\partial u(t,0)}{\partial x}=\sqrt{\frac{c\rho}{k}}\frac{d^{1/2}}{dt^{1/2}}u(t,0)$$

Returning from $u(t,x)$ to $T(t,x)$ we get

$$k\frac{\partial T(t,0)}{\partial x}=\sqrt{c\rho k}\frac{d^{1/2}}{dt^{1/2}}(T_{surf}\{t\}-T_0)\,.$$

The term $k\frac{\partial}{\partial x}T(t,0)=Q(t)$, be termed as heat flux, flowing through the thermocouple wire at the interface of the furnace wall and point of contact (the origin). Therefore the heat flux expression is

$$Q(t)=\sqrt{c\rho k}\frac{d^{1/2}}{dt^{1/2}}\left(T_{surf}(t)-T_0\right)=\frac{k}{\sqrt{\frac{k}{c\rho}}}\frac{d^{1/2}}{dt^{1/2}}\left(T_{surf}(t)-T_0\right)=\frac{k}{\sqrt{\alpha}}\,{}_aD_t^{1/2}T_b(t)$$

The normal integer order diffusion equation contained half (fractional) derivative that was demonstrated in Chapter 1, the similar half derivative appeared in this case too.

3.3 Single Thermocouple Junction Temperature in Measurement of Heat Flux

From the derivation as in above section we can write a general heat flow equation relating the heat-flux conducted through a semi-infinite conductor of heat to the temperature at the origin as time varying constitutive relation:

$$Q_i(t)=\frac{k}{\sqrt{\alpha}}\,{}_aD_t^{1/2}T_b,$$
$${}_aD_t^{1/2}\equiv\frac{d^{1/2}}{[d(t-a)]^{1/2}}$$
$$\alpha=\frac{k}{c\rho}$$

The semi-derivative is shown for initial time point a. When initial forcing conditions (states) are zero, the operator then is

$$ {}_0D_t^{1/2} \equiv \frac{d^{1/2}}{dt^{1/2}}, $$

here α is thermal diffusivity; T_b is the body temperature, at the point of contact of thermocouple to the furnace wall.

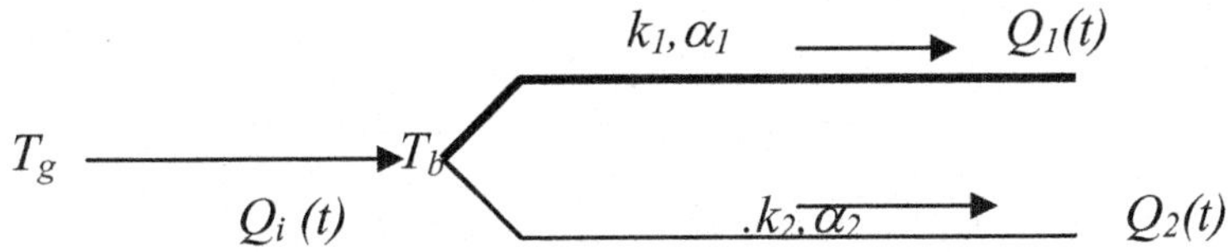

Fig. 3.2 Thermocouple junction for temperature (heat flux) measurement

The following equations define the time domain behavior.

Input heat flux to the thermocouple from steam temperature to the tip of the thermocouple junction $Q_i = hA(T_g(t) - T_b(t))$. At the tip of the thermocouple this input heat flux flows into two thermocouple wires as shown in Figure 3.2. Thus

$$ Q_i(t) - Q_1(t) - Q_2(t) = mc\frac{dT_b}{dt}. $$

Converting this expression to integral form we obtain the thermocouple node temperature is related two heats fluxes as,

$$ T_b(t) = \frac{1}{mc}\,{}_aD_t^{-1}(Q_i(t) - Q_1(t) - Q_2(t)). $$

The two semi-infinite heat conductors have constitutive equations in semi-differential form as derived for Figure 3.1, as

$$ Q_1(t) = \frac{k_1}{\sqrt{\alpha_1}}\,{}_aD_t^{1/2}T_b(t), \text{ and } Q_2(t) = \frac{k_2}{\sqrt{\alpha_2}}\,{}_aD_t^{1/2}T_b(t). $$

Where *hA*, is product of convective heat transfer coefficient and surface area, and *mc* is product of the mass and specific heat. The constitutive equation is obtained by substituting values of $Q's$, as:

$$ hA\left(T_g(t) - T_b(t)\right) - \frac{k_1}{\sqrt{\alpha_1}}\,{}_aD_t^{1/2}T_b(t) - \frac{k_1}{\sqrt{\alpha_2}}\,{}_aD_t^{1/2}T_b(t) = mc\frac{dT_b(t)}{dt}, $$

after taking Laplace Transforms of the constitutive equations we have the following expression:

$$\left(mcs+\frac{k_1}{\sqrt{\alpha_1}}s^{1/2}+\frac{k_2}{\sqrt{\alpha_2}}s^{1/2}+hA\right)T_b(s)=hAT_g(s)$$

The transfer function is as follows:

$$\frac{T_b(s)}{T_g(s)}=\frac{1}{\left(\frac{mc}{hA}\right)s+\frac{1}{hA}\left(\frac{k_1}{\sqrt{\alpha_1}}+\frac{k_2}{\sqrt{\alpha_2}}\right)s^{1/2}+1}=\frac{\left(\frac{1}{mc}\right)}{s+b\sqrt{s}+C}hA$$

$$b=\frac{1}{mc}\left(\frac{k_1}{\sqrt{\alpha_1}}+\frac{k_2}{\sqrt{\alpha_2}}\right);\quad C=\frac{hA}{mc}$$

The value of fractional calculus is clearly demonstrated in this analysis. Conventional approaches require the solution of two simultaneous partial differential equations with ordinary integer order differential equation. The Bode plots show two distinct asymptotes one of slope –10 db/decade (corresponds to semi-pole $s^{1/2}$ behavior) and the next one as –20 db/decade at higher frequency. The diagram is shown in Figure 3.3 with $(mc/hA)=0.005$, and

$$\frac{1}{hA}\left(\frac{k_1}{\sqrt{\alpha_1}}+\frac{k_2}{\sqrt{\alpha_2}}\right)=5.0\,.$$

One more observation may be drawn from this analysis. In order to estimate heat-flux in any thermal system, classical method of utilizing two thermocouples, can be replaced by one thermocouple and from the temperature values obtained as function of time instantaneous semi-differential equation as indicated above may be solved to estimate flowing heat-flux.to estimate flowing heat-flux.

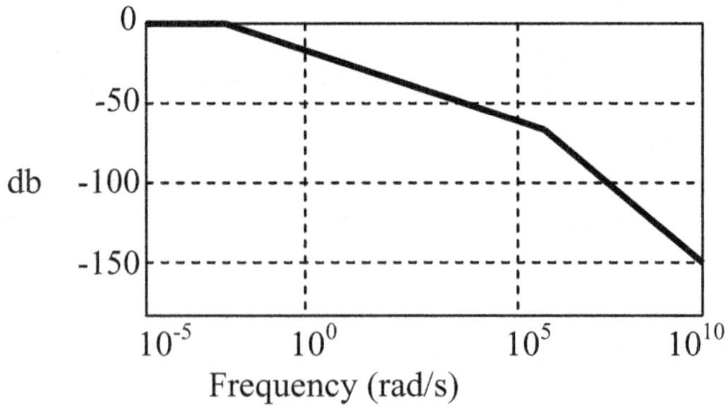

Fig. 3.3 Frequency response amplitude frequency Bode plot

Observing the transfer function of this system, $T_b(s)/T_g(s)$, obtained earlier, it appears as per integer order calculus theory that is a first order system. The first order system responses to a step input is a damped output, with out any oscillation and overshoot. The presence of fractional order terms makes somewhat anomalous, argument (which is detailed in Chapter 10). Though the system appears to be of first order yet the presence of the half order term, in denominator, may give a system response, to a step input as oscillatory with overshoot. Therefore definition of the system order for fractional order system is different, than that in integer order calculus.

The time evolution of $T_b(t)$ can be obtained for Heaviside step input, $T_g(t) = H(t)$ thus $T_g(s) = (1/s)$ put into the obtained transfer function and with help of partial fraction and R -function; $\mathcal{L}\{R_{q,v}(a,0,t)\} = (s^{-v})/(s^q - a)$ as:

$$T_b(s) = \frac{hA}{mc}\left(\frac{\frac{1}{\beta_2-\beta_1}}{\sqrt{s}+\beta_1} - \frac{\frac{1}{\beta_2-\beta_1}}{\sqrt{s}+\beta_2}\right)T_g(s); \quad \beta_1 = -\frac{b}{2}+\frac{1}{2}\sqrt{b^2-4C} \quad \beta_2 = -\frac{b}{2}-\frac{1}{2}\sqrt{b^2-4C}$$

$$T_b(t) = \frac{hA}{mc(\beta_2-\beta_1)}\left[R_{(1/2),-1}(-\beta_1,0,t) - R_{(1/2),-1}(-\beta_2,0,t)\right]$$

The output, $T_b(t)$ to a ramp input $T_g(t) = t$, thus $T_g(s) = (1/s^2)$ similarly is:

$$T_b(t) = \frac{hA}{mc(\beta_2-\beta_1)}\left[R_{(1/2),-2}(-\beta_1,0,t) - R_{(1/2),-2}(-\beta_2,0,t)\right]$$

3.4 Heat Transfer

System identification is the part of control practice, in which the parameters that enter into mathematical model of the system are determined. This is especially important in thermal system with convection, because of presence of heat transfer coefficient, which is never really known with great exactitude, and may vary with time due to physical or chemical changes at the heat transfer surface. There are also other parameters like the thermal capacity of the conductive body, its surface area, and its thermal diffusivity that have to be determined.

Rather than find each of the parameters separately it is more practical to estimate non-dimensional expression that best fits the observed data. In this heat transfer example, consider the cooling (or heating) of a one-dimensional plane wall of thickness L, with spatial uniform initial temperature T_i. There is convective heat transfer coefficient h from one wall to fluid at temperature T_∞. The exact solution with partial differential equation is as follows with the boundary and initial conditions:

The temperature field is $T(x,t)$, where x the coordinate is measured from the wall and t is the time. The transient heat conduction equation in the wall is given by:

$$\frac{\partial T(x,t)}{\partial t} = \alpha \frac{\partial^2 T(x,t)}{\partial x^2},$$

Where α is thermal diffusivity.

The initial and boundary conditions are:

$$\frac{\partial T(x,t)}{\partial x} = 0 \text{ at } x = 0$$

$$k\frac{\partial T(x,t)}{\partial x} + h(T\{x,t\} - T_{\infty}) = 0 \text{, at } x = L$$

$$T(x,t) = T_i \text{, at } t = 0$$

Where k is the thermal conductivity of wall material. With change of variable to make dimensionless equation we get the following transformed unit less variables as:

$$\xi = \frac{x}{L}, \tau = \frac{t\alpha}{L^2}$$

and unit less temperature as

$$\theta = \frac{T(x,t) - T_{\infty}}{T_i - T_{\infty}},$$

we obtain dimensionless equation as:

$$\frac{\partial \theta}{\partial \tau} = \frac{\partial^2 \theta}{\partial \xi^2},$$

With $\frac{\partial \theta}{\partial \xi} = 0$, at $\xi = 0$ and $\frac{\partial \theta}{\partial \xi} + B_i \theta = 0$ at $\xi = 1$, and $\theta = 1$ for $\tau = 0$.

Where the 'size-factor' $B_i = \frac{hL}{k}$, is called Biot's-number.

Solution to this dimensionless equation, which is exact representation of heat transfer, is:

$$\theta(x,\tau) = \sum_{n=1}^{\infty} C_n \exp\left(-\lambda_n^{\,2}\tau\right)\cos\left(\lambda_n x\right),$$

where

$$C_n = \frac{4\sin\lambda_n}{2\lambda_n + \sin(2\lambda_n)}$$

and λ_n are positive roots of transcendental expression $\lambda_n \tan\lambda_n = B_i$.

The dimensionless mean temperature is

$$\bar{\theta}(\tau) = \int_0^1 \theta(\xi,\tau)d\xi .$$

The exact solution is obtained above, by considering various Biot's number (0.1-10). However the first way to approximate this heat transfer phenomena is by having spatial average of the temperature i.e.

$$\bar{T}(t) = \frac{1}{L}\int_0^L T(x,t)dx$$

be taken as dependent variable. In terms of this average temperature the heat balance equation is

$$\frac{d\bar{T}(t)}{dt} + \frac{h}{\rho c}\left(\bar{T} - T_\infty\right) = 0,$$

where ρ is the density and c is specific heat of the wall material. The convective heat transfer initial and boundary conditions are:

$$\frac{\partial T}{\partial x} = 0, \text{ at } x = 0, \text{ and } k\frac{\partial T}{\partial x} + h(T - T_\infty) = 0, \text{ at } x = 1$$

have been used to derive the above average expression as approximation. The equation $T = T_i$ at $t = 0$ gives $\bar{T}(0) = 1$. Using dimensionless variables as done for exact solution case one obtains:

$$\frac{d\bar{\theta}}{d\tau} + B_i\bar{\theta} = 0,$$

with $\bar{\theta}(0) = 1$. Now numerical experiments points toward an interesting observation that the solution with $B_i = 0.1$, the exact solution and this approximation match closely, where as for $B_i = 10$, the deviations are large.

An improvement to above approximation is to write a fractional order differential equation as:

$$\frac{d^q\bar{\theta}}{d\tau^q} + pB_i\bar{\theta} = 0,$$

with $\bar{\theta}(0) = 1$. Here q and p be varied to minimize

$$E = \int_0^{\tau_{\max}} e(\tau)^2 d\tau,$$

where $e(\tau)$ is difference between the exact solution and approximate solution of $\theta(\tau)$. $\tau_{\max}$ is the maximum value of τ to which integration is carried out.

Here the effect of B_i is to be discussed. The fractional order $q \to 1$ and the multiplier of the Biot's number $p \to 1$, as $B_i \to 0$.

This example works well with dynamic system modeling where measurements enable simultaneous time-dependent system identification as well as provide an error signal for feed back controls. In the present heat transfer example, for the wall with simple geometry, a lumped parameter energy balance in which temperature of the system is assumed to be spatially uniform is commonly used to model transient conductive systems exposed to convective heat fluxes at their boundaries. It is simple to fit experimental data and the integer order (in this case first order) differential equation that is easy to solve. For larger size (Biot's number) there is difference between actual temperature field and the spatial average used in lumped model. This necessitates the solution of the partial differential equations for the transient heat conducting in the body which is made difficult by shape, heterogeneity, or the unknown nature of the convective heat transfer coefficient at the boundaries.

This argument is true for any realistic process with transport phenomena. Here a practical method to describe the system with fractional order differential equation aims at reality in system identification and control, as real systems are distributed phenomena.

3.5 Driving Point Impedance of Semi-infinite Lossy Transmission Line

Assuming a lossy *RC* line the boundary value problem can be defined in terms of the current or voltage variables. Since a semi-infinite line is considered, the measurable inputs or outputs are @ $x = 0$, at the left while right end $x = \infty$ is at finite value. In terms of the voltage variable the equations can be written as:

$$\frac{\partial v(x,t)}{\partial x} = -i(x,t)R$$
$$\frac{\partial i(x,t)}{\partial x} = -C\frac{\partial v(x,t)}{\partial t}$$

R, C are resistance and capacitance per unit length.

Differentiating first with respect to x and then substituting second in the first one we get

$$\frac{\partial^2 v}{\partial x^2} = -R\frac{\partial i}{\partial x} = RC\frac{\partial v}{\partial t},$$

choosing $1/RC$ as α we get the problem formulation as:

$$\frac{\partial v(x,t)}{\partial t} = \alpha\frac{\partial^2 v(x,t)}{\partial x^2}, v(0.t) = v_I(t), v(\infty, t) = 0.v(x,0)$$

Given with $i(x,t) = -\frac{1}{R}\frac{\partial v(x,t)}{\partial x}$

In this formulation (v) is the voltage, (i) is the current, (v_i) and is a time-dependent input variable. At $x = \infty$, the condition is of short circuit. A classical solution using iterated Laplace Transform is used to solve this problem.

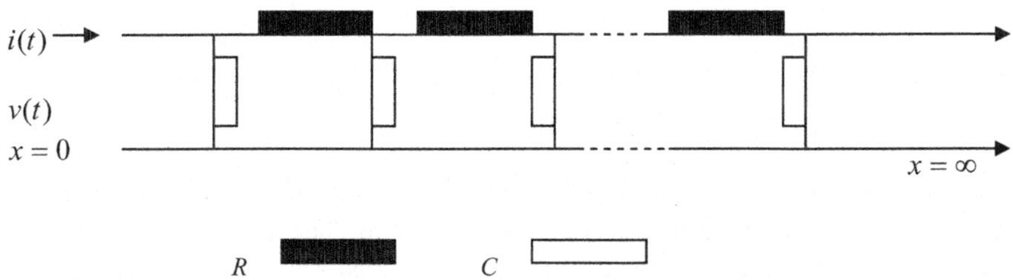

Fig. 3.4 Semi-infinite lossy transmission line

Taking Laplace transform with respect to time and using s as temporal Laplace variable gives:

$$sV(x,s) - v(x,0) = \alpha \frac{d^2V(x,s)}{dx^2},$$

$$\text{with } V(0,s) = V_I(s), V(\infty,s) = 0, I(x,s) = -\frac{1}{R}\frac{dV(x,s)}{dx}.$$

Then taking the Laplace transform with respect to spatial position x, and using p as the 'spatial' Laplace variable gives:

$$\frac{s}{\alpha}V(p,s) - \frac{1}{\alpha}V(p,0) = p^2V(p,s) - pV(0,s) - \left[\frac{dV(0,s)}{dx}\right].$$

Substituting

$$\left[\frac{dV(0,s)}{dx}\right] = V^*,$$

this equation can be manipulated to give $V(p,s)$ as:

$$\left[p^2 - \frac{s}{\alpha}\right]V(p,s) = -\frac{1}{\alpha}V(p,0) + pV(0,s) + V^*(0,s),$$

or

$$V(p,s) = \left[\frac{1}{p^2 - \frac{s}{\alpha}}\right]\left[-\frac{1}{\alpha}V(p,0) + pV(0,s) + V^*(0,s)\right]$$

The first term of the transform, here, of the initial spatial distribution $V(p,0)$ is problem-dependent. After rearrangement and partial fraction the above expression can be expressed as:

$$V(p,s)=\left[\frac{1}{2\sqrt{\frac{s}{\alpha}}\left(p-\sqrt{\frac{s}{\alpha}}\right)}-\frac{1}{2\sqrt{\frac{s}{\alpha}}\left(p+\sqrt{\frac{s}{\alpha}}\right)}\right]\left[-\frac{1}{\alpha}V(p,0)\right]+\left[\frac{1}{p^2-\frac{s}{\alpha}}\right]\left[pV(0,s)+V^*(0,s)\right]$$

Here the first term represents the effect of any initial spatial voltage distribution, while the second term represents the voltage and current present at $x=0$ end of the line. The first term is now inverse-Laplace-transformed with respect to the variable p using convolution, and the second is inverse-Laplace transformed using standard transform pairs.

$$V(x,s)=\int_0^x\frac{1}{2\sqrt{\frac{s}{\alpha}}}e^{+(x-\lambda)\sqrt{\frac{s}{\alpha}}}\left[-\frac{1}{\alpha}v(\lambda,0)\right]d\lambda-\int_0^x\frac{1}{2\sqrt{\frac{s}{\alpha}}}e^{-(x-\lambda)\sqrt{\frac{s}{\alpha}}}\left[-\frac{1}{\alpha}v(\lambda,0)\right]d\lambda+$$
$$V(0,s)\cosh\left(x\sqrt{\frac{s}{\alpha}}\right)+\frac{V^*(0,s)}{\sqrt{\frac{s}{\alpha}}}\sinh\left(x\sqrt{\frac{s}{\alpha}}\right)$$

Equivalently:

$$V(x,s)=\int_0^x\frac{1}{2\sqrt{\frac{s}{\alpha}}}e^{+x\sqrt{\frac{s}{\alpha}}}e^{-\lambda\sqrt{\frac{s}{\alpha}}}\left[-\frac{1}{\alpha}v(\lambda,0)\right]d\lambda-$$
$$\int_0^x\frac{1}{2\sqrt{\frac{s}{\alpha}}}e^{-x\sqrt{\frac{s}{\alpha}}}e^{+\lambda\sqrt{\frac{s}{\alpha}}}\left[-\frac{1}{\alpha}v(\lambda,0)\right]d\lambda+$$
$$\frac{V(0,s)}{2}\left[e^{+x\sqrt{\frac{s}{\alpha}}}+e^{-x\sqrt{\frac{s}{\alpha}}}\right]+\frac{V^*(0,s)}{2\sqrt{\frac{s}{\alpha}}}\left[e^{+x\sqrt{\frac{s}{\alpha}}}-e^{-x\sqrt{\frac{s}{\alpha}}}\right]$$

Collecting the like exponentials gives the following:

$$V(x,s)=\frac{e^{+x\sqrt{\frac{s}{\alpha}}}}{2}\left[V(0,s)+\frac{V^*(0,s)}{\sqrt{\frac{s}{\alpha}}}-\frac{1}{\alpha\sqrt{\frac{s}{\alpha}}}\int_0^x e^{-\lambda\sqrt{\frac{s}{\alpha}}}v(\lambda,0)d\lambda\right]+$$

$$\frac{e^{-x\sqrt{\frac{s}{\alpha}}}}{2}\left[V(0,s)-\frac{V^*(0,s)}{\sqrt{\frac{s}{\alpha}}}+\frac{1}{\alpha\sqrt{\frac{s}{\alpha}}}\int_0^x e^{+\lambda\sqrt{\frac{s}{\alpha}}}v(\lambda,0)d\lambda\right]$$

It should be recognized that the coefficients multiplying the two exponential functions are unknowns. Although the integral and either $V(0,s)$ or $V^*(0,s)$ are given in the problem statement, the other condition (V^* or V, respectively) at $x=0$ is determined as a response to these two given terms. Imposing the boundary condition at $x=\infty$, allows the determination of a relationship between these three terms at $x=0$, and thus allows the impedance and initial condition response of the system.

It is required to evaluate the above equation in the limit $x\to\infty$. In this limit second term in the above equation go to zero due to exponential behavior, however the integral inside the bracket will diverge. That is

$$\lim_{x\to\infty}\left(e^{-x\sqrt{\frac{s}{\alpha}}}\right)\left(\frac{1}{\alpha\sqrt{\frac{s}{\alpha}}}\int_0^x e^{+\lambda\sqrt{\frac{s}{\alpha}}}v(\lambda,0)d\lambda\right)=0.\infty$$

We are thus left with indeterminate form and this can be solved by L'Hopital's rule as follows (after rearrangement):

$$\lim_{x\to\infty}\frac{\frac{1}{2\alpha\sqrt{\frac{s}{\alpha}}}\int_0^x e^{+\lambda\sqrt{\frac{s}{\alpha}}}v(\lambda,0)d\lambda}{e^{+x\sqrt{\frac{s}{\alpha}}}}$$

The L'Hopital rule says that this ratio has the same value as the ratio of the derivatives (with respect to x) of the numerator and denominator. Differentiating the denominator is easy, but differentiating the numerator with respect to x requires Leibniz's rule. Performing the differentiation gives:

$$\lim_{x\to\infty}\frac{\left(\frac{1}{2\alpha\sqrt{\frac{s}{\alpha}}}\frac{d}{dx}\int_0^x e^{+\lambda\sqrt{\frac{s}{\alpha}}}v(\lambda,0)d\lambda\right)}{\left(\sqrt{\frac{s}{\alpha}}e^{+x\sqrt{\frac{s}{\alpha}}}\right)}$$

Combining the leading constants, and applying the Leibniz's rule to the numerator gives:

$$\lim_{x\to\infty}\frac{\frac{1}{2s}e^{+x\sqrt{\frac{s}{\alpha}}}v(x,0)}{e^{+x\sqrt{\frac{s}{\alpha}}}}.$$

It can now be seen that the exponential terms cancel, which leaves the result

$$\lim_{x\to\infty}\frac{1}{2s}v(x,0)=\frac{v(\infty,0)}{2s}.$$

The problem statement however requires that the boundary condition $v(\infty,t)=0$ be satisfied for all time. Thus it is shown that the first term of the main equation equals zero for $x\to\infty$.

From above limit derivations:

$$V(\infty,s)=\frac{e^{+\infty\sqrt{\frac{s}{\alpha}}}}{2}\left[V(0,s)+\frac{V^*(0,s)}{\sqrt{\frac{s}{\alpha}}}-\lim_{x\to\infty}\frac{1}{\alpha\sqrt{\frac{s}{\alpha}}}\int_0^x e^{-\lambda\sqrt{\frac{s}{\alpha}}}v(\lambda,0)d\lambda\right]=0$$

Dropping the limit notation we have

$$V(0,s)+\frac{V^*(0,s)}{\sqrt{\frac{s}{\alpha}}}-\frac{1}{\alpha\sqrt{\frac{s}{\alpha}}}\int_0^x e^{-\lambda\sqrt{\frac{s}{\alpha}}}v(\lambda,0)=0.$$

Remembering that the current anywhere in the line is related to the voltage, then at $x=0$

$$I(0,s)=-\frac{1}{R}\frac{dV(0,s)}{dx}=-\frac{V^*(0,s)}{R}$$

and solving for voltage in terms of source current gives:

$$V(0,s)=\frac{RI(0,s)}{\sqrt{\frac{s}{\alpha}}}+\frac{1}{\alpha\sqrt{\frac{s}{\alpha}}}\int_0^x e^{-\lambda\sqrt{\frac{s}{\alpha}}}v(\lambda,0)d\lambda$$

In evaluating the integral on the right it is now recognized that this term is equivalent to a Laplace transform integral with

$$s \to q = \sqrt{\frac{s}{\alpha}}.$$

Thus the Laplace transform tables can simplify the evaluation of this term as follows:

$$\frac{1}{\alpha\sqrt{\frac{s}{\alpha}}}\int_0^x e^{-\lambda\sqrt{\frac{s}{\alpha}}} v(\lambda,0)d\lambda = \frac{1}{\alpha\sqrt{\frac{s}{\alpha}}}\left[V(q,0)\right]_{q=\sqrt{\frac{s}{\alpha}}}$$

The notation here on the right hand side of this equation is used to indicate the evaluation procedure. First the initial spatial distribution $v(x,0)$ is Laplace transformed with respect to the spatial Laplace variable p to give $V(p,0)$ The integral on the left side of the above equation here is then easily calculated by replacing spatial variable p with

$$q = \sqrt{\frac{s}{\alpha}}.$$

The voltage equation thus becomes

$$V(0,s) = \frac{RI(0,s)}{\sqrt{\frac{s}{\alpha}}} + \frac{1}{\alpha\sqrt{\frac{s}{\alpha}}}\left[V(p,0)\right]_{p=\sqrt{\frac{s}{\alpha}}}.$$

Notice that this contains driving point impedance function $Z(s)$, which is obtained by setting the initial condition, terms to zero.

$$Z(s) = \frac{V(0,s)}{I(0,s)} = \frac{R}{\sqrt{\frac{s}{\alpha}}}$$

or as $\alpha = 1/RC$, the impedance is

$$Z(s) = \frac{V(0,s)}{I(0,s)} = \sqrt{\frac{R}{C}}\frac{1}{\sqrt{s}}.$$

Note that in the impedance expression of $Z(s)$ there are two parts, the forced response due to $I(0,s)$ and the initial condition response due to the initial voltage distribution in the lossy line. The final expression of voltage anywhere in the line as function of the applied voltage at the terminal $V_I(s)$ and the initial condition on the line, as:

$$V(x,s) = \frac{e^{+x\sqrt{\frac{s}{\alpha}}}}{2\alpha\sqrt{\frac{s}{\alpha}}}\left[[V(p,0)]_{p=\sqrt{\frac{s}{\alpha}}} - \int_0^x e^{-\lambda\sqrt{\frac{s}{\alpha}}} v(\lambda,0)d\lambda\right] +$$

$$\frac{e^{-x\sqrt{\frac{s}{\alpha}}}}{2\alpha\sqrt{\frac{s}{\alpha}}}\left[2V_I(s) - \frac{[V(p,0)]_{p=\sqrt{\frac{s}{\alpha}}}}{\alpha\sqrt{\frac{s}{\alpha}}} + \frac{1}{\alpha\sqrt{\frac{s}{\alpha}}}\int_0^x e^{+\lambda\sqrt{\frac{s}{\alpha}}} v(\lambda,0)d\lambda\right]$$

Furthermore the current at any point in the line can be determined directly from the above equation as

$$I(x,s) = -\frac{1}{R}\frac{dV(x,s)}{dx}$$

By applying inverse Laplace transforms, for the driving point impedance expression the voltage-current behavior (with zero initial condition) is obtained as:

$$i(t) = \frac{1}{R\sqrt{\alpha}}\frac{d^{1/2}v(t)}{dt^{1/2}}$$

$$v(t) = R\sqrt{\alpha}\frac{d^{-1/2}i(t)}{dt^{-1/2}}$$

More compactly the voltage-current relation with the initial condition expressed as:

$$v(t) = R\sqrt{\alpha}\frac{d^{-1/2}i(t)}{dt^{-1/2}} + \phi_1(t),$$

$$\phi_1(t) = \mathcal{L}^{-1}\left[\frac{1}{\alpha\sqrt{\frac{s}{\alpha}}}[V(p,0)]_{p=\sqrt{\frac{s}{\alpha}}}\right] = \mathcal{L}^{-1}\left(\frac{1}{\alpha\sqrt{\frac{s}{\alpha}}}\int_0^x e^{-\lambda\sqrt{\frac{s}{\alpha}}} v(\lambda,0)d\lambda\right)$$

or

$$i(t) = \frac{1}{R\sqrt{\alpha}}\frac{d^{1/2}v(t)}{dt^{1/2}} + \phi_2(t)$$

$$\phi_2(t) = -\frac{1}{R\sqrt{\alpha}}\frac{d^{1/2}\phi_1(t)}{dt^{1/2}}$$

3.5.1 Practical Application of the Semi-infinite Line in Circuits

3.5.1.1 Semi-integrator Circuit

The circuit shown in the Figure 3.5 performs the function of semi-integration of the input voltage $v_i(t)$. The half order element (semi-infinite lossy line) is based on one dimensional diffusion equation

$$\frac{\partial v}{\partial t} = \alpha \frac{\partial^2 v}{\partial x^2},$$

which is depicted by a ladder of discrete resistance, and capacitance as shown in Figure 3.4 and its connection is shown in Figure 3.5 in an operational amplifier circuit. The terminal characteristic or the driving point impedance as obtained is described as

$$v(t) = r\sqrt{\alpha}\frac{d^{-1/2}i(t)}{dt^{-1/2}} + \phi_1(t)$$

or

$$i(t) = \frac{1}{r\sqrt{\alpha}}\frac{d^{1/2}v(t)}{dt^{1/2}} + \phi_2(t)$$

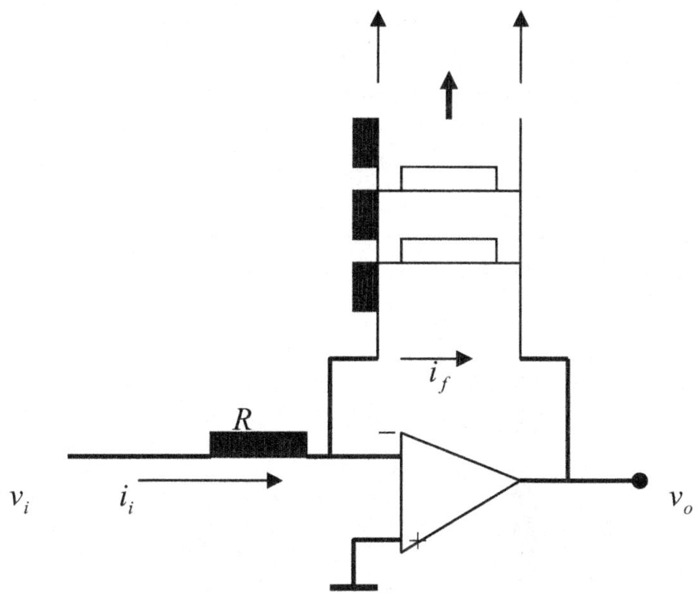

Fig. 3.5 Semi-Integrator

Here $v(t)$ and $i(t)$ are the voltage and current respectively, at the terminal element, r is the resistance per unit length and α is the product of r and c (the capacitance per unit length of the line). The initial condition functions are determined by the initial state of charge and voltage or current that exists on the infinite array of elements. For operational amplifier negative feed back configuration:

$$v_i(t) - 0 = i_i(t)R$$

$$0 - v_0(t) = r\sqrt{\alpha}\frac{d^{-1/2} i_f(t)}{dt^{-1/2}} + \phi_I(t) = r\sqrt{\alpha}\,{}_cD_t^{-1/2} i_f(t)$$

$$i_i(t) = i_f(t)$$

Solving for $v_0(t)$

$$v_0(t) = -r\sqrt{\alpha}\,{}_cD_t^{-1/2}\left\{\frac{1}{R}v_i(t)\right\}$$

$$v_0(t) = -\frac{r\sqrt{\alpha}}{R}\,{}_cD_t^{-1/2} v_i(t)$$

Note the symbolism change from small case differential operator to uppercase one, where the initialization function got included. This will be taken in detail while elaborate explanation of the initialization of fractional differintegrals in Chapter 6.

This is the basis of semi-integrator computing element. The equivalent (uninitialized) impedance form may also be calculated as $Z_f = r\sqrt{\alpha}/s^{1/2}, Z_i = R$.

The transfer function (uninitialized) form is thus is

$$\frac{v_0(s)}{v_i(s)} = -\frac{r\sqrt{\alpha}}{Rs^{1/2}}$$

3.5.1.2 Semi-Differentiator Circuit

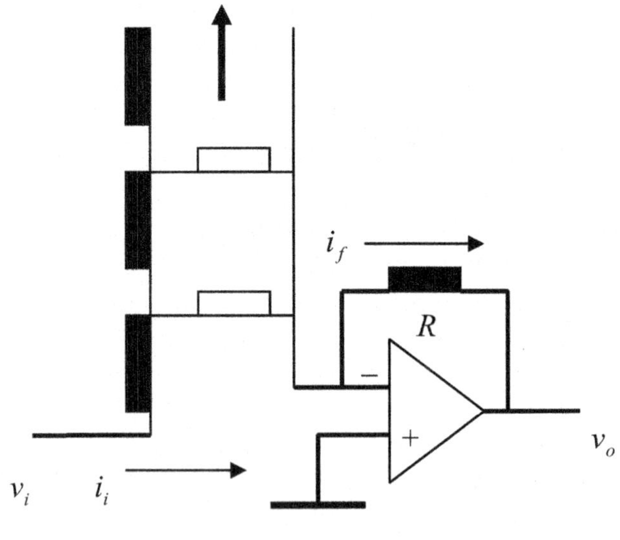

Fig. 3.6. Semi-differentiator

For the circuit in Figure 3.6 the negative feedback configuration gives:

$$i_i(t) = \frac{1}{r\sqrt{\alpha}}\,{}_cD_t^{1/2}(v_i(t) - 0)$$

$$0 - v_o(t) = Ri_f(t)$$

$$i_i(t) = i_f(t)$$

$$v_o(t) = -Ri_f(t) = -\frac{R}{r\sqrt{\alpha}}\,{}_cD_t^{1/2}v_i(t)$$

This formation with the leading coefficients specialized to one, is the basis of semi-differential computing element. Figure 3.7 gives a practical circuit for semi-integration with operational amplifiers.

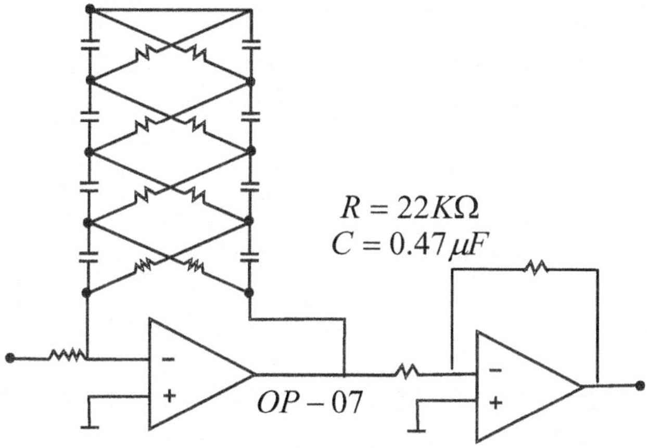

Fig. 3.7 Practical circuit for semi-integrator

The Figure 3.7 circuit is to realize the fractional order PID analog control system. In this circuit the offset adjustment parts are not explicitly shown. The semi integral control will have transfer function as

$$\frac{V_0(s)}{V_i(s)} = \frac{Z_f}{Z_i} = \frac{\sqrt{R\dfrac{1}{Cs}}}{R} = \frac{K}{\sqrt{s}}.$$

By replacing s with $\mathrm{j}\omega$ one gets the relation as

$$\frac{V_0(\mathrm{j}\omega)}{V_i(\mathrm{j}\omega)} = \frac{K}{\sqrt{\omega}}e^{-\mathrm{j}\pi/4}.$$

This circuit behaves as constant phase element of angle -45^0, meaning to a sinusoidal input the circuit will give a constant phase lag to the output by 45^0. By using values of the impedances the transfer function constant

$$K = \frac{\sqrt{\frac{22\times 10^3}{0.47\times 10^{-6}}}}{22\times 10^3} = 9.8,$$

and the transfer function is

$$\frac{V_0(\mathrm{j}\omega)}{V_i(\mathrm{j}\omega)} = 9.8\omega^{-0.5} e^{-\mathrm{j}\pi/4}.$$

The practical results are tabulated below in Table 3.1, where almost a constant phase is demonstrated (around -55^0). The circuit is excited by sinusoidal voltage and the phase lag was recorded along with the peak-peak amplitude.

Table 3.1 Practical Results from semi-integrator circuit measurement

Input frequency (Hz) f	Input frequency (radian) ω	Phase Angle (degree)	V_i Volt	V_0 Volt	$G = \frac{V_0}{V_i}$	$K = G\sqrt{\omega}$	$20\log(G)$ dB
50	314	-50.4	3.8	2.0	0.5263	9.32	-5.57
100	628	-45.0	3.8	1.5	0.3947	9.89	-8.07
150	942	-56.25	3.8	1.2	0.3158	9.69	-10.01
200	1257	-55.40	3.8	1.00	0.2632	9.33	-11.59
250	1571	-60.00	3.8	0.80	0.2105	8.19	-13.53
400	2513	-41.50	3.8	0.75	0.1974	9.89	-14.09
450	2827	-49.10	3.8	0.70	0.1842	9.79	-14.69
600	3770	-51.40	3.8	0.65	0.1710	10.49	-15.34
700	4398	-52.94	3.8	0.60	0.1578	10.46	-16.03
750	4712	-56.25	3.8	0.58	0.1526	10.47	-16.32
900	5655	-69.23	3.8	0.56	0.1473	11.07	-16.63
950	5969	-60.00	3.8	0.54	0.1421	10.9	-16.95

3.5.2 Application of Fractional Integral and Fractional Differentiator Circuit in Control System

Analog or digital realization can give a control system design for fractional order control system. Figure 3.8 gives block diagram representation of a classical integer order system (DC-Motor) being controlled by a fractional order feedback controller.

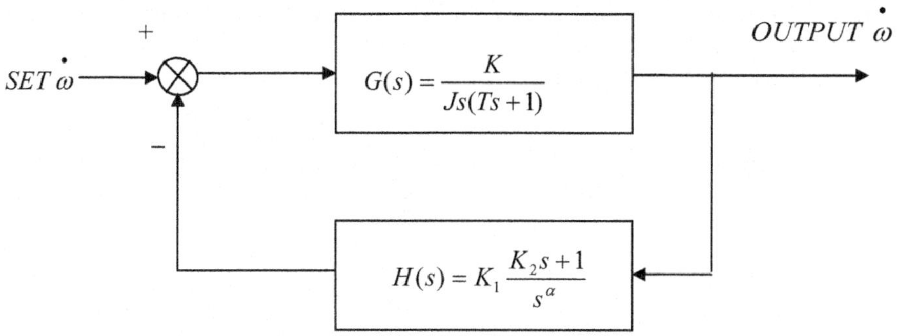

Fig. 3.8 Block diagram of fractional order control system

System transfer function of DC motor is

$$G(s) = \frac{K}{Js(Ts+1)},$$

J is payload (inertia). Phase margin of the controlled system is $\Phi_m = \arg|G(j\omega)H(j\omega)| + \pi$, the controller characteristics is

$$H(s) = K_1 \frac{K_2 s + 1}{s^\alpha}.$$

Here chose $K_2 = T$ Note that $H(s)$ is composed of a differentiator of fractional order $(1-\alpha)$ and an integral controller of order α. This gives constant phase margin as:

$$\Phi_m = \arg|G(j\omega)H(j\omega)| + \pi$$

$$\Phi_m = \arg\left[\frac{K_1 K / J}{(j\omega)^{1+\alpha}}\right] + \pi = \arg\left[(j\omega)^{-(1+\alpha)}\right] + \pi = -(1+\alpha)\frac{\pi}{2} + \pi = \frac{1}{2}(\pi - \pi\alpha)$$

The close loop transfer function is

$$G_c = \frac{G(j\omega)H(j\omega)}{1 + G(j\omega)H(j\omega)} = \frac{KK_1/J}{(s^{1+\alpha} + KK_1/J)}.$$

The step input response will be:

$$y(t) = \mathcal{L}^{-1}\left[\frac{KK_1/J}{s\left(s^{1+\alpha} + KK_1/J\right)}\right] = \left(\frac{KK_1}{J}\right) t^{1+\alpha} E_{1+\alpha,2+\alpha}\left(-\frac{KK_1}{J} t^{1+\alpha}\right)$$

This is "iso-damping" meaning, that the overshoot is same for various payloads (inertia), to have this type of control system is robust and efficient. Figure 3.9 gives the concept of isodamping.

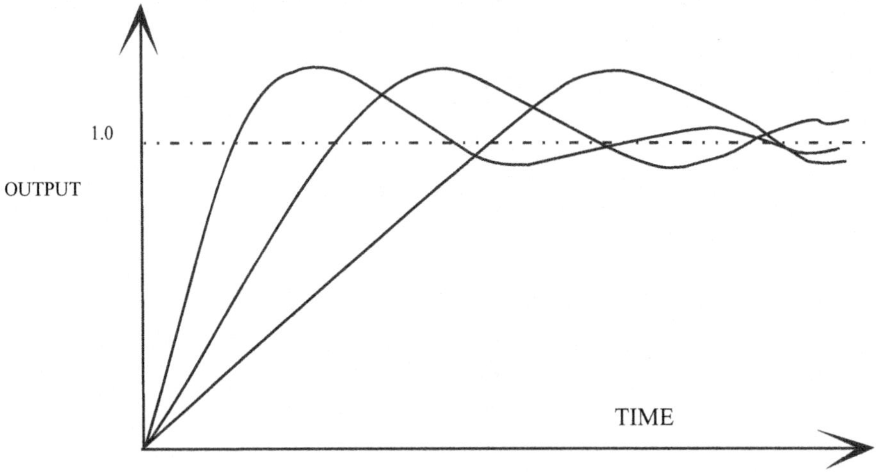

Fig. 3.9 Isodamping in fractional order controlled system.

This feature is of remarkable in field of control science indicating that a system need not be of fractional order to have a fractional order controller. An integer order system gets a robust and efficient feed back control if fractional calculus applied in the field of control science. This concept is dealt in detail in Chapter 9.

3.5.3 Bode's Integrals

The transfer function is a frequency domain representation and is convenient to be described as

$$G(s) = |G(s)| e^{j\angle G(s)},$$

in magnitude and phase angle ($\angle G(s) = \arg G(s)$) in complex frequency domain. For plotting the Bode's magnitude and phase diagrams, log magnitude and angle versus log of frequency, we put $s = j\omega$, where the angular frequency (radians/s) is ω. Thus we have $\ln G(j\omega) = \ln|G(j\omega)| + j\angle G(j\omega)$. The derivative of this log of transfer function that is

$$\frac{d \ln G(j\omega)}{d\omega} = \frac{1}{G(j\omega)} \frac{dG(j\omega)}{d\omega} = \frac{d \ln|G(j\omega)|}{d\omega} + j\frac{d\angle G(j\omega)}{d\omega},$$

is important in control systems design. This derivative of transfer function has two parts, derivative of magnitude and derivative of phase angle, called Bode's integrals.

The feature of iso-damping described in Figure 3.9 requires shaping of the Bode's phase curve, to have constant phase over desired band of frequencies around gain cross over points. The method uses classical Bode's integral formulas, to shape the phase curves. The relation between the phase and amplitudes of a stable minimum phase system has been derived by Bode. The result is based on Cauchy's residue theorem. The first integral is derivative of amplitude. The minimum phase transfer function $G(\mathrm{j}\omega)$ and the phase of system $\angle G(\mathrm{j}\omega)$ at any ω_0 is defined as Bode's first integral formula:

$$\angle G(\mathrm{j}\omega_0) = \frac{1}{\pi}\int_{-\infty}^{+\infty} \frac{d\ln|G(\mathrm{j}\omega)|}{du} \ln\coth\frac{|u|}{2} du \text{ , where } u = \ln\frac{\omega}{\omega_0} .$$

Since $\ln\coth\frac{|u|}{2}$ decreases rapidly as ω deviates from ω_0, the integral mostly depends on

$$\frac{d\ln|G(\mathrm{j}\omega)|}{du},$$

that is slope of Bode magnitude plot near ω_0. This ω_0 could be even gain cross over frequency ω_{gc} .We assume that slope of Bode's plot is almost constant in the neighborhood of ω_0 , and then $\angle G(\mathrm{j}\omega_0)$ is approximated as:

$$\angle G(\mathrm{j}\omega_0) \approx \frac{1}{\pi}\left[\frac{d\ln|G(\mathrm{j}\omega)|}{du}\right]_{\omega_0} \int_{-\infty}^{+\infty} \ln\coth\frac{|u|}{2} du = \frac{1}{\pi}\left[\frac{d\ln|G(\mathrm{j}\omega)|}{du}\right]_{\omega_0} \times \frac{\pi^2}{2} = \frac{\pi}{2}\left[\frac{d\ln|G(\mathrm{j}\omega)|}{du}\right]_{\omega_0}$$

This property is often used in loop shaping of control system. Here measured phase of a system at ω_0 is used to determine approximately the slope of Bode's magnitude plot.

$$s_a(\omega_0) = \left[\frac{d\ln|G(\mathrm{j}\omega)|}{du}\right]_{\omega_0} = \omega_0\left[\frac{d\ln|G(\mathrm{j}\omega)|}{d\omega}\right]_{\omega_0} \approx \frac{2}{\pi}\angle G(\mathrm{j}\omega_0)$$

The derivative of phase is the second formula of Bode's integral. The logarithm of stable system amplitude, at frequency ω_0, given by Bode is:

$$\ln|G(\mathrm{j}\omega_0)| = \ln|k_g| - \frac{\omega_0}{\pi}\int_{-\infty}^{+\infty} \frac{d\left(\angle G(\mathrm{j}\omega)/\omega\right)}{du} \ln\coth\frac{|u|}{2} du ,$$

k_g is the static gain of system at $\omega = 0$.Assume $(\angle G(j\omega_0)/\omega)$ is linear in the neighborhood of ω_0, and then the above integral is approximated as:

$$\ln|G(j\omega_0)| \approx \ln|k_g| - \frac{\omega_0}{\pi}\left[\frac{d(\angle G(j\omega)/\omega)}{du}\right]_{\omega_0} \times \frac{\pi^2}{2}$$

$$\ln|G(j\omega_0)| \approx \ln|k_g| - \frac{\pi\omega_0^{\ 2}}{2}\left(\frac{1}{\omega_0}\left[\frac{d\angle G(j\omega)}{d\omega}\right]_{\omega_0} - \frac{\angle G(j\omega_0)}{\omega_0^{\ 2}}\right)$$

This gives slope of phase as

$$s_p(\omega_0) = \omega_0\left[\frac{d\angle G(j\omega)}{d\omega}\right]_{\omega_0} \approx \angle G(j\omega_0) + \frac{2}{\pi}\left[\ln|k_g| - \ln|G(j\omega_0)|\right]$$

These are useful formulations for shaping the phase curve to get desired response, and are is further demonstrated in Chapter 9.

3.6 Semi Infinite Lossless Transmission Line

Above discussions in earlier sections elaborates on semi-differentiation and semi-integration obtained for driving point impedance of semi-infinite lossy line. A loss less transmission line constitutes of L and C distributed through out its length. In the Figure 3.4 the element L will replace R. The line considered here is the semi-infinite lossless line whose impedance is constant or an operator of zero order. The problem is written as:

$$\frac{\partial^2 v(x,t)}{\partial t^2} = \frac{1}{LC}\frac{\partial^2 v(x,t)}{\partial x^2}, v(0,t) = v_1(t), v(\infty,t) = 0, v(x,0)\ \&\ v'(x,0),$$

is wave equation.

Given with

$$\frac{\partial i(x,t)}{\partial x} = -C\frac{\partial v(x,t)}{\partial t}\ \&\ \frac{\partial v(x,t)}{\partial x} = -L\frac{\partial i(x,t)}{\partial t},$$

where v is the voltage i is the current $v_I(t)$ is time-dependent input variable, L is inductance per unit length and C is capacitance per unit length. A classical solution to this problem is obtained through iterated Laplace transforms as done for semi-infinite lossy line in section 3.5. The main results are given below:

$$V(0,s) = \frac{-V^*(0,s)}{\sqrt{LC}s} + \frac{1}{\sqrt{LC}}[V(p,0)]_{p=s\sqrt{LC}} + \frac{1}{\sqrt{LC}s}[V'(p,0)]_{p=s\sqrt{LC}}$$

$$V^*(0,s) = \frac{dV(0,s)}{dx}\ \&\ V' = \frac{dv(x,0)}{dt}$$

This contains transfer function of driving point (not impedance) as

$$\frac{V(0,s)}{V^*(0,s)} = -\frac{1}{\sqrt{LC}s}$$

and in time domain

$$v(0,t) = -\frac{1}{\sqrt{LC}}\int\frac{dv(0,t)}{dx}dt + \phi_1(t)$$

$\phi_1(t)$ is time dependent initial condition. The transfer function consists of two parts, the forced response due to $V^*(0,s)$ and the initial condition response due to the initial voltage distribution in the loss less line. Using current expression as given the driving point impedance is obtained as follows:

$$V(0,s) = \sqrt{\frac{L}{C}}I(0,s) - \sqrt{\frac{L}{C}}\frac{[I(0,0)]}{s} + \frac{1}{\sqrt{LC}}[V(p,0)]_{p=s\sqrt{LC}} + \frac{1}{\sqrt{LC}s}[V'(p,0)]_{p=s\sqrt{LC}}$$

Notice that the voltage is composed of two parts, the forced response due to $I(0,s)$, and the initial condition response due to the initial voltage distribution in the loss less line. Considering only the first term it can be seen that the impedance looking into this line is thus:

$$Z(s) = \frac{V(0,s)}{I(0,s)} = \sqrt{\frac{L}{C}}$$

which is simply a constant. Mathematically the impedance expressed in time domain as

$$v(0,t) = \sqrt{\frac{L}{C}}i(0,t) + \phi_2(t),$$

has a time dependent initial condition response due to initial voltage and current distribution and can be obtained by Laplace inverse of the last three terms of equation showing $V(0,s), I(0,s)$ relationship i.e.

$$V(0,s) = \sqrt{\frac{L}{C}}I(0,s) - \sqrt{\frac{L}{C}}\frac{[I(0,0)]}{s} + \frac{1}{\sqrt{LC}}[V(p,0)]_{p=s\sqrt{LC}} + \frac{1}{\sqrt{LC}s}[V'(p,0)]_{p=s\sqrt{LC}}$$

Thus it can be seen that a simple constant gain operator (zero-order operator) can also have time-varying initial condition terms. Figure 3.10 gives the diagram of a loss less semi-infinite transmission line (a zero-order element). Though the order of operation is zero i.e. it returns the input function (variable) unaltered (except gain or attenuation) yet in the theory of generalized calculus the initial distributed charges and voltage stored will be returned to the output. This initial function is time varying into future. The initial conditions on the distributed L and C, along the infinite line that gives rise to initialization functions (of time). Note that this particular element (of zero-order) does not call for differintegrations but the initial conditions ϕ associated with this distributed characteristics is very important to generalized theory of initialized (fractional) calculus. Operational amplifier circuit realized with zero-order distributed element will give practical understanding for generalized (initialized) calculus, is dealt in detail in Chapter 6 and 7.

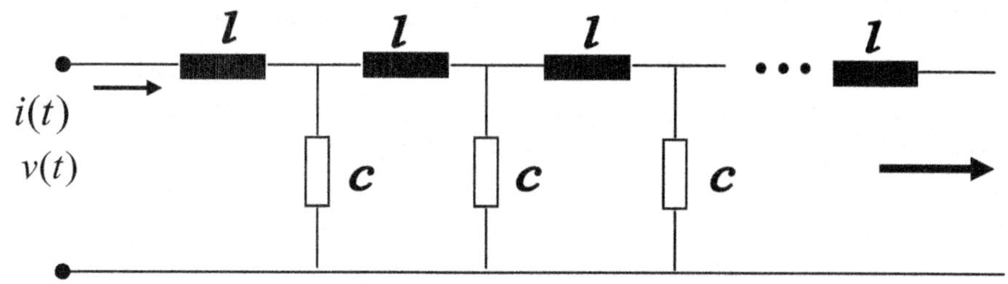

Fig. 3.10 Semi-infinite loss less transmission line

In the Figure 3.6, the input element is a lumped resistor R and the feed back element is a lumped capacitor C. Then this circuit configuration gives lumped integrator circuit. The input voltage $v_i(t)$ to the circuit, let be switched on at some time a, before that time $t=a$ the voltage is zero, and we start the circuit process (of integration) at time $t=c>a$. This implies that the capacitor is pre-charged with $q(c)$ Coulombs, from time a to c with initial voltage $v_o(c)$. This constant is thus initial condition for this lumped element integrator circuit.

The describing equations for this configuration are as follows:

$$v_i(t)-0=i_f R$$

$$0-v_o(t)=\frac{1}{C}\int_{t=c}^{t} i_f(t)dt+\frac{1}{C}\int_{t=a}^{t=c} i_f(t)dt=\frac{1}{C}\int_c^t i_f(t)dt+\frac{q(c)}{C}=\frac{1}{C}\int_c^t i_f(t)dt+\left[0-v_o(c)\right]$$

$$=\frac{1}{C}\,{}_cD_t^{-1}i_f(t)\,.$$

The ${}_cD_t^{-1}$ is integer order 'one' integration process starting from time $t=c$ which includes the initialization, process that is charging of the capacitor C from time $t=a$ to $t=c$ represented as:

$$\psi'(t)=\int_a^c i_f dt=\frac{1}{C}q(c)=-v_o(c)$$

Therefore, the total process is, un-initialized integration starting from time $t = c$ that is:

$$_{c}d_{t}^{-1}i_{f} = \int_{c}^{t} i_{f}dt$$

plus initialization integration process from a to c that is:

$$_{a}d_{c}^{-1}i_{f} = \int_{a}^{c} i_{f}dt = \psi'(t)$$

These equations yield the final result by putting $i_{i}(t) = i_{f}(t)$, as:

$$v_{o}(t) = -\frac{1}{RC}\int_{c}^{t} v_{i}(t)dt + v_{o}(c) = -\frac{1}{RC}\,{}_{c}D_{t}^{-1}v_{i}(t)$$

With $\psi(t) = -RCv_{o}(t)$. This is classical integer order calculus, with initialization as constant.

In the circuit of Figure 3.5 now we replace the input element with semi-infinite loss less (LC) transmission line, a zero order element, and the feed back element with lumped capacitor C_{f}. The transmission line terminal equation is re-written as:

$$i(t) = \sqrt{\frac{C}{L}}v(t) + \phi(t)$$

with $\phi(t)$ as initial charge distribution on the distributed element.

The defining equations of this circuit are:

$$i_{i}(t) = \sqrt{\frac{C}{L}}\left[v_{i}(t) - 0\right] + \phi(t)$$

$$0 - v_{o}(t) = \frac{1}{C_{f}}\int_{t=c}^{t} i_{f}(t)dt + \left[-v_{o}(c)\right]$$

as done for the lumped integrator case above.

$$i_{i}(t) = i_{f}(t)$$

Therefore solving for $v_{o}(t)$ we obtain:

$$v_{o}(t) = -\left[\frac{1}{C_{f}}\sqrt{\frac{C}{L}}\right]\int_{t=c}^{t} v_{i}(t)dt - \frac{1}{C_{f}}\int_{t=c}^{t}\phi(t)dt + v_{o}(c) = -\left[\frac{1}{C_{f}}\sqrt{\frac{C}{L}}\right]{}_{c}D_{t}^{-1}v_{i}(t)$$

where,

$$\psi(t)=\sqrt{\frac{L}{C}}\int_c^t \phi(t)dt + C_f\sqrt{\frac{L}{C}}v_o(c)$$

Here the initialization function is not a constant but a function of time.

This is also expression is similar to classical integer order integrator with lumped parameters as obtained earlier. Here the integrator is realized with distributed element. The important difference in the values of initialization function. For a distributed element integrator, the effect of past history is contained, not only in a constant $v_o(c)$, which is charge on the capacitor C_f, but also carried in the remainder of the $\psi(t)$ function, which accounts for the distributed charge along the semi-infinite line. It is also observed here that the zero order input element since is a wave equation

$$\frac{\partial^2 v(x,t)}{\partial t^2}=\frac{1}{LC}\frac{\partial^2 v(x,t)}{\partial x^2},$$

will simply propagate any perturbations in $v_i(t)$ along the semi-infinite line, never returning, thus never seen again. The only effect being proportional variations in the $i_i(t)$. This behavior is true for terminal charging. However for side charging (arbitrary charging with voltages on the distributed line) an additional time function may return to the circuit output, which is dependent on initial voltage distribution on the line.

The circuit of Figure 3.6 when configured with input element as lumped capacitor C and the feedback element as lumped resistance R behaves as integer order differentiator. The constituent equations are:

$$v_i(t)-0=\frac{1}{C}\int_{t=c}^{t} i_i(t)dt+\frac{1}{C}\int_{t=a}^{t=c} i_i(t)dt=\frac{1}{C}\int_c^t i_i(t)dt+\frac{q(c)}{C}=\frac{1}{C}\int_c^t i_i(t)dt+v_i(c)=\frac{1}{C}\,{}_cD_t^{-1}i_i(t)$$

$$0-v_o(t)=i_f R$$

$$i_i(t)=i_f(t)$$

This gives:

$$v_o(t)=-RC\left[\frac{d}{dt}\left(v_i(t)-v_i(c)\right)\right]=-RC\,{}_cD_t^1 v_i(t)=-RC\left[{}_cd_t^1 v_i(t)+\psi(t)\right]$$

The initialization term

$$\psi(t)=\frac{d}{dt}v_i(c)$$

is taken normally as zero. However presence of initial charge in the input capacitor gives an impulse output at the start of differentiation process at $t = c$.

Modifying the circuit of Figure 3.6 with input element as capacitor C_i and the feedback element with distributed LC zero-order element gives the integer order differentiator transfer character, with the concept of initialization function and generalized calculus. The defining equations are:

$$v_i(t) - 0 = \frac{1}{C_i}\int_c^t i_i(t)dt + v_i(c) = \frac{1}{C_i}\,{}_cD_t^{-1} i_i(t)$$

For the distributed feedback zero order elements the expression in the circuit is:

$$0 - v_o(t) = \sqrt{\frac{L}{C}} i_f(t) + \phi(t) = \sqrt{\frac{L}{C}} i_f(t) + \sqrt{\frac{L}{C}}\psi(t)$$

Putting $i_i(t) = i_f(t)$, yields the final result as:

$$v_o(t) = -\sqrt{\frac{L}{C}}\left[C_i \frac{d}{dt}\left(v_i(t) - v_i(c)\right) + \psi(t)\right] = -C_i\sqrt{\frac{L}{C}}\left[\frac{d}{dt} v_i(t) + \frac{1}{C_i}\psi(t)\right] = -C_i\sqrt{\frac{L}{C}}\,{}_cD_t^1 v_i(t)$$

The generalized differentiation requires an initialization function. However for terminal charging case for integer order differentiation this initialization is zero but for side charged transmission line, an additional time function will be returned to the circuit output.

A simple gain (memory) less zero order operators is realized by configuring the circuit of Figure 3.6 with R_i as lumped resistor at input leg, and R_f as lumped resistor at the feedback. The transfer characteristics will be then:

$$v_o(t) = -\frac{R_f}{R_i} v_i(t) = -\frac{R_f}{R_i}\,{}_cD_t^0 v_i(t) = -\frac{R_f}{R_i}\left[{}_cd_t^0 v_i(t) + \psi(t)\right], \text{ with } \psi(t) = 0,$$

Clearly this circuit has no memory.

Zero order circuit may me realized by employing semi infinite distributed loss less transmission lines at input leg and one lumped resistor R at feedback, of circuit of Figure 3.6.

The input leg equation with LC line is:

$$i_i(t) = \sqrt{\frac{C}{L}}\left[v_i(t) - 0\right] + \phi_i(t) = \sqrt{\frac{C}{L}}\,{}_cD_t^0[v_i(t) - 0]$$

The feedback leg equation is:

$$0 - v_o(t) = Ri_f(t)$$

and $i_i(t) = i_f(t)$ gives:

$$v_o(t) = -Ri_f(t) = -R\sqrt{\frac{C}{L}}\,{}_cD_t^0 v_i(t) = -R\sqrt{\frac{C}{L}}\left\{{}_c d_t^0 v_i(t) + \psi_i(t)\right\}$$

$$\text{where } \psi_i(t) = \sqrt{\frac{L}{C}}\phi_i(t)$$

This zero order operation in general returns the input function $v_i(t)$ (with amplification or attenuation), but also provides the extra time function (associated with the memorized charges on the distributed element). This zero order circuit has memory.

3.7 Partial Differential Equations and Operational Calculus

In earlier discussions what we have seen elaborate detailed derivation of partial differential equations (PDE) leading to half derivative and thus fractional differential equations. The early theory was developed by Oliver Heaviside (1871) for these 'semi-infinite' systems-described by partial differential equations-one dimensional diffusion equation is given as:

$$\frac{\partial^2 u}{\partial x^2} = a^2\frac{\partial u}{\partial t}$$

with parameters,

$$a^2 = \frac{c_p\rho}{k}; \quad a^2 = RC$$

for semi-infinite heat conductor and for semi-infinite lossy transmission lines as described in previous sections of this chapter. The equation is standard Fick's diffusion equation in one dimension with diffusing quantity as: $u(x,t)$, variable of position and time. The initial condition is: $u(x,0) = 0, x > 0$, and the boundary condition is: $u(0,t) = u_0$.

Let the operator $s \equiv \partial/\partial t$, (choice by O. Heaviside to solve Differential equations). Putting this operator in the PDE we obtain differential equation in position variable only as:

$$\frac{\partial^2 u}{\partial x^2} - a^2 su = 0$$

The roots of the above linear differential equation are: $m = \pm a\sqrt{s}$, giving standard solution as:

$$u(x,s) = A\exp(-a\sqrt{s}x) + B\exp(+a\sqrt{s}x)$$

Putting the initial condition we get: $x \to \infty, u(x,0) = 0, B = 0$, and putting the boundary condition we get: $x \to 0, u(0,t) = u_0 = A$; giving solution as: $u(x,s) = u_0 \exp(-a\sqrt{s}x)$ .We expand this exponential solution as power series to get:

$$u(x,s) = u_0 + u_0 \sum_{n=1}^{\infty} \frac{(-ax\sqrt{s})^n}{n!} = u_0 + \sum_{n=1}^{\infty} \frac{(-ax)^n (s)^{\frac{n}{2}}}{n!} u_0$$

Segregating odd and even terms and with re-arrangement we get:

$$u(x,s) = u_0 - \sum_{m=0}^{\infty} \frac{(ax)^{2m+1}}{(2m+1)!} s^m (s^{1/2} u_0) + \sum_{n=0}^{\infty} \frac{(ax)^{2n}}{(2n)!} s^n u_0$$

Use the operator(as above) and use half (time) derivative of u_0 as a constant as $u_0 / \sqrt{\pi t}$, and recognizing the fact that n -th integer derivative of constant u_0 is zero then thereby putting the values of identities as:

$$d^{1/2} u_0 \to s^{1/2} u_0 \equiv \frac{u_0}{\sqrt{\pi t}}; d^n u_0 \to s^n u_0 = 0 \text{ ,with } s^n \to d^n / dt^n \; n \in \mathbb{Z}^+$$

we obtain the solution:

$$u(x,t) = u_0 - \frac{u_0}{\sqrt{\pi}} \sum_{m=0}^{\infty} \frac{(ax)^{2m+1}}{(2m+1)!} \left[\frac{d^m}{dt^m} t^{-1/2} \right] = u_0 - \frac{u_0}{\sqrt{\pi}} \sum_{m=0}^{\infty} \frac{(ax)^{2m+1}}{(2m+1)!} \frac{\Gamma(-\frac{1}{2}+1)}{\Gamma(-\frac{1}{2}-m+1)} t^{-\frac{1}{2}-m}$$

$$u(x,t) = u_0 - \frac{u_0}{\sqrt{\pi}} \sum_{m=0}^{\infty} \frac{(ax)^{2m+1}}{(2m+1)!} \frac{\Gamma\left(1/2\right)}{\Gamma\left(\frac{1}{2}-m\right)} \frac{1}{t^{m+1/2}}$$

Using the property of Gamma function as

$$\Gamma(-m + 1/2) = \frac{[-4]^m m! \sqrt{\pi}}{(2m)!},$$

we simplify the solution as:

$$u(x,t) = u_0 - \frac{u_0}{\sqrt{\pi}} \sum_{m=0}^{\infty} \frac{(-1)^m}{m!} \frac{(ax)^{2m+1}}{(2m+1)2^{2m} t^{m+\frac{1}{2}}}$$

Writing

$$\frac{(ax)^{2m+1}}{(2m+1)2^{2m}t^{m+\frac{1}{2}}} \equiv 2\int_0^{y=\frac{ax}{2\sqrt{t}}} y^{2m}dy$$

and using

$$\sum_{m=0}^{\infty}\frac{(-1)^m}{m!}(x^2)^m \equiv \exp(-x^2),$$

we get exact solution as:

$$u(x,t)=u_0-\frac{2u_0}{\sqrt{\pi}}\int_0^{ax/2\sqrt{t}}\exp(-y^2)dy = u_0\left[1-\mathrm{erf}\left(\frac{ax}{2\sqrt{t}}\right)\right]$$

This is work of Heaviside and his Operational Calculus on PDE defining diffusion.

3.8 Fick's Diffusion Discussion

In all the above sections we have derived the expressions for fractional ½ order or integer order differential equations, relating 'through variables' with 'across variables'; by use of Fick's law of diffusion. Here we elaborate few salient observations regarding classical Fick's law, and then in the next section we try to modify the same.

For $C(x,t)$ the diffusion quantity into media and $J(x,t)$ denoting the flux of the diffusing quantity, we relate first a continuity equation, (also called Fick's second law) as:

$$\frac{\partial}{\partial t}C(x,t) = -\frac{\partial}{\partial x}J(x,t),$$

followed by constitutive equation, (also called Fick's first law-similar to Fourier's Heat conduction law) as:

$$J(x,t) = -\mathbb{D}\frac{\partial}{\partial x}C(x,t).$$

Combining these two equations we obtain the Fick's Diffusion Equation as:

$$\frac{\partial}{\partial t}C(x,t) = \mathbb{D}\frac{\partial^2}{\partial x^2}C(x,t)$$

This is the equation we have often used in the previous sections of this chapter. The Fourier Laplace Transforms are used to analyze the partial differential equations, which are noted as:

$$C(k,s) = \Im\left\{\mathcal{L}\left\{C(x,t);t \to s\right\};x \to k\right\}$$

The Fourier transformed diffusion equation using

$$\Im\left\{\partial_{xx}^{2} C(x,t)\right\} = (\mathrm{i}k)^2 C(k,t) = -k^2 C(k,t)$$

is $\frac{\partial}{\partial t} C(k,t) = -\mathbb{D}k^2 C(k,t)$

The individual modes (wave numbers k) decays or relaxes as

$$C(k,t) = \exp(-\mathbb{D}k^2 t) .$$

The Laplace Transformed of this is

$$C(k,s) = \frac{1}{s + \mathbb{D}k^2} C_0(k) .$$

From this one can calculate mean squared displacement (MSD) as:

$$\left\langle \Delta x^2 \right\rangle = \mathcal{L}^{-1}\left[\lim_{k \to 0}\left\{-\left(d^2 / dk^2\right) C(k,s)\right\}\right].$$

For this $C(k,s)$, with initial condition as $C(x,0) = \delta(x)$, $C_0(k) = 1$ the following steps demonstrates how to get MSD

$$\lim_{k \to 0}\left[-\frac{d^2}{dk^2} C(x,t)\right] = \left(\frac{2\mathbb{D}}{s^2}\right),$$

the subsequent inverse Laplace gives

$$\left\langle \Delta x^2 \right\rangle = \mathcal{L}^{-1}\left(2\mathbb{D}s^{-2}\right) = (2\mathbb{D})t$$

Well, the normal Brownian motion also has the same MSD, as linearly growing, that is:

$$\left\langle \Delta x^2 \right\rangle \approx t .$$

This Fick's diffusion is valid in continuous time random walk (CTRW) approach where the diffusing species (particles, walker) in random Brownian motion, has finite average wait time (or no wait time) as well as finite jump lengths with well defined variance or standard deviation that is, wait time statistics give finite average wait time and jump length statistics give well defined finite MSD (variance and standard deviation). However, if any of these wait time and or jump length statistics are having diverging (not finite) average and or standard deviation we will have anomalous diffusion. These cases either lead to sub-diffusion or super-diffusion phenomena described by fractional diffusion equations. These issues will be elaborated in next chapters.

The Fick's diffusion equation

$$\frac{\partial}{\partial t}C(x,t) = \mathbb{D}\frac{\partial^2}{\partial x^2}C(x,t)\,,$$

when put in integral form is expressed as:

$$C(x,t) = \delta_{x0} + \mathbb{D}\int_0^t \Delta C(x,t')dt'\,,$$

Where $\Delta \equiv \partial^2 / \partial x^2$, Laplacian operator. This integral is very reminiscent of the integral equation of CTRW. In CTRW one imagines a random walker that starts at $x = 0$ and at time $t = 0$, and proceeds by successive jumps. The probability density for a time interval of length t between two consecutive jumps is 'wait-time' denoted by $w(t)$ and the probability density of displacement by a vector x, in a single jump is denoted by $\lambda(x)$; then the integral equation of CTRW reads:

$$f(x,t) = \delta_{x0}\chi(t) + \int_0^t w(t-t')\int_{-\infty}^{\infty}\lambda(x-x')f(x',t')dx'dt'\,,$$

where $\chi(t)$ is the probability that the walker survives at the origin for a time of length t. Here, the walker is assumed to be prepared in its initial position, from which it develops according to $w(t)$. The survival probability $\chi(t)$ is related to 'wait-time' density by

$$\chi(t) = 1 - \int_0^t w(t')dt'$$

The formal similarity between the probability evolution equation $f(x,t)$ and the integral representation of Fick's diffusion suggests there exists a relation between them. If one normalizes the observation $C(x,t)$ to unity at $C(0,0) = 1$ then, they are the same $C(x,t) \equiv f(x,t)$. Substituting the $\chi(t)$, survival probability into the CTRW expression, we obtain:

$$f(x,t) = \delta_{x0}\left[1 - \int_0^t w(t')dt'\right] + \int_0^t w(t-t') \int_{-\infty}^{\infty} \lambda(x-x') f(x',t') dx'dt'$$

$$f(x,t) - \int_0^t w(t-t') \int_{-\infty}^{\infty} \lambda(x-x') f(x',t') dx'dt' = \delta_{x0}\left[1 - \int_0^t w(t')dt'\right]$$

For the above expression the double integral in the LHS is Fourier-Laplace convolution, which when transformed returns multiplicative terms.

Taking Laplace-Fourier of the above expression, with $\Im\{\delta_{x0}\} = 1$, $\mathcal{L}\{1\} = s^{-1}$ we get:

$$F(k,s)[1 - w(s)\lambda(k)] = \frac{1}{s} - \frac{w(s)}{s} ,$$

which gives

$$F(k,s) = C(k,s) = \frac{1 - w(s)}{s[1 - w(s)\lambda(k)]} .$$

This expression will be used in Chapter 4.

For initial delta function excitation $C(x,0) = \delta(x)$ and for natural semi-infinite boundary conditions $C(|x| \to \infty, t) = 0$, the solution is obtained as Gaussian solution, namely:

$$C(x,t) = \frac{1}{\sqrt{4\pi \mathbb{D} t}} \exp\left(-\frac{x^2}{4\mathbb{D} t}\right)$$

This solution (or distribution function) states that for a very small time $t \to 0^+$, that is just after the delta function was placed at the origin for excitation, a finite amount of diffusing quantity is available far from origin! That is, that there exists 'infinite velocity' of propagation for the diffusing element in the media!

The Diffusion equation can be written in vector form by use of Laplacian operator as:

$$\frac{\partial}{\partial t} C(\bar{X}, t) = \mathbb{D} \nabla^2 C(\bar{X}, t) .$$

For isotropic case this can be written as:

$$\frac{\partial}{\partial t} C(r,t) = \mathbb{D} r^{1-d} \frac{\partial}{\partial r} r^{1-d} \frac{\partial}{\partial r} C(r,t)$$

With d representing integer order Euclidian dimensions, say for one-dimension case $d = 1$ the diffusion equation is

$$\frac{\partial}{\partial t}C(r,t) - \mathbb{D}\frac{\partial^2}{\partial r^2}C(r,t) = 0\,,$$

for spherical three-dimensional case ($d = 3$) the diffusion equation is

$$\frac{\partial}{\partial t}C(r,t) = \mathbb{D}\frac{1}{r^2}\frac{\partial}{\partial r}\left[r^2\frac{\partial}{\partial r}C(r,t)\right] = \mathbb{D}\frac{2}{r}\frac{\partial}{\partial r}C(r,t) + \mathbb{D}\frac{\partial^2}{\partial r^2}C(r,t)$$

For two dimensional cases $d = 2$ (Cylinder Geometry) the diffusion equation is:

$$\frac{\partial}{\partial t}C(r,t) = \mathbb{D}\frac{1}{r}\left[\frac{\partial}{\partial r}r\frac{\partial}{\partial r}C(r,t)\right] = \mathbb{D}\frac{1}{r}\frac{\partial}{\partial r}C(r,t) + \mathbb{D}\frac{\partial^2}{\partial r^2}C(r,t)$$

In terms of parameter say geometry g, we may re-write the same as:

$$\frac{\partial}{\partial t}C(r,t) - \mathbb{D}\frac{\partial^2}{\partial r^2}C(r,t) - \frac{2g\mathbb{D}}{r}\frac{\partial}{\partial r}C(r,t) = 0$$

The above three cases will be recovered for planer one dimensional with $g = 0$, for spherical three dimensional with $g = 1$, and with g as ½ we get two dimensional cylindrical case.

In simple way let us ask a question. Whether Euclidian geometrical parameters what we have considered should always be positive integers of dimensional values 1, 2, 3? Well can the geometrical parameters in the Laplacian operator be of only 0, 1, and ½ for the integer order geometry? The answer to this is very well yes; these dimensional number d, and g can be arbitrary real numbers.

Let us give physical reasoning to our hypothesis that fractional Euclidian dimensions or fractional geometrical parameters may be possible in reality. Consider the diffusing species (charges, temperature, particles, solution concentration etc.) travel in a media with lot of heterogeneous microscopic obstacles or attractors. The diffusing species may get to see smeared out media in presence of these inhomogeneous obstacles, and may thus diffuse at faster rate, or it may so happen that the obstacles are traps (singularities) and the diffusing species fall into these and do take longer time to diffuse through. This reasoning thus opens up the possibility that the dimension numbers can be of fraction or even geometric number that can be arbitrary opening up possibility of having fractional order differential diffusion equation for Fick's equation. Note that in previous sections what we have derived as ½ order differential equations is for all semi-infinite diffusing media, without any heterogeneous traps or obstacles. The anomalous case was discussed in chapter 1 with anomalous coefficient d_w, which was one way to describe anomalous case.

3.9 Cattaneo Diffusion

The diffusing flux is allowed some relaxation time τ in this formulation (1948). This is in order to give 'finite velocity' to the diffusing element. What we had observed in Fick's diffusion is that the velocity of the species assumed to be infinite at the start time of diffusing process. The flux equation (Fick's first law) gets modified as:

$$J(x,t)+\tau\frac{\partial}{\partial t}J(x,t)=-\mathbb{D}\frac{\partial}{\partial x}C(x,t),$$

writing this with the continuity equation

$$\frac{\partial}{\partial t}C(x,t)=-\frac{\partial}{\partial x}J(x,t),$$

we obtain the Catteneo Equation as:

$$\frac{\partial}{\partial t}C(x,t)+\tau\frac{\partial^2}{\partial t^2}C(x,t)=\mathbb{D}\frac{\partial^2}{\partial x^2}C(x,t)$$

This equation is a hyperbolic equation, compared to Fick's equation which is a parabolic equation. This Catteneo equation is a 'damped wave equation' also called as Telegrapher's equation. Here the propagation velocity is $v=\sqrt{\mathbb{D}/\tau}$ and is finite.

In a media with memory, the diffusing flux $J(x,t)$ is related to the previous history of diffusing species (density, concentration, temperature etc.) $C(x,t)$ through relaxation function (memory kernel) as:

$$J(x,t)=-\int_0^t K(t-t')\frac{\partial C(x,t')}{\partial x}dt'.$$

This is similar to Boltzmann's superposition law, and Memory Integrals. Put this convolution expression in Catteneo's equation, that is

$$J(x,t)+\tau\frac{\partial}{\partial t}J(x,t)=-\mathbb{D}\frac{\partial}{\partial x}C(x,t),$$

to obtain

$$-\int_0^t K(t-t')\frac{\partial C(x,t')}{\partial x}dt'-\tau\frac{\partial}{\partial t}\int_0^t K(t-t')\frac{\partial C(x,t')}{\partial x}dt'=-\mathbb{D}\frac{\partial}{\partial x}C(x,t)$$

$$\left(1+\tau\frac{\partial}{\partial t}\right)\int_0^t K(t-t')\frac{\partial}{\partial x}C(x,t')dt'=\mathbb{D}\frac{\partial}{\partial x}C(x,t)$$

Applying Leibniz's rule for differentiation of an Integral i.e.

$$D\left[D^{-1}f\right]=D^{-1}\left[Df\right]+f(0)$$

we get:

$$\tau K(0)\frac{\partial}{\partial x}C(x,0)+\int_0^t\left[\tau\frac{\partial}{\partial t}K(t-t')+K(t-t')\right]\frac{\partial}{\partial x}C(x,t)dt'=\mathbb{D}\frac{\partial}{\partial x}C(x,t)$$

We can take from above $\tau K(0)=\mathbb{D}$ and $\tau\frac{\partial}{\partial t}K(t)+K(t)=0$; that is done by equalizing the like terms of LHS and RHS. Above sets give solution for memory kernel as:

$$K(t)=\left(\frac{\mathbb{D}}{\tau}\right)\exp\left(-\frac{t}{\tau}\right)$$

This is relaxation function or Memory Kernel of Catteneo diffusion.

The objective now is to embed by similar computation a generalize Catteneo equation with power-law type memory kernel, like:

$$K(s)=\frac{\mathbb{D}_K}{\tau^{\alpha}}\frac{s^{-1}}{(1+\tau^{-\alpha}s^{-\alpha})}$$

in Laplace domain, or

$$K(t)=\frac{\mathbb{D}_K}{\tau^{\alpha}}E_{\alpha,1}\left[-\left(\frac{t}{\tau}\right)^{\alpha}\right]$$

in Mittag-Leffler functional form in time domain.

This new (power-law) Kernel gives rise to:

$$\frac{\partial}{\partial t}C(x,t)+\tau^{\alpha}\frac{\partial^{\alpha-1}}{\partial t^{\alpha-1}}\frac{\partial^2}{\partial t^2}C(x,t)=\mathbb{D}_K\frac{\partial^{\alpha-1}}{\partial t^{\alpha-1}}\frac{\partial^2}{\partial x^2}C(x,t).$$

A Fractional Differential Equation for diffusion, with power law Kernel for long time decay as $K(t)\sim t^{-\alpha}$, $(0<\alpha<1)$. The generalized Catteneo equation, for flux relaxation of Mittag-Leffler type is thus:

$$J(x,t)+\tau^{\alpha}\frac{\partial^{\alpha}}{\partial t^{\alpha}}J(x,t)=-\mathbb{D}_K\frac{\partial^{\alpha-1}}{\partial t^{\alpha}}\frac{\partial}{\partial x}C(x,t).$$

Few generalizations to anomalous diffusion are considered in next section.

3.10 Anomalous Diffusion

The anomalous diffusion situation can arise if the two equations that is continuity and constitutive equations in Fick's law do not hold at the same time. Either one or the other or both ought to be generalized. First we generalize the continuity equation as:

$$\frac{\partial^\alpha}{\partial t^\alpha} C(x,t) = -\frac{\partial}{\partial x} J(x,t), \text{ with } \alpha < 1$$

This means that the system is in a situation where the number densities of diffusing species are not conserved. The constitutive equation of Fick's law remains the same. This generalization gives generalized diffusion equation as:

$$\frac{\partial^\alpha}{\partial t^\alpha} C(x,t) = \mathbb{D} \frac{\partial^2}{\partial x^2} C(x,t)$$

In the second case we consider that the only equation that differs from ordinary diffusion is the constitutive equation, that is

$$J(x,t) = -\mathbb{D} \frac{\partial^{1-\alpha}}{\partial t^{1-\alpha}} \frac{\partial}{\partial x} C(x,t) .$$

This constitutive equation is derived from stochastic scheme (continuous time random walk CTRW) in macroscopic limit. In a random walk the diffusing species (particle, walker) jumps and waits at the lattice position and that waiting times are drawn from a 'long-tailed' power-law distribution. This implies that the waiting times of the diffusing species have no finite average or mean and variance; whereas the jump length variance is kept finite. These issues will be discussed in the following chapters. The effect is that some particles (walker) of diffusing species get stuck for a very long time giving slow diffusion termed as sub-diffusive process. Using this flux expression we obtain:

$$\frac{\partial}{\partial t} C(x,t) = \mathbb{D} \frac{\partial^{1-\alpha}}{\partial t^{1-\alpha}} \left(\frac{\partial^2}{\partial x^2} C(x,t) \right)$$

giving rise to, same as discussed above, that is:

$$\frac{\partial^\alpha}{\partial t^\alpha} C(x,t) = \mathbb{D} \frac{\partial^2}{\partial x^2} C(x,t)$$

A third case of generalization is considered as:

$$J(x,t) = -D \frac{\partial^{\alpha-1}}{\partial t^{\alpha-1}} \frac{\partial}{\partial x} C(x,t) ,$$

with $\alpha < 1$ while continuity equation remains the same. This generalization of constitutive equation with fractional integral is interpreted as memory integral giving rise to non-local transport theory, establishing a relationship of the flux $J(x,t)$ to the previous histories of gradient of concentration $C(x,t)$, with power-law memory kernel. Combining this with the continuity equation we get

$$\frac{\partial C(x,t)}{\partial t} = \mathbb{D}\frac{\partial^{\alpha-1}}{\partial t^{\alpha-1}}\frac{\partial^2 C(x,t)}{\partial x^2}$$

After rearrangement of fractional derivatives we have:

$$\frac{\partial^{2-\alpha}}{\partial t^{2-\alpha}}C(x,t) = \mathbb{D}\frac{\partial^2}{\partial x^2}C(x,t)$$

The above equation is fractional diffuse-wave equation.

A forth case is taken where the diffusion equation is described, with spatial fractional derivative as:

$$\frac{\partial}{\partial t}C(x,t) = \mathbb{D}_\mu \frac{\partial^\mu}{\partial x^\mu}C(x,t)$$

With $1 < \mu < 2$. Here the waiting time statistic is assumed to have finite average wait times but the variance of the jump lengths are diverging (infinite). This jump length distribution is drawn from long tailed statistical distribution; giving that particle (walker) of diffusing species will be executing unrestricted 'long' jumps. This will be discussed in the next chapter. Note, in this case usual units of diffusion constant will not be valid and will be taken as $\mathrm{cm}^\mu \mathrm{s}^{-1}$.

3.11 Truncation of Semi-Infinite System to a Finite System

In this chapter what is observed is that semi-infinite lossy systems give a terminal relation with half order. The examples we have studied are from semi-infinite heat conducting system and semi-infinite lossy transmission line. The terminal relations obtained in detail shows this half derivative (half integral) behavior between a 'through variable' like current (heat-flux) and 'across variable' like voltage (temperature). In electrical engineering terms we call this terminal relation as impedance $Z(\mathrm{j}\omega) = (\mathrm{j}\omega)^{-1/2}$. Electrochemical impedance is widely used to investigate the interfacial bulk properties of materials and measure the relevant physico-chemical parameters. This impedance in electrochemical studies is called Warburg impedance which is of half order.

In all the cases of semi-infinite system Fick's diffusion law is applied and the terminal behavior appeared as 'half-ordered derivative/ integral'. This Fick's diffusion when restricted to 'finite' system, that is drive point at $x = 0$ and the system is terminated by boundary at $x = \mathrm{L}$ the situation is different. Then the terminal relation may have 'different fractional order'. Let us call characteristic frequency of a truncated semi-infinite system as: $\omega_d \equiv \mathbb{D}/\mathrm{L}^2$, where $\mathbb{D}$ is 'diffusion-constant' from Fick's equation and L is the finite dimension (not extending to infinity).

The observation is that if the frequency of excitation at the origin is large, that is $\omega >> \omega_d$, then system will behave as semi-infinite system returning the half order operator. At low frequency (compared to characteristic frequency of the system) $\omega << \omega_d$ and the terminal impedance behavior depends on whether the diffusing species are reflected or extracted at the far end at $x = \mathrm{L}$. The reason for this is that ω_d is the frequency corresponding to transit time for a particle (excitation) injected at origin $x = 0$ to cover distance $x = \mathrm{L}$. For very high frequency (greater than the characteristic frequency) the particles will not see any boundary at $x = \mathrm{L}$ and system will behave as semi-infinite system.

Consider a system of Figure 3.4 but truncated at length $x = \mathrm{L}$. If there is excitation of voltage at $x = 0$ call it $\tilde{V}$ that will create change in charge density at origin. Let the number density of charges be represented by $C(x,t)$. Therefore at origin the linear variation of charge density vis-à-vis voltage change will be approximately, $\tilde{V} = (dV/dC)\tilde{C}$. The tide denotes the changes. If $\tilde{J}(x,t)$ is the flux of number of charges flowing per unit time per unit area, then current which is conducting is: $\tilde{I}(x,t) = qA\tilde{J}(x,t)$. Here q is the charge that crosses the area A. Therefore, conduction current at $x = 0$ corresponds entirely to the flux changes of the diffusing charges that is, $\tilde{I}(0,t) = qA\tilde{J}(0,t)$, again the tide denotes the small changing quantities.

The first set of boundary conditions are at $x = 0$

$$\tilde{V}(0,t) = \left(\frac{dV}{dC}\right)\tilde{C}(0,t) \text{ at } x = 0 \text{ and } \tilde{I}(0,t) = aA\tilde{J}(0,t) \text{ at } x = 0$$

At $x = \mathrm{L}$ there can be two types of boundary conditions they are: (1) Reflecting boundary, where at the boundary the flux of diffusing species goes to zero and (2) Absorbing boundary; meaning that the diffusing species value at boundary is zero. That is:

$$\frac{\partial \tilde{C}(x,t)}{\partial x} = 0$$

at $x = \mathrm{L}$ indicating reflecting boundary condition, whereas $\tilde{C}(x,t) = 0$ at $x = \mathrm{L}$ indicates absorbing boundary condition. One may extend the thought of having a mixed boundary condition at $x = \mathrm{L}$ with

$$\frac{\partial^\alpha}{\partial x^\alpha}\tilde{C}(x,t) = 0 \text{ at } x = \mathrm{L} \text{ with } 0 \le \alpha \le 1 .$$

The fractional derivative index $\alpha = 0$ gives pure absorbing boundary condition, while $\alpha = 1$ gives a pure reflecting boundary condition. In between values gives a mix of reflecting and absorbing boundary definitions.

The Fick's diffusion equation for small fluctuation amplitudes of sinusoidal concentration may be written in Laplace domain as:

$$\frac{\partial^2}{\partial x^2}\tilde{C}(x,s) = \frac{1}{\lambda^2}\tilde{C}(x,s),$$

with $\lambda^2 = \frac{\mathbb{D}}{s}$ indicating that $\lambda(s)$ is frequency dependent.

This diffusion equation has spatial solution as

$$\tilde{C}(x,s) = A_1 \cosh\frac{x}{\lambda} + A_2 \sinh\frac{x}{\lambda}.$$

For absorbing boundary condition at $x = \mathrm{L}$ we have

$$A_1 \cosh\frac{\mathrm{L}}{\lambda} + A_2 \sinh\frac{\mathrm{L}}{\lambda} = 0.$$

This gives relation as

$$A_1 = -A_2 \tanh\frac{\mathrm{L}}{\lambda}$$

The flux is

$$\tilde{J}(x,s) = -\mathbb{D}\frac{\partial C(x,s)}{\partial x} = -\frac{\mathbb{D}A_1}{\lambda}\sinh\frac{x}{\lambda} - \frac{\mathbb{D}A_2}{\lambda}\cosh\frac{x}{\lambda}$$

At $x = 0$ given $\tilde{I}(0,s) = qA\tilde{J}(0,s)$, from above we get:

$$-\frac{\mathbb{D}}{\lambda}A_2 = \tilde{C}(0,s) = \frac{\tilde{I}(s)}{qA}.$$

From above derivations we obtain:

$$\frac{\tilde{C}(0,s)}{\tilde{I}(0,s)} = \frac{\lambda}{qAD}\tanh\frac{\mathrm{L}}{\lambda}.$$

Using this on $\tilde{V}(0,s) = (dV/dC)\tilde{C}(0,s)$ we get:

$$Z(0,s)=\frac{\tilde{V}(0,s)}{\tilde{I}(0,s)}=\frac{(dV/dC)}{qA\mathbb{D}}\lambda\tanh\frac{\mathrm{L}}{\lambda}=\frac{\mathrm{L}}{qA\mathbb{D}}\left(\frac{dV}{dC}\right)\left(\frac{\lambda}{\mathrm{L}}\right)\tanh\frac{\mathrm{L}}{\lambda}$$

Introducing a new parameter having unit of resistance called diffusion resistance as:

$$\mathrm{R}\equiv\frac{\mathrm{L}}{qA\mathbb{D}}\left(\frac{dV}{dC}\right)$$

we get terminal impedance of truncated transmission lossy line as:

$$Z(s)=\mathrm{R}\left(\frac{\lambda}{\mathrm{L}}\right)\tanh\frac{\mathrm{L}}{\lambda},$$

using the definition of characteristic frequency of this truncated system as defined earlier $\omega_d\equiv\mathbb{D}/\mathrm{L}^2$ we have $\lambda=\mathrm{L}(\omega_d/s)^{1/2}$, we obtain the terminal impedance at $x=0$ as: $Z(s)=\mathrm{R}(\omega_d/s)^{1/2}\tanh\left[(s/\omega_d)^{1/2}\right]$.

To analyze this impedance standard graphical technique is employed, where one plots the 'normalized' impedance plot on a complex plane. Putting $s=\mathrm{j}\omega$ the normalized impedance is

$$Z_N(\omega)=\left(Z(\mathrm{j}\omega)/\mathrm{R}\right)=\left(\omega_d/\mathrm{j}\omega\right)^{1/2}\tanh\left[\left(\mathrm{j}\omega/\omega_d\right)^{1/2}\right],$$

which has magnitude and phase angle. The plot of $|Z_N(\omega)|$ and $\angle\left[Z_N(\omega)\right]$ for values of different ω in radians/second gives the impedance plot. Here in this case one finds that as the value of $\omega>>\omega_d$ the terminal impedance is limiting toward $Z_N(\omega)\sim\left(\mathrm{j}\omega\right)^{1/2}$; showing semi-infinite behavior. At lower frequencies the behavior is different from half-order element.

Same steps can be repeated to get terminal impedance at $x=0$ for a reflecting boundary at $x=\mathrm{L}$ as $Z(s)=\mathrm{R}(\omega_d/s)^{1/2}\coth\left[(s/\omega_d)^{1/2}\right]$. Here too impedance-plots show half-order behavior for higher frequencies.

This discussion will be further carried over in explaining electro-chemical and Warburg impedances in Chapter 9; and how truncation of the semi-infinite system with these boundary conditions can give the structure with 'constant-phase-element' (CPE) which behaves as fractional order (other than half order) system.

3.12 Approximating the Half Order by Self Similar Structure and Its Relation to Continued Fraction Expansion

Figure 3.11 a stage-1 shows a simple capacitor element, for which the terminal voltage and current representation is

$$e(t)=\frac{1}{C}\int_0^t i(\xi)d\xi=\frac{1}{C}\frac{d^{-1}}{dt^{-1}}i(t),$$

with initial condition given as

$$i(t\le 0)=0=e(t\le 0).$$

Extending this to make the circuit for stage-2 of Figure 3.11, the circuit equation is obtained by writing the nodal equations as following:

$$i_0(t)=\frac{e_0(t)}{R_0},\ i(t)-i_0(t)=C_0\frac{de_0(t)}{dt},\ i(t)=\frac{e(t)-e_0(t)}{R_1},$$

simplifying this we get:

$$(R_0+R_1)i(t)+R_0R_1C_0\frac{di(t)}{dt}=e(t)+R_0C_0\frac{de(t)}{dt},$$

with static initial condition as $i(0)=0=e(0)$. Taking Laplace of the stage-2 we get, the following:

$$I(s)[R_0+R_1+R_0R_1C_0s]-R_0R_1C_0i(0)=E(s)[1+R_0C_0s]-R_0C_0e(0)$$

$$\frac{E(s)}{I(s)}=\frac{R_0+R_1+R_0R_1C_0s}{1+R_0C_0s},$$

simplifying this we obtain:

$$\frac{E(s)}{R_1I(s)}=1+\frac{\frac{1}{R_1C_0}}{s+\frac{1}{R_0C_0}}$$

For the stage-3 circuit of Figure 3.11, we have:

$$i(t)=\frac{e(t)-e_1(t)}{R_2},\ i(t)-i_1(t)=C_1\frac{de_1(t)}{dt}.$$

On this we obtain Laplace Transformation and after simplification we obtain

$$\frac{E(s)}{R_2I(s)}=1+\frac{\frac{1}{R_2C_1}}{s+\frac{I_1(s)}{C_1E_1(s)}}$$

From the stage-2 circuit we have

$$\frac{E_1(s)}{R_1 I_1(s)} = 1 + \frac{\dfrac{1}{R_1 C_0}}{s + \dfrac{1}{R_0 C_0}}.$$

We now use some substitution as

$$\omega_0 \equiv (\tau_0)^{-1} = (R_0 C_0)^{-1}, \omega_1 \equiv (R_1 C_0)^{-1}, \omega_2 \equiv (R_1 C_1)^{-1}, \omega_3 \equiv (R_2 C_1)^{-1}$$

to get

$$\frac{E(s)}{R_2 I(s)} = 1 + \frac{\omega_3}{s + \dfrac{\omega_2}{1 + \dfrac{\omega_1}{s + \omega_0}}}$$

Like this we can continue for circuit of stage-n of Figure 3.11 and write, by putting $\omega_{2j} = (R_j C_j)^{-1}; \omega_{2j+1} = (R_{j+1} C_j)^{-1}$, to get:

$$\frac{E(s)}{R_n I(s)} = 1 + \frac{\omega_{2n-1}}{s + \dfrac{\omega_{2n-2}}{1 + \dfrac{\omega_{2n-3}}{s + \dots \dfrac{\omega_1}{s + \omega_0}}}} = 1 + \frac{\omega_{2n-1}}{s+} \frac{\omega_{2n-3}}{1+} \frac{\omega_{2n-3}}{s+} \dots \frac{\omega_2}{1+} \frac{\omega_1}{s+} \frac{\omega_0}{1}$$

$$\frac{E(s)}{R_n I(s)} = 1 + \frac{\omega_{2n-1}}{s+} \frac{\omega_{2n-2}}{1+} \frac{\omega_{2n-3}}{s+} \dots\dots \frac{\omega_2}{1+} \frac{\omega_1}{s+} \frac{\omega_0}{1}$$

Put $v_j = \dfrac{\omega_j}{s}$, to get

$$\frac{E(s)}{R_n I(s)} 1 + \frac{v_{2n-1}}{1+} \frac{v_{2n-2}}{1+} \frac{v_{2n-3}}{1+} \dots\dots\dots\dots\dots \frac{v_2}{1+} \frac{v_1}{1+} \frac{v_0}{1}$$

For the case

$$C_0 = C_1 = C_2 = \dots\dots\dots\dots\dots\dots = C_{n-1} = C$$

and

$$R_0 = R_1 = R_2 = \dots\dots\dots\dots\dots\dots = R_{n-1} = R; R_n = \frac{1}{2} R$$

with,

$$v_0 = v_1 = v_2 = \ldots = v_{2n-3} = v_{2n-2} = \frac{1}{RCs} \equiv v\,,$$

then $v_{2n-1} = \dfrac{2}{RCs} = 2v$, we get:

$$\frac{2E(s)}{RI(s)} = 1 + 2\frac{v}{1+}\frac{v}{1+}\frac{v}{1+}\cdots\cdots\cdots\frac{v}{1+}\frac{v}{1}$$

From the theory of continued fractions as can be inductively derived, we have the following expression:

$$\frac{v}{1+}\frac{v}{1+}\cdots\cdots\frac{v}{1+}\frac{v}{1} = \frac{\sqrt{4v+1}}{1+\left[\dfrac{\sqrt{4v+1}-1}{\sqrt{4v+1}+1}\right]^{2n+1}} - \frac{\sqrt{4v+1}}{2} - \frac{1}{2}$$

Combining the above with the expression of stage-n, circuit obtained as $2n$ and numeratorial v's, and then dividing by $2\sqrt{v}$, we obtain:

$$\frac{E(s)}{I(s)}\sqrt{\frac{Cs}{R}} = \sqrt{\frac{4v+1}{4v}}\left[\frac{[\sqrt{4v+1}+1]^{2n+1} - [\sqrt{4v+1}-1]^{2n+1}}{[\sqrt{4v+1}+1]^{2n+1} + [\sqrt{4v+1}-1]^{2n+1}}\right]$$

This is the final result interrelating the transforms of $e(t)$ and $i(t)$

The right hand side of the above expression when graphically plotted, for values of v on the X-axis; for various values of n, gives a choice of number of stages. For large values of n, the plot of function

$$f(v) = \sqrt{\frac{4v+1}{4v}}\left[\frac{[\sqrt{4v+1}+1]^{2n+1} - [\sqrt{4v+1}-1]^{2n+1}}{[\sqrt{4v+1}+1]^{2n+1} + [\sqrt{4v+1}-1]^{2n+1}}\right],$$

gives value close to unity, over wide ranges of v values. In fact if n is in excess of 10, the function $f(v)$ lies within 2% of unity,

$$\frac{E(s)}{I(s)}\sqrt{\frac{Cs}{R}} \approx 1$$

provided that $6 \le v \le \frac{1}{6}n^2$. Recalling the definition of v, this implies

$$E(s) \approx \sqrt{\frac{R}{Cs}} I(s), \text{ for } 6RC \le \frac{1}{s} \le \frac{1}{6} n^2 RC$$

Inverting this we obtain, approximately the following semi-integration relationship

$$e(t) \approx \sqrt{\frac{R}{C}} \{d_t^{-1/2} i(t)\}, \text{ for } 6RC \le t \le \frac{1}{6} n^2 RC$$

Therefore, the Figure 3.11 (stage-n) circuit performs as semi-integration. The lower time limit for this operation to be holding is about six time RC time constant of the single stage, which are repeated n-times. This can be made small by selecting smaller values of the resistors and capacitors. The upper time limit depends on the number of stages and can be made large, for accuracy. One more observation is that transformed voltage current can be approximated by 'continued fraction expansion' for approximating fractional integration/differentiation (in transformed domain).

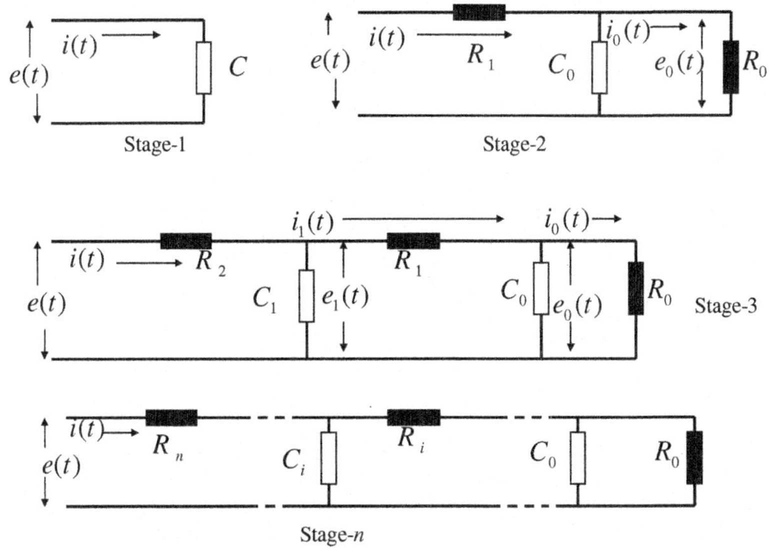

Fig. 3.11 Continued fraction expansion (CFE) with semi-infinite *RC* Transmission Line, and approximation to 'half-order' system

3.13 Dynamics of Chain Network

Consider a Gaussian chain, comprising of N beads connected by spring to form a linear chain. The chain configuration is represented by a set of vectors $\{\overline{r}_n(t)\}$ where

$\{\bar{r}_n(t)\} = \{x_n(t), y_n(t), z_n(t)\}$ that is the position vector of the n th bead at time t, and $n = 0,1,2,\ldots(N-1)$. This model is for study of polymer, as represented by this Gaussian chain, comprising of N monomers connected by harmonic springs. Macro molecular systems in many cases show visco-elastic which contains the feature of ideal solid (elastic) and ideal liquid (viscosity), and this chain dynamics is one of those cases. The potential energy $\mathrm{U}(\{\bar{r}_n(t)\})$ has to account for the elastic contributions and the influence of the external force $\bar{\mathrm{F}}(t)$, let this force act on the bead $n = 0$. Therefore:

$$\mathrm{U}(\{\bar{r}_n(t)\}) = \frac{K}{2}\sum_{n=1}^{N-1}[\bar{r}_n(t) - \bar{r}_{n-1}(t)]^2 - \bar{\mathrm{F}}(t)\bar{r}_0(t)$$

$K = 3T / b^2$ is the (entropic) spring constant, where T is the temperature in units of Boltzmann's constant (k_B) and b is the mean distance between neighboring beads in absence of any external force. The chain network dynamics is dictated by N coupled Langevin equations. Neglecting the hydrodynamic interaction between the beads we have coupled equations as:

$$\zeta\frac{d\{\bar{r}_n(t)\}}{dt} = -\frac{\partial \mathrm{U}\{\bar{r}_n(t)\}}{\partial \bar{r}_n} + f_r(n,t)$$

ζ is the friction constant and $f_r(n,t)$ is the random thermal noise force, which emulates the n th beads interaction (collision) with environment. The noise is Gaussian white noise with zero mean $\langle f_i(n,t)\rangle = 0$ and delta correlated that is, expressed as

$$\langle f_i(n,t) f_j(n,t)\rangle = 2\zeta T\delta_{ij}\delta_{nn'}(t-t'),$$

where i, j denotes the components of the force vector that is $i, j = x, y, z$. Taking the variable n to be continuous that is considering the chain as elastic string and taking the force acting on X-direction as $\bar{\mathrm{F}}(t) = \{F(t), 0, 0\}$, it follows from the potential energy balance equation and applying Langevin equations, the three x, y, z components are obtained as follows (derived later just below, this set of following diffusion equations):

$$\zeta\frac{\partial x_n}{\partial t} = K\frac{\partial^2 x_n}{\partial n^2} + \delta_{n0}F(t) + f_x(n,t)$$

$$\zeta\frac{\partial y_n}{\partial t} = K\frac{\partial^2 y_n}{\partial n^2} + f_y(n,t)$$

$$\zeta \frac{\partial z_n}{\partial t} = K \frac{\partial^2 z_n}{\partial n^2} + f_z(n,t)$$

The above are one dimensional diffusion equation, whose Green's function or modulus is Gaussian and its width increases with time (proportionately with $\sqrt{t}$). At the end of chain network the boundary condition is called Rouse boundary condition given as:

$$\left.\frac{\partial x_n(t)}{\partial n}\right|_{n=0,N} = \left.\frac{\partial y_n(t)}{\partial n}\right|_{n=0,N} = \left.\frac{\partial z_n(t)}{\partial n}\right|_{n=0,N} = 0$$

For the above derivation consider x-component and the equation of potential energy thus is

$$\mathrm{U}(\{x_n(t)\}) = \frac{K}{2}\sum_{n=1}^{N-1}[x_n(t) - x_{n-1}(t)]^2 - \bar{\mathrm{F}}(t)x_0(t),$$

now we do, partial differentiation with respect to x_n as required by the Langevin equation to get $\partial \mathrm{U}(x_n(t))/\partial x_x$, for $n = 0,1,2..N$. We get

$$\frac{\partial U(x_n(t))}{\partial x_n} = \frac{K}{2}\sum_{n=1}^{N-1}\frac{\partial}{\partial x_n}\left[(x_n - x_{n-1})^2\right] = \frac{K}{2}\frac{\partial}{\partial x_n}\left[\ldots + (x_n - x_{n-1})^2 + (x_{n+1} - x_n)^2 + \ldots\right]$$
$$= \frac{K}{2}[2(x_n - x_{n-1}) - 2(x_{n+1} - x_n)] = -K\left[x_{n+1} - 2x_n + x_{n-1}\right]$$

We write double derivative of a function as

$$f^{(2)}(x) = \lim_{h\to 0}\frac{f(x+2h) - 2f(x+h) + f(x)}{h^2}.$$

Here if we let $n = h$ and recognizing that $n = 1$, we may thus cast the $\partial \mathrm{U}(x_n(t))/\partial x_n$ as

$$\partial \mathrm{U}(x_n(t))/\partial x_n = -K\partial^2 x_n(t)/\partial n^2,$$

and this is used in above set of one dimensional equations for chain dynamics. The $\delta_{n0}F(t)$ indicates the use of external force acting on the bead $n = 0$ only acting in X-direction; as $\delta_{n0} = 1$, for $n = 0$ and $\delta_{n0} - 0$ for $n \neq 0$.

Since y, z components are force independent we restrict our self to x component. Converting the x component one dimension diffusion equation obtained above by Fourier transformation we get:

$$\frac{\partial x(p,t)}{\partial t}=-\frac{p^2}{\tau_R}x(p,t)+\frac{1}{N\zeta}F(t)+\frac{1}{\zeta}\varphi_x(p,t)$$

This is obtained by writing $x_n(t)$ as $x(p,t)$, as Fourier transformed to p (Fourier spatial frequency) coordinates, by replacing $\partial/\partial n$ by $-ip$ where $\mathrm{i}=\sqrt{-1}$. Here, $p=0,1,2,.......$ is (space) Fourier variable. Here τ_R denotes the Rouse time and is expressed as $\tau_R=\zeta b^2N^2/3\pi^2T$. Physically, this Rouse time indicates longest internal relaxation time of the harmonic chain. $\varphi_x(p,t)$ denotes the Fourier transform of the thermal noise that is,

$$\varphi_x(p,t)=\frac{1}{N}\int_0^N dn\cos\left(\frac{p\pi n}{N}\right)f_x(n,t)$$

The time series solution $x_n(t)$ of displacement can be expressed in p coordinates as Fourier series that is,

$$x_n(t)=x(0,t)+2\sum_{p=1}^{\infty}x(p,t)\cos\left(\frac{p\pi n}{N}\right),\quad p=0,1,2,...$$

denotes Fourier frequency (p) coordinates and

$$x(p,t)=\frac{1}{N}\int_0^N dn\cos\left(\frac{p\pi n}{N}\right)x_n(t)\,.$$

The Green's function solution for the homogeneous equation

$$\frac{\partial x(p,t)}{\partial t}+\frac{p^2}{\tau_R}x(p,t)=0$$

is $x(p,t)=\exp(-p^2t/\tau_R)$, assuming $x(p,0)=1$ as the initial value. With this Green's function and its convolution with the external forcing functions

$$f^{\mathrm{ext}}(t)=\frac{1}{\zeta}\varphi_x(p,t)+\frac{1}{N\zeta}F(t),$$

one can get particular solution, which is exactly done in following discussion.

Assume that chain is at thermal equilibrium at $t=0$, and a constant force is switched on at time $t=0$, so $F(t)=F_0H(t)$, with $H(t)$ as Heaviside's unit step function. The P coordinate differential-equation thus has solution for this step excitation as:

$$x(p,t)=\frac{1}{\zeta}\int_{-\infty}^{t}dt'\varphi_x(p,t)\exp\left(-p^2(t-t')/\tau_R\right)+\frac{F_0}{N\zeta}\int_0^t dt'\exp\left(-p^2(t-t')/\tau_R\right)$$

The X-component of the trajectory of centre of mass (CM) is given by the 0 th Fourier-frequency (p) component that is,

$$x(0,t)=x_{\text{CM}}(t)=\frac{1}{N}\int_0^N dn[x_n(t)]\,.$$

Putting $p=0$ in above solution we have $\langle x_{\text{CM}}(t)\rangle = F_0 t/N\zeta$ that is, the chain drifts as a whole with constant velocity given as, $v_{\text{CM}}(t)=F_0/N\zeta$ assuming the average contribution of the thermal noise acting on the chain is zero (that is the first term in the above convolution integral). This also implies that friction constant of overall chain is $N\zeta$ that is, sum of friction constants of individual beads.

Now let us consider partial motion of few conglomerates of beads that is, motion of tagged bead say $n=0$. Putting this $n=0$ in Fourier series

$$x_n(t)=x(0,t)+2\sum_{p=1}^{\infty}x(p,t)\cos\left(\frac{p\pi n}{N}\right);$$

we obtain $x_0(t)=x(0,t)+2\sum_{p=1}^{\infty}x(p,t)$.

In this expression putting the solution $x(p,t)$, obtained for step Forcing function we get:

$$\overline{x}_0(t)=\frac{F_0 t}{N\zeta}+\frac{2F_0}{N\zeta}\sum_{p=1}^{\infty}\int_0^t dt'\exp\left(-p^2(t-t')/\tau_R\right)$$

The above expression has a term corresponding to $p=0$, which is the first term, and that corresponds to motion of the tagged bead $n=0$, with the centre of mass of the system that is $x_{\text{CM}}(t)$. The tagged bead will follow this only when time is large (say $t>>\tau_R$). The second term in above expression is the fluctuation component (with non-zero Fourier spatial frequency); will be the motion of the tagged bead at small times $t<<\tau_R$. We thus neglect the first term in the above obtained expression to find average displacement at small times, for $t<<\tau_R$.

$$\overline{x}_0(t)=\frac{2F_0}{N\zeta}\sum_{p=1}^{\infty}\int_0^t dt'\exp\left(-p^2(t-t')/\tau_R\right)=\frac{2F_0}{N\zeta}\sum_{p=1}^{\infty}\left(\frac{\tau_R}{p^2}\right)\int_0^{\frac{p^2 t}{\tau_R}}dx\exp(-x)$$

where $p^2(t-t')/\tau_R = x$. Simplifying the above, with approximating the sum over p with integral gives the following approximate expression

$$\bar{x}_0(t) = \frac{2F_0}{N\zeta}\sum_{p=1}^{\infty}\left(\frac{\tau_R}{p^2}\right)\left(1-\exp\left(-\frac{p^2 t}{\tau_R}\right)\right) \cong \frac{2F_0\tau_R}{N\zeta}\int_{p=1}^{\infty} dp\frac{(1-\exp(-p^2 t/\tau_R))}{p^2}.$$

The approximation used is $\sum_{k=a}^{b} f(k) = \int_{[a,b]} f(\mu)d\mu$, and for a decreasing function we can write the sum over p approximately as:

$$\int_{s=a}^{b} f(s)ds \le \sum_{i=a}^{b} f(i) \le \int_{s=(a-1)}^{b} f(s)ds .$$

Using the change of variable as $p^2 t/\tau_R = y^2$, and taking the lower limit of integration to zero, yields the following approximate expression.

$$\bar{x}_0(t) \cong \frac{2F_0\tau_R}{N\zeta}\int_{y=\sqrt{t/\tau_R}}^{\infty}\sqrt{\frac{\tau_R}{t}}dy\frac{t}{\tau_R}\frac{(1-\exp(-y^2))}{y^2} \cong \frac{2F_0\sqrt{\tau_R}\sqrt{t}}{N\zeta}\left[\int_0^{\infty} dy\frac{1-e^{-y^2}}{y^2}\right]$$

The bracketed integration expression has a value approximately as 1.76, (evaluated numerically) which can be approximated to $\sqrt{\pi}$; along with expression of τ_R taken from definition of Rouse time we get the average displacement of the tagged bead $n=0$, for small times, $t \ll \tau_R$, as:

$$\bar{x}_0(t) = \langle x_0(t)\rangle \cong \frac{2bF_0}{\sqrt{3\pi\zeta T}}(t)^{1/2} = \frac{2}{\sqrt{\pi}}\frac{F_0}{\sqrt{\zeta K}}(t)^{1/2}$$

In the long time regime $t \gg \tau_R$ one finds the second term of the equation in the order τ_R/t smaller than the first term thus one has for $t \gg \tau_R$, $\langle x_0(t)\rangle \cong F_0 t/N\zeta$, which is CM's motion that is collectively the beads move as a chain.

Consider the obtained relation for 'tagged' bead's motion that is,

$$\langle x_0(t)\rangle = 2F_0(t)^{1/2}/\sqrt{\pi}\sqrt{\zeta K},$$

as obtained above, for a step input of 'force'. Recognize that semi-integration of constant F_0 is $d_t^{-1/2}[F_0] = 2F_0(t)^{1/2}/\sqrt{\pi}$. Using this we get semi integral equation, relating the displacement of a tagged bead in linear chain network and arbitrary force $F(t)$ as:

$$x_0(t) = \frac{1}{\sqrt{\zeta K}} \frac{d^{-\frac{1}{2}}}{dt^{-\frac{1}{2}}} F(t)$$

This is a nice way to express chain dynamics as fractional differential equation. For a very long chain N that is very large, Rouse time obeys $\tau_R \propto N^2$, one can describe the linear chain network dynamics as half order integral equation. This can be further expressed, in terms of causal convolution and in terms of Boltzmann's superposition law as:

$$x_0(t) = \frac{1}{\sqrt{\zeta K}} \frac{d^{-\frac{3}{2}}}{dt^{-\frac{3}{2}}} \left[\frac{dF(t)}{dt} \right] = \frac{1}{\sqrt{\zeta K}} \frac{1}{\Gamma(3/2)} \int_{-\infty}^{t} dt'(t-t')^{\frac{3}{2}-1} \frac{dF(t')}{dt'}$$

This explanation of tagged bead moving collectively and the relation with time is proportional to $\sqrt{t}$ is also explained by 'scaling' argument. When the force is applied at time $t = 0$, the number $g(t)$ of the beads (indicating group of beads) which move collectively with the tagged one $n = 0$ increase with time. This number follows the Rouse 'relaxation' time associated with g numbers of beads (of sub-chain) that is $\tau_g \approx \zeta b^2 g^2 / T$, from this we get $g(t) \approx \sqrt{T}(t)^{\frac{1}{2}} / b\sqrt{\zeta}$, describing the short time behavior ($t << \tau_R$). Whereas for longer times, $t >> \tau_R$; $g(t) \approx N$ that is the whole chain moves collectively. The mobility of set of g beads decreases with time as, $\mu(t) \cong (\zeta g(t))^{-1}$ .The average velocity in X-direction $\langle v_x \rangle$ of the tagged bead is given by velocity of set of beads moving together with it, and the average displacement corresponding to blob (group) of beads can be estimated from corresponding blob (group) of $g(t)$ number of collective beads that is $\langle x_0(t) \rangle = \langle v_x \rangle t = (\zeta g(t))^{-1} F_0 t$.

For $t << \tau_R$, $\langle x_0(t) \rangle \cong bF_0(t)^{\frac{1}{2}} / \sqrt{\zeta T}$ and for large times $t >> \tau_R$ the chain drifts as a whole with $\langle x_0(t) \rangle \cong (N\zeta)^{-1} F_0 t$, that is $\langle x_{\mathrm{CM}}(t) \rangle$.

Well for large N, say $N \to \infty$, the short time explanation holds. The case is then like semi-infinite chain of beads connected by spring, and a force acts on the first bead, gives displacement relation as, $\langle x_0(t) \rangle \cong bF_0(t)^{\frac{1}{2}} / \sqrt{\zeta T}$ or $x_0(t) \cong (\zeta K)^{-1/2} D_t^{-\frac{1}{2}} F(t)$. Well semi differintegration is again a natural terminal relation for this system which is semi-infinite!

3.14 Dynamics of Charged Chain Network in Electric Field

Continuing the previous sections discussion let us try and modify the chain network, where the chain comprises of beads (molecules), but all beads are charged. Consider a case where all beads (molecules) are charged with set $\{q_n\}$ Coulombs, $n = 0,1,2,...N$ and the charge distribution is 'uncorrelated' that is, bead to bead charge interaction is weak and correlation is averaged as $\langle q_n q_m \rangle = q^2 \delta_{nm}$. This implies that $\delta_{nm} = 1$, for $n = m$ else $\delta_{nm} = 0$ for $n \neq m$. This is also termed as weak coupling limit, where interaction between the bead charges is neglected. In this scenario the average total charge is calculated as:

$\langle Q_{tot}^2 \rangle = \sum_{n=0}^{N} q_n^2$, this comes from self-correlation, as cross correlation is zero. If all charges are same as q, of same sign then $\langle Q_{tot}^2 \rangle = q^2(N+1)$, for large N, we can write $Q_{tot} \cong q\sqrt{N} = q(N)^{1/2}$.

For a strong interaction of charges between the beads that is cross correlation between the charges of beads too is present along with (obvious) self-correlation, then $\langle Q_{tot}^2 \rangle = \sum_{n=0}^{N} \sum_{m=0}^{N} q_n q_m$, will give $\langle Q_{tot}^2 \rangle = q^2(N+1)^2 \cong q^2(N)^2$, for same sign charges on the beads that are q. We can generalize the system of charges depending on correlation between the beads as:

$\langle Q_{tot}^2 \rangle = q^2 N^{2\gamma}$, when $\gamma = 1/2$, represents (weak interaction) uncorrelated charges, and when $\gamma = 1$, represents strong correlation (strong-interaction) of all the bead charges. Well the case also exits if $\gamma = 0$, meaning that beads have strong interaction and correlation, with arrangement of alternating positive and negative charge distribution. So in general, $0 < \gamma < 1$ for a random distribution of charges on bead, of N bead network, connected with 'spring constant' K, having equivalent average total charge as $\langle Q_{tot}^2 \rangle = q^2 N^{2\gamma}$. While $\gamma > 1/2$ tells positively correlated system of charges, $\gamma < 1/2$ represents negatively correlated system of charges, and $\gamma = 1/2$, represents uncorrelated system of charges.

Let this system of charges on bead network be subjected to an electric field $\bar{\mathrm{E}}(t)$. Then the energy equation of previous section becomes:

$$\mathrm{U}(\{\bar{r}_n(t)\}) = \frac{K}{2}\sum_{n=1}^{N-1}[\bar{r}_n(t) - \bar{r}_{n-1}(t)]^2 - \bar{\mathrm{E}}(t)\sum_{n=1}^{N-1} q_n \bar{r}_n(t)$$

Consider the electric field switching on at time zero, having constant magnitude in X-direction as $\bar{\mathrm{E}}(t) = \{E,0,0\}$, then by the developed arguments as in preceding section we get the X-direction equation (the other two Y and Z direction equations are same) as

$$\zeta \frac{\partial x_n}{\partial t} = K \frac{\partial^2 x_n}{\partial n^2} + q_n E(t) + f_x(n,t)$$

Let us develop the equation of average motion in X-direction, by scaling arguments as described in the previous section, for time $t << \tau_R$. Consider a single bead; the excess charge Q of the collectively moving set of beads (group) $g(t)$ grows as $\langle Q^2 \rangle \cong q^2 (g(t))^{2\gamma}$. The explanation is same as we developed the generalized total correlated charges, to have average charge on the N beads, where the mobility decreases with time as $\mu \approx (\zeta g(t))^{-1}$. The average velocity of the tagged bead in X-direction v_X equals the velocity of the collectively moving set around it, hence

$$\hat{v}_X^2 (g(t)) \cong \mu^2 \langle Q^2 \rangle E^2 \cong q^2 E^2 (g(t))^{2\gamma-2},$$

where, $g(t) \cong \sqrt{T}(t)^{1/2} / b\sqrt{\zeta}$,(is discussed in previous section), for $t << \tau_R$ and $g(t) \cong N$ for $t >> \tau_R$.

The average displacement $\bar{x}_0(t)$ of single bead follows from group displacement of neighbors that is;

$$\bar{x}_0^{\,2}(t) \cong \hat{v}_X^2 t^2 \cong q^2 E^2 \zeta^{-2} (g(t))^{2\gamma-2} t^2,$$

by putting $g(t) \cong \sqrt{T}(t)^{1/2} / b\sqrt{\zeta}$, for $t << \tau_R$, one finds

$$\bar{x}_0^{\,2}(t) \cong \frac{b^{2-2\gamma} q^2 E^2}{\zeta^{1+\gamma} T^{1-\gamma}} t^{1+\gamma}.$$

We see that average displacement scales as power law

$$\bar{x}_0(t) \sim t^{(1+\gamma)/2};$$

from this observation we can write fractional differential equation as done for previous section, relating arbitrary electric field to average tagged displacement as:

$$\bar{x}(t) \approx \frac{b^{1-\gamma} q}{\zeta^{(1+\gamma)/2} T^{(1-\gamma)/2}} \frac{d^{-(1+\gamma)} E(t)}{dt^{-(1+\gamma)}}$$

For uncorrelated charge distribution $\gamma = 1/2$, the average square displacement is

$$\bar{x}^2(t) = \frac{bq^2 E^2}{\zeta^{3/2} T^{1/2}} t^{3/2},$$

giving, average displacement scaling as

$$\bar{x}(t) = \frac{b^{1/2} qE}{\zeta^{3/4} T^{1/4}} t^{3/4},$$

power law. This uncorrelated charge distribution system will have corresponding differential equation as:

$$\bar{x}(t) \approx \frac{b^{1/2} q}{\zeta^{3/4} T^{1/4}} \frac{d^{-3/4} E(t)}{dt^{-3/4}}$$

For large times $t >> \tau_R$, the chain as a whole drifts as $\langle x_0(t) \rangle \cong (N\zeta)^{-1} F_0 t$, derived from previous section. For uncorrelated system of charges we can write $F_0 = \langle Q_{tot} \rangle E = qE\sqrt{N}$, gives $\langle x_0(t) \rangle \cong (qE / \zeta\sqrt{N})t$.

Well, this section gives the fractional order relation other than semi-differintegral equation; which was common for all discussed semi-infinite systems so far. This sub-diffusive phenomenon for chain with charges (uncorrelated) is due to the fact that all the beads of N threaded chain acts to the external electric field. The situation is different from semi-infinite systems where only at one end the excitation acts, the other end is at infinity. However, the above discussion leads to generalized fractional differential equation system.

3.15 Concluding Comments

The practical examples in this chapter demonstrated the reality of the existence of fractional order differentiation and integrations, in natural description of systems. Interesting observations obtained from analysis of semi-infinite systems; heat flow and current flow in lossy lines and for linear chained network dynamics indicate the existence of semi-differintegration operations needed to describe transfer characteristics. Also in realization of the transfer characteristics is possible by circuit synthesis and to have control system with robustness measure, independent of the gain. The 'ifs and buts' regarding the definition of the order of the system for fractional differential equation is an open issue and cannot be directly related to integer order theory definitions. For example a first order system having say highest order of differentiation as unity may show (anomalous response) under damped response to the step input excitation. Now the system looking as first order system but with fractional order differentiation too, will behave as though having some resonance, (anomalously). This behavior speaks that though the system may look classically first order yet due to fractional order terms presence the behavior changes, so does the definition of the order. In this chapter the examples points to the fact that distributed parameters and connectivity as in chain network do point towards fractional order system description and in reality the parameters are indeed distributed, also fractional (half) order differintegration is natural with all the classical theories of physics.

Chapter 4
Concept of Fractional Divergence and Fractional Curl

4.1 Introduction

Fractional kinetic equations of the diffusion are useful approach for the description of transport dynamics in complex systems, which are governed by anomalous diffusion and non-exponential relaxation patterns. The anomalous diffusion can be modeled by fractional differential equation in time as well as space. For the spatial part use of fractional divergence modifies the anomalous diffusion expression, in the modified Fick's law. Application of this fractional divergence is bought out in Nuclear reactor neutron flux definition. When anomalous diffusion is observed in time scale, the modification suggests use of Fractional kinetic equations. The evolution of Fractional Difference Equation, with reference to Fractional Brownian motion and the anomalous diffusion is also discussed in this chapter. Fractional curl operators will play perhaps role in electromagnetic theory and Maxwell equations. Here example in Electromagnetic is taken to have a feel how the fractional curl operator can map E and H fields in between the dual solutions of Maxwell equation.

4.2 Concept of Fractional Divergence for Particle Flux

Because of relative simplicity and widespread use, the basis of local theory is discussed first. The local theory makes use of ADE (advection diffusion equation) as:

$$\frac{\partial C}{\partial t} = \nabla \bullet \left(-vC + \mathbb{D}\nabla C\right) \tag{4.1}$$

where C is the solute concentration and $\mathbb{D}$ and v are local dispersion and velocity tensors respectively. The ADE is based on classical definition of divergence of a vector field. The divergence is defined as the ratio of total flux through a closed surface to the volume enclosed by the surface when the volume shrinks towards zero.

$$\text{div}J \equiv \lim_{V \to 0} \frac{1}{V} \int_S J \bullet n dS \tag{4.2}$$

Where J is flux vector, V is an arbitrary volume enclosed by surface S, and n is unit vector normal to the surface.

The (4.1) becomes

$$\frac{\partial C(x,t)}{\partial t} + v \frac{\partial C(x,t)}{\partial x} = \mathbb{D} \frac{\partial^2}{\partial x^2} C(x,t)$$

for 1D case with constant v, the advection drift velocity. For initial concentration $C(x,0) = \delta(x)$ we get 'Galilei shifted Gaussian distribution' as

$$C(x,t) = \frac{1}{\sqrt{4\pi \mathbb{D} t}} \exp\left(-\frac{(x - vt)^2}{4\mathbb{D}t} \right),$$

for natural boundary condition $C\left(|x| \to \infty, t\right) = 0$.

While the advection velocity is zero, the case is Fick's law and its Gaussian solution to the sharp initial condition (Chapter 3).

This is valid only if the flux is indeed a 'point' vector quantity relative to the scale of observation, (for example heat flow in homogeneous material). Then the limit exists and the operator reduces the familiar dot product with gradient vector $[\partial / \partial x, \partial / \partial y, \partial / \partial z]$. Solute dispersion is counter example since it is primarily due to the velocity fluctuations that arise only as an observation space grows larger, invalidating the limit. The solute flux is due to combined effects of mean velocity (advection) and velocity fluctuation (dispersion). The dispersive fluxes for a given volume are typically averaged in some fashion (volumetric, statistical) and approximated by Fick's first law. Since velocity itself is a variable function of space, as control volume shrinks (as divergence requires), the velocity fluctuations and the dispersive flux disappear. Therefore true divergence of the macroscopic solute flux cannot contain a macroscopic dispersive term.

Because of the limit in (4.2), the classical Gauss divergence theorem discounts macro dispersion until a point vector can approximate the dispersive flux. These calls for separation of scales: The scale of the transport process must be much larger than some finite volume at which the ratio in (4.2) becomes seemingly constant. For these things to happen, the dispersive flux must not increase as volume passes some largest size. This representative elementary volume (REV) for dispersion is point at which the deviations in the velocity field are negligible. The divergence is associated with a non-zero volume and is given by the first derivative of total surface flux to volume (Figure 4.2) rather than the limit of the derivative at zero volume. The dispersion coefficient does not grow (scale), if the ratio of the surface flux to volume is constant over some range of volume (Figure 4.1 and stepped solid lines of Figure 4.2b making piece wise constant slope of Figure 4.2a). Therefore at some larger scale of observation (non-zero), the ratio of the total surface flux to volume is constant over large range of arbitrary volumes and relatively constant first derivative (the de facto divergence) allows assignment of dispersion coefficient. Both volume averaging and ensemble averaging concepts are based on this idea of separation (or distinction) of scale.

At the field scale at least two problems occur that make it difficult to rely on the REV method. First, even if there is a distinct hierarchy, the act of measurement involves a volume integration, which impacts the dispersion coefficient. Second, there is a long-standing and growing body of evidence that real solute materials have evolving heterogeneity. If that is the case then there will be no separation of scales and the Figure 4.2b dashed line will represent the flux, as continuous one instead of stepped one.

An integer order divergence theorem forces a scaling parameter, since ratio of flux to volume is scale dependent. Rather than use a step function approximation (Figure 4.2b), of the growth of dispersive flux with scale, which forces $\mathbb{D}$ to take on increasing values, one might try to describe the evolving dashed line (Figure 4.2b). Non-local including convolutional theories does this by integrating cumulative effects of dispersion over any length scale and or time scale. A subset of these uses the mathematical tools of fractional calculus, which are non local operators for fractal functions.

4.3 Fractional Kinetic Equation

The fundamental laws of physics are written as equations for the time evolution of a quantity $X(t)$ with $dX(t)/dt = -AX$, where this could be Maxwell's equation, Schrodinger's equation or could be Newton's law of motion or Geodesic equations. The mathematical solution for the linear operator A is $X(t) = X(0)\exp\{-At\}$; putting $\tau = A^{-1}$ the standard exponential relaxation (Maxwell-Debye) law is:

$$X(t) = X(0)\exp\{-t/\tau\}. \tag{4.3}$$

Complex systems and investigations of their structural and dynamical properties have established on the physics agenda. These structures with variations are characterized through

a) A large diversity of elementary units.
b) Strong interactions between the units.
c) A non-predictable or anomalous evolution in course of time.

Complex systems and their study play a dominant role in exact and life sciences, embracing a richness of systems such as glasses, liquid crystals, polymers, proteins, biopolymers, or even eco systems. In general the temporal evolution of, and within, such systems deviates from the corresponding standard laws. With the development of higher resolution experiments, these deviations have become more prominent.

One can have stretched exponential behavior as:

$$X(t) = X(0)\exp\left\{-(t/\tau)^{\alpha}\right\} \tag{4.4}$$

With $0 < \alpha < 1$, or one can visualize the asymptotic power law as:

$$X(t) = X(0)(1+t/\tau)^{-n}, \text{ with } n > 0 \tag{4.5}$$

Similarly the diffusion process in various complex systems usually no longer follow Gaussian statistics, and thus Fick's second law fails to describe the related transport behavior (Anomaly of Fick's diffusion law is discussed in Chapter 3, with Catteneo's diffusion and Fractional Diffusion Equation's evolution). Especially one observes deviations from the linear dependence of mean squared displacement. The mean squared displacement is given by

$$\langle \Delta x_j(t)\rangle^2 = \langle x_j(t) - x_j(0)\rangle^2,$$

where the $x_j(t) - x_j(0)$ is the vector distance traveled by a molecule; particle; flux (walker) over some time interval of length t its magnitude is averaged over many such intervals t. If no other walker is encountered then the distance traveled would be proportional to the time interval that is, $\langle x\rangle \propto vt$, where v is walker velocity. In this case (without hindrance) the mean squared displacement would increase as square of time that is, $\langle \Delta x_j\rangle^2 \propto t^2$. In the presence of hindrance or we may say dense phases the t^2 behavior holds only for very short time ($t \to 0$) of the order of collision time. Beyond this time motion is better described as 'random-walk' for which mean squared displacement is proportional to time (6). Mean Squared Displacement is thus

$$\langle x(t)\rangle^2 = \left\langle \frac{1}{N}\sum_{j=0}^{N}\left(x_j(t) - x_j(0)\right)^2 \right\rangle$$

$$< x^2(t) > \approx K_1 t \tag{4.6}$$

This is characteristic of Brownian motion (BM), and such a direct consequence of central limit theorem and the Markovian nature of the underlying stochastic process. A point is mentioned here regarding the graph of this classical Brownian motion where $|\Delta x(t)| \approx \sqrt{(\Delta t)}$ is characteristic scaling; as demonstrated in Chapter 2 box-dimensions of $d_B = 3/2$, and Hurst exponent as $H = 1/2$. Instead anomalous diffusion is found in a wide diversity of systems, its hallmark being non-linear growth of the mean squared displacement in course of time, following power law:

$$< x^2(t) > \approx K_q t^q \tag{4.7}$$

This is ubiquitous to a diverse number of systems. There exists variety of other patterns such as logarithmic time dependence etc. The anomalous diffusion behavior manifests in Eq. (4.7) is intimately connected with break down of the central limit theorem, caused by either broad long tailed distributions or long-range correlations. These broad spatial jumps or waiting times' distributions lead to non-Gaussian and possibly non-Markovian time evolution way of diffusion, manifesting into non-local temporal phenomena. Note that the unit of diffusion coefficient in (4.7) is having unit $[K_q] \equiv \text{cm}^2\text{s}^{-q}$, according to anomalous diffusion exponent q. This exponent if $0 < q < 1$ defines sub-diffusive transport, and defines super-diffusive phenomena for $1 < q < 2$. For $q = 1$ the transport phenomena is normal integer order and Fickian. The sub diffusive regime includes transport of charged carriers in amorphous semiconductor, nuclear magnetic resonance (NMR) diffusometry in percolative and porous systems, Rouse (chain) dynamics in polymers systems, transport in fractal geometries, and dynamics of beads in polymeric networks. The super diffusion systems are rotating flows, turbulent diffusion, transport in heterogeneous rocks, Quantum optics, single molecule spectroscopy, transport in turbulent plasma, and bacterial motions.

Standard integer order kinetic equation when integrated gives:

$$X_i(t) - X_0 = -c\int_0^t X(t) = -c\,{}_0D_t^{-1}X(t) \tag{4.8}$$

${}_0D_t^{-1}$, is the standard Riemann-Liouville integral operator. The number density of the species $i, X_i = X_i(t)$ is a function of time and $X_i(t=0) = X_0$ is the number density of species i at time $t = 0$. If we drop the index i in (4.8) and replace c by c^q, then solution of the generalized fractional order diffusion equation:

$$X(t) - X_0 = -c^q\,{}_0D_t^{-q}X(t) \tag{4.9}$$

is:

$$X(t) = X_0\sum_{k=0}^{\infty}\frac{(-1)^k(ct)^{kq}}{\Gamma(kq+1)} \tag{4.10}$$

can be written as compact form by use of Mittag-Leffler function as:

$$X(t) = X_0 E_q(-c^q t^q) \tag{4.11}$$

The rate of growth of mean squared displacement depends on how often the walker suffers collision. At higher density it will take longer time to diffuse a given distance.

The limiting slope of mean squared displacement considered for time intervals sufficiently long for it to be in linear region is related to diffusing constant $\mathbb{D}$ (appearing in Fick's law) and in (4.1). The relation is Einstein diffusion relation and is:

$$\lim_{t\to\infty}\frac{d}{dt}\left\langle \Delta x_j(t)\right\rangle^2 = 2d\mathbb{D},$$

where d is dimension of space (1, 2, 3) Euclidian or Topological dimension, in normal Fick's diffusion. This relation of MSD and diffusion constant is derived for Fick's diffusion in Chapter 3, by use of Fourier-Laplace transformation. This number could be fractional too indicating anomalous diffusion in fractal landscape with heterogeneity.

4.4 Discrete Difference and Continuum Limit and Differential Operator in Random Walk Context

4.4.1 Integer Order Discrete Difference and Continuum Limit and Differential Operator

Let the random variable $X(t)$ denote position of walker at times t, which increments to a new position as the time changes in steps of τ .Taylor expansion of walker's position at time $t+\tau$ is given as:

$$X(t+\tau) = X(t) + \tau DX(t) + \frac{\tau^2}{2!}D^2X(t) + \ldots + \frac{\tau^n}{n!}D^nX(t) + \ldots = [e^{\tau D}]X(t)$$

where D is derivative operator. We can define an up-shift operator as:

$$E_\tau \equiv E_\tau\left(X\{t\}\right) = X(t+\tau)$$

Comparing this up-shift operator with the above mentioned Taylor expansion we get

$$E_\tau X(t) = \left\{e^{\tau D}\right\}X(t)$$

The symbolic equivalence between up-shift operator and derivative operator is $E_\tau = e^{\tau D}$, and we write a new relation as: $E_\tau \equiv 1-\Delta_{(-)} = e^{\tau D}$; where we define right difference operator as: $\Delta_{(-)} \equiv 1-E_\tau = 1-e^{\tau D}$, which is expanded as following:

$$\Delta_{(-)} = 1-E_\tau = 1-e^{\tau D} = 1-\left(1+\tau D+\frac{(\tau D)^2}{2}+\ldots\right) \approx -\tau D$$

In a way we have defined a derivative operator as following expression:

$$\lim_{\tau\to 0}\left(\frac{\Delta_{(-)}}{\tau}\right)=-D$$

It is clear that D is 'time-derivative' operator and when applied to continuous functions of time yields:

$$\frac{d}{dt}X(t)=-\lim_{\tau\to 0}\frac{\Delta_{(-)}X(t)}{\tau}$$

As done above, similarly we can define down-shift operator as $E_{\tau}^{-1}X(t)\equiv X(t-\tau)$ and then from corresponding Taylor's series expansion of $X(t-\tau)$ we obtain symbolic representation as $E_{\tau}^{-1}\equiv 1-\Delta_{(+)}=e^{-\tau D}$, with $\Delta_{(+)}=1-E_{\tau}^{-1}=1-e^{-\tau D}$, as left-difference operator and the following expression for derivative as:

$$\lim_{\tau\to 0}\left(\frac{\Delta_{(+)}}{\tau}\right)=D$$

The derivative operators obtained by left-difference operator and right-difference operator are equivalent mathematical expression of derivative. To be on safe side we express the derivative operator considering the average (of left and right difference) and express in symmetric form as:

$$\lim_{\tau\to 0}\frac{\Delta_{(+)}-\Delta_{(-)}}{2\tau}=D=\frac{d}{dt}$$

4.4.2 Fractional Order Discrete Difference and Continuum Limit and Fractional Differential Operator

Generalizing the limit of finite difference

$$\Delta_{(+)}=1-E_{\tau}^{-1}=\left(1-e^{-\tau D}\right)=\left(1-e^{-\tau\frac{d}{dt}}\right),$$

we obtain

$$\Delta_{(+)}^{\alpha}=(1-E_{\tau}^{-1})^{\alpha}=\left(1-e^{-\tau\frac{d}{dt}}\right)^{\alpha}.$$

So that we can write for left-sided and right-sided fractional derivative as following generalizations:

$$\lim_{\tau\to 0}\frac{\Delta^{\alpha}_{(+)}}{\tau^{\alpha}}X(t)=\lim_{\tau\to 0}\frac{\left(1-e^{-\tau\frac{d}{dt}}\right)^{\alpha}}{\tau^{\alpha}}X(t)=\frac{d^{\alpha}}{dt^{\alpha}}X(t)$$

$$\lim_{\tau\to 0}\frac{\Delta^{\alpha}_{(-)}}{\tau^{\alpha}}X(t)=\lim_{\tau\to 0}\frac{\left(1-e^{\tau\frac{d}{dt}}\right)^{\alpha}}{\tau^{\alpha}}X(t)=\left(-\frac{d}{dt}\right)^{\alpha}X(t)$$

$$D^{\alpha}_{(\pm)}X(t)=\lim_{\tau\to 0}\frac{\Delta^{\alpha}_{(\pm)}}{\tau^{\alpha}};\alpha>0$$

Consider the right sided fractional difference between $X(t)$ and $X(t-\tau)$ for arbitrary τ that is, $\Delta^{\alpha}_{(-)}X(t)=\left(1-E_{\tau}^{-1}\right)^{\alpha}X(t)$. For integer α and using binomial expansion of this we obtain $\left(1-E_{\tau}^{-1}\right)^{\alpha}=\sum_{k-0}^{\alpha}\binom{\alpha}{k}(-1)^{k}E_{\tau}^{-k}$. If we consider the $E_{\tau}^{-1}=L$ a lag-operator then same can be expressed as

$$(1-L)^{\alpha}=\sum\nolimits_{k=0}^{\alpha}\binom{\alpha}{k}(-L)^{k}.$$

For positive $\alpha>0$ (non-integer) we generalize as

$$\binom{\alpha}{k}=\frac{\alpha!}{k!(\alpha-k)!}=\frac{\Gamma(\alpha+1)}{\Gamma(k+1)\Gamma(\alpha+1-k)},$$

since $\binom{\alpha}{k}=0$ for $k>\alpha$, the binomial series can be extended to infinity (as generalization), and is written as:

$$\Delta^{\alpha}_{(-)}X(t)=\sum_{k=0}^{\infty}\binom{\alpha}{k}(-1)^{k}X(t-k\tau)$$

which rather depends on the values of function in vicinity of local time instant t that is, this fractional difference depends on entire history.

In terms of lag-operator we have

$$(1-L)^{\alpha}=\sum\nolimits_{k=0}^{\infty}\binom{\alpha}{k}(-L)^{k}$$

We have obtained a fractional differencing operator which is defined as an infinite binomial series expansion of power series of the ordinary back shift operator. This gives Autoregressive Fractional Integral Moving Average (ARFIMA); a work horse in Fractional Order Signal Processing. ARFIMA will be also discussed in Chapter 5

4.4.3 *Fourier Representation of Fractional Difference and Derivative*

Fourier series representation of a function that is periodic over an interval 2π is

$$X(t) = \sum_{\omega=-\infty}^{\omega=+\infty} X(\omega)e^{i\omega t}$$

and the Fourier coefficients are defined by

$$\Im[X(t)] = X(\omega) = \frac{1}{2\pi}\int_0^{2\pi} X(t)e^{-i\omega t}\,dt$$

Applying this definition to the following obtained definitions of fractional difference and fractional derivatives

$$\Delta^{\alpha}_{(-)}X(t) = \sum_{k=0}^{\infty}\binom{\alpha}{k}(-1)^k X(t-k\tau)$$

$$D^{\alpha}_{(-)}X(t) = \lim_{\tau\to 0}\frac{\Delta^{\alpha}_{(-)}}{\tau^{\alpha}}$$

We get:

$$\Im\left[D^{\alpha}_{(-)}X(t)\right] = \lim_{\tau\to 0}\frac{1}{2\pi\tau^{\alpha}}\int_0^{2\pi} e^{-i\omega t}\Delta^{\alpha}_{(-)}X(t)\,dt$$

$$\Im\left[D^{\alpha}_{(-)}X(t)\right] = \lim_{\tau\to 0}\frac{1}{\tau^{\alpha}}\sum_{k=0}^{\infty}\binom{\alpha}{k}\frac{(-1)^k}{2\pi}\int_0^{2\pi} e^{-i\omega t}\sum_{\omega'=-\infty}^{\omega'=+\infty} X(\omega')e^{i\omega'(t-k\tau)}\,dt$$

When the integral over time yields delta function so that integral is zero if $\omega \neq \omega'$ and

$$\Im\left[D^{\alpha}_{(-)}X(t)\right] = \lim_{\tau\to 0}\frac{1}{\tau^{\alpha}}\sum_{k=0}^{\infty}\binom{\alpha}{k}(-1)^k e^{-i\omega k\tau}X(\omega),$$

when $\omega = \omega'$. This expression can now be expressed as binomial expansion and following can be written:

$$\Im\left[D^{\alpha}_{(-)}X(t)\right] = \lim_{\tau\to 0}\frac{1}{\tau^{\alpha}}\left(1-e^{-i\omega\tau}\right)^{\alpha} X(\omega)$$

More generally for right and left fractional derivative we get:

$$\Im\left[D^{\alpha}_{(\mp)}X(t)\right] = \lim_{\tau\to 0}\frac{1}{\tau^{\alpha}}\left(1-e^{\mp \mathrm{i}\omega\tau}\right)^{\alpha} X(\omega)$$

In the limit $\tau \to 0$ we expand the exponential ($e^{\mp \mathrm{i}\omega\tau}$) so that the linear terms cancel and we obtain exact result for Fourier representation of Fractional Derivatives as:

$$\Im\left[D^{\alpha}_{(\mp)}X(t)\right] \approx (\pm \mathrm{i}\omega)^{\alpha} X(\omega)$$

Thus we see that effect of fractional derivative operating on function is to multiply the Fourier amplitude of that function by $(\mathrm{i}\omega)^{\alpha}$ or with its complex conjugate. Therefore we obtain useful identity as:

$$D^{\alpha}_{\mp}X(t) = \sum_{\omega=-\infty}^{\omega=+\infty} (\pm \mathrm{i}\omega)^{\alpha} X(\omega)e^{\mathrm{i}\omega t},$$

where it is clear that for non-integer α fractional derivative operators are non-local in time (space); as against locality in integer order counterparts.

4.4.4 Stochastic Fractional Difference Equations

Random time series are traditionally modeled in the physical science using simple random walks. If ξ_j is a random variable intended to represent a step taken at discrete time j then the random walk variable that denotes the total distance travelled in time $t = N\tau$ or after N such steps is

$$X(t) = \sum_{j=1}^{N} \xi_j .$$

Alternatively, $X(t) - X(t-\tau) = \xi_N$.

We simplify notation by setting $\tau = 1$ and define the down-shift operator for unit interval without subscripts; so we can re-write the right-side difference using discrete indices for arbitrary step j as: $(1-E^{-1})X_j = \xi_j$. This is discrete analog of Brownian motion (BM), which is physicists guide to world of stochastic process.

Therefore $(1-E^{-1})^{\alpha} X_j = \xi_j$, with α as non-integer is discrete analog of Fractional Brownian motion (FBM). We solve this case by inverting this basic expression as:

$$X_j = \left(1-E^{-1}\right)^{-\alpha} \xi_j$$

Using the expansion $\left(1-E_{\tau}^{-1}\right)^{-\alpha}=\sum_{k=0}^{\infty}\binom{-\alpha}{k}(-1)^{k}E_{\tau}^{-k}$, with $\tau=1$, and

$$\binom{-\alpha}{k}=\binom{k+\alpha-1}{k}=\frac{(k+\alpha-1)!}{k!(\alpha-1)!}$$

as generalization of Binomial coefficients we obtain:

$$X_{j}=\sum_{k=0}^{\infty}\frac{(k+\alpha-1)!}{k!(\alpha-1)!}E^{-k}\xi_{j}=\sum_{k=0}^{\infty}\frac{(k+\alpha-1)!}{k!(\alpha-1)!}\xi_{j-k},$$

and simplifying with use of Gamma function (that is to extend this to non-integer values) we obtain:

$$X_{j}=\sum_{k=0}^{\infty}\frac{\Gamma(k+\alpha)}{\Gamma(k+1)(\alpha-1)!}\xi_{j-k}$$

We wish to obtain the asymptotic expression for

$$X_{j}=\sum_{k=0}^{\infty}\frac{\Gamma(k+\alpha)}{\Gamma(k+1)(\alpha-1)!}\xi_{j-k},$$

using Stirling's approximations of Gamma functions ratio as:

$$\frac{\Gamma(z+\alpha)}{\Gamma(z+\beta)}\approx z^{\alpha-\beta},\ \left|\arg(z+\alpha)\right|<\pi \text{ and } z\to\infty, \text{ we obtain:}$$

$$X_{j}\approx\sum_{k=0}^{\infty}\frac{k^{\alpha-1}}{(\alpha-1)!}\xi_{j-k}, \text{ as } k>>\alpha; k\to\infty$$

Indicating the strength of contributions to X_j decrease asymptotically with increasing time lag (k) as inverse power-law as long as $0<\alpha<1$.

Spectrum can be obtained for the above expansion

$$X_{j}=\sum_{k=0}^{\infty}\frac{(k+\alpha-1)!}{k!(\alpha-1)!}\xi_{j-k}=\sum_{k=0}^{\infty}\theta(k)\xi_{j-k},$$

by use of discrete Fourier Transform, and averaging the square of modulus of the Fourier amplitude over ensemble of realizations of random fluctuations driving the fractional difference equation $\hat{X}_{\omega}=\hat{\theta}_{\omega}\hat{\xi}_{\omega}$, where $\Im\left(\theta_{k}\right)=\hat{\theta}_{\omega}$.

That is, discrete convolution of two fluctuations is the product of their Fourier coefficients. Spectrum (specifically power spectrum) is defined as

$$S(\omega)=\left\langle\left|\hat{X}_{\omega}\right|^{2}\right\rangle,$$

average over ensemble of realizations of ξ fluctuations. Assuming random fluctuations have a white noise spectrum of unit strength and delta correlated in time ($\hat{\xi}_{\omega}=1$), we get $S(\omega)=\left|\hat{\theta}_{\omega}\right|^{2}$, where

$$\hat{\theta}_{\omega}=\sum_{k=0}^{\infty}\frac{(k+\alpha-1)!}{k!(\alpha-1)!}e^{-ik\omega}=\left(1-e^{-i\omega}\right)^{-\alpha},$$

gives

$$\left|\hat{\theta}_{\omega}\right|^{2}=\left(1-e^{-i\omega}\right)^{-2\alpha}=S(\omega).$$

Rearranging this and using trigonometric identities, as in following steps we get:

$$\begin{aligned}
S(\omega)=(1-e^{-i\omega})^{-2\alpha}&=\left(1-\frac{1}{e^{i\omega}}\right)^{-2\alpha}=\left(\frac{e^{i\omega}}{e^{i\omega}-1}\right)^{2\alpha}\\
&=\left(\frac{e^{i\omega}}{e^{i\omega/2}e^{i\omega/2}-e^{i\omega/2}e^{-i\omega/2}}\right)^{2\alpha}\\
&=\left\{\frac{(e^{i\omega})}{e^{i\omega/2}\left[e^{i\omega/2}-e^{-i\omega/2}\right]}\right\}^{2\alpha}\\
&=\left[\frac{e^{i\omega/2}}{\frac{2i}{2i}(e^{i\omega/2}-e^{-i\omega/2})}\right]^{2\alpha}=\left[\frac{e^{i\omega/2}}{2i\sin(\omega/2)}\right]^{2\alpha}\\
&=\frac{e^{i\alpha\omega}}{\left[2\sin(\omega/2)\right]^{2\alpha}}
\end{aligned}$$

The modulus of the spectrum is thus:

$$\left|S(\omega)\right|=\frac{1}{\left[2\sin\left(\omega/2\right)\right]^{2\alpha}}\approx\frac{1}{\omega^{2\alpha}},$$

for low frequencies $\omega\to 0$.

Above is the power-spectrum of Fractional Difference Process driven by white-noise. Therefore we have obtained inverse power-law (scale-invariant fractal) $S(\omega)\cong(\omega)^{-2\alpha}$ as $\omega\to 0$, for the right-sided fractional difference process driven by white noise. We also observe for infinite series representation of right-sided fractional difference process the statistics X_j are the same as that of ξ_j since $X_j = (1-E^{-1})^{-\alpha}\xi_j$ is linear equation relating the two. Thus, since statistics of ξ_j are Gaussian so too are of X_j. However, whereas spectrum of ξ is flat that is, characteristics of white noise; the X spectrum is inverse power law characteristic of fractional stochastic process, a 'fractal noise'.

Let us find the auto correlation of the time series generated by the above discussed stochastic 'fractional difference process'; from the obtained power spectrum. The $R_{XX}(k)$ represents auto correlation as $R_{XX}(k) = \langle X_j X_{j+k} \rangle$; the $\langle ... \rangle$ represents the averaging over a time period. Therefore,

$$R_{XX}(k) = \langle X_j X_{j+k} \rangle = \frac{1}{2\pi}\int_0^{2\pi} d\omega S(\omega)\cos(k\omega)$$

which is inverse cosine Fourier Transform of power spectrum. This is inverse of 'Wiener-Khintchine' theorem which states that power spectrum of noise is the Fourier Transform of the auto-correlation function of noise. In above integral we write, $\phi = \omega/2$ and $S(\omega) = [2\sin(\omega/2)]^{-2\alpha}$, to get

$$R_{XX}(k) = \frac{1}{2^{2\alpha-1}\pi}\int_0^{\pi/2} d\phi(\sin\phi)^{-2\alpha}\cos(2k\phi).$$

From the (standard) table of integral series and products, and using definition of Beta function as

$$\mathrm{B}(p,q) = \Gamma(p)\Gamma(q)/\Gamma(p+q);$$

and $\cos(k\pi) = (-1+\mathrm{j}0)^k = (-1)^k$ we write the integral as

$$\int_0^{\pi/2} d\phi(\sin\phi)^{-2\alpha}\cos(2k\phi) = \frac{2^\alpha \pi\cos(k\pi)}{(1-\alpha)\mathrm{B}(1+k-\alpha, 1-k-\alpha)} = \frac{2^\alpha \pi(-1)^k\Gamma(1-\alpha)}{\Gamma(1+k-\alpha)\Gamma(1-k-\alpha)}$$

In above use generalization of Binomial coefficient as in Chapter 2 that is,

$$\binom{-\alpha}{k} = \frac{\Gamma(1-\alpha)}{k!\Gamma(1-\alpha-k)} = \frac{(-1)^k\Gamma(\alpha+k)}{k!\Gamma(\alpha)},$$

from this we write

$$\Gamma(1-\alpha-k) = \frac{\Gamma(1-\alpha)\Gamma(\alpha)}{(-1)^k\Gamma(\alpha+k)},$$

and substitute this in above integral to get

$$\int_0^{\pi/2} d\phi(\sin\phi)^{-2\alpha}\cos(2k\phi) = \frac{(2^{\alpha}\pi)\Gamma(\alpha+k)}{\Gamma(1+k-\alpha)\Gamma(\alpha)}.$$

In this $(-1)^{2k} = 1$ is used.

Put this in $R_{XX}(k)$ to get

$$R_{XX}(k) = 2^{2(1-\alpha)} \frac{\Gamma(\alpha+k)}{\Gamma(1+k-\alpha)\Gamma(\alpha)}.$$

For asymptotic explanation use Stirling's approximation of ratio of Gamma function as

$$\Gamma(k+\alpha)/\Gamma(k+\beta) \approx k^{\alpha-\beta},$$

and write $R_{XX}(k) \cong (2^{2(1-\alpha)}/\Gamma(\alpha))|k|^{2\alpha-1}$. So as the lag increases the auto correlation of the power series obtained by the fractional stochastic difference decays as power law as $R_{XX}(\tau) \propto \tau^{2\alpha-1}$.

4.4.5 Random Walker with Memory Concept of Persistence and Anti-persistence Walk with Long Memory and Short Term Memory

The theory of influence of long time memory on stochastic phenomena is what is described in earlier chapters where the MSD is non-linear, $\langle X(t)^2 \rangle \propto t^{2H}$ with, $H \neq 0.5$. The model of these phenomena is relying heavily on the work-horse of statistical physics, the random walk model. In continuum for a simple random walk can be written in a stochastic differential equation

$$\frac{d}{dt}X(t) = \xi(t)$$

where $\xi(t)$ is a stochastic quantity. In above discussion we have made use of $\xi(t)$ as white noise, driving the, stochastic 'fractional difference equation'; generating power spectrum as 'fractal noise' or fractional noise as $S(f) \propto (1/f^{2\alpha})$. The time series as generated by this process on integration will give a motion 'with memory'.

Fractional Gaussian Noise is defined by 'power-spectral' density of the form $S(f) \propto f^{-\beta}$, for a signal or random variable say position which scales as $X(t) = c^{-H}X(ct)$. H, is called Hurst exponent and $\beta = 2H-1$, $\beta = 2\alpha$ and valid for uni-fractal series. With $H = 1$ the time series $\{X(t)\}$ is 'pink' noise with power spectral density as $S(f) \cong (1/f)$. This type of $1/f$ noise is important since it is kind of threshold between persistent stable noise where $0.5 < H < 1$, with $0 < \beta < 1$ and non stationary noise $\beta > 1$.

A white noise (random noise) is characterized by $\alpha = 0$, $\beta = 0$ and $H = 0.5$. The noise characterized by $0 < H < 0.5$ is called anti-persistent, and characterized by $0.5 < H < 1$ is called persistent.

Whereas a random walk or Brownian Motion (BM) is characterized by power spectral density $S(f) \cong f^{-2}$ that is, $\alpha = 1$ and $\beta = 2$. Its Fourier Inverse is $X^2(\tau) \cong \tau$ giving indication of Mean Squared Displacement scales proportional to time. In the similar way as for persistent and anti-persistent noise the walks may be classified as a persistent walk if the spectral density index or the fluctuation is given by $1 < \beta < 2$ and anti-persistent walk given by $2 < \beta < 3$. We call these walks for fluctuating characteristic exponent $1 < \beta < 3$, because these walks may be obtained by 'integrating' the fractal noise defined by $0 < H < 1$.

Here we discuss the step taken by random walk process (with memory). That will give concept of 'persistent' and 'anti persistent' walks. From the analytical results discussed in previous section where spectrum was derived for process defined by right-sided fractional difference stochastic equation; is analogous to 'fractional Brownian Motion', but not the same. The process generates $1/f^{2\alpha}$' fractal-noise'. Note for $\alpha = 1$ the spectral density is of Brownian motion, and for $\alpha = 0$ the process is white-noise. Both processes are random processes 'without memory'; but one signifies motion while other signifies noise. Integrating noise (white noise or fractal noise), we get the motion Brownian or Fractional Brownian Motion (with or without memory respectively).

The analogy is complete if we set $\alpha = H - (1/2)$, so that the spectrum equation is:

$$S(\omega) \approx 1/(\omega^{2\alpha}) = 1/(\omega^{2H-1}) = \omega^{-(2H-1)}.$$

The H parameter is called Hurst index, signifies the degree of 'long-range dependency' (LRD) of process. Where $0 < H < 1$ and $-0.5 < \alpha < 0.5$.

When, $\alpha = 0$ the spectrum is of white noise, (here $H = +0.5$). When, $\alpha = +0.5$ the spectrum is of $S(\omega) \cong (1/\omega)$, the process is discrete $1/f$ noise, (here $H = 1$).

Using the Fourier pair, for a power function that is, $\left|\Im\left\{t^{\alpha}\right\}\right| = \Gamma(1+\alpha)\omega^{-(\alpha+1)}$ and $\angle\Im\left\{t^{\alpha}\right\} = -(\alpha+1)\pi/2$, for a function $f(t) = t^{\alpha}$, $t \geq 0$ and $f(t) = 0$, for $t < 0$, the inverse Fourier is $\Im^{-1}\{|S(\omega)|\} = \Im^{-1}\{\omega^{-(2H-1)}\} = R_{XX}(\tau) \cong \tau^{2H-2}$. This is derived in previous section. Where, $R_{XX}(t) = \langle X(t)X(0)\rangle$ with $\langle ... \rangle$ is average over time interval, gives the auto-correlation function of the process $X(t)$ .This $R_{XX}(t)$ should not be confused with Mean Squared Displacement (MSD). For 'white-noise', $\alpha = 0$ the auto-correlation decays with time that is, for $H = 1/2$, the $R_{XX}(\tau) \propto \tau^{-1}$, and has $|S(\omega)| \cong 1$, that is frequency independent.

In the language of random walks, the above inverse power law

$$(S(\omega) \approx \omega^{-2\alpha} = \omega^{-(2H-1)})$$

and, for $0.5 \le H \le 1$ or $0 \le \alpha \le 0.5$ implies random walker has tendency to continue in direction where he/she is going. This means that there is 'persistence' to the process, given a step in a particular direction that step is 'remembered' and likelihood that next step is also in same direction is greater than reversing the direction. The process is having long tail, lingering memory. For example in this range of Hurst exponent, consider a power-spectrum set for processes as,

$$\{S(\omega)\} = \{\omega^{-0}, \omega^{-0.25}, \omega^{-0.5}, \omega^{-0.75}, \omega^{-1}\}.$$

This set will have corresponding auto-correlation function $R_{XX}(\tau)$ as

$$\{R_{XX}(\tau)\} = \{\tau^{-1}, \tau^{-0.75}, \tau^{-0.5}, \tau^{-0.25}, \tau^{-0}\}.$$

This range of Hurst exponent ($0.5 \le H \le 1$) is most useful range, for LRD processes, spanning from one extreme Brownian case or white-noise process (that is without memory) to other end pure $1/f$ noise process. Note that the auto correlation function is decaying 'slowly' with time, signifying that in this range of Hurst exponent, the process is LRD with lingering long tailed memory.

Analogy for $0 < H < 0.5$ or $-0.5 < \alpha < 0$ implies that random walker prefers to change his/her mind after each step. This is 'anti-persistence' walk that is, in a process a given step in a particular direction is 'remembered' and likelihood that next step being in the same direction is less than that of reversing the step. This is 'short time memory' and process decays monotonically and hyperbolically to zero. Consider, for example set of processes in this range of Hurst exponent ($0 < H < 0.5$) let the power-spectrum set be, $\{S(\omega)\} = \{\omega^{0}, \omega^{0.25}, \omega^{0.5}, \omega^{0.75}, \omega^{1}\}$. These processes will have corresponding auto correlation $R_{XX}(\tau)$, as

$$\{R_{XX}(\tau)\} = \{\tau^{-1}, \tau^{-1.25}, \tau^{-1.5}, \tau^{-1.75}, \tau^{-2}\},$$

indicating these processes decay very fast hyperbolically to zero, case of 'short ranged processes'. The exponent H is Hurst's exponent gives nature of 'fractal' time series defining anomalous diffusion, with memory or sub diffusion less than normal diffusion rate. At $H = 1/2$ the rate of diffusion is Fick's diffusion that is walker, walks without memory.

4.5 Nuclear Reactor Neutron Flux Description

The neutron balance description in Nuclear Reactor is defined by Transport-Theory. The basic transport equations are then approximated by several coupled differential equations. One of the simplified approximation of the reactor representation is given to engineers is the neutron diffusion equation sets in multi energy group or single energy group, In all these diffusion equations the leakage term has Fick's law of diffusion, where the neutron flux is assumed to be a point quantity. For larger reactor representation several of these diffusion equations are formed and modeled by region to region coupling coefficients. Engineering science then proceeds on these approximates to obtain reactor transfer function model and then various control system analysis are done. For complex systems the integer models of the reactor may not suffice and thus a fractional order model for obtaining flux profile or kinetics may describe the complex reality better. The argument is similar to as described for heat transfer model of Chapter 3, where distributed and complex parametric spreads and size factor gets described better by fractional transient heat transfer equation. The point kinetic coupled differential equation is listed as:

$$\frac{1}{v_c}\frac{\partial}{\partial t}\phi(x,t) = \mathbb{D}\nabla\phi(x,t) + \left(\gamma\Sigma_f - \Sigma_a\right)\phi(x,t) + \lambda C(x,t)$$

$$\frac{\partial}{\partial t}C(x,t) = \beta\gamma\Sigma_f\phi(x,t) - \lambda C(x,t)$$

The parameters are:

v_c : Neutron velocity
ϕ : Neutron flux.

C : Density of precursors
$\mathbb{D}$: Neutron diffusion coefficient

γ: Average number of neutrons produced per fission
Σ_f : Macroscopic fission cross-section
Σ_a : Macroscopic absorption cross-section
λ : Radioactive decay constant
β : Delayed neutron fraction that is, fraction of the fission neutrons that are delayed

This is classical integer order expression, the solution to this with boundary and initial conditions give the neutron flux profile inside the reactor and the temporal growth of neutron population.

4.6 Classical Constitutive Neutron Diffusion Equation

In classical sense, the constitutive equation assumes point neutron flux, with v as the average speed of the neutrons passing through area with n neutrons per unit volume as neutron density. The vector quantity representing the neutron flux is J. Following will elucidate the classical statements.

$$\vec{J} = n\,\vec{v}$$
$$\phi = n\,v$$
$$\vec{J} = -\mathbb{D}\,\nabla\,\phi$$

Consider a closed volume; the loss of neutrons from the closed surface is given as surface integral of neutron current. $J \bullet dS$. The loss occurring in the volume by absorption is given by absorption cross-section and then taking volume integral of, $\Sigma_a \phi dV$. This total loss, when equated to the source term gives classical constitutive neutron diffusion equation, as depicted below.

$$\int_S J \bullet dS + \int_V \Sigma_a \phi dV = \int_V S dV$$

The above integral form when converted to volume integral look as below:

$$\int_V (\nabla \bullet J + \Sigma_a \phi - S)dV = 0 \quad \text{or} \quad \nabla \bullet J + \Sigma_a \phi - S = 0 \text{ for equilibrium}$$

Using the expression of $J = -\mathbb{D}\nabla\phi$ we obtain the following

$$-\mathbb{D}\nabla^2\phi + \Sigma_a\,\phi - S = 0$$

In the steady state the RHS of above constitutive equation is zero and if there is time changing flux, then that is put in the RHS as

$$\mathbb{D}\nabla^2\phi - \Sigma_a\,\phi + S = \frac{\partial n}{\partial t} = \frac{1}{v}\frac{d\phi}{dt}$$

4.6.1 Discussion on Classical Constitutive Equations

The classical neutron diffusion constitutive equation as described is based on classical divergence of the divergence of a vector field. The divergence is defined as ratio of total flux through a closed surface to the volume enclosed by the surface, when volume shrinks towards zero.

$$\text{div}J = \lim_{V \to 0} \frac{1}{V} \int_S J \bullet dS$$

Where J is the flux vector, V is an arbitrary volume enclosed by surface S. The dot product of vector J with the surface dS is obvious this is valid only if the flux is indeed a "point" quantity relative to the scale of observation. Neutron diffusion is counter example since primarily due to velocity fluctuations (even at constant energy/temperature) that arise only as observation space grows larger, invalidating the limit. Also the neutrons are no longer in homogeneous medium. The dispersive fluxes for a given volume are typically averaged in some fashion (volumetric, statistical) and are approximated by Fick's first law as we have obtained in deriving the classical constitutive equation for neutron diffusion. $\vec{J} = -\mathbb{D}\nabla\phi$. As the control volume shrinks to zero, the velocity fluctuations and the dispersive flux disappear. Therefore in a true sense the classical divergence theorem discounts the real effects of macroscopic in-homogeneity and the fluctuations associated with neutron diffusion in a reactor.

Because of the limit in the divergence definition, the classical Gauss divergence theorem discounts the effect of large volume until the dipersive; flux can be approximated by a point quantity.

4.6.2 Graphical Explanation

Refer Figure 4.1. In the Figure 4.1(a) it is shown that surface flux with respect to volume of the observation space is of constant slope line. The Figure 4.1(b) plots the ratio of the surface flux with respect to the control volume (first derivative of Figure 4.1a).

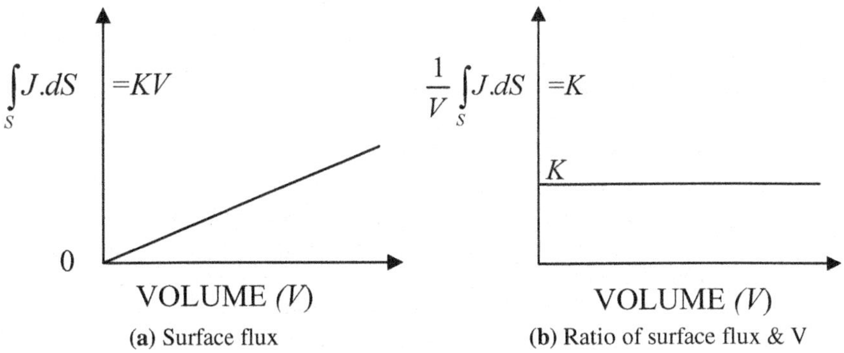

Fig. 4.1 Classical definition of divergence of flux vector.

The Figure 4.1 show, in simplistic manner if the surface flux of neutrons with average constant velocity grows in linear fashion with respect to the volume of the observation space then in this case the ratio of the surface flux to the control volume remains fixed. In this particular (ideal) case making the control volume shrink to zero will yield ideal definition of the divergence of the vector flux (neutron current density). This simplistic picture neglects the effect of inhomogeneous medium and macroscopic dispersion, fluctuating velocity effects and effects due to neighborhood, neutron currents.

Figure 4.2 is extension of Figure 4.1 showing the macroscopic effects of surface flux manifestaion as the control volume is enlarged. The observation space when enlarged as shown in Figure 4.2(b) captures dspersive effect of neutrons as magnified by stair case type of ratio of surface flux to the volume figure. The effect can be seen as surface flux gets manifested as some power of observation space (volume).

Figure 4.2(b) is the first derivative of the Figure 4.2(a) and shows that at quasi large observation space (control volume) one gets seemingly constant ratio of surface flux to volume, therefore yielding a non-local divergence of the nuetron flux vector. This definition of non-local divergence is what contradicts classical divergence where control volume is made to shrink to zero.Therefore, the larger observation space does captures the reality of associated with neutron diffusion process.

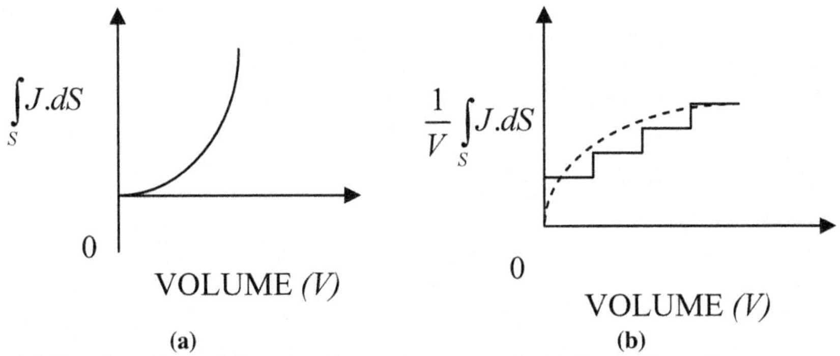

Fig. 4.2 Showing effect of dipersive flux and neutron velocity fluctuation with macroscopic scale of observation extension.

4.6.3 About Surface Flux Curvature

Refer Figure 4.2(a). The curvature is concave is nature as the observation space (control volume) is made bigger. Contention could have been that why the curvature is taken as concave instead of convex. Here some practical reasoning will elucidate the nature of curve shown in Figure 4.2. For a very small observation space area the surface flux is product neutron current and that area. As the area is made larger the neighbouring neutrons does effect the neutron current in the wider area of measurement. This gives the larger value of the neutron current for newer area considered. This increment in the neutron current is what gets integrated in the surface integral giving the concave shape of the Figure 4.2(a).

This is elaborated in the Figure 4.3. Let the observation surface area for measurement of neutron surface flux be divided into squares as shown. Assume that each centre of the square is having one neutron. If all the neutrons are at rest with out any velocity fluctutaions then there will not be any finite probabilty that it may jump across to the next adjacent boxes. However the case is not so, as there always is finite probabilty of having neutrons designated for a particular box finding into adjacent box. However if the area of observation is very small as depicted by smaller circles inside each box the fluctutaion effect of neutron velocity will not be observed. Therefore with the smaller circles in the observation space measures a smaller neutron current (solely due to presence of its own neutron in the squrae box). However, the observation area is made into larger circles as shown in the Figure 4.3. Here we see that with enlarged area the effect of neutrons in the adjacent square will enhance the neutron current compared to the first smaller area.Also this bigger circle will catch the effect of velocity fluctuations, and therefore will show larger magnitude of neutron current. This simplistic explanation is justified for the observation that the shape of the surface flux with respect to the observation space (Figure 4.2) is concave and not convex.

4.6.4 Statistical and Geometrical Explanation for Non-local Divergence

The Figure 4.3 divides the space into grids. The fluctuations in velocity causing the violation of classical limit of volume shrinkage towards zero, for classical divergence also elucidated by the fact that at a particular space the neutrons will have spatial long tailed distributions. The effect of this long tailed statistical probability distribution will thus gets enhanced by the use of non-local divergence and this reality effect will thus be shown with avoidance of volume shrinkage to zero. Also in reality the coupling between various zones in the reactor take place. The non-local divergence with the principle of non-zero volume therefore is the apt tool for constitutive equation for neutron diffusion equation for reactor description.

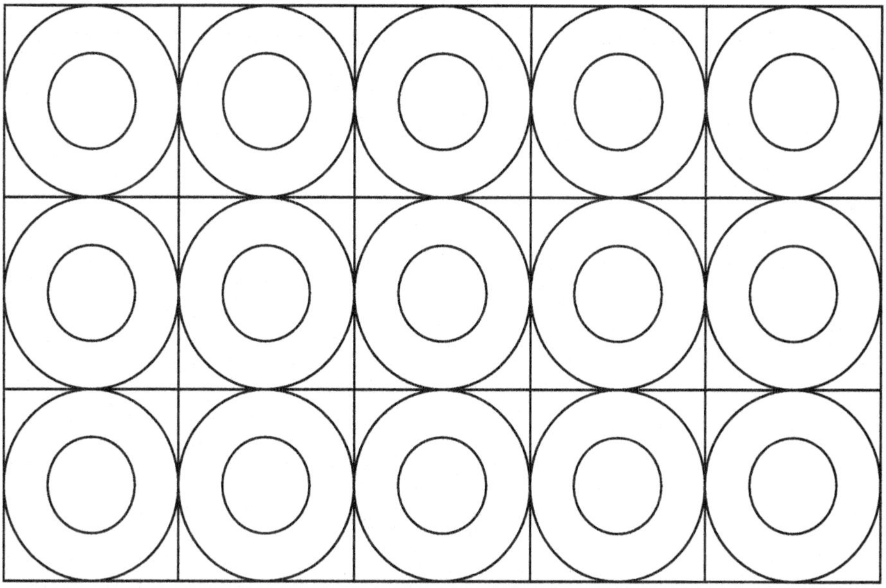

Fig. 4.3 Showing the effect of growing observation space modifying the neutron current

Refer the classical neutron diffusion equation and Figure 4.2. The surface flux is $\int_S J.dS$ then the ratio of the surface flux to the volume is taken to consider the leakage through surface term as $\nabla.J$, at volume shrinking towards zero. Then with Fick's law we get the leakage of neutrons through a closed surface as $\nabla.(-\mathbb{D}\nabla\phi) = -\mathbb{D}\nabla^2\phi$. This term in one dimension case is

$$-\mathbb{D}\frac{d^2\phi}{dx^2}$$

Examining the above and relating this to Figure 4.2, one may say that divergence is slope of the surface flux (Figure 4.2a). This curvature if has square law variation in the shape then the double derivative will be constant and we have the entire curvature captured in that constant. If the curvature of the surface flux (Figure 4.2a) is not having $\approx x^2$ variation, but has say $\approx x^{1.5}$ variation then double derivative will not capture the information about the curvature in a constant.

4.6.5 Point Kinetic Equation in Heterogeneous Background

Consider Figure 4.3, the reactor matrix is heterogeneous. This comprises of lumped fuel central circle, encircled by moderator and the matrix gets repeated. Consider collisions of neutrons in this heterogeneous reactor with characteristic length scales of migration of neutrons, where fuel material is dispersed in lumps within the moderator m and fuel f, with all the neutrons with same speed v and assuming the angular flux is only linearly anisotropic. Then the conservation equation for the neutron collision and reactor process as well as initial internal boundary conditions are.

$$\frac{1}{v}\frac{\partial}{\partial t}\phi_l + \nabla\bullet\vec{J}_l + \Sigma_{al}\,\phi_l = S_l\,,$$

and

$$\frac{1}{v}\frac{\partial}{\partial t}\vec{J}_l + \frac{1}{3}\nabla\phi_l + \Sigma_{trl}\,\vec{J}_l = 0$$

with initial condition as

$\phi_l(r,0) = \phi_{l0}(r)$. The subscript l is allowed to play the role of m moderator and f fuel. The inner boundary condition at the mf interface,

$$-\vec{\mathrm{n}}_{mf}\bullet\mathbb{D}_m\nabla\phi_m = -\vec{\mathrm{n}}_{mf}\bullet\mathbb{D}_f\nabla\phi_f$$

and $\phi_m = \phi_f$. Here $\vec{\mathrm{n}}$ is unit vector normal at the interface which is directed from moderator toward fuel. The cross-sections are $\Sigma_{al} = \Sigma_{tl} - \Sigma_{sl}$ and $\Sigma_{tr} = \Sigma_t - \bar{\mu}_0\,\Sigma_s$, where $\bar{\mu}_0$ is average cosine angle and Σ_{tl}, Σ_{sl} and Σ_{trl} are total, scattering and transport cross-sections, respectively, for region l. The $\phi(r,t), \vec{J}(r,t)$ $\Sigma(r,t)$ and $S(r,t)$ variables are space and time dependent, generally. Typical consideration is to neglect, time derivative in neutron current equation, i.e. $v^{-1}\partial\vec{J}_l/\partial t$, if the time variation rate of neutron current density is much slower than the collision frequency, that is $v\Sigma_{tl}$.

Meaning that,

$$\left|\vec{J}_l\right|^{-1}(\partial\vec{J}_l/\partial t) << v\Sigma_{tl}\,.$$

For thermal light water reactors the $v\Sigma_{tl} \approx 10^{-5}\mathrm{s}^{-1}$, the above simplifies to Fickian case with $\vec{J}_l = -\mathbb{D}_l\nabla\phi_l$ where diffusion constant is

$$\mathbb{D}_l = (3\Sigma_{trl})^{-1} = 3[\Sigma_{tl} - \bar{\mu}_0\,\Sigma_{sl}]^{-1}$$

is also a function of space and time.

In the above discussion we assumed firstly that; absorption is much less likely than scattering. This is valid for the moderating and structural materials but not for fuel and control rods. Secondly, a linear spatial variation of the neutron distribution is assumed.

This satisfied a few mean free paths away from the boundary, for large homogeneous medium with uniform source distribution. Thirdly, isotropic scattering is assumed, this is satisfied for scattering from heavy atomic mass nuclei.

However, the Fickian assumptions exhibit highly anomalous diffusion phenomena due to highly heterogeneous configuration, especially in the presence of strong neutron absorbers, in the form of fuel control rods, shut off rods and chemical shims present in the coolant or poisons in moderator. The dynamics of these absorbers radically change the local energy generation, giving new method of description of point kinetics, with non-Fickian approximations. One type of modification is to include relaxation times and rewrite the neutron current equation as given by Catteneo (1948) dealt in Chapter 3.

For times scales in which flux varies rapidly, the condition

$$\left|\vec{J}_l\right|^{-1}(\partial\vec{J}_l/\partial t) << v\Sigma_{tl}$$

is not satisfied; then we modify the neutron current equation as:

$$\tau_l\frac{\partial}{\partial t}\vec{J}_l + \vec{J}_l = -\mathbb{D}_l\nabla\phi_l\,,$$

where

$$\tau_l = \frac{1}{v\Sigma_{trl}} = \frac{3\mathbb{D}_l}{v}\,,\ \text{when}\ t << \tau_l\,.$$

Putting this Catteneo's diffusion in

$$\frac{1}{v}\frac{\partial}{\partial t}\phi_l + \nabla\bullet\vec{J}_l + \Sigma_{al}\,\phi_l = S_l\,,$$

and rearranging we obtain.

$$\frac{\partial^2 \phi_l}{\partial t^2} + \left(v \Sigma_{al} + \frac{1}{\tau_l} \right) \frac{\partial \phi_l}{\partial t} - v_{ph}^2 \nabla^2 \phi_l = \frac{v}{\tau_l} [S_l - \Sigma_{al} \phi_l] + v \frac{\partial S_l}{\partial t},$$

The propagation phase velocity of 'neutron wave' is

$$v_{ph} = \pm \sqrt{v \mathbb{D}_l / \tau_l}\,.$$

When time scales of $\left|\vec{J}_l\right| (\partial \vec{J}_l / \partial t) << v \Sigma_{tl}$ is satisfied then $\vec{J}_l = -\mathbb{D}_l \nabla \phi_l$ Fick's case is recovered with above obtained equation simplified as

$$\frac{1}{v} \frac{\partial}{\partial t} \phi_l + \Sigma_{al} \phi_l = \nabla \bullet \mathbb{D}_l \nabla \phi_l\,.$$

(without source term).

For a neutron as diffusion species, if there is non-linear scaling of mean squared displacement $\langle r^2(t) \rangle \approx t^q$, where $q > 1$ implies super diffusion, and $0 < q < 1$ representing sub diffusion we can generalize the Catteneo diffusion of neutrons as, fractional model of neutron balance; only for moderator region m as:

$$\tau_m^q \frac{1}{v} \frac{\partial^{q+1} \phi_m}{\partial t^{q+1}} + \tau_m^q \Sigma_{am} \frac{\partial^q \phi_m}{\partial t^q} + \frac{1}{v} \frac{\partial \phi_m}{\partial t} + \Sigma_{am} \phi_m = \mathbb{D}_m \nabla^2 \phi_m\,,$$

and the current equation as

$$\tau_m^q \frac{\partial^q \bar{J}_m}{\partial t^q} + \bar{J}_m = -\mathbb{D}_m \nabla \phi_m\,.$$

Suitable interface conditions are then

$$\tau_m^q \frac{\partial^q \vec{J}_m}{\partial t^q} + \vec{n}_{mf} \bullet \mathbb{D}_m \nabla \phi_m = \tau_f^q \frac{\partial^q \vec{J}_f}{\partial t^q} + \vec{n}_{fm} \bullet \mathbb{D}_f \nabla \phi_f\,,$$

where $J_m = \vec{n}_{mf} \bullet \vec{J}_m$ and $J_f = \vec{n}_{fm} \bullet \vec{J}_f$. When 'fractional relaxation time' $\tau_m^q \to 0$, with $q \to 1$, normal Fickian equations are recovered. The conservation of

$$\tau_m^q \frac{\partial^q \vec{J}_m}{\partial t^q} + \vec{n}_{mf} \bullet \mathbb{D}_m \nabla \phi_m = \tau_f^q \frac{\partial^q \vec{J}_f}{\partial t^q} + \vec{n}_{fm} \bullet \mathbb{D}_f \nabla \phi_f$$

equation is valid as

$$\frac{1}{v} \frac{\partial}{\partial t} \phi_m + \Sigma_{am} \phi_m = \nabla \bullet \mathbb{D}_m \nabla \phi_m$$

represents balance to be met categorically. But the current law is modified as with fractional relaxation times. The time relaxation may stem from possible 'obstacles' and 'traps' that delay the neutrons to move, and thus introduce 'memory' effect into the motion without affecting the neutron conservation.

However, the consideration of high heterogeneity may not be sufficient condition for Fractional Order Generalization. So far we have assumed average constant velocity of drift of neutrons that is, thermal speed. However, there are neutrons from 10MeV to less than 0.01MeV of energy and the cross-section is velocity (energy) dependent. In that case the detailed group treatment is required. Heterogeneity effect of a reactor system with fast neutrons is less sensitive than heterogeneity effect with thermal neutrons due to mean free path of neutrons, varying in fuel as well as moderator (for thermal reactors).

4.6.6 Revisiting the Realm of Brownian Motion

Refer Figure 4.3 and consider it to be homogeneous lattice with lattice constant Δx. A typical Brownian walk invented by Perrin considers a walker (particle neutron) jumps at each time step $t = 0, \Delta t, 2\Delta t, ...n\Delta t, ...$ to a randomly selected direction, thereby covering the distance Δx, the lattice constant. In discrete time steps of span Δt the walker (particle-neutron) is assumed to jump to one of its nearest neighbor sites. Such process can be modeled by a master equation:

$$N_j(t+\Delta t) = \frac{1}{2}N_{j-1}(t) + \frac{1}{2}N_{j+1}(t)$$

In the above equation, the index denotes the position on the underlying one-dimensional lattice. The equation defines the probability distribution function (PDF), or concentration/ number of neutron particles at position j at time $t+\Delta t$ in dependence of population of neutron of the two adjacent sites $j \pm 1$ at time t. The constants ½ expresses that jump direction are isotropic; meaning that there is equal probability of jumping toward, forward or backward direction. In continuum limit $\Delta t \to 0$ and $\Delta x \to 0$, the Taylor expansion in time and space give:

$$N_j(t+\Delta t) = N_j(t) + \Delta t \frac{\partial N_j}{\partial t} + \mathrm{R}_1\left([\Delta t]^2\right)$$

$$N_{j\pm 1}(t) = N(x,t) \pm \Delta x \frac{\partial N}{\partial x} + \frac{(\Delta x)^2}{2}\frac{\partial^2 N}{\partial x^2} + \mathrm{R}_2\left([\Delta x]^3\right)$$

Using these two expressions and putting into master equation, recognizing the notations $N(x,t) = N_j(t)$; neglecting the remainders R_1 and R_2, leads to Fick's diffusion equation:

$$\frac{\partial N(x,t)}{\partial t} = \mathbb{D}\frac{\partial^2}{\partial x^2}N(x,t),$$

where $\mathbb{D}$ is diffusion constant with units cm^2s^{-1} and defined in the limits as:

$$\mathbb{D} \equiv \lim_{\Delta x \to 0, \Delta t \to 0} \frac{(\Delta x)^2}{2\Delta t},$$

the same as obtained in Chapter 3. The solution $N(x,t)$ for sharp initial condition is called propagator and is Gaussian in shape. The Fourier-Laplace transform of this Fick's equation is,

$$N(k,s) = N_0(k)\left[s + \mathbb{D}k^2\right]^{-1}$$

as demonstrated in Chapter 3.

The diffusion equation is one of the most fundamental equations in physics and engineering. This is direct consequence of 'central limit theorem'. Under the condition that the first two moments of $N(x,t)$ or the probability distribution function (pdf), describing the appropriately normalized distance covered in a jump event and the variance, $\overline{x} = \sum_i x_i$ and $< \Delta x^2 >$, as well as a mean time span Δt between any two individual jump event exists, and is finite. The CLT assures that the random walk process is characterized by a mean velocity $\overline{v} = \overline{x} / \Delta t$, and diffusion a constant $\mathbb{D}$ as defined above. Furthermore, for long times i.e. after a large enough number of steps, the $N(x,t)$ or the pdf of being at a certain position x at time t, is governed by above diffusion equation, and is given by Gaussian shape.

4.6.7 The Continuous Time Random Walk (CTRW) Model

In discussion of previous section, the particle neutrons jumps at equal lattice spacing, and spends zero time (or a constant time) there that is, each step jump has equal length (the lattice distance) and nil waiting time at the lattice. We may also have situation that neutron particles can jump different site lengths and can even wait for some time at those lattice positions. Assume that the neutrons are disciplined and they move by not spending infinitely long times at sites and they do not jump infinite (long) lattice lengths still the Fick's law will be applicable. Meaning if the waiting times are having finite average and the jump lengths are having finite variance from a mean jump length, then too the governing law will be Fick's diffusion. This will be demonstrated in this section.

The CTRW model is based on the idea that the lengths of a given jump, as well as the waiting time elapsing between two successive jumps are drawn from a joint pdf $\psi(x,t)$ which is called 'jump pdf'. This pdf is composed of two parts the 'jump length pdf' $\lambda(x)$ and 'waiting time pdf' $w(t)$. The simple situation is that these pdf are decoupled that is, jump length and waiting times are independent random variable; then $\psi(x,t) = \lambda(x)w(t)$. These pdf have corresponding Laplace and Fourier Transforms as, $w(s)$, $\lambda(k)$ and $\psi(k,s)$. Different types of CTRW processes can be categorized as average waiting time between jumps as

$$T = \int_0^{\infty} dt(t)w(t)$$

and jump length variance

$$\sigma^2 = \int_{-\infty}^{\infty} dx(x^2)\lambda(x)\,,$$

being finite or diverging, respectively.

The integral

$$\chi(t) = 1 - \int_0^{t} dt' w(t')$$

represents the cumulative probability that a neutron number density (pdf) when created at time $t = 0$ say $N(x,0) = N_0(x)$ initially will not wait from time $t' = 0$ to $t' = t$. Let $N_0(k)$ be the Fourier Transformed initial condition, and $N(k,s)$ denotes Fourier-Laplace Transformed of the pdf the neutron number density at position x and at time t that is, $N(x,t)$. The Laplace transformed $\chi(t)$ is

$$\chi(s) = \frac{1 - w(s)}{s}\,.$$

At the position x and time t there will constitution of neutron number density that is purely stationary at that position minus the fraction of neutron number density that will have probability of jumping from these location and time with jump probability $\psi(x,t)$.

Thus the consolidated number density that will be present at position x and at time t is $N(k,s) - \psi(k,s)N(k,s)$. The Fourier-Laplace transformed parameters are used, since in this k and s domain the multiplication explanation is easier, else we have to explain via convolution-integrals. This consolidated number density obtained is result of the initial neutron number density (pdf) which has manifested as result of 'no wait' from initial time to present time $t' = t$, that is given as $\chi(s)N_0(k)$ and equating the two we have

$$N(k,s) - \psi(k,s)N(k,s) = \chi(s)N_0(k)$$

giving:

$$N(k,s) = \frac{1 - w(s)}{s}\frac{N_0(k)}{1 - \psi(k,s)}$$

This expression was obtained by CTRW expression in Chapter 3.

Consider $N_0(k)=1$ that is, Fourier Transformed of unit delta-function $\delta(x)$ as initial neutron number density. Consider the $w(t)=\tau^{-1}\exp(-t/\tau)$, a Poisson's process for the wait-time pdf. Thus average wait time between each successive jump is finite that is, $T=\tau$. Consider together with wait-time pdf, an independent jump length pdf as Gaussian i.e.

$$\lambda(x)=(4\pi\sigma^2)^{-1/2}\exp(-x^2/4\sigma^2)$$

leading to finite variance as $2\sigma^2$. The jump pdf is decoupled that is, $\psi(k,s)=\lambda(k)w(s)$. The corresponding Laplace and Fourier transformed asymptotic expansions, for wait-time and jump length pdf are: $w(s)\sim 1-s\tau+\mathrm{R}_1(\tau^2)$ and $\lambda(k)\sim 1-\sigma^2k^2+\mathrm{R}_2(k^4)$; with higher order remainder terms.

In fact any pdf with finite average and finite variance leads to this same result, to lowest orders; and therefore, the above expansions are asymptotic representations for late (long) time limits. Putting these we have:

$$N(k,s)=\frac{1-w(s)}{s}\frac{N_0(k)}{1-\psi(k,s)}=\frac{1-(1-s\tau)}{s}\frac{1}{1-(1-s\tau)(1-\sigma^2k^2)}\cong\frac{1}{s+\mathbb{D}_1k^2}\text{, with}$$

$$\mathbb{D}_1\equiv\frac{\sigma^2}{\tau}.$$

We have got a Fourier-Laplace expression for finite average wait-time and finite variance of jump-length as $sN(k,s)+\mathbb{D}_1k^2N(k,s)=1$ re-writing this as $[sN(k,s)-1]+\mathbb{D}_1k^2N(k,s)=0$ and using $\mathcal{L}\{\partial N(x,t)/\partial t\}=sN(x,s)-N_0(x)$ and $\Im\{\partial^2N(x,t)/\partial x^2\}=-k^2N(k,t)$ with $N_0(k)=1\equiv\Im\{\delta(x)\}$ we recover the Fick's diffusion equation

$$\frac{\partial N(x,t)}{\partial t}=\mathbb{D}_1\frac{\partial^2}{\partial x^2}N(x,t).$$

Note the diffusion constants $\mathbb{D}$ and $\mathbb{D}_1$ are same. The subscript is used to distinguish this constant with the constants of diffusion appearing in fractional diffusion equation (in next sections).

4.7 Diffusion with Long Rests

Consider the situation where the characteristic waiting time $T=\infty$ that is, diverges; but the jump-length variance $\sigma^2<\infty$, finite. This diverging average is character of a pdf of power-law, long-tailed type distribution. Let the waiting-time pdf be $w(t) \sim (t/\tau)^{1+\alpha}$, for $0 < \alpha < 1$. The corresponding Laplace asymptote of this power-law "fractal' waiting-time pdf is $w(s) \sim 1-(s\tau)^{\alpha}$. Let the jump-length pdf be Gaussian as done in previous section giving finite variance of jump length as: $\lambda(k) \sim 1-\sigma^2 k^2$. Putting these in the expression for $N(k,s)$ we get:

$$N(k,s) = \frac{1-w(s)}{s}\frac{N_0(k)}{1-\psi(k,s)} = \frac{1-\{1-(s\tau)^{\alpha}\}}{s}\frac{N_0(k)}{1-[1-(s\tau)^{\alpha}][1-\sigma^2 k^2]} \cong \frac{[N_0(k)/s]}{1+\mathbb{D}_{\alpha} s^{-\alpha} k^2},$$

where $D_{\alpha} \equiv \sigma^2/\tau^{\alpha}$ with dimensions of $\mathrm{cm}^2\mathrm{s}^{-\alpha}$.

Employing

$$\mathcal{L}\left\{{}_0D_t^{-q} f(x,t)\right\} = s^{-q} F(x,s), \; q \geq 0$$

one gets, fractional integral equation

$$N(x,t) - N_0(x) = {}_0D_t^{-\alpha}\mathbb{D}_{\alpha}\frac{\partial^2}{\partial x^2}N(x,t)$$

Differentiating (the above) both sides once w.r.t. time we have

$$\frac{\partial}{\partial t}N(x,t) = {}_0D_t^{1-\alpha}\mathbb{D}_{\alpha}\frac{\partial^2}{\partial x^2}N(x,t) \; ,$$

a fractional differential equation of diffusion. Fractionally differentiating by order α the fractional integral equation derived above we get alternate form as:

$${}_0D_t^{\alpha}N(x,t) - \frac{t^{-\alpha}}{\Gamma(1-\alpha)}N_0(x) = \mathbb{D}_{\alpha}\frac{\partial^2}{\partial x^2}N(x,t)$$

This integrodifferential nature of Riemann-Liouvelli fractional operator ${}_0D_t^{1-\alpha}$ with integral kernel $K(t) \propto t^{\alpha-1}$ ensures the non-Markovian nature of this 'sub-diffusive' process defined above. The calculation of MSD from the relation

$$\left\langle \Delta x^2 \right\rangle = \lim_{k\to 0}\left\{-(d^2/dk^2)N(k,s)\right\},$$

gives MSD as

$$\left\langle \Delta x^2 \right\rangle = [2\mathbb{D}_{\alpha}/\Gamma(1+\alpha)]t^{\alpha}.$$

For $\alpha = 1$ we get MSD for integer order Fick's law diffusion

$$\left\langle \Delta x^2 \right\rangle = 2\mathbb{D}_1 t$$

(Chapter 3). Note that the fractional index α appears naturally in this MSD calculation. We have termed as sub-diffusion, as the spread, quantified by MSD appears slower as compared to ideal Fick's case, or Brownian motion. Note that in the fractional diffusion equation obtained above, the single modes (k) decays in accordance to Mittag-Leffler pattern, that is, with

$$N(k,t) \cong E_\alpha(-\mathbb{D}_\alpha k^2 t^\alpha)$$

the asymptotic power-law behaviour

$$N(k,t) \sim \left(D_\alpha k^2 t^\alpha \Gamma(1-\alpha)\right)^{-1}.$$

This typical Mittag-Leffler behavior of the mode relaxation replaces the exponential decay

$$N(k,t) \cong \exp(-\mathbb{D}_1 k^2 t)$$

occurring for normal diffusion (Chapter 3).

From the expression obtained

$$N(k,s) = \frac{\left[N_0(k,s)/s\right]}{1+\mathbb{D}_\alpha s^{-\alpha} k^2},$$

for $N_0(k) = 1$ as initial condition, one can first Fourier invert to get:

$$N(x,s) = \frac{1}{2} s^{\frac{\alpha}{2}-1} \exp\left(-|x| s^{\alpha/2}\right).$$

Then expand the exponential term in its Taylor series, and Laplace invert term-by-term, to get power series solution. The solution to such diffusion equations is examined in Chapter 11.

4.8 Diffusion with Long Jumps

The opposite case from the above process is that characteristic waiting times $T < \infty$ is finite but the jump-length variance $\sigma^2 = \infty$ diverges. We can have Poisson's process from which waiting-times are drawn, and a Levy distribution for jump length pdf. The Fourier representation of Levy distribution is

$$\lambda(k) = \exp(-\sigma^{\mu} |k|^{\mu}) \sim 1 - \sigma^{\mu} |k|^{\mu}$$

For $1 < \mu < 2$, corresponding to the asymptotic behavior as

$$\lambda(x) \sim \sigma^{\mu} |x|^{-1-\mu} \text{ for } |x| \gg \sigma .$$

The wait-time asymptote approximation is $w(s) \sim 1 - s\tau$. Putting these two values, we obtain, for $N_0(k) = 1$, the following Fourier-Laplace relation.

$$N(k,s) = \frac{1-w(s)}{s} \frac{N_0(k)}{1-\psi(k,s)} = \frac{1-(1-s\tau)}{s} \frac{N_0(k)}{1-(1-s\tau)(1-\sigma^{\mu}|k|^{\mu})} \cong \frac{1}{s+\mathbb{D}_{\mu}|k|^{\mu}}$$

Using

$$\Im\left\{ {}_{-\infty}D_x^{\mu} f(x) \right\} = -|k|^{\mu} f(k),$$

the modified definition from, usual definition that is,

$$\Im\left\{ {}_{-\infty}D_x^{\mu} f(x) \right\} = (-\mathrm{i})^{\mu} |k|^{\mu} f(k),$$

that is, by suppressing the imaginary unit we get:

$$\frac{\partial}{\partial t} N(k,t) = \mathbb{D}_{\mu} [{}_{-\infty}D_x^{\mu} N(x,t)]$$

Here the generalized diffusion constant is

$$\mathbb{D}_{\mu} \equiv \sigma^{\mu} / \tau$$

with unit of $\mathrm{cm}^{\mu}\mathrm{s}^{-1}$. The Laplace inverse of $N(k,s)$ gives

$$N_0(k,t) = \exp(-\mathbb{D}_{\mu} |k|^{\mu} t) .$$

This situation of 'long-jumps' is Markovian process in time, with no 'temporal' memory. Refer Chapter 2 for memory integrals. However, one may argue that due to fractional operators appearing in space variable, the process shall be perhaps called non-Markovian in space; having spatial memory kernel. This process remembers its spatial past! Space-time is linked and is different sides of the same coin!

Due to the asymptotic property of this pdf for long jumps

$$\lambda(x) \sim \sigma^{\mu} |x|^{-1-\mu} ,$$

having infinite variance, very long jumps may occur with significant probability than exponentially decaying pdf like formerly employed ‘Gaussian’ jump-length pdf. The scale invariance nature of pdf as used for the jump-length leads to clustering nature of the neutron distribution spatially, i.e. local motion is occasionally interrupted by long sojourns, on all length scales that is, one finds clusters of local motion within clusters. Here one can assign a fractal dimension $d_B = \mu$ with $1 < \mu < 2$, and can analyze the system with normal integer order Fick’s law. The asymptotic solution to the spatial fractional diffusion equation for the long jumps case is

$$N(x,t) \sim \frac{\mathbb{D}_\mu t}{|x|^{1+\mu}}, \text{ with } \mu < 2$$

Due to this the MSD diverges that is, $\langle \Delta x \rangle^2 = \infty$. Clearly we can have argument that massive particle as neutron cannot jump infinitely far. For such massive particles finite velocity of propagation exists making instantaneous long jumps impossible. But in reality we have nuclear reactors where the dimensions are large especially for high power reactors. The neutrons do have ‘coupling’ between spatially distributed ‘point’ reactors. These power reactors are having dimensions much larger than the average diffusion lengths of neutron and are called coupled core reactors. The spatial heterogeneity in small scales do manifest as fractal dimensions in space which makes the anomalous transport of neutrons contrary to belief that it can only reside and ‘walk’ as local Brownian motion. Long-jumps can therefore take place for these ‘massive’ neutrons in a ‘fractal’ heterogeneous spatial backdrop.

In the derivation of both the sub-diffusive case and diffusion with long-jumps the diffusion limit $(k,s) \to (0,0)$ was drawn. Equivalently, the diffusion limit corresponds to choosing $(\sigma,\tau) \to (0,0)$ with $\mathbb{D}_\alpha = \sigma^2 / \tau^\alpha$ a constant or $\mathbb{D}_\mu = \sigma^\mu / \tau$ a constant, which matches the limit

$$\mathbb{D}_1 = \lim_{\Delta x \to 0, \Delta t \to 0} (\Delta x)^2 / (2\Delta t)$$

drawn in the Brownian case. In the sense these Fractional Diffusion Equations obtained are valid in the diffusion limit $t \gg \tau$.

The above generalization methods makes us to think a random walk characterized by broad tailed pdf for both waiting-times and jump-lengths with infinite average wait-time T and infinite jump-length variance σ^2 has constitutive relation as:

$$\frac{\partial}{\partial t} N(x,t) = {}_0D_t^{1-\alpha} (\mathbb{D}_{\alpha,\mu}) \nabla^\mu N(x,t),$$

along with

$$\mathbb{D}_{\alpha,\mu} \equiv \frac{\sigma^\mu}{\tau^\alpha},$$

having unit of $\text{cm}^\mu \text{s}^{-\alpha}$.

We can define pseudo MSD as

$$\left\langle x^2(t)\right\rangle_P \sim t^{2\alpha/\mu}$$

or by

$$\left\langle x^2(t)\right\rangle_P \sim t^{2+\alpha-\mu}$$

for $1<\mu<2$, $0<\alpha<1$, and

$$\left\langle x^2(t)\right\rangle_P \sim t^{3-\mu} \text{ for } 1<\mu<2\,,\alpha>1\,,$$

for these anomalous diffusion cases. In these sections we have also observed the explanation of several cases by different diffusion constants, appropriately selected as 'new transport coefficients' those are $\mathbb{D}$, $\mathbb{D}_\alpha$ $\mathbb{D}_\mu$ and $\mathbb{D}_{\alpha,\mu}$ with different units.

Table 4.1 Phase table for the Fractional Diffusion Equation

Temporal Fractional Order α	Spatial Fractional Order μ	Type of Walk	Average Waiting Time T	Jump-Length Variance σ^2	Nature of Diffusion
$0<\alpha<1$	$0<\mu<2$	Long-Jump	∞	∞	Non-Markovian
$\alpha\ge 1$	$0<\mu<2$	Long-Jump	$<\infty$	∞	Markovian
$0<\alpha<1$	$\mu\ge 2$	Sub-diffusion	∞	$<\infty$	Non-Markovian
$\alpha\ge 1$	$\mu\ge 2$	Brownian	$<\infty$	$<\infty$	Markovian

Table 4.1 presents the summary of the Fractional Diffusion Equation, obtained above as

$$\frac{\partial}{\partial t}N(x,t) = {}_0D_t^{1-\alpha}(\mathbb{D}_{\alpha,\mu})\frac{\partial^\mu}{\partial x^\mu}N(x,t)\,,$$

for different phases with values of spatial-temporal orders of fractional derivative order. The different 'phases' are four distinct domains which can be characterized according to the diverging or finite characteristic (average) waiting times and the jump-lengths variance. If the parameters α and μ become greater than one and two respectively, the corresponding diffusion CTRW process locks onto process dominated by finite T and finite σ^2 (Brownian case). For diverging variance of jump-length $\sigma^2=\infty$, one observes that either the traditional Markovian process for $\alpha\ge 1$, or the non-Markovian process (for $\alpha<1$) with single modes decaying as Mittag-Leffler pattern is evolved.

4.9 Fractional Divergence in Neutron Diffusion Equations

The neutron leakage through closed surface enclosed by finite control volume may be changed as

$$-\mathbb{D}\frac{d^{\beta}\phi}{dx^{\beta}}=\frac{\partial^{\alpha}}{\partial x^{\alpha}}(-\mathbb{D}\frac{d\phi}{dx}).$$

The bracket quantity is Fick's first law. Here β is between 1 and 2 and α is between 0 and 1, non-integer fractional real numbers. In the divergence formulations, the vector,

$$[\frac{\partial^{\alpha}}{\partial x^{\alpha}},\frac{\partial^{\alpha}}{\partial y^{\alpha}},\frac{\partial^{\alpha}}{\partial z^{\alpha}}]\rightarrow\nabla^{\alpha}$$

the fractional divergence operator, which acts on the neutron current vector J.

The constitutive equation in fractional divergence form may thus be as follows:

$$\nabla^{\alpha}J+\Sigma_a\phi-S=0$$
$$\nabla^{\alpha}(-\mathbb{D}\nabla\phi)+\Sigma_a\phi-S=0$$
$$\mathbb{D}\nabla^{1+\alpha}\phi-\Sigma_a\phi+S=0$$

Converting to one-dimensional form to get neutron flux profile in a reactor or neutron flux profile near a source in scattering medium we will look at solutions of the form as follows:

$$\mathbb{D}\frac{d^{1+\alpha}\phi}{dx^{1+\alpha}}-\Sigma_a\phi+S=0,\ldots 0<\alpha<1$$
$$\mathbb{D}\frac{d^{\beta}\phi}{dx^{\beta}}-\Sigma_a\phi+S=0,\ldots 1<\beta<1$$

One may interpret the simplified form of $\nabla^{\alpha}.J$ is that a fractional divergence operator is applied to Fickian dispersion term. For illustration of how fractional derivatives relate to the definition of divergence in neutron transport, consider two simple functions $f(x)=x^2$ and $f(x)=x^{1.5}$. Recalling the fundamental of the derivatives we recall that the derivatives of function give the idea of the curvature contained in the curve. Successive derivatives will strip of the independent variable and subsequently will show up the curvature contents. Take the example $f(x)=x^2$ the first and second derivatives are $f^{(1)}(x)=2x$ and $f^{(2)}(x)=2$ respectively. In this example the second derivative contains all the information about the function and that in constant (i.e.2). This argument favors well for well-behaved integer order (power) functions. Now let us change the function and take any real order function say $f(x)=x^{1.5}$, the first and second derivative is $f^{(1)}(x)=1.5x^{0.5}$ and

$f^{(2)}(x) = 0.75x^{-0.5}$. In both the derivatives for this real ordered function the derivatives vary with the independent variable. Refer Figure 4.2b; the effect of growing control volume is approximated as steps. Each step signifies growing value of $\mathbb{D}$. The shape emerged as dotted curve in the Figure 4.2b, approximates the first derivative of the Figure 4.2a. Here the first derivative or any other higher order integer derivative fails to contain the curvature information.

In the functions with real order (other than integer order) if the concepts of fractional calculus are applied then we can contain the curvature information of $f(x) = x^{1.5}$ by taking 1.5 derivative of the $f(x)$ symbolically $d^{1.5}x^{1.5} / dx^{1.5} = 1.33$.

$$\frac{d^{\alpha}}{dx^{\alpha}} x^{u} = \frac{\Gamma(u+1)}{\Gamma(u-\alpha+1)} x^{u-\alpha}$$

$$\frac{d^{1.5}}{dx^{1.5}} x^{1.5} = \frac{\Gamma(1.5+1)}{\Gamma(1.5-1.5+1)} x^{1.5-1.5} = \Gamma(2.5) = 1.33$$

Therefore if the fractional differential operator is chosen in which the fractional order of differentiation matches the power law scaling of the function, then the curvature is reduced to a constant and all the scaling information is contained in the order of the derivative and that constant. If the neutron plume is traveling through material with evolving heterogeneity, then a fractional divergence might account for the increased dispersive flux over larger range of measurement scale.

This argument sets the stage for writing neutron diffusion equation with fractional calculus. The above derivation and discussion completes the description of neutron diffusion equations in fractional calculus.

In deriving the fractional differential equations for neutron diffusion in an enclosed volume we argued the basis of taking a larger observation space for defining divergence. The non-local formation of the divergence thus gives the effect of macroscopic effects caused by velocity fluctuations, the coupling effect of nearby zonal neighborhood neutrons (refer Figure 4.3). This therefore is making the constitutive descriptive equations closer to reality. The classical integer order constitutive differential equations are approximations to everything as "point" quantity in time or in space. The classical integer order methods do not thus take into account the space history or time history and therefore cannot represent the natural laws close to reality. Fractional calculus does take of all these reality and therefore more appropriate for representation of natural phenomena. Refer Figure 4.3, as an out side observer let us try to visualize space squares as depicted in the figure. The squares without any neutrons in them look differently to an outside observer as compared to the squares with the neutrons. While observer sitting in the same squares will not notice the difference in the squares with or without neutrons. So the observer in the same space will apply point quantity and will try to describe the neutron balance by classical integer order calculus. Whereas to the outside observer the squares or the space will appear transformed with or without presence of neutrons. The outside observer thus will apply this space transformation correction factor and obtain some different results, and that result will be close to reality.

4.9.1 Solution of Classical Constitutive Neutron Diffusion Equation (Integer Order)

This section will serve as revision to simple classical solution of the diffusion equation. Then in the next section we will solve the fractional differential equation obtained. This we will demonstrate for the space variables in one dimension for simplicity.

$$\mathbb{D}\nabla^2\phi - \Sigma_a\phi + S = \frac{dn}{dt} = \frac{1}{v}\frac{d\phi}{dt}$$
$$S = k_\infty \Sigma_a \phi$$
$$\mathbb{D}\nabla^2\phi + (k_\infty - 1)\Sigma_a\phi = \frac{1}{v}\frac{d\phi}{dt}$$

The flux term is variable of space and time. The source term multiplication law governs S. The separation of variables will give following for the flux term which can be substituted in the basic constitutive equation and following expressions will emerge.

$$\phi = \phi(r)e^{-\Lambda t}$$
$$\mathbb{D}\nabla^2\phi(r) + (k_\infty - 1)\Sigma_a\phi(r) = \frac{-\Lambda}{v}\phi(r)$$

Λ: is positive for sub critical and negative for super critical and zero for critical equilibrium reactor. The space coordinate r, we replace by x and with substitution of B as geometric buckling we get following simple form. The temporal solution is avoided for simplicity.

$$B^2 = \frac{(k_\infty - 1)\Sigma_a + \frac{\Lambda}{v}}{\mathbb{D}}$$
$$\frac{d^2\phi(x)}{dx^2} + B^2\phi(x) = 0$$

Here we can apply standard Laplace method with initial conditions at $x = 0$ at the center point of the reactor geometry having constant flux and at the walls at $x = a$ zero flux. General Laplace formula for derivative of function is indicated below and is applied to have polynomial form. Here p is space-Laplace variable.

$$p^2\Phi(p) - \sum_{k=0}^{1} p^k \frac{d^{2-k-1}\phi(x)}{dx^{2-k-1}}]_{@\,x=0} + B^2\Phi(p) = 0$$
$$p^2\Phi(p) - \frac{d\phi(x)}{dx}]_{@\,x=0} - p\phi(x)]_{@\,x=0} + B^2\Phi(p) = 0$$

$$\frac{d\phi(x)}{dx}]_{@\,x=0} = 0 \text{ and } \phi(x)]_{@\,x=0} = C$$

The above initial condition gives simple equation as:

$$p^2\Phi(p) - pC + B^2\Phi(p) = 0$$

$$\Phi(p) = \frac{pC}{p^2 + B^2}$$

Taking inverse $\phi(x) = C\cos Bx$.

4.9.2 Solution of Fractional Divergence Based Neutron Diffusion Equation (Fractional Order)

With the extension of the above method we try to solve the fractional differential equation:

$$\frac{d^\beta \phi(x)}{dx^\beta} + B^2\phi(x) = 0$$

$$1 < \beta < 2$$

$$\mathcal{L}\left(\frac{d^\alpha f(x)}{dx^\alpha}\right) = p^\alpha F(p) - \sum_{k=0}^{n-1} p^k \frac{d^{\alpha-k-1} f(x)}{d^{\alpha-k-1}}]_{@\,x=0}$$

$$p^\beta \Phi(p) - \frac{d^{\beta-1}\phi(x)}{dx^{\beta-1}}]_{@\,x=0} - p\frac{d^{\beta-2}\phi(x)}{dx^{\beta-2}}]_{@\,x=0} + B^2\Phi(p) = 0$$

The above is Laplace Transformation for LHD definition of the fractional derivative Note that in the expression fractional derivative of the flux as initial condition appear, which currently is tough to visualize and relate physically. Let us try to make use of Laplace transformation of RHD Caputo definition, as given below:

$$\mathcal{L}\left(\frac{d^\alpha}{dx^\alpha} f(x)\right) = p^\alpha F(p) - \sum_{k=0}^{n-1} p^{\alpha-k-1} \frac{d^k}{dx^k} f(x)_{@\,x=0}$$

$$p^\beta \Phi(p) - p^{\beta-1}\phi(x)_{@\,x=0} - p^{\beta-2}\frac{d}{dx}\phi(x)_{@\,x=0} + B^2\Phi(p) = 0$$

The above expression we relate physically to the earlier initial condition taking second term as C and third term as zero as done earlier. Here the integer order derivative comes as initial condition, therefore physically realizable from measurements and observations.

$$\Phi(s) = \frac{p^{\beta-1}C}{p^{\beta} + B^2}$$

$$\phi(x) = C\mathcal{L}^{-1}\left[\frac{p^{\beta-1}}{p^{\beta} + B^2}\right]$$

The solution of the fractional differential equation for the constitutive neutron balance equation therefore is with Laplace identity

$$\mathcal{L}\left\{E_{\alpha}(-\lambda t^{\alpha})\right\} = \frac{p^{\alpha-1}}{p^{\alpha} + \lambda},$$

we obtain:

$$\phi(x) = CE_{\beta}(-B^2 x^{\beta}) = C\sum_{k=0}^{\infty}\frac{(-B^2 x^{\beta})^k}{\Gamma(\beta k + 1)}$$

$$\phi(x) = C + C\frac{(-B^2 x^{\beta})}{\Gamma(\beta + 1)} + C\frac{(B^4 x^{2\beta})}{\Gamma(2\beta + 1)} + C\frac{(-B^6 x^{3\beta})}{\Gamma(3\beta + 1)} + \ldots$$

The above is flux mapping obtained by the solution of fractional order neutron constitutive equations obtained by concept of fractional divergence.

Let us see what classical flux pattern and fractional order flux pattern are same when we take the fractional order equal to 2 the integer order.

Solution in the classical form is cosine function and series representation of the same is:

$$\cos(x) = 1 - \frac{x^2}{2!} + \frac{x^4}{4!} - \frac{x^6}{6!} + \ldots$$

$$\beta = 2, \phi_{\beta=2}(x) = CE_{\beta}(-B^2 x^2) = C\sum_{k=0}^{\infty}\frac{(-B^2 x^2)^k}{\Gamma(2k + 1)} = C\left[1 - \frac{(B^2 x^2)}{\Gamma(3)} + \frac{(B^2 x^2)^2}{\Gamma(5)} - \frac{(B^2 x^2)^3}{\Gamma(7)} + \ldots\right]$$

$$\Gamma(n + 1) = n!$$

$$\phi_{\beta=2}(x) = C\left[1 - \frac{(Bx)^2}{2!} + \frac{(Bx)^4}{4!} - \frac{(Bx)^6}{6!} + \ldots\right] \approx C.\cos(Bx)$$

Therefore when the fractional order equals the integer order we get classical flux profile. This is proof of our assumption that indeed neutron flux being not a point quantity be represented as fractional divergence of order less than unity.

4.9.3 Fractional Geometrical Buckling and Non-point Reactor Kinetics

The above concept of fractional divergence gave a deviation from ideal flux map (cosine). The term geometrical buckling is indicative of the flux profile of neutron flux inside the reactor. Measuring the actual flux distribution and then controlling the power of reactor is one mode of reactor control. Now if the control computer is kept with a map of cosine table and the neutron spatial detectors are mapping in each control cycle a deviation then the unwarranted correction cycles will keep the control devises moving. Actually the correction may not be called for if the control computer is programmed with actual fractional geometrical buckling data. The fractional divergence has given the new thought of 'fractional geometrical buckling', which in turns when, used with basic multiplying factor k_∞ gives rise to a concept of fractional criticality. The describing reactor kinetics with fractional divergence will give concept of non-point kinetic description. The concept of fractional criticality is described in following section.

4.9.4 Fractional Reactor Kinetic Equation

We assume that the diffusion process is carried out exclusively in the moderator m, which has characteristic length l_m (distance between the two fuel rods of Figure 4.3); and say the characteristic length of the system is L_s (example fuel assembly length). Then the Fickian and non-Fickian cases are valid as long as length scale restrictions are met that is, $l_m << L_s$. In the light water systems mean free path of thermal neutrons is 1cm, comparable to fuel pin diameter and on the orders of centimeters for fast neutrons. The relaxation times are $\tau_m \equiv 3v^{-1}\mathbb{D}_m$ and order of magnitude of characteristic time constant may be defined as $t_d \equiv L_\phi^2 / v\mathbb{D}_m$, where L_ϕ^2 are characteristic lengths associated with 'changes' in the neutron flux. It can be observed that $(\tau_m / t_d) \cong 3\mathbb{D}_m^2 / L_\phi^2$. Since $\mathbb{D}_m^2 / L_\phi^2 << 1$, the time scale restriction is thus $(\tau_m / t_d) << 1$, in fractional form we may write $(\tau_m / t_d)^\alpha << 1$. In generalizing the neutron diffusion in heterogeneous medium with relaxation times included, we have obtained following type of equation as:

$$\tau_m^\alpha \frac{1}{v} \frac{\partial^{\alpha+1}\phi_m}{\partial t^{\alpha+1}} + \tau_m^\alpha \Sigma_{am} \frac{\partial^\alpha \phi_m}{\partial t^\alpha} + \frac{1}{v}\frac{\partial \phi_m}{\partial t} + \Sigma_{am}\phi_m = \mathbb{D}_m \nabla^2 \phi_m .$$

Including source term as

$$S + \tau^k \frac{\partial^q S}{\partial t^q},$$

putting in above and dropping the subscripts m, we write generalized equation as:

$$\frac{\tau^q}{v} \frac{\partial^{q+1}\phi}{\partial t^{q+1}} + \tau^q \Sigma_a \frac{\partial^q \phi}{\partial t^q} + \frac{1}{v}\frac{\partial \phi}{\partial t} = S - \Sigma_a \phi + \mathbb{D}\nabla^2\phi + \tau^q \frac{\partial^q S}{\partial t^q},$$

the six group source term is approximated as

$$S = (1-\beta)k_\infty \Sigma_a \phi + \sum_{i=1}^{6} \lambda_i C_i \,,$$

where the first term represents the production of prompt neutrons and second term is total rate of formation of delayed neutrons. Using this source (rather internal source) in the above fractional generalized equation we obtain.

$$\frac{\tau^\alpha}{v}\frac{\partial^{\alpha+1}\phi}{\partial t^{\alpha+1}} + \tau^\alpha[\Sigma_a + (1-\beta)k_\infty \Sigma_a]\frac{\partial^\alpha \phi}{\partial t^\alpha} + \frac{1}{v}\frac{\partial \phi}{\partial t}$$
$$= [(1-\beta)k_\infty \Sigma_a - \Sigma_a]\phi + \mathbb{D}\nabla^2\phi + \sum_{i=1}^{6}\lambda_i C_i + \tau^\alpha \sum_{i=1}^{6}\left(\lambda_i \frac{\partial^\alpha C_i}{\partial t^\alpha}\right)$$

The Laplacian $\nabla^2\phi$ can be replaced by Geometrical Buckling $-B_g^2\phi$, above can be expressed as

$$\frac{\tau^\alpha}{v}\frac{\partial^{\alpha+1}\phi}{\partial t^{\alpha+1}} + \tau^\alpha[\Sigma_a + (1-\beta)k_\infty \Sigma_a]\frac{\partial^\alpha \phi}{\partial t^\alpha} + \frac{1}{v}\frac{\partial \phi}{\partial t}$$
$$= \left[(1-\beta)k_\infty \Sigma_a - \Sigma_a - \mathbb{D}B_g^2\right]\phi + \sum_{i=1}^{6}\lambda_i C_i + \tau^\alpha \sum_{i=1}^{6}\left(\lambda_i \frac{\partial^\alpha C_i}{\partial t^\alpha}\right)$$

We can use separation of variables and write $\phi \equiv \phi(t)\psi(r)$ and $C_i \equiv C_i(t)\psi(r)$, where $\psi(r)$ represents fundamental spatial mode of steady state system. Using $\phi(t) = vn(t)$, we write time dependent neutron population dynamics as:

$$\tau^\alpha \frac{d^{\alpha+1}n(t)}{dt^{\alpha+1}} + \tau^\alpha[\Sigma_a + (1-\beta)k_\infty \Sigma_a]v\frac{d^\alpha n(t)}{dt^\alpha} + \frac{dn(t)}{dt}$$
$$= \left[(1-\beta)k_\infty \Sigma_a - \Sigma_a - \mathbb{D}B_g^2\right]vn(t) + \sum_{i=1}^{6}\lambda_i C_i + \tau^\alpha \sum_{i=1}^{6}\left(\lambda_i \frac{d^\alpha C_i}{dt^\alpha}\right)$$

Using, standard reactor physics notations as:

$$L^2 \equiv \mathbb{D}/\Sigma_a \,, l \equiv \left[v\Sigma_a(1+L^2B_g^2)\right]^{-1}, \Lambda \equiv 1/(k_\infty \Sigma_a v)\,, k_{eff} \equiv k_\infty/(1+L^2B_g^2)\,,$$

$$\rho \equiv (k_{eff}-1)/k_{eff}$$

We get,

$$\tau^\alpha \frac{d^{\alpha+1}n}{dt^{\alpha+1}} + \tau^\alpha\left[\frac{1}{l} + \frac{(1-\beta)}{\Lambda}\right]\frac{d^\alpha n}{dt^\alpha} + \frac{dn}{dt} = \frac{\rho-\beta}{\Lambda}n + \sum_{i=1}^{6}\lambda_i C_i + \tau^\alpha\left(\sum_{i=1}^{6}\lambda_i \frac{d^\alpha C_i}{dt^\alpha}\right)$$

Here ρ is reactivity, l is prompt neutron lifetime for finite media, Λ is neutron generation time, β is fraction of delayed neutron precursors, λ_i is radioactive-decay constant for delayed neutron precursor, C_i is concentration of delayed neutron precursor, and τ is relaxation time. The fractional-order $0<\alpha<1$ gives sub-diffusion and $1<\alpha<2$ gives super diffusion. While $\tau^{\alpha} \to 0$ recovers classical reactor point kinetics

$$\frac{dn}{dt}=\frac{\rho-\beta}{\Lambda}n+\sum_{i=1}^{6}\lambda_i C_i+\tau^{\alpha}\left(\sum_{i=1}^{6}\lambda_i\frac{d^{\alpha}C_i}{dt^{\alpha}}\right).$$

The fractional order term $d^{\alpha+1}n/dt^{\alpha+1}$ is important contribution for rapid changes in neutron density, $n(t)$ for example in case of reactor trip, turbine trip, while $d^{\alpha}n/dt^{\alpha}$ contributes when changes in neutron density is relatively slow example during start up, and power maneuvering by control rods. The term $d^{\alpha}C_i/dt^{\alpha}$ becomes significantly of importance while in case of shut-down process, and also to understand ADS (Accelerator Driven Sub-criticality systems); that is sub-critical state, characterized by low fraction of delayed neutron and by 'small' Doppler reactivity coefficients.

This is a kind of Fractional Differential Equation, obtained by generalizing the Catteneo's diffusion where relaxation times are taken. However, there can be other type of generalization considering either infinite jump lengths of neutrons, or by considering infinite wait times between collisions as were discussed.

The previous section gave boundary value problem, and thereby obtained was spatial flux distribution inside the nuclear reactor. Here we modify the point kinetic equations and call them Fractional coupled kinetic equation; the solution to this gives the time evolution of neutron population. Following is a case where infinite wait-time statistic gives fractional rate change in point kinetic, which is solved. The reactor fractional kinetic equations are given by following pair of fractional differential equations.

$$\frac{1}{v_c}\frac{\partial^{\alpha}}{\partial t^{\alpha}}\phi(x,t)=\mathbb{D}\nabla\phi(x,t)+\left(\gamma\Sigma_f-\Sigma_a\right)\phi(x,t)+\lambda C(x,t)$$

$$\frac{\partial^{\alpha}}{\partial t^{\alpha}}C(x,t)=\beta\gamma\Sigma_f\phi(x,t)-\lambda C(x,t)$$

Where $0<\alpha<1/2$, and initial condition is $\phi(x,0)=\phi_0(x)$

The parameters are:

v_c : Neutron velocity

ϕ : Neutron flux.

C : Density of precursors

$\mathbb{D}$: Neutron diffusion coefficient

γ : Average number of neutrons produced per fission

Σ_f : Macroscopic fission cross-section

Σ_a : Macroscopic absorption cross-section
λ : Radioactive decay constant
β : Delayed neutron fraction that is fraction of the fission neutrons that are delayed

Rewriting the above two equations by replacing with $B = \beta\gamma\Sigma_f$ and $\Sigma = \gamma\Sigma_f - \Sigma_a$, we get:

$$\frac{1}{v_c}\frac{\partial^\alpha}{\partial t^\alpha}\phi(x,t) = \mathbb{D}\nabla^2\phi(x,t) + \Sigma\phi(x,t) + \lambda C(x,t)$$

$$\frac{\partial^\alpha}{\partial t^\alpha}C(x,t) = B\phi(x,t) - \lambda C(x,t)$$

Rearranging from the first equation of above we obtain $C(x,t)$ as

$$C(x,t) = \frac{1}{\lambda v_c}\frac{\partial^\alpha}{\partial t^\alpha}\phi(x,t) - \frac{\mathbb{D}}{\lambda}\nabla^2\phi(x,t) - \frac{\Sigma}{\lambda}\phi(x,t)$$

and then using this in second equation we get:

$$\frac{\partial^{2\alpha}}{\partial t^{2\alpha}}\phi(x,t) = \mathbb{D}v_c\frac{\partial^\alpha\left(\nabla^2\phi(x,t)\right)}{\partial t^\alpha} + \Sigma v_c\frac{\partial^\alpha\phi(x,t)}{\partial t^\alpha} - \lambda\frac{\partial^\alpha\phi(x,t)}{\partial t^\alpha}$$
$$+ + \lambda B v_c\phi(x,t) + \lambda\mathbb{D}v_c\nabla^2\phi(x,t) + \lambda v_c\Sigma\phi(x,t)$$

The solution of this is demonstrated in Chapter 11, by decomposition technique, to get analytical approximate solution as series

$$\phi(x,t) = \sum_{i=0}^{\infty}\phi_i(x,t),$$

the components are:

$$\phi_0(x,t) = \phi(x,0) = \phi_0(x)$$

$$\phi_1(x,t) = M_1(x)\frac{t^\alpha}{\Gamma(\alpha+1)} + M_2(x)\frac{t^{2\alpha}}{\Gamma(2\alpha+1)}$$

$$\phi_3(x,t) = N_1(x)\frac{t^{2\alpha}}{\Gamma(2\alpha+1)} + N_2(x)\frac{t^{3\alpha}}{\Gamma(3\alpha+1)} + N_3(x)\frac{t^{4\alpha}}{\Gamma(4\alpha+1)}$$

$$\phi_3(x,t) = G_1(x)\frac{t^{3\alpha}}{\Gamma(3\alpha+1)} + G_2(x)\frac{t^{4\alpha}}{\Gamma(4\alpha+1)} + G_3(x)\frac{t^{5\alpha}}{\Gamma(5\alpha+1)} + G_4(x)\frac{t^{6\alpha}}{\Gamma(6\alpha+1)}$$

and so on.

Here the functions of x appearing as above are listed as follows:

$$M_1(x) = \mathbb{D}v_c\nabla^2\phi_0(x) + \Sigma v_c\phi_0(x) - \lambda\phi_0(x)$$
$$M_2(x) = \lambda Bv_c\phi_0(x) + \mathbb{D}\lambda v_c\nabla^2\phi_0(x) + \lambda v_c\Sigma\phi_0(x)$$
$$N_1(x) = \mathbb{D}v_c\nabla^2 M_1(x) + \Sigma v_c M_1(x) - \lambda M_1(x)$$
$$N_2(x) = \mathbb{D}v_c\nabla^2 M_2(x) + \Sigma v_c M_2(x) - \lambda M_2(x) + \lambda Bv_c M_1(x) + \mathbb{D}\lambda v_c\nabla^2 M_1(x) + \lambda v_c\Sigma M_1(x)$$
$$N_3(x) = \lambda Bv_c M_2(x) + \mathbb{D}\lambda v_c\nabla^2 M_2(x) + \lambda v_c\Sigma M_2(x)$$
$$G_1(x) = \mathbb{D}v_c\nabla^2 N_1(x) + \Sigma v_c N_1(x) - \lambda N_1(x)$$
$$G_2(x) = \mathbb{D}v_c\nabla^2 N_2(x) + \Sigma v_c N_2(x) - \lambda N_2(x) + \lambda Bv_c N_1(x) + \mathbb{D}\lambda v_c\nabla^2 N_1(x) + \lambda v_c\Sigma N_1(x)$$
$$G_3(x) = \mathbb{D}v_c\nabla^2 N_3(x) + \Sigma v_c N_3(x) - \lambda N_3(x) + \lambda Bv_c N_2(x) + \mathbb{D}\lambda v_c\nabla^2 N_2(x) + \lambda v_c\Sigma N_2(x)$$
$$G_4(x) = \lambda Bv_c N_3(x) + \mathbb{D}\lambda v_c\nabla^2 N_3(x) + \lambda v_c\Sigma N_3(x)$$

The solution in time for flux growth is summation of all these flux components obtained.

4.9.5 Growth of Neutron Flux with Time for Different Values of Fractional Orders and Fractional Criticality

The solution of the fractional reactor point kinetics is shown in the previous section as series solution (as initial value problem). In the present analysis we shall use the above obtained series solution to present the nature of the neutron flux for different values of fractional order.

For $v_c = 220{,}000\text{cm}/\text{s}$, $B = 000735\text{cm}^{-1}$, $\mathbb{D} = 0.356\text{cm}^2\text{s}^{-\alpha}$, $\lambda = 0.08\text{s}^{-1}$, $\Sigma = 0.005\text{cm}^{-1}$, with initial condition as: $\phi(x,0) = \phi(x) = 1.0$, the following Table 4.1 gives the values with respect to time.

Table 4.2 Neutron flux $\phi(t)$, growth for different fractional order values

Time (s)	$\alpha = 0.1$	$\alpha = 0.2$	$\alpha = 0.3$	$\alpha = 0.4$	$\alpha = 0.5$
0.00010	1.04535×10^8	6.21599×10^6	3.58370×10^5	2.00988×10^4	1.13743×10^3
0.00039	1.59622×10^8	1.42279×10^7	1.22266×10^6	1.01416×10^5	8.23638×10^3
0.00068	1.89937×10^8	1.99782×10^7	2.02159×10^6	1.97079×10^5	1.86944×10^4
0.00097	2.12264×10^8	2.48265×10^7	2.78898×10^6	3.01545×10^5	3.16326×10^4
0.00126	2.3039×10^8	2.91405×10^7	2.78898×10^6	3.01545×10^5	3.16326×10^4
0.00155	2.45852×10^8	3.3087×10^7	4.26719×10^6	5.29060×10^5	6.34629×10^4
0.00184	2.59449×10^8	3.67595×10^7	4.98668×10^6	6.50071×10^5	8.19238×10^4
0.00213	2.71652×10^8	4.02168×10^7	5.69632×10^6	7.75079×10^5	1.01890×10^5
0.00242	2.8277×10^8	4.34989×10^7	6.39764×10^6	9.03656×10^5	1.23257×10^5
0.00271	2.93014×10^8	4.66344×10^7	7.09176×10^6	1.03547×10^6	1.45939×10^5
0.00300	3.02536×10^8	4.96447×10^7	7.77958×10^6	1.17024×10^6	1.69866×10^5

From observations of Table 4.2, the inference is drawn that neutron flux multiplication can be obtained at fractional order values at $\alpha > 0.3$. The neutron flux growths for the lower values of $\alpha < 0.3$ have saturation tendency with time. The rate of flux growth tends to decrease with time. Whereas for fractional order values $\alpha > 0.3$ the rate of neutron flux growth is either positive or increases with time. This implies that there is possibility of nuclear reactor achieving the desired neutron multiplication factor k_∞ or criticality at fractional values indicating concept of 'fractional criticality'.

4.10 Concept of Fractional Curl in Electromagnetics

Fractional curl operator has been utilized to find the new set of solutions to Maxwell's equations by fractionalizing the principle of duality. New set of solutions are named as fractional dual solutions to the Maxwell's equation.

4.10.1 Concept of Chirality

Chirality is 'handedness', the term chiral describes an object, especially a molecule which has or produces a non-superimposeable mirror image of it itself. In chemistry, such a molecule is called an enantiomer or is said to exhibit 'chirality' enatiomerism. The term "chiral" comes from the Greek word for the human hand, which itself exhibits such non-superimposeability of the left hand precisely over the right. Due to the opposition of the fingers and thumbs, no matter how the two hands are oriented, it is impossible for both hands to exactly coincide-Helices, chiral characteristics (properties), chiral media, order, and symmetry all relate to the concept of left- and right-handedness.

Wave propagation as handedness is wave polarization and described in terms of helicity (occurs as a helix). Electrical polarization occurs perpendicular to the direction of propagation. If during the wave propagation through material the electric vector rotates clockwise then the material is right-handed, and if rotation occurs counterclockwise then material is left-handed. Historically, the orientation of a polarized electromagnetic wave has been defined for light by the orientation of the electric vector, and for radio waves, by the orientation of magnetic vector.

4.10.2 Duality of Solutions

The electromagnetic theory is based on principle of duality. Dual solutions of electromagnetic means any solution to a problem containing electric source can be converted into a dual solution to the problem containing a magnetic source. Duality does arise in circuit theory as circuit of Thevenin voltage source can be converted to Norton current source. In electromagnetic the set $(E, H, D, B, \mu, \varepsilon)$ has dual solution with set $(H, -E, B, -D, \varepsilon, \mu)$.

This can be demonstrated by example as replacing $E \rightarrow H$, $H \rightarrow -E$, $\mu \rightarrow \varepsilon$, and $\varepsilon \rightarrow \mu$ in $\nabla \times E = -\mathrm{j}\omega H$ and $\nabla \times H = \mathrm{j}\omega E$, keeps the two Maxwell equations same. The complex notation $\mathrm{j} = \sqrt{-1}$, in $\mathrm{j}\omega$ is used for representing time-rate change, whereas, complex notation $\mathrm{i} = \sqrt{-1}$, in $\mathrm{i}k$ is used for representing 'spatial rate of change' specifically spatial harmonics.

4.10.3 Fractional Curl Operator

Fractional curl operator has been utilized to find the new set of solutions to Maxwell's equations by fractionalizing the principle of duality. In electromagnetic the principle of duality states that if $(E, \eta H)$ is one set solutions (original) to Maxwell equations, then other set of solutions (dual to original) is $(\eta H, -E)$, where η is the impedance of the medium. The solution that may be regarded as intermediate step between the original and dual to the original solutions may be obtained using the following relationship.

$$E_{fd} = \frac{1}{(\mathrm{j}k)^{\alpha}} (\nabla \times)^{\alpha} E \text{ and } \eta H_{fd} = \frac{1}{(\mathrm{j}k)^{\alpha}} (\nabla \times)^{\alpha} H$$

Where $(\nabla \times)^{\alpha}$ means the fractional curl operator and $k = 2\pi f \sqrt{\mu\varepsilon} = \omega\sqrt{\mu\varepsilon}$ is the wave number of the medium. The subscripted E_{fd}, H_{fd} notations mean fractional dual solutions. In the above definition of fractional curl operator the factor $(\mathrm{j}k)^{\alpha}$ is used to normalize the result, since fractional derivatives of periodic (harmonic) signals will return this after derivative operation (Chapter 5 and10 gives detailed account of the fractional derivative of periodic signals).

4.10.4 Wave Propagation in Unbounded Chiral Medium

In a chiral medium electric flux and magnetic flux densities is composite quantity and represented as

$$\mathbf{D} = \varepsilon\left[E + \beta(\nabla \times E)\right] \text{ and } \mathbf{B} = \mu\left[H + \beta(\nabla \times H)\right]$$

Therefore the D (and B) at any point x depends on the electric field at other point x. This spatially dispersive property is non-local property and here the fractional calculus is used. The factor β is the chirality property of the media.

Consider a uniform plane wave propagating in z-direction in an unbounded loss less isotropic chiral medium. According to field decomposition field quantities E and H may be thought of two plane waves i.e. (E_{+}, H_{+}) and (E_{-}, H_{-}), that are called 'wave-fields'. The electric fields corresponding to two-wave fields are:

$$E_{\pm}(z) = E_{\pm}(0)\exp(ik_{\pm}z),$$

where $k_{\pm} = k(1 \pm \kappa_r)$ is the wave number of the two wave fields. $k = \omega\sqrt{\mu\varepsilon}$ and $\kappa_r = \kappa\sqrt{\mu_0\varepsilon_0 / \mu\varepsilon}$. ($\kappa$ is 'chirality' parameter). Using the following relation

$$\eta_{\pm}H_{\pm}(z) = \pm iE_{\pm}(z)$$

Corresponding magnetic field may be obtained. In the above expression

$$\eta_{\pm} = \sqrt{\frac{\mu_{\pm}}{\varepsilon_{\pm}}} = \eta .$$

This means each wave field sees media with equivalent constitutive parameters as (ε_+, μ_+) and (ε_-, μ_-). Also, each wave-field sees 'chiral medium' as 'achiral medium', with equivalent parameters as (ε_+, μ_+) and (ε_-, μ_-).

Medium parameters of the equivalent isotropic media are related to the parameters of this chiral medium by

$$\varepsilon_{\pm} = \varepsilon(1 \pm \kappa_r) \text{ and } \mu_{\pm} = \mu(1 \pm \kappa_r)$$

Simple expressions of wave fields can be written as

$$E_+ = \frac{1}{2}(E - j\eta H)$$

$$E_- = \frac{1}{2}(E + j\eta H)$$

$$H_+ = \frac{1}{2}\left(H + j\frac{E}{\eta}\right)$$

$$H_- = \frac{1}{2}\left(H - j\frac{E}{\eta}\right)$$

The total field in chiral medium, the original field is:

$$E(z) = E_+^{i}(0)\exp(ik_+z) + E_-^{i}(0)\exp(ik_-z)$$
$$\eta H(z) = i[E_+^{i}(0)\exp(ik_+z) - E_-^{i}(0)\exp(ik_-z)] .$$

Fractionalizing the electric field $E_+(z)$ *and* $E_-(z)$ we get

$$E_{fd+}(z) = \frac{1}{(ik_+)^{\alpha}}(\nabla\times)^{\alpha}E_+(z)$$

$$= \frac{1}{(\mathrm{i}k_+)^\alpha}\left\{(z\times)^\alpha E_+^{\mathrm{i}}(0)\right\}\left\{\frac{d^\alpha}{dz^\alpha}\exp(\mathrm{i}k_+ z)\right\}$$

$$= E_+^{\mathrm{i}}(0)\exp\left\{\mathrm{i}\left(k_+ z + \frac{\alpha\pi}{2}\right)\right\}$$

Similarly

$$E_{fd-}(z) = E_-^{\mathrm{i}}(0)\exp\left\{\mathrm{i}\left(k_- z - \frac{\alpha\pi}{2}\right)\right\}$$

Fractional dual fields corresponding to the original field may be written as

$$E_{fd}(z) = E_+^{\mathrm{i}}(0)\exp\left\{\mathrm{i}\left(k_+ z + \frac{\alpha\pi}{2}\right)\right\} + E_+^{\mathrm{i}}(0)\exp\left\{\mathrm{i}\left(k_- z - \frac{\alpha\pi}{2}\right)\right\}$$

$$\eta H_{fd}(z) = \mathrm{i}\left[E_+^{\mathrm{i}}(0)\exp\left\{\mathrm{i}\left(k_+ z + \frac{\alpha\pi}{2}\right)\right\} - E_-^{\mathrm{i}}(0)\exp\left\{\mathrm{i}\left(k_- z - \frac{\alpha\pi}{2}\right)\right\}\right]$$

It is obvious that for $\alpha = 0$ $E_{fd}(z) = E(z)$ and $\eta H_{fd}(z) = \eta H(z)$.

For $\alpha = 1$

$$E_{fd}(z) = \eta H(z) \text{ and } \eta H_{fd}(z) = -E(z)$$

which is consistent with electromagnetic principle of duality. For any value in between solutions may be regarded as intermediate between the original and dual to the original solutions.

4.10.5 Reflection in Chiral Medium

Consider a plane wave hits a boundary normally a chiral to chiral interface, at $z = 0$, the intrinsic impedance before the interface is η and after the interface is η_1. The total electric field of the wave fields before the interface is:

$$E_\pm(z) = E_\pm^{\mathrm{i}}(0)\exp\left(\mathrm{i}k_\pm z\right) + R_{\mp\pm}E_\pm^{\mathrm{i}}(0)\exp\left(\mathrm{i}k_\mp z\right)$$

Consider $R_\mp$ as the reflection coefficient for the negative incidence wave field and positive reflected wave field; and $R_\pm$ is the reflection coefficient for positive incidence and negative reflected wave field. For reciprocal chiral medium the reflection coefficient becomes

$$R_{\mp\pm} = R = \frac{\eta_1 - \eta}{\eta_1 + \eta}.$$

So we can write:

$$\begin{aligned} E_\pm(z) &= E_\pm^{\mathrm{i}}(0)\exp\left(\mathrm{i}k_\pm z\right) - RE_\pm^{\mathrm{i}}(0)\exp\left(-\mathrm{i}k_\mp z\right) \\ &= E_\pm^{\mathrm{i}}(0)\left\{\exp(\mathrm{i}k_\pm z) + R.\exp(-\mathrm{i}k_\mp z)\right\} \end{aligned}$$

The corresponding magnetic field is:

$$\eta H_\pm(z) = \pm \mathrm{i}E_\pm^{\mathrm{i}}(0)\left\{\exp(\mathrm{i}k_\pm z) - R.\exp(-\mathrm{i}k_\mp z)\right\}$$

Using these expressions the total field is written as:

$$\begin{aligned} E(z) &= E_+^{\mathrm{i}}(0)\left\{\exp(\mathrm{i}k_+ z) + R.\exp(-\mathrm{i}k_- z)\right\} + E_-^{\mathrm{i}}(0)\left\{\exp(\mathrm{i}k_- z) + R.\exp(-\mathrm{i}k_+ z)\right\} \\ &= E_1 + E_2 + E_3 + E_4 \\ \eta H(z) &= \mathrm{i}E_+^{\mathrm{i}}(0)\left\{\exp(\mathrm{i}k_+ z) - R.\exp(-\mathrm{i}k_- z)\right\} - \mathrm{i}E_-^{\mathrm{i}}(0)\left\{\exp(\mathrm{i}k_- z) - R.\exp(-\mathrm{i}k_+ z)\right\} \\ &= \eta H_1 + \eta H_2 + \eta H_3 + \eta H_4 \end{aligned}$$

The fractional dual solution to each of the field is given as:

$$E_{1fd} = \frac{1}{\left(\mathrm{i}k_+\right)^\alpha}(\nabla\times)^\alpha\left\{E_+^{\mathrm{i}}(0)\exp(\mathrm{i}k_+ z)\right\} = (\mathrm{i})^\alpha E_+^{\mathrm{i}}(0)\exp(\mathrm{i}k_+ z)$$

The other components are:

$$\begin{aligned} E_{2fd} &= -(\mathrm{i})^\alpha(-1)^\alpha E_+^{\mathrm{i}}(0)\exp(-\mathrm{i}k_- z) \\ E_{3fd} &= (-\mathrm{i})^\alpha E_-^{\mathrm{i}}(0)\exp(\mathrm{i}k_- z) \\ E_{4fd} &= -(-\mathrm{i})^\alpha(-1)^\alpha E_-^{\mathrm{i}}\exp(-\mathrm{i}k_+ z) \\ \eta H_{ifd} &= \mathrm{i}(\mathrm{i})^\alpha E_+^{\mathrm{i}}(0)\exp(\mathrm{i}k_+ z) \\ \eta H_{2fd} &= \mathrm{i}(\mathrm{i})^\alpha(-1)^\alpha E_+^{\mathrm{i}}(0)\exp(-\mathrm{i}k_- z) \\ \eta H_{3fd} &= -\mathrm{i}(-\mathrm{i})^\alpha E_-^{\mathrm{i}}(0)\exp(\mathrm{i}k_- z) \\ \eta H_{4fd} &= -\mathrm{i}(-\mathrm{i})^\alpha(-1)^\alpha E_-^{\mathrm{i}}(0)\exp(\mathrm{i}k_+ z) \end{aligned}$$

Using these expressions in the total field expressions of $E(z)$ and $\eta H(z)$ we get:

$$E_{fd}(z)=(\mathrm{i})^{\alpha}E_{+}^{\mathrm{i}}(0)\Big[\exp\{\mathrm{i}(k_{+}z)\}+R(-1)^{\alpha}\exp\{-\mathrm{i}(k_{-}z)\}\Big]$$
$$+(-\mathrm{i})^{\alpha}E_{-}^{\mathrm{i}}(0)\Big[\exp\{\mathrm{i}(k_{-}z)\}+R(-1)^{\alpha}\exp\{-\mathrm{i}(k_{+}z)\}\Big]$$
$$\eta H_{fd}(z)=\mathrm{i}(\mathrm{i})^{\alpha}E_{+}^{\mathrm{i}}(0)\Big[\exp\{\mathrm{i}(k_{+}z)\}-R(-1)^{\alpha}\exp\{-\mathrm{i}(k_{-}z)\}\Big]$$
$$-\mathrm{i}(-\mathrm{i})^{\alpha}E_{-}^{\mathrm{i}}(0)\Big[\exp\{\mathrm{i}(k_{-}z)\}-R(-1)^{\alpha}\exp\{-\mathrm{i}(k_{+}z)\}\Big]$$

On simplification we have:

$$E_{fd}(z)=\exp\left(\frac{\mathrm{i}\alpha\pi}{2}\right)\left\{\begin{array}{l}(\mathrm{i})^{\alpha}E_{+}^{\mathrm{i}}(0)\left[\exp\left\{\mathrm{i}\left(k_{+}z-\alpha\dfrac{\pi}{2}\right)\right\}+R.\exp\left\{-\mathrm{i}\left(k_{-}z-\alpha\dfrac{\pi}{2}\right)\right\}\right]\\ +(-\mathrm{i})^{\alpha}E_{-}^{\mathrm{i}}(0)\left[\exp\left\{\mathrm{i}\left(k_{-}z-\alpha\dfrac{\pi}{2}\right)\right\}+R.\exp\left\{-\mathrm{i}\left(k_{+}z-\alpha\dfrac{\pi}{2}\right)\right\}\right]\end{array}\right\}$$

$$\eta H_{fd}(z)=\mathrm{i}\exp\left(\frac{\mathrm{i}\alpha\pi}{2}\right)\left\{\begin{array}{l}(\mathrm{i})^{\alpha}E_{+}^{\mathrm{i}}(0)\left[\exp\left\{\mathrm{i}\left(k_{+}z-\alpha\dfrac{\pi}{2}\right)\right\}-R.\exp\left\{-\mathrm{i}\left(k_{-}z-\alpha\dfrac{\pi}{2}\right)\right\}\right]\\ -(-\mathrm{i})^{\alpha}E_{-}^{\mathrm{i}}(0)\left[\exp\left\{\mathrm{i}\left(k_{-}z-\alpha\dfrac{\pi}{2}\right)\right\}-R.\exp\left\{-\mathrm{i}\left(k_{+}z-\alpha\dfrac{\pi}{2}\right)\right\}\right]\end{array}\right\}$$

For $\alpha=0$ we get $E_{fd}(z)=E(z)$ and $\eta H_{fd}(z)=\eta H(z)$ which is original set of solutions. For $\alpha=1$ we get $E_{fd}(z)=\eta H(z)$ and $\eta H_{fd}(z)=-E(z)$ which is dual to the original set of solutions. Above expressions for fields with $0<\alpha<1$ may be regarded as fractional dual solutions corresponding to original set of solutions.

4.10.6 Transverse Wave Impedance

Transverse wave impedance is defined as

$$Z_{fdxy}=E_{fdx}/H_{fdy}=-E_{fdy}/H_{fdx},$$

where

$$E_{fdx}(z)=\left(e^{\mathrm{i}\alpha\pi/2}\right)\left\{\begin{array}{l}\left[(\mathrm{i})^{\alpha}E_{+}^{\mathrm{i}}(0)\left\{e^{\mathrm{i}(k_{+}z-\alpha(\pi/2))}+R.e^{-\mathrm{i}(k_{-}z-\alpha(\pi/2))}\right\}\right]\\ +\left[(-\mathrm{i})^{\alpha}E_{-}^{\mathrm{i}}(0)\left\{e^{\mathrm{i}(k_{-}z-\alpha(\pi/2))}+R.e^{-\mathrm{i}(k_{+}z-\alpha(\pi/2))}\right\}\right]\end{array}\right\}$$

$$\eta H_{fdy}(z)=\left(e^{\mathrm{i}\alpha\pi/2}\right)\left\{\begin{array}{l}\left[(\mathrm{i})^{\alpha}E_{+}^{\mathrm{i}}(0)\left\{e^{\mathrm{i}(k_{+}z-\alpha(\pi/2))}-R.e^{-\mathrm{i}(k_{-}z-\alpha(\pi/2))}\right\}\right]\\ +\left[(-\mathrm{i})^{\alpha}E_{-}^{\mathrm{i}}(0)\left\{e^{\mathrm{i}(k_{-}z-\alpha(\pi/2))}-R.e^{-\mathrm{i}(k_{+}z-\alpha(\pi/2))}\right\}\right]\end{array}\right\}$$

Using $k_{+}=k+k\kappa_{r}$ and $k_{-}=k-k\kappa_{r}$, we get

$$E_{fdx}(z) = \left(e^{\mathrm{i}\alpha\pi/2}\right)\left[e^{\mathrm{i}(kz-\alpha(\pi/2))} + R.e^{-\mathrm{i}(kz-\alpha(\pi/2))}\right]\left\{(\mathrm{i})^{\alpha} E_{+}^{\mathrm{i}}(0)e^{\mathrm{i}k\kappa_r z} + (-\mathrm{i})^{\alpha} E_{-}^{\mathrm{i}}(0)e^{-\mathrm{i}k\kappa_r z}\right\}$$

$$H_{fdy}(z) = \left(\frac{1}{\eta}e^{\mathrm{i}\alpha\pi/2}\right)\left[e^{\mathrm{i}(kz-\alpha(\pi/2))} - R.e^{-\mathrm{i}(kz-\alpha(\pi/2))}\right]\left\{(\mathrm{i})^{\alpha} E_{+}^{\mathrm{i}}(0)e^{\mathrm{i}k\kappa_r z} + (-\mathrm{i})^{\alpha} E_{-}^{\mathrm{i}}(0)e^{-\mathrm{i}k\kappa_r z}\right\}$$

Substituting these in the expression for transverse impedance we get

$$Z_{fdxy} = \eta\frac{\left[e^{\mathrm{i}(kz-\alpha(\pi/2))} + R.e^{-\mathrm{i}(kz-\alpha(\pi/2))}\right]}{\left[e^{\mathrm{i}(kz-\alpha(\pi/2))} - R.e^{-\mathrm{i}(kz-\alpha(\pi/2))}\right]}$$

For Perfect Electric Conductor (PEC) interface, above impedance expression yields $Z_{fdxy} = \mathrm{i}\eta\tan(kz - \alpha(\pi/2))$, and at $z = 0$, $Z_{fdxy} = \mathrm{i}\eta\tan(\alpha\pi/2)$
Similarly

$$Z_{fdyx} = -E_{fdy}/H_{fdx} = \mathrm{i}\eta\tan(kz - \alpha(\pi/2))$$

at $z = 0$, $Z_{fdyx} = -\mathrm{i}\eta\tan(\alpha\pi/2)$.

From above derivations, it is clear that for $\alpha = 0$, $Z_{fd} = 0$ at $z = 0$, which describes a situation as if PEC interface is placed at $z = 0$. For $\alpha = 1$, the transverse wave impedance is placed at $z = 0$ which gives impedance $Z_{fd} = \infty$, describing a situation as if Perfect Magnetic Conductor (PMC) interface is placed at $z = 0$. Hence, for $0 < \alpha < 1$, the wave impedance describes a surface located at $z = 0$, where impedance is in between PEC and PMC.

Consider a chiral slab which is in between two different chiral media, the front media with intrinsic impedance η_1 and rear media with impedance η_2. The width of middle chiral slab is L and its intrinsic impedance is η_3. The coordinate $z = 0$ is the front end of slab and $z = -L$ is the rear end of the slab. The transverse impedance for the fractional dual solution is:

$$Z_{1fdxy} = \eta_1\frac{\left[e^{\mathrm{i}(k_1 z-\alpha(\pi/2))} + R_1.e^{-\mathrm{i}(k_1 z-\alpha(\pi/2))}\right]}{\left[e^{\mathrm{i}(k_1 z-\alpha(\pi/2))} - R_1.e^{-\mathrm{i}(kz-\alpha(\pi/2))}\right]}, \text{ with } R_1 = \frac{Z_{2fdxy} - \eta_1}{Z_{2fdxy} + \eta_1}$$

$$Z_{2fdxy} = \eta_1\frac{\left[e^{\mathrm{i}(k_2 z-\alpha(\pi/2))} + R.e^{-\mathrm{i}(k_2 z-\alpha(\pi/2))}\right]}{\left[e^{\mathrm{i}(k_2 z-\alpha(\pi/2))} - R.e^{-\mathrm{i}(k_2 z-\alpha(\pi/2))}\right]}, \text{ with } R_2 = \frac{\eta_3 - \eta_2}{\eta_3 + \eta_2}$$

For a value $\alpha = 2k_2L/\pi$, the situation is as if there is no chiral slab. The above derivation is extended, for a case of chiral slab backed by PEC interface the result is:

$$Z_{1fdxy} = \eta_1 \frac{\left[e^{i(k_1 z-\alpha(\pi/2))} + R_1.e^{-i(k_1 z-\alpha(\pi/2))}\right]}{\left[e^{i(k_1 z-\alpha(\pi/2))} - R_1.e^{-i(kz-\alpha(\pi/2))}\right]}, \; R_1 = \frac{Z_{2fdxy} - \eta_1}{Z_{2fdxy} + \eta_1},$$

where $Z_{2fdxy} = i\eta_2 \tan(k_2 L - (\alpha\pi/2))$. It may be observed here, that for $\alpha = 0$, it is situation of slab backed by PEC interface, while for $\alpha = 1$, the slab is backed by a PMC interface. With $0 < \alpha < 1$ deals with fractional dual solution.

4.10.7 Propagation of Electromagnetic Waves in Bi-isotropic Medium

Solution to the Maxwell's Equations, for bi-isotropic media gives two circularly polarized waves. One of them is Right Circularly Polarized (RCP) and other is Left Circularly Polarized (LCP); traveling with different wave numbers k_+ and k_-, obviously with different phase velocities $v_{ph} = \omega / k$. Propagation of a plane wave through a plane parallel structure of bi-isotropic medium can be analyzed in terms of two non-interacting scalar Transmission Lines, with two eigen solutions waves, in the same way as simple isotropic media. The main difference is that in the case of bi-isotropic medium re-ected waves has different wave numbers k, different from the incident waves. This implies that LCP wave will become RCP wave upon re-ection from the interface, and vice-versa. Thus incident and re-ected waves will see different effective medium, therefore the corresponding Transmission Line becomes non-symmetric.

Thus for two circularly polarized TEM eigen waves depending only upon z component, the following are the Maxwell's Equations

$$\text{curl}\{\bar{E}_\pm(z)\} = -i\omega\mu_\pm \bar{H}_\pm(z),$$

and $\text{curl}\{\bar{H}_\pm(z)\} = i\omega\varepsilon_\pm \bar{E}_\pm(z)$, we can re-write these as

$$u_z \times E'_\pm(z) = -i\omega\mu_\pm H_\pm(z),$$

and $u_z \times H'_\pm(z) = i\omega\varepsilon_\pm E_\pm(z)$. The primed symbol, that are, $E'_\pm, H'_\pm$ represents the differentiation with respect to z. In terms of Circularly Polarized (CP) unit vector $u_\pm$ satisfying $u_z \times u_\pm = \pm i u_\pm$, we may write $E\pm = u_\pm E_\pm$ and $H\pm = u_\pm H_\pm$. Using these values, we may re-write the Maxwell Equations as:

$$E'_\pm(z) = -i\omega\mu_\pm\{\mp i H_\pm(z)\} \text{ and } \mp i H'_\pm(z) = -i\omega\varepsilon_\pm E_\pm(z)$$

The above represents Transmission Line Equations as $V'(z) = -i\omega L I(z)$ and $I'(z) = -i\omega C V(z)$, by identifying $V_\pm \equiv E_\pm$ and $I_\pm \equiv \mp i H_\pm$.

4.10.8 Fractional Non-symmetric Transmission Line

The positive and negative wave fields are represented by their respective voltage and current components. These components are added to get total voltage and current. For a boundary with reflection coefficient $R_{\pm\mp}$, we can write the wave fields as,

$$V_{\pm}(z) = V_{\pm}(0)e^{ik_{\pm}z} + R_{\mp\pm}V_{\pm}(0)e^{-ik_{\pm}z}$$

$$ZI_{\pm}(z) = V_{\pm}(0)e^{ik_{\pm}z} - R_{\mp\pm}V_{\pm}(0)e^{-ik_{\mp}z}$$

For reciprocal chiral medium

$$R_{\mp\pm} = R = \frac{Z_L - Z}{Z_L + Z},$$

so we write

$$V_{\pm}(z) = V_{\pm}(0)\{e^{ik_{\pm}z} + Re^{-ik_{\pm}z}\}$$

and

$$ZI_{\pm}(z) = V_{\pm}(0)\{e^{ik_{\pm}z} - Re^{-ik_{\mp}z}\},$$

now voltage and current can be written as

$$V_{fd}(z) = e^{i\alpha\pi}V_{+}(0)\left[e^{i(k_{+}z-\alpha\pi)} + R.e^{-i(k_{-}z-\alpha\pi)}\right] + e^{i\alpha\pi}V_{-}(0)\left[e^{i(k_{-}z-\alpha\pi)} + R.e^{-i(k_{+}z-\alpha\pi)}\right]$$

$$ZI_{fd}(z) = e^{i\alpha\pi}V_{+}(0)\left[e^{i(k_{+}z-\alpha\pi)} - R.e^{-i(k_{-}z-\alpha\pi)}\right] + e^{i\alpha\pi}V_{-}(0)\left[e^{i(k_{-}z-\alpha\pi)} - R.e^{-i(k_{+}z-\alpha\pi)}\right]$$

Well, for $\alpha = 0$, $V_{fd}(z) = V(z)$ and $ZI_{fd}(z) = ZI(z)$. For

$$\alpha = (1/2)\ V_{fd}(z) = ZI(z)$$

and $ZI_{fd}(z) = V(z)$, which is original and dual to the original set of solutions for a Transmission Line. For $0 < \alpha < (1/2)$, solution sets V_{fd} and ZI_{fd} may be regarded as intermediate step between the original and dual to the original solutions of Transmission Line. The Transmission Line corresponding to solution set V_{fd} and ZI_{fd} may be termed a fractional dual transmission line.

4.10.9 Input Impedance of Terminated Fractional Non-symmetric Line

The impedance is;

$$Z_{fd} = V_{fd}(z)/I_{fd}(z),$$

using $k_{+} = k + k\kappa_r$ and $k_{-} = k - k\kappa_r$, we write $e^{ik_{+}z} = e^{ikz} \times e^{ik\kappa_r}$ and $e^{ik_{-}z} = e^{ikz} \times e^{-ik\kappa_r}$. Using these values and putting in V_{fd} and ZI_{fd} obtained above we get:

$$V_{fd}(z) = \frac{1}{(Z_L + Z)} e^{i\alpha\pi} \left[V_{+}(0) e^{ik\kappa_r z} + V_{-}(0) e^{-ik\kappa_r z} \right] \times 2\cos(kz - \alpha\pi) \left[Z_L + iZ\tan(kz - \alpha\pi) \right]$$

$$ZI_{fd}(z) = \frac{1}{(Z_L + Z)} e^{i\alpha\pi} \left[V_{+}(0) e^{ik\kappa_r z} + V_{-}(0) e^{-ik\kappa_r z} \right] \times 2\cos(kz - \alpha\pi) \left[iZ_L \tan(kz - \alpha\pi) + Z \right]$$

$$Z_{fd} = Z \frac{Z_L + iZ\tan(kz - \alpha\pi)}{Z + iZ_L \tan(kz - \alpha\pi)},$$

for $\alpha = 0$

$$Z_{fd} = Z \frac{Z_L + iZ\tan(kz)}{Z + iZ_L \tan(kz)};$$

which is relation of input impedance at any point for a Transmission Line having characteristic impedance Z. We consider it as a original Transmission Line. Now for $\alpha = (1/2)$, we get

$$Z_{fd} = Z \frac{Z_L + iZ\cot(kz)}{Z + iZ_L \cot(kz)},$$

which is relation for input impedance at any point for dual to original transmission line.

It may be noticed that at a particular point along the original Transmission Line input admittance becomes the corresponding input impedance of the dual Transmission Line. So for $\alpha = 0$, Z_{fd} expression represents input impedance of the original Transmission Line and $\alpha = (1/2)$ represents the dual to the original Transmission Line, and when $0 < \alpha < (1/2)$ represents impedance of fractional dual Transmission Line.

Variation in value of α along the fractional Transmission Line corresponds to variation in the observation point along the original Transmission Line. So, the behavior along the Transmission Line may be studied by changing the fractional parameter along the corresponding fractional Transmission Line, while keeping the observation point constant.

Setting the observation point $z = 0$ then

$$Z_{fd} = Z \frac{Z_L - iZ\tan(\alpha\pi)}{Z - iZ_L\tan(\alpha\pi)},$$

it is obvious that

$$Z_{fd}\Big|_{\alpha=0} = Z_L,\ Z_{fd}\Big|_{\alpha=(1/2)} = Z^2 / Z_L \text{ or } \bar{Z}_{fd}\Big|_{\alpha=0} = \bar{Z}_L$$

and

$$Z_{fd}\Big|_{\alpha=(1/2)} = 1/\bar{Z}_L = \bar{Y}_L .$$

This means if original Transmission Line that is $\alpha = 0$ deals with Transmission Line which is terminated by Z_L load, then $\alpha = 0.5$ deals with Transmission Line which is terminated by a load, having impedance Y_L. For $0 < \alpha < 0.5$ represents a new line terminated by a load, which may be regarded as in between Z_L and Y_L.

Now consider further special cases. If $Z_L = 0$ then $Z_{fd}\Big|_{\alpha=0} = 0$, and $Z_{fd}\Big|_{\alpha=0.5} = \infty$, and if $Z_L = \infty$, then $Z_{fd}\Big|_{\alpha=0} = \infty$ and $Z_{fd}\Big|_{\alpha=0.5} = 0$. For $0 < \alpha < 0.5$, if load Z_L is inductive, for an original load of Z_L short circuit, and Z_L is capacitive if originally the Z_L is open circuit.

4.11 Concluding Comments

This chapter extends the argument about reality of having the concept of fractional order calculus to describe the Nuclear Reactor and application in Electromagnetic Theory. With regards to the electromagnetic theory the application of the fractional curl will see extension of its usage in description of Left Handed Materials (LHM) or Metamaterials where the electromagnetic gets reversed. Reversal of Snell's law, reversed Doppler Effect and superlumilaty are terms associated with LHM utilized for presently to have perfect focusing of beams by straight surfaces. Future will see the geometrical interpretation of the concept of fractional curl and its use in formation of turbulence in flow of fluids and electromagnetic. The concept of fractional divergence as introduced in reactor description will in future lead to development of reactor criticality concepts based on fractional geometrical buckling, and fractional criticality or fractional multiplication factor. This enables to describe the reactor flux profile more closely to actual and maintain efficient correction and control. The fractional divergence will be used to describe several anomalous effects presently observed in diffusion experiments, which, is presently ode to non-linearity effects and its explanation through integer order theory or by probabilistic methods. The concept of fractional stochastic difference equations and anomalous transport described here by fractional diffusion equation, to explain the long tailed time-series behavior, are also observed in various physiological processes of non-linear dynamics in biology and other self-similar 'fractal' phenomena. The use of slope, in log-log plot of the asymptotic frequency spectra for physiological phenomena having fractal property is an indicator of unhealthiness or healthiness. The time series of healthy individual's heart beat that is interbeat interval increment shows 'anti- persistence' behavior as described in random walk phenomena can be obtained from spectral analysis. Also the method described here for anomalous transport, can be applied to studies of economics more specifically econo-physics. The fluctuations of stock market or ruin theory of a company or say distribution of wealth indeed are classical cases to study by theories of anomalous transport developed here to describe neutron diffusion.

Chapter 5
Fractional Differintegrations Insight Concepts

5.1 Introduction

This chapter describes the geometric and physical interpretation of fractional integration and fractional differentiation. As a start point the Riemann-Liouville (RL) fractional integration is taken. Briefly existence of fractional differintegration is discussed along with useful tricks to obtain the fractional differintegration. The geometric interpretation is developed first for RL integration process along with concept of transformed time scales, and in-homogeneous time axis. Thereafter the RL definition is geometrically explained by convolution of the power function and the integrand, and as area under shape changing curve is demonstrated. The concept of delay is developed for Grunwald-Letnikov differintegration process and this is converted into the specific definition of short-memory principle, used for computer applications. The GL differintegration is also explained as in the classical calculus by considering infinitesimal quantities for the independent variable and the function, and explained graphically. The GL definition is expanded with binomial coefficients and its application to numerical regression. These methods are advance algorithms to get digital realization for fractional order controllers. The application to solve fractional differential equation numerically is demonstrated. Small introduction is made regarding definitions of Local Fractional Derivatives (LFD) for continuous but nowhere differentiable functions. These LFD (Kolwankar-Gangal K-G definition's) utility is extended to measure critical point behaviors of physical system and its relation to 'fractal' dimension. The demonstration is made to have fractional integration and fractional differentiation, for fractal distributed quantities; thus, line, surface and volume integration can be performed when the measurable quantities are distributed in fractal form, Thereby generalizing the Gauss's and Stroke's law for fractal distributed quantities.

5.2 Calculating Fractional Integral

Our ability to calculate explicitly the fractional integral of $f(x)$ depends on proficiency in performing integration

$$\int_0^x (x-u)^{v-1} f(u)du \, , v>0 \, .$$

However because of nature of 'singular' kernel $(x-u)^{v-1}$, it is possible to develop certain analytical tricks that allow us to calculate the fractional integral. Even for explicit calculations of fractional derivatives one has to fractionally integrate the function. This trick can be used for large classes of well behaved functions, and analytical functions.

5.2.1 Existence of Fractional Differeintegration

We take our clue from Integer Order Calculus, that the class of differentiegrable function is a classically 'well-behaved' function. Off course most of the classical functions of mathematical physics and engineering are 'well behaved' and thus we can go ahead for fractional differentiation and integration. Say the speedometer reading of a moving car or odometer reading of moving vehicle (function with respect to time) is well behaved, and on this reading we can calculate the differentiation and or integration to derive automatic control functions. If they were not the control action will be exceptional.

Candidate function is defined on a closed interval $[a \le x \le y]$ that is bounded everywhere in half open interval $a < x \le y$ and is better behaved at the 'lower-limit terminal' a; than is $(x-a)^{-1}$. Better behaved than $(x-a)^{-1}$ means that

$$\lim_{x \to a} \left[(x-a) f(x) \right] = 0 \, .$$

Physically, it is the area under the curve near the lower-limit terminal a; which indeed should be zero. That is,

$$\lim_{(a+) \to a} \int_a^{a+} f(x)dx = 0 \, .$$

We define class of 'differintegrable series' to be all finite sums of functions each of which may be represented as product of power of $(x-a)$ and analytic function of $(x-a)^{1/n}$, where n is positive integer. For, $a_0 \ne 0$ we represent the differintegrable series as

$$f(x) = (x-a)^p \sum_{k=0}^{\infty} a_k (x-a)^{k/n} \, .$$

For say $n = 2$, the series expansion is:

$$f(x)=a_0(x-a)^p+a_1(x-a)^{p+\frac{1}{2}}+a_2(x-a)^{p+1}+........$$

$$(x-a)f(x)=a_0(x-a)^{p+1}+a_1(x-a)^{p+\frac{3}{2}}+a_2(x-a)^{p+2}+......$$

Then

$$\lim_{x\to a}(x-a)f(x)=0\,,$$

only if $p+1>0$ that is, $p>-1$. Meaning better behaved than $(x-a)^{-1}$, the function is differintegrable (fractionally).

In addition to the bounded example the functions

$$f(x)=(x-a)^{-1/2}\,,\ f(x)=(x-a)^{-3/4}\left[\sin\sqrt{x-a}\right],\ f(x)=(x-a)\,,\log x\,,\ f(x)=0$$

$$f(x)=\frac{1}{2}\sin\left[\pi(x-a)\right]-1\,,$$

Heaviside's unit step, Square Wave function, is fractionally differinitegrable. The functions which are not differintegrable are like

$$f(x)=(x-a)^{-1}\,,\ f(x)=\sqrt{\frac{9}{4}-(x-a)^2}\ .$$

5.2.2 *Useful Procedure for Calculating Fractional Integral*

The procedure is that we have to express the fractional integral of a power of x, times a function $f(x)$ in terms of fractional integral of $f(x)$. This is demonstrated below:

$$D^{-v}\left[xf(x)\right]=\frac{1}{\Gamma(v)}\int_0^x(x-u)^{v-1}\left[uf(u)\right]du\,,\ v>0$$

If we replace the term in brackets in the integrand by $\left[x-(x-u)\right]f(u)$, the integral becomes:

$$\begin{aligned}D^{-v}\left[xf(x)\right]&=\frac{1}{\Gamma(v)}\int_0^x(x-u)^{v-1}\left\{\left[x-(x-u)\right]f(u)\right\}du\\&=\frac{x}{\Gamma(v)}\int_0^x(x-u)^{v-1}f(u)du-\frac{v}{v\Gamma(v)}\int_0^x(x-u)^{v}f(u)du\\&=xD^{-v}f(x)-vD^{-(v+1)}f(x)\end{aligned}$$

So for

$$D^{-v}\left[xe^{ax}\right]=xD^{-v}e^{ax}-vD^{-(v+1)}e^{ax}=xE_x(v,a)-vE_x(v+1,a),$$

similarly we can use this trick for

$$D^{-v}\left[x\cos ax\right]=xD^{-v}\cos ax-vD^{-(v+1)}\cos ax=xC_x(v,a)-vC_x(v+1,a),$$

and for

$$D^{-v}\left[x\sin ax\right]=xS_x(v,a)-vS_x(v+1,a)$$

The above method can be generalized if a function is multiplied by x^p, where p is nonnegative integer. The procedure is demonstrated as:

$$D^{-v}\left[x^p f(x)\right]=\frac{1}{\Gamma(v)}\int_0^x(x-u)^{v-1}\left[u^p f(u)\right]du\text{, With } v>0.$$

Replace $u^p\equiv[x-(x-u)]^p=\sum_{k=0}^{p}(-1)^k\binom{p}{k}x^{p-k}(x-u)^k,$

then integral becomes:

$$D^{-v}\left[x^p f(x)\right]=\frac{1}{\Gamma(v)}\sum_{k=0}^{p}(-1)^k\binom{p}{k}x^{p-k}\int_0^x(x-u)^{v+k-1}f(u)du$$

$$=\frac{1}{\Gamma(v)}\sum_{k=0}^{p}(-1)^k\binom{p}{k}\Gamma(v+k)x^{p-k}D^{-(v+k)}f(x)$$

Using

$$\binom{-z}{n}=\frac{\Gamma(1-z)}{n!\Gamma(1-z-n)}=(-1)^n\frac{\Gamma(z+n)}{n!\Gamma(z)}=(-1)^n\binom{z+n-1}{n},$$

we write:

$$D^{-v}\left[x^p f(x)\right]=\sum_{k=0}^{p}\binom{-v}{k}\left[D^k x^p\right]\left[D^{-(v+k)}f(x)\right]$$

For example

$$D^{-v}\left[x^{p}e^{ax}\right]=\frac{1}{\Gamma(v)}\sum_{k=0}^{p}(-1)^{k}\binom{p}{k}\Gamma(v+k)x^{p-k}E_{x}(v+k,a)$$

5.2.3 Calculating Fractional Integral with Non-zero Lower Limit

For calculating the fractional integral when lower terminal is not zero.

$${}_{c}D_{x}^{-v}f(x)=\frac{1}{\Gamma(v)}\int_{c}^{x}(x-u)^{v-1}f(u)du\ ,\ v>0$$

and $0\le c<x$. With change of variable as $u=x(1-y)$, we write:

$${}_{c}D_{x}^{-v}f(x)=\frac{x^{v}}{\Gamma(v)}\int_{0}^{\chi}y^{v-1}f(x-xy)dy\,,\ \chi=\frac{x-c}{x}$$

For example take $f(x)=x^{\mu}$, $\mu>-1$ and substituting in above we have:

$${}_{c}D_{x}^{-v}x^{\mu}=\frac{x^{\mu+v}}{\Gamma(v)}\int_{0}^{\chi}y^{v-1}(1-y)^{\mu}dy=\frac{x^{\mu+v}}{\Gamma(v)}\mathrm{B}_{\chi}(v,\mu+1)\,,\ \chi=\frac{x-c}{x}$$

(in terms of incomplete Beta function). In this if we let $c=0$, then we get

$${}_{0}D_{x}^{-v}x^{\mu}=\frac{\Gamma(\mu+1)}{\Gamma(\mu+v+1)}x^{\mu+v}$$

With this observation we can also write

$${}_{c}D_{x}^{-v}e^{ax}=e^{ac}E_{x-c}(v,c)\,,$$

and

$${}_{c}D_{x}^{-v}\cos ax=(\cos ac)C_{x-c}(v,a)-(\sin ac)S_{x-c}(v,a)$$

5.2.4 Fractional Integral for Analytical Function

Let $f(x)$ be analytical function in interval $[0,X]$ and we try to obtain its fractional integral expression.

$$
{}_0D_x^{-v} f(x) = \frac{1}{\Gamma(v)} \int_0^x (x-u)^{v-1} f(u)du \,, \; v > 0 \,, \; 0 < x \leq X
$$

$$
= \frac{1}{\Gamma(v)} \int_0^x y^{v-1} f(x-y)dy
$$

For analytic function, for all $y = a$, $\in [0, X]$ we have series expansion as:

$$
f(x-y) = f(x) + \sum_{k=1}^{\infty} (-1)^k \frac{D^k}{k!} y^k ,
$$

converges for all y in interval $[0, x]$. Substituting this analytical expansion in the integral we get:

$$
{}_0D_x^{-v} f(x) = \frac{f(x)}{\Gamma(v)} \int_0^x y^{v-1} dy + \frac{1}{\Gamma(v)} \int_0^x y^v \left[\sum_{k=1}^{\infty} (-1)^k \frac{D^k f(x)}{k!} y^{k-1} \right] dy
$$

We may change the order of summation and integration to obtain:

$$
{}_0D_x^{-v} f(x) = \frac{1}{\Gamma(v)} \sum_{k=0}^{\infty} \frac{(-1)^k D^k}{k!(v+k)} x^{v+k} = \frac{1}{\Gamma(v)} \left[\frac{f(x)}{v} x^v - \frac{f'(x)}{(v+1)} x^{v+1} + \frac{f''(x)}{2!(v+2)} - \ldots \right]
$$

Thus we have expressed fractional integral of analytical function in terms of ordinary derivatives of that function.

Recalling the expression

$$
D_x^{-(v+k)} [1] = \frac{1}{\Gamma(v+k+1)} x^{v+k} ,
$$

and using the same we may write the above expansion as:

$$
{}_0D_x^{-v} f(x) = \sum_{k=0}^{\infty} (-1)^k \binom{v+k-1}{k} \left[D^k f(x) \right] \left[D^{-(v+k)} (1) \right] = \sum_{k=0}^{\infty} \binom{-v}{k} \left[D^k f(x) \right] \left[D^{-(v+k)} (1) \right]
$$

5.3 Fractional Differintegration of Product of Two Functions

The rule for differentiating a product of two functions in integer order calculus is:

$$
\frac{d^n [fg]}{dx^n} = \sum_{j=0}^{n} \binom{n}{j} \frac{d^{n-j} [f]}{dx^{n-j}} \frac{d^j [g]}{dx^j} ,
$$

and off course restricted to nonnegative integer n.

This finite sum could well be extended to infinite sum, recognizing the fact that $\binom{n}{j}=0$, for $j>n$. From above product rule we generalize as:

$$D^{\mu}[fg]=\sum_{k=0}^{\infty}\binom{\mu}{k}D^{k}[g]D^{\mu-k}[f],\ \mu>0.$$

Applying this rule to $[fg]=[t^{p}f(t)]$, we get:

$$D^{\mu}\left[t^{p}f(t)\right]=\sum_{r=0}^{p}\binom{\mu}{r}\left[D^{r}t^{p}\right]\left[D^{\mu-r}f(t)\right],\ \text{for}\ p=0,1,2,\ldots\ \text{and}\ \mu>0,$$

From here if $[fg]=[tf(t)]$, then $D^{\mu}\left[tf(t)\right]=tD^{\mu}f(t)+\mu D^{\mu-1}f(t)$, for $\mu>0$.

It is interesting to note that if $0<\mu<1$, and then the term $D^{\mu-1}f(t)$ is fractional integral and not fractional derivative!

The same expression we apply to get:

$$D^{\mu}\left[tE_{t}(w,a)\right]=tE_{t}(w-\mu,a)+\mu E_{t}(w-\mu+1,a)$$

The product rule for integer order integration comes from repeated integration by parts as:

$$\frac{d^{-n}[fg]}{[d(x-a)]^{-n}}=\sum_{j=0}^{\infty}\binom{-n}{j}\frac{d^{-n-j}[f]}{[d(x-a)]^{-n-j}}\frac{d^{j}[g]}{[d(x-a)]^{j}},$$

for n, nonnegative integer.

We thus expect the product rule for general arbitrary $v>0$, as:

$$D^{-v}[fg]=\sum_{k=0}^{\infty}\binom{-v}{k}\left[D^{k}g\right]\left[D^{-(v+k)}f\right]$$

The generalized product rule called as Leibniz's rule for fractional differintegration is:

$$\frac{d^{q}[fg]}{[d(x-a)]}=\sum_{j=0}^{\infty}\binom{q}{j}\frac{d^{q-j}[f]}{[d(x-a)]^{q-j}}\frac{d^{j}[g]}{[d(x-a)]^{j}},\ \text{for any}\ q.$$

Take a fractional differential equation

$$tD^{1/2}y(t) - y(t) = 0\,,$$

we apply $D^{1/2}$ on this and write the same as:

$$D^{1/2}\left[tD^{1/2}y(t)\right] = D^{1/2}y(t)\,.$$

Here we apply Leibniz's rule to get:

$$tD^{1/2}\left[D^{1/2}y(t)\right] + \frac{1}{2}D^{-1/2}\left[D^{1/2}y(t)\right] = D^{1/2}y(t)\,,$$

which is

$$tDy(t) + \frac{1}{2}y(t) = D^{1/2}y(t)\,.$$

But from the initial equation we have

$$D^{1/2}y(t) = {y(t)}/{t}\,,$$

substituting this on above obtained equation we have now a linear ordinary integer order differential equation as:

$$tDy(t) + \frac{1}{2}y(t) = \frac{y(t)}{t}\,,$$

its solution is by simple integration, for $t > 0$, $y(t) = Kt^{-1/2}e^{-1/t}$, with K as constant of integration.

Let us consider another similar equation:

$$ty(t) - \sqrt{\pi}D^{-1/2}y(t) = 0\,,$$

which is a fractional integral equation. Operating with $D^{1/2}$ and rearranging we have:

$$D^{1/2}\left[ty(t)\right] = \sqrt{\pi}D^{1/2}\left[D^{-1/2}y(t)\right] = \sqrt{\pi}y(t)\,.$$

Now apply Leibniz's rule and we get:

$$tD^{1/2}[y(t)] + \frac{1}{2}D^{-1/2}[y(t)] = \sqrt{\pi}y(t)\,,$$

but initial equation is

$$D^{-\frac{1}{2}}y(t)=\frac{t}{\sqrt{\pi}}y(t),$$

and substituting, this on above we obtain:

$$tD^{\frac{1}{2}}y(t)+\frac{t}{2\sqrt{\pi}}y(t)=\sqrt{\pi}y(t)$$

Taking the ordinary derivative of original equation

$$ty(t)-\sqrt{\pi}D^{-\frac{1}{2}}y(t)=0,$$

we get:

$$tDy(t)+y(t)=\sqrt{\pi}D^{\frac{1}{2}}y(t).$$

Eliminating $D^{\frac{1}{2}}y(t)$, from this equation and previous derived one we have

$$t^2Dy(t)+\left(\frac{3}{2}t-\pi\right)y(t)=0,$$

an ordinary differential equation and its solution for $t>0$ is obtained by integration as $y(t)=Kt^{-\frac{3}{2}}e^{-\frac{\pi}{t}}$ with K as constant for integration.

5.4 Symbol Standardization and Description for Differintegration

Mathematicians have used several notations since the birth of fractional calculus. As mentioned in the Introduction chapter (Historical Development of Fractional Calculus) several contemporary notations for fractional differentiation and fractional integration. Here attempt will be made to standardize the notations as differintegrals. The same operator used as integrator when index is negative and differentiator when index is positive. Separate notation will be used to indicate initialized differintegral operator and un-initialized operator. However the difference in notations is made clear as 'un-initialized' and 'initialized' differintegrals, the concept of initialization function $\psi(f,q,a,c,t)$ is dealt in details in the next chapters.

${}_cD_t^q f(t)$ Represents initialized q-th order differintegration of $f(t)$ from start point c to t. ${}_cd_t^q f(t)$ Represents un-initialized generalized (or fractional) *q-th* order differintegral. This is also same as

$$\frac{d^q f(t)}{[d(t-c)]^q} \equiv {}_c d_t^q f(t),$$

shifting the origin of function at start of the point from where differintegration starts. This un-initialized operator can also be short formed as $d^q f(t)$. The initialization function (not a constant) is represented as $\psi(f,q,a,c,t)$ meaning that this is function of independent variable *t*, and is for differintegral operator of order *q*, for the function $f(t)$ born at $t=a$ (before that the function is zero), and differintegral process starting at $t=c$ This initialization function can be short formed as $\psi(t), \psi(f,q,t)$. Therefore the expression between initialized differintegral and un-initialized one is:

$${}_c D_t^q f(t) = {}_c d_t^q f(t) + \psi(f,q,a,c,t)$$

The notation contains lower limit of the process at the front subscript and the order of the process at the tail superscript, with independent variable with respect to what is being differintegrated.

For Caputo derivative we will describe ${}_c^C d_t^q f(t)$ as the uninitialized one and ${}_c^C D_t^q f(t)$ for initialized case. For a generalized case of fractional derivative of order $0<\alpha<1$ with type parameter $0 \le \beta \le 1$ the fractional initialized derivative representation is ${}_a^\beta D_t^\alpha f(t)$. For $\beta = 1$, we have Caputo derivative and for $\beta = 0$, we get Riemann-Liouvelli (RL) derivative. For local fractional derivative at a point, t_0 we describe the symbol as $\mathbf{D}_t^q f(t_0)$.

5.5 Riemann-Liouville Fractional Differintegral

5.5.1 Scale Transformation

The integration in fractional calculus is embedded part of the fractional differintegration. The RL definition is described as:

$${}_0 D_t^{-\alpha} f(t) = \frac{1}{\Gamma(\alpha)} \int_0^t f(\tau)(t-\tau)^{\alpha-1} d\tau$$

Take the function g as which is basically scaling the time τ variable to the function $g(\tau)$, is scale transformation concept. Described as:

$$g(\tau)=\frac{1}{\Gamma(\alpha+1)}\left\{t^{\alpha}-(t-\tau)^{\alpha}\right\}$$

$$\tau \rightarrow g(\tau)$$

$$dg(\tau)=\frac{(t-\tau)^{\alpha-1}}{\Gamma(\alpha)}$$

$$_0D_t^{-\alpha}f(t)=\int_0^t f(\tau)dg(\tau)$$

Therefore the fractional integration of the function is area under the curve for the plot of $f(\tau)$ and $g(\tau)$, from $0-t$. Let us take three axes $\tau, g(\tau), f(\tau)$, making a cubic room with floor comprising of plane $\tau, g(\tau)$. We plot the function as from $0<t<\tau$, in the floor

$$g(\tau)=\frac{1}{\Gamma(\alpha+1)}\left\{t^{\alpha}-(t-\tau)^{\alpha}\right\}.$$

This is depicted in Figure 5.1. Along the obtained curve (on the floor) we build a fence of varying height $f(\tau)$, so the top edge of the fence is 3D line. Points are $\tau, g(\tau), f(\tau)$ for $0<\tau<t$.

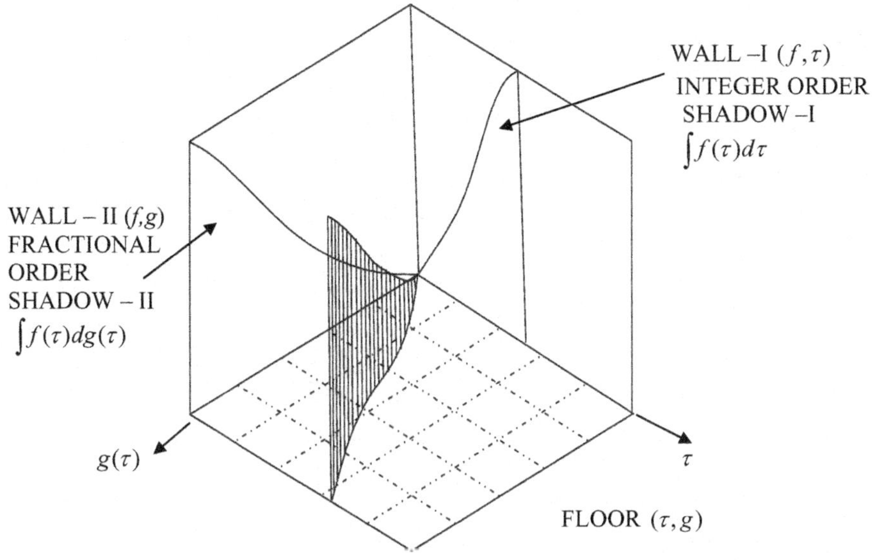

Fig. 5.1 The 3D representation of RL fractional integration.

The shadow on the wall $(\tau, f(\tau))$ as shown in Figure 5.1, is well known area under the curve and is normal integer order integration

$$ {}_0D_t^{-1} f(t) = \int_0^t f(\tau) d\tau . $$

The second shadow on the wall $(g(\tau), f(\tau))$ is geometric interpretation of the fractional integration of $f(t)$, i.e.

$$ {}_0D_t^{-\alpha} f(t) = \int_0^t f(\tau) dg(\tau) $$

for fixed t that is, area under shadow of wall-II. The observation from Figure 5.1 is that what happens when t is changing (namely growing), the fence changes simultaneously. Its length and in certain sense its shape changes. The area under shape changing curve (shadow on wall-II) when differentiated by integer order m such that integer $m > \alpha$, will give fractional derivative of order $m - \alpha$. That is

$$ D_t^{\mu} f(t) = D_t^m \int_0^t f(\tau) dg(\tau) , $$

where $\mu = m - \alpha$, and $g(\tau)$ is power function defined as above, with power α, translating uniform time scale $\tau \to g(\tau)$.

The wall $(f(\tau), g(\tau))$, depicting shadow growing as $t = 9 \to 10$ depicting fractional integration is shown in Figure 5.2. Area under shape changing curve, is fractional integration

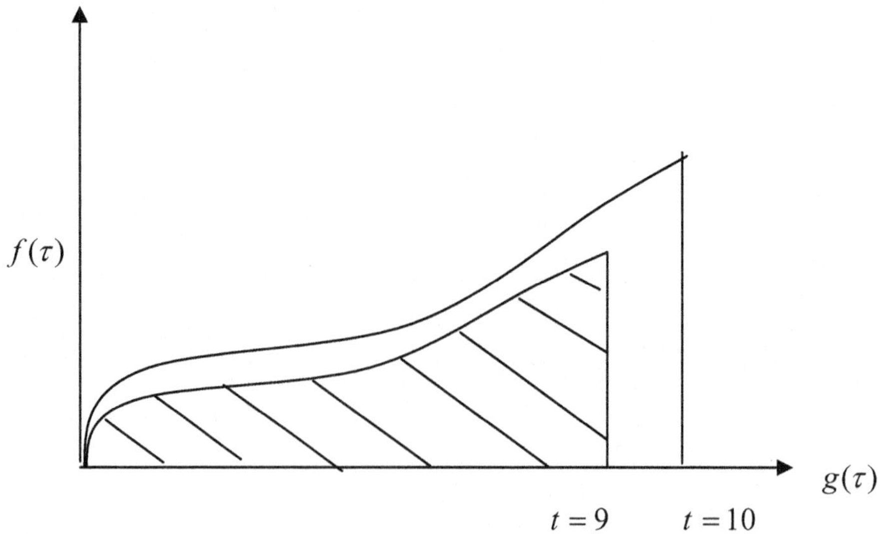

Fig. 5.2 Shadow on the wall showing fractional integration as t grows.

5.5.2 Changing Shape of Curve While Obtaining Fractional Integration and Differentiation

The explanation about, difference in concept of area under a constant curve-integer order integration and area under changing shape of curve-fractional integration is proved here. The generalized integrationof Riemann-Liouvelli type, of any real number q-folds is

$$ {}_0d_t^{-q} = \frac{1}{\Gamma(q)} \int_0^t (t-\tau)^{q-1} f(\tau) d\tau $$

In the above expression let us put

$$ \tau = t - x^{\frac{1}{q}}, $$

giving

$$ d\tau = -\frac{1}{q} x^{\frac{1-q}{q}} dx, $$

substituting this change of variable the generalized (fractional) order integration relation is:

$$ {}_0d_t^{-q} = \frac{1}{\Gamma(q)} \int_0^t (t-\tau)^{q-1} f(\tau) d\tau = \frac{1}{\Gamma(q)} \int_{t^q}^{0} \left(t - t + x^{\frac{1}{q}} \right)^{q-1} \left(-\frac{1}{q} x^{\frac{1-q}{q}} \right) f(t - x^{\frac{1}{q}}) dx $$

$$ = \frac{1}{q\Gamma(q)} \int_{t^q}^{0} - \left(x^{\frac{1}{q}} \right)^{q-1} x^{\frac{1-q}{q}} f(t - x^{\frac{1}{q}}) dx $$

$$ = \frac{1}{\Gamma(q+1)} \int_0^{t^q} f(t - x^{\frac{1}{q}}) dx $$

Therefore, the fractional integration of a function as defined above, by convolution integral is also visualized geometrically, as an area under the 'changing curve' as time (t) grows, that is $f(t - x^{1/q})$ and the limit of integration is from $x = 0$ to $x = t^q$. This is the major difference as geometrical interpretation from integer order calculus, where the integration (one-fold) means as the time grows, and the integration is area under the same function f from $x = 0$ to $x = t$. Whereas, in fractional integration the function itself keeps on changing its shape from time-to-time and new area be evaluated at each time, under a new function transformed as: $f(x) \rightarrow f(t - x^{1/q})$, with limits of integration as transformed from $(0-t)$ to $(0-t^q)$ that is compressed (or expanded scale). This geometrical explanation is available in a simple function

$f(x)=(x+2)^2$, in integer order integration yields the area under the curve $(x+2)^2$ from say $x=0$, to $x=t=t_1$ for any time the function is same. While changing the integration to any other limit, one has to add the incremental area under the same curve, $(x+2)^2$ from, time $x=t_1$ to $x=t_2$, to that obtained earlier. For the case of fractional integration it is not so. The same function $(x+2)^2$ when fractionally integrated by say half order then the curve keeps on changing the shape. When $t=0.001$, the area under the curve is for the function $f(x)=(0.001-x^2+2)^2 \approx (4-4x^2+x^4)$, from $x=0$ to $x=\sqrt{0.001}$. For, $t=1$ the semi-integration will yield area under a different function that is $f_1(x)=(1-x^2+2)^2=9-6x^2+x^4$, from $x=0$ to $x=\sqrt{1}$. For $t=3$, a different curve is integrated that is $f_2(x)=25-10x^2+x^4$, from $x=0$ to $x=\sqrt{3}$, and so on. Obviously, after evaluating all these integrations with the changing functions as time variable grows, a scalar multiplication by a constant that is $(1/0.5\Gamma(0.5))$ will give correct semi-integration of the function. The fractional integration is embedded in the differentiation process too, and thus as opposed to integer order calculus, the fractional differentiation is non-local quantity. The differentiation in fractional order is non-local process, and requires lower terminal and upper terminal for the interval under consideration, for the fractional differentiation process.

For $t=10$, τ is varied from 0-10. $g(\tau)$, is formed and then $f(\tau)$ is plotted. The Figure 5.2 shows the change in the shape of the curve from $t=9$ to $t=10$ and the integration under the new shape. The difference in the integer order integration is that the new shape of the curve from 0-10 as compared to old one 0-9. When t changes the entire preceding time interval changes as well. The changing area under the shape changing curve, when differentiated by integer order m just greater than the fractional order α used for integration, will yield fractional differentiation of order $m-\alpha$.

5.5.3 Homogeneous and Heterogeneous Scales in Fractional Integration/Differentiation

Let us consider a moving object fitted with speedometer and a clock. The observer in the moving object records the speed at the end of each second (time-interval) as $v_1, v_2, v_3, \ldots v_{10}$. At the end of 10 seconds the observer in the moving object calculates the distance, $S_N = v_1+v_2+v_3+\ldots v_{10}$, where the observer takes by the local clock each interval of time as 'one-second'. Now the observer stationary at the fixed frame of reference knows that the clock of the mobile observer is running slow. The actual time at the end of each one-second is 1, 2, 4, 8, 16, 32, 64 .... Then the actual distance traveled as seen by stationary observer (call cosmic observer) as per the actual time (call cosmic time) is $S_o = v_1+2v_2+4v_3+8v_4+16v_5+32v_4+64v_5+\ldots$, is much more than S_N. The integration of local velocity (speed) with local time is integer order integration and the integration of the speed recorded with respect to the transformed time is fractional order integration. These phenomena when moving body changes its position in space-time, the gravitation field in the entire space-time changes due to movement of the object. As a consequence, the 'cosmic-time-interval', which corresponds to history of the movement of the moving object, changes. $S_N, v(\tau), \tau$, is individual, distance, speed, and time, of the observer moving with the object. $S_o, v_o(t), t$, distance, speed and time as recorded by the observer outside on fixed frame of reference (cosmic). They are related as

$$S_N(t) = {}_0D_t^{-1}v(t) = \int_0^t v(\tau)d\tau\,,\ S_o(t) = {}_0D_t^{-\alpha}v(t) = \int_0^t v(\tau)dg(\tau)\,,\ v(t) = {}_0D_t^{\alpha}S_o(t)$$

is individual velocity of the local speedometer as related to cosmic-distance, and

$$v_o(t) = \frac{d}{dt}S_o(t) = \frac{d}{dt}\,{}_0D_t^{-\alpha}v(t) = {}_0D_t^{1-\alpha}v(t)\,.$$

The first derivative of cosmic distance is the cosmic velocity, and the cosmic velocity is fractional derivative of order $(1-\alpha)$ of the local velocity. Figure 5.3 gives the two kind of time homogeneous (local time) and the heterogeneous (transformed time). In above, example the variable t is used as notation for both the observers N and O.

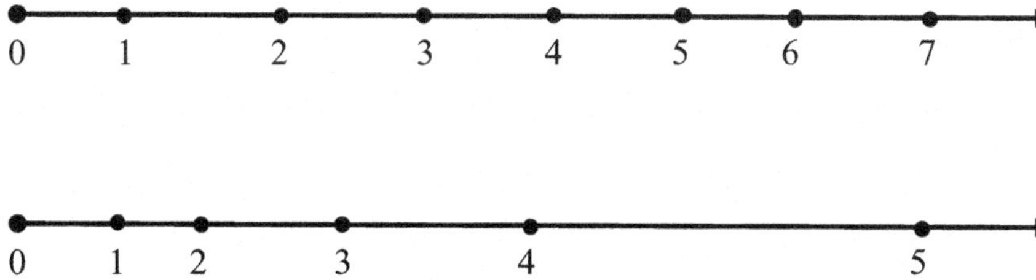

Fig. 5.3 Homogeneous and heterogeneous time.

The $g(\tau)$ describes in-homogeneous time scales, which depends not only on the τ, but also on the parameter t, representing the last measured value of the individual time (of the moving object). Fractional integration in time means transformation of the local time to cosmic-time.

5.5.4 Convolution Example

The Riemann-Liouville definition for the fractional integral is:

$$\frac{d^{-q}}{\left[d(t-a)\right]^{-q}}f(t) = \frac{1}{\Gamma(q)}\int_a^t (t-\tau)^{q-1}f(\tau)d\tau\,,\ q \ge 0$$

This definition is extended to fractional differentiation with m as integer as:

$$\frac{d^{m-q}}{[d(t-a)]^{m-q}} f(t) \equiv \frac{1}{\Gamma(q)} \frac{d^m}{dt^m} \int_a^t (t-\tau)^{q-1} f(\tau) d\tau, \; q \geq 0, m > q$$

The RL integral can be viewed as convolution of the integrand function with power function, when both the functions are causal can be expressed as:

$$\frac{d^{-q}}{[d(t-0)]} f(t) = f(t) * h(t) = f(t) * \left(\frac{1}{\Gamma(q) t^{-q+1}} \right) = \frac{1}{\Gamma(q)} \int_0^t \frac{f(\tau) d\tau}{(t-\tau)^{-q+1}}$$

Where

$$h(t) = \frac{1}{\Gamma(q) t^{-q+1}}.$$

Causal functions mean that no convolution response can obtained before the function $f(t)$ is applied Figure 5.4 demonstrates the convolution process. The function is

$$f(t) = \cos\left(\frac{2\pi}{5} t \right),$$

the order of fractional integration is half $q = 1/2$. The Figure a) show $h(\tau)$ vs. τ, the Figure b) show the $f(t), t > 0$. The Figure c) show $h(-\tau)$. The curve is obtained for the value $t = 5$, and the d) show the plot of $h(t-\tau) @ t = 5$ i.e. $h(5-\tau)$ vs. τ. The part e) of the figure show the full integrand for $t = 5$. Now moving this $h(t-\tau)$ for several continuous values of t from 0-10, repeating the graphs d) and e) and obtaining the value of the integral of the product (for several values of t) the final graph f) is obtained. For $t = 5$ the graph e) show full integrand as $h(5-\tau) f(\tau)$. The integral of this product becomes the $t = 5$, value of the semi-integral of $f(t)$.

In this convolution example the convolution of power function $t^{q-1}/\Gamma(\alpha)$ is carried out with $f(t)$. This is Laplace convolution, represented symbolically as ${}_0d_t^{-q} f(t) = \{f(t)\} * \{t^{q-1}/\Gamma(q)\}$.

There is Fourier convolution of a power function that is given as: $\{f(x)\} \otimes \{x^{\mu-1}/\Gamma(\mu)\} = {}_{-\infty}d_x^{-q} f(x)$, which is fractional integral starting from $-\infty$.

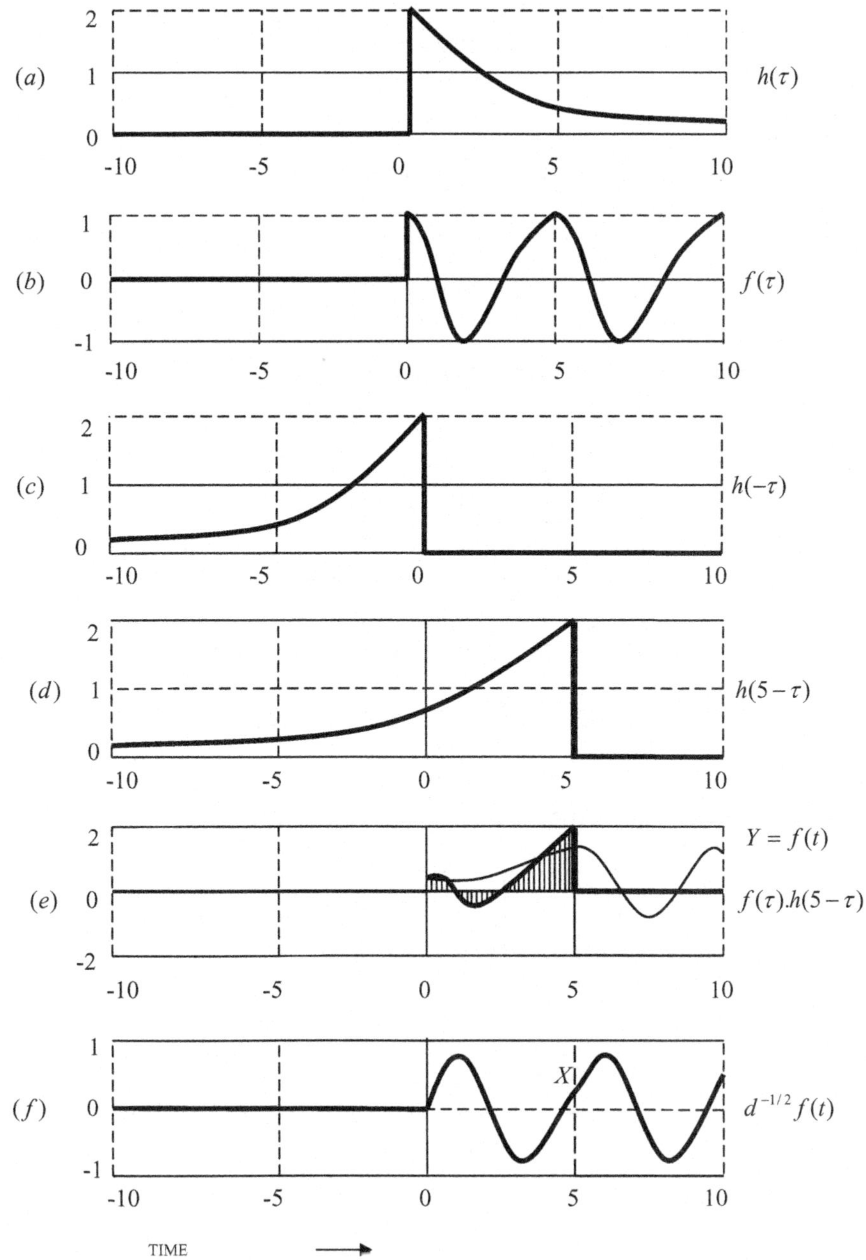

Fig. 5.4 RL integral interpreted as convolution.

In the Figure 5.4f, the point X is

$$\int_0^5 f(\tau)h(5-\tau)d\tau\,,$$

definite value of the integration.

Figure 5.5 demonstrates the several $h(t-\tau)$ for $t = 1, 2, 3, 4, 5, 6, \ldots$

Figure 5.6 demonstrates the semi-derivative of

$$f(t) = \cos\left(\frac{2\pi}{5}t\right),$$

is obtained from differentiating once the RL semi integral graph.

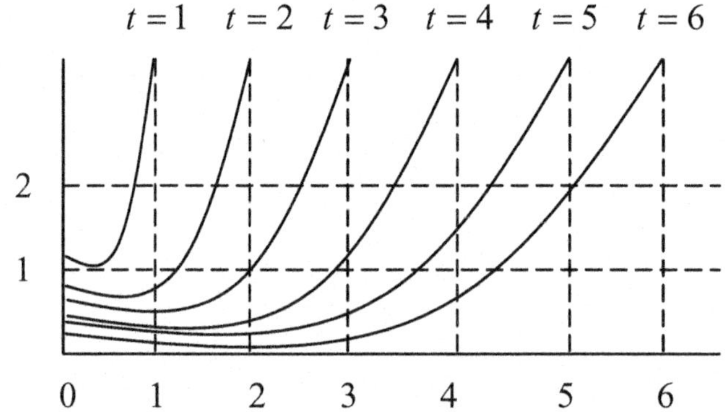

Fig. 5.5 Convoluting function *h* (t) for several *t*.

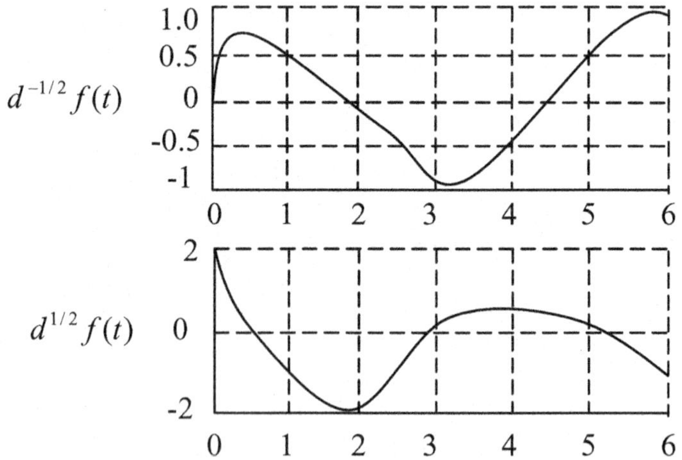

Fig. 5.6 Semi-derivative of function. Differentiating once the RL semi-integral of *f* (t).

5.5.5 Practical Example of RL Differitegration in Electrical Circuit Element Description

These types of intermediate devises are becoming reality in electrical circuits as evident from patent US 20060-267595 of November 2006. We shall start with a resistoductance alone, which is a linear circuit element whose behavior is intermediate between that of an inductor element and ohmic resistor element. The term resistoductance is combination of pure resistance and pure inductor. As integer order equations the fractional order requires fractional derivatives (or integrals) as initial conditions. How does then one relate to initial condition expressed in terms of fractional differintegrals? The constitutive equation of such an element is:

$$v(t) = K\,{}_0D_t^{\alpha} i(t) \text{ or } i(t) = \frac{1}{K}\,{}_0D_i^{-\alpha} v(t)$$

Here $v(t)$ is across variable i.e. the voltage across the circuit element and $i(t)$ is the through variable i.e. current through the circuit element. If $\alpha = 0$, the circuit is purely resistive and $K = R$ ohms and if $\alpha = 1$, the circuit is purely inductive and $K = L$ henrys.

For a step input of voltage, $v(t) = V_0$ applied at $t = 0$. The current is described as fractional integration of the forcing function i.e.

$$i(t) = \frac{1}{K}\,{}_0D_t^{-\alpha} v(t).$$

In terms of convolution definition (as forcing function being causal) and

$$\phi(t) = \frac{t^{\alpha-1}}{\Gamma(\alpha)},$$

the power function, the current is obtained as:

$$i(t) = \frac{1}{K}\left[\phi(t) * v(t)\right] = \int_0^t \frac{(t-\tau)^{\alpha-1}}{K\Gamma(\alpha)} V_0 d\tau = \frac{V_0}{K\Gamma(\alpha)} \int_t^0 -(x)^{\alpha-1} dx$$

$$= \frac{V_0}{K\Gamma(\alpha)}\left[-\frac{x^{\alpha-1+1}}{\alpha-1+1}\right]_t^0 = \frac{V_0}{K\Gamma(\alpha)}\left[-\frac{x^{\alpha}}{\alpha}\right]_t^0 = \frac{V_0}{K\alpha\Gamma(\alpha)} t^{\alpha} a$$

Using $\Gamma(\alpha+1) = \alpha\Gamma(\alpha)$ we obtain the current to step input as:

$$i(t) = \frac{V_0}{K\Gamma(\alpha+1)} t^{\alpha}$$

For pure resistance $\alpha = 0$, and $i(t) = V_0 / K$ and for pure inductance $\alpha = 1$, and $i(t) = (V_0 / K)t$.

The initial value of the current vanishes i.e. there is no instantaneous current, only retarded response. However, the first ordinary derivative of $i(t) = K_1 t^{\alpha}$ is unbounded, so that a finite though undefined current can be reached in arbitrary small time interval. The change of $i(t)$ is described by the fractional differential equation as: ${}_0D_t^{\alpha} i(t) = V_0 / K$.

In accordance with the theory of fractional differential in terms of RL derivatives, an initial condition involving ${}_0D_t^{\alpha-1} i(t)$ is thus required. Physically this initial condition has no representation and cannot be directly obtained from measurement. This condition can be found by taking the first order integral of the constitutive equation. This process relates the fractional (immeasurable) initial condition to something of reality and measurable as:

$$\left[{}_0D_t^{\alpha-1} i(t)\right]_{t\to 0} = \left[{}_0D_t^{-1} (V_0 / K)\right]_{t\to 0}$$

In the case under consideration, voltage stress is finite at all times hence

$$\left[{}_0D_t^{-1} V_0\right]_{t\to 0} = 0,$$

which leads to condition of zero initial condition involving fractional differintegral, namely:

$$\left[{}_0D_t^{\alpha-1} i(t)\right]_{t\to 0} = 0.$$

The same consideration applies to general finite voltage $v(t)$, and equation be solved is ${}_0D_t^{\alpha} i(t) = v(t) / K$, and same zero initial condition be attached.

Now for the impulse input voltage at time 0 i.e. $v(t) = B\delta(t)$ at $t = 0$, the current expression is again

$$i(t) = \frac{1}{K} {}_0D_t^{-\alpha} v(t)$$

using the convolution definition and

$$\phi(t) = \frac{t^{\alpha-1}}{\Gamma(\alpha)},$$

the current expression is:

$$i(t) = \frac{1}{K}\left[\phi(t) * v(t)\right] = \frac{1}{K}\left[\phi(t) * B\delta(t)\right] = \frac{B}{K}\phi(t) = B\frac{t^{\alpha-1}}{K\Gamma(\alpha)}$$

This is obtained as the convolution of function with impulse at $t = 0$, returns the function itself. The power function with gain B is retuned. This is property of the convolution. As observed from the derived current expression for the impulse voltage, the initial voltage-stress singularity give rise to lower order current singularity, since resistoductance cannot respond instantaneously.

The impulse response is mathematical convenience to evaluate transfer characteristics, is seldom used in practice, because it is even more problematic to apply homogeneous impulse voltage on circuit element, than to apply step. However investigating the impulse response will follow the same reasoning as for the step.

For the impulse voltage excitation for $t > 0$, the fractional differential equation is, ${}_0D_t^{\alpha} i(t) = 0$. In accordance with the theory of fractional differential equations with RL derivatives, an initial condition involving $[{}_0D_t^{\alpha-1} i(t)]_{t\to 0}$ is required. This can be found through integration of constitutive equation as:

$$\left[{}_0D_t^{\alpha-1} i(t)\right]_{t\to 0} = \left[{}_0D_t^{-1}(v(t)/K)\right]_{t\to 0} = B/K$$

Which gives the initial condition in terms of fractional differintegral as

$$\left[{}_0D_t^{\alpha-1} i(t)\right]_{t\to 0} = B/K\ .$$

This fractional differintegral initial condition is non-zero well defined and bounded, whereas both current and its integer order derivatives are unbounded, and its first order integral is zero so that a meaningful initial condition expressing the loading conditions cannot be obtained using integer order derivatives.

In above example of resistoductance, it is possible to attribute physical meaning to initial condition expressed in terms of fractional differintegral. Expressing initial condition in terms of fractional derivative of a function $u(t)$ is not a problem, because it does not require a direct experimental evaluation of these fractional derivatives. Instead one should consider its counterpart (in separable twin), $v(t)$ via basic physical law, and measure (or consider) its initial values.

Similarly other intermediate models can be considered as resistocaptance. Resistocaptance will be similar in nature to resistoductance, where the circuit element will be intermediate between pure resistance and pure capacitance. The constitutive part will be

$$v(t) = \frac{1}{K}\,{}_0D_t^{-\alpha} i(t)$$

Or $i(t) = K\,{}_0D_t^{\alpha} v(t)$, where for $\alpha = 1$ the element is pure capacitor with $K = C$ farads, and for $\alpha = 0$ the element will be pure conductance $K = G$ mho.

These intermediate models can explain the behavior of 'time-constant' dispersion effects in the circuit behavior when the relaxation observations cannot be explained by single $\tau = R/L$ or RC time constant.

5.6 Grunwald-Letnikov Fractional Differinteration

The basic definition of Grunwald-Letnikov (GL) is:

$$\frac{d^q f(t)}{[d(t-a)]^q} \equiv \lim_{N\to\infty} \frac{\left(\frac{t-a}{N}\right)^{-q}}{\Gamma(-q)} \sum_{j=0}^{N-1} \frac{\Gamma(j-q)}{\Gamma(j+1)} f\left(t - j\left[\frac{t-a}{N}\right]\right)$$

When the index q is negative the above process is fractional integration and when q is positive fraction the process tends to fractional differentiation. For understanding the recursive formulation of GL method consider small example with $q = 1/2, N = 4$. The GL expansion for the four terms is:

$$\frac{d^{1/2} f(t)}{[d(t-a)]^{1/2}} \equiv \frac{\left(\frac{t-a}{N}\right)^{-1/2}}{\Gamma\left(-\frac{1}{2}\right)} \left\{ \begin{array}{c} \frac{\Gamma(-1/2)}{\Gamma(1)} f(t) + \\ \frac{\Gamma(1/2)}{\Gamma(2)} f\left(t - \left(\frac{t-a}{4}\right)\right) + \\ \frac{\Gamma(3/2)}{\Gamma(3)} f\left(t - 2\left(\frac{t-a}{4}\right)\right) + \\ \frac{\Gamma(5/2)}{\Gamma(4)} f\left(t - 3\left(\frac{t-a}{4}\right)\right) \end{array} \right\}$$

The process explanation is that the function $f(t)$ is first multiplied by a constant then time shifted by amount $((t-a)/N)$ for $(N-1)$ many times and each shifted term gets a weight multiplication ($\Gamma(j-q)/\Gamma(-q)\Gamma(j+1)$) before summing and then scaled by a factor

$$\left(\frac{t-a}{N}\right)^{-q}.$$

Figure 5.7 shows the diagrammatic representation of weighted addition for the four terms as derived for semi-differentiation of the function $f(t)$.

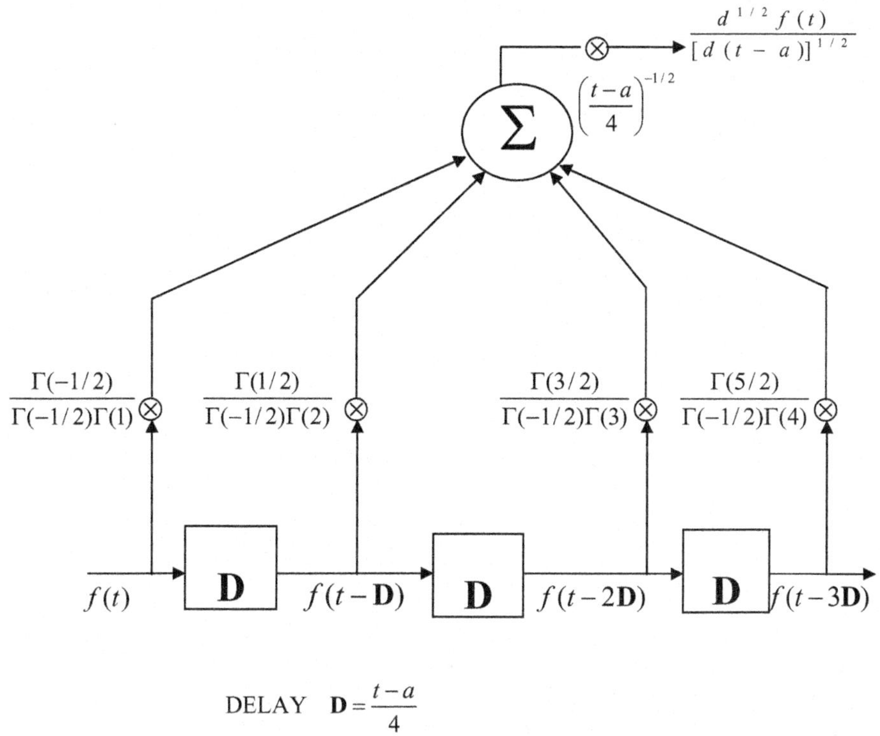

Fig. 5.7 Time delay weighted summation of semi-derivative (four-terms)

Thus the observation is that semi-derivative evaluation for a function by GL method is seen to be a summation of progressively delayed evaluation of $f(t)$ multiplied by progressively decreasing constant weights, and finally multiplied by $\left(\frac{t-a}{N}\right)^{-q}$. In reality choosing the value $N=4$ is crude approximation. The definition states $N \to \infty$, so take large value as $N = 10{,}000$. The following case is considered for $q = 1/2, N = 10{,}000$:

$$\frac{d^{1/2}f(t)}{[d(t-a)]^{1/2}} \equiv \frac{\left(\frac{t-a}{10,000}\right)^{-1/2}}{\Gamma\left(-\frac{1}{2}\right)} \left\{ \begin{array}{c} \frac{\Gamma(-1/2)}{\Gamma(1)} f(t)+ \\ \frac{\Gamma(1/2)}{\Gamma(2)} f\left(t-\left(\frac{t-a}{10,000}\right)\right)+ \\ \frac{\Gamma(3/2)}{\Gamma(3)} f\left(t-2\left(\frac{t-a}{10,000}\right)\right)+ \\ * \\ \frac{\Gamma(j-1/2)}{\Gamma(j+1)} f\left(t-j\left(\frac{t-a}{10,000}\right)\right)+ \\ * \\ \frac{\Gamma(9997.5)}{\Gamma(9999)} f\left(t-9998\left(\frac{t-a}{10,000}\right)\right)+ \\ \frac{\Gamma(9998.5)}{\Gamma(10,000)} f\left(t-9999\left(\frac{t-a}{10,000}\right)\right) \end{array} \right\}$$

By comparison it is seen that the four-term expansion and the 10,000-term expansion of the semi-derivative in GL method Gamma function based coefficients are the same for the first N terms. The time shift factor (incremental delay) is off course very much smaller and indeed approaches zero as the number of terms are made towards infinity.

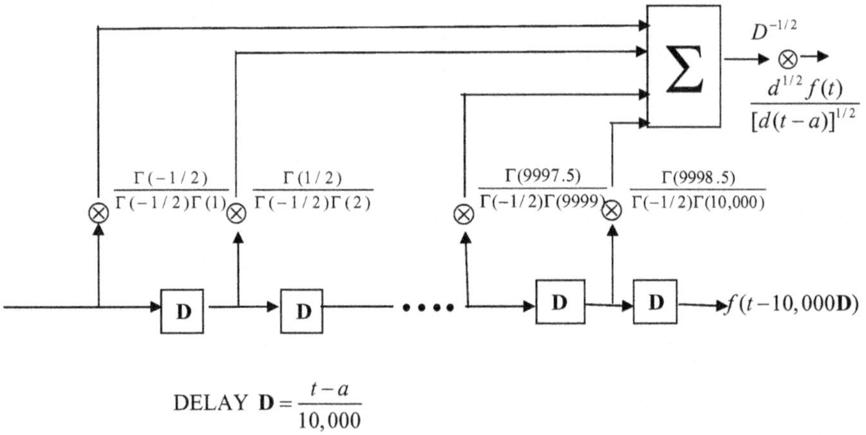

Fig. 5.8 Time delay weighted summation of semi-derivative (large number terms) higher order approximation of GL.

The diagram of Figure 5.8 shows the higher order approximation with large number of terms. The diagrams of Figure 5.7 and Figure 5.8 are similar except the delay steps are smaller in the higher order approximation. One observation is made as the time (independent variable) grows the delay between the consecutive terms increases as

$$\mathbf{D} = \Delta T = \frac{t-a}{N},$$

is the delay. Meaning that for fixed N, as the time grows the resolution between the samples of the function decreases. This is overcome by spreading the sampling instrument N in number uniformly in one interval keeping ΔT same will rewrite the GL approximate as:

$${}_aD_t^q f(t) = \lim_{\Delta T \to 0} \frac{(\Delta T)^{-q}}{\Gamma(-q)} \sum_{j=0}^{N-1} \frac{\Gamma(j-q)}{\Gamma(j+1)} f(t - j(\Delta T))$$

5.7 Unification of Differintegration through Binomial Coefficients

The coefficient in the Figure 5.7 and 5.8, for GL approximation in series form however is through Gamma functions. Practically it is tough to evaluate Gamma function especially when the argument is a large number. Also for electronics and real time computer realization the use of Gamma function is tough. Here the unification approach is presented by use of binomial coefficients. First integer order unification of repeated differentiation and repeated folded integration is presented. The same can be generalized by use of Letnikov theorem to any arbitrary order (not in the scope here).

The repeated integer order differentiation is as follows:

$$f^{(1)}(t) = \frac{df}{dt} = \lim_{h\to 0} \frac{f(t) - f(t-h)}{h}$$

$$f^{(2)}(t) = \frac{d^2 f}{dt^2} = \lim_{h\to 0} \frac{f^{(1)}(t) - f^{(1)}(t-h)}{h} = \lim_{h\to 0} \frac{1}{h}\left\{\frac{f(t)-f(t-h)}{h} - \frac{f(t-h)-f(t-2h)}{h}\right\}$$

$$= \lim_{h\to 0} \frac{f(t) - 2f(t-h) + f(t-2h)}{h^2}$$

$$f^{(3)} = \frac{d^3 f}{dt^3} = \lim_{h\to 0} \frac{f(t) - 3f(t-h) + 3f(t-2h) - f(t-3h)}{h^3}$$

By induction:

$$f^{(n)} = \frac{d^n f}{dt^n} = \lim_{h\to 0} \frac{1}{h^n} \sum_{r=0}^{n} (-1)^r \binom{n}{r} f(t-rh)$$

The binomial coefficient

$$\binom{n}{r} = \frac{n(n-1)(n-2)...(n-r+1)}{r!},$$

for $r > 1$.

The above derivative expression is generalized for any p integer where $p \le n$, is:

$$f_h^{(p)}(t) = \frac{1}{h^p}\sum_{r=0}^{n}(-1)^r\binom{p}{r}f(t-rh)$$

and obviously

$$\lim_{h\to 0} f_h^{(p)}(t) = f^{(p)}(t) = \frac{d^p f}{dt^p},$$

because in such a case as follows, all the coefficients after $\binom{p}{p}$ are zero.

For integration case the index is negative, for convenience sake let us define (like binomial coefficient) the following:

$$\begin{bmatrix} p \\ r \end{bmatrix} = \frac{p(p+1)(p+2)...(p+r-1)}{r!}.$$

Call this as pseudo-binomial coefficient.

Then

$$\binom{-p}{r} = \frac{-p(-p-1)(-p-2)...(-p-r+1)}{r!} = (-1)^r\begin{bmatrix} p \\ r \end{bmatrix}.$$

Using this we have for integration (differentiation with negative index) as:

$$f_h^{(-p)}(t) = \frac{1}{h^{-p}}\sum_{r=0}^{n}\begin{bmatrix} p \\ r \end{bmatrix}f(t-rh) = h^p\sum_{r=0}^{n}\begin{bmatrix} p \\ r \end{bmatrix}f(t-rh)$$

Taking interval $t-a$ and dividing them equally by n write the limit as:

$$\lim_{\substack{h\to 0\\ nh=t-a}} f_h^{(-p)} = {}_aD_t^{(-p)}f(t)$$

Take $p=1$, then $f_h^{(-1)}(t) = h\sum_{r=0}^{n} f(t-rh)$. Taking into account $t-nh=a$,

$$\lim_{\substack{h\to 0\\ nh=t-a}} f_h^{(-1)}(t) = {}_aD_t^{-1} f(t) = \int_0^{t-a} f(t-z)dz = \int_a^t f(\tau)d\tau$$

Take $p=2$, then

$$\begin{bmatrix} 2 \\ r \end{bmatrix} = \frac{2.3.4...(2+r-1)}{r!} = r+1,$$

giving

$$f_h^{(-2)}(t) = h^2 \sum_{r=0}^{n} (r+1) f(t-rh).$$

Rearranging and putting $t+h=y$ we get

$$f_h^{(-2)}(t) = h\sum_{r=0}^{n} [(rh)f(t-rh) + f(t-rh)] = h\sum_{r=1}^{n+1} (rh) f(y-rh)$$

and taking $h\to 0$ we get

$$\lim_{\substack{h\to 0\\ nh=t-a}} f_h^{(-2)}(t) = {}_aD_t^{(-2)} f(t) = \int_0^{t-a} zf(t-z)dz = \int_a^t (t-\tau) f(\tau)d\tau.$$

Take $p=3$, then $\begin{bmatrix} 3 \\ r \end{bmatrix} = \frac{3.4...(3+r-1)}{r!} = \frac{(r+1)(r+2)}{1.2}$ we have

$$f_h^{(-3)}(t) = \frac{h}{1.2}\sum_{r=0}^{n} (r+1)(r+2)h^2 f(t-rh) \to f_h^{(-3)}(t) = \frac{h}{1.2}\sum_{r=1}^{n+1} r(r+1)h^2 f(y-rh),$$

here also $t+h=y$ is substituted. Expressing the above by rearranging:

$$f_h^{(-3)}(t) = \frac{h}{1.2}\sum_{r=1}^{n+1} (rh)^2 f(y-rh) + \frac{h^2}{1.2}\sum_{r=1}^{n+1} (rh) f(y-rh),$$

taking $h\to 0$ we obtain:

$${}_aD_t^{(-3)} f(t) = \frac{1}{2!}\int_0^{t-a} z^2 f(t-z)dz = \int_a^t (t-\tau)^2 f(\tau)d\tau.$$

Because $y\to t$, as $h\to 0$ and

$$\lim_{\substack{h\to 0\\ nh=t-a}} \frac{h^2}{1.2}\sum_{r=1}^{n+1} rhf(y-rh) = \lim_{\substack{h\to 0\\ nh=t-a}} h\int_a^t (t-\tau)f(\tau) = 0$$

Generally this process suggests that:

$${}_aD_t^{-p} f(t) = \lim_{\substack{h\to 0\\ nh=t-a}} h^p \sum_{r=0}^{n} \begin{bmatrix} p \\ r \end{bmatrix} f(t-rh) = \frac{1}{(p-1)!}\int_a^t (t-\tau)^{p-1} f(\tau)d\tau \,.$$

Repeated p folds integration.

Applying the Letnikov theorem above expressions obtained can be generalized for any arbitrary order of differentiation and integration process with binomial coefficients expressed as weights as indicated below:

For arbitrary differentiation of real order the coefficient approximation for

$$(-1)^k \binom{\alpha}{k} \text{ is}$$

$w_0^{(\alpha)} = 1$ and $w_k^{(\alpha)} = \left(1 - \frac{1+\alpha}{k}\right) w_{k-1}^{(\alpha)}$; used for recursive computation, these are weights.

For arbitrary integration of real order the coefficient approximation for $\begin{bmatrix} \alpha \\ k \end{bmatrix}$ is

$$w_0^{(-\alpha)} = 1 \text{ and } w_k^{(-\alpha)} = \left(1 - \frac{1-\alpha}{k}\right) w_{k-1}^{(-\alpha)};$$

used for recursive computation, these are weights.

A caution is put here the factorials in the binomial (for differentiation) and pseudo binomial (for integration) coefficients get generalized by Gamma functions as indicated in the Introduction chapter, for arbitrary order. However these weights are approximates and are helpful in recursive formulation for computation and real time applications, obviously with error.

5.8 Short Memory Principle- A Moving Start Point Approximation and Its Error

For $t >> a$, the number of addends in fractional differintegral approximates becomes enormously large. However, it follows from the expressions for GL definitions and unification arguments in the preceding sections, that for large t the role of "history" of the behavior of the function $f(t)$ near the lower terminal $t = a$, (the start point of the differintegral) process, can be neglected, under certain assumptions. Those observations lead us to the formulation of the "short-memory" principle, which means taking into account only the "recent past" of the function behavior. That is in the interval $[t-L,t]$, where L is memory length (in unit of time). Therefore the approximation is:

$$ {}_aD_t^{\alpha} f(t) \approx {}_{t-L}D_t^{\alpha} f(t), \; (t > a+L) $$

In other words, according to the short-memory principle the fractional differintegrational with lower limit *a*, is approximated by fractional differintegration with "moving lower limit", $(t-L)$. With this approximation the addends in the GL process is always not greater than $[L/h]$. In the selection of number of addend the rule followed is thus, by this short memory principle (for $N(t)$) is

$$ N(t) = \min\left\{\left[\frac{t}{h}\right], \left[\frac{L}{h}\right]\right\}, $$

and where h is step size.

Of course, for this simplification penalty is paid in terms of accuracy. Following rule will explain this. We consider the function in the interval (a,b) to be bounded by $f(t) \le M$ and have the error value as per required accuracy as ε. Then, following estimate is the rule:

Error in short-memory principle expressed as:

$$ \Delta(t) = \left| {}_aD_t^{\alpha} f(t) - {}_{t-L}D_t^{\alpha} f(t) \right| \le \frac{ML^{-1}}{\left|\Gamma(1-\alpha)\right|} $$

Where $(a+L \le t \le b)$ and $f(t) \le M$ for $a \le t \le b$.

This inequality rule can be used for determining the "memory-length" (in unit of time) provided the required accuracy is met i.e. $\Delta(t) \le \varepsilon$ and $a+L \le t \le b$ therefore:

$$ L \ge \left(\frac{M}{\varepsilon \left|\Gamma(1-\alpha)\right|} \right)^{1/\alpha} $$

Summarizing the approximates for differintegration process we write:

$$ {}_aD_t^{\alpha} = \begin{cases} \dfrac{d^{\alpha}}{dt^{\alpha}} & \Re e(\alpha) > 0 \\ 1 & \Re e(\alpha) = 0 \\ \int (d\tau)^{-\alpha} & \Re e(\alpha) < 0 \end{cases} $$

$$
{}_{(t-L)}D_t^{\pm\alpha} f(t) \approx h^{-(\pm\alpha)} \sum_{j=0}^{N(t)} \omega_j^{\pm\alpha} f(t-jh)
$$

$$
N(t) = \min\left\{\left[\frac{t}{h}\right], \left[\frac{L}{h}\right]\right\}
$$

$$
\omega_0^{\pm\alpha} = 1, \omega_j^{\pm\alpha} = \left(1 - \frac{1 \pm \alpha}{j}\right)\omega_{j-1}^{\pm\alpha}
$$

5.9 Matrix Approach to Discretize Fractional Differintegration and Weights

The weights or the coefficients for approximation of fractional differintegration as described in the preceding sections are:

For differentiation:

$$
w_k^{\alpha} = (-1)^k \binom{\alpha}{k}, \text{ for } k = 0, 1, 2, 3...N .
$$

This can be written as:

$$
w_0^{\alpha} = 1; w_k^{\alpha} = \left(1 - \frac{1+\alpha}{k}\right), \text{ for } k = 1, 2, 3...N .
$$

For integration:

$$
w_k^{-\alpha} = \begin{bmatrix} \alpha \\ k \end{bmatrix}, \text{ for } k = 0, 1, 2, 3, ...N .
$$

This can be written as:

$$
w_0^{-\alpha} = 1, w_k^{-\alpha} = \left(1 - \frac{1-\alpha}{k}\right), \text{ for } k = 1, 2, 3, ..N .
$$

Both are same formulation. However in system identification the most appropriate value of order α must be found; this means that various values of α are considered and for each α we have to calculate w_k^{α}. In such cases above recursive method is not easy instead Fast Fourier Transform (FFT) is used. The weights w_k^{α} can be considered (for differentiation) as coefficient of the series function $(1-z)^{\alpha}$. The power series is expanded as:

$$(1-z)^{\alpha}=\sum_{k=0}^{\infty}(-1)^{k}\binom{\alpha}{k}z^{k}=\sum_{k=0}^{\infty}w_{k}^{(\alpha)}z^{k}$$

Substituting $z=e^{-j\varphi}$ we have

$$(1-e^{-j\varphi})^{\alpha}=\sum_{k=0}^{\infty}w_{k}^{(\alpha)}e^{-j\varphi}$$

and the coefficients $w_{k}^{(\alpha)}$ are expressed in FFT as

$$w_{k}^{(\alpha)}=\frac{1}{j2\pi}\int_{0}^{2\pi}f_{\alpha}(\varphi)e^{jk\varphi}d\varphi\,,$$

Where $f_{\alpha}(\varphi)=(1-e^{-j\varphi})^{\alpha}$. $j=\sqrt{-1}$

$w_{k}^{(\alpha)}$, can be computed using FFT. Since in this case we always obtain a finite number of coefficients the FFT can be used with short memory principle.

The approximation to fractional derivative can be written as fractional difference approach:

$$ {}_{a}D_{t_k}^{\alpha}f(t)\approx\frac{\Delta^{\alpha}f(t_k)}{h^{\alpha}}=h^{-\alpha}\sum_{j=0}^{k}(-1)^{j}\binom{\alpha}{j}f_{k-j},\ \text{for}\ k=1,2,3,\ldots N$$

Following formulation of the above in matrix notation is helpful for coefficient evaluation:

$$\begin{bmatrix} h^{-\alpha}\Delta^{\alpha}f(t_0)\\ h^{-\alpha}\Delta^{\alpha}f(t_1)\\ *\\ h^{-\alpha}\Delta^{\alpha}f(t_{N-1})\\ h^{-\alpha}\Delta^{\alpha}f(t_N)\end{bmatrix}=\frac{1}{h^{\alpha}}\begin{bmatrix} w_0^{(\alpha)} & 0 & 0 & * & 0\\ w_1^{(\alpha)} & w_0^{(\alpha)} & 0 & * & 0\\ * & * & * & 0 & 0\\ w_{N-1}^{(\alpha)} & w_{N-2}^{(\alpha)} & * & w_0^{(\alpha)} & 0\\ w_N^{(\alpha)} & w_{N-1}^{(\alpha)} & * & w_1^{(\alpha)} & w_0^{(\alpha)}\end{bmatrix}\begin{bmatrix} f_0\\ f_1\\ *\\ f_{N-1}\\ f_N\end{bmatrix}$$

In the above coefficient matrix the coefficients are symbolically are:

$$w_{j}^{(\alpha)}=(-1)^{j}\binom{\alpha}{j},\ j=0,1,2,\ldots N\,.$$

With the argument as indicated above, for obtaining them from FFT we have a generating polynomial and whose coefficients will the Triangular matrix with $w_{j}^{(\alpha)}$, with truncation as described by:

$$Q(z)=\sum_{k=0}^{\infty}w_{k}^{(\alpha)}z^{k}\leftrightarrow\text{trunc}_{N}(Q(z))\overset{\text{def}}{=}\sum_{k=0}^{N}\omega_{k}^{(\alpha)}z^{k}=Q_{N}(z)\,.$$

The matrix notation we write in short as

$$\left[h^{-\alpha}\Delta^{\alpha} f(t_k)\right] = [\mathrm{B}].[f_k],$$

where matrix $\left[\mathrm{B}\right]$ includes the $1/h^{\alpha}$. We call

$$\mathrm{B}_N^{(\alpha)} = \left[\mathrm{B}\right] = \beta_{\alpha}(z) = h^{-\alpha}(1-z)^{\alpha}.$$

The approximation for fractional integration follows from B_N by doing inverse operation. Define $\mathrm{I}_N^{(\alpha)} = (\mathrm{B}_N^{(\alpha)})^{-1}$; the generating polynomial representation will be for integration as:

$$\mathrm{I}_N^{(\alpha)} \leftrightarrow \varphi_N = \mathrm{trunc}_N(\beta_{\alpha}^{-1}(z)) = \mathrm{trunc}_N(h^{\alpha}(1-z)^{-\alpha})$$

For integration operation the coefficient matrix is:

$$\mathrm{I}_N^{(\alpha)} = h^{\alpha}\begin{bmatrix} w_0^{(-\alpha)} & 0 & 0 & 0 & 0 \\ w_1^{(-\alpha)} & w_0^{(-\alpha)} & 0 & 0 & 0 \\ * & * & w_0^{(-\alpha)} & 0 & 0 \\ w_{N-1}^{(-\alpha)} & w_{N-2}^{(-\alpha)} & * & * & 0 \\ w_N^{(-\alpha)} & w_{N-1}^{(-\alpha)} & * & w_1^{(-\alpha)} & w_0^{(-\alpha)} \end{bmatrix}$$

The integral matrix representation is in short:

$$\left[h^{\alpha}\Delta^{-\alpha} f(t_k)\right] = \left[\mathrm{I}_N^{(\alpha)}\right].[f_k], k = 0,1..N.$$

The weights for integration symbolically are:

$$w_j^{(-\alpha)} = (-1)^j \binom{-\alpha}{j} = \begin{bmatrix} \alpha \\ j \end{bmatrix}, \text{ for } j = 0,1,2,3,\ldots N$$

5.10 Use of Discrete Fractional Order Differintegration in Fractional Order Signal Processing

Above discussion on weights and previous discussion (Chapter 4) on the fractionally differenced process generating Autoregressive Fractional Integral Moving Average (ARFIMA) as,

$$(1-L)^{\alpha} = \sum_{k=0}^{\infty} \binom{\alpha}{k} (-L)^k$$

(where L is the lag operator), gives insight to some very important properties of ARFIMA in context of fractional order signal processing subjects. ARIFMA is a generalization of Autoregressive Moving Average Method (ARMA), used in control theory for system identification.

These fractionally differenced processes (ARFIMA) exhibits long term memory (LRD) persistence, or anti-persistence short term memory.

The fractional order, α of ARFIMA process is between (- ½ to ½) and this is most useful range.

The spectral density, as derived in Chapter 4, is

$$S(\omega) \cong [2\sin(\omega/2)]^{-2\alpha} \cong 1/\omega^{2\alpha}.$$

For $\alpha = 1$, the spectral density is of Brownian Motion (BM). Brownian Motion is output of an integrator driven by a white noise.

When, $\alpha = 0$ ARFIMA process is white noise. For $-(1/2) < \alpha < 0$ the ARFIMA process is with 'short memory' anti-persistence and autocorrelation of the process decays monotonically and hyperbolically to zero. For range $0 < \alpha < +(1/2)$, the ARFIMA process is with long memory; the auto correlation of this time series (which is long range dependent) decays slowly (with heavy tail) as power law function of time. This range is most important for the processes exhibiting Long Range Dependence (LRD). For, $\alpha = +(1/2)$ the spectral density of ARFIMA is like discrete $1/f$ noise. For $\alpha = -(1/2)$ the process is not invertible. These were elaborated in Chapter 4.

The discussion brings out concept of $1/f$ noise for $\alpha = +(1/2)$; this was developed by Bernamont in 1937. The spectral density of $1/f$ noise is $C/|f|^{\beta}$; $0 < \beta < 1$. The $1/f$ noise is typical process that has long memory also termed as 'pink' noise and 'flicker' noise. The basic $1/f$ is output of a fractional order integrator system whose input is white noise. we have developed this concept in Chapter 4. It appears in widely different systems such as radioactive decay, chemical systems, biology, fluid dynamics, astronomy, electronic devices, optical systems, network traffic and economics. In this range of $0 < \alpha < +(1/2)$, the process is Long Range Dependent (LRD), and most useful for identification of system with lingering memory.

Also we can consider $1/f$ type of noise as output of Fractional Integrator with Transfer Function $H(s) = 1/(s)^q$, whose impulse response is $h(t) = t^{q-1}/\Gamma(q)$ for $0 < q < 1$. For $q = 1$, the output of (integer) integrator to white noise input $x(t) = \delta(t)$ will have spectral density as $1/\omega^2 \equiv 1/f^2$; yielding Brownian motion, $q = 1$. The auto-correlation function of the output, $y(t)$ of the Fractional Integrator is $R_{yy}(t) \propto t^{2q-1}$.

Fractional Gaussian Noise (FGN) is another kind of $1/f$ noise. FGN can be seen as stationary increments of self-similar process called Fractional Brownian Motion

(FBM). The FBM plays fundamental role in modeling and identifying Long Range Dependence (LRD). The increment in time series is

$$G_H(k) = B_H(k) - B_H(k-1)\,, k \in \mathbb{Z}\,;$$

of the FBM process, the B_H are called FGN. The scaling properties of 'fractal noise', FGN are characterized by Hurst exponent H. The scaling is $Y(t) = c^{-H} Y(ct)$, (where is c a constant). The FGN is defined by spectral density $S(f) \cong f^{-\beta}$. The model for long range dependence was introduced by Mandelbrot and Van Ness (1968). The value of this LRD lays in financial data, communications networks data, and video traffic bio-corrosion data. Formally let us have stationary time series $Y(k) = \{Y(k)\}$, with zero mean. The time series Y is LRD if autocorrelation

$$R_{yy}(k) = \langle Y(k)Y(0)\rangle \cong c_Y |k|^{-\gamma}\,,\ k \to \infty$$

with, $0 < \gamma < 1$ and spectral density is $S_y(\omega) = c_\omega |\omega|^{-\beta}, 0 < \beta < 1$. The asymptotic exponent, β giving self affine property of the time series, is related to Hurst exponent as $\beta = 2H - 1$, theoretically valid for an infinitely long 'mono-fractal' noise; Fractional Gaussian Noise (FGN).

Here we relate the Hurst exponent, H with γ, β and fractional differencing number as defined for AFRIMA above α. The Hurst exponent was introduced in Chapter 2 and 4 is an index that also characterizes the LRD, and this has to be $0.5 < H < 1$, for process to be LRD. The relationships are $\beta = 1 - \gamma$, $H = (1+\beta)/2$ and $H = \alpha + (1/2)$.

This Hurst exponent gives degree of LRD, for a process; that is autocorrelation of the time series and takes place at $0.5 < H < 1$, where the value $0 < H < 0.5$ indicates a time series with negative autocorrelation (that is a decrease between the values will be followed by an increase; anti-persistent); and $0.5 < H < 1$ indicates a time series with positive autocorrelation (that is an increase between the values will probably followed by another increase, persistence). A value $H = 0.5$ ($\alpha = \beta = 0, \gamma = 1$) indicates white-noise and random walk, where it is equally likely that decrease or increase will follow from any particular value that is the time series has no previous memory. For LRD processes, past events have a decaying effect on the future.

With $H = 1$ the series is 'pink' noise. This type of time-series of pink noise is important because this is a kind of threshold between the persistent stable noise ($0.5 < H < 1,\quad 0 < \beta < 1$), and the non stationary noise $\beta > 1$. Note that common random white noise is characterized by $\beta = 0$ and $H = 0.5$, whereas a random walk or Brownian motion is characterized by $\beta = 2$. The noise characterized by $0 < H < 0.5$ is called anti-persistent while $0.5 < H < 1$ is persistent noise.

In the same way we may call anti-persistent walk given by fluctuation with, $1<\beta<2$ and persistent walk given by fluctuation with $2<\beta<3$. The fluctuation characteristic by exponent $1<\beta<3$ we call 'walks' because they may be obtained by integrating fractal noises characterized by $0<H<1$.

5.11 Infinitesimal Element Geometrical Interpretation of Fractional Differintegrations

The GL definition is described below:

$$
{}_aD_t^q f(t) = \lim_{N\to\infty} \frac{\left(\frac{t-a}{N}\right)^{-q}}{\Gamma(-q)} \sum_{j=0}^{N-1} \frac{\Gamma(j-q)}{\Gamma(j+1)} f\left(t-j\left(\frac{t-a}{N}\right)\right) = \lim_{\Delta T\to 0} \sum_{j=0}^{N-1} \frac{\Gamma(j-q)}{\Gamma(-q)\Gamma(j+1)} \frac{f(t-j\Delta T)}{\Delta T^q}
$$
$$
\Delta T = (t-a)/N, N\to\infty, \Delta T\to 0
$$

The nature of the definition may be explored with the j-th and $j+1$-th term. In a general sense if the terms are additive and $q<0$, then integration is in effect. If the terms are differenced and $q>0$ then differentiation is suggested. Then

$$
{}_aD_t^q f(t) = \lim_{\Delta T\to 0} \left\{ \ldots + \frac{\Gamma(j-q)}{\Gamma(-q)\Gamma(j+1)} \frac{f(t-j\Delta T)}{\Delta T^q} + \frac{\Gamma(j+1-q)}{\Gamma(-q)\Gamma(j+2)} \frac{f(t-(j+1)\Delta T)}{\Delta T^q} + \ldots \right\}
$$

Dividing through out by the coefficients of j-th term, and combining the j-th and ($j+1$)-th term gives:

$$
{}_aD_t^q f(t) = \lim_{\Delta T\to 0} \left\{ \ldots + \alpha\left(\frac{f(t-j\Delta T) + \beta f(t-(j+1)\Delta T}{\Delta T^q} \right) + \ldots \right\}, j = 1,2,3\ldots
$$
$$
\beta = \frac{\Gamma(j+1-q)\Gamma(j+1)}{\Gamma(j-q)\Gamma(j+2)}, \alpha = \frac{\Gamma(j-q)}{\Gamma(-q)\Gamma(j+1)}
$$

5.11.1 Integration

Now when $q=-1$, then $\beta=1$ and $\alpha=1$. The GL equation simplifies for j and $j+1$ term as:

$$
{}_aD_t^{-1} f(t) = \lim_{\Delta T\to 0} \left\{ \ldots + \Delta T\left[f(t-j\Delta T) + f(t-(j+1)\Delta T) \right] + \ldots \right\}, j = 1,2,3\ldots
$$

Which is conventional integration. For simplicity take $q=1/2$, then

$$ {}_aD_t^{-1/2}f(t) = \lim_{\Delta T \to 0}\left\{ ... + \alpha \Delta T^{1/2}\left[f(t - j\Delta T) + \beta_{-1/2} f(t-(j+1)\Delta T \right] + ... \right\},\ j = 1,2,3... $$

$$ \beta_{-1/2} = \frac{j+\frac{1}{2}}{j+1} $$

For all j greater than equal to zero β is positive. The subscripted symbol indicates semi-integration process. For the general case the expression is:

$$ {}_aD_t^{-q}f(t) = \lim_{\Delta T \to 0}\left\{ ... + \alpha \Delta T^{q}\left[f(t - j\Delta T) + \beta_{-q} f(t-(j+1)\Delta T \right] + ... \right\},\ j = 1,2,3... $$

Thus α and β are always positive, when q is between 0 and –1, and the above summation is seen to be integration process and indeed fractional. A geometric approximation to this integration is given in the Figure 5.9.

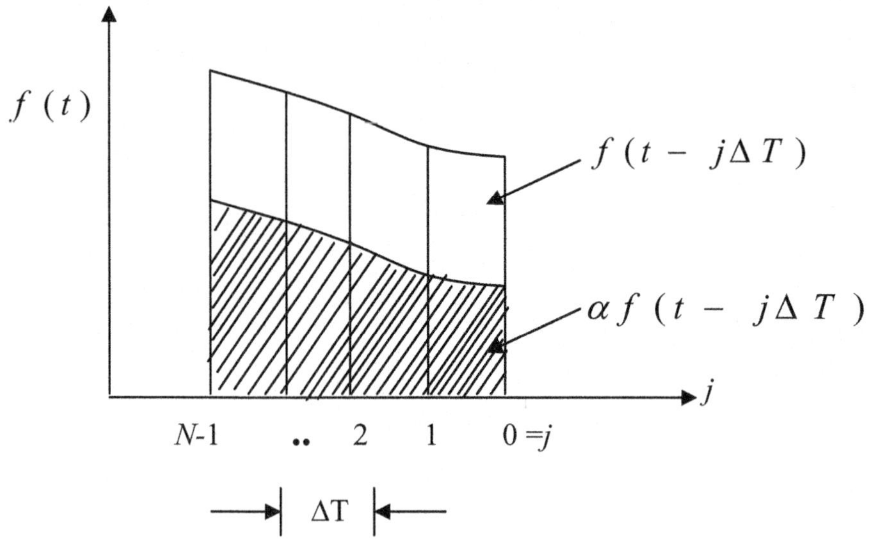

Fig. 5.9 Geometric interpretation if infinitesimal incremental fractional integration.

If $q = -1$, then $\sum \alpha f(t - j\Delta T)\Delta T$ is an area represented below the curve $\alpha f(t - j\Delta T)$. If $q = -2$ (Outside the domain of consideration here as q is restricted to –1), then $\sum \alpha f(t - j\Delta T)\Delta T^2$ is a volume. Then the series $\sum \alpha f(t - j\Delta T)\Delta T^q$ for $0 \ge q > -1$ may be thought of as "fractional-area". In Figure 5.9 fractional integral is area under the $\alpha f(t - j\Delta T)$ curve multiplied by ΔT^{q-1}, a fractional scaled version of the shaded area or full area.

5.11.2 Differentiation

For $q = 1, \alpha = 0$ for all *j* except for $j = 0$. The GL expansion is:

$$_aD_t^1 f(t) = \lim_{\Delta T \to 0} \left\{ \frac{f(t) - f(t - \Delta T)}{\Delta T} \right\}$$

Which is conventional integer order differentiation.

For general case we take $q = 1/2$, then

$$\beta_{\frac{1}{2}} = \frac{\left(j - \frac{1}{2}\right)}{(j+1)}$$

here for all $j \geq 1 \to \beta_{\frac{1}{2}} > 0$. It can be seen that β_q will be positive for all q when $j \geq 1$. It is also true that $0 > \alpha \geq -1$, for all q in the range. Therefore:

$$_aD_t^{\frac{1}{2}} f(t) = \lim_{\Delta T \to 0} \left\{ ...\alpha \Delta T^{-\frac{1}{2}} \left[f(t - j\Delta T) + \beta_{\frac{1}{2}} f(t - (j+1)\Delta T) \right] + ... \right\}$$

For general q the expression is:

$$_aD_t^q f(t) = \lim_{\Delta T \to 0} \left\{ ...\alpha \Delta T^{-q} \left[f(t - j\Delta T) + \beta_q f(t - (j+1)\Delta T) \right] + ... \right\}$$

So after the first $j = 0$ term, it is seen that all terms are a direct sum of negatively weighted functions, again integration process. However the effectiveness of the weighing ΔT^{-q} is changed as $q > 0$. The first $j = 0$ term for all q is $\Delta T^{-q} f(t)$, thus considering the first two terms we have:

$$_aD_t^q f(t) = \lim_{\Delta T \to 0} \left\{ \frac{f(t) - qf(t - \Delta T)}{\Delta T^q} + ... \right\}$$

This brings in an effective differentiation (for $q > 0$) though scaled by ΔT^q instead of ΔT, as in the case of one-order differentiation.

If the q is taken as 1 then this returns to rate of change, like velocity (slope). For value 2 i.e. outside the range it yields acceleration. For 0-1 therefore the expression may be called as fractional rate of change of function. Figure 5.10, shows the $j = 0$ and $j = 1$ points of a geometric approximation to the q th derivative. The slope between the curves multiplied by ΔT^{-q+1}, is loosely a geometric interpretation for this part of the fractional derivative or fractional rate. The remaining terms after the first two are like integration terms may be interpreted in as in Figure 5.9, but meaningless at limit $\Delta T \to 0$.

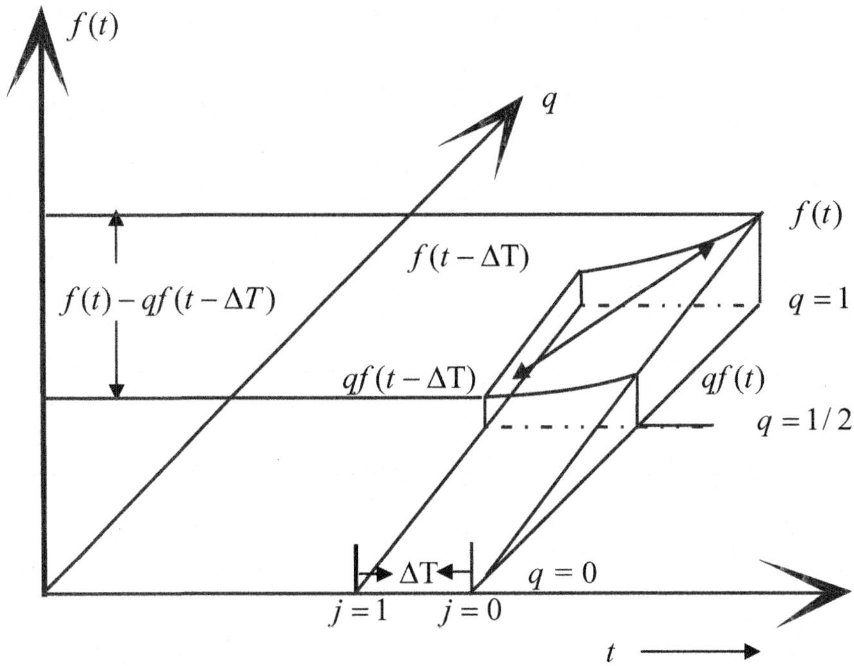

Fig 5.10 Geometric interpretation of fractional differentiation for j=0 to j=1.

5.12 Local Fractional Derivatives (LFD)

Fractals and multifractals functions and corresponding curves or surfaces are found in numerous places in non-linear and non-equilibrium phenomenon. For example low viscous turbulent fluid, Brownian motion. These phenomena give occurrence of continuous but highly irregular (non-differentiable) curves. However the precise nature of connection between the dimensions of the graph of fractal curve and fractional differentiability property was only recognized recently. One way to express LFD is by limit of first derivative of fractional integral of order $1-\alpha$ of a function $f(x)$ subtracted from an offset that is value of function at x_0 that is $f(x_0)$ when $x \to x_0$, and is expressed as follows:

The fractional integral is

$$I(x) = \frac{1}{\Gamma(1-\alpha)} \int_{x_0}^{x} (x-x')^{-\alpha} \left[f(x') - f(x_0) \right] dx',$$

the first derivative of this fractional integral while $x \to x_0$ is LFD at point.

$$\mathbf{D}^{\alpha} \left[f(x) \right]_{x=x_0} = \lim_{x \to x_0} I'(x)$$

The way of classical definitions of fractional derivatives (forward derivative) do indicate that the derivative at a point is determined by all the states that the function had been through in the past till that point. This is non-local character of the fractional derivative in classical sense. Well if a function is defined in interval $[a,b]$ then if one has to find fractional derivative at a point $a < x_0 < b$, then one has to know not only entire character of the function from a to x_0 but also to calculate backward derivative. One also needs to know the 'future' of the function (from x_0 to b) and only when these forward and backward fractional derivatives are equal then we can say that fractional derivative at local point x_0 exists. This puts a demanding restriction on this approach.

K.M. Kolwankar and Anil D. Gangal introduced a new notion of Local Fractional Derivative (LFD). This method will be termed as KG Local Fractional Derivative.

In this section only definition of this LFD will be introduced. The classical definition of fractional derivatives of Riemann-Liouvelli (RL), Grunwald-Letnikov (GL) discussed in detail in this book, makes it a non-local property. Also as discussed the RL approach of fractional derivative of a constant makes it non-zero. These two features make extraction of scaling information somewhat difficult.

5.12.1 KG- LFD for Order Less than Unity

LFD tries to overcome these issues by having a neighborhood point approach as defined for a function $f:[0,1] \to \mathbb{R}$ (Function defined for values 0 to one). Why this domain is chosen, the reason is the basic definition of box-dimension where the geometrical method assumes that function is normalized between 0 and one; refer Chapter 2.

$$\mathbf{D}^q f(x_0) = \lim_{x \to x_0} \frac{d^q \left(f(x) - f(x_0)\right)}{d(x - x_0)^q}$$

If this limit exists and is finite then we say the LFD of order q, for $0 < q < 1$ at $x = x_0$ exists. In this definition the lower limit x_0 is treated as a constant. The subtraction of $f(x_0)$ corrects for the fact that fractional derivative of a constant is not zero, for Riemann-Liouvelli fractional derivative. Whereas, limit $x \to x_0$, is taken to remove non-local contents. Advantage of defining LFD in this manner lies in its local nature and hence allowing the study of point wise behavior of functions.

Say the, point of interest is $x_0 = 0$, and $q = 1$ then the LFD is slope at point zero for function $f(x)$. That is:

$$\mathbf{D}^1 f(0) = \lim_{x \to 0} \frac{d}{dx} f(x)$$

The logic in this approach is to get 'fractional-slope' at a point $x = x_0$, the constant offset gets subtracted. Due to non-local character of RL definition and due to the fact that Fractional Derivative of a constant is not zero, addition of constant changes the magnitude of Fractional Derivative. This KG definition of LFD is offsetting the effect of constant in RL derivative by shifting the point of interest x_0 and $f(x_0)$ to the origin (0, 0). Thereby, nullifying the effect of constant by subtracting $f(x_0)$, the lower limit is treated as a constant.

5.12.2 KG- LFD for Order Greater than Unity

The method of obtaining KG-LFD for fractional order greater than one is as follows. The idea is to expand the function in Taylor series at the point $x = x_0$, by choosing an integer N such that $N < q \le N+1$ and follow the same procedure of subtracting this value $f(x_0)$ from the function and then take the Fractional Derivative in the limit of $x \to x_0$, this is formulated as follows; for a function $f:[0,1] \to \mathbb{R}$ (Function defined for values 0 to one).

$$\mathbf{D}^q f(x_0) = \lim_{x \to x_0} \frac{d^q}{[d(x-x_0)]^q}\left(f(x) - \left[f(x_0) + f^{(1)}(x_0)(x-x_0) + \frac{f^{(2)}}{2!}(x-x_0)^2 + \ldots + \frac{f^{(N)}}{N!}(x-x_0)^N \right]\right)$$

$$\mathbf{D}^q f(x_0) = \lim_{x \to x_0} \frac{d^q \left(f(x) - \sum_{n=0}^{N} \frac{f^{(n)}}{\Gamma(n+1)}(x-x_0)^n \right)}{[d(x-x_0)]^q}$$

If this limit exits and is finite where N is the largest integer for which N^{th} derivative of $f(x)$ exists at x_0 and is finite, then we say LFD of order q with $N < q \le N+1$ at x_0 exists.

5.12.3 Critical Order of a Function and Its Relation to the Box Dimension

Definition of critical order is α at $x = x_0$ as supremum (of q) the order of LFD, of order less than q exists at $x = x_0$.

Let us take a function $f(x) = a + bx + c|x|^{\beta}$ with $\beta > 1$, say $\beta = 1.5$, then we have $f(0) = a$ and $f^{(1)}(0) = b$ at $x = x_0 = 0$ with $N = 1$ for this case we get:

$$D^q f(0) = \frac{d^q}{[d(x-0)]^q}\left[f(x) - f(0) - f^{(1)}(0)x \right] = \left[\frac{d^q}{dx^q} c|x|^{\beta} \right]_{x=0} = \left[c\frac{\Gamma(\beta+1)}{\Gamma(\beta-q+1)} x^{\beta-q} \right]_{x=0}$$

Clearly at the origin for $q = \beta$, fractional derivative exists and the fractional derivatives diverge at origin for $q > \beta$. The critical order for this example is thus β.

Let us take example of an irregular function and find out its Critical Order. The example is of Weierstrass Function

$$W_\lambda(x) = \sum_{k=1}^{\infty} \lambda^{(s-2)k} \sin \lambda^k x,$$

with $\lambda > 1$ and $1 < s < 2$; and $x \in \mathbb{R}$ (real number). The box dimension is $d_B = s$, for this function. The Hurst exponent is thus $H = 2 - d_B = 2 - s$, for this function. This Weierstrass function is continuously fractionally differentiable locally for orders $q < (2-s)$ and not for any fractional order q between $(2-s)$ and 1. This we discuss here in this section. Note that $W_\lambda(0) = 0$. Now we fractionally differentiate this:

$$\frac{d^q}{dx^q} W_\lambda(x) = \sum_{k=1}^{\infty} \lambda^{(s-2)k} \frac{d^q}{dx^q} \sin \lambda^k x$$

Use scaling property of the Fractional Differintegration as:

$$\frac{d^q}{dx^q} f(\beta x) = \beta^q \frac{d^q}{d(\beta x)^q} f(\beta x)$$

to write

$$\frac{d^q \sin \lambda^k x}{dx^q} = \left(\lambda^k\right)^q \frac{d^q \sin \lambda^k x}{d\left(\lambda^k x\right)^q},$$

thus:

$$\frac{d^q}{dx^q} W_\lambda(x) = \sum_{k=1}^{\infty} \lambda^{(s-2+q)k} \frac{d^q}{d(\lambda x)^q} \sin \lambda^k x$$

For

$$0 < q < 1, \frac{d^q}{dx^q} \sin x = \frac{d^q}{dx^q} \int_0^x \cos t dt = \frac{d^{q-1}}{dx^{q-1}} \cos x,$$

and putting this above we get:

$$\frac{d^q}{dx^q}W_\lambda(x) = \sum_{k=1}^{\infty} \lambda^{(s-2+q)k} \frac{d^{q-1}}{d(\lambda x)^{q-1}} \cos(\lambda^k x)$$

The fractional integral of $\cos\left(\lambda^k x\right)$, of order $1-q$ is bounded uniformly for all values of $\lambda^k x$. Here since $q<1$, the operator d^{q-1} is actually fractional integration operation. This implies that RHS of the obtained expression for fractional derivative of $W_\lambda(x)$ will converge if $s-2+q<0$ or $q<2-s$. This justifies taking the fractional derivative operator inside the sum. The fractional integral of $\cos x$ as $x \to 0$ is zero ($D_x^{-v} \cos ax = C_x(v,a)$, and $C_0(v,a)=0$).

Therefore,

$$\lim_{x\to 0} \frac{d^q W_\lambda(x)}{dx^q} = 0 \text{; for } q < 2-s \text{.}$$

For any other point, other than zero i.e., $x = \chi \neq 0$ we use $x' = x - \chi$, and; $\tilde{W}_\lambda(x') = W_\lambda(x'+\chi) - W_\lambda(\chi)$, so that $\tilde{W}_\lambda(0) = 0$.

We have:

$$\tilde{W}_\lambda(x') = \sum_{k=1}^{\infty} \lambda^{(s-2)k} \sin \lambda^k (x'+\chi) - \sum_{k=1}^{\infty} \lambda^{(s-2)k} \sin\lambda^k \chi$$

$$= \sum_{k=1}^{\infty} \lambda^{(s-2)k} \left(\cos\lambda^k \chi \sin\lambda^k x' + \sin\lambda^k \chi(\cos\lambda^k x' - 1)\right)$$

Taking Fractional Derivative of $\tilde{W}_\lambda(x')$ w.r.t. x' and following the same procedure as above we can show that Fractional Derivative (LFD) for Weierstrass function of order $q<2-s$ exists at all points, and for $q>2-s$ the LFDs diverge (does not exist).

The critical order is therefore $\alpha = 2-s$ for Weierstrass function. Therefore the Weierstrass function is not differentiable by any integer order 1, 2... though is Fractionally Differentiable at all points. This is one classic example of a function which is continuous everywhere but nowhere differentiable, like graph of stock-index fluctuations!

We therefore write the rule relating the critical order to the box dimensions as for a function $f:[0,1] \to \mathbb{R}$ to be a continuous one, if $\mathbf{D}^q f(x_0) = 0$ for $q<\alpha$ where $0<q,\alpha<1$ for all x_0 then box dimension of the function is $d_B \le 2-\alpha$. Also if there exist a sequence of x_n in close proximity of x_0 and as $n\to\infty$ such that $\mathbf{D}^q f(x_0) = \pm\infty$ for $q>\alpha$ for all x_0 then $d_B \ge 2-\alpha$.

5.12.4 Information Content in LFD

As mentioned in the Chapter 4 the derivative gives the measure of curvature of a function. In order to see the information contained in the LFD we consider the 'fractional' Taylor series, with remainder term for a real function $f(x)$. Let

$$F(x_0, x-x_0; q) = \frac{d^q}{\left[d(x-x_0)\right]^q}\left(f(x)-f(x_0)\right),$$

it is therefore clear that $\mathbf{D}^q f(x_0) = F(x_0, 0; q)$. Now for $0<q<1$, using Riemann-Liouvelli fractional integration definition we write from above formulation of F, by use of integration by parts, the following expression.

$$\begin{aligned} f(x)-f(x_0) &= \frac{1}{\Gamma(q)}\int_0^{x-x_0}(x-x_0-\xi)^{q-1}F(x_0,\xi;q)d\xi \\ &= \frac{1}{\Gamma(q)}\left[F(x_0,\xi;q)\int(x-x_0-\xi)^{q-1}d\xi\right]_0^{x-x_0} \\ &\quad +\frac{1}{\Gamma(q)}\int_0^{x-x_0}\frac{dF(x_0,\xi;q)}{d\xi}\frac{(x-x_0-\xi)^q}{q}d\xi \end{aligned}$$

provided the second term in the above expression exists. Therefore we obtain:

$$f(x)-f(x_0) = \frac{\mathbf{D}^q f(x_0)}{\Gamma(q+1)}(x-x_0)^q + \frac{1}{\Gamma(q+1)}\int_0^{x-x_0}\frac{dF(x_0,\xi;q)}{d\xi}(x-x_0-\xi)^q d\xi$$

That is the expression for 'fractional' Taylor series is as follows:

$$f(x) = f(x_0) + \frac{\mathbf{D}^q f(x_0)}{\Gamma(q+1)}(x-x_0)^q + R_q(x,x_0)$$

The remainder is given as:

$$R_q(x,x_0) = \frac{1}{\Gamma(q+1)}\int_0^{x-x_0}\frac{dF(x_0,\xi;q)}{d\xi}(x-x_0-\xi)^q d\xi$$

This expansion of a function at a given point is fractional Taylor series expansion involving only the lowest leading terms. The further expansion may be carried on to higher terms provided the corresponding remainders are defined too. We note that

LFD gives a coefficient A where $A = \mathbf{D}^q f(x_0)$, the LFD at point; for approximating a function $f(x)$ in vicinity of x_0; by

$$f(x) \sim f(x_0) + A(x - x_0)^q / \Gamma(q+1).$$

This coefficient is non trivial at $q = \alpha$, the critical order. One may generalize the Taylor series for higher order derivatives N integer and fractional number $0 < q < 1$ as:

$$f(x) = \sum_{n=0}^{N} \frac{f^{(n)}(x_0)}{\Gamma(n+1)} \Delta^n + \frac{\mathbf{D}^q f(x_0)}{\Gamma(q+1)} \Delta^q + R_q(x_0, \Delta); \; \Delta = x - x_0 > 0$$

We note that the function is differentiable up to N with existence of first N derivatives, the Taylor expansion provides approximation of $f(x)$ in vicinity of x_0.

For $q = 1$, the expansion is

$$f(x) = f(x_0) + \frac{\mathbf{D}^1 f(x_0)}{\Gamma(2)}(x - x_0) + R_1(x, x_0) \approx f(x_0) + \{f^{(1)}(x_0)\}(x - x_0)$$

This is the equation of straight line. That is, the curves passing through x_0 and having the same slope (tangent) forms an equivalence class of functions (approximated by linear functions). Analogously all the curves (functions) with the same 'critical order' α and the same LFD $\mathbf{D}^\alpha$ will form an equivalence class approximated and represented by power law i.e., x^α. This fractional Taylor series expansion may be taken as geometric interpretation of fractional derivatives generalizing the tangent concept. This is a useful observation when we try to approximate an irregular function (at a point) by piecewise smooth scaling function (as power law).

The information that LFD gives is about strength of singularity. Consider a case of single isolated singularity as $f(x) = ax^\alpha$ with $0 < \alpha < 1$ and $x > 0$. At the origin the critical order gives the 'order of singularity', whereas the LFD value at $x = 0$ i.e. $\mathbf{D}^\alpha f(0) = a\Gamma(\alpha+1)$, gives the 'strength of singularity'. If there are multiple singularities we can use the LFD to detect weaker singularity masked by stronger one. Consider a function $f(x) = ax^\alpha + bx^\beta$ with $x > 0$ and $0 < \alpha < \beta < 1$. The LFD of order of this function at $x = 0$ by using KG definition is

$$\mathbf{D}^\alpha f(0) = \left[a\Gamma(\alpha+1) + b\frac{\Gamma(\beta+1)}{\Gamma(\beta-\alpha+1)} x^{\beta-\alpha} \right]_{x=0} = a\Gamma(\alpha+1).$$

Using this value of stronger singularity we write a function $G(x;\alpha)$ as:

$$G(x;\alpha) = f(x) - f(0) - \frac{\mathbf{D}^{\alpha} f(0)}{\Gamma(\alpha+1)} x^{\alpha},$$

which for the function $f(x) = ax^{\alpha} + bx^{\beta}$, $0 < \alpha < \beta < 1$ and $x > 0$ is:

$$G(x;\alpha) = ax^{\alpha} + bx^{\beta} - f(0) - \frac{\mathbf{D}^{\alpha} f(0)}{\Gamma(\alpha+1)} x^{\alpha}$$

$$= ax^{\alpha} + bx^{\beta} - 0 - a\frac{\Gamma(\alpha+1)}{\Gamma(\alpha+1)} x^{\alpha} = bx^{\beta}$$

$$\frac{d^q G(x;\alpha)}{dx^q} = b\frac{\Gamma(\beta+1)}{\Gamma(\beta-q+1)} x^{\beta-q}$$

At $q = \beta$ the above expression is finite and thus critical order of the function $G(x;\alpha)$ is β at the origin. This gives a method of using LFD to extract the secondary singularity of the function masked by the primary singularity of higher strength.

The concept of fractional differentiation especially LFD allows us to quantify the loss of differentiability of 'uni-fractal' or 'multi-fractal' irregular and rough functions (signals). The larger the irregularity of the function the smaller is the extent of differentiability and smaller is the value of Holder exponent. Bigger the fractal dimension the smaller is the extent of differentiability. Local Taylor series expansions or fractional Taylor approximation provides way to approximate irregular rough functions by piece wise scaling functions. In particular Holder exponents (box-dimensions) are related to the critical order of the corresponding Fractional (Local) derivative. This is useful in study of irregular signals, image processing where one can characterize singularities with image data. This LFD can extract information about the dimension of irregular graph, say the data of mortality rate from epidemic outburst; the fluctuations in stock market. The change in dimension of graph from say about 1.5 to a lower value approximately 1.0 gives the indication about setting in of regular process from the random Brownian case (white-noise) to a regular behavior. A signal arrival point in time buried in large white noise can be estimated by looking at the point where dimension of the graph changes. Thus arrival time estimation of first seismic wave (buried in high noise background) can be estimated by LFD and its relation to the 'fractal-dimension'.

For example an abrupt phase change of a process at point $x_0 = 0$ is approximated as shown in Figure 5.11.

$$\left|f(x) - P_n(x - x_0)\right| = C\left|x - x_0\right|^h$$
$$x_0 = 0; P(x) = 2 + 3x$$
$$f(x) = (2 + 3x) \pm 4\left|x\right|^{1/2}$$

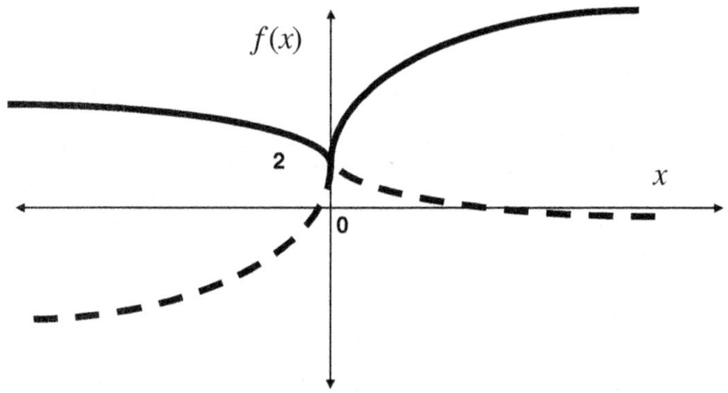

Fig. 5.11 Non-Differentiability at critical point zero is approximated by function

Non-differentiability irregularity, at point $x_0 = 0$ in this case of Figure 5.11 has value $f(x) = 2$, is approximated by a polynomial $P(x - x_0) = 2 + 3x$ with non-linear power law

$$C\left|x - x_0\right|^h = 4\left|x\right|^{0.5}.$$

Response functions of several processes diverge algebraically near critical point, for example Vander Wall's equation at critical point, Stress-strain curve near yield point, Magnetic Currie points etc. This way one can get 'continuous phase-transition' explanations. The enlarging of the differentiability of the Figure 5.11 approximation at critical point is shown in Figure 5.12. The function is continuous at critical point $x_0 = 0$ from differentiation of order $q = 0$ till $q < 0.5$. Beyond that $q > 0.5$ order the function diverges. The critical order is half, and the value at half order is $8\Gamma(1.5)$. Say in this phase transition 'loss of integer order differentiability' represents curve of specific heat with temperature, then at this phase transition point this finite value of fractional derivative $8\Gamma(1.5)$, may be termed as 'Fractional Latent Heat'. This way one can have LFD to expand the 'critical-points' or 'non-differentiable points' or 'irregular points' of physical systems.

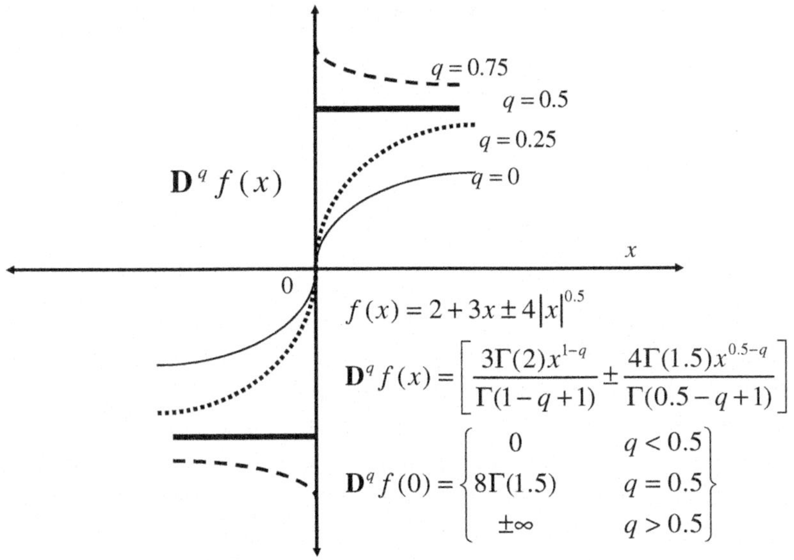

Fig. 5.12 Observation of fractional differentiability for a non-differentiable function

5.12.5 Finding Holder Exponent for Singularity at a Point

We have made use of concept of Holder exponent at the 'critical-point' where differentiability gets lost, in earlier sections. The Holder exponent of a singularity can be evaluated using the Wavelet Transform (WT). The WT makes use of scaling functions that have the property of being 'localized' in both time and in frequency. A scaling coefficient s characterizes and measures the width of a wavelet. Continuous Wavelet Transform (CWT) is defined for a given signal $f(x)$ as

$$W_{w,x_0}(f) = \int_{-\infty}^{+\infty} \frac{1}{w}\psi\left(\frac{x - x_0}{w}\right) f(x)dx,$$

where $\psi(u)$, with $u = (x - x_0)/w$ as the 'kernel' is also called 'wavelet filter', at origin $u = 0$, with 'unit window' $w = 1$. The WT can be used to determine singularity because $\psi(u)$, the kernel can be chosen in such a way as to be an orthogonal to polynomials up to degree n that is,

$$\int_{-\infty}^{+\infty} \frac{1}{w}\psi\left(\frac{x - x_0}{w}\right) x^m dx = 0;$$

for all m such that $0 \le m \le n$.

If function $f(x)$ full fills the condition

$$\left|f(x)-P_n(x-x_0)\right|\le C\left|x-x_0\right|^{h(x_0)},\ h(x_0)>n;$$

its WT is given as

$$W_{w,x_0}(f)=C\left|w\right|^{h(x_0)}\int_{-\infty}^{+\infty}\left|u\right|^{h(x_0)}\psi(u)du\propto\left|w\right|^{h(x_0)},$$

where $u=(x-x_0)/w$. Therefore at least theoretically a Holder exponent of singularity can be evaluated as scaling exponent of WT coefficient for $w\to 0$.

One of the kernel useful in this particular application is Mexican hat function (second derivative of Gaussian) that is, $\psi(u)=(1-u^2)\exp(-u^2/2)$. The Mexican hat wavelet integrates to zero polynomial biases up to degree $n=1$. Finally, due to exponential convergence of this wavelet to zero, in this Mexican hat function for large $|u|$, we may assume that this wavelet explores a 'window size' a ten times the scale s. These above arguments are true if there exists no other singularity in near neighborhood of x_0. Therefore calculate the $\left|W_{w,x_o}(f)\right|$ at a point x_0, with window of the kernel w varying from 0 to say 100, and then one can use linear regression for finding Holder exponent at x_0 that is, $h(x_0)$ by using the

$$\log\left|W_{w,x_0}(f)\right|=h(x_0)\log\left|w\right|+C,$$

the slope of this as $|w|\to 0$.

5.13 Numerical Solution of Fractional Order Differential Equation by Use of Grunwald-Letnikov Technique

5.13.1 The Algorithm

In this algorithm, the fractional differential term is directly replaced by the numerical approximation definition given by Grunwald-Letnikov. But in order to do that, we have to discretize the time with a sampling period satisfying the Nyquist criterion.

Thus, we have $t = h*n$ and $f(t)$ is denoted by $f(n)$ where $f(n)$ is the discretized function. Using Grunwald's approximation,

$$D_t^{\alpha}f(Kh)=\lim_{h\to 0}\frac{1}{h^{\alpha}}\sum_{j=0}^{n}(-1)^j\frac{\Gamma(\alpha+1)}{\Gamma(\alpha-j+1)\Gamma(j+1)}f(Kh-jh) \tag{5.1}$$

where K is the number of data points available.

Consider a system described by the fractional order differential equation given by:

$$a\frac{d^n y(t)}{dt^n} + by(t) = u(t)\text{ , with }(0 < n < 1) \tag{5.2}$$

Using approximation in (5.1),

$$a\frac{1}{h^n}\sum_{j=0}^{n}(-1)^j \frac{\Gamma(n+1)}{\Gamma(n-j+1)\Gamma(j+1)} y(Kh-jh) + by(Kh) = u(Kh) \tag{5.3}$$

System output is given by

$$y(Kh) = \frac{u(Kh) - \frac{1}{h^n}\sum_{j=0}^{n}(-1)^j \frac{\Gamma(n+1)}{\Gamma(n-j+1)\Gamma(j+1)} y(Kh-jh)}{b + \frac{a}{h^n}} \tag{5.4}$$

5.13.2 Obtaining the Step Response

Equation (5.4) directly gives the solution for a simple n-th order FO system. Step response of the system represented by equation (5.5) is studied for various values of n.

$$a\frac{d^n}{dt^n} y(t) + by(t) = u(t)\text{ with }0 < n < 1 \tag{5.5}$$

For all the plots, the total number of points taken for evaluation is 1000, shown in Figure 5.13 to Figure 5.16.

5.13.3 Fractional Order System and Integer Order System Comparision

5.13.3.1 Order of the FOS-*n*

The step response of different values of n is plotted. It can be seen that the time constant of the system increases as the value of n reaches from 0 to 1.

The step response for values of n greater than one is interesting. Since the order of the system is more, the response is more of an oscillating kind tending toward second order system with damping. This observation of oscillation relaxation is made for a non-Newtonian fluid under compression, where the order of fractional differential equation $n = 1.5$ gives a very good curve fitting on experimental data.

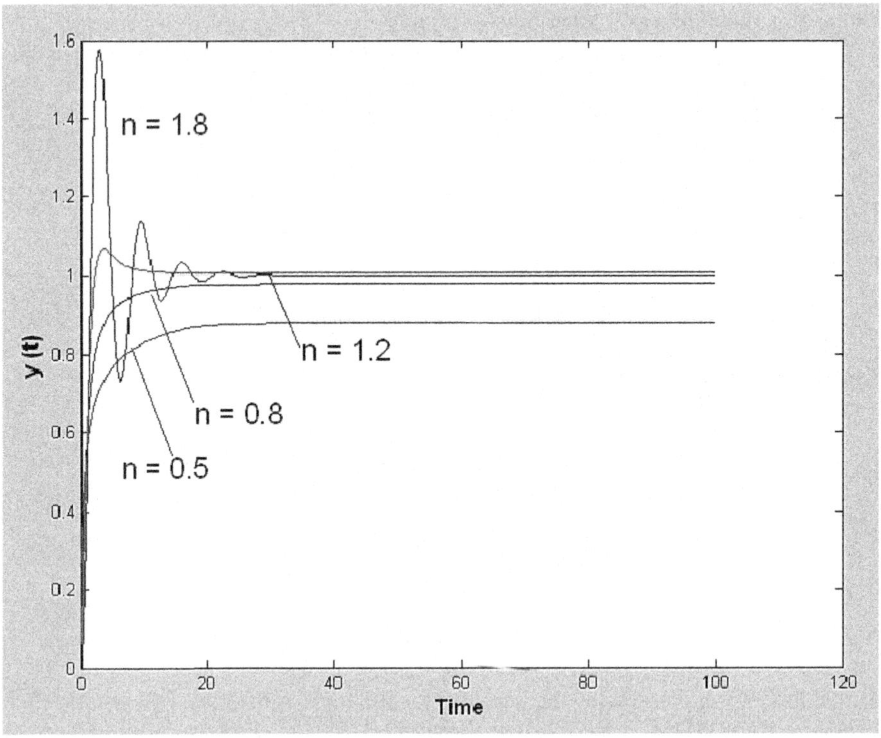

Fig 5.13 Step response of the system for different values of n using $a=b=1$ and $y\ (0)=0$

5.13.3.2 Significance of Parameters *a* and *b*

Now the values of the coefficients a and b are changed for $n = 1.2$. It can be observed that these parameters control the amount of damping experienced by the system as well as the steady state value of the solution.

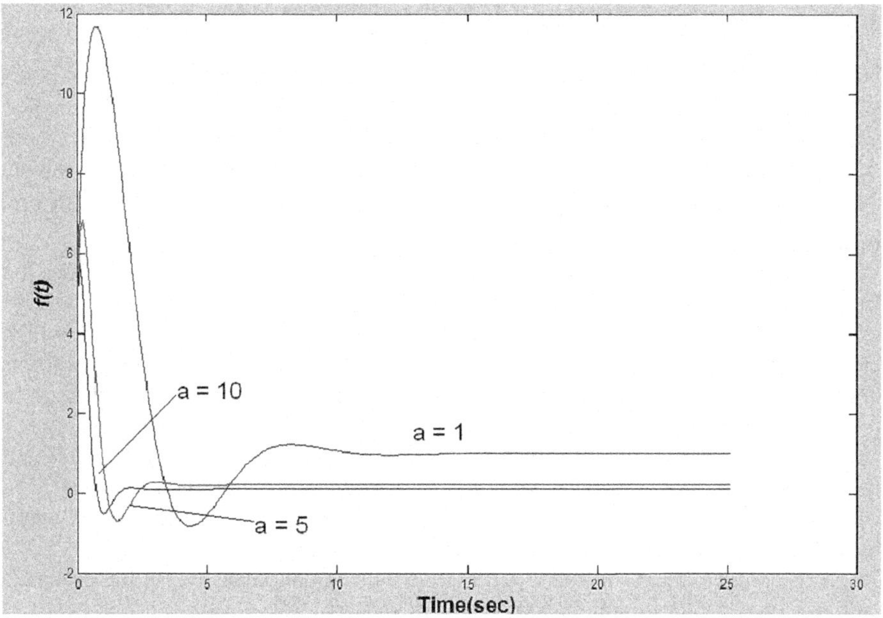

Fig 5.14 Step response for different values of *a*, depicting increased damping for greater values of *a*.

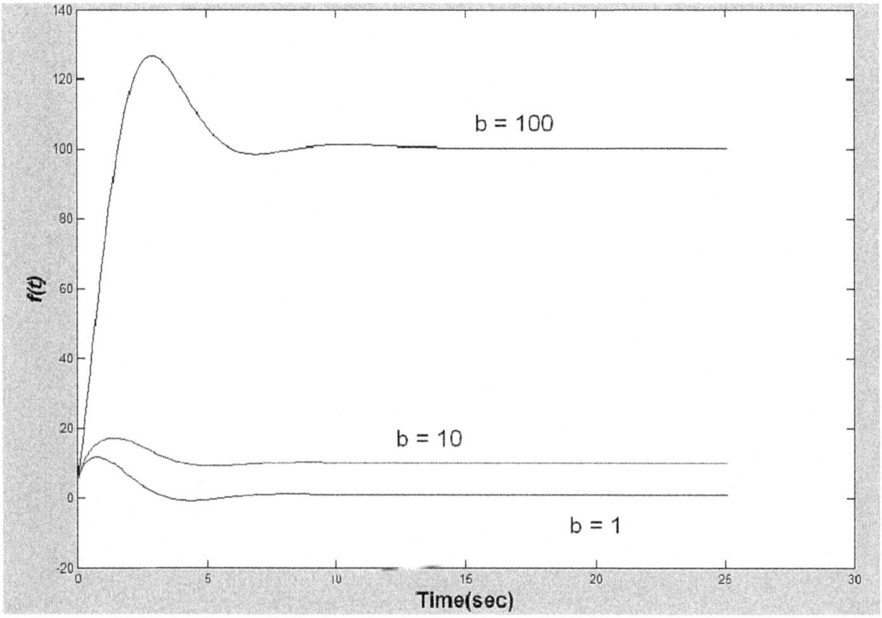

Fig 5.15 Step response for different values of parameter *b*.

5.13.3.3 Effect of Initial Conditions

Unlike any integral differentiator operator, fractional differential operator is not a local operator. For integer derivative at a point *P*, only the knowledge of function values in the neighborhood of *P* is sufficient. When *n* is not an integer, it is not sufficient just to know the functional values of neighborhood of *P* in order to evaluate $D_t^n y(t)$. Rather we need to have information about the entire history of the function from initial instant t_0 to *t*.

Ordinary and fractional derivatives differ in several ways. The signal to which a fractional derivative is applied starts at $t = 0$. Before that, the signal is always assumed to be zero. If that is not the case, $t = 0$ must be moved to a time before which the signal does not differ from zero. The fractional derivative of a constant is not zero because the signal change from zero to a finite value at $t = 0$ gives a contribution to the fractional derivative at all later times. A derivative of integer order also assumes a value different from zero at the step at $t = 0$, which we call a delta functional, but there is no contribution at later times since the delta functional is very short. The fractional derivative is not zero even if the signal is zero because the signal might have differed from zero at an earlier time.

The plot given below compares the memory for initial values for an integer order system with n=1 and a fractional order system with n=1.75.

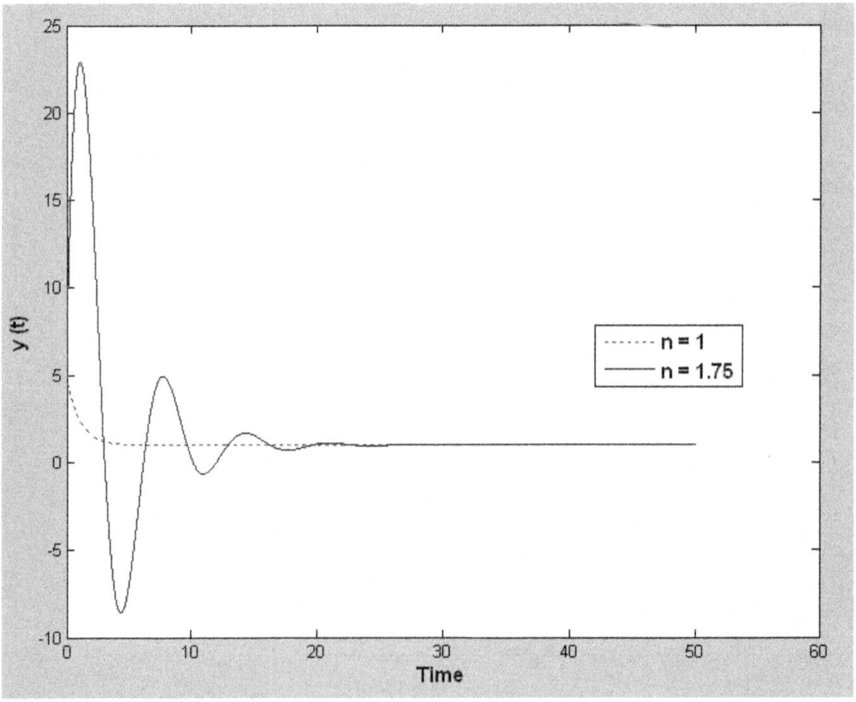

Fig. 5.16 Effect of initial conditions on a system with n =1, 1.75 for the step response

From the above plot, it can be concluded that the effect of initial conditions in case of integer order systems dies out very fast when compared to that of fractional order systems. For n=1.75, the effect of initial condition is present feebly even after 25 sec while this effect dies down in 3 sec for the integer order system.

5.14 Line, Surface and Volume Integration of Fractal Distributions

Here in this section a method is described to get Fractal Distribution represented by 'Fractional Continuum Medium' and then to perform surface or volume integration. The fractional integrals are considered as an approximation of integrals on fractals. This type of new approach is applicable in processes where the fractal features of the processes or the medium impose the necessity of using nontraditional tools in 'regular' smooth physical equations. Smoothening of microscopic characteristics over the physically infinitesimal volume/area transforms the initial fractal distribution into fractional continuous model. The order of fractional integration is taken as equal to the fractal dimension of the distribution. Therefore fractional integrals may be considered as an approximation of integrals on fractals. Say the number of particles N enclosed in a volume with characteristic size R satisfies the scaling law $N(R) \sim R^d$;(d is real number) whereas for a regular n-dimensional Euclidian object we have $N(R) \sim R^n$ (n is 1, 2, and 3 integer). The difference between the real fractal medium and fractional continuous medium model is analogous to the difference between the real atomic structure (fractal medium) and usual continuous models of those medium.

The RL definition of Fractional Integration is:

$$ {}_0D_x^{-\alpha} f(x) = \frac{1}{\Gamma(\alpha)} \int_0^x (x-u)^{\alpha-1} f(u)du \text{ , } v > 0 \text{ .} $$

Consider the RHS and say the element du is changed to elementary volume element dV_3, in the Euclidian topological space, comprising of $dV_3 = (dx)(dy)(dz)$. Then the volumetric integration of order 3α is expressed as:

$$ {}_0D_{x,y,z}^{-3\alpha} f(x,y,z) = \frac{1}{\Gamma^3(\alpha)} \int_0^x \int_0^y \int_0^z (x-u)^{\alpha-1} (y-v)^{\alpha-1} (z-w)^{\alpha-1} f(u,v,w) du dv dw $$

Noting $dV_3 = dudvdw$ and writing $d = 3\alpha$ we can re-cast the above integral as:

$$ {}_0D_{x,y,z}^{-d} f(x,y,z) = \frac{1}{[\Gamma(d/3)]^3} \int_0^x \int_0^y \int_0^z (x-u)^{\frac{d}{3}-1} (y-v)^{\frac{d}{3}-1} (z-w)^{\frac{d}{3}-1} f(u,v,w) dV_3 $$

The LHS of above is 'volume' integration on fractional volume V_d with elemental volume dV_d, symbolically.

$$ {}_0D_{x,y,z}^{-d} f(x,y,z) = \int_V f(x,y,z)(dV_d) . $$

Equating the two with simplification we get:

$$ \int_V f(x,y,z)(dV_d) = \int_V \frac{1}{\Gamma^3(d/3)}(x-u)^{\frac{d}{3}-1}(y-v)^{\frac{d}{3}-1}(z-w)^{\frac{d}{3}-1} f(u,v,w)dV_3 $$

We infer that fractional volume element dV_d is related to 3D- Euclidian elemental volume element dV_3 by a function, call it

$$ K_3(x,y,z,d) = \frac{(xyz)^{\frac{d}{3}-1}}{\Gamma^3(d/3)}, \text{ as: } dV_d = K_3(x,y,z,d)dV_3 . $$

Here we can approximate

$$ \Delta x = \Delta y = \Delta z = \Delta r , $$

with $x \approx y \approx z = r$ and in spherical coordinates we write:

$$ K_3(r,d) \cong \frac{r^{d-3}}{\Gamma^3(d/3)}, \text{ for } 2 < d < 3 $$

Consider the ball of region W such that $r \le R$ and spherically symmetric distribution of particles with number density as $n(r,t) = n(r)$. Then number of particles enclosed in a fractal volume V_D, will be

$$ N_d(r) = \int_W n(r,t)dV_d . $$

The number of particles in a 3D volume is

$$ N(r) = \int_W n(r,t)dV_3 . $$

The elemental volume is

$$ dV_3 = 4\pi r^2 dr $$

in spherical geometry and considering $n(r,t) = n_0$, as a constant we have

$$N(r) = \int_0^R 4\pi r^2 n_0 dr = \frac{4}{3}\pi R^3 n_0$$

Similarly we have

$$N_d(r) = \int_0^R 4\pi r^2 n_0 dV_d = \frac{4\pi n_0}{\Gamma^3(d/3)}\int_0^R r^2 r^{d-3} dr = \frac{4\pi n_0}{\Gamma^3(d/3)}\frac{R^d}{d} \sim R^d$$

We have used the derived approximate expression

$$dV_d = \frac{r^{d-3}}{\Gamma^3(d/3)} dV_3 = \frac{r^{d-3}}{\Gamma^3(d/3)}\left(4\pi r^2\right)dr\,,$$

to calculate $N_d(r)$, in above problem.

With the similar arguments as above we can have relation between fractional infinitesimal area (surface) dS_d and Euclidian infinitesimal area (surface) dS_2 to have surface integration of fractal surfaces. The relation is approximately:

$$dS_d = K_2(r,d)dS_2\,,$$

with

$$K_2(r,d) = \frac{r^{d-2}}{\Gamma^2(d/2)}\,,$$ for $1 < d < 2$. Note that $K_2(r,2) = 1$ for $d = 2$.

This relation can be helpful in finding 'flux' through a fractal surface. The usual definition of flux is 'surface' integral given as:

$$\phi_{(S)} \equiv \int_S (J \bullet dS_2)\,,$$

where $J(r,t)$ denotes a flowing quantity. Say q are charges flowing with number density of $n = n(r,t)$ with velocity $v = v(r,t)$, and then $J = qnv$ is current density of the flowing quantity. The fractional generalization of the flux for a fractal distribution can be:

$$\phi_{S_d} = \int_S (J(r,t) \bullet dS_d)\,,$$

where

$$dS_d = \frac{r^{d-2}}{\Gamma^2(d/2)} dS_2 .$$

From the above discussion we can have the relation for line integration for fractal (line) distribution, with relating the elemental line elements as:

$$dL_d = K_1(r,d) dL_1 \text{, for } 0 < d < 1 ,$$

with

$$K_1(r,d) = \frac{r^{d-1}}{\Gamma(d)} .$$

5.15 Fractional Generalization of Gauss's Law and Stroke's Law

The Gauss's law is following, giving the relation of surface integral and volume integral for region (volume) W, bounded by a surface $S = \partial W$. That is closed surface integral of a vector $J(r,t)$ is equal to the volume integral of divergence of the vector $J(r,t)$.

$$\oint_{\partial W} (J(r,t) \bullet dS_2) = \int_W \text{div}[J(r,t)] dV_3 = \int_W \nabla \bullet J(r,t) dV_3 ,$$

where $J(r,t)$ is a vector field say in x direction so that,

$$\text{div}(J) = \partial J / dr = \partial J / \partial x .$$

Using the above derived relation we write

$$dS_d = K_2(r, d_2) dS_2 ,$$

with

$$K_2(r,d_2) = \frac{r^{d_2-2}}{\Gamma^2(d_2/2)} ,$$

and we get

$$\oint_{\partial W} (J(r,t) \bullet dS_d) = \oint_{\partial W} K_2(r,d_2)(J(r,t) \bullet dS_2) .$$

Using the usual Gauss's law as described above, the RHS becomes:

$$\oint_{\partial W} (J(r,t) \bullet dS_d) = \int_W \text{div}[K_2(r,d_2) J(r,t)] dV_3$$

Using the expression obtained

$$dV_d = K_3(r,d_3)dV_3$$

Where

$$K_3(r,d_3) = r^{d_3-1}\Gamma^{-3}(d_3/3),$$

we write the fractional generalization of Gauss's law as:

$$\oint_{\partial W} \left(J(r,t) \bullet dS_d\right) = \int_W \left(K_3(r,d_3)\right)^{-1} \mathrm{div}\left[K_2(r,d_2)J(r,t)\right] dV_d$$

Analogous to the above Gauss's law for fractal distribution we can generalize the Stroke's law as:

$$\oint_L (E \bullet dL_d) = \int_S \left(K_2(r,d_2)\right)^{-1} \left[\mathrm{curl} K_1(r,d_1)E\right] dS_d$$

where $K_1(r,d_1) = r^{d_1-1}\Gamma^{-1}(d_1)$; as obtained in the previous section.

In this chapter we mention the difference between the real fractal medium structures and replacing with by a fractional continuous mathematical model to have approximate line, surface and volume integrals. These tools are helpful in generalizing the laws of physics as charge conservation laws , gravitational potentials and integral Maxwell equations to obtain approximate representation where the distribution of charges or masses are fractal in nature. The application of this technique in analyzing dusty-plasma will be helpful.

5.16 Concluding Comments

The physical and geometric interpretation of fractional differintegration process has shown some insight of the mathematics, which lays in-between complete integration and complete differentiation. The elaborate block diagrams provide understanding for computation of these fractional processes. The up coming field of local fractional derivatives is just introduced, which is a tool for description of fractal process. The concepts of minimizing computation effort by digital signal processing fundamentals have given a direction to evolve efficient algorithms for digital control science applications. The concepts are evolving even today, and future will see much more insight into the concepts of fractional differ integrations. The use of fractional calculus to magnify the physical critical points non-differentiability has been examined along with its use to regularize the irregular behavior of functions. The use of fractional calculus to model a rough irregular system of charges lead to different sets of Maxwell equations in electrodynamics; this method could be of use even to represent irregular boundaries for fractal boundary value problems.

Chapter 6
Initialized Differintegrals and Generalized Calculus

6.1 Introduction

This chapter demonstrates the need for a non-constant initialization for the fractional calculus. Here basic definitions are formed for the initialized fractional differintegrals (differentials and integrals). Here two basic popular definitions of fractional calculus are considered, those are Riemann-Liouville (RL) and Grunwald-Letnikov (GL). Two forms of initialization methods are prevalent, the 'terminal initialization' and the 'side initialization'. The issue of initialization has been an essentially a neglected subject in the development of the fractional calculus. Liouville's choice of lower limit as $-\infty$ and Riemann's choice as c were in fact were issues related to the same initialization. Ross and Caputo maintained that to satisfy the composition of the fractional differintegrals, the integrated function and its integer order derivatives must be zero, for all times up to and including the start of fractional differintegration. Ross stated, "The greatest difficulty in Riemann's theory is the interpretation of complimentary function. The question of existence of complimentary function caused much of confusion. Liouvelli was led to error and Riemann became inextricably entangled in his concept of a complimentary function." The complimentary function issue is raised here because an initialization function, 'which accounts for effect of history', of the function, for fractional derivatives and integrals, will appear in the definitions of this chapter. The form of initialization function is kept similar to what Riemann has used as complimentary function $\psi(x)$ however it's meaning and use is different. Constant initialization of the past is insufficiently general, the widely used contemporary equations for the Laplace transform for differintegrals based on that assumption also lacks generality. Therefore the generalized form is presented here. In solution of fractional differential equations with assumed history, the set of initializing constants representing the values of fractional differ integrals at $t = 0$, are ineffective, will be deliberated in this chapter. Therefore required is 'non-constant initialization' for generalized concept of integration and differentiation. The Caputo definition vis-à-vis Riemann-Liouvelli definition with initialization is compared and the difficulty in physically making them compatible is too probed into. A brief discussion on criteria and properties of generalized calculus as given by Ross (1974) is mentioned and then simple examples provided for getting the gist of fractional calculus, with importance given to initialization. In this chapter a section is included to evaluate fractional derivative and fractional integral of periodic functions, with lower terminal initialized to minus infinity and also with a finite start point in differintegration. The solution of Fractional Advection Dispersion Equation is solved in context of boundary condition, and concept of Levy stable law is demonstrated, as Long Range Dependency (LRD). This Long Range Dependence is extended to identify spiky delay dynamics of computer based system. A new aspect of identification of random delays (processes) of spiked nature is developed vis-à-vis fractional Langevin's equation, driven by shot noise; the concept is general and can be generalized to other fractional stochastic processes, say market economy or say price fluctuation and any other fluctuating dynamics of random nature. These two dynamic problems are taken in view of initial condition and boundary condition for solution of Fractional Differential equation System.

6.2 Notations of Differintegrals

Mathematicians have used several notations since the birth of fractional calculus. As mentioned in the Introduction chapter (Historical Development of Fractional Calculus) several contemporary notations for fractional differentiation and fractional integration. Here attempt will be made to standardize the notations as differintegrals. The same operator used as integrator when index is negative and differentiator when index is positive. Separate notation will be used to indicate initialized differintegral operator and un-initialized operator.

${}_cD_t^q f(t)$ Represents 'initialized' q-th order differintegration of $f(t)$ from start point c to t. ${}_cd_t^q f(t)$ Represents 'un-initialized' generalized (or fractional) q-th order differintegral. This is also same as

$$\frac{d^q f(t)}{[d(t-c)]^q} \equiv {}_cd_t^q f(t),$$

shifting the origin of function at start of the point from where differintegration starts. This un-initialized operator can also be short formed as $d^q f(t)$. The initialization function (not a constant) is represented as $\psi(f,q,a,c,t)$ meaning that this is function of independent variable t, and is for differintegral operator of order q, the function born at $t=0$ (before that the function is zero), and differintegral process starting at c. This initialization function can be short formed as $\psi(t), \psi(f,q,t)$. Therefore the expression between initialized differintegral and un-initialized one is:

$${}_cD_t^q f(t) = {}_cd_t^q f(t) + \psi(f,q,a,c,t)$$

The notation contains lower limit of the process at the front subscript and the order of the process at the tail superscript, with independent variable with respect to what is being differintegrated.

6.3 Requirement of Initialization

In real applications, it is usually the case that the problem to be solved is in some way isolated from past that is, it should not be necessary to retreat to $-\infty$ in time to start the analysis. Usually, the analyst desires to start the analysis at some time $t = t_0$, with the knowledge (or assumption) of all values of the function and its derivatives. Specifically, $f(t_0), f^{(1)}(t_0), f^{(2)}(t_0).....f^{(n)}(t_0)$, in the case of integer order calculus. In modern parlance this collection is called 'state' and contains the effect of all the past history. One way the behavior of the semi-infinite transmission line can be described is in terms of its input behavior (impedance) at the open end of the line, that is, as semi-differential equation. However, to practically use such fractional order differential equation requires additional function of time. In terms of the physics this time-function relates back to the initial voltage distribution (distributed initialization) on the semi-infinite lossy line. From Chapter 3 the input terminal behavior of the same is:

$$V(0,s) = \frac{rI(0,s)}{\sqrt{\frac{s}{\alpha}}} + \frac{1}{\alpha\sqrt{\frac{s}{\alpha}}} \int_0^{\infty} e^{\sqrt{\frac{s}{\alpha}}\lambda} V(\lambda,0) d\lambda ,$$

where λ is, the dummy variable of integration and the variable s is Laplace variable. The above expression in Chapter 3 was obtained by iterated Laplace transformation technique applied to the basic diffusion equation. However, an attraction of the fractional calculus is the ability to express the behavior of the line, (a distributed system or mathematically partial differential equation) as part of the system of distributed equations using fractional differential equations. Such a fractional differential equation for the semi-infinite lossy line is:

$$\frac{d^{1/2}v(t)}{dt^{1/2}} = r\sqrt{\alpha}i(t) ,$$

assuming $v(x,0) = 0$. To initialize this distributed system a function of time $\psi(t)$ must be added to account for the integral term of the obtained expression for $V(0,s)$ written above. With this fractional differential equation is:

$$\frac{d^{1/2}v(t)}{dt^{1/2}} + \psi(t) = r\sqrt{\alpha}i(t) .$$

The focus in this chapter will be on $\psi(t)$. Clearly, one can addend such terms in ad hoc way to the fractional differential equations which are being solved, the formal approach to evaluate this function is presented in detail in this chapter. If the analyst is constrained that the initial function value and all its derivatives are zero, the range of applicability for this entire class of problems, which includes eventually all distributed systems, will greatly be limited. Therefore all fractional ordinary differintegral equations require initialization terms to be associated with each fractional differintegration term, in order to complete the description. This requirement is a generalization to the requirement of a set of initialization constants 'states', in integer order ordinary differential equations. Fundamentally it is the information to start the integration process of the differential equations while properly accounting for the effect of the past.

6.4 Initialization Fractional Integration (Riemann-Liouville Approach)

This non-constant initialization function $\psi(t)$, which shall be elucidated clearly brings out the past history, also brings to the definition of the fractional integral the effect of past namely, effect of fractionally integrating the function from its birth. This added effect will also be influencing the process after time *t*, the start of integration process.

Consider fractional order q integration of the $f(t)$, the first starting at $t = a$, and second starting at $t = c > a$

$$ {}_a d_t^{-q} f(t) = \frac{1}{\Gamma(q)} \int_a^t (t-\tau)^{q-1} f(\tau) d\tau \tag{6.1}$$

$$ {}_c d_t^{-q} f(t) = \frac{1}{\Gamma(q)} \int_c^t (t-\tau)^{q-1} f(\tau) d\tau \tag{6.2}$$

Assume that function was born at $t = a$, that is $f(t) = 0$ for all time less than equal to a. i.e. $f(t) = 0, t \le a$. Then the time period between a and c may be considered as history. The assumption is that the integral $\left({}_c d_t^{-q} f(t)\right)$, is properly initialized so that it should function as continuation of integral starting at $t = a$. To this therefore an initialization must be added (to ${}_c d_t^{-q} f(t)$), so that fractional integral starting at $t = c$ should be identical to the result starting at $t = a$ for $t > c$. We call what Riemann proposed as complimentary function as initialization function as ψ.

We have for the above argument the following:

$$ {}_c d_t^{-q} f(t) + \psi = {}_a d_t^{-q} f(t), t > c \text{. Then, } \psi = {}_a d_t^{-q} f(t) - {}_c d_t^{-q} f(t), t > c$$

Therefore,

$$\psi = \frac{1}{\Gamma(q)} \int_a^c (t-\tau)^{q-1} f(\tau) d\tau \equiv {}_a d_c^{-q} f(t) . \qquad t > c \tag{6.3}$$

Here ψ is a function of independent variable *t*, thus is 'non-constant'. For integer order integration we put $q = 1$, and see that

$$\psi = \int_a^c f(\tau)d\tau = K ,$$

a constant. Because of increased complexity of the initialization relative to the integer order calculus, it is important to formalize the initialization process. This formalization will include the initialization term into the definition of these fundamental fractional order calculus operators.

Two types of initialization are considered. The terminal initialization, where it is assumed that the differintegral operator can be initialized (charged) by effectively differintegrating prior to the start time, $t = c$; and the side-initialization where fully arbitrary initialization may be applied to the differintegral operator at time $t = c$. These are in contemporary terms may be stated as terminal charging and side charging. First we restrict to RL type of differintegrals for formalizing these definitions. This initialization function ψ has the effect of allowing the function $f(t)$ and its derivatives to start at a value other than zero, namely ${}_aD_c^{-q} f(t)_{@t=c}$, and continues to contribute to differintegral response after $t = c$. That is, a function of time is added to the uninitialized integral, (not just a constant at $t = c$).

6.4.1 Terminal Initialization

The standard contemporary definition of fractional integral (RL) is accepted if an only if the differintegrand $f(t) = 0$ for all $t \le a$. The initialization period (or space) is region $a \le t \le c$. The fractional integration takes place for $t > c \ge a$. Further the fractional integration starts at $t = c$ (i.e. point of initialization).

$${}_aD_t^{-q} f(t) \equiv \frac{1}{\Gamma(q)} \int_a^t (t-\tau)^{q-1} f(\tau)d\tau \, , \; q \ge 0. \& .t > a \tag{6.4}$$

subject to $f(t) = 0$ for all $t \le a$.

The following definition of fractional integration will apply generally (at any $t > c$):

$${}_cD_t^{-q} f(t) \equiv \frac{1}{\Gamma(q)} \int_c^t (t-\tau)^{q-1} f(\tau)d\tau + \psi(f,-q,a,c,t), \tag{6.5}$$

$$q \ge 0, t > a, c \ge a \; \& \; f(t) = 0 \; @ \; t \le a$$

The function $\psi(f,-q,a,c,t)$ is called the initialization function and will be chosen such that

$${}_aD_t^{-q} f(t) = {}_cD_t^{-q} f(t) . \qquad t > c \tag{6.6}$$

This condition gives:

$$\frac{1}{\Gamma(q)}\int_a^t (t-\tau)^{q-1} f(\tau)d\tau = \frac{1}{\Gamma(q)}\int_c^t (t-\tau)^{q-1} f(\tau)d\tau + \psi(f,-q,a,c,t)\,. \tag{6.7}$$

Since, $\int_a^t g(\tau)d\tau = \int_a^c g(\tau)d\tau + \int_c^t g(\tau)d\tau$.

Therefore we get

$$\psi(f,-q,a,c,t) = {}_aD_c^{-q} f(t) = \frac{1}{\Gamma(q)}\int_a^c (t-\tau)^{q-1} f(\tau)d\tau \tag{6.8}$$
$$t > c, \&, q > 0$$

This expression for $\psi(t)$ gives 'terminal initialization', and also brings out in the definition of fractional integral the effect of the past 'history', namely the effect of fractionally integrating the $f(t)$ from a to c. This effect is also called terminal charging.

6.4.2 Side-Initialization

When ψ, is arbitrary and terminal initialization equation is not valid then the effect is called 'side-initialization', or side charging. Figure 6.1 demonstrates concept initialization as block diagram, as a signal flow graph.

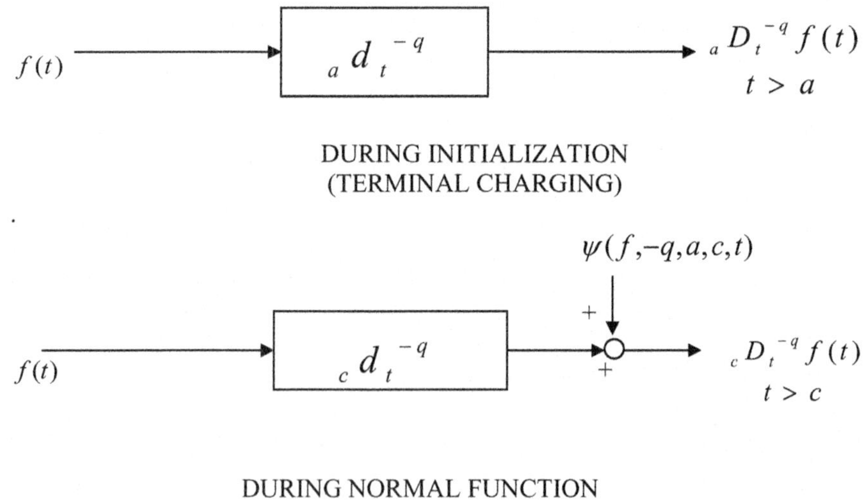

Fig. 6.1 Signal flow graph for demonstrating initialization of fractional integration

Example 1. Let function $f(t)=t$ for $t>0$ and $f(t)=0$ for $t<0$. The semi-integral process with initialization is demonstrated below from start point of integration @ $t=1$. By applying RL formulations for fractional integration we obtain the following:

$$_0D_t^{-1/2}t=\frac{1}{\Gamma(1/2)}\int_0^t(t-\tau)^{0.5-1}\tau.d\tau=\frac{1}{\Gamma(1/2)}\int_t^0\frac{(t-x)(-dx)}{x^{1/2}}=\frac{t^{3/2}}{\frac{3}{2}\cdot\frac{1}{2}\Gamma(1/2)}=\frac{4}{3\sqrt{\pi}}t^{3/2}$$

$$_1D_t^{-1/2}t=\frac{1}{\Gamma(1/2)}\int_{t-1}^0\frac{(t-x)(-dx)}{x^{1/2}}+\psi(t,-0.5,0,1,t)=\frac{2}{3\sqrt{\pi}}\left[(t-1)^{1/2}(2t+1)\right]+\psi(t,-1/2,0,1,t)$$

$$t>1,\&,\Gamma(0.5)=\sqrt{\pi}$$

$$\psi(t,-1/2,0,1,t)=\frac{2}{3\sqrt{\pi}}\left[2t^{3/2}-(2t+1)(t-1)^{1/2}\right]$$

Note may be taken about the nature of initialization function, which is semi-integration of the function from 0-1, and is function of *t*. Also as time passes by the function decays to zero, i.e. 'the effect of history is forgotten as future grows!'; $\lim_{t\to\infty}\psi(t,-0.5,0,1,t)=0$, at least in this particular example. This is shown in the Figure 6.2:

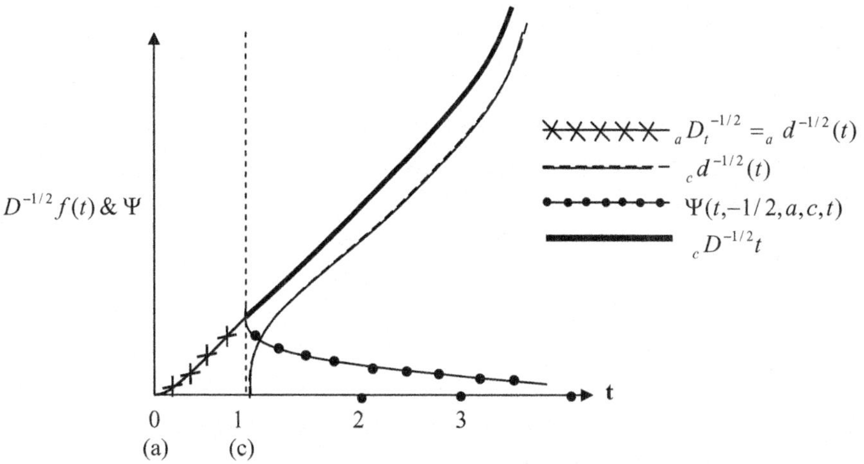

Fig. 6.2 Graphical representation of initialized differintegration for *f (t)=t*

6.5 Initializing Fractional Derivative (Riemann-Liouvelle Approach)

This is contrary to integer order derivative, which is a point and also local quantity (property) where initialization is not called for. However the definitions of fractional derivatives do contain fractional integration and thus fractional derivative of a function is not a point quantity. For fractional derivative initialization process is called for. Fractional derivative is non-local operator and therefore has history. In solution of differential equations the initialization constants, which sets the initial values of the derivatives, really have the effect of accounting for the integration of the derivative from $-\infty$ to starting time of the integration (of the differential equation). In the fractional calculus initialization for derivatives are also required for handling the effect of 'distributed initialization', in actual system. Distributed initialization means initial voltage/charge profile in semi-infinite distributed transmission line (lossy of order ½ or lossless of order zero), or in case of initial strain distribution in elastic semi-infinite bar (order one). Extending generalization concept, the integer order derivative also calls for initialization in 'fractional context'. Thus, a generalized integer order differentiation is defined as (with initialization)

$$ {}_cD_t^m f(t) \equiv \frac{d^m}{dt^m} f(t) + \psi(f, m, a, c, t) \cdot \tag{6.9}$$

$t > c$. Here m is integer and $\psi(f, m, a, c, t)$ is an initialization function. Now bare or un-initialized fractional derivative is defined as:

$$ {}_aD_t^q f(t) \equiv {}_aD_t^m \; {}_aD_t^{-p} f(t) \tag{6.10}$$

$q \geq 0$, $t > a$, and $f(t) = 0$ at $t \leq a$; $q = m - p$ Meaning m is the integer just greater than the fractional order q, by amount p. The function is born at $t = a$ and before that the value is zero. The differentiation starts at $t > c$. Now as in fractional integration case, $\psi(f, -p, a, a, t) = 0$. Further consider the $h(t) = {}_aD_t^{-p} f(t)$ i.e. fractional integral of function starting at a with initialized term $\psi(h, m, a, a, t) = 0$; the initialized fractional derivative looks and defined as for $q \geq 0, t > c \geq a$.

$$ {}_cD_t^q f(t) \equiv {}_cD_t^m f(t) \, {}_cD_t^{-p} f(t) \, . \tag{6.11}$$

6.5.1 Terminal Initialization

The definition and concept is similar to that obtained as terminal initialization for fractional integrals. The requirement is also same as for the fractional integrals, that is

$$ {}_cD_t^q f(t) = {}_aD_t^q f(t) \tag{6.12}$$

for all $t > c \geq a$. Specifically this requires compatibility of the derivatives starting at $t = a$ and $t = c$ for all $t > c$. Therefore it follows that:

$$ {}_cD_t^m \; {}_cD_t^{-p} f(t) = {}_aD_t^m \; {}_aD_t^{-p} f(t) \tag{6.13}$$

Expanding the fractional integrals with initialization we obtain:

$$ {}_cD_t^m\left(\frac{1}{\Gamma(p)}\int_c^t(t-\tau)^{p-1}f(\tau)d\tau+\psi(f,-p,a,c,t)\right)={}_aD_t^m\left(\frac{1}{\Gamma(p)}\int_a^t(t-\tau)^{p-1}f(\tau)d\tau+\psi(f,-p,a,a,t)\right) \tag{6.14}$$

For, $t>c$ and $\psi(f,-p,a,a,t)=0$. Using the definition of generalized integer order derivative as defined above we get:

$$\begin{aligned}&\frac{d^m}{dt^m}\left\{\frac{1}{\Gamma(p)}\int_c^t(t-\tau)^{p-1}f(\tau)d\tau+\psi(f,-p,a,c,t)\right\}+\psi(h_1,m,a,c,t)=\\&\frac{d^m}{dt^m}\frac{1}{\Gamma(p)}\int_a^t(t-\tau)^{p-1}f(\tau)d\tau+\psi(h_2,m,a,a,t)\end{aligned} \tag{6.15}$$

where $h_1={}_cD_t^{-p}f(t)$ and $h_2={}_aD_t^{-p}f(t)$ the integer order derivative is initialized at $t=a$, thus $\psi(h_2,m,a,a,t)=0$. After rearranging the integrals we get:

$$\psi(h_1,m,a,c,t)=\frac{d^m}{dt^m}\left(\frac{1}{\Gamma(p)}\int_a^c(t-\tau)^{p-1}f(\tau)d\tau-\psi(f,-p,a,c,t)\right). \tag{6.16}$$

This is the expression and 'the requirement for the initialization for the derivative in general'.

Under condition of terminal charging of the fractional integral

$$\psi(f,-p,a,c,t)=\frac{1}{\Gamma(p)}\int_a^c(t-\tau)^{p-1}f(\tau)d\tau,$$

is the initialization function of fractional integration as defined and derived earlier (6.8). Therefore the $\psi(h_1,m,a,c,t)=0$, a very important result is seen, that is 'integer order differentiation cannot be initialized through the terminal (terminal charging)'.

6.5.2 Side-Initialization

Refer the expression (6.16) and the requirement for initialization for general derivative as obtained in the terminal charging case. If side charging is employed meaning that the function $\psi(f,-p,a,c,t)$ is arbitrary. Thus it can be inferred from the requirement equation (6.16) that $\psi(f,-p,a,c,t)$ or $\psi(h_1,m,a,c,t)$ can be arbitrary but not both together, but also should then satisfy the requirement expression derived above.

The generalized expression for the side charging case can be stated as:

$$_cD_t^q f(t) = {_cD_t^m}\left\{\frac{1}{\Gamma(p)}\int_c^t (t-\tau)^{p-1} f(\tau)d\tau + \psi(f,-p,a,c,t)\right\}, (t.>c) \quad (6.17)$$

m, is positive integer $> q$, with $q = m - p$

$$_cD_t^q f(t) = \frac{d^m}{dt^m}\frac{1}{\Gamma(p)}\int_c^t (t-\tau)^{p-1} f(\tau)d\tau + \frac{d^m}{dt^m}\psi(f,-p,a,c,t) + \psi(h,m,a,c,t) \quad (6.18)$$

where, $h(t) = {_aD_t^{-p}} f(t)$. Here both the initialization terms are arbitrary and thus may be considered as a single (arbitrary) term, namely

$$\psi(f,q,a,c,t) \equiv \frac{d^m}{dt^m}\psi(f,-p,a,c,t) + \psi(h,m,a,c,t). \quad (6.19)$$

In case of terminal charging the fractional integral initialization part

$$\psi(f,-p,a,c,t) = \frac{1}{\Gamma(p)}\int_a^c (t-\tau)^{p-1} f(\tau)d\tau \text{ for } t > c \quad (6.20)$$

Figure 6.3 demonstrates the initialization concept for fractional derivative

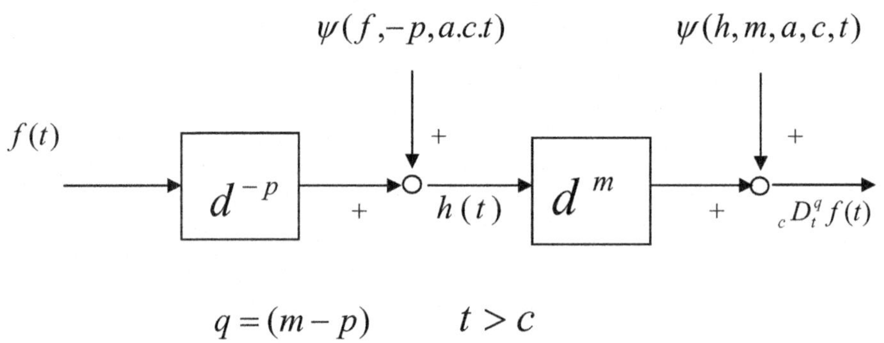

Fig. 6.3 Initialization of fractional derivative

6.6 Initializing Fractional Differintegrals (Grunwald-Letnikov Approach)

Here in this approach too, take function's starting time as a, and the differinteration process starts at $t = c$. An initialization (notation same as for RL-approach) is introduced to account for past history and goes back to $t = a$, with $f(t) = 0$ @ all time before $t = a$. Then differintegration with arbitrary order q is:

$$ {}_aD_t^q f(t) = \frac{d^q f(t)}{[d(t-a)]^q} \equiv \lim_{N\to\infty} \frac{\left(\frac{t-a}{N}\right)^{-q}}{\Gamma(-q)} \sum_{j=0}^{N-1} \frac{\Gamma(j-q)}{\Gamma(j+1)} f\left(t - j\left\{\frac{t-a}{N}\right\}\right) \tag{6.21} $$

$$ t > a \;\&\; f(t) = 0 \;@\; t \le a $$

Grunwald-Letnikov (GL) definition for differintegrals will generally apply as:

$$ {}_cD_t^q f(t) \equiv \frac{d^q f(t)}{[d(t-c)]^q} + \psi(f,q,a,c,t). \tag{6.22} $$

and $f(t) = 0, (at\ t \le a)$ and $c \ge a$. Here again $\psi(f,q,a,c,t)$ is selected such that ${}_cD_t^q f(t)$ will produce the same result as ${}_aD_t^q f(t)$ for $t > c$. Expressing as

$$ {}_cD_t^q f(t) = {}_cd_t^q f(t) + \psi(f,q,a,c,t) = {}_aD_t^q f(t) \tag{6.23} $$

will be self explanatory, for all $t > c$, and $f(t) = 0$ for all $t \le a$. Therefore, $\psi(f,q,a,c,t) = {}_aD_t^q f(t) - {}_cd_t^q f(t)$ or identifying that ${}_aD_t^q f(t) \to {}_ad_t^q f(t)$ i.e. un-initialized differintegral, as per standard notation we write the same as: $\psi(f,q,a,c,t) = {}_ad_t^q f(t) - {}_cd_t^q f(t)$. In this substituting GL series we obtain the following:

$$ \psi(f,q,a,c,t) = \lim_{N_1\to\infty} \left\{ \frac{\left(\frac{t-a}{N_1}\right)^{-q}}{\Gamma(-q)} \sum_{j=0}^{N_1-1} \frac{\Gamma(j-q)}{\Gamma(j+1)} f\left(t - j\frac{t-a}{N_1}\right) \right\} $$

$$ - \lim_{N_2\to\infty} \left\{ \frac{\left(\frac{t-c}{N_2}\right)^{-q}}{\Gamma(-q)} \sum_{j=0}^{N_2-1} \frac{\Gamma(j-q)}{\Gamma(j+1)} f\left(t - j\frac{t-c}{N_2}\right) \right\} \tag{6.24} $$

For all $t > c$ and $f(t) = 0$ for $t < a$.

After considerable manipulations by adjusting delay element as equal, that is $N_2 = ((t-c)/(t-a))N_1$, and adjusting with

$$ \Delta T = (t-a)/N_1 \;\&\; N_3 = ((c-a)/(t-a))N_1 $$

$$\psi(f,q,a,c,t)=\lim_{N_1\to\infty}\left\{\frac{\Delta T^{-q}}{\Gamma(-q)}\sum_{j=0}^{N_3-1}\frac{\Gamma(N_1-1-q-j)}{\Gamma(N_1-j)}f\left(t-[N_1-1-j]\Delta T\right)\right\} \tag{6.25}$$

6.7 Properties and Criteria for Generalized Differintegrals

One of the fundamental problems of fractional calculus is the requirement that the function and its derivatives be identically equal to zero at the start of initialization (i.e. start of differintegration process) at time $t=c$. This needed to assure composition or index law holds implying that

$$_cD_t^v\,{}_cD_t^u f(t) = {}_cD_t^u\,{}_cD_t^v f(t) = {}_cD_t^{u+v} f(t)\,.$$

It is difficult in engineering sciences, to always require that the functions and its derivatives be at zero (rest) at initialization instants. This fundamentally implies that 'there can be no initialization or composition is lost'. Thus it is not in general true that

$$f-\frac{d^{-Q}}{dt^{-Q}}\frac{d^{Q}f}{dt^{Q}}=0\,.$$

Therefore while solving a fractional differential equation of the form

$$\frac{d^{Q}f}{dt^{Q}}=F\;,$$

additional terms must be added like:

$$f-\frac{d^{-Q}}{dt^{-Q}}\frac{d^{Q}f}{dt^{Q}}=C_1t^{Q-1}+C_2t^{Q-2}+\ldots C_mt^{Q-m}$$

to achieve the most general solution:

$$f=\frac{d^{-Q}F}{dt^{-Q}}+C_1t^{Q-1}+C_2t^{Q-2}+\ldots+C_mt^{Q-m}\,.$$

These issues describe says that the index law or the composition law is inadequate.

Minimal set criteria have been thought fit to be applied for fractional (or generalized) calculus. They are listed as follows and are called Ross (1974) criteria:

(i) If $f(z)$ is analytic function of the complex variable z, the differintegral ${}_cD_z^v f(z)$ is analytic function of z and v.

(ii) The operator ${}_cD_x^{v}f(x)$ must produce the same result of differentiation, when v is a positive integer.

(iii) If v is negative integer (say $v=-n$) then ${}_cD_x^{-n}f(x)$ must produce the same result of n-fold integration of function $f(x)$ and ${}_cD_x^{-n}f(x)$ must vanish along with $f^{(1)}, f^{(2)}, \dots f^{(n-1)}$ all the $(n-1)$ derivatives at $x=c$.

(iv) 'Zero' operation leaves the function unchanged.

$$ {}_cD_x^{0}f(x)=f(x) $$

(v) Linearity of fractional (generalized) differintegral operator:

$$ {}_cD_x^{-q}\left[af(x)+bg(x)\right]=a\,{}_cD_x^{-q}f(x)+b\,{}_cD_x^{-q}g(x) $$

(vi) The law of exponents for arbitrary order is holding:

$$ {}_cD_x^{-u}\,{}_cD_x^{-v}f(x)={}_cD_x^{-u-v}f(x)={}_cD_x^{-v-u}f(x) $$

The above notations are used by Ross.

It should be noted that there is a minor conflict contained in these criteria. Also clear notation explanation be given as the ${}_cD_x^{q}f(x)$ in the above criteria is un-initialized differintegrals. It is correct as the function itself starts at c, and before that the same is zero. So

$$ @\, t=c\,,\; {}_cD_x^{q}f(x)={}_cd_x^{q}f(x)\,. $$

The criteria (ii) and (iii) calls for backward compatibility and criteria (vi) calls for index law to be holding vis-à-vis integer order calculus.

The fundamental theorem of integer order calculus violates this "zero law" as:

$$ d^{-m}d^{m}f(x)\neq d^{0}f(x)=f(x)\,, $$

for all $f(x)$ and for all m (integer). The fundamental theorem states that

$$ \int_c^t f'(t)=f(t)-f(c)\,, $$

and can be thus observed that reversal of differentiation and integration differs from $f(t)$ by $f(c)$, that is by the initialization (constant in integer order calculus). This failure in backward compatibility & index law is handled in the integer order calculus by constant of integration and by complimentary function for solution of differential equations (in ad-hoc manner). The law of exponents (index-law) is demonstrated in Figure 6.4.

The composition law of index we summarize, with simplified example as:

1. $D^{-\mu}\left[D^{-v}f(t)\right]=D^{-(\mu+v)}f(t)=D^{-v}\left[D^{-\mu}f(t)\right]$, for $\mu,v>0$, implies commutation and computation (addictiveness) is valid for fractional integration.
2. Integer derivative of fractional integration of a function is not equal to fractional integration of integer derivative of the function. They are related by

$$D\left[D^{-v}f(t)\right]=D^{-v}\left[Df(t)\right]+t^{v-1}\left[\Gamma(v)\right]^{-1}f(0);$$

that is initial condition. Here $0<v<1$. This follows from fundamental theorem of calculus-a generalization of the same. Thus Riemann-Liouvelli and Caputo derivatives are not same unless the initial conditions are zero (static)
3. Fractional integration of order $(v-1)$ of integer order derivative, of a function is not same as fractional integration of order $0<v<1$, but again related by initial condition as

$$D^{-v-1}\left[Df(t)\right]=D^{-v}f(t)-t^{v}\left[\Gamma(v+1)\right]^{-1}f(0).$$

4. Arbitrary composition is invalid that is:

$$D^{u}\left[D^{v}f(t)\right]\neq D^{u+v}f(t),$$

for any $u,v\in\mathbb{R}$, this composition and commutation is only valid for fractional integrations. Let us take $f(t)=\sqrt{t}$ with $u=1/2$ and $v=3/2$. Then $D^{u}(\sqrt{t})=D^{1/2}(t^{1/2})=(\sqrt{\pi})/2$,
Similarly $D^{v}(\sqrt{t})=D^{3/2}t^{1/2}=0$.
Therefore $D^{u}\left[D^{v}(\sqrt{t})\right]=0$.
Now calculate as: $D^{v}\left[D^{u}(\sqrt{t})\right]=D^{3/2}(\sqrt{\pi}/2)=-(1/4\sqrt{t^{3}})$.
Now calculate $D^{u+v}(\sqrt{t})=D^{2}(\sqrt{t})=-(1/4\sqrt{t^{3}})$.
This calculation demonstrates that
$D^{u}\left[D^{v}f(t)\right]\neq D^{v}\left[D^{u}f(t)\right]$ and $D^{u}\left[D^{v}f(t)\right]\neq D^{u+v}f(t)$
This demonstrates that $D^{u}\left[D^{v}f(t)\right]$and $D^{v}\left[D^{u}f(t)\right]$both exist but they are not equal. Things could be worse that is one of them may not exist. Say take $f(t)=t^{-1/2}$ and let $u=-1/2$ and $v=1$, then $D^{v}\left[D^{u}f(t)\right]=0$,
But $D^{v}f(t)=-(1/2)t^{-3/2}$, and $D^{u}\left[D^{v}f(t)\right]$does not exist since

$\int_{0}^{t}(t-u)^{-1/2}u^{-3/2}du$ diverges.

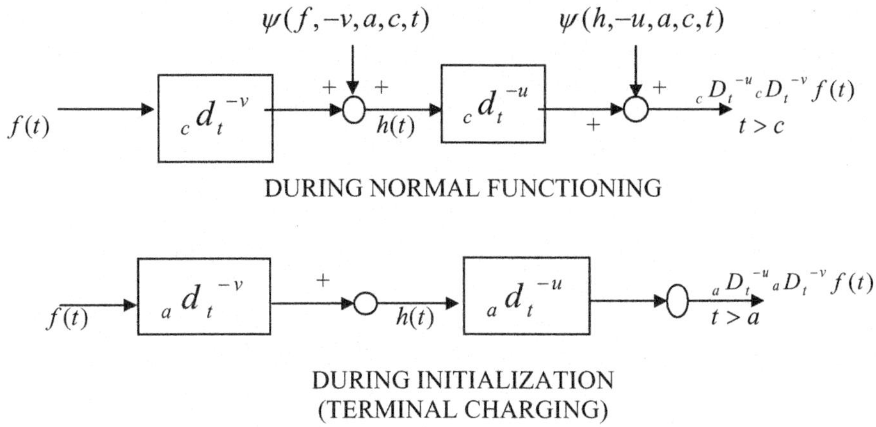

Fig. 6.4 Demonstration of index law

The discussion in all the differintegrations is limited to the real domain. Under the condition of terminal charging of the u-th and v-th differintegrations,

$$ {}_cD_t^u\, {}_cD_t^v f(t) = {}_cD_t^v\, {}_cD_t^u f(t) = {}_cD_t^{u+v} f(t) \quad \text{for } t > 0 $$

Under the following condition:

a) $u < 0, v < 0$ for continuous $f(t)$
b) $u > 0, v > 0$ For $f(t)$ are m times differentiable. ${}_aD_t^m f(t)$ exists and is non-zero continuous function of t for $t > a$, where m an integer larger than integer part $[u]$ or $[v]$.
c) $u < 0, v > 0$ same as (b).

6.7.1 Terminal Charging

Under the conditions of terminal charging the above properties and criteria holds, this provides credibility to the initialized fractional (generalized) calculus. Some conditions are however imposed say on linearity of fractional integrals

$$ {}_cD_t^{-v}(bf(t) + kg(t)) = b\,{}_cD_t^{-v} f(t) + k\,{}_cD_t^{-v} g(t), (t > c) $$

holds only if

$$ \psi(bf + kg, -v, a, c, t) = b\psi(f, -v, a, c, t) + k\psi(g, -v, a, c, t) $$

Relative to the criteria of backward compatibility with the integer order calculus, the addition of the initialized function is clearly a generalization relative to integer order calculus. In a strict sense $\psi(t) \neq 0$, violates the criteria 2; however, we are looking for generalization of integer order calculus and it is clear that this generalization (i.e. addition of initialization function) will be very useful in many applications.

Relative to the criteria of zero order property holds for terminal charging.

Relative to linearity holds for the terminal charging subject to above said rule. Relative to composition rule the above a) b) c) should follow.

It is noted that the, $f^{(k)}(c)=0$ for all *k*, no longer exists. This constraint has effectively been contained (shifted to) the requirement $f(t)=0$ for all $t \leq a$. This allows initialization of fractional differential equations.

In summary terminal charging case is backward compatible with integer order calculus and satisfies the applicable criteria established by Ross.

6.7.2 Side-Charging

The case for side charging is less definitive. Criteria for backward compatibility are the same as the terminal charging case. Relative to zero property the condition

$$\psi(f,-p,a,c,t)=\frac{1}{\Gamma(p)}\int_a^c (t-\tau)^{p-1} f(\tau)d\tau = 0 = \psi(h,m,a,c,t)$$

is required for side charging since ψ is arbitrary. When this conditions are not met the zero order operation on $f(t)$ will return $f(t)+g(t)$, i.e. the original function with extra time function $((g(t))$, the effect of initialization. Relative to linearity the side charging demands additional requirements about initialization.

These are not so much of issue as it appears for practical applications. In the solution of fractional differential equations $\psi(t)$ will be chosen in the much the same manner as initialization are currently chosen for ordinary differential equations in integer order. This will imply nature of $f(t)$ from *a* to *c*. The new aspect is that to achieve a particular initialization for a given composition now requires attention to the initialization of the composing elements.

6.8 Initialization with Caputo Derivative and Its Difficulties

In real applications, it is usually the case that the problem to be solved is in some way isolated from past. That is, it should not be necessary to retreat to $-\infty$ in time; that is time immemorial, to start the analysis. Usually, the analyst desires to start the analysis at some time $t=t_0$, with the knowledge (or assumption) of all values of the function and its derivatives. Specifically, $f(t_0), f^{(1)}(t_0), f^{(2)}(t_0).....f^{(n)}(t_0)$, in the case of integer order calculus. In modern parlance, this collection is called 'state' and contains the effect of all the history. However, in integer order calculus these 'initial-conditions' are constants. A non-constant initialization or function (of time) appears, as history in initializing differentiation (anti differentiation), as required for the fractional calculus, is well-documented for Riemann-Liouvelli (RL) and Grunwald-Letnikov's (GL) formulations, as described in the earlier sections of this chapter. However, the requirement of fractional initialization, conditions of type, $f(t_0), f^{(1/2)}(t_0), f^{(-1/2)}(t_0)...f^{(n-\alpha)}(t_0)$ are required which some-what (seems to be) physically unrealizable for those formulations, however great effort may be required to relate them physically (explained in Chapter 5).

Caputo derivative (1967) definition is made to attach integer order initializations and making the fractional derivative of a constant as zero. However, these may be mathematically pure as requirement in Caputo formulation requirement is integer order initial conditions, yet the mathematical purity gives hard physical reality. A popular belief that Caputo derivative formulation of fractional derivative properly accounts for the initialization effect is not generally true when applied to fractional differential equations. Constant initialization of the past is insufficiently general, the widely used contemporary equations for the Laplace transform for differintegrals based on that assumption also lacks generality. Therefore, the generalized form requires an initialization function. In solution of fractional differential equations with assumed history, the set of initializing constants representing the values of fractional differintegrals at $t=0$, is ineffective, thus required is 'non-constant initialization' for generalized concept of integration and differentiation as explained in earlier sections of this chapter. The requirement of this initialization function for initializing generalized derivatives in RL formulations and difficulty in Caputo formulation, to get equivalent history function is demonstrated in this section; by solving few examples. Attempt is made to clarify, that initialization function, or assumed history function, required for Caputo formulation can give a drastic condition tracing to $-\infty$, which is much away from the actual functional behavior of the function being considered. Whereas, in RL formulation with terminal initialization, the continuity and the nature of the function is maintained throughout the interval. In this section, attempt is made through simple examples to point out the difficulties in initializing Caputo formulation, to physical reality.

6.8.1 Relation between Caputo and Rieman-Liouvelli (RL) Fractional Derivative and Issues Relating to Initialization

The generalized integration (RL), of any real number q -folds is:

$$ {}_0d_t^{-q} = \frac{1}{\Gamma(q)}\int_0^t (t-\tau)^{q-1} f(\tau)d\tau $$

In the above, expression let us put $\tau = t - x^{\frac{1}{q}}$, giving

$$ d\tau = -\frac{1}{q} x^{\frac{1-q}{q}} dx , $$

substituting this change of variable the generalized (fractional) order integration relation is:

$$
\begin{aligned}
{}_0d_t^{-q} &= \frac{1}{\Gamma(q)}\int_0^t (t-\tau)^{q-1} f(\tau)d\tau = \frac{1}{\Gamma(q)}\int_{t^q}^{0}\left(t-t+x^{\frac{1}{q}}\right)^{q-1}\left(-\frac{1}{q}x^{\frac{1-q}{q}}\right)f(t-x^{\frac{1}{q}})dx \\
&= \frac{1}{q\Gamma(q)}\int_{t^q}^{0} -\left(x^{\frac{1}{q}}\right)^{q-1} x^{\frac{1-q}{q}} f(t-x^{\frac{1}{q}})dx \\
&= \frac{1}{\Gamma(q+1)}\int_0^{t^q} f(t-x^{\frac{1}{q}})dx
\end{aligned}
$$

Therefore, the fractional integration of a function as defined above, by convolution integral can also be visualized geometrically, as an area under the 'changing curve' as time (t) grows that is, $f(t-x^{1/q})$ and the limit of integration is from $x=0$ to $x=t^q$. This is detailed in Chapter 5. The Rieman-Liouvelli (RL) derivative, and its initialization issues, covered in detail in earlier sections in this chapter. In this section, the Caputo derivative and RL derivative difference is pointed out, then following is the demanding initialization issues for the Caputo derivative is elaborated by one simple example, in following section. After that demonstration, generalization is made in subsequent sections. The symbolism of the initialized derivative, un-initialized derivative and the initialization function is maintained as elucidated in earlier sections of this chapter.

6.8.2 Un-initialized Derivatives RL and Caputo

To understand the basic issues of differences between Caputo and RL derivative let us consider a simple function:

$$
\begin{aligned}
f(t) &= (t+2)^2, t \ge -2 \\
f(t) &= 0, t < -2
\end{aligned}
$$

This means that system $f(t)$ is at rest before time $t=-2$. Our aim is to differentiate by order α (we take as $\alpha = 0.5$), from time $t \ge 0$, where the function value is $f(t=0)=4$, by forward differentiation method. For demonstration, we take the order of differentiation as less than one. Therefore, the highest integer order in this case is $m = 1 > 1/2 = \alpha$, as required by Caputo and RL definitions. The rule of fractionally differentiable is maintained as per existence criteria (Chapter 5) for this function that is, the candidate function is defined and bounded in the interval and at the lower terminal $a=-2$, is 'better' behaved than $(t-a)^{-1}$.

By Caputo definition, choosing integer order derivative as unity, and then obtaining semi integration of that derivative of the function from $a=0$, we get:

$$ {}_0^C d_t^{1/2}(t+2)^2 = {}_0 d_t^{-(1/2)}\left[\frac{d}{dt}(t+2)^2\right] = {}_0 d_t^{-(1/2)}(2t) + {}_0 d_t^{-(1/2)}\{(4)\} $$
$$ = \frac{1}{\Gamma(1/2)}\left[2\int_0^t (t-\tau)^{\frac{1}{2}-1}\tau d\tau + 4\int_0^t (t-\tau)^{\frac{1}{2}-1} d\tau\right] $$

With the change of variable $(t-\tau) = dx$, we get $d\tau = -dx, \tau = t - x$, substituting this and integrating with changed limits, we have:

$$ {}_0^C d_t^{1/2}(t+2)^2 = \frac{1}{\Gamma(1/2)}\left[\frac{8}{3}t^{3/2} + 8t^{1/2}\right] \tag{6.26} $$

Note that, in (6.26) there is no singularity term. Note also that Caputo definition of fractional derivative is 'fractional integration of a derivative of a function'.

The un-initialized RL derivative is as follows:

$$ {}_0 d_t^{1/2}(t+2)^2 = \frac{d}{dt}\left[{}_0 d_t^{-(1/2)}(t+2)^2\right] $$

The un-initialized semi-integration is

$$ {}_0 d_t^{-(1/2)}(t+2)^2 = \frac{1}{\Gamma(1/2)}\int_0^t (t-\tau)^{\frac{1}{2}-1}(\tau+2)^2 d\tau $$

Changing variable as $(t-\tau) = x$ gives:

$$ {}_0 d_t^{-(1/2)}(t+2)^2 = \frac{1}{\Gamma(1/2)}\int_t^0 x^{-\frac{1}{2}}(t-x+2)^2(-dx) $$
$$ = \frac{1}{\Gamma(1/2)}\int_t^0 \{-(t+2)^2 x^{-\frac{1}{2}}dx + 2(t+2)x^{\frac{1}{2}} - x^{\frac{3}{2}}\}dx $$

Integration of this expression leads to:

$$ {}_0 d_t^{-(1/2)}(t+2)^2 = \frac{1}{\Gamma(1/2)}\left[2(t+2)^2 t^{\frac{1}{2}} - \frac{4}{3}(t+2)t^{\frac{3}{2}} + \frac{2}{5}t^{\frac{5}{2}}\right] $$

Simplifying the above yields:

$$ {}_0 d_t^{-(1/2)}(t+2)^2 = \frac{1}{\Gamma(1/2)}\left[\frac{16}{15}t^{\frac{5}{2}} + \frac{16}{3}t^{\frac{3}{2}} + 8t^{\frac{1}{2}}\right]. $$

Differentiating by one integer order we obtain the RL un-initialized semi-derivative of the function as:

$$ {}_0d_t^{1/2}(t+2)^2 = \frac{d}{dt}\left[\frac{1}{\Gamma(1/2)}\left\{\frac{16}{15}t^{\frac{5}{2}} + \frac{16}{3}t^{\frac{3}{2}} + 8t^{\frac{1}{2}}\right\}\right] = \frac{1}{\Gamma(1/2)}\left[\frac{8}{3}t^{\frac{3}{2}} + 8t^{\frac{1}{2}} + 4t^{-1/2}\right] \quad (6.27) $$

Note that the above expression (6.27) has a singularity at the start point of differentiation $t = 0$. Note also that, the RL fractional derivative is 'a derivative of fractionally integrated function'.

From the above expression (6.26) and (6.27) we find that the relation between un-initialized RL derivative and Caputo derivative is,

$$ \left[{}_0d_t^{1/2} f(t)\right]_{RL} = {}_0^C d_t^{1/2} f(t) + \frac{f(0)}{\Gamma(1-1/2)} t^{-1/2} \quad (6.28) $$

Noting, $f(t=0) = 4$, and the expression (6.28) equate the two definitions by a singularity at start point of differentiation.

We can therefore write a general expression for $\alpha < 1$ order RL and Caputo derivative as

$$ \left[{}_a d_t^{\alpha} f(t)\right]_{RL} = {}_a^C d_t^{\alpha} f(t) + \frac{f(a)}{\Gamma(1-\alpha)}(t-a)^{-\alpha} \quad (6.29) $$

These two derivatives are equal only if at the start point of fractional differentiation the function is at rest, or the start point of differentiation coincides with the start point of the process (demonstrated below). Generally, if start point of differentiation differs from the start point of process, these two derivatives are not equal. The Caputo derivative is added by a function, which is singular at the start point, of the differentiation process. However, at steady state these two derivatives converge.

For generalization proof of (6.29) let us consider $a = 0$, and observe the effect of Differentiation of integral, as follows:

From definition of fractional integral as discussed above, we have the equivalence as:

$$ {}_0d_t^{-q} f(t) = \frac{1}{\Gamma(q)} \int_0^t (t-\tau)^{q-1} f(\tau)\, d\tau = \frac{1}{\Gamma(q+1)} \int_0^{t^q} f(t - x^{\frac{1}{q}})\, dx $$

To differentiate the above we use Leibniz's rule, of differentiation of an integration (of several variables) namely:

$$ \frac{d}{dt}\int_0^t G(x,t) = G(t,t) + \int_0^t \frac{\partial G}{\partial t} dx, $$

the above formulation is function in (x,t)

Put $x^{\frac{1}{q}} = y$, and we obtain $x = y^q$, and $dx = qy^{q-1}dy$ and obtain

$$G(t, y) = f(t - y)qy^{q-1},$$

the fractional integral expression becomes:

$${}_0d_t^{-q} f(t) = \frac{1}{\Gamma(q)} \int_0^t (t-\tau)^{q-1} f(\tau)d\tau = \frac{1}{\Gamma(q+1)} \int_0^{t^q} f(t - x^{\frac{1}{q}})dx = \frac{1}{\Gamma(q+1)} \int_0^t G(t, y)dy$$

Differentiation of above expression, with Leibniz's rule one obtains:

$$d_t^1[{}_0d_t^{-q}] = \frac{1}{\Gamma(q+1)} G(t,t) + \frac{1}{\Gamma(q+1)} \int_0^{t^q} \frac{\partial}{\partial t} f(t - x^{\frac{1}{q}})dx$$

The first term is $G(t,t) = f(0)qt^{q-1}$ and the second term, is obtained by putting $t - x^{\frac{1}{q}} = \xi$

$$\frac{1}{\Gamma(q+1)} \int_0^{t^q} \frac{\partial}{\partial t} f(t - x^{\frac{1}{q}})dx = \frac{1}{\Gamma(q)} \int_0^t (t-\xi)^{q-1} \frac{d}{dt} f(\xi)d\xi = {}_0d_t^{-q}[d_t^1 f(t)]$$

Substituting these, and simplifying we obtain, the derivative of (fractional) integral operation as:

$$d_t^1[{}_0d_t^{-q}] = {}_0d_t^{-q}[d_t^1 f(t)] + \frac{t^{q-1}}{\Gamma(q)} f(0),$$

putting $q = 1 - \alpha$, and applying definition of RL and Caputo derivatives we get (6.29) that is:

$$[{}_0d_t^{\alpha}]_{RL} = {}_0^Cd_t^{\alpha} f(t) + \frac{t^{-\alpha}}{\Gamma(1-\alpha)} f(0)$$

6.8.3 Evaluation of RL and Caputo Derivative from the Start Point of the Function

Let us now evaluate fractional derivative Caputo and RL from the start point of function itself ($a = -2$).

The Caputo derivative is:

$$
\begin{aligned}
{}_{-2}^{C}d_t^{1/2}(t+2)^2 &= {}_{-2}d_t^{-1/2}(2t) + {}_{-2}d_t^{-1/2}(4) \\
&= \frac{1}{\Gamma(1/2)}\left[2\int_{-2}^{t}(t-\tau)^{\frac{1}{2}-1}\tau d\tau + 4\int_{-2}^{t}(t-\tau)^{\frac{1}{2}-1}d\tau\right] \\
&= \frac{1}{\Gamma(1/2)}\left[2\int_{t+2}^{0}x^{-\frac{1}{2}}(t-x)(-dx) + 4\int_{t+2}^{0}x^{-\frac{1}{2}}(-dx)\right] \\
&= \frac{1}{\Gamma(1/2)}\left[4t(t+2)^{\frac{1}{2}} - \frac{4}{3}(t+2)^{\frac{3}{2}} + 8(t+2)^{\frac{1}{2}}\right] \\
&= \frac{1}{\Gamma(1/2)}(t+2)^{\frac{1}{2}}\left[4t - \frac{4}{3}(t+2) + 8\right] \\
&= \frac{8}{3\Gamma(1/2)}(t+2)^{\frac{3}{2}}
\end{aligned}
$$

The RL derivative is:

$$
\begin{aligned}
{}_{-2}d_t^{1/2}(t+2)^2 &= \frac{d}{dt}\left\{{}_{-2}d_t^{-(1/2)}(t+2)^2\right\} = \frac{1}{\Gamma(1/2)}\frac{d}{dt}\int_{-2}^{t}(t-\tau)^{\frac{1}{2}-1}(\tau+2)^2 d\tau \\
&= \frac{1}{\Gamma(1/2)}\frac{d}{dt}\int_{t+2}^{0}x^{-\frac{1}{2}}(t-x+2)^2(-dx) \\
&= \frac{1}{\Gamma(1/2)}\frac{d}{dt}\int_{t+2}^{0}x^{-\frac{1}{2}}\left[(t+2)^2 - 2(t+2)x + x^2\right](-dx) \\
&= \frac{1}{\Gamma(1/2)}\frac{d}{dt}\left[\int_{t+2}^{0}-(t+2)^2x^{-\frac{1}{2}}dx + \int_{t+2}^{0}2(t+2)x^{\frac{1}{2}}dx - \int_{t+2}^{0}x^{\frac{3}{2}}dx\right] \\
&= \frac{1}{\Gamma(1/2)}\frac{d}{dt}\left[2(t+2)^2(t+2)^{\frac{1}{2}} - \frac{4}{3}(t+2)(t+2)^{\frac{3}{2}} + \frac{2}{5}(t+2)^{\frac{5}{2}}\right] \\
&= \frac{1}{\Gamma(1/2)}\frac{d}{dt}\left[\frac{16}{15}(t+2)^{\frac{5}{2}}\right] \\
&= \frac{8}{3\Gamma(1/2)}(t+2)^{\frac{3}{2}}
\end{aligned}
$$

The above example states that the Caputo derivative and the RL derivative are equal when the differentiation process starts with the birth point of function or process itself.

6.8.4 Initialization of Caputo Derivative

Let $\psi(t)$ be the time dependent initialization function taking into account history of the system, function $f(t)$, which starts at time $t=a$ and the differentiation process starts at time $t=c$, where $a<c<t$, and $f(t)=0$, for $t\le a$.

For un-initialized fractional integration of order q, the RL formulation is:

$${}_a d_t^{-q} f(t) = \frac{1}{\Gamma(q)} \int_a^t (t-\tau)^{q-1} f(\tau) d\tau \text{, for } t > a \text{.}$$

and

$${}_c d_t^{-q} f(t) = \frac{1}{\Gamma(q)} \int_c^t (t-\tau)^{q-1} f(\tau) d\tau \text{, for } t > c$$

The time between a and c is considered history of the fractional integral ${}_c d_t^{-q} f(t)$, an-initialization function.

Thus the expression, for entire integration should be, for $t \geq c$, is:

$${}_c d_t^{-q} f(t) + \psi(t) = {}_a d_t^{-q} f(t)$$

The initialization function in terminal initialization sense is fractionally integrating the function $f(t)$ from time a to c. Thus, initialization function is:

$$\psi(t) = \frac{1}{\Gamma(q)} \int_a^c (t-\tau)^{q-1} f(\tau) d\tau$$

and for this terminal initialization the integral can be initialized prior to the start time $t = c$, of the differintegration process. In the standard symbols, one may thus express the initialization as:

$${}_c D_t^{-q} f(t) \equiv {}_c d_t^{-q} f(t) + \psi(f, -q, a, c, t)$$

The meaning of $\psi(f, -q, a, c, t)$ is fractionally integrating(by order q)the function f from time a to c, for getting a time varying function ψ, for purpose of initializing the fractional integration process for $t > c$.

We can therefore generalize the fractional derivative (as done for fractional integral) for RL definition as:

$${}_c D_t^{q} f(t) \equiv {}_c D_t^{m} \, {}_c D_t^{-p} f(t)$$

Where m is an integer such that $(m-1) < q < m$ and $p = m - q, q \geq 0, t > c \geq a$.

The above expression is for initialized derivative, and converting the same by expansion with un-initialized derivative and initialization function (as done for fractional integration case) we get:

Take as $h(t) = {}_a D_t^{-p} f(t)$, then

$$ {}_cD_t^m h(t) = {}_cd_t^m \{h(t)\} + \psi(h,m,a,c,t) $$

$$ {}_cD_t^q f(t) = \frac{d^m}{dt^m}\left\{ {}_cd_t^{-p} f(t) + \psi(f,-p,a,c,t) \right\} + \psi(h,m,a,c,t) $$

For terminal initialization for integer order derivative (m), the initialization is zero that is: $\psi(h,m,a,c,t) = 0$,

For terminal initialization the $\psi(t)$ is the operation on the function itself from time a to c. For side initialization the initialization function $\psi(t)$ could be anything, such as Dirac's delta function or Heaviside's step function or the operation of obtaining initialization function is on any other function other than $f(t)$, call it $f_i(t) \neq f(t)$.

For the considered example $f(t) = (t+2)^2$ take,

$$ m = 1, \alpha = 1/2, 0 < \alpha < 1, c = 0, a = -2 $$

Then

$$ \begin{aligned} {}_0D_t^\alpha f(t) &= {}_0D_t^1 \left\{ {}_0D_t^{-(1-\alpha)} f(t) \right\} \\ &= \frac{d}{dt}\left\{ {}_0D_t^{-(1-\alpha)} f(t) \right\} + \psi(h,1,a,0,t) \\ &= \frac{d}{dt}\left\{ {}_0d_t^{-(1-\alpha)} f(t) + \psi(f,-[1-\alpha],a,0,t) \right\} + 0 \\ &= \frac{d}{dt}\left\{ \frac{1}{\Gamma(1-\alpha)} \int_0^t (t-\tau)^{-\alpha} f(\tau)d\tau \right\} + \frac{d}{dt}\left\{ \frac{1}{\Gamma(1-\alpha)} \int_a^0 (t-\tau)^{-\alpha} f(\tau)d\tau \right\} \end{aligned} $$

We will use above formulation for initialized fractional semi-derivative for RL formulation, for terminal initialization case.

$$ {}_0D_t^{1/2}(t+2)^2 = \frac{d}{dt}\left\{ \frac{1}{\Gamma(1/2)} \int_0^t (t-\tau)^{\frac{1}{2}-1} (\tau+2)^2 d\tau \right\} + \frac{d}{dt}\left\{ \frac{1}{\Gamma(1/2)} \int_{-2}^0 (t-\tau)^{\frac{1}{2}-1} (\tau+2)^2 d\tau \right\} $$

For the first term in RHS of above we have already calculated as un-initialized semi derivative as above:

$$ {}_0d_t^{1/2}(t+2)^2 = \frac{d}{dt}\left[\frac{1}{\Gamma(1/2)} \left\{ \frac{16}{15} t^{\frac{5}{2}} + \frac{16}{3} t^{\frac{3}{2}} + 8t^{\frac{1}{2}} \right\} \right] = \frac{1}{\Gamma(1/2)} \left[\frac{8}{3} t^{\frac{3}{2}} + 8t^{\frac{1}{2}} + 4t^{-1/2} \right] $$

The second term we evaluate as total initialization, for terminal initialization case as:

$$\frac{d}{dt}\left\{\frac{1}{\Gamma(1/2)}\int_{-2}^{0}(t-\tau)^{\frac{1}{2}-1}(\tau+2)^2 d\tau\right\}$$

$$=\frac{d}{dt}\int_{t+2}^{t}\frac{1}{\Gamma(1/2)}x^{-\frac{1}{2}}(t-x+2)^2(-dx)$$

$$=\frac{d}{dt}\left[\frac{1}{\Gamma(1/2)}\left\{\begin{array}{l}-2(t+2)^2 x^{\frac{1}{2}}]_{t+2}^{t}\\ +\frac{4}{3}(t+2)x^{\frac{3}{2}}]_{t+2}^{t}-\frac{2}{5}x^{\frac{5}{2}}]_{t+2}^{t}\end{array}\right\}\right]$$

$$=\frac{1}{\Gamma(1/2)}\left[\frac{8}{3}(t+2)^{\frac{3}{2}}-\frac{8}{3}t^{\frac{3}{2}}+8t^{\frac{1}{2}}+4t^{-\frac{1}{2}}\right]$$

From above derivation, what we get as initialized RL semi-derivative as:

$${}_0D_t^{1/2}(t+2)^2=\frac{1}{\Gamma(1/2)}\frac{8}{3}(t+2)^{\frac{3}{2}}$$

This is same as RL (and Caputo) derivative un-initialized start point at -2, meaning equal to (as derived earlier)

$${}_{-2}D_t^{1/2}(t+2)^2={}_{-2}d_t^{1/2}(t+2)^2={}_{-2}^{C}d_t^{1/2}(t+2)^2=\frac{1}{\Gamma(1/2)}\frac{8}{3}(t+2)^{\frac{3}{2}}$$

as, is obvious in this case $\psi(t)$ the required initialization will be zero that is:

$$\frac{d}{dt}\psi(\{t+2\}^2,-1/2,-2,-2,t)=\frac{d}{dt}\psi(f,-(1-\alpha),a,a,t)=0,$$

That is at the start point of the function the initialization is zero, the two definitions, RL and Caputo, and derivatives are equal.

We have seen the relation between un-initialized Caputo and RL derivative as (6.29):

$${}_0d_t^{\alpha}f(t)={}_0^{C}d_t^{\alpha}+f(0)\frac{t^{-\alpha}}{\Gamma(1-\alpha)},$$

For case $0<\alpha<1$, to this we add on both the sides the initialization function expression of the fractional derivative and obtain:

$${}_0d_t^{\alpha}f(t)+\frac{d}{dt}\psi(f,-(1-\alpha),a,0,t)={}_0^{C}d_t^{\alpha}f(t)+f(0)\frac{t^{-\alpha}}{\Gamma(1-\alpha)}+\frac{d}{dt}\psi(f,-(1-\alpha),a,0,t)$$

$${}_0D_t^{\alpha}f(t)={}_0^{C}d_t^{\alpha}f(t)+f(0)\frac{t^{-\alpha}}{\Gamma(1-\alpha)}+\frac{d}{dt}\psi(f,-(1-\alpha),a,0,t)$$

(6.30)

The Caputo derivative will be equal to initialized RL derivative if we make for (6.30):

$$\frac{d}{dt}\psi(f,-(1-\alpha),a,0,t)=-f(0)\frac{t^{-\alpha}}{\Gamma(1-\alpha)},$$

In our example, considered here, we can thus write this in form:

$$\frac{d}{dt}\psi(f,-1/2,-2,0,t)=-4\frac{t^{-1/2}}{\Gamma(1-1/2)}=\frac{-4t^{-1/2}}{\Gamma(1/2)} \tag{6.31}$$

In the above case this initialization is equal to differentiating a constant function by order half $f(t)=-4$, so that

$$\frac{d^{1/2}}{dt^{1/2}}(-4)t^{0}=-4\frac{\Gamma(0+1)}{\Gamma(0-1/2+1)}t^{0-\frac{1}{2}}=-\frac{4t^{-1/2}}{\Gamma(1/2)}$$

In this chosen example, one can choose the history function, $f(0-)=4$ as a constant to have Caputo derivative same as RL initialized derivative, and that function is semi-differentiated and then subtracted from Caputo to make them RL initialized derivative same as Caputo. However, choosing this constant as -4 will make the function $f(t)$ discontinuous at $t=0$, as $f(0-)=-4\neq f(0+)=4$. But the argument suggests, that in order to make the RL initialized derivative and the Caputo process same, one has to use a constant function (at least in this chosen example) and the concept of terminal initialization will not work. This calls for side-initialization that is having initialization function $\psi(t)$, derived from any other function other than the function $f(t)$; call it $f_i(t)\neq f(t)$.

Let the initialization function be formed for Caputo derivative by a history function, which is analytical at $t=c=0$, is represented by Taylor series, for $t\leq 0$ as:

$$f_i(\tau)=\sum_{n=0}^{\infty}\frac{f_i^{(n)}(0-)}{n!}\tau^n,\tau<0$$

For Caputo initialization we have

$$\frac{d}{dt}\left\{\sum_{n=0}^{\infty}\frac{f_i^{(n)}(0-)}{n!}\frac{1}{\Gamma(1-\alpha)}\int_a^0(t-\tau)^{-\alpha}\tau^n d\tau\right\}=f(0+)\frac{t^{-\alpha}}{\Gamma(1-\alpha)}$$

For $n=0$ we get,

$$f_i(0-)\frac{d}{dt}\int_a^0(t-\tau)^{-\alpha}d\tau=f_i(0-)\left[(t-a)^{-\alpha}-t^{-\alpha}\right]=-t^{-\alpha}f(0+).$$

For this identity to be valid then $a \to -\infty$, gives $f_i(0-) = f(0+)$ for $-\infty < t < 0$. In our example therefore to have Caputo and RL derivative initialized and equated we have to make the function $f(t) = (t+2)^2$ abruptly take a constant value that is, value of the function at $t = 0$ and be a constant at value $f(0) = 4$, from start point $t = -\infty$, till $t = 0$. In this argument though the continuity of the function at $t = 0$, is maintained though, the function looses its differentiability at this point. Therefore, the history function for Caputo derivative in this example is a constant Heaviside's step function, multiplied by constant of value $f(0) = 4$ starting at $t = -\infty$. This is though not at all a backward continuation of the function being considered for derivative process. It is different from original function form from $-\infty$ to the start point of differentiation, process i.e. $t = 0$.

This condition seems to be physically demanding. Say the function represents, velocity that started at some time -2, and we are obtaining fractional acceleration by this process of half derivative, for obtaining forces acting. Now the situation is that if we want to initialize with Caputo formulation then we have to consider that system has constant velocity from time immemorial to time zero, and then suddenly takes our defined trajectory of the function. This seems to be unrealistic physically, at least in this case.

In the terms of RL initialization let us take $f_i(t) = f(0+) = 4$, as history function from $-2 \le t \le 0$. In case of terminal initialization as demonstrated above the history function for RL was taken as $f_i(t) = f(t) = (t+2)^2$ for $-2 \le t \le 0$. Evaluating the initialization term with revised history function $f_i(t) = 4$ we obtain:

$$
\begin{aligned}
\frac{d}{dt}\psi\{4,-1/2,-2,0,t\} &= \frac{d}{dt}\int_{-2}^{0}\frac{1}{\Gamma(1/2)}(t-\tau)^{\frac{1}{2}-1}4d\tau \\
&= \frac{4}{\Gamma(1/2)}\frac{d}{dt}\int_{t+2}^{t}x^{-\frac{1}{2}}(-dx) \\
&= \frac{d}{dt}\{-\frac{8}{\Gamma(1/2)}t^{\frac{1}{2}}+\frac{8}{\Gamma(1/2)}(t+2)^{\frac{1}{2}}\} \\
&= -\frac{4t^{-\frac{1}{2}}}{\Gamma(1/2)}+\frac{4(t+2)^{-\frac{1}{2}}}{\Gamma(1/2)}
\end{aligned}
$$

The un-initialized RL derivative as obtained earlier is

$$
{}_0d_t^{1/2}(t+2)^2 = \frac{8t^{\frac{3}{2}}}{3\Gamma(1/2)}+\frac{8t^{\frac{1}{2}}}{\Gamma(1/2)}+\frac{4t^{-\frac{1}{2}}}{\Gamma(1/2)} = {}_0^Cd_t^{1/2}(t+2)^2+\frac{4t^{-\frac{1}{2}}}{\Gamma(1/2)}
$$

Therefore the initialized RL derivative is:

$$ {}_0D_t^{1/2}(t+2)^2 = {}_0d_t^{1/2}(t+2)^2 + \frac{d}{dt}\psi\{4,1/2,-2,0,t\} = \frac{8t^{\frac{3}{2}}}{3\Gamma(1/2)} + \frac{8t^{\frac{1}{2}}}{\Gamma(1/2)} + \frac{4(t+2)^{-\frac{1}{2}}}{\Gamma(1/2)} $$

The observation from above is that the singularity is shifted at, $t=-2$ from $t=0$, by taking this initialization function as constant of (value, 4) from start point of the function ($t=-2$) to start point of the fractional derivative ($t=0$) process. However pushing the start point at minus infinity i.e., taking $t \to -\infty = -2$, makes the initialized RL derivative equal to Caputo derivative, exactly i.e.

$$ {}_0D_t^{1/2}(t+2)^2 = {}_0d_t^{1/2}(t+2)^2 + \frac{d}{dt}\psi\{4,1/2,-\infty,0,t\} = \frac{8t^{\frac{3}{2}}}{3\Gamma(1/2)} + \frac{8t^{\frac{1}{2}}}{\Gamma(1/2)} = {}_0^Cd_t^{1/2}(t+2)^2 $$

The exactness in RL and Caputo is thus achieved by pushing the singularity to $-\infty$, at least for this example.

The value of the initialization function is:

$$ \psi_i(t) = \frac{d}{dt}\{\psi(f_i(t)=4,1/2,-\infty,0,t)\} = -4\frac{t^{-\frac{1}{2}}}{\Gamma(1/2)} = -f(0+)\frac{t^{-\alpha}}{\Gamma(1-\alpha)} $$

that is, as though using a constant function and from time immemorial ($-\infty$) and putting the formulation of terminal initialization of RL derivative for obtaining initialization function to equate Caputo formulation, to cancel the singularity function which appears as difference of RL un-initialized and Caputo un-initialized derivatives. However, it is observed that this history function makes the first derivative at $t=0$ as discontinuous. This restriction may not be physically viable for physical processes, though mathematically pure.

6.8.5 Generalization of RL and Caputo Formulations

The expression (6.29) and (6.30) may be extended by the above example for cases $1<\alpha<2$, the just greatest integer for RL and Caputo formulation is $m=2$ the initialization expression is:

$$ {}_0D_t^{\alpha} f(t) = {}_0^Cd_t^{\alpha} f(t) + \frac{(t-0)^{-\alpha} f(0)}{\Gamma(1-\alpha)} + \frac{(t-0)^{1-\alpha} f'(0)}{\Gamma(2-\alpha)} + \frac{d^2}{dt^2}\psi\{f,-(2-\alpha),a,0,t\}, \quad t>0 \ldots.. \tag{6.32} $$

For RL derivative initialized case to be equal to Caputo derivative initialization needs be the following expression:

$$ \frac{(t-0)^{-\alpha} f(0)}{\Gamma(1-\alpha)} + \frac{(t-0)^{1-\alpha} f'(0)}{\Gamma(2-\alpha)} + \frac{d^2}{dt^2}\psi\{f,-(2-\alpha),a,0,t\} = 0 \tag{6.33} $$

Again considering the initialization function as analytic for $t \le 0$, (Taylor-Riemann series)

$$f_1(t) = \sum_{n=0}^{\infty} \frac{f_1^{(n)}(0-)}{n!} t^n$$

We obtain the condition for initialization as:

$$\frac{1}{\Gamma(2-\alpha)} \frac{d^2}{dt^2} \int_a^0 (t-\tau)^{2-\alpha-1} \sum_{n=0}^{\infty} \frac{f_1^{(n)}(0-)}{n!} d\tau = -\frac{t^{-\alpha}}{\Gamma(1-\alpha)} f(0+) - \frac{t^{1-\alpha}}{\Gamma(2-\alpha)} f'(0+) \quad (6.34)$$

Equating the $n=0$ and $n=1$ in this case gives the history function that will make the RL derivative and Caputo derivative equal, will have condition $f_1(0-) = f(0+)$, and $f_i'(0-) = f'(0+)$, making history function as $f_1(t) = f(0+) + tf'(0+)$ for $-\infty \le t \le 0$.

For the case $f(t) = (t+2)^2$ at $t=0$, and say the $\alpha = 1.5$ the history function will be $f_1(t) = 4t + 4$, for $t \le 0$.Clearly this history function will have second derivative discontinuous at $t = 0$. Again, this possess a physical restriction, as to look at the function from time immemorial as a very different from the process itself.

We therefore can generalize (6.29), (6.30) and (6.31) by induction, for any integer m so those fractional RL and Caputo derivatives are initialized, for $(m-1) < \alpha < m$.

$${}_0D_t^{\alpha} f(t) = {}_0^C d_t^{\alpha} f(t) + \sum_{n=0}^{m-1} \frac{t^{n-\alpha}}{\Gamma(n-\alpha+1)} f^{(n)}(0+) + \frac{d^m}{dt^m} \psi\{f, -(m-\alpha), a, 0, t\}, \ t > 0$$

To have RL derivative and Caputo the same, for $t > 0$ we have:

$$\frac{d^m}{dt^m} \psi\{f, -(m-\alpha), a, 0, t\} = -\sum_{n=0}^{m-1} \frac{t^{n-\alpha}}{\Gamma(n-\alpha+1)} f^{(n)}(0+)$$

$$\frac{d^m}{dt^m} \frac{1}{\Gamma(m-\alpha)} \int_a^0 (t-\tau)^{m-\alpha-1} f_1(\tau) d\tau = -\sum_{n=0}^{m-1} \frac{t^{n-\alpha}}{\Gamma(n-\alpha+1)} f^{(n)}(0+)$$

Putting Taylor expression for $f_1(t)$ the history function for $-\infty \le t \le 0$, the expression is:

$$\frac{1}{\Gamma(m-\alpha)} \frac{d^m}{dt^m} \int_a^0 (t-\tau)^{m-\alpha-1} \sum_{n=0}^{\infty} \frac{f_1^{(n)}(0-)}{n!} d\tau = -\sum_{n=0}^{m-1} \frac{t^{n-\alpha}}{\Gamma(n-\alpha+1)} f^{(n)}(0+),$$

will thus be valid for: $f_1^{(i)}(0-) = f^{(i)}(0+)$, for $i = 0, 1, 2, \ldots (m-1)$. Therefore, the generalized history function to have initialized RL and Caputo derivative same for fractional derivative is a polynomial of order $(m-1)$ for $-\infty \le t \le 0$, represented as:

$$f_1(t) = f^{(m-1)}(0)t^{m-1} + f^{(m-2)}(0)t^{m-2} + \ldots,$$

This initialization law gives a discontinuity for derivatives higher than $(m-1)$ for the function with history at $t = 0$

6.8.6 Observations Regarding Difficulties in Caputo Initialization and Demanding Physical Conditions vis-à-vis RL Initialization Conditions and Relation to Physics in Solving Fractional Order Differential Equations

The above discussion, in the sections, suggests that Caputo inferred history has discontinuous integer order derivatives at the start point of fractional differentiation process. While, the RL initialization is smooth continuation of function being fractionally differentiated, for terminal initialization case; there is difference in the two derivatives RL initialized with Caputo initialized, and is profound at large times, however the functional forms of the two derivatives are identical, (that is) power function with time for times significantly larger than the start point of fractional differentiation. The RL un-initialized derivative of the function starting at the start point of the function itself is a smooth backward continuation of initialized fractional RL derivative starting other than the start point of the function, for terminal initialization. It is also noted that RL fractional derivative, using Caputo inferred history for initialization for $-\infty \le t \le c = 0$, is same as Caputo fractional derivative. The two definitions of derivatives are exactly same if and only if the initialization part is zero or the derivative process for RL and Caputo starts at the birth of the function ($f(t) = f'(t) = 0$ at $t \le a = -2$). Else, the two derivatives are equated by a polynomial that is singular at the start point of differentiation ($t = c = 0$).

Say we have a differential equation

$$y(t) = {}_0^C d_t^{\alpha} x(t)$$

Using Caputo inferred history may be acceptable if (a) it is found that the history acceptable to physics of the problem and (b) it is acceptable to have discontinuity of the integer order derivative of $x(t)$ of the order $m-1$, where $(m-1) < \alpha < m$.

Suppose we consider a slightly complicated fractional order differential equation as:

$$y(t) = {}_0^C d_t^{\frac{3}{2}} x(t) + {}_0^C d_t^{\frac{1}{2}} x(t) + x(t),$$

and suppose we assume (for simplicity) that both the fractional derivative terms have the same history i.e. $x_1(t)$, for $t \le 0$. For this example, the orders of the terms lay between different integer orders, for the first term of fractional derivative, the order $1 < \alpha = 1.5 < 2$, and for the second fractional derivative, the order is between,

$0 < \alpha = 0.5 < 1$. From our discussion on generalization of the initialization, the history for the first term is $x_1(t) = x(0) + tx'(0)$, and the second term the history will be $x_1(t) = x(0)$ for $-\infty \le t \le 0$, clearly only history that will satisfy both the fractional terms is $x_1(t) = x(0)$, for $-\infty \le t \le 0$, with $x'(0) = 0$ (which is not the case, as with this history the first derivative will be discontinuous at $t = 0$).

If on the other hand each have separate histories that means each fractional differential terms in the equation is disconnected from one other, and being acted on by their own individual history, in negative time. Then at time zero the entire individual position, velocity, have the same value at time zero, under this condition and scenario the initialization of each term reverts to the case, as discussed for, the case $y(t) = {}_0^C d_t^\alpha x(t)$.

6.9 Fractional Differintegrations for Periodic Signals

6.9.1 Fractional Derivative/Integral of Generalized Periodic Function

The function $f_{PER}(t)$, which can be expressed as sum of complex conjugated exponential functions as:

$$f_{PER}(t) = \sum_{k=1}^{\infty} A_k \exp\left(\frac{2\pi jkt}{T}\right) + \bar{A}_K \exp\left(-\frac{2\pi jkt}{T}\right)$$

where A_k and $\bar{A}_k$ are conjugate constants, T is the period of the function, and then using the differentigration expression for exponential as:

$$\frac{d^q}{dt^q}\exp\left(\pm\frac{2\pi jkt}{T}\right) = \left[\pm\frac{2\pi jkt}{T}\right]^q \frac{d^q}{d(\pm 2\pi jkt/T)^q}\exp\left(\pm\frac{2\pi jkt}{T}\right)$$
$$= t^{-q}\exp\left(\frac{\pm 2\pi jkt}{T}\right)\gamma^*\left(-q, \frac{\pm 2\pi jkt}{T}\right)$$

gives complete description of fractional derivative of a periodic function. After that, use of expansion of incomplete Gamma function for small and large values of t one gets the asymptotic expansions. [Abramowitz and Stegun]

For small values of t the power series of incomplete Gamma function when used, we get:

$$\frac{d^q}{dt^q}\exp\left(\frac{\pm 2\pi jkt}{T}\right) = t^{-q}\sum_{m=0}^{\infty}\frac{[\pm 2\pi jkt/T]^m}{\Gamma(m-q+1)}.$$

Therefore in the limit of small t, the fractional differintegration of periodic function is:

$$\frac{d^q}{dt^q} f_{PER}(t \to 0) = \sum_{k=1}^{\infty} t^{-q} \frac{[A_k + \bar{A}_k]}{\Gamma(1-q)} + \sum_{k=1}^{\infty} t^{1-q} \frac{\mathrm{j}k[A_k - \bar{A}_k]}{T\Gamma(2-q)} + ...\sum_{k-1}^{\infty} ....$$

The asymptotic formulation for large t for incomplete Gamma function gives using the fact $(\pm \mathrm{j})^q$ a complex number $\exp(\pm\pi \mathrm{j} q/2)$ yields the following:

$$\frac{d^q}{dt^q}\exp\left(\frac{\pm 2\pi \mathrm{j}kt}{T}\right) \approx \left[\frac{\pm 2\pi \mathrm{j}k}{T}\right]^q \exp\left(\frac{\pm 2\pi kt}{T}\right) - \sum_{m=0}^{\infty} t^q \frac{[\pm 2\pi \mathrm{j}kt/T]^{-1-m}}{\Gamma(-q-m)}$$

Yielding:

$$\frac{d^q}{dt^q} f_{PER}(t \to \infty) = \sum_{k=1}^{\infty}\left[\frac{2\pi k}{T}\right]^q \left\{A_k \exp\left(2\pi \mathrm{j}\left[\frac{kt}{T} + \frac{q}{4}\right]\right) + \bar{A}_k \exp(-2\pi \mathrm{j}\left[\frac{kt}{T} + \frac{q}{4}\right]\right\}$$

$$+\sum_{k=1}^{\infty}\left\{t^{-1-q}\frac{\mathrm{j}[A_k - \bar{A}_k]T}{2\pi k\Gamma(-q)} + t^{-2-q}\frac{[A_k + \bar{A}_k]T^2}{4\pi^2\Gamma(-q-1)} + \ldots\right\}$$

In the above expression, the terms grouped inside the first summation are periodic. They show that the effect of differintegration of order q is to change the amplitude of each component by $[2\pi k/T]^q$ and to change the phase by an angle $q\pi/2$. Within the second summation, the terms are aperiodic; provided $q > -1$, as the condition for standard fractional differentiability, they are representing transients, which die down, and become insignificant at large times.

One can use this and can write:

$$\frac{d^q}{dx^q}\cos(x) = \cos\left(x + q\frac{\pi}{2}\right) + \frac{x^{-2-q}}{\Gamma(-q-1)} - \frac{x^{-4-q}}{\Gamma(-q-3)} + \ldots$$

$$\frac{d^q}{dx^q}\sin(x) = \sin\left(x + q\frac{\pi}{2}\right) + \frac{x^{-1-q}}{\Gamma(-q)} - \frac{x^{-3-q}}{\Gamma(-q-3)} + \ldots$$

This above described method is one of the methods to obtain the fractional differintegrals of periodic function. However, other methods are available, as to fractional derivatives of the trigonometric functions may be represented by higher transcendental functions or transcendental trigonometric functions in terms of Mittag-Leffler functions Miller Ross functions etc, some were discussed earlier in Chapter 1 and Chapter 2.

6.9.2 Fractional Derivative of Periodic Function with Lower Terminal Not at Minus Infinity

The ${}_{-\infty}D_t^q \sin\lambda t = \lambda^q \sin\left(\lambda t + q\frac{\pi}{2}\right)$, meaning that if the derivative (or anti derivative process is started from time immemorial then we get a steady state condition, that the periodic function gets phase shifted either leading , or lagging by $q\frac{\pi}{2}$. But the situation is different if the lower limit of differentiation (integration) is shifted from time immemorial to some other time (say time $t = c$). In that case, we write initialized differintegral of a function as uninitialized part plus initialization function as:

$$ {}_cD_t^q f(t) = {}_ad_t^q f(t) + \psi(f, q, a, c, t) \, , $$

where the function $f(t) = 0$, for $t < a$, and the initialization function is result of differintegrating the function from time a to c by order q , for terminal charging (terminal-initialization)case and can be chosen as arbitrary as per requirement for side charging (side-initialization) case. The uninitialized differentigral is:

$$ {}_ad_t^q f(t) = \frac{d^q}{d[t-a]^q} f(t) \, . $$

The differintegration of sinusoidal function can be represented as:

$$ \frac{d^q}{d[t-a]^q}\sin(t) = \sin\left(t + q\frac{\pi}{2}\right) + \frac{[t-a]^{-1-q}}{\Gamma(-q)} - \frac{[t-a]^{-3-q}}{\Gamma(-q-2)} + ... $$

This expression suggests that fractional differintegration of sinusoidal periodic function gives a phase change in the periodic part and there are transients associated with the non-periodic part which dies down at large times, becoming insignificant.

Putting $a = 0$ we get;

$$ {}_0d_t^q \sin(t) = \sin\left(t + q\frac{\pi}{2}\right) + \frac{t^{-1-q}}{\Gamma(-q)} - \frac{t^{-3-q}}{\Gamma(-q-2)} + ... $$

Let us put $q = \frac{1}{2}$, and $a \neq 0$, the semi-differentiation is:

We get,

$$ {}_ad_t^{1/2} \sin(t) = \sin\left(t + \frac{\pi}{4}\right) + \frac{[t-a]^{-3/2}}{\Gamma(-1/2)} - \frac{[t-a]^{-7/2}}{\Gamma(-5/2)} + ... $$

Taking the $a=-\infty$, the polynomial of fractional order above to zero gives:

$$_{a=-\infty}d_t^{1/2}\sin(t)=\sin\left(t+\frac{\pi}{4}\right)$$

Putting $q=-\dfrac{1}{2}$, and $a\neq 0$, we have semi-integration as:

$$_a d_t^{-1/2}\sin(t)=\sin\left(t-\frac{\pi}{4}\right)+\frac{[t-a]^{-1/2}}{\Gamma(1/2)}-\frac{[t-a]^{-5/2}}{\Gamma(-3/2)}+\ldots$$

In this case also putting $a=-\infty$, gives, higher order terms in the remainder polynomial as zero yielding:

$$_{a=-\infty}d_t^{-1/2}\sin(t)=\sin\left(t-\frac{\pi}{4}\right)$$

Using scaling law $\dfrac{d^q}{dx^q}f(\lambda x)=\lambda^q\dfrac{d^q}{d(\lambda x)^q}f(\lambda x)$, we get:

$$_0d_t^q\sin(\lambda t)=\lambda^q\sin\left(t+q\frac{\pi}{2}\right)+\lambda^q\frac{t^{-1-q}}{\Gamma(-q)}-\lambda^q\frac{t^{-3-q}}{\Gamma(-q-2)}+\ldots$$

Putting $q=1$ in the expression

$$_0d_t^q\sin(t)=\sin\left(t+q\frac{\pi}{2}\right)+\frac{t^{-1-q}}{\Gamma(-q)}-\frac{t^{-3-q}}{\Gamma(-q-2)}+\ldots,$$

we get:

$$_0d_t^1\sin(t)=\sin\left(t+\frac{\pi}{2}\right)+\frac{t^{-2}}{\Gamma(-1)}-\frac{t^{-4}}{\Gamma(-3)}+\ldots.=\sin\left(t+\frac{\pi}{2}\right)+0=\cos(t)=\frac{d}{dt}\sin(t).$$

Here we have used properties of reciprocal Gamma function which is zero at all negative integer points.

Putting $q=-1$ in the expression

$$_0d_t^q\sin(t)=\sin\left(t+q\frac{\pi}{2}\right)+\frac{t^{-1-q}}{\Gamma(-q)}-\frac{t^{-3-q}}{\Gamma(-q-2)}+\ldots$$

we get

$$ {}_0d_t^{-1}\sin(t)=\sin\left(t-\frac{\pi}{2}\right)+\frac{t^0}{\Gamma(1)}-\frac{t^{-2}}{\Gamma(-1)}+....=\sin\left(t-\frac{\pi}{2}\right)+1=-\cos(t)+1=\int_0^t \sin(t)dt $$

The above expression shows that the expression for fractional differentigration of $\sin(t)$ is correct for one complete integer order differentiation and one complete integration of integer order.

In above derivation we have assumed the arbitrary initialization function as zero i.e. $\psi(\sin(t),q,a,c,t)=0$, and the $f(t)=\sin t(t)=0$, for $t<a=0$.

6.10 Fractional Advection Dispersion Equation and Its Solution

An example of fractional differential equation is taken without going into thought of earlier sections of dichotomy of using RL or Caputo formulation in the fractional derivative part. This example is taken and also its physical significance is discussed. This is boundary value problem and the spatial fractional derivatives are Fourier transformed, and therefore the limit of integration is from $-\infty$ to $+\infty$. Thus in this case RL or Caputo initialization issues are not considered. Fractional Advection Dispersion Equation, which is generalization of Advection Dispersion Equation (ADE) also called Fokker Plank Equation (FPE). This originates from theory of conservation of probability (Chapter 1) and describes transport of solute concentration $C(x,t)$ in medium. The generalized equation is:

$$ \frac{\partial C}{\partial t}=-v\frac{\partial C}{\partial x}+\frac{1}{2}(1+\beta)\mathbb{D}_\alpha\frac{\partial^\alpha C}{\partial x^\alpha}+\frac{1}{2}(1-\beta)\mathbb{D}_\alpha\frac{\partial^\alpha C}{\partial(-x)^\alpha},\ \alpha<2 $$

Here we solve the boundary value problem for instantaneous injection of a unit Dirac delta function spike of solute (that is Green's function) via Fourier transformation. The boundary condition is $C\left(|x|\to\infty,t\right)=0$. The unit of $\mathbb{D}_\alpha$ is $\mathrm{cm}^\alpha/\mathrm{s}$ is diffusion (dispersion constant). For $\alpha=2$, the normal ADE is recovered as

$$ d^2/dx^2=d^2/d(-x)^2. $$

Transforming the above equation by Fourier transform we get:

$$ \frac{d}{dt}\hat{C}(k,t)=-v(\mathrm{i}k)\hat{C}(k,t)+\frac{1}{2}(\mathrm{i}k)^\alpha\mathbb{D}_\alpha(1+\beta)\hat{C}(k,t)+\frac{1}{2}(-\mathrm{i}k)^\alpha\mathbb{D}_\alpha(1-\beta)\hat{C}(k,t) $$

We have used $\Im\left\{D_\pm^\alpha f(x)\right\}=(\pm \mathrm{i}k)^\alpha\hat{f}(k)$, with $\mathrm{i}=\sqrt{-1}$; that is generalized Fourier identity (Chapter 1). The solution in k Fourier-coordinates is thus:

$$ \hat{C}(k,t)=\exp\left[\frac{1}{2}(1-\beta)(-\mathrm{i}k)^\alpha\mathbb{D}_\alpha t+\frac{1}{2}(1+\beta)(\mathrm{i}k)^\alpha\mathbb{D}_\alpha t-\mathrm{i}kvt\right] $$

Let us simplify, the above obtained solution with $\mathrm{B} = |\cos(\pi\alpha)/2|\mathbb{D}_\alpha$ and identities $\mathrm{i} = e^{\mathrm{i}\pi/2}$ and $e^{\mathrm{i}\theta} = \cos\theta + \mathrm{i}\sin\theta$ to get:

$$\hat{C}(k,t) = \exp\left\{-\mathrm{B}t|k|^\alpha \left[1 + \mathrm{i}\beta\,\mathrm{sgn}(k).\tan(\pi\alpha/2)\right] - \mathrm{i}kvt\right\}$$

This above expression does not have closed form inverse. Hence putting it in the form of characteristic function (substituting k with $-k$), the density $\hat{C}(-k,t)$, can be manipulated in canonical forms of α -stable densities that is:

$$\hat{C}(-k,t) = \exp\left\{-\mathrm{B}t|k|^\alpha \left[1 - \mathrm{i}\beta\,\mathrm{sgn}(k).\tan(\pi\alpha/2)\right] + \mathrm{i}kvt\right\}$$

where $\sigma = (\mathrm{B}t)^{1/\alpha}$ indicates a stable density that is shifted by a mean $\mu = vt$ and is invariant upon scaling by $t^{1/\alpha}$. For $\beta = 0$ the solution implies symmetrical Fractional Fokker Plank System or symmetrical fractional Advection Dispersion equation, given by vector equation as:

$$\frac{\partial C}{\partial t} = -v \bullet \nabla C + \mathbb{D}_\alpha \nabla^\alpha C\,,$$

with $\alpha < 0$, and symmetrical solution in k -Fourier coordinate is

$$\hat{C}(k,t) = \exp\left(-\mathrm{B}t|k|^\alpha - \mathrm{i}kvt\right).$$

It is simple matter to show that for any $\alpha < 2$ the stable variables have infinite variance (as for power-law), since one or both tails decay as $f_\alpha(x) \sim c|x|^{-1-\alpha}$. However a finite sampling of the density (that is plume) yields a finite sample variance. Since, the density is scale invariant with $t^{1/\alpha}$, the sample variance must grow proportional to the square of this. So the density (plume) undergoing this Levy motion, would have sample variance that grows proportional to $t^{2/\alpha}$, or always equal to or faster than the Fickian growth. The solution to the classical ADE with continuous source initial condition is generally written in closed form using the error function.

$$\mathrm{erf}(x) = \frac{2}{\sqrt{\pi}} \int_0^x \exp(-y^2)\,dy\,.$$

The step function, of constant C_0 boundary value problem using the classical ADE is given by

$$C = \frac{C_0}{2}\left[1-\mathrm{erf}\left(\frac{x-vt}{2\sqrt{\mathbb{D}t}}\right)\right].$$

For the symmetric fractional ADE, ($\beta = 0$) the solution for the similar step function uses generalized error function as

$$C = \frac{C_0}{2}\left[1-\mathrm{serf}\left(\frac{x-vt}{(\mathrm{B}t)^{1/\alpha}}\right)\right],$$

where $\mathrm{serf}_\alpha(x) = 2\int_0^x f_\alpha(y)dy$.

The function $f_\alpha(x)$ is standard symmetric α-stable density, defined with $\mu = vt$, $\sigma = (\mathrm{B}t)^{1/\alpha}$ and $f_\alpha(k) = \exp\left(-\sigma^\alpha |k|^\alpha + ik\mu\right)$. This $\alpha-$ stable density will be elaborated in following section.

6.11 Identification of Random Delays

We might expect that network traffic would best be represented, modeled or simulated by having some random source and random sized packets in a network. Following this line of thinking, the distribution of delay times on network access, would be Poisson distribution. As it turns out this as naïve model for network traffic seems to be wrong. Network traffic is best modeled by a process which displays non-random nature of Hurst parameter H which gives Long Ranged Dependence (LRD) with lingering memory tail, and a non-Gaussian distribution. 'Indeed Nature's prediction toward LRD has been well documented in hydrology, metrology and geophysics'. Here in this section LRD is established for 'stochastic delay' of computer based network system.

6.11.1 Random Delay a Stochastic Behavior

Dynamics of delay in any systems demonstrate the stochastic behavior. The delay of random nature has wide spikes and if a statistics be taken, it is like a power law, with pronounced tail. For example, effect of network delay in control system is very widely researched topic, and has practical relevance to modern computer control industry. A Brownian motion to model the stochastic process of 'random delay dynamics' is proposed, through fractional equivalent of Langevin equation driven by 'shot-noise'. A shot noise results when a 'memory-less filter' is excited by train of impulses derived from a homogeneous Poisson's Point Process (PPP). For simplicity we assume the impulse response of the filter as $\varphi(t) = \exp(-t/\eta)$ for $t \geq 0$. Therefore we can represent a shot-noise as

$$F(t)=\sum_{k=1}^{n} a_k \varphi(t-t_k) .$$

We consider the fractional form of Langevin type stochastic differential equation, and replace standard 'white noise' Gaussian stochastic driving excitation force, by 'shot-noise'

$$F(t)=\sum_{k=1}^{n} a_k \varphi(t-t_k)$$

whose each pulse has randomized amplitude, a; the pdf of the amplitude $P(a)$, has characteristic function, $W(k)=\exp\{-b^{\alpha}|k|^{\alpha}\}$, with $0<\alpha\leq 2$.

The force is acting on a delay generating block where the Fractional equivalent of Langevin equation is dynamic representation of the system as:

$$D_t^q \tau(t)=\lambda\tau(t)+F(t),$$

with $0<q\leq 1$, and τ_0 representing fractional initial condition of the delay generating dynamic system. This equation is detailed in next subsection. The driving force is train of pulses, will give a delay function of time $\tau(t)$, which also may be called fractional stochastic variable, from this above dynamic system as $\tau(t;\tau_0,F)$. The fluctuation of this fractional stochastic variable is, $\tau-\tau(t;\tau_0,F)$. This fluctuating delay function then generates spiked delay as $\delta(\tau-\tau(t;\tau_0,F))$.

The physics of the delay generating system can be explained in following manner that is, say for one moment we have demanded a particular data (information), say the computer gives that data (information) after time of 500mS; the next moment a different data (information) is called for and the new data (information) now gets available say at 5000mS. The type of data or information signifies the amplitude, a_k of shot-noise pulse, which after exciting the delay generating system, makes the particular data or information available. So these 500mS and 5000mS duration becomes the stochastic variable $\tau(t)$ with height at the data availability time, gives spike that is, $\delta(\tau(t))$. So a plot of $\tau-\tau(t;\tau_0,F)$ and its delay time $\delta(\tau-\tau(t;\tau_0,F))$, will give spike nature of fluctuating delay dynamics. These spikes are really spiky in nature, deviate and fluctuates a lot. The first type of data again when asked may take 10mS or even may take 10,000mS; that depends on the load on network traffic. Therefore, the fluctuations are really spiky in real terms.

The classical method of fluctuation dynamics is by Gaussian assumption of the random behavior, and dynamics of the same applied to fluctuations in financial assets is widely used in mathematical finance because of simplification it provides in analytical calculations. This gives integer order differential equation formulations giving Gaussian solutions.

Mandelbrot who introduced the term 'fractal' observed that in addition to being non-Gaussian, the stochastic process of financial returns shows interesting property of 'self-similarity' that is, the statistical dependencies of 'random phenomena like financial returns, Brownian motion, have similar functional form as discussed earlier in this book, for various time increments Δt. Motivated by 'pronounced tails and the stable functional forms at different time scales, Mandelbrot proposed that distributions of 'such' fluctuation is consistent with α -stable Levy distributions that is the fluctuating processes can be modeled as α -stable Levy process. Thus from the point of view of the fractal concept here in this section, attempt is made to provide fractality concept for random fluctuating delay dynamics of general (computer, finance) system. It is well known that the trajectories of Brownian and Levy stochastic processes are fractals as discussed in earlier chapter, it means that they are non-differentiable, self similar curves whose fractal dimensions are different from topological dimensions.

Here in this section, developed a new extension of fractality concept for dynamics of random delay. It is proposed here that a possible fractional calculus approach to model the evolution of stochastic dynamics of random delay. The proposed fractional dynamic stochastic approach allows obtaining the probability distribution function (pdf) of the modeled random delay. As an application of the developed general approach we derive the equation of pdf of increments of random delay $\Delta\tau$, as a function of increment of time Δt, $\Delta\tau(\Delta t)=\tau(t+\Delta t)-\tau(t)$ where the value of random delay generating system at any time is $\tau(t)$. Statistical properties of incremental delay of computer control system play important role in understanding the control system for example its stability measures with gain variations and its robustness. The theoretically predicted pdf of increments of delay $\Delta\tau$ as a function of increment of time Δt is to be experimentally verified, for calibration of numerical parameters of the proposed dynamical model for random delay.

6.11.2 About Levy Distribution

A random variable X is said to have an α -stable distribution was demonstrated in previous section; if and only if its characteristic function has the following form:

$$f(x)=\exp\left\{i\delta x-\gamma|x|^{\alpha}\left[1+i\beta\,\mathrm{sgn}(x)w(x,\alpha)\right]\right\},\ i=\sqrt{-1}$$

$$\text{With } w(x,\alpha)=\begin{cases}\tan\dfrac{\alpha\pi}{2}; & \alpha\neq 1\\[2mm] \dfrac{2}{\pi}\log|x|; & \alpha=1\end{cases},\ \text{and } \mathrm{sgn}(x)=\begin{cases}1; & x>0\\ 0; & x=0\\ -1 & x<0\end{cases}$$

$$0<\alpha\leq 2; -1\leq\beta\leq 1; \gamma>0; -\infty<\delta<\infty$$

This was also discussed in brief in Chapter 1. Therefore, an α -stable distribution can be completely determined by four parameters. **1.** The characteristic exponent α. It is the shape parameter which specifies the thickness of the tail of the probability density function. Lesser the value more pronounced is the tail, indicating strong lingering memory. In other words, α changes the level of spikiness in the distribution, the larger the value of α, the less likely it is to observe random variable that is distant from its central location. For a normal distribution $\alpha = 2$, where the tail decays exponentially fast from central 'dome', indicating that there is less likelihood of presences or a random variable at far places, from 'mean'. This parameter is also called 'stability' parameter. For $\alpha < 2$ the distribution of the random variable has no finite variance, and for $\alpha < 1$, the finite mean does not exist. Generally, $0 < \alpha \le 2$. **2.** The skewness index β. Positive values for β make the distribution skewed toward the right tail and negative values make it skewed toward the tail on the left hand side. For normal distribution or any symmetric distributions $\beta = 0$. Generally $-1 \le \beta \le +1$. 3. The variable γ is called the scale parameter or dispersion parameter and it expresses the dispersion of the distribution. For normal distribution γ is equal to standard deviation. For non-normal distribution this γ has a finite non-negative value but it is not the same as standard deviation. For non-normal stable distribution the standard deviation is infinite. For all cases $\gamma > 0$ 4. The variable δ is called the location parameter and it is an expression of the mean or median of the entire distribution. This is also termed as measure of centrality or mean. For a normal distribution and other stable distributions for which $\alpha > 1$; δ is same as the mean value of distribution. When $\alpha = 1$ mean value is not defined, here the value, δ of the distribution is not the same as mean value. Generally $-\infty < \delta < +\infty$.

The α -stable distributions obey two major properties. 1. The stability property, which states that the sum of weighted independent α -stable random variables is still stable with the same characteristic exponent α. 2. The generalized central limit theorem which states that the sum of a number of independently and identically distributed (i.i.d) random variables, can only be a stable distribution. The generalized central limit theorem defines the randomness as a result of cumulative effects and these effects are distributed with heavy-tailed probability density.

A symmetric characteristic function for Levy distribution is

$$f(\theta) = \exp(-b^{\alpha} |\theta|^{\alpha}) .$$

The pdf of Levy stable distribution can be obtained by performing numerically Fourier cosine transform giving

$$f(x) = \frac{1}{\pi} \int_0^{\infty} dk \exp(-|bk|^{\alpha}) \cos kx .$$

The delays of computer control system or any other dynamics of fluctuation like in finance, follows a random behavior, could be explained by the heavy tailed distributions of such type. The delays in computer control system arise due to forces of random nature are inbuilt into the software delays caused by 'multi-tasking programming methods', and the randomness in 'Network Traffic', 'data bus traffics', hardware 'arbitration logic' of share resources, 'decision making algorithms' etc.

6.11.3 Fractional Stochastic Dynamic Model

Let us take an example of delays in computer systems. Here it is proposed to describe the dynamics of random delay of computer control system $\tau(t)$ by the fractional stochastic differential equation, representing Brownian motion like system.

$$\frac{d^q}{dt^q}\tau(t) = \lambda\tau(t) + F(t) \tag{6.35}$$

where, $0 < q \leq 1$.

We have 'fractional stationary' condition, as initial condition as:

$$\left.\frac{d^{q-1}}{dt^{q-1}}\tau(t)\right|_{t=0} = \tau_0 \tag{6.36}$$

where λ is a constant, expected delay rate (like Poisson's process formulation) this value could be zero too that case is discussed later. $F(t)$, is the random forcing function and d^q / dt^q is Riemann-Liouvelli (RL) fractional derivative, of order q. Using the definition of RL fractional integration, d^{-q} / dt^{-q} of a function $f(t)$ that is:

$${}_0I_t^q = {}_0D_t^{-q} \equiv \frac{d^{-q}}{dt^{-q}} f(t) = \frac{1}{\Gamma(q)}\int_0^t dt' \frac{f(t')}{(t-t')^{1-q}},$$

for $0 < q \leq 1$, yields solution to (6.35) as:

$$\tau(t;\tau_0,F) = t^{q-1}E_{q,q}\left(\lambda t^q\right)\tau_0 + \int_0^t dt' F(t')(t-t')^{q-1}E_{q,q}\left(\lambda(t-t')^q\right) \tag{6.37}$$

$$E_{\alpha,\beta}(z) = \sum_{k=0}^{\infty} z^k / \Gamma(\alpha k + \beta)$$

is two parameter Mittag-Leffler function.

The fractality index q is related to Mandelbrot's self-similarity parameter H, also called Hurst Index (was introduced in Chapter 4), where, $q = H + (1/2)$.

The mathematical motivation for applying the fractional stochastic problem (6.35) and (6.36) is following. It is easy to see when $q = 1$, the Eq. (6.35) reduces to the standard (integer order) Langevin equation with integer order initial condition $\tau(t)|_{t=0} = \tau_0$, and Eq. (6.37) gives standard well-known stochastic Poisson's process problem.

$$\tau(t;\tau_0,F) = e^{\lambda t}\tau_0 + \int_0^t dt' F(t') e^{\lambda(t-t')},$$

because of $E_{1,1}(z) = e^z$.

Thus we see that the fractional stochastic initial value problem (6.35) and (6.36) seems to be fractional generalization of well known Langevin approach to fluctuating phenomena. We define the probability distribution function (pdf) $P_q(\tau,t)$ of the fractional stochastic variable $\tau(t)$ in the following way, by Fourier integral of Dirac's delta function.

$$
\begin{aligned}
P_q(\tau,t) &= \left\langle \delta(\tau-\tau(t;\tau_0,F)) \right\rangle \\
&= \left\langle \frac{1}{2\pi}\int_{-\infty}^{\infty} d\omega \exp\{[\mathrm{i}\omega(\tau-\tau(t;\tau_0,F)\}\right\rangle \\
&= \left\langle \frac{1}{2\pi}\int_{-\infty}^{\infty} d\omega \exp\left\{[\mathrm{i}\omega(\tau-t^{q-1}E_{q,q}(\lambda t^q)\tau_0)]-\mathrm{i}\omega\int_0^t dt' F(t')(t-t')^{q-1}E_{q,q}\left(\lambda(t-t')^q\right)\right\}\right\rangle \\
&= \frac{1}{2\pi}\int_{-\infty}^{\infty} d\omega \exp\{\mathrm{i}\omega(\tau-t^{q-1}E_{q,q}(\lambda t^q)\tau_0)\} \\
&\quad \times \left\langle \exp\left\{-\mathrm{i}\omega\int_0^t dt' F(t')(t-t')^{q-1}E_{q,q}\left(\lambda(t-t')^q\right)\right\}\right\rangle
\end{aligned}
\tag{6.38}
$$

where the brackets $\langle ... \rangle$ means the averaging over the all possible realizations of the random force $F(t)$. Thus the 'averaging operator' $\langle ... \rangle$ is applied to second term, as the first term in (6.38) is force free. Here, $\mathrm{i}=\sqrt{-1}$. In (6.38) we have used the Fourier identities, with $2\pi\xi=\omega$, for $\hat{\delta}(\xi)=\int_{-\infty}^{+\infty} dx\{\exp(-\mathrm{i}2\pi x\xi)\delta(x)\}=1$. The inverse Fourier as

$$
\int_{-\infty}^{+\infty} 1 e^{\mathrm{i}2\pi x\xi} d\xi = \delta(x),
$$

and orthogonality of Fourier kernel as

$$
\int_{-\infty}^{+\infty} e^{\mathrm{i}2\pi\xi_1 t}\left[e^{\mathrm{i}2\pi\xi_2 t}\right]^* dt = \int_{-\infty}^{+\infty} e^{-\mathrm{i}2\pi(\xi_2-\xi_1)t} dt = \delta(\xi_1-\xi_2)
$$

Let the stochastic force $F(t)$ be a generalized shot noise as defined in statistical communication theory as follows, and explained in previous subsection as:

$$
F(t)=\sum_{k=1}^{n} a_k \varphi(t-t_k) \tag{6.39}
$$

Here, a_k are the random amplitudes, $\varphi(t)$ is the 'impulse' response function of 'memory-less linear filter', and t_k are the homogeneously distributed on time interval $[0,T]$ moments of time, the number n obeys the Poisson law. We guess that, defined by Eq. (6.39), random force $F(t)$ describes the influence of the different fluctuating factors on the 'delay generation dynamics'. These forces of random nature are inbuilt into the software delays caused by 'multi-tasking programming methods', and the randomness in 'Network Traffic', 'data bus traffics', hardware 'arbitration logic' of share resources, 'decision making algorithms' etc. A single-shot-noise pulse $a_k\varphi(t-t_k)$ describes the influence of a piece of information which has become available at the random moment t_k on the decision-making process at a later time t. The amplitude a_k responds to the magnitude of pulse $\varphi(t-t_k)$; it will depend on type of information and will, therefore, be subjected to probability distribution. For simplicity we assume that each pulse has the same functional form or, in other words, one general response function φ can be used to describe the delay process; in our case it is $\exp(-t/\eta)$. Thus, the averaging procedure includes three statistically independent averaging techniques, which is used to calculate, the average of (6.38) that is:

$$\left\langle \exp\left\{-\mathrm{i}\omega\int_0^t dt' F(t')(t-t')^{q-1}E_{q,q}\left(\lambda(t-t')^q\right)\right\}\right\rangle$$
$$=\left\langle \exp\left\{-\mathrm{i}\omega\int_0^t dt'[\sum_{k=1}^{n} a_k\varphi(t'-t_k)](t-t')^{q-1}E_{q,q}\left(\lambda(t-t')^q\right)\right\}\right\rangle$$

Let

$$\int_0^t dt'\varphi(t-t'')(t-t')^{q-1}E_{q,q}\left(\lambda(t-t')^q\right)=R_q$$

a function of t' and λ which will be derived later. Using this we have simplified above expression as

$$\left\langle \exp\left\{-\mathrm{i}\omega\int_0^t dt' F(t')(t-t')^{q-1}E_{q,q}\left(\lambda(t-t')^q\right)\right\}\right\rangle=\left\langle \exp\left\{-\mathrm{i}\omega R_q\sum_{k=1}^{n} a_k\right\}\right\rangle$$

Now we demonstrate term by term averaging over all possible parameters of force.

1. Averaging over random amplitudes, a_k that is $\langle ...\rangle_{a_k}$,

$$\langle f_1\rangle_{a_k}=\int da_1...da_n P(a_1,...,a_n)(f_1) \tag{6.40}$$

where $P(a_1,...,a_n)$ is the probability distribution of amplitudes a_k. Using (6.40) we get

$$\left\langle \exp\left\{-\mathrm{i}\omega\int_0^t dt' F(t')(t-t')^{q-1} E_{q,q}\left(\lambda(t-t')^q\right)\right\}\right\rangle_{a_k} = \int da_1 da_2 ... da_n P(a_1 ... a_n) \exp\left\{-\mathrm{i}\omega R_q \sum_{k=1}^{n} a_k\right\}$$

Define a function $W(\zeta)$ is characteristic function of the probability distribution $P(a)$, as

$$W(\zeta) = \int_{-\infty}^{\infty} dae^{-i\zeta a} P(a),$$

and pdf $P(a)$ is distribution function of the amplitudes of the shot-noise pulses. Also for simplicity assume that;

$$P(a_1,...,a_n) = \prod_{k=1}^{n} P_1(a_k) = (P(a))^n .$$

Using this simplification we factorize and write the following, for averaging over a_k, as:

$$\begin{aligned}
&\left\langle \exp\left\{-\mathrm{i}\omega\int_0^t dt' F(t')(t-t')^{q-1} E_{q,q}\left(\lambda(t-t')^q\right)\right\}\right\rangle_{a_k} \\
&\quad = \int da_1 P(a_1) e^{-\mathrm{i}\omega R_q a_1} da_2 P(a_2) e^{-\mathrm{i}\omega R_q a_2} ... da_n P(a_n) e^{-\mathrm{i}\omega R_q a_n} \\
&\quad = \int \left[daP(a) e^{-\mathrm{i}\omega R_q a}\right]^n = \left[W(\omega R_q)\right]^n
\end{aligned}$$

2. Averaging over t_k on time interval T.

$$\langle f_2 \rangle_T = \frac{1}{T}\int_0^T dt_1 ... \frac{1}{T}\int_0^T dt_n (f_2) \tag{6.41}$$

Above obtained average expression, we now use (6.41) and write to average over t_k

$$\begin{aligned}
\left\langle \exp\left\{-\mathrm{i}\omega\int_0^t dt' F(t')(t-t')^{q-1} E_{q,q}\left(\lambda(t-t')^q\right)\right\}\right\rangle_{a_k, t_k} &= \frac{1}{T}\int_0^T dt_1 \frac{1}{T}\int_0^T dt_2 ... \frac{1}{T}\int_0^T dt_n \left[W(\omega R_q)\right]^n \\
&= \frac{1}{T^n}\left[\int_0^T dt W(\omega R_q)\right]^n
\end{aligned}$$

3. Averaging over random numbers n of time moments t_k

$$\langle f_3 \rangle_n = \sum_{n=0}^{\infty} \frac{\bar{n}^n}{n!} e^{-\bar{n}} f_3 \tag{6.42}$$

where $\bar{n} = \nu T$ and ν is the density of points t_k on time interval $[0,T]$. Using (6.42), we get

$$\left\langle \exp\left\{ -\mathrm{i}\omega \int_0^t dt' F(t')(t-t')^{q-1} E_{q,q}\left(\lambda(t-t')^q\right) \right\} \right\rangle_{a_k,t_k,n} = \sum_{n=0}^{\infty} \frac{\bar{n}^n}{n!} e^{-\bar{n}} \frac{1}{T^n} \left[\int_0^T dt W(\omega R_q) \right]^n$$

$$= e^{-\nu T} \sum_{n=0}^{\infty} \frac{\nu^n T^n}{n!} \frac{1}{T^n} \left[\int_0^T dt W(\omega R_q) \right]^n$$

Let us calculate the RHS of above

$$e^{-\nu T} \sum_{n=0}^{\infty} \frac{\nu^n T^n}{n!} \frac{1}{T^n} \left[\int_0^T dt W(\omega R_q) \right]^n = e^{-\nu T} \sum_{n=0}^{\infty} \frac{1}{n!} \left[\nu \int_0^T dt \{W(\omega R_q)\} \right]^n$$

$$= e^{-\nu \int_0^T dt} \times e^{\nu \int_0^T dt \{W(\omega R_q)\}}$$

$$= e^{-\nu \int_0^T dt [1 - W(\omega R_q)]}$$

$$= e^{-J_q(\omega,t)}$$

with $J_q(\omega,t) = \nu \int_0^T dt \left[1 - W(\omega R_q) \right]$.

Taking into the account of Eq. (6.38) and performing the averaging in accordance with (6.40)-(6.41), as described and derived above we obtain, with the time interval $[0,T]$ is changed as $[0,t]$.

$$P_q(\tau,t) = \frac{1}{2\pi} \int_{-\infty}^{\infty} d\omega \exp\left\{ \mathrm{i}\omega(\tau - t^{q-1} E_{q,q}(\lambda t^q)\tau_0 \right\} \exp\left\{ -J_q(\omega,t) \right\} \tag{6.43}$$

where the following notation is introduced as derived from above averaging method

$$J_q(\omega,t) = \nu \int_0^t dt' \left[1 - W\left(\omega \int_0^t dt'' \varphi(t''-t')(t-t'')^{q-1} E_{q,q}\left(\lambda(t-t'')^q\right) \right) \right] \tag{6.44}$$

Here the function $W(\zeta)$ is characteristic function of the probability distribution $P_1(a)$,

$$W(\zeta) = \int_{-\infty}^{\infty} da e^{-i\zeta a} P_1(a)$$

and pdf $P_1(a)$ is a "one-particle" distribution to be introduced into consideration because of simple assumption that is, we consider a delay dynamics when probability distribution $P(a_1,...,a_n)$ is factorized as product of n equal "one-particle" distribution $P_1(a)$, Eq. (6.45).

$$P(a_1,...,a_n) = \prod_{k=1}^{n} P_1(a_k) \tag{6.45}$$

To evaluate the integrals in Eq. (6.43) and (6.44) we should know the type of response function $\varphi(t)$ and pdf $P_1(a)$. Let us choose an exponential response function,

$$\varphi(t) = \begin{cases} e^{\left(-\frac{t}{\eta}\right)}; & t \ge 0 \\ 0; & t < 0 \end{cases} \tag{6.46}$$

This means, that the impact has a characteristic-time (time constant) as a number η. Evaluate the integral over dt' in Eq. (6.44),

$$\begin{aligned} &\int_0^t dt' \varphi(t-t'')(t-t')^{q-1} E_{q,q}\left(\lambda(t-t')^q\right) \\ &= \int_{t''}^t dt' e^{-(t'-t'')/\eta} (t-t')^{q-1} \sum_{k=0}^{\infty} \frac{\lambda^k (t-t')^{qk}}{\Gamma(qk+q)} \end{aligned} \tag{6.47}$$

Expanding in series the $e^{-(t'-t'')/\eta}$ and using the formula

$$\int_{t''}^t dt'(t-t')^{a-1}(t'-t'')^{b-1} = (t-t'')^{a+b-1} \frac{\Gamma(a)\Gamma(b)}{\Gamma(a+b)},$$

makes RHS of Eq. (6.47) as

$$\int_{t''}^t dt' e^{-(t'-t'')/\eta} (t-t')^{q-1} \sum_{k=0}^{\infty} \frac{\lambda^k (t-t')^{qk}}{\Gamma(qk+q)} = (t-t'')^q \sum_{k=0}^{\infty} \left(-\frac{t-t''}{\eta}\right)^k E_{q,q+k+1}\left(\lambda(t-t'')^q\right).$$

The function $J_q(\omega,t)$ as in Eq. (6.44) then is:

$$J_q(\omega,t) = v \int_0^t dt' \left[1 - W(\omega R_q(t';\lambda)\right] \tag{6.48}$$

with

$$R_q(t';\lambda) = (t')^q \sum_{k=0}^{\infty} \left(-\frac{t'}{\eta}\right)^k E_{q,q+k+1}(\lambda t'^q) \tag{6.49}$$

As a second step choose the Levy α-stable distribution $P_1(a)$ as "one-particle" probability distribution function:

$$P_1(a) = \frac{1}{2\pi} \int_{-\infty}^{\infty} d\zeta e^{i\zeta a} W(\zeta), \text{ and } W(\zeta) = \exp\left\{-b^\alpha |\zeta|^\alpha\right\},\ 0 < \alpha \le 2 \tag{6.50}$$

where b is scale parameter of Levy α-stable distribution.

Thus in accordance with Eq. (6.42) the new general equation for the pdf of the fractional stochastic delay process $\tau(t)$ described by Eq. (6.34) can be rewritten as:

$$P_q(\tau,t) = \frac{1}{2\pi} \int_{-\infty}^{\infty} d\omega \exp\left\{i\omega(\tau - t^{q-1} E_{q,q}\left(\lambda t^q\right)\tau_0)\right\} \times \exp\left\{-v \int_0^t dt' \left[1 - \exp\left\{-b^\alpha \left|\omega R_q(t';\lambda)\right|^\alpha\right\}\right]\right\} \tag{6.51}$$

with $R_q(t')$ defined as in Eq. (6.49).

Note that putting $q = 1$ in Eq. (6.49) we get $R_1(t') = \left(\eta / (\lambda\eta + 1)\right)\left(e^{\lambda t'} - e^{-t'/\delta}\right)$.

6.11.4 *Fractional Delay Dynamics*

Let us apply the developed general approach to derive the analytical expression for the pdf of increments $\Delta\tau$ of the delay generating system as a function of time increment Δt, $\Delta\tau = \tau(t+\Delta t) - \tau(t)$, where the value of the instantaneous delay of the delay generating system is $\tau(t)$. The pdf of delay increments fluctuation plays an important role in understanding the delay dynamics, and in overall system engineering to plant control by computers. As usual we define the pdf $P_q(\Delta\tau, t, \Delta t; \alpha)$ of the increments $\Delta\tau(\Delta t) = \tau(t+\Delta t) - \tau(t)$ of the delay dynamics system of delay $\tau(t)$ as a function of incremental time Δt by following expression:

$$P_{\alpha,q}(\Delta\tau, t, \Delta t) = \left\langle \delta\left(\Delta\tau - \left\{\tau(t+\Delta t;\tau_0,F) - \tau(t;\tau_0,F)\right\}\right)\right\rangle \tag{6.52}$$

where $\langle ... \rangle$ means averaging over the all possible realizations of random force $F(t)$ in accordance with Eq. (6.40)-(6.42). Repeating the same steps used above for derivation of Eq. (6.51), we find for the pdf $P_{\alpha,q}(\Delta\tau, t, \Delta t)$ as:

$$P_{\alpha,q}(\Delta\tau, t, \Delta t)$$
$$= \frac{1}{\pi}\int_0^\infty d\omega \cos\omega\{\Delta\tau - \left((t+\Delta t)^{q-1} E_{q,q}\left(\lambda (t+\Delta t)^q\right)\right)$$
$$- t^{q-1} E_{q,q}\left(\lambda t^q\right)\tau_0\} \exp\left\{-L_{\alpha,q}(\omega, t, \Delta t; \lambda)\right\} \quad (6.53)$$

with

$$L_{\alpha,q}(\omega, t, \Delta t; \lambda) = v\int_0^t dt' \left(1 - \exp\left\{-b^\alpha \omega^\alpha \left|R_q(t' + \Delta t; \lambda) - R_q(t'; \lambda)\right|^\alpha\right\}\right),$$

where $R_q(t'; \lambda)$ is by Eq. (6.49). Equation (6.53) presents a new general expression for the fractional pdf of increments of delay $\Delta\tau = \tau(t+\Delta t) - \tau(t)$ fluctuations, when delay $\tau(t)$ is described by fractional stochastic differential equations (6.35).

For a case when $\lambda = 0$ the Eq. (6.53) can be rewritten as

$$P_{\alpha,q}(\Delta\tau, t, \Delta t)$$

$$= \frac{1}{\pi}\int_0^\infty d\omega \cos\omega\left\{\Delta\tau - \frac{1}{\Gamma(q)}\left((t+\Delta t)^{q-1} - t^{q-1}\right)\tau_0\right\} \times \exp\left\{-L_{\alpha,q}(\omega, t, \Delta t)\right\} \quad (6.54)$$

where

$$L_{\alpha,q}(\omega, t, \Delta t) - v\int_0^t dt' \left(1 - \exp\left\{-b^\alpha \omega^\alpha \left|r_q(t' + \Delta t) - r_q(t')\right|^\alpha\right\}\right),$$

and $r_q(t')$ is obtained from Eq. (6.49) by placing $\lambda = 0$, that is

$$r_q(t') = R_q(t'; \lambda = 0) = (t')^q E_{1,1+q}\left(-t'/\eta\right) \quad (6.55)$$

Further in limit for large time the asymptotic pdf we obtain as:

$$P_{\alpha,q}(\Delta\tau, \Delta t) = \lim_{t\to\infty} P_{\alpha,q}(\Delta\tau, t, \Delta t) = \frac{1}{\pi}\int_0^\infty d\omega \cos(\omega\Delta\tau) \times e^{-L_{\alpha,q}(\omega, \Delta t)} \quad (6.56)$$

Here $L_{\alpha,q}(\omega, \Delta t)$ is defined as:

$$L_{\alpha,q}(\omega,\Delta t)=\lim_{t\to\infty}L_{\alpha,q}(\omega,t,\Delta t)=v\int_0^\infty dt'\left(1-\exp\left\{-b^\alpha\omega^\alpha\left|r_q(t'+\Delta t)-r_q(t')\right|^\alpha\right\}\right) \quad (6.57)$$

In Eq. (6.57) $0<q\le 1$ and $1<\alpha\le 2$.

The limiting pdf $P_{\alpha,q}(\Delta\tau,\Delta t)$ is characterized by the fractality index q and the Levy index α. Thus, it is shown how general fractional dynamic approach developed here; one derives the expression for pdf $P_{\alpha,q}(\Delta\tau,\Delta t)$ of increments of delay dynamics. The new pdf $P_{\alpha,q}(\Delta\tau,\Delta t)$ allows studying any statistical and scaling dependencies of the fluctuating dynamics and developing the new general approach to deal with random delays in computer controls, and evaluate its risks and robustness.

In special case of integer order delay dynamics case with $q=1$ the general Eq. (6.57) can be expressed as:

$$L_{\alpha,1}(\omega,\Delta t)=v\int_0^\infty dt'\left\{1-\exp\left[-b^\alpha\left|\omega\right|^\alpha\eta^\alpha\left|e^{-t'/\eta}(1-e^{-\Delta t/\eta})\right|^\alpha\right]\right\} \quad (6.58)$$

where we have used $\Gamma(1)=1$ and $E_{1,2}(z)=\left(e^z-1\right)/z$. Then Eq. (6.56) leads to

$$P_{\alpha,1}(\Delta\tau,\Delta t)=\frac{1}{\pi}\int_0^\infty d\omega\cos(\omega\Delta\tau)$$

$$\times\exp\left\{-v\int_0^\infty dt'\left\{1-\exp\left[-b^\alpha\omega^\alpha\eta^\alpha\left(e^{-t'/\eta}(1-e^{-\Delta t/\eta})\right)^\alpha\right]\right\}\right\}$$

Substituting with u instead of t' we have $u=b\omega e^{-t'/\eta}(1-e^{-\Delta t/\eta})$, $du=-dt'(u/\eta)$ we get,

$$P_{\alpha,1}(\Delta\tau,\Delta t)=\frac{1}{\pi}\int_0^\infty d\omega\cos\left(\omega\Delta\tau\right)\exp\left\{-v\eta\int_0^{b\eta(1-e^{-\Delta t/\eta})\omega}\frac{du}{u}(1-e^{-u^\alpha})\right\} \quad (6.59)$$

The pdf in Eq. (6.59) is α-generalization of Eq. (6.40) and (6.41), the Gaussian case will be obtained for $\alpha=2$, for Eq. (6.59) that is, $P_{2,1}(\Delta\tau,\Delta t)$.

Obtain the delay values for large number of time, and say we form a set $\{\tau_i\}=\{\tau(t_1),\tau(t_2),\tau(t_3).......\}$. In this set we take differential delay and make the set as $\{\Delta\tau_i\}=\{\Delta\tau_1=\tau(t_2)-\tau(t_1),\Delta\tau_2=\tau(t_3)-\tau(t_2),\Delta\tau_3=\tau(t_4)-\tau(t_3),.......\}$. This new set of data has value zero, positive and negative; showing incremental delay spread.

Simply filling of empirical values for data by ordinary symmetrical Levy α-stable is by

$$P_{\text{LEVY}}(\Delta\tau / c) = \frac{1}{\pi c}\int_0^\infty dk \cos(k\Delta\tau / c)\exp(\gamma k^\alpha),$$

where α is Levy index and γ is scale factor be performed as designed by Famma and Roll in 1972. This we will detail in Chapter 10. Here, c is number in order to normalize the data set. In our delay case this could be one.

For numerical calculations of (6.57) that is:

$$L_{\alpha,q}(\omega,\Delta t) = \lim_{t\to\infty} L_{\alpha,q}(\omega,t,\Delta t) = v\int_0^\infty dt'\left(1-\exp\left\{-b^\alpha\omega^\alpha\left|r_q(t'+\Delta t)-r_q(t')\right|^\alpha\right\}\right)$$

Write as:

$$L_{\alpha,q}(\chi,\Delta y) = K\int_0^\infty dz\left(1-\exp\left\{-S_q^\alpha\chi^\alpha\left|s_q(z+\Delta y)-s_q(z)\right|^\alpha\right\}\right),$$

where $\Delta y = \Delta t / \omega$ is new dimensionless parameter, signifying change in time, the other dimensionless parameters are $K = v\eta$ and $S_q = b\eta^q / c$, and the function is $s_q(z) = z^q E_{1,1+q}(z)$. Then plot the pdf as:

$$P_{\alpha,q}(\Delta\tau / c,\Delta y) = \frac{1}{\pi c}\int_0^\infty d\chi \cos(\chi.\Delta\tau / c)\times\exp\{-L_{\alpha,q}(\chi,\Delta y)\},$$

for various values of α and q starting with $\alpha = 2$, $q = 1$ and decreasing the same. By varying these fractality parameters empirically fitting of the data of delay, the pdf can be seen.

6.11.5 The Random Dynamics of Computer Control System

From the basic signals and systems point of view, the random delay of computer control system can be regarded as an output signal of "a (delay generating) dynamic system" driven by some known signal say 'shot-noise'. If we know? better about the generating 'dynamic system', can we do better job in compensating it. Well, the answer is yes. This is a new research frontier for exploration which is especially fundamental and important in the understanding of randomized delay.

The same effect as above cause randomness of the time variant of $\tau(t)$. This implies that the distribution of such stochastic process follows a stable model. The variation of $\tau(t)$, with the bursts (spikes) implies the heavy tailed behavior in the distribution. α-stable distribution is different from Gaussian in a way that there only exists moments of order less than characteristic exponent α. Hence the variance of a stable (non-Gaussian) distribution is divergent (infinite) unless $\alpha = 2$, in which case the distribution becomes a Gaussian. This is called the 'Fractional Lower Order Moment' statistics. It has been established that the fundamental solutions of fractional order diffusion equations generate probability density functions (pdf) evolving in time or varying in space related to stable distributions, earlier in this book. Thus the randomization dynamics is well defined by fractional differential equations.

Thus basic research question is

1. How to use Fractional Calculus to best characterize the Random Dynamics of physics of delay?
2. How to make use of the acquired knowledge of random dynamics to design a better controller, possibly also a fractional order controller, to accommodate the randomness in $\tau(t)$ influence on control performance of computer control system.

Here we have elaborated a fractality concept and fractional differential equation in mathematics to represent randomized computer control delay. Our main assumption is that the fluctuating delay can be adequately described by means of fractional calculus, non-Gaussian, long range dependence, heavy tailed stochastic process. To describe the dynamics of the random delay, we have introduced a new fractional stochastic differential equation, driven by a random forcing function 'shot-noise', for which each pulse has random amplitude with α -stable Levy distribution. As a result we have obtained the general expression for the fractional pdf of incremental delays with time. Future studies and experiments on computer control system delay data is required to calibrate the parameters obtained in this description of pdf. The new fractional pdf has fractality parameters q and α. The parameter q describes the dynamical memory effects in the delay stochastic evolution, while the Levy index α describes the long range dependencies of external impacts on the delay dynamics, like 'multi-tasking programming methods', and the randomness in 'Network Traffic', 'data bus traffics', hardware 'arbitration logic' of share resources, 'decision making algorithms' etc. More research and experimental data are indeed required to quantify these effects described in these random delay phenomena of computer control system. In Chapter 10 identification algorithm will be discussed for estimating H and important parameters for the Levy stable distribution from the data series of the experiment.

6.12 Concluding Comments

Strong motivation exists for the study, development of the fractional calculus. This may be readily verified and validated by large number of applications discussed in preceding chapters, where need for fractional calculus and initialization in particular is pointed. Fractional integration and fractional differentiation processes require the use of non-constant initialization function. This initialization function is the behavior of the function before the process of differintegration is taken up, and makes up the process continuous in future. Whereas, for Riemann-Liouvelli (RL) formulation, with terminal initialization the continuity and the nature of the function is maintained throughout the interval, which is not so simple to initialize Caputo formulation. Though Caputo formulation requires integer order initial states for solving compared to fractional order initial conditions for RL formulations, yet the generalized theory of initialization function is far from simple while dealing with Caputo formulation. Some future developments in theory of fractional differential equation solution should be able to merge these two diverging definitions; and provide ease in tackling generalized dynamic systems. The initialization function is history or memory associated with process of differintegration of the function since its birth. This history or memory fades away as the time passes by and appears as output of fractional differintegrator as an added function. The initialized differintegration process is generalization of the total calculus theory, for fractional order systems or integer order systems. The unifying concepts and notation of fractional calculus provide a significant benefit that greatly simplifies the solution of certain differential equations (distributed systems). Perhaps the strongest motivation to develop the fractional calculus is the belief that a wide variety of physical problems that have resisted compact (and first principles) description when using integer order calculus will yield to the methods of fractional calculus, otherwise major recourse was probabilistic methods. A new extension of a fractality concept and fractional calculus for random delay dynamics in computer control system as well as solution of Fractional Advection Dispersion Equation is described. These examples show that Long Ranged Dependence (LRD) of the solution in these processes with fractality has long tailed power law stable distributions, different from normal Gaussian distributions. These processes have fluctuation dynamics which deviates larger than normal fluctuation stochastic processes, like Brownian Paths. These anomalous processes well suited to be described by fractional differintegration be it Riemann-Liouvelli type be it Caputo formulation type; the engineers and scientists would like them to have solution with close physical picture relating to initial values or initialization functions; and they would like to relate them to experimental observations; irrespective of type of fractional differintegrations used.

Chapter 7
Generalized Laplace Transform for Fractional Differintegrals

7.1 Introduction

Differential equations of fractional order appear more and more frequently in various research areas of science and engineering. An effective method for solving such equations is needed. The method of Laplace transforms technique gives almost unified approach to solve the fractional differential equations. Also generalization of the same in view of initial conditions appropriately put (terminal/side charging) gives unified generalized approach. Also the fundamental fractional order differential equation concept is touched; its solution is the fundamental time response, whose combination provides solution to complicated systems. From this transfer function is constructed with fractional pole, which is the transfer function of the fundamental fractional differential equation, and is fundamental building block for more complicated fractional order systems. In this chapter scalar initialization and vector initialization problem is taken to describe the approaches developed for initialization function. These problems give insight into fractional "state" variable concepts and general system description of fractional order systems, and controls. For fractional order control system stability analysis transformed Laplace $s^q \to w$ plane (wedge) is introduced. The pole-placement and its properties for control system stability for fractional order systems are carried on in this w-plane. The realization of fractional Laplace operator by rational function approximation is also introduced in this chapter. Here generalized stationary conditions are discussed and idea is developed as generalized Laplace transform to define Riemann-Liouvelli and/ or Caputo derivative (or even a derivative having the mix of two!).

7.2 Recalling Laplace Transform Fundamentals

Let us recall some basic facts about Laplace transforms, developed for classical integer order calculus. The function $F(s)$ of complex variable (frequency) s is defined by:

$$F(s)=\mathcal{L}\{f(t)\}=\int_0^{\infty} e^{-st} f(t)dt$$

is called Laplace transform of the function $f(t)$, which is called the original. For Laplace transform to exist, $f(t)$ must be of an exponential order α. In other words the function $f(t)$, must not grow faster than a certain exponential function when t tends to infinity. The original $f(t)$ can be restored from Laplace transform $F(s)$ with the help of inverse transform:

$$f(t)=\mathcal{L}^{-1}\{F(s)\}=\frac{1}{j2\pi}\int_{\gamma-j\infty}^{\gamma+j\infty} e^{st}F(s)ds .$$

Where $\gamma=0$ for all singularities of $F(s)$ if in the left half of s-plane (LHP) i.e. all $F(s)$ poles at LHP.

Laplace transform of $\int_b^t f(t)dt$ is given by

$$\mathcal{L}\left\{\int_b^t f(t)dt\right\}=\frac{1}{s}F(s)+\frac{1}{s}\int_b^0 f(t)dt .$$

Constructing this for initialized case, the generalized integer order Laplace transformation for integration operation can be expressed as:

$$\mathcal{L}\left\{{}_bD_t^{-1}f(t)\right\}=\mathcal{L}\left\{{}_bd_t^{-1}f(t)+\psi(f,-1,a,b,t)\right\}=\mathcal{L}\left\{\int_b^t f(t)+\psi(f,-1,a,b,t)\right\}$$

$$\mathcal{L}\left\{{}_bD_t^{-1}f(t)\right\}=\frac{1}{s}F(s)+\frac{1}{s}\int_b^0 f(t)dt+\mathcal{L}\{\psi(f,-1,a,b,t)\} .$$

However in the most general case the ψ is selected arbitrarily, if chosen as constant K, then since $\mathcal{L}\{K\}=K/s$, it is clear that this term contains initialization effect of the second term on the RHS of the above Laplace expression for initialized integer order integration. Hence it is not necessary to include such terms that redundantly bring in the effect of initialization from the integer order calculus. Therefore:

$$\mathcal{L}\left\{{}_bD_t^{-1}f(t)\right\}=\frac{1}{s}F(s)+\mathcal{L}\{\psi(f,-1,a,b,t)\}$$

For multiple integer order integrals the Laplace Transform derivation is as follows:

$$\mathcal{L}\{g(t)\} = \mathcal{L}\left\{\int_b^t\int_b^{t_1}\int_b^{t_2}......\int_b^{t_{n-1}} f(t_n)dt_n dt_{n-1}....dt_2 dt_1\right\},$$

is to be evaluated. For convenience write

$$g(t_j) = \int_b^{t_j} g(t_{j+1})dt_{j+1},$$

for $j = 1,2,...n-2$, and

$$g(t_{n-1}) = \int_b^{t_{n-1}} f(t_n)dt_n .$$

Then starting from out side we have:

$$\mathcal{L}\{g(t)\} = \mathcal{L}\left\{\int_b^t g(t_1)\,dt_1\right\} = \frac{1}{s}\mathcal{L}\{g(t_1)\} + \frac{1}{s}\int_b^0 g(t_1)\,dt_1 = \frac{1}{s}\mathcal{L}\{g(t_1)\} + \frac{c_1}{s},$$

where constant $c_1 = \int_b^0 g(t_1)dt_1$. Going one level inside i.e. replacing the $g(t_1)$ with $\int_b^{t_1} g(t_2)dt_2$ we get:

$$\mathcal{L}\{g(t)\} = \frac{1}{s}\left[\mathcal{L}\left\{\int_b^{t_1} g(t_2)dt_2\right\}\right] + \frac{c_1}{s} = \frac{1}{s}\left[\frac{1}{s}\mathcal{L}\{g(t_2)\} + \frac{c_2}{s}\right] + \frac{c_1}{s} = \frac{1}{s^2}\mathcal{L}\{g(t_2)\} + \frac{c_2}{s^2} + \frac{c_1}{s}$$

Repeating this n times we get the result as:

$$\mathcal{L}\left\{\int_b^t\int_b^{t_1}\int_b^{t_2}......\int_b^{t_{n-1}} f(t_n)dt_n dt_{n-1}...dt_2 dt_1\right\} = \frac{1}{s^n}F(s) + \sum_{i=1}^{n}\frac{c_i}{s^i}, n = 1,2,3,.....,$$

where coefficients c_i is given as $c_i = \int_b^0 g(t_i)dt_i$

Thus

$$\mathcal{L}\{D^{-3}f(t)\} = \frac{1}{s^3}F(s) + \frac{c_1}{s} + \frac{c_2}{s^2} + \frac{c_3}{s^3}.$$

With this observation can we write the Laplace of fractional integral as:

$$\mathcal{L}\left\{{}_0D_t^{-2.3} f(t)\right\} = \frac{1}{s^{2.3}} F(s) + \frac{c_1}{s} + \frac{c_2}{s^2};$$

with choice of integer order integration index greater but just less than the fractional order integration index. That is, 2 is the integer order index, just less than 2.3 the fractional order index.

Consider the integer order derivative and its Laplace transform as

$$\mathcal{L}\left\{\frac{d}{dt} f(t)\right\} = \mathcal{L}\left\{f^1(t)\right\} = sF(s) - f(0^+).$$

For the initialized calculus case the formulation is

$$\mathcal{L}\left\{{}_0D_t^1 f(t)\right\} = \mathcal{L}\left\{\frac{df(t)}{dt} + \psi(f,1,a,0,t)\right\}.$$

Substituting the first definition into the second one we get:

$$\mathcal{L}\left\{{}_0D_t^1 f(t)\right\} = sF(s) - f(0^+) + \mathcal{L}\left\{\psi(f,1,a,0,t)\right\}.$$

Again in the most general case the ψ is selected arbitrarily. If it is chosen as value of the function $f(t)$ at $t = 0^+$, represented by Dirac delta function as: $\psi = -f(t)\delta(t-0^+)$ then,

$$\mathcal{L}\{\psi\} = \mathcal{L}\left\{-f(t)\delta(t-0^+)\right\} = -f(0^+).$$

This term then contains the initialization effect brought in by the integer order calculus Laplace expansion. It is again not necessary to include the redundant term.

The notation $f(0^+)$ may be changed to $f(0)$ with understanding that $t = 0$ will be actually $t = 0^+$. For repeated derivative the Laplace Transform gives:

$$\mathcal{L}\left\{f^{(n)}(t)\right\} = s^n F(s) - s^{n-1} f(0) - s^{n-2} f^{(1)}(0) - \ldots.. - f^{(n-1)}(0)$$

Therefore

$$\begin{aligned}\mathcal{L}\left\{{}_0D_t^3 f(t)\right\} &= s^3 F(s) - s^2 f(0) - sf'(0) - f''(0) = \\ &= s^3 F(s) - D^2\left\{f(t)\right\}_{@t=0} - D\left\{f'(t)\right\}_{@t=0} - \left\{f''(t)\right\}_{@t=0}\end{aligned}$$

With similar argument we can write Laplace of fractional derivative as:

$$\begin{aligned}\mathcal{L}\left\{{}_0D_t^{2.3} f(t)\right\} &= s^{2.3} F(s) - s^{1.3} f(0) - s^{0.3} f'(0) - s^{-0.7} f''(0) \\ &= s^{2.3} F(s) - D^{1.3}\left\{f(t)\right\}_{@t=0} - D^{0.3}\left\{f'(t)\right\}_{@t=0} - D^{-0.7}\left\{f''(t)\right\}_{@t=0}\end{aligned}$$

Here the nearest integer order derivative to 2.3 is chosen to be 3.

For semi-derivative the Laplace transform, may be expressed as:

$$\mathcal{L}\left\{{}_0D_t^{0.5}f(t)\right\} = s^{0.5}F(s) - s^{-0.5}f(0) = s^{0.5}F(s) - D^{-0.5}\left\{f(t)\right\}_{@t=0}$$

Here the nearest integer order derivative to 0.5 is chosen to be 1.

The observation is for Laplace Transform of fractional derivative, the requirement is of fractional initial states, like $f^{(2.3)}(0), f^{(1.3)}(0), f^{(-0.5)}(0)$, and they need be finite too. Experimentally these fractional states are hard to visualize, though one may try to correlate these with the integer order states with some effort and physical reasoning. For sake of solution with this extension of integer order calculus for Laplace of fractional order differentiation and fractional order integration, assume that these fractional initial states exist and are finite.

Let us take a fractional integral equation as:

$$ty(t) = \int_0^t (t-\tau)^{-\frac{1}{2}} y(\tau)d\tau$$

The above is form of semi-integration, of Riemann-Liouvelli (RL) definition; except Gamma of ½ is missing from the expression. Putting that constant multiplier of reciprocal of Gamma of ½ to both the sides we rearrange the integral equation and write as

$$ty(t) - \sqrt{\pi}\,{}_0D_t^{-\frac{1}{2}}y(t) = 0$$

$$\left[t\,{}_0D_t^0 - \sqrt{\pi}\,{}_0D_t^{-\frac{1}{2}}\right]y(t) = 0$$

We use the identity $\mathcal{L}\{1\} = s^{-1}$, with $\mathcal{L}\{t\} = s^{-2} = -D_s^1\left[\mathcal{L}\{1\}\right]$, and $\mathcal{L}\left\{tD^0y(t)\right\} = -D_s^1Y(s)$, to get Laplace of the above integral equation as:

$$D_s^1Y(s) + \sqrt{\pi}s^{-\frac{1}{2}}Y(s) = 0$$

$$\frac{dY}{Y} = -\sqrt{\pi}s^{-\frac{1}{2}}ds$$

Integrating both the sides and with integration constant k, we get:

$\ln Y(s) = -2\sqrt{\pi s} + k$, giving $Y(s) = Ke^{-2\sqrt{\pi s}}$, using inverse Laplace the solution, for $t > 0$ is:

$$y(t) = Kt^{-\frac{3}{2}}e^{-\frac{\pi}{t}}$$

Let us apply this observation to solve, a fractional differential equation

$$tD^{1/2}y(t) - y(t) = 0$$

We use the identity $\mathcal{L}\{1\} = s^{-1}$, with $\mathcal{L}\{t\} = s^{-2} = -D_s^1\left[\mathcal{L}\{1\}\right]$, and get

$$\mathcal{L}\left\{tD^{1/2}y(t)\right\} = -D\left[\mathcal{L}\left\{D^{1/2}y(t)\right\}\right] = -D_s^1\left[s^{1/2}Y(s) - D^{-1/2}y(0)\right],$$

using this in the fractional differential equation, we obtain:

$$-D_s^1\left[s^{1/2}Y(s) - D^{-1/2}y(0)\right] - Y(s) = 0$$

Differentiating (with respect to s) we obtain:

$$-D_s^1\left[s^{1/2}Y(s)\right] + 0 - Y(s) = 0$$

$$-\left\{\left(D_s^1 s^{1/2}\right)Y(s) + s^{1/2}D_s^1 Y(s)\right\} - Y(s) = 0$$

$$-\frac{1}{2}s^{-1/2}Y(s) - s^{1/2}D_s^1 Y(s) - Y(s) = 0$$

Dividing the above expression by ½ and rearranging we obtain

$$D_s^1 Y(s) + \left(\frac{1}{2}s^{-1} + s^{-1/2}\right)Y(s) = 0$$

which is a first order Linear Differential Equation on $Y(s)$, which is rearranged for solving by integration as:

$$\frac{dY}{Y} = \left(-\frac{1}{2}s^{-1} - s^{-1/2}\right)ds$$

Integrating both the sides we obtain the solution with k as constant of integration as:

$$\ln Y = -\frac{1}{2}\ln s - 2\sqrt{s} + k$$

$$Y(s) = Ks^{-1/2}e^{-2\sqrt{s}}$$

Using standard Laplace Inverse we get the solution for $t > 0$ as:

$$y(t) = \mathcal{L}^{-1}\{Y(s)\} = Kt^{-1/2}e^{-1/t}$$

This example demonstrates use of Laplace Transforms to solve Fractional order, integral and differential equation vis-à-vis integer order calculus. However, these examples are of Fractional Differential/Integral equations with non-constant coefficients and non-linear fractional order systems. Nevertheless, these examples could be solved by Laplace Technique to arrive at analytical close form solutions. Thus strong motivation exists to generalize the Laplace Transformation and apply the same for solution of Fractional Order Differential Equations.

7.3 Laplace Transform of Fractional Integrals

Consider starting point of integration as $c = 0$, for simplicity. The Laplace transform of the initialized fractional integral looks like:

$$\mathcal{L}\left\{{}_0D_t^{-q} f(t)\right\} = \int_0^{\infty} e^{-st}\left(\int_0^t \frac{(t-\tau)^{q-1}}{\Gamma(q)} f(\tau)d\tau + \psi(f,-q,a,0,t)\right)dt \tag{7.1}$$

$$q > 0, t > 0 .$$

Apply convolution for Laplace transforms i.e.

$$h(t) * g(t) = \int_0^t h(\tau)g(t-\tau)d\tau \leftrightarrow H(s)G(s) .$$

For the above fractional integral take

$$h(t) = f(t); g(t) = \frac{t^{q-1}}{\Gamma(q)}; \mathcal{L}\left\{g(t)\right\} = \frac{1}{s^q} .$$

The Laplace of the fractional integral therefore is

$$F(s)G(s) = \frac{1}{s^q}\mathcal{L}\left\{f(t)\right\}, \ q > 0 .$$

This gives the result as:

$$\mathcal{L}\left\{{}_0D_t^{-q} f(t)\right\} = \frac{1}{s^q}\mathcal{L}\left\{f(t)\right\} + \mathcal{L}\left\{\psi(f,-q,a,0,t)\right\} , q > 0 \tag{7.2}$$

Here the observation is that the initialization function $\psi(f,-q,a,0,t)$ may be thought to have equivalent (compound) effect of initialization to create ${}_0D_t^{-q} f(t)$. For fractional case ${}_0D_t^{-q} f(t)$ can be composed in infinite ways as opposed to integer order calculus, where only, integer order combinations are possible for composition, as indicated below for integer order n:

$$\mathcal{L}\left\{f^{(n)}(t)\right\} = s^n F(s) - s^{n-1} f(0) - s^{n-2} f^{(1)}(0) ... + f^{(n-1)}(0)$$

This implies that $f^{(n)}$ is decomposed into n separate differentiations of unity order; conversely $f^{(n)}$ is composed of n separate discrete differentiation. But for fractional order this composition is not discrete but can thus be decomposed in infinite ways. This is described next.

7.3.1 Decomposition of Fractional Integral in Integer Order

For generalized integration (with initialization) the composition law holds. The decomposition of the following Laplace of initialization function is demonstrated by integration by parts, for $q > 1$. The initialization function $\psi(f,-q,a,0,t)$ is q-th order integration of $f(t)$ from a to zero and is function of t, defined as in the Chapter 6 of initialized differintegrals. The function $f(t) = 0$, for $t < a$ and Laplace is from $t = 0 > a$

$$\psi(f,-q,a,0,t) = \frac{1}{\Gamma(q)} \int_a^0 (t-\tau)^{q-1} f(\tau) d\tau \qquad q > 1 \tag{7.3}$$

$$\mathcal{L}\{\psi(f,-q,a,0,t)\} = \int_0^\infty e^{-st} \psi(f,-q,a,0,t) dt \tag{7.4}$$

Take $u = \psi(f,-q,a,0,t), dv = e^{-st} dt$, then

$$du = \frac{d}{dt}\psi(f,-q,a,0,t)dt = \psi^{(1)} dt$$

$$v = -\frac{e^{-st}}{s}$$

yields

$$\mathcal{L}\{\psi(f,-q,a,0,t)\} = \left[-\frac{e^{-st}}{s}\psi(f,-q,a,0,t)\right]_0^\infty + \frac{1}{s}\int_0^\infty e^{-st}\psi^{(1)}(f,-q,a,0,t)dt \tag{7.5}$$

$$\mathcal{L}\{\psi(f,-q,a,0,t)\} = 0 + \frac{1}{s}\psi(f,-q,a,0,t)_{@t=0} + \frac{1}{s}\mathcal{L}\left\{\psi^{(1)}(f,-q,a,0,t)\right\} \tag{7.6}$$

Repeating this same another time gives:

$$\mathcal{L}\{\psi(f,-q,a,0,t)\} = \frac{1}{s}\psi(f,-q,a,0,t)_{@t=0} + \frac{1}{s}\left[\frac{1}{s}\psi^{(1)}(f,-q,a,0,t)_{@t=o} + \frac{1}{s}\mathcal{L}\left\{\psi^{(2)}(f,-q,a,0,t)\right\}\right] \tag{7.7}$$

Now repeating the process a total number of n times where n is integer such that $n+1>q>n$ gives, the expression for the equivalent initialization function. For $q=2.3>1$, then $n=2$.

$$\mathcal{L}\{\psi(f,-q,a,0,t)\}=\frac{1}{s^n}\mathcal{L}\{\psi^{(n)}(f,-q,a,0,t)\}+\sum_{j=1}^{n}\frac{1}{s^j}\psi^{(j-1)}(f,-q,a,0,t)_{@t=0} \tag{7.8}$$

Putting this, the Laplace transform of the fractional integral yields:

$$\mathcal{L}\{{}_0D_t^{-q}f(t)\}=\frac{1}{s^q}\mathcal{L}\{f(t)\}+\frac{1}{s^n}\mathcal{L}\{\psi^{(n)}(f,-q,a,0,t)\}+\sum_{j=1}^{n}\frac{1}{s^j}\psi^{(j-1)}(f,-q,a,0,t)_{@t=0} \tag{7.9}$$

The expression states that the q-th differintegral is composed (or can be decomposed) of n order 1 integer integrations and a fractional integration of order $q-n$ refer figure 7.1. For $q=2.3$, meaning the composition have $n=2$ full integer order integration and $q-n=0.3$ order fractional integration. Further the order 1 integrations are each initialized by a constant $\psi^{(j-1)}{}_{@t=0}$, (the integer order integrations are initialized via constants).

The fractional integration of order 2.3, with integer order as $n=2$ is thus:

$${}_0D_t^{-2.3}f(t)={}_0d_t^{-1}\left\{{}_0d_t^{-1}\left\{{}_0d_t^{-0.3}+\psi^{(2)}(t)\right\}+\psi^{(1)}(0)\right\}+\psi(0)$$

The Laplace is

$$\begin{aligned}\mathcal{L}\left\{{}_0D_t^{-2.3}f(t)\right\}&=\frac{1}{s}\left[\frac{1}{s}\left[\frac{1}{s^{0.3}}F(s)+\mathcal{L}\left\{\psi^{(2)}(t)\right\}\right]+\mathcal{L}\left\{\psi^{(1)}(0)\right\}\right]+\mathcal{L}\{\psi(0)\}\\&=\frac{1}{s^{2.3}}F(s)+\frac{1}{s^2}\mathcal{L}\left\{\psi^{(2)}(t)\right\}+\frac{1}{s}\left\{\psi^{(1)}(0)\right\}+\mathcal{L}\{\psi(0)\}\end{aligned}$$

For $\psi(0),\psi^{(1)}(0)$ is chosen to be constants K's, to initialize integer order integrations. With, Laplace of constant as $s^{-1}K$, the initialized Laplace expression for 2.3 fractional integration with $\psi(t)=\psi(f,-2.3,a,0,t)$ as:

$$\mathcal{L}\left\{{}_0D_t^{-2.3}f(t)\right\}=\left[\frac{1}{s^{2.3}}F(s)+\frac{1}{s^2}\mathcal{L}\left\{\psi^{(2)}(t)\right\}\right]+\frac{1}{s^2}\psi^{(1)}(0)+\frac{1}{s}\psi(0)$$

Laplace for semi-integration of a function is generalized from Figure 7.1 decomposition as:

$${}_0D_t^{-0.5}f(t)={}_0d_t^{-0.5}f(t)+\psi(f,-0.5,a,0,t)$$

$$\mathcal{L}\left\{{}_0D_t^{-0.5} f(t)\right\} = \frac{1}{s^{0.5}} F(s) + \mathcal{L}\{\psi(f,-0.5,a,0,t)\}$$

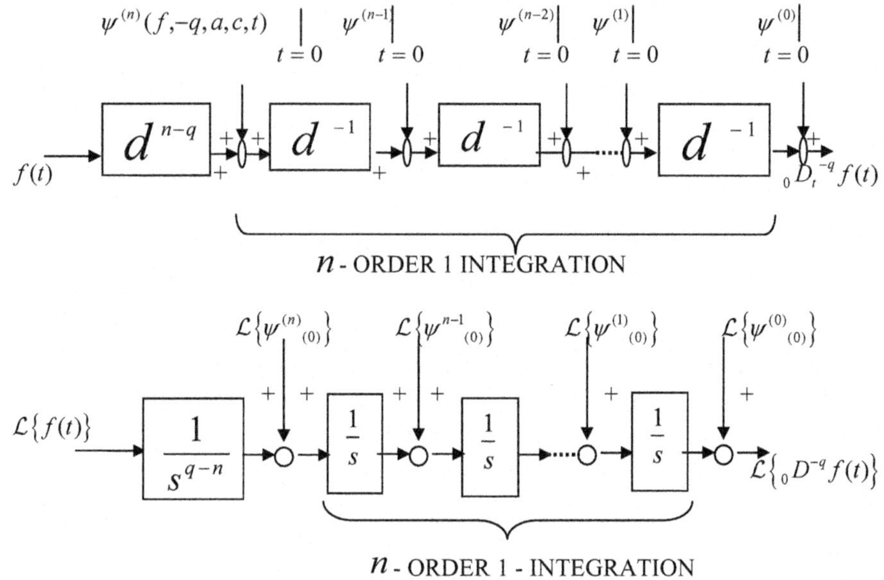

$$\mathcal{L}\left\{\psi^{(n)}{}_{(0)}\right\} = L\left\{\psi^{(n)}(f,-q,a,c,t)_{t=0}\right\}$$

Decomposition for $q > 1$

Fig. 7.1 Block diagram representing integer order decomposition of fractional integral (time and frequency domain)

Following we describe for sake of comparison the Laplace transform for multiple Integer order Integrals. The compatibility is observed for (7.9) and (7.10) for $q = n = 1, 2, 3 \ldots$ and properly choosing ψ.

$$\mathcal{L}\{g(t)\} = \mathcal{L}\left\{{}_b^t\!\int {}_b^{t1}\!\int {}_b^{t2}\!\int \ldots\ldots {}_b^{tn-1}\!\int f(t_n)dt_n dt_{n-1} \ldots\ldots dt_2 dt_1\right\} = \frac{1}{s^n}\mathcal{L}\{f(t)\} + \frac{c_1}{s} + \frac{c_2}{s^2} + \ldots + \frac{c_n}{s^n}$$
$$= \frac{1}{s^n}\mathcal{L}\{f(t)\} + \sum_{j=1}^{n} \frac{c_j}{s^j} \tag{7.10}$$

The coefficients are $c_j = \int_b^0 g(t_j)dt_j$.

In the Figure 7.1, the integer order integrations could just be replaced by general (fractional) integrations each also of order 1. Each of these then will allow non-constant initialization function. Figure 7.2 is ${}_0D_t^{-q}f(t)$ decomposed into n generalized order 1 integrations and a fractional integration of order $-q+n$. Where n is greatest integer less than q, with Further the order 1 integrations are each initialized by a constant $q>0$.The break up expression for Figure 7.2 is:

$$\begin{aligned}&{}_0D_t^{-q}f(t)=x_n(t)=\psi_1+{}_0d_t^{-1}x_{n-1}(t)=\psi_1+{}_0d_t^{-1}(\psi_2+{}_0d_t^{-1}x_{n-2}(t))=\\&\psi_1+{}_0d_t^{-1}\psi_2+{}_0d_t^{-2}(\psi_3+{}_0d_t^{-1}x_{n-3}(t))=\psi_1+{}_0d_t^{-1}\psi_2+{}_0d_t^{-2}\psi_3+{}_0d_t^{-3}x_{n-3}(t)\end{aligned}\tag{7.11}$$

Observing the fact that $x_1(t)={}_0D_t^{-q+n}f(t)=\psi(f,-q+n,a,0,t)+{}_0d_t^{-q+n}f(t)$, repeating the above iteration n times we have the following:

$${}_0D_t^{-q}f(t)={}_0d_t^{-q}f(t)+{}_0d_t^{-n}\psi(f,-q+n,a,0,t)+\sum_{j=1}^{n}{}_0d_t^{-(j-1)}\psi_j(x_j,-1,a,0,t)\tag{7.12}$$

as the mathematical representation of Figure 7.2. Taking the Laplace transforms we get.

$$\mathcal{L}\left\{{}_0D_t^{-q}f(t)\right\}=\frac{1}{s^q}\mathcal{L}\{f(t)\}+\frac{1}{s^n}\mathcal{L}\{\psi(f,-q+n,a,0,t)\}+\sum_{j=1}^{n}\frac{1}{s^{(j-1)}}\mathcal{L}\{\psi_j(x_j,-1,a,0,t)\}\tag{7.13}$$

This expression is generalization of the expressions of Figure 7.1, where the initialization effect was done by constants Further the order 1 integrations are each initialized by a constant K's, values of $\psi^{(n)}(t)_{@t=0}$. In this case, of Figure 7.2 the initialization effect is carried out by functions of time instead (i.e. ψ_n), Here for $q=2.3$, the integer $n=2$, meaning the composition with three integer order integration and one fractional order $-q+n=-0.3$ integration.

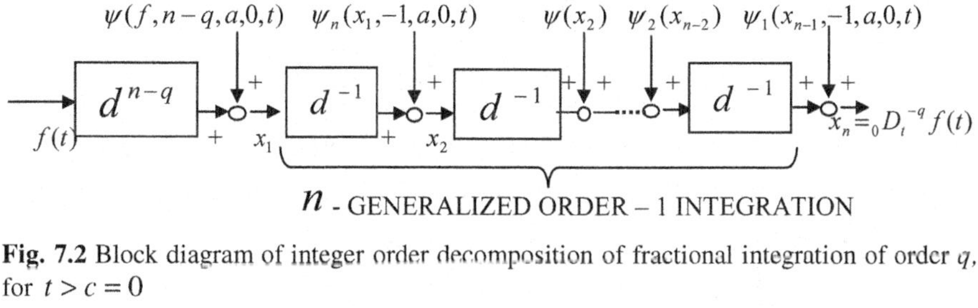

Fig. 7.2 Block diagram of integer order decomposition of fractional integration of order q, for $t>c=0$

For q as 2.3, then Laplace from figure 7.2 is composed, with $n=2$ as:

$$\mathcal{L}\left\{{}_0D_t^{-2.3} f(t)\right\} = \frac{1}{s^{2.3}} F(s) + \frac{1}{s^2} \mathcal{L}\{\psi(f,-0.3,a,0,t)\} + \frac{1}{s} \mathcal{L}\{\psi_2(x_2,-1,a,0,t)\} + \mathcal{L}\{\psi_1(x_1,-1,a,0,t)\}$$

The fractional integral composition is:

$${}_0D_t^{-2.3} f(t) = [{}_0d_t^{-2.3} f(t) + {}_0d_t^{-2} \psi(f,-0.3,a,0,t)] + \psi_1(x_1,-1,a,0,t) + {}_0d_t^{-1} \psi_2(x_2,-1,a,0,t)$$

The semi-integration is composed, with $n = 0$ as:

$${}_0D_t^{-0.5} f(t) = {}_0d_t^{-0.5} f(t) + \psi(f,-0.5,a,0,t),$$

and its Laplace as:

$$\mathcal{L}\left\{{}_0D_t^{-0.5} f(t)\right\} = \frac{1}{s^{0.5}} F(s) + \mathcal{L}\{\psi(f,-0.5,a,0,t)\}$$

7.3.2 Decomposition of Fractional Order Integral in Fractional Order

Here the decomposition of ${}_0D_t^{-q} f(t)$ is not limited to only integer order integral elements. Refer figure 7.3 where this decomposition is indicated. The mathematics is explained below (for convenience subscript is dropped i.e. ${}_0D_t^{-q} \to D^{-q}$)

$${}_0D_t^{-q} f(t) = x_{n+1}(t) = D^{-q_n} D^{-q_{n-1}} \ldots D^{-q_2} D^{-q_1} x_1(t) \qquad t > 0. \& .q_j > 0 \tag{7.14}$$

where $q = \sum_{i=1}^{n} q_i$.

Then starting from inside:

$$\mathcal{L}\{x_2(t)\} = \mathcal{L}\left\{{}_0d_t^{q_1} x_1(t)\right\} + \mathcal{L}\{\psi_1(x_1,-q_1,a,0,t)\},$$

can be written with simplified symbols as

$$\mathcal{L}\{x_2\} = \mathcal{L}\{d^{-q_1} x_1\} + \psi_1(s) = s^{-q_1} \mathcal{L}\{x_1\} + \psi_1(s)$$

$$\mathcal{L}\{x_3\} = \mathcal{L}\{d^{-q_2} x_2\} + \psi_2(s)$$

$$\mathcal{L}\{x_3\} = s^{-q_2}\{s^{-q_1} \mathcal{L}\{x_1\} + \psi_1(s)\} + \psi_2(s) = s^{-q_2-q_1} \mathcal{L}\{x_1\} + s^{-q_1} \psi_1(s) + \psi_2(s)$$

$$\mathcal{L}\{x_4\} = \mathcal{L}\{d^{-q_3} x_3\} + \psi_3(s) = s^{-q_3} \mathcal{L}\{x_3\} + \psi_3(s) = s^{-q_3-q_2-q_1} \mathcal{L}\{x_1\} + s^{-q_3-q_2} \psi_1(s) + s^{-q_3} \psi_2(s) + \psi_3(s)$$

$$\mathcal{L}\{x_4\} = s^{-q_3-q_2-q_1} \mathcal{L}\{x_1\} + s^{-q_3-q_2} \psi_1(s) + s^{-q_3} \psi_2(s) + \psi_3(s)$$

Repeating this process till x_{n+1}, we arrive at the following:

$$\mathcal{L}\{x_{n+1}\} = s^{-q_1-q_2\cdots\cdots-q_n}\mathcal{L}\{x_1\} + s^{-q_2-q_3\cdots\cdots-q_n}\psi_1(s) + s^{-q_3-q_4\cdots\cdots-q_n}\psi_2(s) + \ldots + s^{-q_{n-1}-q_n}\psi_{n-2}(s) + s^{-q_n}\psi_{n-1}(s) + \psi_n(s)$$

Defining B_a as

$$B_a = \sum_{i=a}^{n} q_i \,,$$

we can have general form of decomposition expression as, where $1 \le a \le n$;

$$\mathcal{L}\{x_{n+1}\} = s^{-B_1}\mathcal{L}\{x_1\} + s^{-B_2}\psi_1(s) + s^{-B_3}\psi_2(s) + \ldots + s^{-B_n}\psi_{n-1}(s) + \psi_n(s) \quad (7.15)$$

Summarizing this in compact form we get:

$$\mathcal{L}\{{}_0D_t^{-q}f(t)\} = s^{-q}\mathcal{L}\{f(t)\} + \psi_n(s) + \sum_{j=1}^{n-1} s^{-B_{j+1}}\psi_j(s) \quad B_a = \sum_{i=a}^{n} q_i \quad 1 \le a \le n \quad (7.16)$$

The effective initialization here in this case is:

$$\mathcal{L}\{\psi_{eff}\} = \psi_n(s) + \sum_{j=1}^{n-1} s^{-B_{j+1}}\psi_j(s)$$

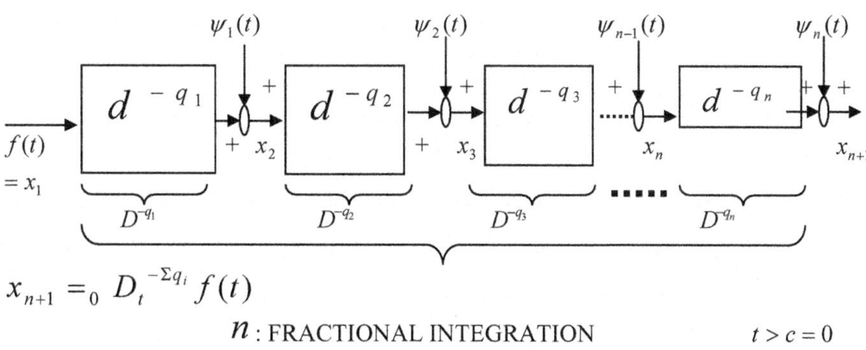

Fig. 7.3 Decomposition of fractional order integral in fractional order

Let us decompose the fractional integration of order 2.3 by fractional integration of order 0.7 three times and followed by fractional integration of 0.2. This means $n = 4$, as per Figure 7.5. The composition for Laplace following the Figure 7.3 will be:

$$X_1(s) = F(s)\,,\ \psi_1(s) = \mathcal{L}\{\psi_1(x_1, -0.7, a, 0, t)\}$$

$$X_2(s) = \frac{1}{s^{0.7}}X_1(s) + \psi_1(s)$$

$$X_3(s)=\frac{1}{s^{0.7}}X_2(s)+\psi_2(s)$$

$$X_4(s)=\frac{1}{s^{0.7}}X_3(s)+\psi_3(s)$$

$$X_5(s)=\frac{1}{s^{0.2}}X_4(s)+\psi_4(s)=\mathcal{L}\left\{{}_0D_t^{-2.3}f(t)\right\}$$

From Figure 7.3 the decomposition of the fractional integral of order 2.3 by fractional integrals of order 0.7 three numbers plus fractional integral of order 0.2 is:

$${}_0D_t^{-2.3}f(t)=\left[{}_0d_t^{-0.2}\left({}_0d_t^{-0.7}\left({}_0d_t^{-0.7}\left({}_0d_t^{-0.7}f(t)+\psi_1(t)\right)+\psi_2(t)\right)+\psi_3(t)\right)+\psi_4(t)\right]$$
$${}_0D_t^{-2.3}f(t)={}_0d_t^{-2.3}f(t)+{}_0d_t^{-1.6}\psi_1(t)+{}_0d_t^{-0.9}\psi_2(t)+{}_0d_t^{-0.2}\psi_3(t)+\psi_4(t)$$

The Laplace decomposition of fractional integral of order 2.3 is:

$$\mathcal{L}\left\{{}_0D_t^{-2.3}f(t)\right\}=\frac{1}{s^{2.3}}F(s)+\frac{1}{s^{1.6}}\psi_1(s)+\frac{1}{s^{0.9}}\psi_2(s)+\frac{1}{s^{0.2}}\psi_3(s)+\psi_4(s)$$

7.4 Laplace Transformation of Fractional Derivatives

The derivative starting point is taken as $c=0$ for simplicity and the function $f(t)$ is born at a, before $t=a$ the function is zero. Here we will find

$$\mathcal{L}\left\{{}_0D_t^u f(t)\right\}=\mathcal{L}\left\{{}_0D_t^m\,{}_0D_t^{-p}f(t)\right\}$$

With $u>0$, and m is the least integer greater than u, such that $u=m-p$. For $u=2.3$, the integer $m=3$. In the case of integration decomposition the integer n is chosen as the greatest integer less than the fractional order of integration q. Using the definitions and expanding we get (for $t>0$)

$$\mathcal{L}\left\{{}_0D_t^u f(t)\right\}=\mathcal{L}\left\{{}_0D_t^m\,{}_0D_t^{-p}f(t)\right\}=\mathcal{L}\left\{\frac{d^m}{dt^m}\left(\int_0^t\frac{(t-\tau)^{p-1}}{\Gamma(p)}f(\tau)d\tau\right)\right\}+\mathcal{L}\left\{\frac{d^m}{dt^m}\psi(f,-p,a,0,t)\right\}+$$
$$+\mathcal{L}\left\{\psi(h,m,a,0,t)\right\} \tag{7.17}$$

where $h(t)={}_0d_t^{-p}f(t)$. Defining equivalent initialization function as $\psi_{eq}(f,u,a,0,t)$:

$$\psi_{eq}(f,u,a,0,t)=\left\{\frac{d^m}{dt^m}\psi_{eq}(f,-p,a,0,t)\right\}+\left\{\psi(h,m,a,0,t)\right\} \tag{7.18}$$

Now consider the first term of the expanded expression i.e.

$$\mathcal{L}\left\{\frac{d^m}{dt^m}\left(\int_0^t \frac{(t-\tau)^{p-1}}{\Gamma(p)} f(\tau)d\tau\right)\right\} = \int_0^\infty e^{-st}\frac{d}{dt}\left(\frac{d^{m-1}}{dt^{m-1}}\int_0^t \frac{(t-\tau)^{p-1}}{\Gamma(p)} f(\tau)d\tau\right)dt \tag{7.19}$$

Apply integration by parts by selecting $u = e^{-st}$ and

$$dv = \frac{d}{dt}\left(\frac{d^{m-1}}{dt^{m-1}}\int_0^t \frac{(t-\tau)^{p-1}}{\Gamma(p)} f(\tau)d\tau\right)dt$$

Therefore, $du = -se^{-st}dt$ and $v = \frac{d^{m-1}}{dt^{m-1}}\int_0^t \frac{(t-\tau)^{p-1}}{\Gamma(p)} f(\tau)d\tau$, yields

$$= \left[e^{-st}\frac{d^{m-1}}{dt^{m-1}}\int_0^t \frac{(t-\tau)^{p-1}}{\Gamma(p)} f(\tau)d\tau\right]_0^\infty + s\int_0^\infty e^{-st}\frac{d}{dt}\left(\frac{d^{m-2}}{dt^{m-2}}\int_0^t \frac{(t-\tau)^{p-1}}{\Gamma(p)} f(\tau)d\tau\right)dt \tag{7.20}$$

The first term is zero after putting the end values; therefore we get the Laplace expression as;

$$\mathcal{L}\left\{\frac{d^m}{dt^m}\left(\int_0^t \frac{(t-\tau)^{p-1}}{\Gamma(p)} f(\tau)d\tau\right)\right\} = s\mathcal{L}\left\{\frac{d}{dt}\left(\frac{d^{m-2}}{dt^{m-2}}\int_0^t \frac{(t-\tau)^{p-1}}{\Gamma(p)} f(\tau)d\tau\right)\right\} \tag{7.21}$$

Repeating the process m times yields:

$$\mathcal{L}\left\{\frac{d^m}{dt^m}\int_0^t \frac{(t-\tau)^{p-1}}{\Gamma(p)} f(\tau)d\tau\right\} = s^m\mathcal{L}\left\{{}_0d_t^{-p} f(t)\right\} \tag{7.22}$$

using this result the Laplace expression for fractional derivative is:

$$\mathcal{L}\left\{{}_0D_t^u f(t)\right\} = \mathcal{L}\left\{{}_0D_t^m\,{}_0D_t^{-p} f(t)\right\} = s^m\mathcal{L}\left\{{}_0d_t^{-p} f(t)\right\} + \mathcal{L}\left\{\frac{d^m}{dt^m}\psi(f,-p,a,0,t)\right\} + \mathcal{L}\left\{\psi(h,m,a,0,t)\right\} \tag{7.23}$$

or

$$\mathcal{L}\left\{{}_0D_t^u f(t)\right\} = s^m\mathcal{L}\left\{{}_0d_t^{-p} f(t)\right\} + \mathcal{L}\left\{\psi_{eq}(f,u,a,0,t)\right\} \tag{7.24}$$

here applying integral Laplace result for Laplace of ${}_0d_t^{-p}$, we get:

$$\mathcal{L}\left\{{}_0D_t^u f(t)\right\} = s^{m-p}\mathcal{L}\left\{f(t)\right\} + \mathcal{L}\left\{\psi_{eq}(f,u,a,0,t)\right\} = s^u\mathcal{L}\left\{f(t)\right\} + \mathcal{L}\left\{\psi_{eq}(f,u,a,0,t)\right\} \tag{7.25}$$

This is most general form of Laplace Transform of the fractional derivative, and is similar to what is obtained for Laplace of fractional integration. As the case with fractional integrals the fractional derivatives also can be decomposed (or is composed off) infinite ways, thus several possible formulations exist for $\psi_{eq}(f,u,a,c,t)$.

7.4.1 Decomposition of Fractional Order Derivative in Integer Order

Equivalent form for $\mathcal{L}\{\psi(f,u,a,0,t)\}$ of (7.25) is considered. Consider the expression appeared in the previous derivation (7.17) i.e.

$$\mathcal{L}\left\{\frac{d^m}{dt^m}\psi(f,-p,a,0,t)\right\}=\int_0^\infty e^{-st}\frac{d}{dt}\left(\frac{d^{m-1}}{dt^{m-1}}\psi(f,-p,a,0,t)\right)dt$$

For integrating by parts take $u=e^{-st}$, and

$$dv=\frac{d}{dt}\frac{d^{m-1}}{dt^{m-1}}\psi(f,-p,a,0,t)dt\,.$$

So $du=-se^{-st}dt$ and

$$v=\frac{d^{m-1}}{dt^{m-1}}\psi(f,-p,a,0,t)\,.$$

Therefore

$$\mathcal{L}\left\{\frac{d^m}{dt^m}\psi(f,-p,a,0,t)\right\}=\left[e^{-st}\frac{d^{m-1}}{dt^{m-1}}\psi(f,-p,a,0,t)\right]_0^\infty+s\int_0^\infty e^{-st}\frac{d^{m-1}}{dt^{m-1}}\psi(f,-p,a,0,t)dt$$

$$=\left[-\frac{d^{m-1}}{dt^{m-1}}\psi(f,-p,a,0,t)\right]_{t=0}+s\mathcal{L}\left\{\frac{d^{m-1}}{dt^{m-1}}\psi(f,-p,a,0,t)\right\}$$

(7.26)

These results thus use to get:

$$\mathcal{L}\left\{\frac{d}{dt}\frac{d^{m-2}}{dt^{m-2}}\psi(f,-p,a,0,t)\right\}=-\left(\frac{d^{m-2}}{dt^{m-2}}\psi(f,-p,a,0,t)\right)_{t=0}+s\mathcal{L}\left\{\frac{d^{m-2}}{dt^{m-2}}\psi(f,-p,a,0,t)\right\}$$

(7.27)

Therefore,

$$\mathcal{L}\left\{\frac{d^m}{dt^m}\psi(f,-p,a,0,t)\right\}=-\left[\frac{d^{m-1}}{dt^{m-1}}\psi(f,-p,a,0,t)\right]_{t=0}$$
$$+s\left[-\left(\frac{d^{m-2}}{dt^{m-2}}\psi(f,-p,a,0,t)\right)_{t=0}+s\mathcal{L}\left\{\frac{d^{m-2}}{dt^{m-2}}\psi(f,-p,a,0,t)\right\}\right] \tag{7.28}$$

Repeating this process m times and writing:

$$\frac{d^k}{dt^k}\psi(f,-p,a,0,t)=\psi^{(k)}(f,-p,a,0,t)$$

gives:

$$\mathcal{L}\left\{\frac{d^m}{dt^m}\psi(f,-p,a,0,t)\right\}=s^m\mathcal{L}\{\psi(f,-p,a,0,t)\}-\sum_{j=1}^{m}s^{j-1}\psi^{m-j}(f,-p,a,0,t)]_{t=0} \tag{7.29}$$

Substituting the above obtained expression into equation for Laplace transform of fractional derivative we obtain Laplace transform of generalized derivative decomposed into integer differentiations with $\psi^{m-j}(f,-p,a,c,t)_{@t=0}$, a constant initialization, as:

$$\mathcal{L}\left\{{}_0D_t^u f(t)\right\}=\mathcal{L}\left\{{}_0D_t^m\,{}_0D_t^{-p} f(t)\right\}$$
$$=s^m\mathcal{L}\left\{{}_0d_t^{-p}f(t)\right\}+s^m\mathcal{L}\{\psi(f,-p,a,0,t)\}-\left(\sum_{j=1}^{m}s^{j-1}[\psi^{m-j}(f,-p,a,0,t)]_{t=0}\right)$$
$$+\mathcal{L}\{\psi(h,m,a,0,t)\} \tag{7.30}$$

Here

$${}_0d_t^{-p}f(t)=\frac{1}{\Gamma(p)}\int_0^t(t-\tau)^{p-1}f(\tau)d\tau=h(t),$$

the un-initialized fractional integration starting at $t=c=0$.

$$\mathcal{L}\left\{{}_0D_t^u f(t)\right\}=\mathcal{L}\left\{{}_0D_t^m\,{}_0D_t^{-p} f(t)\right\}$$
$$=s^u\mathcal{L}\{f(t)\}-\left(\sum_{j=1}^{m}s^{j-1}(\psi^{m-j}(f,-p,a,0,t)]_{t=0}\right)+\mathcal{L}\{\psi(h,m,a,0,t)\} \tag{7.31}$$

Figure 7.4 gives the illustration of the decomposition.

The composition of Laplace of fractional derivative of order 2.3 with integer order derivative (order 3) and then fractional order integration of order 0.7 is:

$$\begin{aligned}\mathcal{L}\left\{{}_0D_t^{2.3}f(t)\right\} &= \mathcal{L}\left\{{}_0D_t^3\,{}_0D_t^{-0.7}f(t)\right\}\\ &= s^{2.3}F(s)+s^3\mathcal{L}\left\{\psi(f,-0.7,a,0,t)\right\}\\ &\quad-\begin{pmatrix}\left[\psi^{(2)}(f,-0.7,a,0,t)\right]_{t=0}+s\left[\psi^{(1)}(f,-0.7,a,0,t)\right]_{t=0}\\ +s^2\left[\psi(f,-0.7,a,0,t\right]_{t=0}\end{pmatrix}\\ &\quad+\mathcal{L}\left\{\psi({}_ad_t^{-0.7}f(t),3,a,0,t)\right\}\end{aligned}$$

Comparing with integer order, Laplace of third derivative of a function is given as:

$$\mathcal{L}\left\{{}_0D_t^3f(t)\right\}=s^3F(s)-\left(f^{(2)}(t)_{t=0}+sf^{(1)}(t)_{t=0}+s^2f(t)_{t=0}\right)$$

The similarity in these two cases is observed, and is explained as above.

The semi-derivative of a function gets decomposed as:

$$\mathcal{L}\left\{{}_0D_t^{0.5}f(t)\right\}=\mathcal{L}\left\{{}_0D_t^1\,{}_0D_t^{-0.5}f(t)\right\}=s^{0.5}F(s)-\psi(f,-0.5,a,0,t)+\mathcal{L}\left\{\psi([{}_ad_t^{-0.5}f(t)],1,a,0,t\right\}$$

Yields, similarity with Laplace of first order integer derivative:

$$\mathcal{L}\left\{{}_0D_t^1f(t)\right\}=sF(s)-f(t)_{t=0}$$

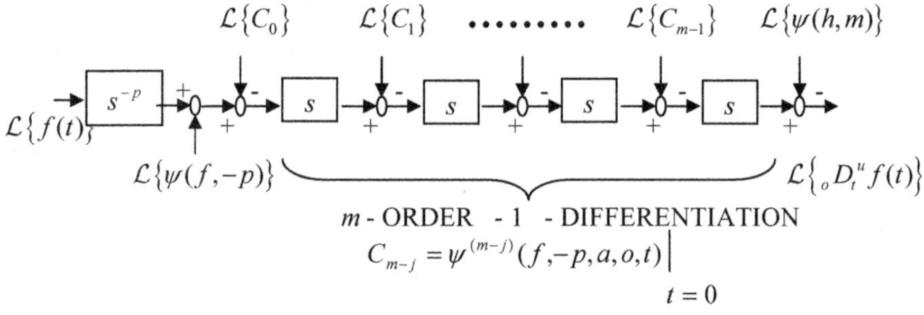

Fig. 7.4 Decomposition of fractional derivative into integer order

Note what has been done here is to decompose the integer order derivative part of ${}_0D_t^uf(t)$ namely into m integer order 1 differentiations. For integer order Laplace transform of derivative we have

$$\mathcal{L}\left\{f^{(m)}(t)\right\}=s^m\mathcal{L}\left\{f(t)\right\}-\sum_{j=1}^{m}s^{j-1}f^{(m-j)}(0)\,.$$

In equation (7.31) put $p = 0$. This becomes zero-order operation then set $\psi(h,m,a,0,t) = 0$, as discussed for terminal charging case for integer order initialization in Chapter 6, gives a specialized case with

$$\psi^{(m-j)}(f,0,a,0,t)\ _{@t=0} = f^{(m-j)}(0).$$

This will give the above result of integer order repeated derivative Laplace relation as indicated above.

In context of equation (7.31) the Laplace expression of total initialized system is:

$$\mathcal{L}\{\psi(f,u,a,0,t)\} = -\sum_{j=1}^{m} s^{j-1}[\psi^{(m-j)}(f,-p,a,0,t)]_{@t=0} + \mathcal{L}\{\psi(h,m,a,0,t)\}.$$

Further generalization is possible with (7.31). From equation (7.17) consider the term

$$\mathcal{L}\left\{\frac{d^m}{dt^m}\psi(f,-p,a,0,t)\right\}$$

and for complete arbitrary initialization let,

$$\psi(f,-p,a,0,t) = {}_0d_t^{1-m}\psi_1(t) + {}_0d_t^{2-m}\psi_2(t) + \ldots\ldots {}_0d_t^{m-m}\psi_m(t) = \sum_{j=1}^{m} {}_0d_t^{j-m}\psi_j(t).$$

Then

$$\mathcal{L}\left\{\frac{d^m}{dt^m}\psi(f,-p,a,0,t)\right\} = \mathcal{L}\left\{\sum_{j=1}^{m}\frac{d^j\psi_j(t)}{dt^j}\right\} = \sum_{j=1}^{m} s^j\mathcal{L}\{\psi_j(t)\}.$$

Here redundant terms have been dropped. Therefore the equation (7.31) is further generalized as:

$$\mathcal{L}\{{}_0D_t^u f(t)\} = \mathcal{L}\{{}_0D_t^m\,{}_0D_t^{-p} f(t)\} = s^u\mathcal{L}\{f(t)\} + \sum_{j=1}^{m} s^j\mathcal{L}\{\psi_j(t)\} + \mathcal{L}\{\psi(h,m,a,0,t)\},$$

the difference with (7.31) is that, the orders of 1 derivatives are each initialized by time varying functions. Meaning the order 1 differentiations is now generalized order 1 differentiation. The derivative initialization of (7.17) i.e. $\mathcal{L}\{\psi(h,m,a,0,t)\}$ can also be similarly decomposed, to have similar effect, distributed over the order 1 derivative term.

For derivative of order 2.3 the Laplace composition is:

$$\mathcal{L}\left\{{}_0D_t^{2.3}f(t)\right\} = \mathcal{L}\left\{{}_0D_t^3\,{}_0D_t^{-0.7}f(t)\right\} =$$
$$= s^{2.3}F(s) + s\mathcal{L}\left\{\psi_1(t)\right\} + s^2\mathcal{L}\left\{\psi_2(t)\right\} + s^3\mathcal{L}\left\{\psi_3(t)\right\} + \mathcal{L}\left\{\psi({}_ad_t^{-0.7}f(t),3,a,0,t)\right\}$$

The Laplace composition in this case with semi derivative is:

$$\mathcal{L}\left\{{}_0D_t^{0.5}f(t)\right\} = \mathcal{L}\left\{{}_0D_t^1\,{}_0D_t^{-0.5}f(t)\right\} = s^{0.5}F(s) + s\mathcal{L}\left\{\psi_1(t)\right\} + \mathcal{L}\left\{\psi({}_ad_t^{-0.5}f(t),1,a,0,t)\right\}$$

7.4.2 Decomposition of Fractional Derivative in Fractional Order

Figure 7.5 demonstrates this concept as signal flow graph.

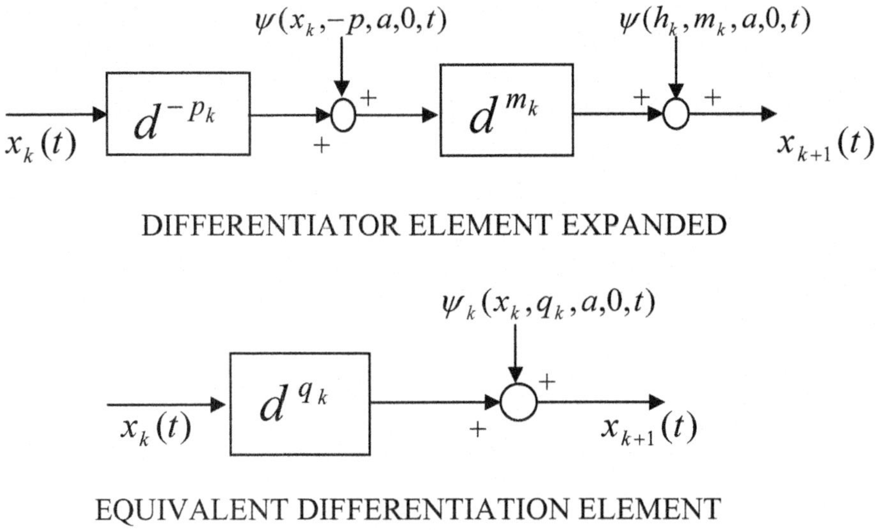

Fig. 7.5 Decomposition of fractional derivative (fractional differentiation element)

As the decomposition of fractional integration was done by breaking into several fractional elements $-q_k$ in the integration section, similarly by replacing all the fractional integration components $-q_k = r_k$, this process is carried out.

The result of the figure 7.5 is

$$\mathcal{L}\left\{{}_0D_t^q f(t)\right\} = \mathcal{L}\left\{{}_0D_t^{B_1}x_1(t)\right\} = \mathcal{L}\left\{x_{n+1}(t)\right\} = s^{B_1}\mathcal{L}\left\{x_1(t)\right\} + \psi_n(s) + \sum_{j=1}^{n-1} s^{B_{j+1}}\psi_j(s)$$

$$B_a = \sum_{i=a}^{n} r_i \tag{7.32}$$

$$r_k \geq 0 \text{ and } 1 \leq a \leq n,$$

with initialization as:

$$\psi_k(x_k, r_k, a, 0, t) = \frac{d^{m_k}}{dt^{m_k}} \psi(x_k, -p_{k-1}, a.0, t) + \psi(h_k, m_k, a, 0, t),$$

where m_k is the least integer such that $r_k = m_k - p_k$ and $h_k = {}_0d_t^{-p_k} x_k(t)$.

Let us decompose the fractional differentiation of order 2.3 by fractional differentiation of order 0.7 three times and followed by fractional differentiation of 0.2, composing this derivative by four number of fractional order derivatives $n = 4$, for this case. The composition for Laplace following the Figure 7.5 will be:

$$x_1 = f(t), \text{ with } r_1 = r_2 = r_3 = 0.7; r_4 = 0.2 \text{ and } m_1 = m_2 = m_3 = m_4 = 1$$

The initialization functions for each stage is:

$$\psi_1 = \psi_1(x_1, r_1, a, 0, t) = \frac{d}{dt}\left[\psi(x_1, -0.3, a, 0, t)\right] + \psi({}_0d_t^{-0.3} f(t), 1, a, 0, t)$$

$$\psi_2 = \psi_2(x_2, r_2, a, 0, t) = \frac{d}{dt}\left[\psi(x_2, -0.3, a, 0, t)\right] + \psi({}_0d_t^{-0.3} x_2, 1, a, 0, t)$$

$$\psi_3 = \psi_3(x_3, r_3, a, 0, t) = \frac{d}{dt}\left[\psi(x_3, -0.3, a, 0, t)\right] + \psi({}_0d_t^{-0.3} x_3, 1, a, 0, t)$$

$$\psi_4 = \psi_4(x_4, r_4, a, 0, t) = \frac{d}{dt}\left[\psi(x_4, -0.8, a, 0, t)\right] + \psi({}_0d_t^{-0.3} x_4, 1, a, 0, t)$$

$$\mathcal{L}\left\{{}_0D_t^{2.3} f(t)\right\} = \mathcal{L}\left\{{}_0D_t^{0.7+0.7+0.7+0.2} f(t)\right\} = \mathcal{L}\left\{x_5(t)\right\}$$

$$= s^{2.3}F(s) + \psi_4(s) + s^{1.6}\psi_1(s) + s^{0.9}\psi_2(s) + s^{0.2}\psi_3(s)$$

7.4.3 Effect of Terminal Charging on Laplace Transforms

In all above discussions the initialization terms ψ is taken as completely arbitrary treating as 'side charging' of the differintegrals. The composition for Laplace described above is for any type of initialization method. The effect of 'terminal charging' is readily determined by appropriate substitutions for the ψ terms. That is for integer derivative make $\psi(h, m, a, 0, t) = 0, m = 1, 2 \ldots$, and for the fractional integrations take

$$\psi(f, -p, a, 0, t) = \frac{1}{\Gamma(p)} \int_a^0 (t-\tau)^{p-1} f(\tau) d\tau$$

This part of fractional integration's initialization comes alone as in initialized fractional integration and also for initialized fractional derivative.

Some modest simplifications are done to Laplace transform of the fractional integral the term $\psi^{(k)}$ for terminal charging, for fractional integration decomposition of figure 7.1, as following.

$$\psi^{(k)}(f,-q,a,0,t)]_{t=0}=\left\{\frac{d^{k-1}}{dt^{k-1}}\left(\frac{d}{dt}\int_a^0\frac{(t-\tau)^{q-1}}{\Gamma(q)}f(\tau)d\tau\right)\right\}_{t=0}\quad, t>0$$

Apply Leibniz's rule and we have:

$$\psi^{(k)}(f,-q,a,0,t)]_{t=0}=\left\{\frac{d^{k-2}}{dt^{k-2}}\left(\frac{d}{dt}\int_a^0\frac{(q-1)(t-\tau)^{(q-1)-1}}{\Gamma(q)}f(\tau)d\tau\right)\right\}_{t=0}$$

Substitute

$$\frac{(q-1)}{\Gamma(q)}=\frac{1}{\Gamma(q-1)},$$

then integral is recognized as $(q-1)$-th order differintegral and continuing the process gives

$$\psi^{(k)}(f,-q,a,0,t)]_{t=0}=\psi(f,-q+k,a,0,t)]_{t=0}, \text{ for } t>0$$

In similar fashion a more general term is to be substituted for terminal initial function as:

$$\psi^{(n)}(f,-q,a,0,t)=\psi(f,-q+n,a,0,t)=\frac{1}{\Gamma(q-n)}\int_a^0(t-\tau)^{q-n-1}f(\tau)d\tau$$

$$t>0, \text{ and } (n+1>q>n>0), \text{ for integer, } n$$

For $q=2.3$ fractional integration case with decomposition as per figure 7.1, $n=2$ then:

$$\psi^{(2)}(f,-2.3,a,0,t)=\psi(f,-0.3,a,0,t)=\frac{1}{\Gamma(0.7)}\int_a^0(t-\tau)^{-0.7}f(\tau)d\tau$$

7.5 Start Point Shift Effect

7.5.1 Fractional Integral

Specifically effect of start of function at non-zero a is considered and differintegration start at c (non-zero) is shown. $c>a\geq 0.\&.q>0$.

Then Laplace transform of uninitialized fractional integral is (for $q > 0$ and $a \ge 0$)

$$\mathcal{L}\left\{{}_aD_t^{-q} f(t)\right\} = \mathcal{L}\left\{{}_ad_t^{-q} f(t)\right\} = \mathcal{L}\left\{\frac{1}{\Gamma(q)}\int_a^t (t-\tau)^{q-1} f(\tau)d\tau\right\}$$

By definition $f(t) = 0$ for $t \le a$ thus following can be formulated by unit step $H(t)$ at $t = a$ as $f(t) = H(t-a)f(t)$. The result is

$$\mathcal{L}\left\{{}_aD_t^{-q} f(t)\right\} = \mathcal{L}\left\{{}_ad_t^{-q} f(t)\right\} = \mathcal{L}\left\{{}_0d_t^{-q}(H(t-a)f(t))\right\}$$

7.5.2 Fractional Derivative

Under the same condition as above the un-initialized fractional derivative is considered then

$$\mathcal{L}\left\{{}_ad_t^{q} f(t)\right\} = \mathcal{L}\left\{{}_ad_t^{m}\, {}_ad_t^{-p} f(t)\right\} = \mathcal{L}\left\{\frac{d^m}{dt^m}\frac{1}{\Gamma(p)}\int_a^t (t-\tau)^{p-1} f(\tau)d\tau\right\},$$

as before m is the least integer greater than q, and $q > 0$,with $p = m - q$.

Again by definition $f(t) = H(t-a)f(t)$. Thus

$$\mathcal{L}\left\{{}_ad_t^{q} f(t)\right\} = \mathcal{L}\left\{\frac{d^m}{dt^m}\frac{1}{\Gamma(p)}\int_0^t (t-\tau)^{p-1} H(\tau-a) f(\tau)d\tau\right\} = \mathcal{L}\left\{\frac{d^m}{dt^m}\, {}_0d_t^{-p}(H(t-a)f(t))\right\}$$
$$= \mathcal{L}\left\{{}_0d_t^{q} H(t-a)f(t)\right\}$$

So the final result is

$$\mathcal{L}\left\{{}_ad_t^{q} f(t)\right\} = \mathcal{L}\left\{{}_0d_t^{q} H(t-a)f(t)\right\}$$

7.6 Laplace Transform of Initialization Function

7.6.1 Fractional Integral

From shifting theorem of Laplace transform we have

$$\mathcal{L}\left\{g(t-a)H(t-a)\right\} = e^{-as}\mathcal{L}\left\{g(t)\right\},$$

for $a \ge 0$. Taking $f(t) = g(t-a)$ we have thus for $a \ge 0$

$$\mathcal{L}\{f(t)H(t-a)\} = e^{-as}\mathcal{L}\{f(t+a)\}.$$

Now the fractional integral under terminal charging the desired Laplace transform is:

$$\mathcal{L}\{\psi(f,-q,a,c,t)\} = \mathcal{L}\{{}_a d_t^{-q} f(t)\} - \mathcal{L}\{{}_c d_t^{-q} f(t)\} = \mathcal{L}\{{}_0 d_t^{-q} f(t)H(t-a)\} - \mathcal{L}\{{}_0 d_t^{-q} f(t)H(t-c)\}$$
$$= s^{-q}\mathcal{L}\{f(t)H(t-a)\} - s^{-q}\mathcal{L}\{f(t)H(t-c)\}$$

Finally

$$\mathcal{L}\{\psi(f,-q,a,c,t)\} = s^{-q}\left[e^{-as}\mathcal{L}\{f(t+a)\} - e^{-cs}\mathcal{L}\{f(t+c)\}\right]$$

7.6.2 *Fractional Derivative*

The equivalent initialization function, for fractional derivative of order u ,is:

$$\psi_{eq}(f,u,a,c,t) = \frac{d^m}{dt^m}\psi(f,-p,a,c,t) + \psi(f,-p,a,c,t),$$

where m is the least integer greater than u, and $u = m - p$ For terminal charging it has been shown $\psi(h,m,a,c,t) = 0$, therefore

$$\mathcal{L}\{\psi(f,u,a,c,t)\} = s^m\mathcal{L}\{\psi(f,-p,a,c,t)\}$$

In this, expression applying the shifting theorem, as obtained in previous section of Laplace the final result for is $0 < a \le c$

$$\mathcal{L}\{\psi(f,u,a,c,t)\} = s^u\left[e^{-as}\mathcal{L}\{f(t+a)\} - e^{-cs}\mathcal{L}\{f(t+c)\}\right]$$

7.7 Examples of Initialization in Fractional Differential Equations

Proper initialization is crucial in the solution and understanding of fractional differential equations the examples will elucidate the application of initialized fractional calculus to the solution of differential equations.

Example 1

$${}_0D_t^{1/2} f(t) + bf(t) = 0 \qquad t > 0; \qquad {}_0D_t^{-1/2} f(t)_{@t=0} = C$$

Assuming here the fractional initial state is finite and observable in the experiment.

The notation above is un-initialized semi-derivative and by obtaining Laplace transform will give

$$F(s)=\frac{C}{s^{1/2}+b}.$$

The inverse Laplace gives $f(t)=Ct^{-1/2}E_{0.5,0.5}(-b\sqrt{t})$ and for b=1, the solution is:

$$f(t)=C\left(\frac{1}{\sqrt{\pi t}}-e^{t}\operatorname{erfc}\left(\sqrt{t}\right)\right).$$

Now the initialized approach is given below, considering semi-derivative as initialized.

$${}_0D_t^{1/2}f(t)+bf(t)=0 \qquad t>0$$

with $\psi(f,1/2,a,0,t)$ is arbitrary, for 'side initialization' case. Therefore the equation is written as

$${}_0d_t^{1/2}f(t)+\psi(f,1/2,a,0,t)+bf(t)=0\,,\ t>0$$

With $\psi(f,1/2,a,0,t)$ arbitrary choice.

Laplace transforms gives now,

$$F(s)=\frac{-\psi(f,1/2,a,0,s)}{s^{1/2}+b}=-\frac{\psi(s)}{s^{1/2}+b}$$

This Laplace transform is same as for un-initialized approach (done above). When $\psi(t)=-C\delta(t)$, i.e. 'when impulse at $t=0$ is used to initialize the fractional differential equation'. Now using R function (variant of Mittag-Leffler function) the generalized inverse of the initialized transform by applying

$$R_{q,v}(\alpha,c,t)\equiv\sum_{n=0}^{\infty}\frac{(\alpha)^{n}(t-c)^{(n+1)q-v-1}}{\Gamma((n+1)q-v)};\mathcal{L}\left\{R_{q,v}(\alpha,c,t)\right\}=\frac{s^{v}}{s^{q}-\alpha}$$

With convolution integral definition we obtain most general solution of the form as:

$$f(t)=-\int_{0}^{t}R_{1/2,0}(-b,0,t-\tau)\psi(\tau)d\tau \qquad t>0$$

Now with this arbitrary initialization the above convolution integral is the most general solution. If $\psi(t)=-C\delta(t)$, from the derived general equation we obtain, $f(t)=CR_{1/2,0}(-b,0,t)$, which is identical to the above result of un-initialized case.

For 'terminal initialization' a more useful result would appear using the terminal charging definition as

$$\psi(f,q,a,0,t)=\frac{d^m}{dt^m}\psi(f,-p,a,0,t)+\psi(h,m,a,0,t) \quad q>0 \quad h(t)={}_a d_t^{-p} f(t)$$

In the present case $q=m-p$, $m=1$, $p=1/2$ and $\psi(h,m,a,0,t)=0$. Then

$$\psi(f,1/2,a,0,t)=\frac{d}{dt}\psi(f,-1/2,a,0,t)=\frac{d}{dt}\frac{1}{\Gamma(1/2)}\int_a^0 (t-\tau)^{-1/2} f(\tau)d\tau$$

Example 2

$${}_0D_t^Q f(t)+{}_0D_t^q f(t)=H(t) \qquad \left[{}_0D_t^{Q-1} f(t)+{}_0D_t^{q-1} f(t)\right]_{@t=0}=C$$

The problem uses symbolism as un-initialized fractional differential operator. Two separate ones to identify the most general solution as will replace the initialization of equation:

$${}_0d_t^Q f(t)+\psi_1(f,Q,a,0,t)+{}_0d_t^q f(t)+\psi_2(f,q,a,0,t)=H(t)$$

and taking Laplace transform:

$$s^Q F(s)+s^q F(s)=H(s)-\psi_1(f,Q,a,0,s)-\psi_2(f,q,a,0,s)$$

$$F(s)=\left(\frac{s^{-q}}{s^{Q-q}+1}\right)[H(s)-\psi_1(f,Q,a,0,s)-\psi_2(f,q,a,0,s)]$$

again using generalized R function and the convolution definition we get a most generalized form of solution is.

$$f(t)=\int_0^t R_{Q-q,-q}(-1,0,t-\tau)(H(\tau)-\psi_1(f,Q,a,0,\tau)-\psi_2(f,q,a,0,\tau))d\tau .$$

In this general expression allows having effect of continuing past, as in terminal initialization. Giving these initialization function arbitrary values of say $\psi_1=-C_1\delta(t).\&.\psi_2=-C_2\delta(t)$ and putting $C=C_1+C_2$ the general solution will be

$$f(t)=-CR_{Q-q,-q}(-1,0,t)+\int_0^t R_{Q-q,-q}(-1,0,t-\tau)(H(\tau))d\tau .$$

It is to show that in this case of side charging the effect of continuing past is not shown in the solution.

Example 3

Consider the same example as in 2 i.e.

$$ {}_cD_t^Q f(t) + {}_cD_t^q f(t) = H(t) $$

This may be written as

$$ {}_cd_t^Q f(t) + {}_cd_t^q f(t) = H(t) - \psi_1(f,Q,a,c,t) - \psi_2(f,q,a,c,t) = H(t) - \psi_{eq}(t) $$

The Laplace transform of un-initialized fractional derivative is

$$ \mathcal{L}\left\{ {}_cd_t^q f(t) \right\} = \mathcal{L}\left\{ {}_0d_t^q H(t-c) f(t) \right\} = e^{-cs} s^q \mathcal{L}\{ f(t+c) \}, $$

where $H(t-c)$ is unit step function, at time $t = c$. Thus, the Laplace transform of the equation yield:

$$ \mathcal{L}\{ f(t+c) \} = \frac{H(s) - \psi_{eq}(s)}{e^{-cs} s^q (s^{Q-q} + 1)} = e^{cs} G(s) \left[H(s) - \psi_{eq}(s) \right], $$

where $G(s) = \dfrac{1}{s^q (s^{Q-q} + 1)}$.

Consider two time intervals. First interval $a_1 = c_1 = 0$ and second interval $a_2 = c_2 = 1$. First interval is used as initial period (charging) for the second interval. We take excitation (forcing function) as $H_1(t) = (H(t) - H(t-1))$, i.e. square window of height unity from 0-1. Also $f(t) = 0, at\, t < 0$ which also tells $\psi_1(f,Q,0,0,t) = \psi_2(f,q,0,0,t) = 0$.

In this first interval solution is:

$$ \mathcal{L}\{ f_1(t) \} = \frac{1 - e^{-s}}{s^{q+1}(s^{Q-q} + 1)} = \left(\frac{1 - e^{-s}}{s} \right) G(s) $$

And

$$ f_1(t) = R_{Q-q,-q-1}(-1,0,t) - H(t-1) R_{Q-q,-q-1}(-1,0,t-1) \qquad t > 0 $$

For $0 < t < 1$ then

$$ f_1(t) = R_{Q-q,-q-1}(-1,0,t) $$

In this second interval consider the forcing function $H(t) = 0$, and the interval 1 is the initialization interval. The equation for this interval is

$$ {}_1d_t^Q f_2(t) + {}_1d_t^q f_2(t) = -\psi_{eq}(t)\,. $$

Using Laplace we get:

$$ \mathcal{L}\{f_2(t+1)\} = \frac{-\psi_{eq}(s)}{e^{-s}s^q(s^{Q-q}+1)} $$

The ψ_{eq} is taken based on the historic forcing $H_1(t)$ as was in the first interval.

$$ \psi_{eq} = -H_1(t) = -(H(t) - H(t-1))\,\cdot $$

Hence, $\psi_{eq}(s) = -\frac{1-e^{-s}}{s}$. Substituting these we get

$$ e^{-s}\mathcal{L}\{f_2(t+1)\} = \frac{1-e^{-s}}{s^{q+1}(s^{Q-q}+1)} = \frac{1-e^{-s}}{s}G(s) = \mathcal{L}\{f_1(t)\} $$

Applying the Laplace shift rule we have $\mathcal{L}\{f_2(t)H(t-1)\} = \mathcal{L}\{f_1(t)\}$.

Thus for $t > 1$ we have important result $f_2(t) = f_1(t), t > 1$.

7.8 The Fundamental Fractional Order Differential Equation

The problem to be addressed in this section is the solution of the fundamental linear fractional order differential equation (7.33). This system is considered to be fundamental because its solution is the fundamental time response, whose combination provides the solution of more complicated systems, analogous to the exponential function for the integer order differential equation. The fundamental equation is:

$$ {}_cD_t^q x(t) \equiv {}_cd_t^q x(t) + \psi(x,q,a,c,t) = -ax(t) + bu(t)\,,\ q > 0 \tag{7.33} $$

Where the left side should be interpreted to be q-th derivative of $x(t)$ starting at time c and continuing until time t. Here it will be assumed for clarity that the problem starts at $c = 0$. We also assume temporarily in this section that the initialization function $\psi(x,q,a,c,t) = 0$. Thus we will be concerned only with the forced response for time being. Rewriting the equation (7.33) with these assumptions we get:

$$ {}_0d_t^q x(t) = -ax(t) + bu(t)\,,\ q > 0 \tag{7.34} $$

Ignoring the initialization terms the equation (7.27) can be Laplace transformed as:

$$s^q X(s) = -aX(s) + bU(s) \tag{7.35}$$

This equation can be rearranged to obtain the system transfer function:

$$\frac{X(s)}{U(s)} = G(s) = \frac{b}{s^q + a} \tag{7.36}$$

This is the transfer function of the fundamental linear fractional order differential equation. As such, it contains the fundamental 'fractional-pole' and is fundamental building block for more complicated fractional order systems. As the constant b in equation (7.36) is a constant multiplier, it can be assumed without loss of generality to be unity. Typically transfer functions are used to study various properties of a particular system. Specifically they can be inverse Laplace transformed to obtain the system impulse response, which can then be used with convolution approaches to the problem. Generally, if $U(s)$ is given then the product $G(s)U(s)$ can be expanded using partial fractions, and forced response obtained by inverse transforming each separately. To accomplish these tasks, it is necessary to obtain the inverse of equation (7.36), which is the impulse response, or generalized exponential function, of the fundamental fractional-order system.

7.8.1 The Generalized Impulse Response Function

Although the Laplace transform of equation (7.36) is not contained in standard Laplace transform table, the following transform pair is available:

$$\frac{1}{s^q} = \mathcal{L}\left\{\frac{t^{q-1}}{\Gamma(q)}\right\}, \; q > 0 \tag{7.37}$$

If we expand the right hand side of equation (7.36) in describing powers of s, we can then inverse transform the series term by term and obtain the generalized impulse response. Then expanding the equation (7.36) about $s = \infty$, we get:

$$G(s) = \frac{1}{s^q + a} = \frac{1}{s^q} - \frac{a}{s^{2q}} + \frac{a^2}{s^{3q}} - \ldots = \frac{1}{s^q}\sum_{n=0}^{\infty}\frac{(-a)^n}{s^{nq}} \tag{7.38}$$

This series can be inverse transformed term-by-term using (7.37). The result is:

$$\mathcal{L}^{-1}\{G(s)\} = \mathcal{L}^{-1}\left\{\frac{1}{s^q} - \frac{a}{s^{2q}} + \frac{a^2}{s^{3q}} - \ldots\right\} = \frac{t^{q-1}}{\Gamma(q)} - \frac{at^{2q-1}}{\Gamma(2q)} + \frac{a^2 t^{3q-1}}{\Gamma(3q)} - \ldots \tag{7.39}$$

The right side (7.39) now be collected into summation and used as definition of the generalized impulse response function:

$$F_q[-a,t] \equiv t^{q-1}\sum_{n=0}^{\infty}\frac{(-a)^n t^{nq}}{\Gamma(\{n+1\}q)}, \; q>0 \tag{7.40}$$

We thus have important Laplace identity as

$$\mathcal{L}\{F_q[a,t]\} = \frac{1}{s^q - a}, \; q>0.$$

Also the F function is generalization of exponential function for $q=1$,

$$F_1[-a,t] = \sum_{n=0}^{\infty}\frac{(-at)^n}{\Gamma(n+1)} = e^{-at}$$

This generalization is the basis for solution of most linear fractional order differential equations.

Here we have established the F- function as the impulse response of the fundamental linear differential equations. This function is important because it will allow the creation of concise theory for fractional order systems, which is a generalization of that of integer order systems, and where the F- function generalizes and replaces the usual exponential function.

Several other variants of this F- function is possible for the solution of the fundamental equation (7.34), like Miller-Ross function R function and G-function listed in Chapter 2.

From equations (7.39) and (7.40) we obtained impulse response as:

$$g(t) = \mathcal{L}^{-1}\{G(s)\} = t^{q-1}\sum_{n=0}^{\infty}\frac{(-a)^n t^{nq}}{\Gamma(nq+q)} \equiv F_q[-a,t], \; q>0 \tag{7.41}$$

The function $F_q[a,t]$ is closely related to the Mittag-Leffler function $E_q\left[at^q\right]$, where one-parameter Mittag-Leffler function in series form is defined as:

$$E_q[x] \equiv \sum_{n=0}^{\infty}\frac{x^n}{\Gamma(nq+1)}, \; q>0 \tag{7.42}$$

Letting $x=-at^q$, this becomes:

$$E_q\left[-at^q\right] \equiv \sum_{n=0}^{\infty}\frac{(-a)^n t^{nq}}{\Gamma(nq+1)}, \; F_q[-a,t] \tag{7.43}$$

which is similar to $F_q[-a,t]$ expressed in (7.41), but not same as equation (7.41).

The Laplace transform of this Mittag-Leffler function (7.43) can also be obtained via term-by-term transform of the series expansion as:

$$\mathcal{L}\left\{E_q\left[-at^q\right]\right\}=\mathcal{L}\left\{\frac{1}{\Gamma(1)}-\frac{at^q}{\Gamma(1+q)}+\frac{a^2t^{2q}}{\Gamma(1+2q)}+\ldots\right\}=\frac{1}{s}-\frac{a}{s^{q+1}}+\frac{a^2}{s^{2q+1}}+\ldots \quad (7.44)$$

or compactly:

$$\mathcal{L}\left\{E_q\left[-at^q\right]\right\}=\frac{1}{s}\left[1-\frac{1}{s^q}+\frac{a^2}{s^{2q}}+\ldots\right]=\frac{1}{s}\sum_{n=0}^{\infty}\left(\frac{-a}{s^q}\right)^n=\frac{1}{s}\sum_{n=0}^{\infty}\frac{(-a)^n}{s^{nq}} \quad (7.45)$$

It should be noted that the summation expression (7.45) is similar to equation (7.42). Using equation (7.42) the equation (7.45) can be rewritten as:

$$\mathcal{L}\left\{E_q\left[-at^q\right]\right\}=\frac{1}{s}\left[\frac{s^q}{s^q+a}\right] \quad (7.46)$$

or equivalently:

$$\mathcal{L}\left\{E_q\left[-at^q\right]\right\}=\frac{1}{s}\left[s^q\mathcal{L}\left\{F_q[-a,t]\right\}\right] \quad (7.47)$$

Thus a general result can be expressed as:

$$\mathcal{L}\left\{E_q\left[\pm at^q\right]\right\}=\frac{s^{q-1}}{s^q\mp a},\ q>0 \quad (7.48)$$

From equation (7.47) the relation between Mittag-Leffler (E_q) function and Robotnov-Hartley function (F_q) is:

$$_0d_t^{q-1}F_q\left[a,t\right]=E_q\left[at^q\right] \quad (7.49)$$

This demonstration was to show a method of obtaining solution of the 'fundamental fractional order (linear) differential equation' equation (7.34) by use of Robotnov-Hartley function. This $F-$ function was utilized by Robotnov to study hereditary integrals in solid mechanics. Solution to (7.34) may be obtained in terms of Miller-Ross function and its Laplace transform. Miller-Ross function is fractional derivative of the exponential function defined as

$$E_t(v,a)\equiv {}_0d_t^{-v}\exp(at),$$

whose Laplace transform is

$$\mathcal{L}\left\{E_t(v,a)\right\}=\frac{s^{-v}}{s-a}.$$

Also recent developments to study diffusion and fractional kinetic equations, use more complicated Fox functions in solving of fractional order differential equations.

Extending this developed technique we obtain the solution of the equation (7.34) for a unit step input excitation. This can be obtained via Laplace transforms by transforming the input function $u(t)$, its Laplace is $1/s$, which must be multiplied by the transfer function $G(s)$ equation (7.36) where $b=1$ is taken. We get:

$$X(s)=\frac{1}{s}\left[\frac{1}{s^q+a}\right] \tag{7.50}$$

Manipulating the equation (7.50) we obtain:

$$X(s)=\frac{1/a}{s}\left[\frac{a}{s^q+a}\right]=\frac{1/a}{s}\left[1-\frac{s^q}{s^q+a}\right]=\frac{1/a}{s}-\frac{s^q/a}{s(s^q+a)} \tag{7.51}$$

Using expression (7.48) this equation (7.51) is inverse Laplace transformed, the result is the step response of the system:

$$x(t)=\mathcal{L}^{-1}\left\{\frac{1}{s(s^q+a)}\right\}=\frac{1}{a}\left(1-E_q\left[-at^q\right]\right)=\frac{1}{a}\left(H(t)-E_q\left[-at^q\right]\right) \tag{7.52}$$

Heaviside step is $H(t)=\begin{cases}1 & t\geq 0\\ 0 & t<0\end{cases}$

Taking integer derivative of equation (7.52) gives $F_q\left[-a,t\right]$, the impulse response:

$$F_q\left[-a,t\right]=\frac{1}{a}\frac{d}{dt}\left(H(t)-E_q\left[-at^q\right]\right) \tag{7.53}$$

Referring to equation (7.48), and multiplying the Laplace transforms there by s^{-q} gives:

$$s^{-q}\mathcal{L}\left\{E_q\left[-at^q\right]\right\}=s^{-1}\mathcal{L}\left\{F_q\left[-a,t\right]\right\}=\frac{1}{s(s^q+a)},$$

which is equation (7.50). Inverse transforming this equation using expression (7.52) shows that step response of equation (7.34) with $b=1$, is also $q-$th fractional integral of the Mittag-Leffler function that is:

$$x(t)=\mathcal{L}^{-1}\left\{\frac{1}{s(s^q+a)}\right\}=\frac{1}{a}\left[H(t)-E_q(-at^q)\right]={}_0d_t^{-q}E_q\left[-at^q\right] \tag{7.54}$$

Few more interesting Laplace pairs can be obtained by taking the q-th derivative of the F-function. Taking the un-initialized fractional derivative $({}_0d_t^q)$, and in Laplace domain multiplying by s^q gives:

$$\mathcal{L}^{-1}\left\{\frac{s^q}{s^q+a}\right\} = {}_0d_t^q F_q\left[-a,t\right] \tag{7.55}$$

This expression (7.55) is also the integer derivative of the Mittag-Leffler function:

$$\mathcal{L}^{-1}\left\{\frac{s^q}{s^q+a}\right\} = {}_0d_t^q F_q\left[-a,t\right] = {}_0d_t^q E_q\left[-at^q\right] \tag{7.56}$$

The expression (7.56) can also be re-written as:

$$\mathcal{L}^{-1}\left\{\frac{s^q}{s^q+a}\right\} = \mathcal{L}^{-1}\left\{1-\frac{a}{s^q+a}\right\} = \delta(t) - aF_q\left[-a,t\right] \tag{7.57}$$

Observing the expressions (7.55) and (7.56) we can write the following:

$${}_0d_t^q F_q\left[-a,t\right] = \delta(t) - aF_q\left[-a,t\right] \tag{7.58}$$

This $F_q\left[a.t\right]$ function is generalization of exponential function of the integer order calculus where it demonstrates the 'eigenfunction property'; i.e. returning the same function upon q-th fractional differentiation. Also the above equation (7.58) shows that $x(t) = F_q\left[-a,t\right]$, Robotnov-Hartley function is impulse response of fundamental fractional order differential equation given by equation (7.34), for $u(t) = \delta(t)$ and $b = 1$.

7.9 Problem of Scalar Initialization

Linear scalar constant coefficient fractional differential equation for $0 < q < 1$ is given as ${}_0d_t^q x(t) = Ax(t) + Bu(t)$, assuming causality of the system and the system was at rest meaning $x(t) = 0$ for all $t < a$. The q-th derivative of $x(t)$ starts at time a, and continues at time t. This means

$${}_0d_t^q x(t) \equiv \frac{d}{dt}\left[\frac{1}{\Gamma(1-q)}\int_a^t \frac{x(\tau)}{(t-\tau)^q}d\tau\right].$$

Let the initialization of the scalar starts at some time c after a. Choosing $c = 0$, the differential equation can be written as ${}_0d_t^q x(t) + {}_ad_t^q x(t) = Ax(t) + Bu(t)$. Here the time axis is broken in two parts, one from $a \to 0$ and the other from $0 \to t$. Here as stated $c=0$, is chosen. The term ${}_a d_0^q x(t)$ above represents this initialization response due to behavior of system before $t=c=0$. It should be noticed that the past history of the particular variable that is fractionally differentiated must be known as long as the system has been operated to obtain the correct initialization response. The equation then can be expressed as in terms of initialized fractional derivative as:

$${}_c D_t^q x(t) = {}_0 d_t^q x(t) + \psi(q,x,a,0,t) = Ax(t) + Bu(t),$$

where for terminal charging for $t>0$, the initialization function is:

$$\psi(q,x,a,0,t) \equiv {}_a d_0^q x(t) = \frac{d}{dt}\left[\frac{1}{\Gamma(1-q)}\int_a^0 \frac{x(\tau)}{(t-\tau)^q} d\tau\right],$$

This initialization function is described in Chapter 6, called the initialization function for fractional derivative. Using the Laplace Transforms of the initialized expression we obtain the following:

$$s^q X(s) + \psi(q,x,a,0,s) = AX(s) + BU(s),$$

Where

$$\psi(q,x,a,0,s) = \mathcal{L}\left\{\frac{d}{dt}\left[\frac{1}{\Gamma(1-q)}\int_a^0 \frac{x(\tau)}{(t-\tau)^q} d\tau\right]\right\},$$

is the Laplace transform of the initialized function. Rearranging the above expression we get:

$$X(s) = \frac{B}{s^q - A}U(s) - \frac{1}{s^q - A}\psi(q,x,a,0,s).$$

Taking inverse Laplace transform of this expression we obtain time response:

$$x(t) = \int_0^t F_q[A,\tau]Bu(t-\tau)d\tau - \int_0^t F_q[A,\tau]\psi(q,x,a,0,t-\tau)d\tau.$$

Here F-function is the impulse response of fundamental linear differential equation (as explained in the previous sections), and is defined as

$$F_q[A,t] \equiv t^{q-1}\sum_{n=0}^{\infty}\frac{A^n t^{nq}}{\Gamma(nq+n)}.$$

The solution $x(t)$ represents any forced response due to $u(t)$, and the second term of the solution expresses the initialized response of the system due to past history of $x(t)$, before time $t=0$. Clearly for integer order systems $(q=1)$, this initialization term $\psi(1,x,a,0,s)$ equals a constant and for fractional order systems, this term $\psi(1,x,a,0,s)$ is time varying expression into the future. That implies the past history of $x(t)$ has appearance of a time dependent forcing term into the 'infinite' future.

Let us choose the history as a constant meaning $x(t)=k$, for $-\infty < t \le 0$. Then the initialization function becomes as limit $a \to \infty$. Described as follows:

$$\psi(q,x,-\infty,0,s) = \mathcal{L}\left\{ \lim_{a\to-\infty} \frac{d}{dt}\left[\frac{1}{\Gamma(1-q)} \int_a^0 \frac{k}{(t-\tau)^q} d\tau \right]\right\}$$

$$= \mathcal{L}\left\{ \lim_{a\to-\infty} \frac{d}{dt}\left[\frac{k}{\Gamma(1-q)} \frac{-(t-\tau)^{1-q}}{(1-q)} \right]_{\tau=a}^{\tau=0}\right\}$$

$$= \mathcal{L}\left\{ \lim_{a\to-\infty} \frac{d}{dt}\left[\frac{k}{\Gamma(1-q)} \left(\frac{(t-a)^{1-q}}{(1-q)} - \frac{t^{1-q}}{(1-q)} \right) \right]\right\}$$

$$= \mathcal{L}\left\{ \lim_{a\to-\infty} \frac{k}{\Gamma(1-q)} \left(\frac{1}{(t-a)^q} - \frac{1}{t^q} \right)\right\} = -\mathcal{L}\left\{ \frac{k}{\Gamma(1-q)} \frac{1}{t^q} \right\} = -ks^{q-1}$$

Inserting this back into Laplace expression we obtain:

$$X(s) = \frac{B}{s^q - A} U(s) + \frac{ks^q}{s(s^q - A)}$$

Inverting this we obtain solution:

$$x(t) = B\int_0^t F_q\left[A,\tau\right] u(t-\tau) d\tau + kE_q\left[At^q\right].$$

The first term is convolution of input $u(t)$ with Impulse response (F-function) and second term is initialization function response and in this particular case of history (k) is Mittag-Leffler function. For integer order systems $q=1$ the initialized Laplace term (second term of $X(s)$), in case of the constant is k/s.

7.10 Problem of Vector Initialization

Vector space representation is useful for 'systems of fractional differential equations'. Once the minimal basis value q is chosen the vector representation can be expressed as: ${}_cD_t^q \overline{x}(t) = \mathrm{A}\overline{x}(t) + \mathrm{B}\overline{u}(t)$, the vector $\overline{x}(t)$ is given for $a \le t \le c$ or initialization vector $\bar{\psi}(q,\bar{x},a,c,t)$ given for $t > c$, and the output vector is $\bar{y}(t) = \mathrm{C}\bar{x}(t) + \mathrm{D}\bar{u}(t)$.

The parameters (Matrices) $\mathrm{A}, \mathrm{B}, \mathrm{C}, \mathrm{D}$ are usual state-space representation for systems of differential equations as for integer order systems, representing state matrix, input and output matrix and feed through matrix. The vector fractional order differential equation expressed above can be written as:

$$ {}_c d_t^q \bar{x}(t) + \bar{\psi}(q,\bar{x},a,0,t) = \mathrm{A}\bar{x}(t) + \mathrm{B}\bar{u}(t) $$

where

$$ \bar{\psi}(q,\bar{x},a,0,t) = \left[\psi(q,x_1,a,0,t), \psi(q,x_2,a,0,t) \ldots \psi(q,x_n,a,0,t)\right]^T = \left[\psi_1, \psi_2, \ldots, \psi_n\right]^T $$

At this point it is important to notice that the fractional dynamic variable in the system of vector space equations are not states in true sense of the name state-space control terminology. In usual integer order system theory, the set of state of the system, known at any given point in time along with the system equations, are sufficient to predict the response of the system both forward or backward in time.

The collection of constant numbers $\bar{x}(t_0)$ at time t_0 specify the complete 'state' of the system at that time. Therefore, the system will have unique time response given its 'initial-state'.

Fractional dynamic variables do not represent the state of system at any given time (alone) due to presence of the initialization function vector (history function), carrying information about the history of the elements of the system. Consequently, as the initialization function vector is generally present, the set of elements of the vector $\bar{x}(t)$, evaluated at any point in time, does not specify the entire 'state' of the system. Thus in fractional system setting, the ability to predict the future response of a system requires the set of fractional differential equations along with the initialization function sets. The initialization problem of the vector fractional differential equations can be solved as solved for the scalar case (in the previous section), by using Laplace transformation:

The Laplace transformed vector equation is:

$$ s^q \bar{X}(s) + \bar{\psi}(s) = \mathrm{A}\bar{X}(s) + \mathrm{B}\bar{U}(s) $$

and the output equation is transformed as:

$$ \bar{Y}(s) = \mathrm{C}\bar{X}(s) + \mathrm{D}\bar{U}(s) . $$

After performing matrix algebraic manipulations we obtain:

$$ \left(\mathrm{I}s^q - \mathrm{A}\right)\bar{X}(s) = \mathrm{B}\bar{U}(s) - \bar{\psi}(s) , $$

where I is $n \times n$ Identity Matrix matching dimensions of state matrix A.

Therefore the Laplace vector solution is thus obtained as:

$$\bar{X}(s) = \left(\mathrm{I}s^q - \mathrm{A}\right)^{-1} \mathrm{B}\bar{U}(s) - \left(\mathrm{I}s^q - \mathrm{A}\right)^{-1} \bar{\psi}(s),$$

inserting this in output expression to have output Laplace solution as:

$$\bar{Y}(s) = \left\{\mathrm{C}\left(\mathrm{I}s^q - \mathrm{A}\right)^{-1} \mathrm{B} + \mathrm{D}\right\}\bar{U}(s) - \mathrm{C}\left(\mathrm{I}s^q - \mathrm{A}\right)^{-1} \bar{\psi}(s),$$

Inverting this one we obtain time response as:

$$\bar{y}(t) = \mathrm{C}\int_0^t F_q\left[\mathrm{A}, \tau\right]\left\{\mathrm{B}\bar{u}(t-\tau) - \bar{\psi}(t-\tau)\right\} d\tau + \mathrm{D}\bar{u}(t)$$

or equivalently can be expressed as

$$\bar{y}(t) = \mathrm{C}\int_0^t F_q\left[\mathrm{A}, t-\tau\right]\left\{\mathrm{B}\bar{u}(\tau) - \bar{\psi}(\tau)\right\} d\tau + \mathrm{D}\bar{u}(t)$$

The above solution requires the use of matrix F-function, which can be obtained by the use of its series expansion. Matrix F-function is defined as

$$F_q\left[\mathrm{A}, t\right] \equiv t^{q-1} \sum_{n=0}^{\infty} \frac{\mathrm{A}^n t^{nq}}{\Gamma(nq+n)} \text{ for } q > 0 \text{ where A is } n \times n \text{ system matrix.}$$

Consider the Laplace transformed equation

$$\bar{Y}(s) = \left\{\mathrm{C}(\mathrm{I}s^q - \mathrm{A})^{-1}\mathrm{B} + \mathrm{D}\right\}\bar{U}(s) - \mathrm{C}(\mathrm{I}s^q - \mathrm{A})^{-1}\bar{\psi}(s),$$

derived earlier in this section. In this case the system transfer matrix is

$$\bar{G}(s) = \left\{\mathrm{C}(\mathrm{I}s^q - \mathrm{A})^{-1}\mathrm{B} + \mathrm{D}\right\},$$

is system representation of multivariate system. In this representation thus the system description is obtained as

$$\bar{Y}(s) = \bar{G}(s)\bar{U}(s) - \mathrm{C}(\mathrm{I}s^q - \mathrm{A})^{-1}\bar{\psi}(s).$$

The $\mathrm{A}, \mathrm{B}, \mathrm{C}, \mathrm{D}$ are usual state-space representation matrix namely system state matrix, input matrix, output matrix and feed-through matrix, as per multivariate state-vector representation, of modern control science.

7.11 Laplace Transform $s \to w$ Plane for Fractional Controls Stability

As it is difficult to visualize multiple Riemann sheets, it is useful to perform 'conformal transformation' into another complex plane $s \to w$ plane. System dynamics are described by singularity (pole) location of the transformed transfer function in new w-plane. The transfer function

$$G(s) = \frac{1}{s^q + 1}, (0 < q < 1)$$

does not have any singularity (pole) in anywhere in $s-$plane (primary Riemann sheet), but after crossing branch-cut the secondary Riemann sheet will contain the singularity (pole). For $q = 1/2$ the denominator $(s^{\frac{1}{2}} + 1)$ goes to zero at $s = 1 + \mathrm{j}0 = \exp(\pm \mathrm{j}2\pi)$, which is underneath the $s-$plane negative real axis $\exp(\pm \mathrm{j}\pi)$, entering 'secondary Riemann sheet', giving pole of the function there.

Here in this section basic control theory knowledge is required, in Laplace domain. To understand the behavior and stability property of the fundamental fractional differential equation

$${}_0d_t^q x(t) = -ax(t) + bu(t),$$

it is necessary to analyze the pole-location of the system transfer function i.e.

$$G(s) = \frac{X(s)}{U(s)} = \frac{b}{s^q + a}.$$

For classical control theory $q = 1$, that is integer order system, the pole-location are studied in the complex Laplace plane (s-plane). The stability boundary in the s-plane is the imaginary axis. Any pole lying to the right of the imaginary axis represents an unstable time response. Examining

$$G(s) = \frac{b}{s^q + a},$$

however indicates that the poles of $G(s)$ must now be evaluated in s^q plane. Rather than dealing with fractional powers of s, the analysis is carried out in $w = s^q$ and then the pole location properties will be studied in the new complex w-plane.

To simplify the discussion we limit the fractional order as $0 < q \le 1$. The mapping from $s \to w$ is as follows:

$$w = \rho e^{j\phi} = \alpha + j\beta \text{ and } s = re^{j\theta}.$$

Then defining $w = s^q = \left(re^{j\theta}\right)^q = r^q e^{jq\theta} = \rho e^{j\phi}$, implying $\rho = r^q$ and $\varphi = q\theta$.

With this equation, it is possible to map either lines of constant radius, or lines of constant angle from the s-plane into w-plane. Of particular interest is the image of the s-plane stability boundary (s-plane imaginary axis), that is $s = re^{\pm j\frac{\pi}{2}}$, maps as $w = r^q e^{\pm jq\frac{\pi}{2}}$. This is pair of lines at $\varphi = \pm q\pi/2$. Thus, the right half plane (RHP) of the s-plane maps into a 'wedge' in the w-plane of angle less than $\pm 90q$ degree that is RHP of s-plane maps into $|\varphi| < q\pi/2$.

As an example for semi-differential equation of order ½, the RHP of s-plane maps into the wedge bounded by $|\varphi| < \pi/4$. A half-order system with its w-plane poles in the wedge that is $|\varphi| < \pi/4$, would be unstable and corresponding F-function (impulse response) would grow with out bound.

It is also important to consider the mapping of the negative real axis of s-plane $s = re^{\pm j\pi}$, the mapping is $w = r^q e^{\pm jq\pi}$. Therefore, the entire 'primary-sheet' of the s-plane maps into a w-plane wedge of angle less than $\pm 180q$ degree; while all the 'secondary-sheet' s-plane maps into the remainder of the w-plane. For half-order system the negative real axis of s-plane maps into w-plane lines at ± 90 degree. Figure 7.6 gives pole location properties of w-plane

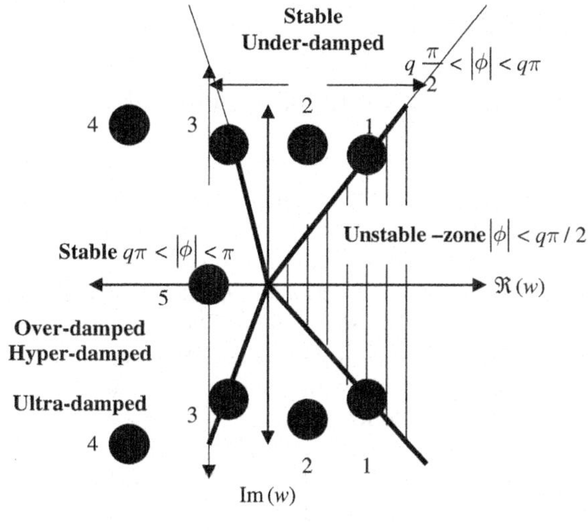

Fig. 7.6 w-Plane stability zones and various pole locations

Table 7.1 Properties of poles in *w*-plane

Pole Number (Fig-7.6)	Region in *w*-plane	Region in *s*-plane	Property of time-response
1	$\lvert\phi\rvert < q\frac{\pi}{2}$	$\Re e(s) > 0$ **(RHP)**	**UNSTABLE**
2	$q\frac{\pi}{2} < \lvert\phi\rvert < q\pi$	$\Re e(s) < 0$ **(LHP)**	**STABLE UNDERDAMPED OSCILLATORY**
3	$\phi = q\pi$	**NEGATIVE REAL AXIS** $s = r\exp(\pm j\pi)$	**STABLE OVERDAMPED**
4	$q\pi < \lvert\phi\rvert < \pi$	**SECONDARY RIEMANN SHEET** $\lvert\theta\rvert > \pi$	**STABLE HYPERDAMPED**
5	$\lvert\phi\rvert = \pi$	**SECONDARY RIEMAN SHEET**	**STABLE ULTRADAMPED**

F-function time response $F_q[a,t]$, depends on both q, and on parameter a, which is the pole location of the system transfer function $G(s) = \dfrac{b}{s^q + a}$. For a fixed value of q the angle φ of the parameter a, as measured from the positive real axis of *w*-plane, determines the type of response to expect. This pole location is depicted in the Table 7.1 with properties of time response.

All usual control system analysis Nyquist, root-locus plot concerning poles or eigenvalues can be used in the *w*-plane, remembering the stability boundary is now the image *s*-plane imaginary axis. Hyper-damped and ultra-damped system will be taken up again in Chapter 10.

7.12 Rational Approximations of Fractional Laplace Operator

There are lots of approaches to get rational approximation of fractional order Laplace operator. The s^q keeps on appearing for fractional order systems and controls. The rational function approximation gives direction to realize this fractional order Laplace operator in circuit impedance and admittance forms. This section gives insight into simple method of approximating the fractional Laplace operator. It is well known that for interpolation or evaluation purposes, rational functions are (sometimes) superior to polynomial fit, because of their ability to model the function with poles and zeros. In other words for evaluation purposes rational approximations frequently converge much more rapidly than power series expansion (PSE), and have wider domain of convergence in the complex plane. These approximations can be viewed in the Laplace domain, as rational approximations of the fractional order operators. Furthermore, these approximations exhibit a common feature, which we observe in all good rational approximations: they have poles and zeros interlaced on the negative real axis of the s plane, and the distance between successive poles and zeros decrease as the approximations is improved by increasing the degree of the numerator and denominator polynomial.

There are number of methods to evaluate rational approximations of the fractional Laplace operator, they are: General continued fractional expansion (CFE) method, Carlson's method, Matsuba's method, Oustaloup's method, Chareff's method, S C Dutta Roy's method, Wang's method and Jones method.

7.12.1 Finding Arbitrary Root of Polynomial Approximation for Fractional Laplace Operator

Here, in this section the Carlson's approach using Newton's approximation to find arbitrary root of polynomial is described. If n th root of a is to be found so that $x^n = a$, or $x = \sqrt[n]{a}$ in functional form, for even or odd root is given as:

$$x_i = x_{i-1}\frac{(n-1)(x_{i-1})^n + (n+1)a}{(n+1)(x_{i-1})^n + (n-1)a}$$

The method is derived from a regular Newton process for iterative approximation of the n th root. The starting point of the method is

The above expression can be used for finding root of a function instead of a number. The above expression gives $G(s) = a$, then for a function.

Say we want to approximate semi-integral Laplace operator $(s)^{-1/2}$ to be realized by this method then take $a = (1/s)$, then square root of $G(s) = (1/s)$ is given iteratively as, with $n = 2; x_0 = H_0(s) = 1$.

We get $x_1 = H_1(s)$ as the first rational approximate for initial function $H_0(s) = 1$ as:

$$H_1(s) = H_0(s)\frac{(n-1)\left(H_0(s)\right)^n + (n+1)G(s)}{(n+1)\left(H_0(s)\right)^n + (n-1)G(s)} = (1)\frac{(1)(1) + (3)\left(\frac{1}{s}\right)}{(3)(1) + (1)\left(\frac{1}{s}\right)} = \frac{s+3}{3s+1}$$

The second approximate, $x_2 = H_2(s)$ is obtained by putting the obtained $H_1(s)$ as:

$$H_2(s) = \left(\frac{s+3}{3s+1}\right)\left[\frac{(1)\left(\frac{s+3}{3s+1}\right)^2 + (3)\left(\frac{1}{s}\right)}{(3)\left(\frac{s+3}{3s+1}\right)^2 + (1)\left(\frac{1}{s}\right)}\right] = \frac{(s+3)\left(s^3 + 33s^2 + 27s + 3\right)}{(3s+1)\left(3s^2 + 27s^2 + 33s + 1\right)}$$

The second approximate for semi integration Laplace operator is:

$$H_2(s) = \frac{s^4 + 36s^3 + 126s^2 + 84s + 9}{3s^4 + 84s^3 + 126s^2 + 36s + 1} \approx \left(\frac{1}{s}\right)^{1/2}$$

For realizing fractional capacitor of transfer function $\sqrt[3]{(1/s)}$, we again use the above iteration expression with $H_0(s) = 1$. The driving point impedance for this is just a shunt resistance of 1 Ohm. The first approximate expression is thus:

$$H_1(s) = \frac{s+2}{2s+1}$$

The driving point impedance of this first approximate is formed by first a shunt resistance of 2 Ohm, then series resistance of 0.6667 Ohm, followed by shunt impedance (capacitor) of $(1/0.750s)$ Ohm. From this first approximation the second approximation is given as:

$$H_2(s) = \frac{s^5 + 24s^4 + 80s^3 + 92s^2 + 42s + 4}{4s^5 + 42s^4 + 92s^3 + 80s^2 + 24s + 1}$$

This second approximate is by ladder realization, and has driven point impedance with: shunt resistance of 4 Ohm, followed by capacitor as series element of impedance $(1/3.375s)$ Ohm, then a shunt resistance of 1.66 Ohm, followed by series element of series combination of resistance of value 0.3113 Ohm with capacitor of impedance $(1/0.2851s)$ Ohm, this again shuntcd by capacitor of value $(1/0.2851s)$ Ohm, followed by series element a resistor of 0.7072 Ohm, shunted by capacitor of value $(1/2.463s)$ Ohm, followed by series element of resistor value of 0.2657 Ohm, and then shunted by resistance of 0.1195 Ohm and lastly terminated (shunted) by inductor impedance of value $0.1003s$ Ohm. A long chain of series shunts with combination of R, L, C ladder.

Physically these approximations steps are to be viewed as impedance (rather immitance), realized by R, L, C network. The fundamentals of circuit synthesis to realize these fractional Laplace operators in network forms, and present day Hybrid Micro Circuit fabrication techniques will give electrical circuit components for fractional order circuit applications, in future.

7.12.2 *Fractional Power Pole and Fractional Power Zero to Approximate Fractional Laplace Operator*

A Fractional slope on the log-log Bode plot has been observed in characterizing a certain type of physical phenomena and is called the fractal system or the fractional power pole, FPP (or Fractional Power Zero FPZ). In order to represent and study its dynamical behavior, a method of singularity function is discussed here, which consists of cascaded branches of a number of poles-zero (negative real) pairs. Moreover, the distribution spectrum of the system can also be easily calculated and its accuracy depends on a prescribed error specified in the beginning.

This method would thereafter be used widely in approximating fractional order transfer functions for the Lead Compensators as well as the $PI^{\alpha}D^{\beta}$ controllers.

7.12.2.1 Singularity Structure for a Single Fractional Power Pole (FPP)

A single FPP system can be modeled in the frequency domain as follows:

$$H(s)=\frac{1}{\left(1+\frac{s}{p_T}\right)^m},\tag{7.59}$$

where $1/p_T$ is the relaxation time constant and $0<m<1$.

As shown in Figure 7.7, the line with slope of $(-20m)\mathrm{dB/decade}$ is approximated by a number of zigzag lines connected together with alternate slopes of $0\mathrm{dB/decade}$ and $-20\mathrm{dB/decade}$. The high-and the low-frequency properties of the magnitude of the transfer function with a single-fractional power pole suggest that the lowest and the highest singularities of the transfer function approximation must be poles.

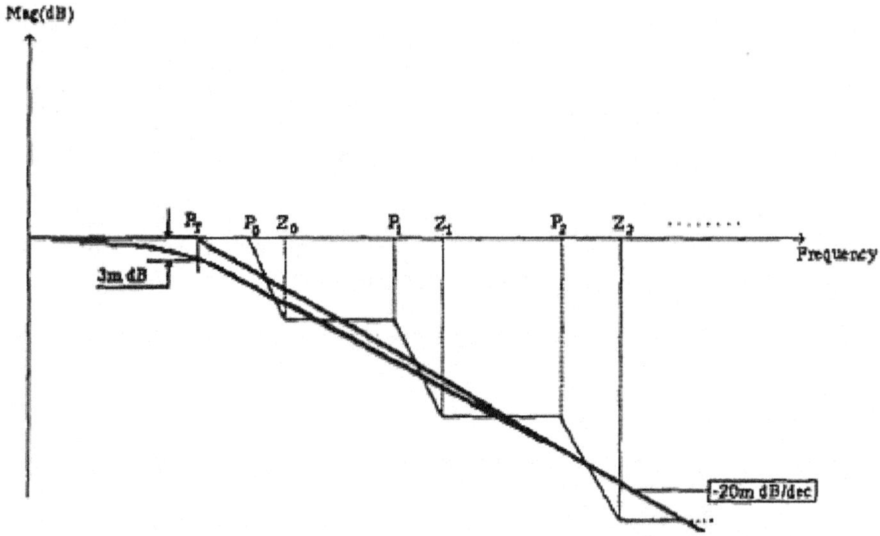

Fig. 7.7 Bode plots of an FPP with slope of -20mdB/decade and its approximation as zigzag straight lines with individual slopes of -20dB/decade and 0dB/dec

Thus, we can rewrite (7.59) as represented by its pole-zero pair as:

$$H(s)=\frac{1}{\left(1+\frac{s}{p_T}\right)^m}=\lim_{N\to\infty}\frac{\prod_{i=0}^{N-1}\left(1+\frac{s}{z_i}\right)}{\prod_{i=0}^{N}\left(1+\frac{s}{p_i}\right)},\tag{7.60}$$

where $(N+1)$ is the total number of the singularities.

Hence, for a finite range of frequency, (7.60) can be truncated to a finite number N, and the approximation becomes:

$$H(s)=\frac{1}{\left(1+\frac{s}{p_T}\right)^m}\approx\frac{\prod_{i=0}^{N-1}\left(1+\frac{s}{z_i}\right)}{\prod_{i=0}^{N}\left(1+\frac{s}{p_i}\right)} \tag{7.61}$$

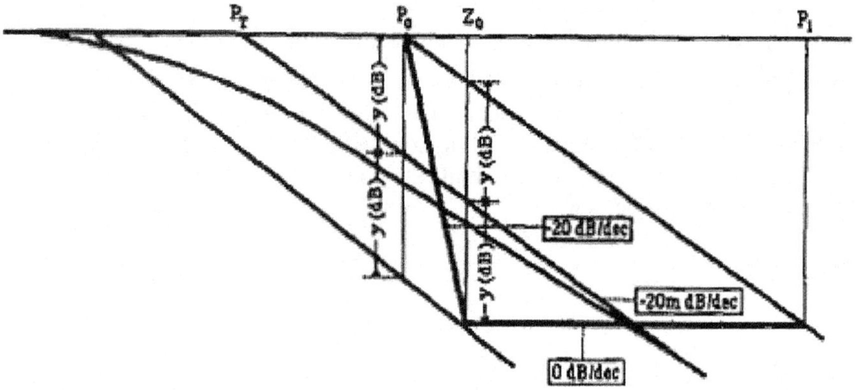

Fig. 7.8 Choosing the singularities for approximation by assuming a constant error between the -20 dB/decade line and the zigzag lines

7.12.2.2 Geometrical Derivation of Recurring Relationship of Fractional Power Pole for Fractional Integration

Refer Figure 7.9 for expansion of fractional power pole with N stage interlaced poles and zeros (Figure 7.8). In this Figure 7.9 only two pairs of pole zeros are shown.

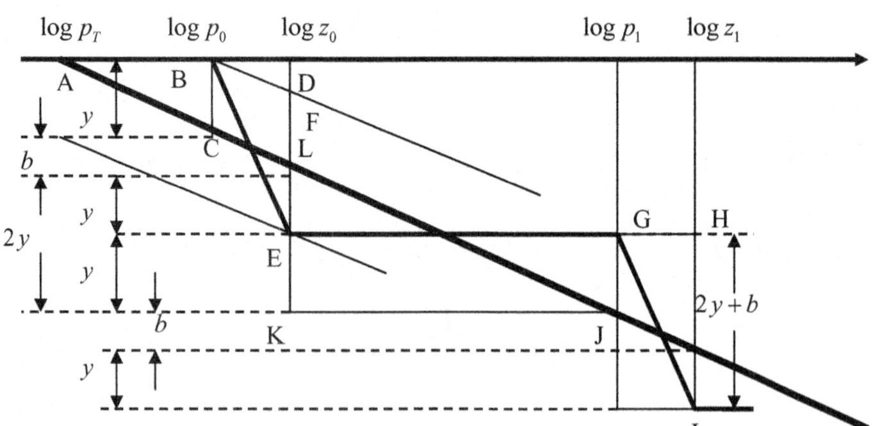

Fig. 7.9 Showing expanded view of shaping of fractinal pole by series of poles and zeros, within y dB error

From the fractional power pole p_T, move in the X-axis to the point when the inclined line of slope $-20m$ is y dB away ($BC = y$), here we get the first integer order pole p_0. That is point B in the X-Axis. From this pole we draw a slope of -20 and stop at point D on the X-axis where the segment BE is at y dB away from the original $20m$ line, that is, ($EL = y$). At this point we place a zero, z_0 on the X-axis that is, at point D. This zero makes the line straight EG, with total slope zero, and we stop at G, which is y dB away from the original $-20m$ line ($EK = y$). The process gets repeated, so that the fractional power pole line of slope $-20m$ dB/decade is y dB away from the zigzag approximated expansion.

The relation between the fractional pole (relaxation time constant) p_T and the first pole is found from triangle ABC.

$$\frac{BC}{AB} = 20m$$

$BC = y$, and $AB = \log p_0 - \log p_T$

Thus

$$\frac{y}{\log p_0 - \log p_T} = 20m$$

Giving $\log p_0 = \log p_T + 20m$. Therefore, $p_0 = p_T 10^{\left[\frac{y}{20m}\right]}$

The relation between p_0 and z_0 is obtained from triangle BDE and triangle BDF. From triangle BDF we get,

$$\frac{DF}{BD} = \frac{b}{\log z_0 - \log p_0} = 20m,$$

giving $b = 20m(\log p_0 - \log z_0)$ .From triangle BDE, we have

$$\frac{ED}{BD} = \frac{2y + b}{\log z_0 - \log p_0} = 20$$

giving $\dfrac{2y}{\log z_0 - \log p_0} + \dfrac{b}{\log z_0 - \log p_0} = 20,$

Here substituting the value of $b = 20m(\log z_0 - \log p_0)$, we get:

$$\frac{2y}{\log z_0 - \log p_0} + 20m = 20\,,$$

giving:

$$\log z_0 - \log p_0 = \frac{2y}{20(1-m)}$$

or

$$z_0 = p_0 10^{\left[\frac{y}{10(1-m)}\right]}$$

The relation between p_1 and z_0 is obtained from triangle JKL, where:

$$\frac{KL}{JK} = \frac{2y}{\log p_1 - \log z_0} = 20\,,$$

giving $p_1 = z_0 10^{\left[\frac{y}{10m}\right]}$.

The relation between z_1 and p_1 is obtained from the triangle GHI, observing the triangle GHI, which is equal to triangle BDE. Thus $GH = BD, HI = DE$.

$$\frac{HI}{GH} = \frac{2y+b}{\log z_1 - \log p_1} = 20\,,$$

gives: $z_1 = p_1 10^{\left[\frac{y}{10(1-m)}\right]}$.

7.12.2.3 Recursive Algorithm for Fractional Power Pole

A way to choose the singularities (the pole-zero pair) for the approximation is developed as follows. Let us assume that the maximum discrepancy or the error between the zigzag lines and the desired line is chosen to be (y)dB dB, as shown in Figures 7.7, 7.8 and 7.9. Then the poles and zeros of the singularity function can be obtained as follows:

The first pole at

$$p_0 = p_T 10^{[y/20m]}$$

The first zero, at

$$z_0 = p_o 10^{[y/10(1-m)]}$$

The second pole, at

$$p_1 = z_o 10^{[y/10m]}$$

The second zero at

$$z_1 = p_1 10^{[y/10(1-m)]}$$

The Nth zero,

$$z_{N-1} = p_{N-1} 10^{[y/10(1-m)]}$$

The $(N+1)^{th}$ pole,

$$p_N = z_{N-1} 10^{[y/10m]}$$

Where p_T is the comer frequency and is determined at a point of $(-3m)$dB from the original transfer function as shown in Figure 7.7, p_0 is the first singularity and is determined by the specified error, y decibel, and p_N is the last singularity and is determined by N.

Now, let

$$a = 10^{[y/10(1-m)]}$$
$$b = 10^{[y/10m]}$$

Then

$$ab = 10^{[y/10m(1-m)]}$$

Therefore, we can obtain the distribution of these poles and zeros as:

$$\frac{z_o}{p_o} = \frac{z_1}{p_1} = \dots\dots\dots\dots\dots\dots \frac{z_{N-1}}{p_{N-1}} = a$$
$$\frac{p_1}{z_o} = \frac{p_2}{z_1} = \dots\dots\dots\dots\dots\dots \frac{p_N}{z_{N-1}} = b \quad (7.62)$$

In addition, the location ratio of a pole to a previous pole is equal to the location ratio of a zero to a previous zero and it is equal to ab; i.e.:

$$\frac{z_1}{z_0} = \frac{z_2}{z_1} = \dots\dots\dots\dots \frac{z_{N-1}}{z_{N-2}} = ab$$
$$\frac{p_1}{p_0} = \frac{p_2}{p_1} = \dots\dots\dots\dots \frac{p_N}{p_{N-1}} = ab \quad (7.63)$$

From the above relation we can also generate these poles and zeros from the first pole p_0 using the following algorithm as:

$$\begin{aligned} p_i &= (ab)^i p_0 \\ z_i &= (ab)^i a p_0 \end{aligned} \tag{7.64}$$

It is interesting to note that both p_i and z_i are in geometrical progression form with ratio equal to ab. The approximated transfer function can be written as follows:

$$H(s) = \frac{1}{\left(1+\frac{s}{p_T}\right)^m} \approx \frac{\prod_{i=0}^{N-1}\left(1+\frac{s}{z_i}\right)}{\prod_{i=0}^{N}\left(1+\frac{s}{p_i}\right)} = \frac{\prod_{i=0}^{N-1}\left(1+\frac{s}{(ab)^i a p_o}\right)}{\prod_{i=0}^{N}\left(1+\frac{s}{(ab)^i p_o}\right)} \tag{7.65}$$

To draw this rational approximation up to maximum frequency $\omega_{\max}$, which can be $100\omega_H$, we need N pairs of interlaced poles and zeros. In the logarithmic scale the frequency spread is $\log\omega_{\max} - \log p_0$. The ratio between subsequent poles is as given in (7.63) is

$$\frac{p_1}{p_0} = \frac{p_2}{p_1} = \ldots = \frac{p_N}{p_{N-1}} = ab\,,$$

Therefore, then the in the logarithmic scale the distance between successive poles are

$$(\log p_1 - \log p_0) = (\log p_2 - \log p_1) = \ldots = (\log p_N - \log p_{N-1}) = \log(ab)\,.$$

Therefore

$$N - 1 = \frac{\log\omega_{\max} - \log p_0}{\log p_N - \log p_{N-1}} = \frac{\log\left(\omega_{\max}/p_0\right)}{\log(ab)}$$

This can be approximated to nearest integer as:

$$N = \text{Integer}\left[\frac{\log\left(\omega_{\max}/p_0\right)}{\log(ab)}\right] + 1$$

7.12.2.4 Singularity Structure for a Single Fractional Power Zero (FPZ)

The singularity structure can be obtained on similar lines as that of an FPP, as:

$$H(s)=\left(1+\frac{s}{p_T}\right)^m \approx \frac{\prod_{i=0}^{N}\left(1+\frac{s}{z_i}\right)}{\prod_{i=0}^{N}\left(1+\frac{s}{p_i}\right)}=\frac{\prod_{i=0}^{N}\left(1+\frac{s}{(ab)^i z_0}\right)}{\prod_{i=0}^{N}\left(1+\frac{s}{(ab)^i a z_0}\right)}$$

Note that the structure has N zeroes and N poles.

Table 7.2 gives rational approximates for several fractional order Laplace operator for fractional integration realization.

Table 7.2 Fractional Operators with approximately 2db error from 0.01-100 rad/sec.

Laplace fractional operator	Rational Approximate
$\frac{1}{s^{0.1}}$	$\frac{220.4s^4+5004s^3+5038s^2 234.5s+0.4840}{s^5+359.8s^4+5742s^3+4247s^2 147.7s+0.2099}$
$\frac{1}{s^{0.2}}$	$\frac{60.95s^4+816.9s^3+582.8s^2+23.24s+0.04934}{s^5+134.0s^4+956.5s^3+383.5s^2+8.953s+0.01821}$
$\frac{1}{s^{0.3}}$	$\frac{23.76s^4+224.9s^3+129.1s^2 4.733s+0.01052}{s^5+64.51s^4+252.2s^3+63.61s^2+1.104s+0.002276}$
$\frac{1}{s^{0.4}}$	$\frac{25.00s^4+558.5s^3+664.2s^2+44.15s+0.1562}{s^5+125.6s^4+840.6s^3+317.2s^2+7.428s+0.02343}$
$\frac{1}{s^{0.5}}$	$\frac{15.97s^4+593.2s^3+1080s^2+135.4s+1}{s^5+134.3s^4+1072s^3+543.4s^2+20.10s+0.1259}$
$\frac{1}{s^{0.6}}$	$\frac{8.579s^4+255.6s^3+405.3s^2 35.93s+0.1696}{s^5+94.22s^4+472.9s^3+134.8s^2+2.639s+0.009882}$
$\frac{1}{s^{0.7}}$	$\frac{5.406s^4+177.6s^3+209.6s^2+9.197s+0.01450}{s^5+88.12s^4 279.2s^3+33.30s^2+1.927s+0.0002276}$
$\frac{1}{s^{0.8}}$	$\frac{5.235s^3+1453s^2+5306s+254.9}{s^4+658.1s^3+5700s^2+658.2s+1}$
$\frac{1}{s^{0.9}}$	$\frac{1.766s^2+38.27s+4.914}{s^3+36.15s^2+7.789s+0.01000}$

These approximate representations are utilized to have impedance on structures for realization of fractional capacitance and thus with operational amplifier circuit technique the fractional order integrator and fractional order differentiator is realized.

7.13 Realization of Constant Phase Element

Any rational polynomial is characterized by its poles and zeros. In the above sections the magnitude asymptotic plot is used to get approximation for fractional operator, by interlacing poles and zeros. In the above sections an algorithm is made where the Bode asymptotic magnitude plot is switched between $0\text{dB}/\text{decade}$ and $\pm 20\text{dB}/\text{decade}$, by alternate placement of real poles and zeros, such that looking at overall frequency range magnitude plot appears as rolling off at slope $\pm 20\alpha\mathrm{dB/decade}$, with α being the fractional order. In this section Bode phase plot is flattened, to achieve 'Constant Phase Element' (CPE).

Irrespective of the method used to obtain rational polynomial the observations are (a) poles and zeros are interlaced alternatively along the negative real axis of the s -plane, (b) ratio of consecutive poles and consecutive zeros is fixed that is $p_{i+1}/p_i = z_{i+1}/z_i = \text{constant}$, $\forall i$, $i = 1, 2, \ldots n-1$ for approximation by n^{th} order rational polynomial and (c) the distance between consecutive pole-zero pairs on $\log\omega$ axis is constant that is,

$$\sqrt{p_{i+1}z_{i+1}} / \sqrt{p_i z_i} = \text{constant.} \ \forall i, \ i = 1, 2, \ldots (n-1).$$

7.13.1 Asymptotic Bode Phase plot

The asymptotic magnitude plot is used to calculate the position of poles and zeros to achieve rational approximation, as described in previous sections. In this section, an algorithm is derived for calculation of poles and zeros using Bode asymptotic phase plot. The phase plot is nonlinear; it is plot of $\log\omega$ vs. $-\tan^{-1}(\omega/p)$ where p is the single pole for which the phase plot is drawn. For the first order pole the phase variation from low frequency to high frequency is 0^0 to -90^0 and at frequency equal to the pole value the phase value is -45^0. For asymptotic phase plot, the entire frequency range $(0,\infty)$ is divided in three parts, low frequency: the phase is constant at 0^0, medium frequency: the phase varies from 0^0 to -90^0 and high frequency: the phase is again constant or equal to -90^0. Hence the bode phase plot for a single pole (p) is approximated by a straight line asymptote in each of the three frequency region as given below,

$$\phi_p(\omega) = \begin{cases} 0^0 & : \ \omega \in (0, p/\lambda) \\ -\dfrac{45^0}{\log\lambda}\log\dfrac{\lambda\omega}{p} & : \ \omega \in (p/\lambda, \lambda p) \\ -90^0 & : \ \omega \in (\lambda p, \infty) \end{cases}$$

where λ is the factor deciding the boundaries between the frequency ranges, and hence the slope of the medium frequency asymptote. For different values of λ, the medium frequency asymptote gives better approximation to actual. Asymptotic plot for $\lambda = 10$ gives minimum RMS error, and gives best approximation. For asymptotic Bode phase plot of single zero all the points discussed are same except the sign of the phase angle which is opposite, i.e., $\phi_z(\omega) = -\phi_p(\omega)$.

7.13.2 Pole Zero Calculation for Constant Phase

The basic idea of getting constant phase element (CPE) is by slope cancellation of asymptotic phase plots for zeros and poles, as they have opposite sign but same value for the slope. The fundamental problem can be stated as: find out the poles and zeros of the rational polynomial such that the rational polynomial will have constant phase around $\phi_{req} = \alpha \times 90^0$, with an error less than ε, for all $\omega \in (\omega_l, \omega_h)$. Any $\phi_{req} \in (0^0, 90^0)$ or $\alpha \in (0,1)$ is possible as the middle frequency asymptote varies from 0^0 to 90^0. The basic idea of the solution is illustrated in Figure 7.10.

Pole p_1 is selected such that its asymptotic plot passes through (ω_l, ϕ_{req}) and calculated using the following

$$\phi_p(\omega) = \begin{cases} 0^0 & : \quad \omega \in (0, p/10) \\ -45^0 \log(\omega / p) - 45^0 & : \quad \omega \in (p/10, 10p), \\ -90^0 & : \quad \omega \in (10p, \infty) \end{cases}$$

Then z_1 and subsequent poles and zeros are selected so as to keep the asymptotic plot constant at, about ϕ_{req}.

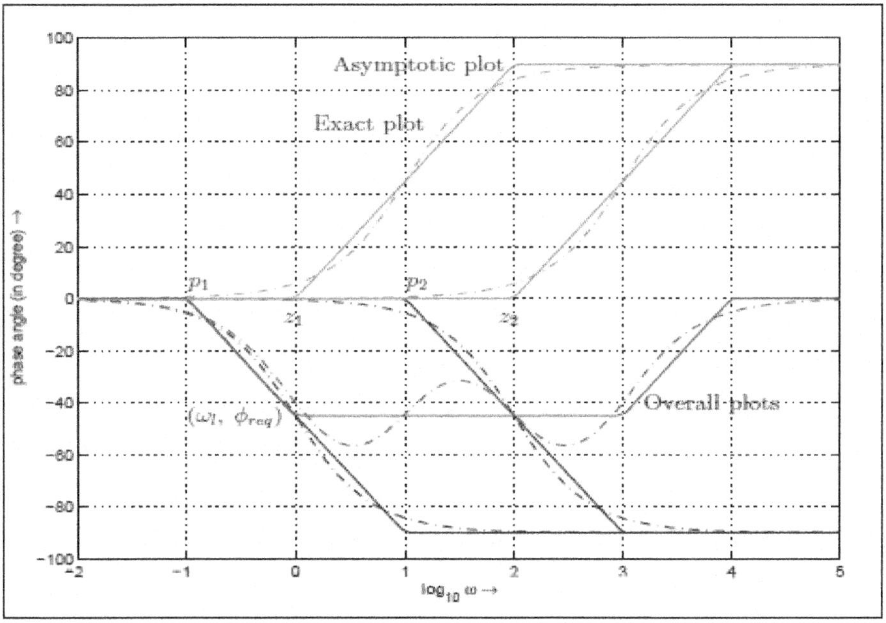

Fig. 7.10 Asymptotic and exact phase plots illustrating the basic idea

Figure 7.10 is drawn for $\phi_{req} = -45^0$, $\omega_l = 1\text{radian}/\sec$, $\omega_h = 1000\text{radians}/\sec$.

With above selection of poles and zeros the asymptotic phase plot is a straight line at ϕ_{req} but actual phase plot is oscillating about asymptotic phase plot, with RMS error (of order 8.5^0), apart from average value of phase angle 46.31^0 is also different from ϕ_{req}. This is achieved by 2 poles-zero pairs over three decades of frequency. These problems can be rectified by increasing the pole-zero pair in the desired frequency band. This is done by pushing $z_1, p_2, z_3, \ldots$ closer toward left, to achieve this design parameter μ is introduced, and interlaced poles and zeros sequence is obtained as:

$$p_1 = 10^{\left[\frac{\phi_{req}+45\log\omega_l}{45}+1\right]},\ z_1 = 10\omega_l,\ p_2 = 10^{[\log(p_1)+2-\mu]},\ z_2 = 10^{[\log(z_1)+2-\mu]} \text{ till } p_n \geq \omega_h$$

The changes because of introduction of $\mu = 0.5$ can be observed as the asymptotic plot is no more straighter; instead changes so as to counter the variation in the actual phase plot. The main purpose to introduce μ is for RMS error reduction (which comes down to 4.85^0), almost half of earlier case. But with this some undesirable effects are observed, such as, the average value is not close to -45^0, it is -57.34^0 with shrinkage in frequency range on either side.

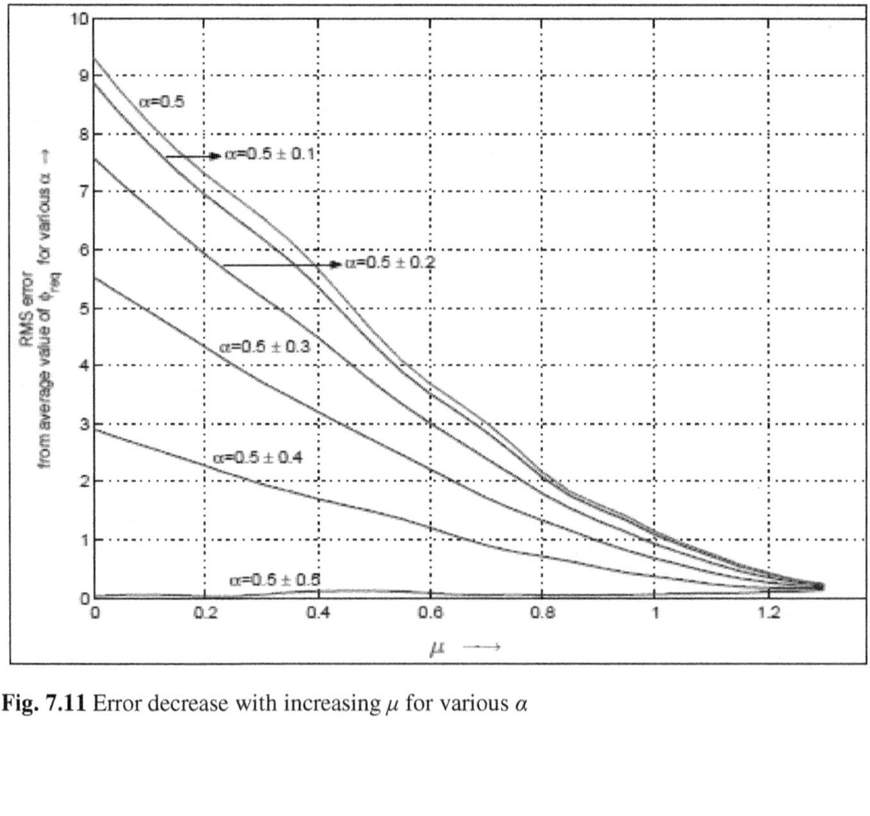

Fig. 7.11 Error decrease with increasing μ for various α

The effect of μ is noted in figure 7.11, where RMS error is plotted with various fractional orders α .The observations are:

1. RMS error decreases as μ increases. The range of μ is $(0,2)$. At $\mu = 2$ the $(i+1)^{th}$ entity is equal to i^{th} entity, hence the termination condition $p_n \geq \omega_h$ will never reach. RMS error monotonically decreases with increase of μ, irrespective of fractional order α; however, the error depends on the angle, maximum at $\alpha = 0.5$, and decreasing on either side.
2. The mean value of the phase increases with increasing μ, though the ripples in phase value are less, but oscillating about a value greater than ϕ_{req} .
3. Relative increase in the mean value of phase (ϕ_{avg}) is same irrespective of value of α that is,

$$\left. \frac{\phi_{avg}\big|_{\mu=1.2}}{\phi_{avg}\big|_{\mu=0.2}} \right|_{\alpha=0.1} = \left. \frac{\phi_{avg}\big|_{\mu=1.2}}{\phi_{avg}\big|_{\mu=0.2}} \right|_{\alpha=0.8} .$$

The increase in the average phase ϕ_{avg} value from the required phase ϕ_{req} value is measured by defining a parameter,

$$\varphi = \frac{\phi_{avg}}{\phi_{req}} = \frac{\phi_{avg}}{90 \times \alpha},$$

it is a factor by which the mean value of phase increases with increasing μ.
4. Frequency band of constant phase shrinks on both the sides, with increasing μ, for constant number of pole-zero pairs.

The inclusion of μ in the pole-zero calculations is fulfilling its primary purpose that is reducing the RMS error in the phase plot. In lieu of this the point 2 and 4 above appears as problem. The point 3 above gives a solution to increase the mean value of required phase. By adjusting the required phase value, ϕ_{req} initially that is, using $\phi_{new} = \phi_{req} / \varphi$, instead of ϕ_{req} for pole-zero calculation will neutralize the increase in mean value, because of μ. The factor φ is the second design parameter.

The problem regarding the frequency band shrinking is tackled by designing the rational function on wider frequency band $\left(\omega_l^*, \omega_h^*\right) = \left(\omega_l / 10^{\epsilon}, 10^{\delta} \omega_h\right)$, such that $\left(\omega_l, \omega_h\right)$ falls well within almost at the centre of the chosen wider frequency band. Select μ and φ such that on $\left(\omega_l, \omega_h\right)$, $\phi_{avg} = \phi_{req}$ and error $e_{RMS} \leq \varepsilon / \sqrt{2}$, where ε is the desired error. This will increase number of pole-zero pairs as wider frequency is used. Hence curtail the frequency over band such that phase remains within $\phi_{req} \pm \varepsilon$ in (ω_l, ω_h). This will reduce the pole-zero pairs and increase value of the lowest entity, hence implementation will be easier. If ω_l' is the frequency at which the phase plot is entering $\phi_{req} \pm \varepsilon$ band, then ω_l' to ω_l is the frequency band that can be removed. The frequency band from ω_l^* to ω_l' is the band required to reach phase angle 0^0 to $\phi_{req} \pm \varepsilon$, see Figure 7.12 and 7.13 drawn for $\phi_{req} = 45^0$ that is, half order differentiator ($\alpha = +0.5$). The algorithm is as follows in the subsequent sections.

7.13.3 Calculation for Pole-Zero Position of Fractional Order Impedance

The impedance of the Fractional Order is realized by using the lumped *RC* network or using the operational amplifier with n passive components. The realizable values for different fractions of the impedances are obtained if the design is firmed around the operating bandwidth and gain of the impedance. This sets the values of phase slope for a given fraction and needs a proper choice of the poles and zeros. These calculations are obtained by using the following algorithm. This particular algorithm gives n^{th} order rational transfer function

$$G(s) = \prod_{i=1}^{n} \left([s + z_i]/[s + p_i]\right),$$

such that

$$\angle G(j\omega) = \phi_{req} \pm \varepsilon, \forall \omega \in (\omega_l, \omega_h).$$

7.13.4 Algorithm

The positioning of the pole-zero is incorporated with the help of following algorithm

Given: α or $\phi_{req} = \alpha \times 90^0$ on (ω_l, ω_h) with error $\leq \varepsilon$.

1. Initialization

   a. $e^*_{RMS} = \kappa \dfrac{\varepsilon}{\sqrt{2}}$, where $\kappa \in (0,1)$ is a factor selected to improve the robustness of the algorithm with respect to error. Nominal value $\kappa = 0.9$

   b. Expand frequency band $(\omega_l^*, \omega_h^*) = (\omega_l / 10^{\epsilon}, 10^{\delta}\omega_h)$ nominal values $\epsilon = 3$ and $\delta = 2$ gives frequency band as $(\omega_l / 10^3, 10^2 \omega_h)$.

   c. Select μ, from Figure 7.11, slightly less than the value corresponding to e_{RMS}. Nominal value $\mu = 0.9 \times \mu(e^*_{RMS})$

   d. Set, $\varphi = 1, \phi_{new} = \phi_{req} / \varphi$

2. Hunt for μ:
   a. Calculate, p_i and z_i using,

   $$p_1 = 10^{\left[\frac{\phi_{req}+45\log\omega_l}{45}+1\right]},\ z_1 = 10\omega_l,\ p_2 = 10^{[\log(p_1)+2-\mu]},$$

   $z_2 = 10^{[\log(z_1)+2-\mu]}$ ...till $p_n \geq \omega_h$ ϕ_{new}, μ, and $\left(\omega_l^*, \omega_h^*\right)$.

   b. Calculate, ϕ_{avg} on $\left(\omega_l, \omega_h\right)$, $\varphi = \phi_{avg} / \phi_{req}$ and $\phi_{new} = \phi_{new} / \varphi$
      i. If $\varphi = 1$, go to 3c
      ii. Else, go to 3a
   c. Calculate, e_{RMS}, RMS error in phase value from ϕ_{avg} on $\left(\omega_l, \omega_h\right)$.
      i. If $e_{RMS} > e_{RMS}^*$, set, $\mu = \mu + 0.05$, go to 3a
      ii. Else go to 4
3. Frequency band Adjustment
   a. Find out the frequency ω_l' at which the first maximum of phase plot occurs after it enters into, $\phi_{req} \pm \varepsilon$ band.
   b. Find out the frequency, ω_h' at which the last minimum of phase plot occurs before it leaves the, $\phi_{req} \pm \varepsilon$ band.
   c. Calculate, $\hat{\omega}_l = \dfrac{\omega_l \omega_l^*}{\omega_l'}$ and $\hat{\omega}_h = \dfrac{\omega_h \omega_h^*}{\omega_h'}$

   i. If $(\omega_l^*, \omega_h^*) \neq (\hat{\omega}_l, \hat{\omega}_h)$, set $(\omega_l^*, \omega_h^*) = (\hat{\omega}_l, \hat{\omega}_h)$

   $$e_{RMS} = \frac{\varepsilon}{\sqrt{2}},\ \text{go to 3.}$$

   ii. Else go to 5.
4. Verification:

   Calculate the poles and zeros finally using

   $$p_1 = 10^{\left[\frac{\phi_{req}+45\log\omega_l}{45}+1\right]},\ z_1 = 10\omega_l,\ p_2 = 10^{[\log(p_1)+2-\mu]},$$

   $z_2 = 10^{[\log(z_1)+2-\mu]}$ ...till $p_n \geq \omega_h$, μ obtained in 3c in latest iteration, ϕ_{new} obtained in 3b in latest iteration and $(\hat{\omega}_l, \hat{\omega}_h)$ obtained in 4c.

   a. If phase plot is not satisfactory go to 2, and adjust, $\kappa = 0.9\kappa, \in = \in + 1, \delta = \delta + 1$, that is the wider band (ω_l^*, ω_h^*).
   b. Else stop.

The κ introduced in step-2 ensures that tightening of frequency band in step-4 will keep $e_{RMS} \leq \varepsilon / \sqrt{2}$. The frequency range can be expanded to any number of decades on either side of the desired band, but it will unnecessarily increase the computational burden, however, practically 2 or 3 decades are sufficiently good to realize a CPE. The phase plot permanently enters into $\phi_{req} \pm \varepsilon$ band at ω_l', then by pushing back the frequency range just by the frequency interval required to bring phase plot from 0^0 to well within $\phi_{req} \pm \varepsilon$ band in step-4, the number of pole-zero pairs is reduced. The frequency overhang between ω_l' and ω_l has to be removed in that. Sometimes the number of pole-zeros may not be very convenient to implement, by relaxing or tightening the step change in μ the number of pole-zero pairs can be reduced or increased, respectively. Also the initial guess in step-2c should be on lower side, and then this algorithm will pull it up to the optimum value. The convergence of this algorithm depends mainly on the step of improvement in μ. Number of pole-zero pairs depends on the permissible error and frequency band of interest. Generally three poles-zero pair per decade gives a constant phase plot within $\pm 1^0$ error, but again depends on the value of α.

7.13.5 Design and Performance of Fractional Order Impedance

Design example for

$$s^{0.5} \approx G(s) = \prod_{i=1}^{6} \left[(s+z_i)/(s+p_i) \right],$$

for $\phi_{req} = \angle G(j\omega) = 45^0 \pm 1^0$

1. Given: $\alpha = +0.5$ or $\phi_{req} = 45^0$ on $(\omega_l, \omega_h) = (100, 1000)\text{radian/sec}$ with error $\varepsilon = 1^0$.
2. Initialization $e^*_{RMS} = 0.5 \times \frac{1}{\sqrt{2}} = 0.3536$, corresponding $\mu = 1.1$ (Figure 7.11), to start with and $(\omega_l^*, \omega_h^*) = (\omega_l/10^3, 10^2\omega_h)$.
3. Finding μ: about 10 iterations are required to get final value of μ; pole-zeros are calculated in each iteration and φ and e_{RMS} are computed to make decisions. The complete procedure is done using Mat lab® program.

Finally using $\mu = 1.15$ and frequency range (100, 10000) rad/sec, the rational approximation for $s^{+0.5}$ is obtained, and the bode plot for the same is shown in Figure 7.12. On the final plot with various parameters are $e_{RMS} = 0.3361$, $\phi_{avg} = 45^0$, minimum value in (ω_l, ω_h) is 44.055^0 and maximum value is 45.675^0, with six pole zero pairs are used for achieving this (Figure 7.13). Table 7.3 gives rational approximated with six poles and zeros for half order differentiator with above algorithm.

Table 7.3 Gives the values of six pole zero pairs for rational approximation of CPE for half order differentiator with one degree accuracy and five decades of frequency

i	1	2	3	4	5	6
z_i	2.2537	15.955	112.95	799.65	5661.1	40078
p_i	6.0406	42.764	302.75	2143.3	15173	107420

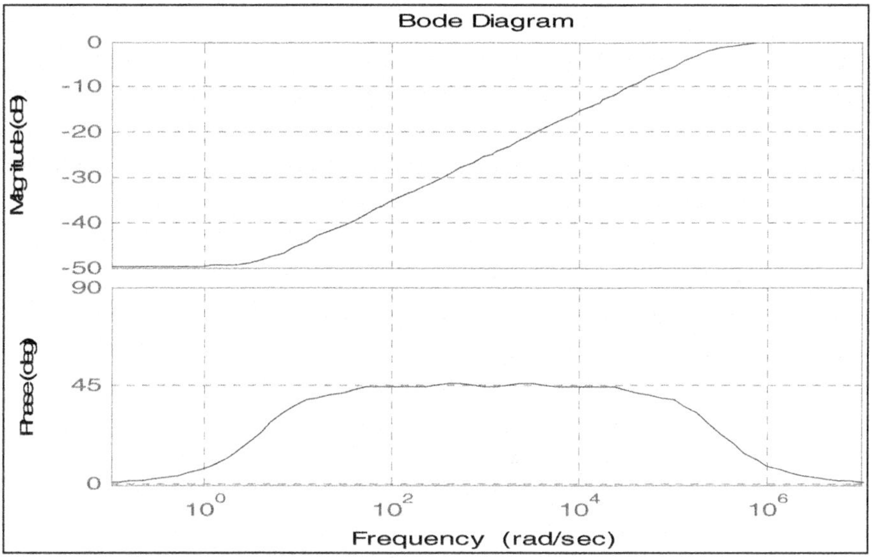

Fig. 7.12 Bode Plot for α=+0.5

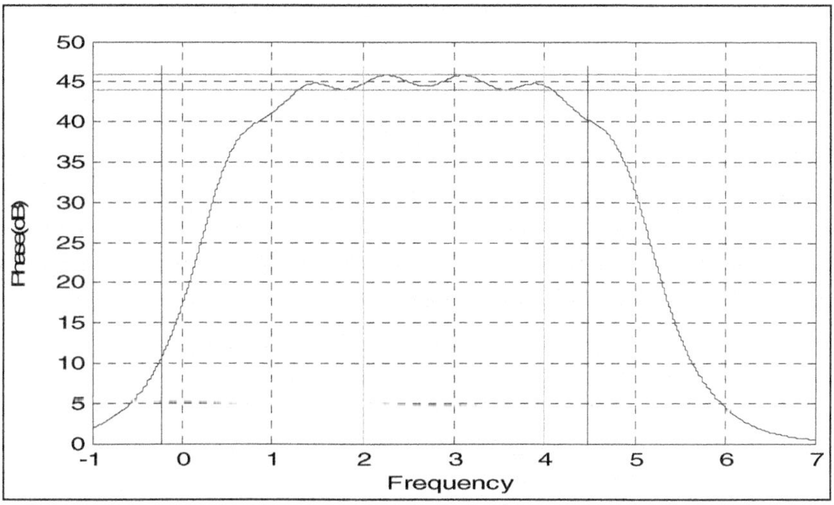

Fig. 7.13 Phase plot for $\phi_{req} = +45^0$ on (100, 10000) radian/sec, with $\varepsilon \leq 1^0$

7.14 Laplace Transform and Charaterization of Type of Fractional Derivative

The meaning of type of fractional derivative is Caputo type and Riemann-Liouvelli type of fractional derivative definition. We have generalized the two definitions in Chapter 1 with type parameter β. The generalized Laplace transform in integer order theory is:

$$\mathcal{L}\left\{\frac{d^n}{dx^n} f(x)\right\} = s^n F(s) - \sum_{k=0}^{n-1} s^k f^{(n-1-k)}(0^+),$$

where $n \in \mathbb{N}$ natural number. Following this we have generalized the above for fractional order α, for Riemann-Liouvelli derivative as:

$$\begin{aligned}\mathcal{L}\left\{{}_0D_x^{\alpha} f(x)\right\} &= \mathcal{L}\left\{\frac{d^n}{dx^n}\, {}_0I_x^{n-\alpha} f(x)\right\} \\ &= s^n \mathcal{L}\left\{{}_0I_x^{n-\alpha} f(x)\right\} - \sum_{k=0}^{n-1} s^k \frac{d^{n-1-k}}{dx^{n-1-k}}\left[{}_0I_x^{n-\alpha} f(0^+)\right] \\ &= s^{\alpha} F(s) - \sum_{k=0}^{n-1} s^k\, {}_0D_x^{\alpha-1-k} f(0^+)\end{aligned}$$

Here $n-1 < \alpha < n$. For Caputo derivative the Laplace can be derived using the above method as:

$$\mathcal{L}\left\{{}_0^CD_x^{\alpha} f(x)\right\} = s^{\alpha} F(s) - \sum_{k=0}^{n-1} s^{\alpha-1-k} f^{(k)}(0^+)$$

The essential difference between the two is that RL requires

$${}_0D_x^{\alpha-1-k} f(0^+), k = 0,1,2...(n-1)$$

number fractional initial states, the Caputo requires $f^{(k)}(0^+)$, $k = 0,1,2,...(n-1)$ integer order initial states. One may generalize the above two as the fractional derivative of solution $g(x)$ of integral equation

$${}_0I_x^{\alpha} g(x) = \frac{1}{\Gamma(\alpha)} \int_0^x (x-u)^{\alpha-1} g(u) du = f(x)$$

in case it exists. Let us denote the solution by ${}_0^*D_x^{\alpha} f(x)$ then following assertions that are equivalent for $n-1 \le \alpha \le n$

a) $s^{\alpha}F(s)=G(s)$

b) $g(x)={}_0^{*}D_x^{\alpha}f(x)$

c) $\frac{d^k\left[{}_0I_x^{n-\alpha}f(x)\right]}{dx^k}$ are equal to ZERO at $x=0$ for $k=0,1,2,...(n-1)$ together with ${}_0D_x^{\alpha}f(x)=g(x)$, the Riemann-Liouvelli derivative.

d) $f^{(k)}(x)$ are equal to ZERO at $x=0$ for $k=0,1,2,3,...(n-1)$, together with ${}_0^CD_x^{\alpha}f(x)=g(x)$, the Caputo derivative.

With type parameter $0\le\beta\le 1$ we have generalized definition in Laplace transform, for fractional order derivative of order $0<\alpha<1$ as:

$$\mathcal{L}\left\{{}_a^{\beta}D_t^{\alpha}f(t)\right\}=s^{\alpha}F(s)-s^{\beta(\alpha-1)}\left[{}_0D_t^{(1-\beta)(\alpha-1)}f(0^+)\right]$$

where ${}_0D_t^{(1-\beta)(\alpha-1)}f(t)$ is Riemann-Liouvelli fractional derivative type $\beta=0$.

For, fractional order $0<\alpha<1$, and for any type, β we define fractional derivative as:

$$ {}_a^{\beta}D_x^{\alpha}f(x)={}_aI_x^{\beta(1-\alpha)}\frac{d}{dx}\,{}_aI_x^{(1-\beta)(1-\alpha)}f(x)$$

As an example consider the differential equation of order $0<\alpha<1$, with type $0\le\beta\le 1$, and with C as a constant. The given initial condition is

$${}_0I_t^{(1-\beta)(1-\alpha)}f(0^+)=f_0\text{, a constant.}$$

$${}_0^{\beta}D_t^{\alpha}f(t)=C$$

Using generalized Laplace identity as obtained above we write the Laplace transformed equation, recognizing that ${}_0I_t^{(1-\beta)(1-\alpha)}\equiv{}_0D_t^{(1-\beta)(\alpha-1)}$; as follows.

$$s^{\alpha}F(s)-s^{\beta(\alpha-1)}f_0=\frac{C}{s}.$$

Giving $F(s)=\frac{C}{s^{\alpha+1}}+\frac{f_0}{s^{\alpha+\beta(1-\alpha)}}$.

Inverting this we get:

$$f(t)=\frac{Ct^{\alpha}}{\Gamma(\alpha+1)}+\frac{f_0 t^{(1-\beta)(\alpha-1)}}{\Gamma[(1-\beta)(1-\alpha)+1]}$$

For RL type with $\beta=0$, we have solution

$$f(t)=\frac{Ct^{\alpha}}{\Gamma(\alpha+1)}+\frac{f_0 t^{(\alpha-1)}}{\Gamma(2-\alpha)}$$

For Caputo type with $\beta=1$, we have solution

$$f(t)=\frac{Ct^{\alpha}}{\Gamma(\alpha+1)}+f_0$$

Consider another type of generalized fractional relaxation equation as:

$${}_0^{\beta}D_t^{\alpha} f(t)=-Cf(t)$$

With given, initial condition as $[{}_0I_t^{(1-\beta)(1-\alpha)} f(t)]_{t=0^+}=f_0$ with C as 'relaxation-constant'. With $0<\alpha<1$ and $0\le\beta\le 1$. Using the generalized Laplace identity as in above example we get:

$$F(s)=\frac{s^{\beta(\alpha-1)} f_0}{s^{\alpha}+C}$$

To get Laplace inverse re-write the above expression as:

$$F(s)=\frac{s^{\alpha-\gamma}}{C+s^{\alpha}}=s^{-\gamma}\frac{1}{Cs^{-\alpha}+1}=\sum_{k=0}^{\infty}(-C)^k s^{-\alpha k-\gamma},$$

where $\gamma=\alpha+\beta(1-\alpha)$.

Using $\mathcal{L}^{-1}\left\{x^{\alpha-1}/\Gamma(\alpha)\right\}=s^{-\alpha}$, and applying this term by term we obtain the response as:

$$f(t)=t^{\gamma-1}\sum_{k=0}^{\infty}\frac{(-Ct^{\alpha})^k}{\Gamma(\alpha k+\gamma)}=f_0 t^{(1-\beta)(\alpha-1)}E_{\alpha,\alpha+\beta(1-\alpha)}(-Ct^{\alpha})$$

Here, we have used the definition of 'two parameter Mittag-Leffler function'

$$E_{a,b}(x)=\sum_{k=0}^{\infty}\frac{x^k}{\Gamma(ak+b)},$$

for all $a > 0$ and $b \in \mathbb{C}$. This is an entire function of order $1/a$, and completely monotone for $0 < a \le 1$ and $b \ge a$.

1) For $C = 0$ the result is

$f(t) = \dfrac{f_0 t^{(1-\beta)(\alpha-1)}}{\Gamma[(1-\beta)(\alpha-1)+1]}$, because $E_{a,b}(0) = \dfrac{1}{\Gamma(b)}$. In this case if; $\beta = 0$ that is, RL case, the solution is $f(t) = f_0 t^{\alpha-1}/\Gamma(\alpha)$.

Take the ${}_0D_t^{\alpha}\left[t^{\alpha-1}\right]$, which is $\Gamma(\alpha)t^{-1}/\Gamma(0)$ zero. For Caputo case with $\beta = 1$ the solution is $f(t) = f_0$, revealing that Caputo fractional derivative of constant is zero.

2) For $\beta = 1$ Caputo derivative type, $f(t) = f_0 E_{\alpha}(-Ct^{\alpha})$, where $E_{\alpha}(x) = E_{\alpha,1}(x)$, denotes the ordinary 'one parameter Mittag-Leffler function'. For RL case with $\beta = 0$, the solution is $f(t) = f_0 t^{\alpha-1} E_{\alpha,\alpha}(-Ct^{\alpha})$.

These examples make us to think about the concept of 'stationarity' in fresh way, in fractional calculus domain.

Let us consider a generalized Fractional Diffusion equation represented as for $0 < \alpha < 1$.

$${}_0^{\beta}D_t^{\alpha} f(r,t) = \mathbb{D}\Delta f(r,t)$$

Given integral initial condition as:

$$I_{0^+}^{(1-\beta)(1-\alpha)} f(r,0^+) = f_{r0} = f_0\delta(r).$$

The Laplacian operator is $\Delta \equiv \partial^2/\partial r^2$, the Fourier of this is $\Im\{\Delta\} = -k^2$; using this and, with $\Im\{f_0\delta(r)\} = f_0$ taking Laplace-Fourier of the given fractional diffusion equation we obtain:

$$F(k,s) = \frac{s^{\beta(\alpha-1)} f_0}{\mathbb{D}k^2 + s^{\alpha}}$$

Taking Laplace inverse we get decay of a mode k, for this fractional diffusion as:

$$f(k,t) = f_0 t^{(1-\beta)(1-\alpha)} E_{\alpha,\alpha+\beta(1-\alpha)}(-\mathbb{D}k^2 t^{\alpha})$$

For $k = 0$, indicates that solution of the fractional diffusion equation cannot be a probability density function 'pdf' except when $\beta = 1$ (the Caputo case). For other cases where $\beta \neq 1$ the spatial integral (initial condition) is time dependent, and then $f(r,t)$ would need to be divided by $t^{(1-\beta)(\alpha-1)}$ to admit probabilistic interpretation.

So with $\beta = 1$ we have:

$$F(k,s) = \frac{s^{\alpha-1}}{\mathbb{D}k^2 + s^\alpha},$$

and from Chapter 3 and 4 we also have

$$F(k,s) = \frac{1-w(s)}{s[1-w(s)\lambda(k)]},$$

with $w(t)$ and $\lambda(x)$ as wait time and jump length probabilities. Making these two $F(k,s)$ equal and re arranging we get:

$$\frac{1-\lambda(k)}{\mathbb{D}k^2} = \frac{1-w(s)}{s^\alpha w(s)}$$

Because the LHS does not depend on s and RHS does not depend on k, the RHS and LHS should be equal to a common constant; and let that constant be τ_0^α. We obtain, jump length densities expression as $\lambda(k) = 1 - \mathbb{D}\tau_0{}^\alpha k^2$ and wait time density expression as

$$w(s) = \frac{1}{1 + \tau_0{}^\alpha s^\alpha}.$$

Note $\mathbb{D}\tau_0{}^\alpha$ is similar to mean square displacement, of single jump $\langle x^2 \rangle$.

The inverse of Laplace is, wait time density

$$w(t) = \frac{1}{\tau_0}\left(\frac{t}{\tau_0}\right)^{\alpha-1} E_{\alpha,\alpha}\left(-\frac{t^\alpha}{\tau_0{}^\alpha}\right).$$

For $\alpha = 1$ the waiting time pdf is from Poisson's process that is,

$$w(t) = \frac{1}{\tau_0} e^{-t/\tau_0},$$

found in normal diffusion equation. This was assumed in Chapter 4 for CTRW models for normal diffusion where one has average wait times.

For $0 < \alpha < 1$, characteristic difference arises from asymptotic behavior for $t \to 0$ and $t \to \infty$, for the above Mittag-Leffler function. The asymptotic behavior for

$t \to 0$ is obtained by noting $E_{\alpha,\alpha}(0)=1$ and hence $w(t) \sim t^{\alpha-1}$. For the case $\alpha<1$, the waiting time density is 'singular' at origin implying a 'statistical abundance' of very short intervals between the jumps, as compared to the Poisson's case where $w(t)=\tau_0^{-1}$ a constant. For large $t \to \infty$, the asymptotic wait time is $w(t) \sim t^{-1-\alpha}$, with long tail.

7.15 Generalized Stationary Conditions

Consider a fractional 'stationary' differential equation with order $0<\alpha<1$ and type $0 \le \beta \le 1$ as ${}_0^{\beta}D_t^{\alpha} f(t)=0$ with initial condition as ${}_0D^{(1-\beta)(\alpha-1)} f(0^+)=f_0$, fractional Riemann-Liouvelli derivative initial condition as stationary. We can re-write it and state fractional integral initial value given as constant that is ${}_0I_t^{(1-\beta)(1-\alpha)} f(0^+)=f_0$. For $\alpha=1$, the conventional definition of stationary condition is recovered as ${}_0^{\beta}D_t^{1} f(t)=0$, for any type β; with initial condition as $f(0^+)=f_0$. The solution of this fractional stationary differential equation is:

$$f(t)=\frac{f_0 t^{(1-\beta)(\alpha-1)}}{\Gamma[(1-\beta)(\alpha-1)+1]}$$

This may be seen by inserting the above $f(t)$ expression into, the generalized definition of fractional derivative as;

$${}_a^{\beta}D_x^{\alpha} f(x)={}_aI_x^{\beta(1-\alpha)} \frac{d}{dx}\, {}_aI_x^{(1-\beta)(1-\alpha)} f(x)\text{, for } 0<\alpha<1 \text{ and using}$$

the fractional integral of power function expression

$${}_aI_x^{\alpha}(x-a)^{\lambda}=\frac{\Gamma(\lambda+1)}{\Gamma(\alpha+\lambda+1)}(x-a)^{\alpha+\lambda}.$$

For RL case with $\beta=0$ the function $f(t)=f_0 t^{\alpha-1}/\Gamma(\alpha)$ and for Caputo case with $\beta=1$ the function is $f(t)=f_0$,a constant.

Note that the integral ${}_0I_t^{(1-\beta)(1-\alpha)} f(t)=f_0$ remains conserved and constant for all t; while the function $f(t)$ itself varies. In particular $\lim_{t\to 0} f(t)=\infty$ and $\lim_{t\to\infty} f(t)=0$. For $\beta=1$ and $\alpha=1$ one recovers $f(t)=f_0$ as usual. Fractional derivative of constant for Caputo type is zero, and integer order derivative of constant of any type is zero.

The new type of stationary states for which a fractional integral (rather than function itself) is constant is arrived at. It seems rather lack of knowledge about fractional stationarity that is particularly responsible for this difficulty of deciding which type of fractional derivative should be used when generalizing traditional equation of motion thermodynamics electrodynamics control system and others. Nevertheless, let us carry on with this dichotomy.

7.16 Concluding Comments

Many of the forms described in this chapter on the Laplace transform have been derived to demonstrate how initialization function generalizes the Laplace transforms (seen in integer order calculus). Here also several possible compositions of fractional integration and fractional derivatives are demonstrated. Reader is advised to imagine further regarding the compositions of these operators, as to adhere to constraints of continuity and index (composition) laws to use these forms. The selection of appropriate initialization function when structuring fractional differential equations will be analogous to, but somewhat more demanding, than selection of constants when structuring integer order systems. Experience working with particular equation or type of physical systems will be required of analyst in selection of initialization function. The use of Laplace transforms and initialization is obtained for scalar and vector initialization examples, will be suitable for fractional order system modeling, for single input single output, as well as multi input multi output systems. The solution of fundamental fractional differential equation contains fractional pole in its transfer function and property of which is important in system analysis. The analysis of stability of the pole (and zero) property is to be carried out in transformed Laplace (w) plane, and the theory and interpretation of control system be mapped here, for future use of fractional order control system. Also the Laplace operator for fractional capacitor is represented as rational function approximation to synthesize circuits for fractional order control system, in analog way. These concepts are very useful for futuristic control systems. The practical case of obtaining constant phase element 'fractional' Laplace operator (for a desired frequency band of interest) is described, which is useful for fractional order integrator/differentiator circuit realization. The concept of generalized 'fractional' initialized states is discussed, in view of Riemann-Liouvelli and Caputo definition, presently let us carry on with these conflicting definitions, and realize the world of fractional order calculus.

Chapter 8
Application of Generalized Fractional Calculus in Electrical Circuit Analysis and Electromagnetics

8.1 Introduction

The fractional calculus is widely popular, especially in the field of viscoelasticity. In this chapter variety of applications are discussed. This chapter is application oriented to demonstrate the fundamental of generalized (fractional) calculus developed earlier, with particular reference to initialization concepts. Here the treatment is to show coupling effect of the initialization functions, and the use of developed Laplace technique. The applications and potential applications of fractional calculus are in diffusion process, electrical science, electrochemistry, material creep, viscoelasticity, control science, electro magnetic theory and several more. In this chapter the fractal distribution effect of charges and its electromagnetic parameters is developed, based on representation of these fractal distributions by fractional differential elemental volume, surface or line. The generalizations of set of Maxwell's equations are carried out for fractal distributions. The chapter also introduces the concept of representing fractional order transfer functions by rational polynomial ratios; also discusses issues of digitizing those fractional order transfer functions, and realize them through circuit techniques. This chapter restricted to electronics and electrical circuit models, and 'fractal' electromagnetic.

8.2 Electronics Operational Amplifier Circuits

8.2.1 Operational Amplifier Circuit with Lumped Components

The operational amplifier is well known in electrical science to provide gain. Figure 8.1 gives simple scheme.

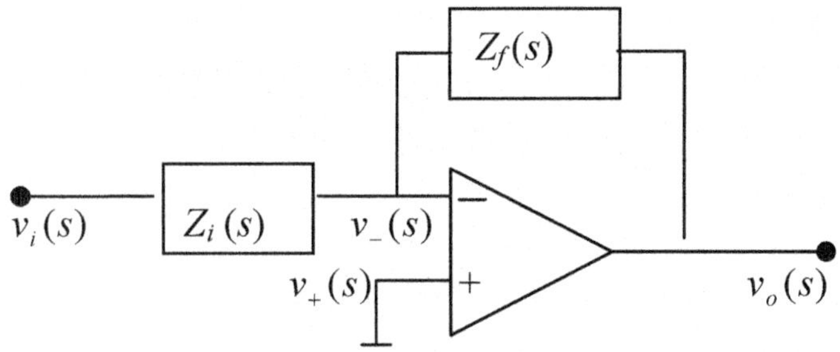

Fig. 8.1 Operational amplifier circuit with lumped impedances.

We revise concept of impedance transfer function. The terms Z_f and Z_i are output and input impedances respectively, will represent general linear circuit together with initialization functions. Therefore not strictly as impedance but some similarity to perhaps memory element!

The Laplace transforms variables:

$$\text{Input side: } v_i(s) - v_-(s) = i_i(s) Z_i(s)$$

$$\text{Output/feedback element: } v_-(s) - v_o(s) = i_f(s) Z_f(s)$$

$$\text{Inverting node: } i_i(s) - i_f(s) - i_-(s) = 0$$

For negative feedback configuration the amplifier will have, from its basic property of high input impedance of NI (+) and I (-) terminals $i_- \cong 0$ therefore $v_- \cong 0$. Putting these values we obtain transfer function (voltage to voltage) as:

$$G(s) = \frac{v_o(s)}{v_i(s)} = -\frac{Z_f(s)}{Z_i(s)}$$

For integer order integration $Z_i(s) = R$ and $Z_f(s) = 1/Cs$, $G(s) = 1/RCs$; interchanging these impedances will give integer order differentiation, with $G(s) = RCs$. However, with these choice of impedances, the circuits are unstable. Therefore for integration if we connect R in parallel with C for $Z_f(s)$, and with $Z_i(s) = R$, this stabilizes the dc-gain, and the transfer function is then

$$G(s) = \frac{v_o(s)}{v_i(s)} = \frac{1}{RCs + 1},$$

and for integer order differentiator $Z_i(s)$ is R parallel with C and $Z_f(s)$ is R series with C and gives transfer function as

$$G(s)=\frac{v_o(s)}{v_i(s)}=\frac{1}{RC}\frac{s}{\left(s+1/RC\right)^2}.$$

With this modified we have the transfer function not equal to pure integer order integrator or differentiator that is, $G(s)\neq s^{\pm 1}$. The integration circuit thus approximates integration action for the frequencies greater than $10/RC$ [radian / s] and the band of frequencies of differentiation is $(0.1/RC)$ [radian / s]. A pure integration or differentiation mathematically is for frequency ranges up to infinity and down to DC that, is zero radian/s. Thus practical integer order differentiation and integration circuits does the operation of integration/differentiation for a band of frequencies; hence are approximates.

This point is worth noting here, though basic circuit elements available as $(Cs)^{-1}$ or Ls have integer order impedance functions, the transfer function $G(s)$ is emulating integer order integration or differentiation on a frequency band and not for all frequencies. This point is a practical point to realize Fractional Order differintegration with these types of impedances, as pole-zero interlacing, which works in the desired frequency band. The Figure 8.1 is still valid for fractional order differintegration, with $G(s)=Ks^{\pm\alpha}$, $\alpha\in(0,1)$.

8.2.2 Operational Amplifier Integrator with Lumped Element

Figure 8.2 describes classical integrator circuit.

The defining equations are:

$$v_i(t)-v_-(t)=i_iR$$

$$v_-(t)-v_o(t)=\frac{1}{C}{}_cD_t^{-1}i_f(t)=\frac{1}{C}\int_c^t i_f(\tau)d\tau+\frac{1}{C}\int_a^c i_f(\tau)d\tau$$

$$=\frac{1}{C}\int_c^t i_f(\tau)d\tau\frac{1}{C}q(c)=\frac{1}{C}\int_c^t i_f(\tau)d\tau+\left[v_-(c)-v_o(c)\right]$$

Above substitution of initial charge stored from time $t=a$ to time $t=c$ as integral of current is done. $q(c)=\int_a^c i_f(\tau)d\tau=C\left[v_-(c)-v_o(c)\right]$ as initial charge on the capacitor at the start of integration process at time $t=c$. For operational amplifier putting $i_f=i_i$ and $v_-(t)=0$ yields.

$$v_o(t)=-\frac{1}{RC}\int_c^t v_i(\tau)d\tau+v_o(c)=-\frac{1}{RC}{}_cD_t^{-1}v_i(t)$$

With $\psi(v_i,-1,a,c,t)=-RCv_o(c)$ as initializing function.

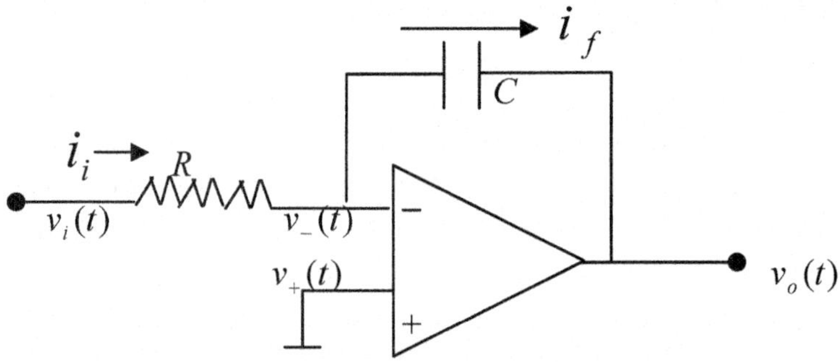

Fig. 8.2 Integrator circuit with lumped element

Then with $RC = 1$, this integrator is represented as $v_o(t) = -{}_cD_t^{-1}v_i(t)$. The initialization function as defined stores the past history (contained in single constant $-v_o(c)$).

When this circuit is used in analog PID controller, the integral reset rate is defined as $T_i = \left|\frac{Z_f}{Z_i}\right| = \frac{1}{RC}$ in unit of per second.

8.2.3 Operational Amplifier Integrator with Distributed Element

The input resistor is replaced by semi-infinite loss less transmission line, with distributed inductance per unit length l and capacitance per unit length c. The semi-infinite loss less line has characteristics equation as wave equation of voltage as:

$$\frac{\partial^2 v(x,t)}{\partial t^2} = \frac{1}{lc}\frac{\partial^2 v(x,t)}{\partial x^2}$$

The terminal characteristics are:

$$i(t) = \sqrt{\frac{c}{l}}v(t) + \phi_1(t) = kv(t) + \phi_1(t)\text{, where } k = \sqrt{c/l}$$

or

$$i(t) = k\,{}_cD_t^0 v(t)\text{, where } \psi(v,0,a,c,t) = \frac{1}{k}\phi_1(t)$$

or

$$v(t) = \frac{1}{k}i(t) + \phi_2(t)$$

or

$$v(t) = \frac{1}{k}\,{}_cD_t^0 i(t) \text{ where } \psi(i,0,a,c,t) = k\phi_2(t)$$

Notice that differentigration operation for voltage-current transfer relationship is of order zero. This may be called a "zero-order" element. The voltage-current relationship of a lumped resistor is also of zero order differintegration. In the loss less transmission line it is the initial condition on the distributed inductor and capacitor along infinite lines gives rise to initial function (of time). Figure 8.3 gives the semi-infinite loss less (zero-order) element representation and Figure 8.4 is operational amplifier circuit based on this loss less line.

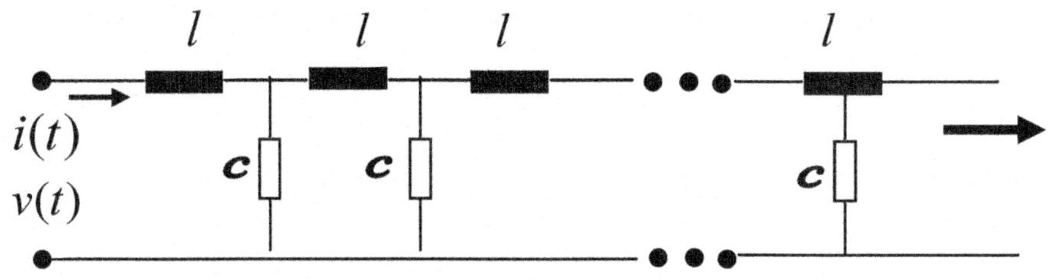

Fig. 8.3 Semi-infinite loss less line (zero-order differintegrator)

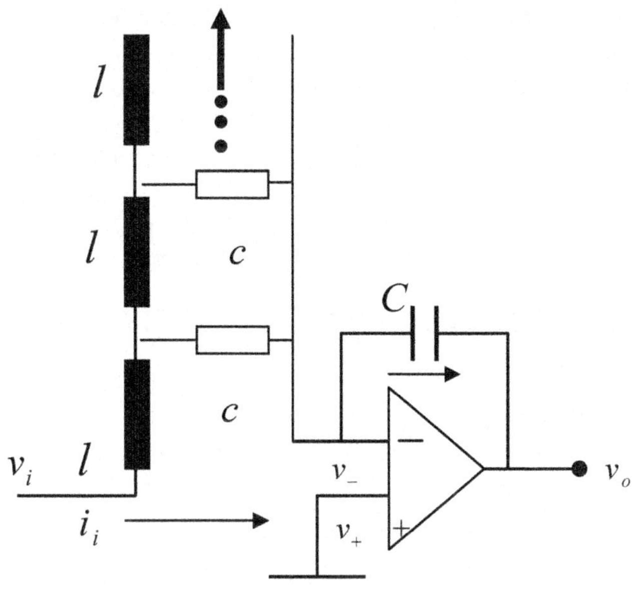

Fig. 8.4 Integrator circuit with distributed elements(one-order-integrator)

The semi-infinite loss less line is connected as input element between input terminal and inverting node. The input element equation is:

$$i_i(t) = k\left[v_i(t) - v_-(t)\right] + \phi_1(t)$$

For feed back element

$$v_-(t)-v_o(t)=\frac{1}{C}\,{}_cD_t^{-1}i_f(t)=\frac{1}{C}\int_c^t i_f(\tau)d\tau-v_o(c)$$

Putting $i_f=i_i$ and $v_-(t)=0$ we have expression for output as:

$$v_o(t)=-\frac{1}{C}\int_c^t[kv_i(\tau)+\phi_1(\tau)]d\tau+v_o(c)=-\left(\frac{1}{C}k\right)\int_c^t v_i(\tau)d\tau-\frac{1}{C}\int_c^t\phi_1(\tau)d\tau+v_o(c)$$

In compact generalized calculus notation we summarize the transfer expression as:

$$v_o(t)=-\left(\frac{1}{C}k\right)\,{}_cD_t^{-1}v_i(t)\quad\text{where}\quad\psi(v_i,-1,a,c,t)=\frac{1}{k}\int_c^t\phi_1(\tau)d\tau+C\frac{1}{k}v_o(c)$$

Selecting the coefficient values and circuit constants as unity we have: $v_o(t)=-{}_cD_t^{-1}v_i(t)$. This is the same expression what is got in classical integrator circuit with lumped elements. The difference exists in the values of the initialization functions. For this integrator with distributed elements the effect of "past history" is contained not in a constant $-v_o(c)$, which is the charge stored in the integrating capacitor C (the lumped element), but is in the remainder of the function $\psi(t)$. This accounts for the distributed charge along the semi-infinite line. The observation is that the "zero-order differintegral" element that is the semi-infinite loss less line will simply propagate the changes in input excitation $(v_i(t))$ along the infinite line and never this perturbation be seen again. The only effect seen being proportional to the variation of the source end current $(i_i(t))$. This is true for 'terminal charging'. For side charging or a 'side charged' line an additional function of time may be impressed at output point of the circuit, which is dependent on the initial voltage distribution on the line.

8.2.4 Operational Amplifier Differential Circuit with Lumped Elements

Figure 8.5 describes the circuit, which is classical differentiator block.

The expressions are for this classical differentiator:

$$v_i(t)-v_-(t)=\frac{1}{C}\,{}_cD_t^{-1}i_i(t)=\frac{1}{C}\int_c^t i_i(\tau)d\tau+\frac{1}{C}\int_a^c i_i(\tau)d\tau$$

$$=\frac{1}{C}\int_c^t i_i(\tau)d\tau+\frac{q(c)}{C}=\frac{1}{C}\int_c^t i_i(\tau)d\tau+v_{i-(-)}(c)$$

$$v_-(t)-v_o(t)=i_f(t)R$$

Putting $i_i(t)=i_f(t)=0$ and $v_-(t)=0$ we have

$$v_o(t)=-Ri_f(t)=-RC\left[\frac{d}{dt}(v_i(t)-v_{i-(-)}(c))\right]$$

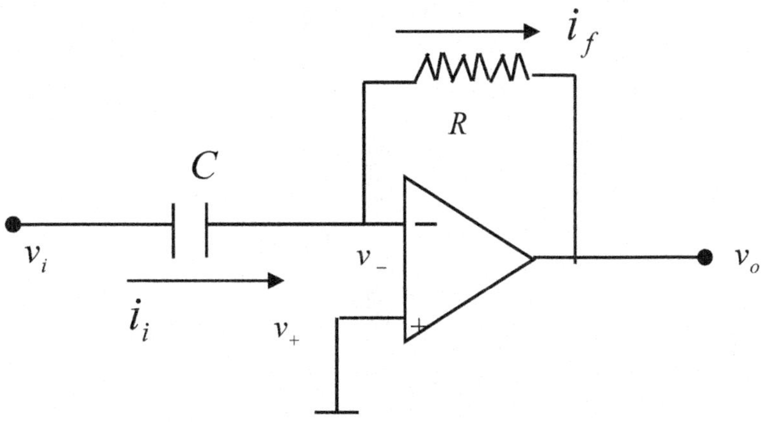

Fig. 8.5 Differentiator circuit with lumped elements

In generalized calculus terms for $t > c$, we get $v_o(t) = -RC\,{}_cD_t^1 v_i(t)$. The initialization function to this generalized derivative is $\psi(v_i, 1, a, 0, t) = \frac{d}{dt} v_{i-(-)}(c)$. This initialization is normally put as zero. However this initialization term can give a "pulse" response at time $t = c$.

8.2.5 Operational Amplifier Differentiator with Distributed Element

Figure 8.6 is the circuit where at the feed back element of circuit of Figure 8.5 is replaced by semi-infinite loss less transmission line (zero order distributed element).

The following is the derivation for the differentiator with distributed zero order elements:

$$v_i(t) - v_-(t) = \frac{1}{C}\,{}_cD_t^{-1} i_i(t) = \frac{1}{C}\int_c^t i_i(\tau)d\tau + v_{i-(-)}(c)$$

The distributed loss less (zero-order) element is connected at feed back path and using its relation we get:

$$v_-(t) - v_o(t) = \frac{1}{k}\,{}_cD_t^0 i_f(t) = \frac{1}{k} i_f(t) + \frac{1}{k}\psi(i_f, 0, a, c, t)\text{, with } k = \sqrt{\frac{c}{l}}$$

Putting $i_i(t) = i_f(t)$ and $v_-(t) = 0$ we have the differentiator expression:

$$v_o(t) = -\frac{1}{k}\left[C\frac{d}{dt}\left(v_i(t) - v_{i-(-)}(c)\right) + \psi(i_f, 0, a, c, t)\right]$$

$$v_o(t) = -C\frac{1}{k}\left[\frac{d}{dt}v_i(t) + \frac{1}{C}\psi(v_i, 1, a, c, t)\right]$$

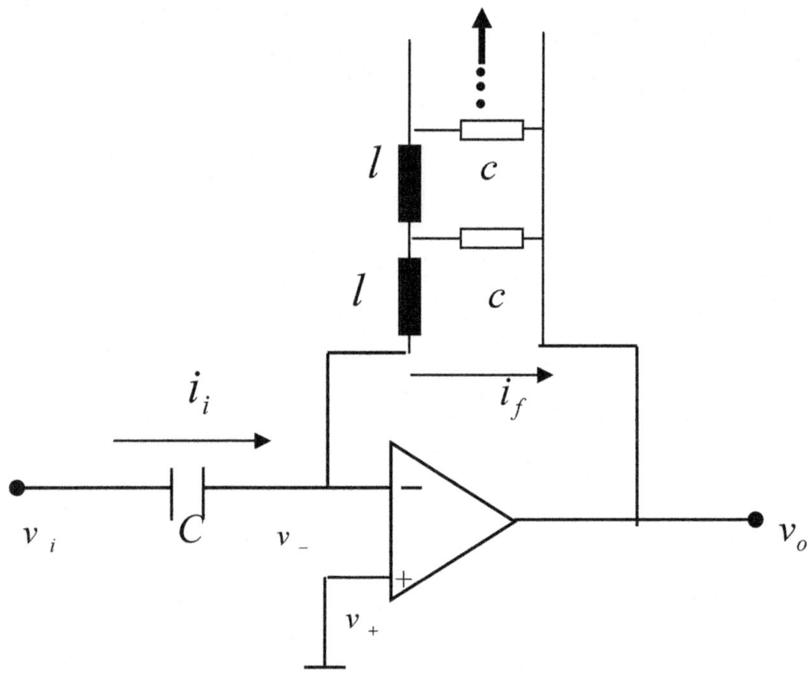

Fig 8.6 Differentiator circuit with distributed element (one-order differentiator)

In terms of generalized calculus notations and concepts the differentiator circuit is described by $v_o(t) = -C\frac{1}{k}\,{}_cD_t^1 v_i(t)$, where initialization function associated with the derivative operator is $\psi(v_i, 1, a, c, t) = -\frac{1}{C}\psi(i_f, 0, a, c, t)$. As in the integrator case with distributed zero order elements, the perturbations at the source terminal end will be propagated away and never return back (reflected). It is noted that for terminally charged integer order differentiator with lumped elements $\psi_{eff} = 0$.

8.2.6 Operational Amplifier as Zero Order Gain with Lumped Components

In Figure 8.1 consider lumped impedances as pure lumped resistances. Therefore the circuit transfer function is

$$v_o(t) = -\frac{R_f}{R_i} v_i(t) = -\frac{R_f}{R_i}\,{}_cD_t^0 v_i(t)$$

Where clearly $\psi(v_i, 0, a, c, t) = 0$.

Meaning that this circuit configuration has no memory. Also is classical zero order operator.

8.2.7 Operational Amplifier as Zero Order Gain with Distributed Elements

Figure 8.7 gives the circuit diagram where feedback element and the input element of Figure 8.1 are constituted by distributed element as semi-infinite loss less line.

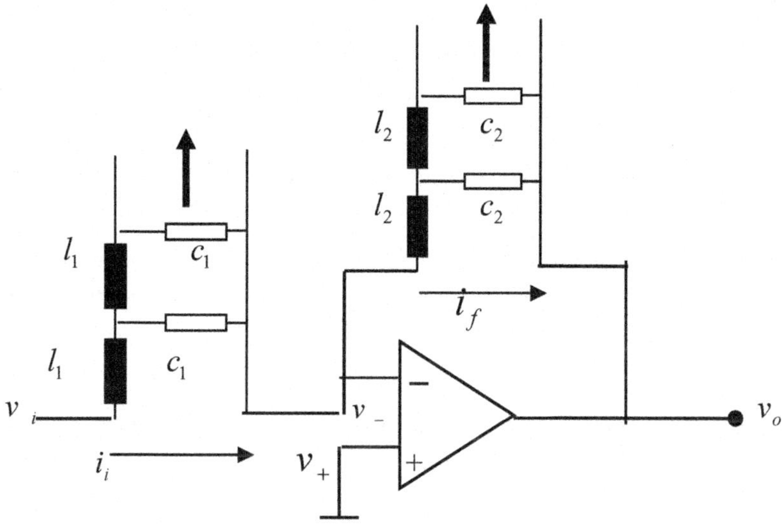

Fig. 8.7 Gain circuit with distributed elements (zero order differintegration)

The describing relations are:

$$i_i(t) = \sqrt{\frac{c_1}{l_1}}\, {}_cD_t^0 \left[v_i(t) - v_-(t) \right],$$

where initialization $\psi(v_i - v_-, 0, a, c, t) = \sqrt{\frac{l_1}{c_1}}\phi_1(t)$

$$v_-(t) - v_o(t) = \sqrt{\frac{l_2}{c_2}}\, {}_cD_t^0 i_f(t),$$

where the initialization $\psi(i_f, 0, a, c, t) = \sqrt{\frac{c_2}{l_2}}\phi_2(t)$

Putting $i_i = i_f$ and $v_-(t) = 0$ we have the expression:

$$
\begin{aligned}
v_o(t) &= -\sqrt{\frac{l_2}{c_2}}\,{}_cD_t^0\left[\sqrt{\frac{c_1}{l_1}}\,{}_cD_t^0 v_i(t)\right] = -\sqrt{\frac{c_1 l_2}{c_2 l_1}}\,{}_cD_t^0 v_i(t) \\
&= -\sqrt{\frac{l_2}{c_2}}\,{}_cd_t^0\left[\sqrt{\frac{c_1}{l_1}}\left\{{}_cd_t^0 v_i(t) + \psi(v_i - v_-, 0, a, c, t) + \psi(i_f, 0, a, c, t)\right\}\right] \\
&= -\sqrt{\frac{c_1 l_2}{c_2 l_1}}\,{}_cd_t^0 v_i(t) - \sqrt{\frac{l_2}{c_2}}\phi_1(t) - \phi_2(t)
\end{aligned}
$$

Summarily: $v_o(t) = -\sqrt{\frac{c_1 l_2}{c_2 l_1}}\,{}_cD_t^0 v_i(t)$ and initialization to this is

$$
\psi(v_i, 0, a, c, t) = \psi(v_i - v_-, 0, a, c, t) + \sqrt{\frac{l_1}{c_1}}\psi(i_f, 0, a, c, t) = \sqrt{\frac{l_1}{c_1}}\phi_1(t) + \sqrt{\frac{c_2 l_1}{c_1 l_2}}\phi_2(t)
$$

Summarizing the above observation leads that zero order gain circuit returns the input function $v_i(t)$ but provides extra time function based on initial distribution of voltages currents in the distributed elements, i.e. $\psi(v_i, 0, a, c, t)$.

8.2.8 *Operational Amplifier Circuit for Semi-differintegration by Semi-infinite Lossy Line*

Figure 8.8 describes the semi-infinite lossy line (a half order element). As opposed to loss less case the voltage current relation is of half order here as it was zero order for loss less case.

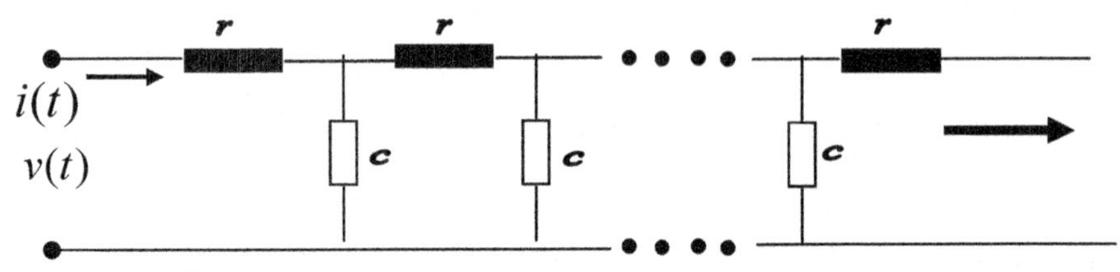

Fig. 8.8 Semi-infinite lossy line (half order element)

The diffusion equation corresponding to this lossy line is $\frac{\partial v(x,t)}{\partial t} = \alpha \frac{\partial^2 v(x,t)}{\partial x^2}$.

In the Figure 8.8 as shown the $v(t)$ and $i(t)$ are the voltage and the current at the source terminal. The per unit length distributed resistance is r and per unit length distributed capacitance is c. The diffusivity coefficient is $\alpha = rc$. The initial state of charge and voltage that exists in infinite array of elements is ψ.

The terminal characteristics is $v(t) = r\sqrt{\alpha}\frac{d^{-1/2}}{dt^{-1/2}}i(t) + \phi_1(t)$ or

$v(t) = r\sqrt{\alpha}\,{}_cD_t^{-1/2}i(t)$, where $\psi(i,-1/2,a,c,t) = \frac{1}{r\sqrt{\alpha}}\phi_1(t)$ or

$$i(t) = \frac{1}{r\sqrt{\alpha}}\frac{d^{1/2}}{dt^{1/2}}v(t) + \phi_2(t)\text{, or}$$

$i(t) = \frac{1}{r\sqrt{\alpha}}\,{}_cD_t^{1/2}v(t)$, where $\psi(v,1/2,a,c,t) = r\sqrt{\alpha}\phi_2(t)$

8.2.9 Operational Amplifier Circuit for Semi-integrator

Figure 8.9 is the circuit for semi-integrator.

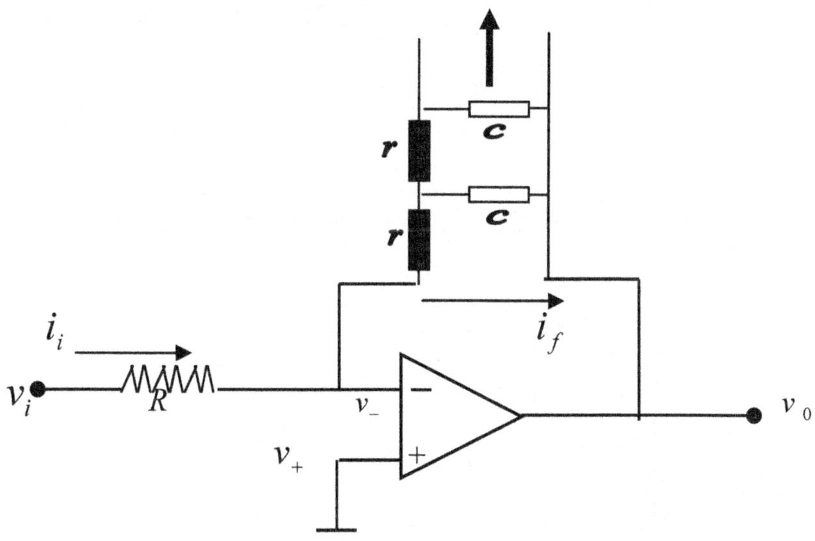

Fig. 8.9 Semi-integrator circuit (half order integrator)

The component equations are the following:

$v_i(t) - v_-(t) = i_i(t)R$

$v_-(t) - v_o(t) = r\sqrt{\alpha}\,{}_cD_t^{-1/2}i_f(t)$, where $\psi(i_f,-1/2,a,c,t) = \frac{1}{r\sqrt{\alpha}}\phi_1(t)$

Putting $i_f = i_i$ and $v_-(t) = 0$ we have the expression for semi integration:

$$v_o(t) = -\frac{r\sqrt{\alpha}}{R}\,{}_cD_t^{-1/2}v_i(t),$$

where $\psi(v_i, t, -1/2, a, c, t) = R\psi(i_f, -1/2, a, c, t) = \frac{R}{r\sqrt{\alpha}}\phi_1(t)$.

Alluding to Figure 8.1 the input element and output element impedances are $Z_f = \frac{r\sqrt{\alpha}}{s^{1/2}}$ and $Z_i = R$, which leads to transfer function in s-domain as $\frac{V_0(s)}{V_i(s)} = -\frac{r\sqrt{\alpha}}{Rs^{1/2}}$. Realizing this with $r = 1K\Omega$ and $c = 1\mu F$ with $R = 22K\Omega$ gives transfer function for the semi-integrator as $\frac{V_0(s)}{V_i(s)} = -1.4374s^{-0.5}$.

Replacing $s \to j\omega$, the transfer function becomes $\frac{V_0(j\omega)}{V_i(j\omega)} = -\frac{1.4374}{\sqrt{\omega}}\exp\left(-j\frac{\pi}{4}\right)$.

Following an inverting amplifier after semi-integration stage of Fig 8.9, one gets fractional (semi) integral control block for analog fractional order PID circuit. This block will behave as constant phase element of -45^0. The controller constant like reset rate of PID is $T_i = \frac{Z_f}{Z_i} = \frac{\sqrt{\frac{r}{c}}}{R} = 1.4374$.

8.2.10 Operational Amplifier Circuit for Semi-differentiator

Figure 8.10 shows a semi-differentiator circuit with semi-infinite lossy element

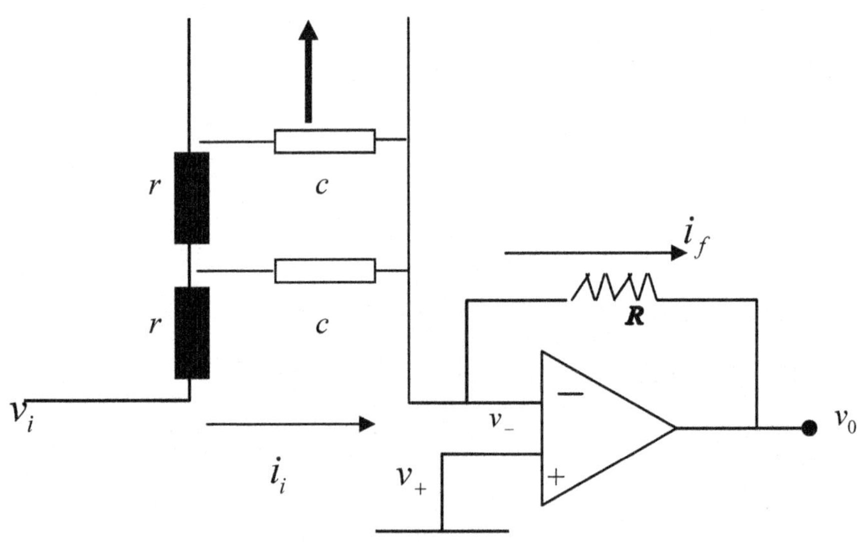

Fig. 8.10 Semi-differentiator circuit.

The expressions for circuit of Figure 8.10 are:

$$i_i(t) = \frac{1}{r\sqrt{\alpha}}\, {}_cD_t^{1/2}\left[v_i(t) - v_-(t)\right]$$

Initializing expression is

$$\psi(v_i - v_-, 1/2, a, c, t) = r\sqrt{\alpha}\phi_2(t)$$

and $v_-(t) - v_o(t) = i_f(t)R$. Putting $i_f = i_i$ and $v_-(t) = 0$ we have:

$$v_o(t) = -i_f(t)R = -\frac{R}{r\sqrt{\alpha}}\, {}_cD_t^{1/2} v_i(t)$$

and $\psi(v_i, 1/2, a, c, t) = \psi(v_i - v_-, 1/2, a, c, t) = r\sqrt{\alpha}\phi_2(t)$

8.2.11 Cascaded Semi-integrators

Figure 8.11 represents two semi-integrator circuits in series, which gives integral operation of order one.

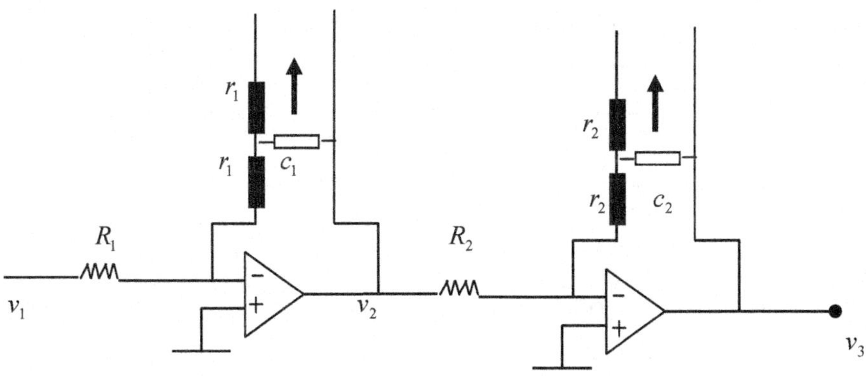

Fig. 8.11 Cascaded semi-integrators.

The circuit of Figure 8.11 has the following expressions:

$v_2(t) = -\frac{r_1\sqrt{\alpha_1}}{R_1}\, {}_cD_t^{-1/2} v_1(t)$, with $\psi_1(v_1, -1/2, a, c, t)$ is first integrator, transfer function $v_3(t) = -\frac{r_2\sqrt{\alpha_2}}{R_2}\, {}_cD_t^{-1/2} v_2(t)$, with $\psi_2(v_2, -1/2, a, c, t)$ is the second integrator expression.

Then combining we have

$$v_3(t) = \frac{r_1 r_2\sqrt{\alpha_1\alpha_2}}{R_1 R_2}\, {}_cD_t^{-1} v_1(t)\,,$$

with

$$\psi(v_1,-1,a,c,t) = {}_c d_t^{-1/2} \psi_1(v_1,-1/2,a,c,t) - \frac{R_2}{r_2\sqrt{\alpha_2}} \psi_2(v_2,-1/2,a,c,t)$$

8.2.12 Semi-integrator Series with Semi-differentiator Circuit

Figure 8.12 gives the schematic of cascaded semi differentiator and semi integrator to have overall zero-order.

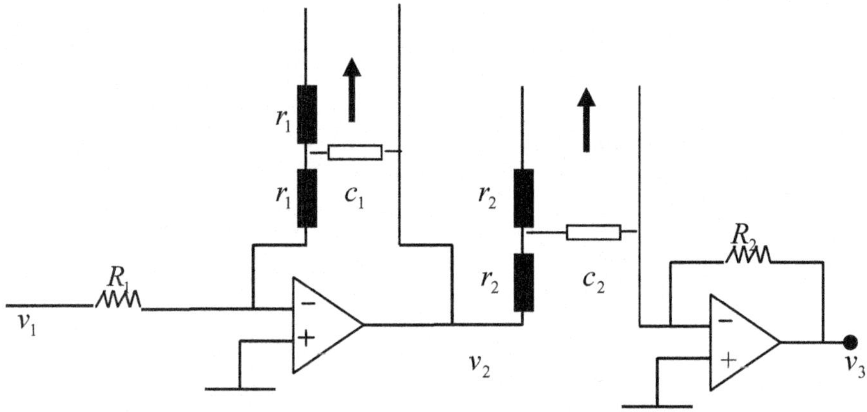

Fig. 8.12 Cascade circuit semi-integrator and semi-differentiator.

$$R_1 = R_2 = r_1 = r_2 = \sqrt{\alpha_1} = \sqrt{\alpha_2} = 1 .$$

The solution is $v_3(t) = -{}_c D_t^{1/2} v_2(t) = -{}_c D_t^{1/2}\left[-{}_c D_t^{-1/2} v_1(t)\right] = {}_c D_t^{0} v_1(t)$, as expected.

Here $v_3(t) = v_1(t)$ only if

$$\psi(v_1,0,a,c,t) = {}_c d_t^{1/2} \psi(v_1,1/2,a,c,t) - \psi(v_2,1/2,a,c,t) = 0$$

Where $\psi(v_1,0,a,c,t)$ is the combined initialization function for the combined operator ${}_c D_t^0 v_1(t)$. This is satisfied under terminal charging. However under side charging conditions it is seen that $v_3(t) = v_1(t) + \psi(v_1,0,a,c,t)$ and the zero-property are not satisfied (not necessarily satisfied), as was indicated in the concept of initialized generalized calculus.

8.3 Battery Dynamics

8.3.1 Battery as Fractional Order System

This section analyzes a simple battery. Electrolytic cell is known to exhibit fractional behavior, typically of half order. The fractional system is electrode-electrolyte interface where diffusion takes place. This diffusion process is called the Warburg impedance (or a constant phase CPE element). The two phases of battery operation is considered, charging and discharging phases. The discharge phase is load drawing (usage). The charging phase takes place from $t = a = 0$, to $t = c$, with actual current flowing i.e. charging occurring for $a = 0 \le t \le b$. Later is the load drawing usage phase. Figure 8.13 gives diagram of battery circuit charging, discharging phase circuit and the charge current profile. Here constant current charging is assumed. The block "W" represents Warburg impedance of electrode-electrolyte interface.

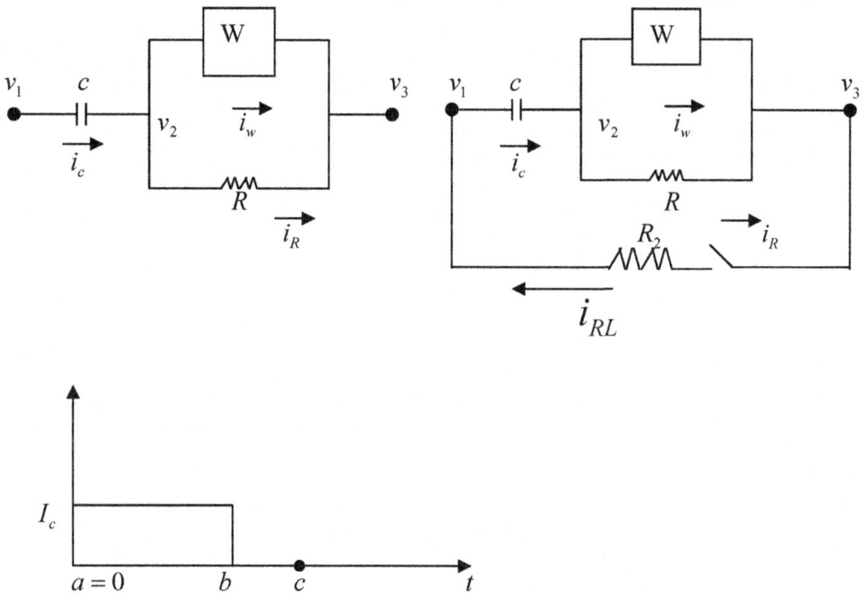

Fig. 8.13 Battery Charging, discharging circuit and charging current profile.

8.3.2 Battery Charging Phase

Time domain equations from Figure 8.13 are:

$$v_1(t) - v_2(t) = \frac{1}{C}\,{}_0D_t^{-1} i_c(t) = \frac{1}{C}\int_0^t i_c(\tau)d\tau + v_{1-2}(0)$$

$$v_2(t) - v_3(t) = i_R(t)R$$

$$i_c(t) = i_R(t) + i_W(t), \text{ and}$$

$$i_W(t) = \frac{1}{B}\,{}_0D_t^{1/2}\left[v_2(t) - v_3(t)\right]$$

Taking Laplace transform yields:

$$V_1(s)-V_2(s)=\frac{1}{Cs}I_c(s)+\frac{v_{1-2}(0)}{s}$$
$$V_2(s)-V_3(s)=I_R(s)R$$
$$I_c(s)=I_R(s)+I_W(s)\text{, and}$$
$$I_W(s)=\frac{1}{B}\left[s^{1/2}\left(V_2(s)-V_3(s)\right)+\psi(v_2-v_3,1/2,0,c,s)\right]$$

Eliminating $I_W(s),I_R(s),V_2(s)$ and rearranging we arrive at following:

$$I_c(s)=\frac{s\left(\frac{1}{B}s^{1/2}+\frac{1}{R}\right)}{s+\frac{1}{CB}s^{1/2}+\frac{1}{RC}}\{V_1(s)-V_3(s)\}+\frac{\frac{1}{B}s\psi(v_2-v_3,1/2,0,c,s)-v_{1-2}(0)\left\{\frac{1}{B}s^{1/2}+\frac{1}{R}\right\}}{s+\frac{1}{CB}s^{1/2}+\frac{1}{RC}}$$

or $$V_1(s)-V_3(s)=\left\{\frac{s+\frac{1}{CB}s^{1/2}+\frac{1}{RC}}{s\left(\frac{1}{B}s^{1/2}+\frac{1}{R}\right)}\right\}I_c(s)-\frac{\frac{1}{B}\psi(s)}{\frac{1}{B}s^{1/2}+\frac{1}{R}}+\frac{v_{1-2}(0)}{s}$$

These Laplace expressions are used to carry out transfer function analysis and to obtain time domain responses. The Laplace transform pairs useful are following:

$$\mathcal{L}^{-1}\left(\frac{B}{\sqrt{s}+\frac{B}{R}}\right)=B\left[\frac{1}{\sqrt{\pi t}}-\frac{B}{R}e^{\left(\frac{B}{R}\right)^2 t}\operatorname{erfc}\left(\frac{B}{R}\sqrt{t}\right)\right]$$

$$\mathcal{L}^{-1}\left[\frac{\frac{1}{C}}{\sqrt{s}\left(\sqrt{s}+\frac{B}{R}\right)}\right]=\frac{1}{C}e^{\left(\frac{B}{R}\right)^2 t}\operatorname{erfc}\left(\frac{B}{R}\sqrt{t}\right)$$

$$\mathcal{L}^{-1}\left[\frac{\frac{B}{RC}}{s\left(\sqrt{s}+\frac{B}{R}\right)}\right]=\frac{1}{C}\left[1-e^{\left(\frac{B}{R}\right)^2 t}\operatorname{erfc}\left(\frac{B}{R}\sqrt{t}\right)\right]$$

Using standard Laplace transforms tables the time response is got for an impulse charging ($I_c(s)=1$) as:

$$v_1(t)-v_3(t)=\frac{1}{C}+B\left\{\frac{1}{\sqrt{\pi t}}-\frac{B}{R}\exp\left(\frac{B^2}{R^2}t\right)\operatorname{erfc}\left(\frac{B\sqrt{t}}{R}\right)\right\},$$

During initialization period here both the initialization functions are taken as zero for $0 = a < t \le c$. The above time expression is for impulse charging and for any general charging this expression is convoluted with the charging current profile i.e. general $i_c(t)$, to get following:

$$v_1(t) - v_3(t) = \int_0^t \left\{ \frac{1}{C} + B \left\{ \frac{1}{\sqrt{\pi\tau}} - \frac{B}{R} e^{\left(\frac{B}{R}\right)^2 \tau} \operatorname{erfc}\left(\frac{B\sqrt{\tau}}{R} \right) \right\} \right\} i_c(t-\tau) d\tau$$

As in the Figure 8.13 the constant current charge case for charging period $i_c(t) = I_c$ for $0 = a < t \le b$, then the step current can be resolved as function of two unit step functions. $i_c(t-\tau) = I_c(H(t-\tau) - H(t-\tau-b))$, and the charging equation thus becomes:

$$v_1(t) - v_3(t) = \int_0^t \left[\frac{1}{C} + B \left\{ \frac{1}{\sqrt{\pi\tau}} - \frac{B}{R} e^{\left(\frac{B}{R}\right)^2 \tau} \operatorname{erfc}\left(\frac{B\sqrt{\tau}}{R} \right) \right\} \right] I_c d\tau$$

$$-H(t-b) \int_0^{t-b} \left[\frac{1}{C} + B \left\{ \frac{1}{\sqrt{\pi\tau}} - \frac{B}{R} e^{\left(\frac{B}{R}\right)^2 \tau} \operatorname{erfc}\left(\frac{B\sqrt{\tau}}{R} \right) \right\} \right] I_c d\tau$$

Using the solved integral formula

$$\int_0^t e^{E\tau} \operatorname{erfc}\left(\sqrt{E\tau}\right) d\tau = \frac{1}{B}\left[e^{Et} \operatorname{erfc}\left(\sqrt{Et}\right) - 1 \right] + \frac{2}{\sqrt{\pi}} \sqrt{Et}$$

We get the charging equation as:

$$v_1(t) - v_3(t) = I_c \left[\frac{t}{C} + 2B\sqrt{\frac{t}{\pi}} - \frac{R}{B} \left\{ e^{\left(\frac{B}{R}\right)^2 t} \operatorname{erfc}\left(\frac{B\sqrt{t}}{R} \right) - 1 + \frac{2}{\sqrt{\pi}} \left(\frac{B\sqrt{t}}{R} \right) \right\} \right] -$$

$$I_c H(t-b) \left[\frac{t-b}{C} + 2B\sqrt{\frac{t-b}{\pi}} - \frac{R}{B} \left\{ e^{\left(\frac{B}{R}\right)^2 (t-b)} \operatorname{erfc}\left(\frac{B\sqrt{t-b}}{R} \right) - 1 + \frac{2}{\sqrt{\pi}} \left(\frac{B\sqrt{t-b}}{R} \right) \right\} \right]$$

Figure 8.14 gives the charging voltage $v_1(t) - v_3(t)$ for charging period of $0 < t < b = 1$, and during relaxation period $t > 1$. The same figure demonstrates the voltage across Warburg element $v_2(t) - v_3(t)$. For simplicity the parameters of battery chosen as $R = B = C = 1.0$.

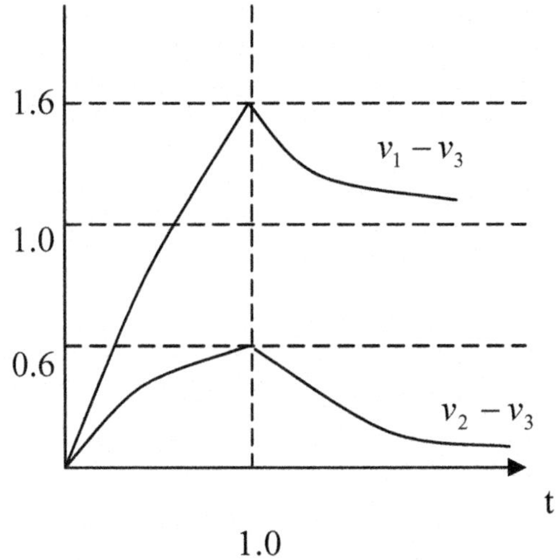

Fig. 8.14 Battery charging till time 1 and thereafter relaxation period for constant current charging current up to time 1.

The above charging expression of total battery includes the effect of both the Warburg element and the capacitor, which is segregated as described in following expressions. Since the $v_{1-2}(0) = 0$ must be zero, the voltage $v_2(t) - v_3(t)$ is determined as:

$$v_2(t) - v_3(t) = \{v_1(t) - v_3(t)\} - \{v_1(t) - v_2(t)\} = \{v_1(t) - v_3(t)\} - \frac{1}{C}\int_0^t i_c(\tau)d\tau$$

$$= \{v_1(t) - v_3(t)\} - \frac{1}{C}\int_0^t I_c\{u(\tau) - u(\tau - b)\}d\tau$$

Using the derived expression for $v_1(t) - v_3(t)$, the Warburg element will have voltage

$$v_2(t) - v_3(t) = \int_0^t B\left\{\frac{1}{\sqrt{\pi\tau}} - \frac{B}{R}e^{\left(\frac{B}{R}\right)^2\tau}\operatorname{erfc}\left(\frac{B\sqrt{\tau}}{R}\right)\right\}I_c d\tau -$$

$$H(t-b)\int_0^{t-b} B\left\{\frac{1}{\sqrt{\pi\tau}} - \frac{B}{R}e^{\left(\frac{B}{R}\right)^2\tau}\operatorname{erfc}\left(\frac{B\sqrt{\tau}}{R}\right)\right\}I_c d\tau$$

The above expression can be good example for mathematical basis for an initialization function for systems of this type with constant current charging. These equations are also valid for $t > c$ with external open circuit. Also if it is desired to initialize at $c = 0$, then axis shifting with $t \to (t - c)$ may be carried out for the expression obtained for Warburg impedance.

Warburg element is semi-derivative constitutive element $i_W = \frac{1}{B}\,{}_0D_t^{1/2}\{v_2(t) - v_3(t)\}$, is assumed is terminally charged. Then the initialization function for the element is defined formally as $\psi(f,u,a,c,t) = \frac{d^m}{dt^m}\psi(f,-p,a,c,t) + \psi(h,m,a,c,t)$, by standard rule described in earlier chapter for initialization concepts. Because of assumption of 'terminal charging', with $\psi(h,1,a,c,t) = 0$, we thus have

$$\psi_W = \frac{1}{\Gamma(1/2)}\frac{d}{dt}\int_a^c (t-\tau)^{\frac{1}{2}-1}\{v_2(\tau) - v_3(\tau)\}d\tau, \text{ for } t > c\,.$$

8.3.3 Battery Discharge Phase

Here in the solution demonstration will be made that to transform $s \to p$ and obtain solution. Referring to the circuit in Figure 8.13 we write the following:

$$v_1(t) - v_2(t) = \frac{1}{C}\,{}_cD_t^{-1}i_c(t) = \frac{1}{C}\int_c^t i_c(t)dt + v_{1-2}(c)$$

$$v_2(t) - v_3(t) = i_R(t)R$$

$$i_c(t) = i_R(t) + i_W(t) = i_{RL}(t)$$

$$v_3(t) - v_1(t) = i_{RL}(t)R_L \text{ , and}$$

$$i_W(t) = \frac{1}{B}\,{}_cD_t^{1/2}\{v_2(t) - v_3(t)\}$$

Laplace transforming the above set we obtain:

$$V_1(s) - V_2(s) = \frac{1}{Cs}I_c(s) + \frac{v_{1-2}(c)}{s}$$

$$V_2(s) - V_3(s) = I_R(s)R$$

$$I_c(s) = I_R(s) + I_W(s) = I_{RL}(s)$$

$$V_3(s) - V_1(s) = I_{RL}(s)R_L$$

$$I_W(s) = \frac{1}{B}\left\{s^{1/2}\left(v_2(s) - v_3(s)\right) + \psi_W(v_2 - v_3, 1/2, 0, c, s)\right\}$$

From charging analysis of equation $V_1(s) - V_3(s)$ and with $I_c(s) = I_{RL}(s) = \{V_3(s) - V_1(s)\}/R_L$ and taking $a = 0, c = 1$

$$V_1(s)-V_3(s)=\left[\frac{R_L s\left(\frac{s^{1/2}}{B}+\frac{1}{R}\right)}{R_L\frac{s^{3/2}}{B}+\left(1+\frac{R_L}{R}\right)s+\frac{1}{CB}s^{1/2}+\frac{1}{RC}}\right]\left[\frac{v_{1-2}(1)}{s}-\frac{\psi_W(v_2-v_3,1/2,0,1,s)}{B\left(\frac{s^{1/2}}{B}+\frac{1}{R}\right)}\right]$$

There is no forced term (forced function is zero) but contains initialization terms from capacitor $v_{1-2}(1)$ and from Warburg impedance ψ_W, the historic past of the element.

Technique to solve the above expression is to convert $s\to p$ domain. By putting $p=s^{1/2}$ so $p^2=s$, $p^3=s^{3/2}$, so that we get:

$$V_1(p)-V_3(p)=\left[\frac{p^2\left(p+\frac{B}{R}\right)}{p^3+\frac{B}{R_L}\left(1+\frac{R_L}{R}\right)p^2+\frac{1}{R_L C}p+\frac{B}{R_L RC}}\right]\left[\frac{v_{1-2}(1)}{p^2}+\frac{\frac{1}{B}\psi(v_2-v_3,1/2,0,1,p^2)}{\frac{p}{B}+\frac{1}{R}}\right]$$

Here appropriate substitution is called for $\psi(p^2)$, and then using partial fractions to achieve p domain response. This obtained p domain response may be transformed back to s domain and then inverse Laplace operation to get time domain answer. To carry out special transform technique is required. The special transformation is 'conformal transformation'.

It is possible to simplify and rewrite the above expression in p-variable as:

$$V_1(p)-V_3(p)=\frac{(p+k_1)(p^2+k_2p+k_3)}{(p+a)(p^3+k_4p+k_4)}$$

This can be partial fractioned and written as, first in p variable and then substituting $p=\sqrt{s}$ as:

$$V_1(p)-V_3(p)=\frac{A}{p+a}+\frac{B}{p+b}+\frac{C}{p+c}+\frac{D}{p+d}=\frac{A}{\sqrt{s}+a}+\frac{B}{\sqrt{s}+b}+\frac{C}{\sqrt{s}+c}+\frac{D}{\sqrt{s}+d}$$

Inverse Laplace transform will give time response as:

$$v_1(t)-v_3(t)=AF_{1/2}[-a,t]+BF_{1/2}[-b,t]+CF_{1/2}[-c,t]+DF_{1/2}[-d,t]$$

Here $F_{1/2}$ is Robotnov-Hartley function, A,B,C,D,a,b,c,d are complex numbers (in general).

It is now possible to do fractional order system analysis and design directly on the transformed plane. To do this it is essential to choose the greatest common fraction q of a particular system. Once this is done, all powers of s^q are replaced by powers of $p = s^q$. Then standard pole-zero analysis is done. This analysis includes root finding, partial fractions, root locus compensation etc. This is what is 'conformal-transformation' $s \to w$ plane discussed in Chapter 7.

8.4 Tracking Filter

The concept of fractional calculus adds another dimension to many applications. The example of fractional order tracking filter will demonstrate the degree of freedom in tuning. This concept is welcome in control science applications where conventional proportional integral and derivative (PID) controller gets more degree of freedom as order control of integration and derivative part is also available. By this we get new type of controller as $PI^{\alpha}D^{\beta}$. The fractional order-tracking filter is just one example.

Here a noisy signal $x(t)$ is to be filtered to yield filtered signal $y(t)$. Filter description is:

$$({}_0D_t^q + a)y(t) = ax(t) \quad \text{or} \quad {}_0d_t^q y(t) + \psi(y,q,0,c,t) + ay(t) = ax(t)$$

Taking Laplace gives:

$$s^q Y(s) + \psi(s) + aY(s) = aX(s)$$

$$Y(s) = \frac{a}{s^q + a} X(s) - \frac{\psi(s)}{s^q + a}$$

The use of $\psi(s)$ is to "pre-charge" the filter to maximize the filter lag, if taken zero then:

$$\frac{Y(s)}{X(s)} = \frac{a}{s^q + a}, \text{ is the Transfer Function of the filter.}$$

For a unit step input X(s), the filter response is given by $Y(s) = \dfrac{a}{s(s^q + a)}$. The full range of freedom is available $0 \le q \le 1$. Taking $q = 1/2$ we have time response $y(t) = 1 - e^{a^2 t}\mathrm{erfc}(at^{1/2})$.

For $q = 1/2$, the forced response is obtained for a Heaviside unit step $H(t)$ (as above) is from listed transform tables. Following steps demonstrates the solution of filter response for any order other than half, for unit step.

Let us write the filter output as

$$Y(s) = \frac{a}{s(s^q + a)}.$$

This we rewrite as follows:

$$Y(s) = a\frac{1/a}{s}\left(\frac{a}{s^q + a}\right) = a\frac{1/a}{s}\left(1 - \frac{s^q}{s^q + a}\right) = \frac{1}{s} - \frac{s^q}{s\left(s^q + a\right)}$$

Using the Laplace pair

$$\mathcal{L}\left\{E_q\left[-at^q\right]\right\} = \frac{1}{s}\left[s^q \mathcal{L}\left\{F_q\left[-at\right]\right\}\right]$$

and

$$\mathcal{L}\left\{E_q\left[at^q\right]\right\} = \frac{s^{q-1}}{s^q - a}, q > 0$$

we get the inverse of Laplace of

$$Y(s) = \frac{1}{s} - \frac{s^q}{s(s^q + a)} = \frac{1}{s} - \frac{s^{q-1}}{s^q + a};$$

to get time response:

$$y(t) = \left[H(t) - E_q(-at^q)\right] = 1 - E_q[-at^q],$$

Where

$$H(t) = \begin{cases} 1 & t \geq 0 \\ 0 & t < 0 \end{cases}$$

is Heaviside step.

For a first order *RC* filter (integer-order filter with order $q = 1$) the Heaviside step response is: $y(t) = 1 - \exp(-at)$, with time constant $\tau = RC = 1/a$. For $q = 1$, the one parameter Mittag-Leffler function $E_q\left[-at^q\right] = E_1\left[-at\right] = \exp(-at)$. Here passing remark is made as for integer order calculus the exponential function appears as solution, for fractional order calculus Mittag-Leffler or its variants say Robotnov-Hartley function, Miller-Ross function etc. are the basis, and appears in solutions. (Refer Figure 8.15).

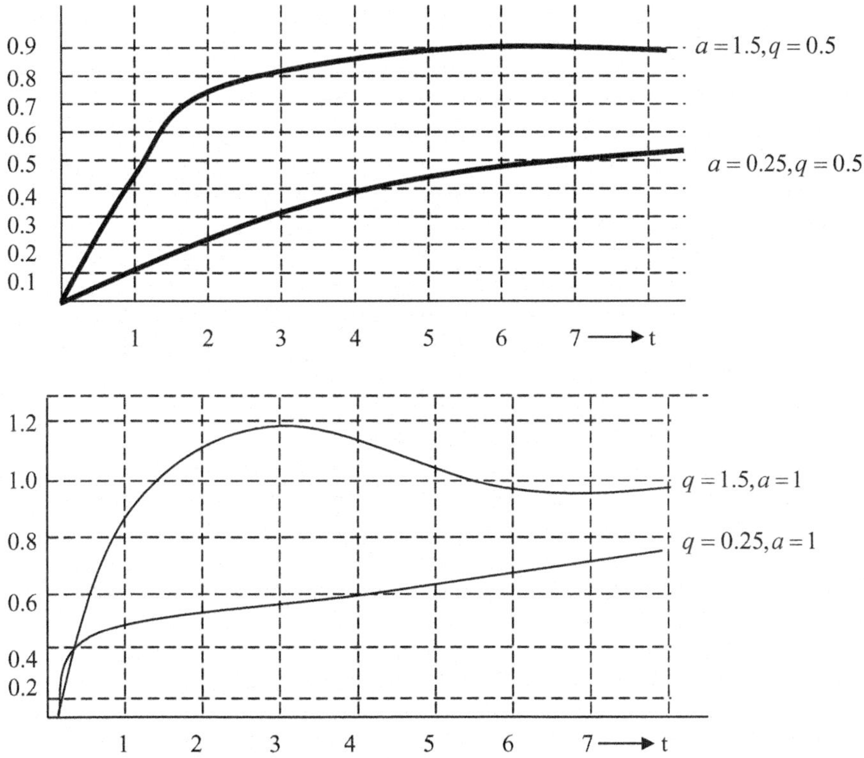

Fig. 8.15 (a) Response to step for varying a at q=0.5
(b) Response to a step for varying q at a=1

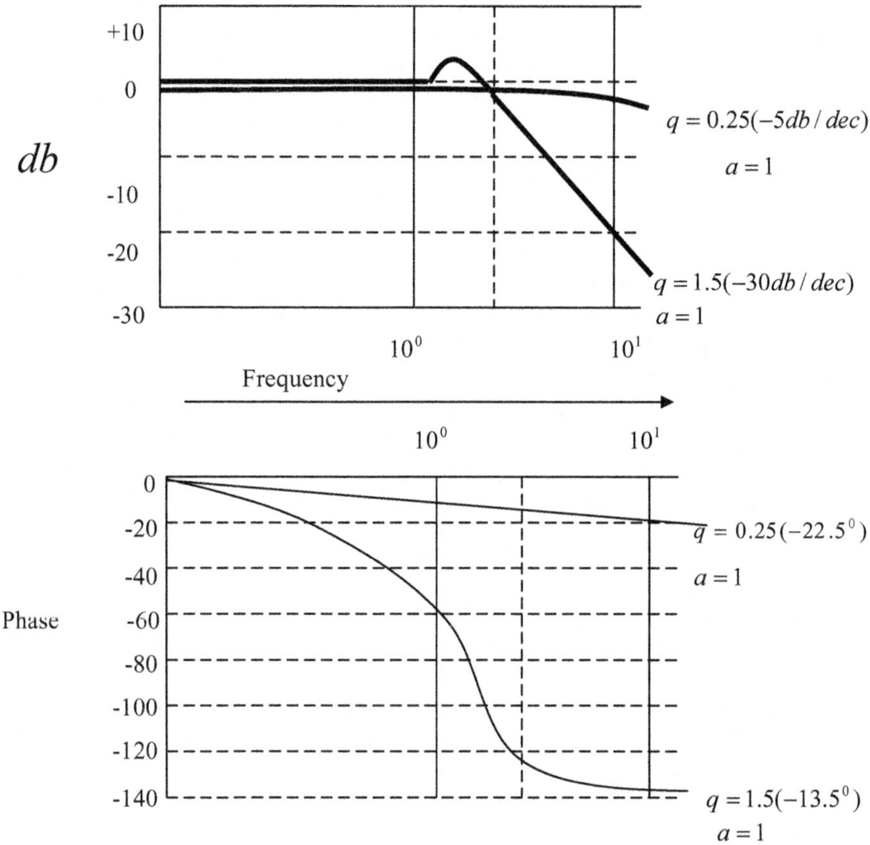

Fig. 8.16 Gain and phase plot with change of order q with constant a

Observations: The tracking of the filter performance with varying q the order as said in Figure 8.16 gives a variable slope of cut-off roll over at the same corner frequency. Also varying the order q, by keeping the corner frequency unchanged, also controls the phase. Conventionally the tuning of filter call for tuning by adjustment of a, to have particular response in time domain, whereas the extra freedom of q gives temporal response adjustment at the user's will. This extra freedom is what the fractional order tuners give i.e. to get temporal or frequency characteristics at user's will. This is the concept of 'fractional pole', which is fundamental to all these fractional order systems.

8.5 Fractional Order State Vector Representation in Circuit Theory

This example in this chapter is chosen as working model for vector space representation. The vector initialization issues have been touched in Chapter 7. This example will elaborate on the same topic and may be therefore extended for any system having set of fractional differential equations as basic phenomena representation. This example pertains to Electrical Science circuit analysis. Figure 8.17 represents the circuit diagram of a voltage source connected by an inductor to a semi-infinite lossy transmission line. As described earlier the lossy semi-infinite line has got terminal characteristics defined by semi-differential element of fractional calculus.

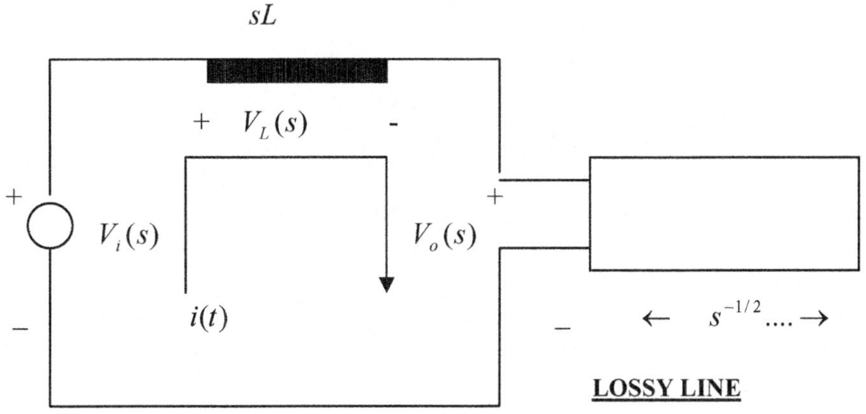

Fig. 8.17 Voltage Source connected to a lossy transmission line through series inductor

Dynamic relationship between current and voltage for inductor is: $v_L(t) = L\left[{}_cD_t^1 i(t)\right]$, where L is the inductance of the coil. For a semi-infinite lossy line the dynamic relationship is: $v_o(t) = \alpha\, {}_cD_t^{-1/2} i(t)$, where α is line constant depends on distribute ohmic resistance per unit length and distributed capacitance per unit length. In the differential form with $c = 0$ (start time) these equations can be expressed as:

$$ {}_0d_t^1 i(t) + \psi(1, i(t), a, 0, t) = \frac{1}{L} v_L(t) \text{ and } {}_0d_t^{1/2} v_o(t) + \psi(1/2, v_o(t), a, 0, t) = \alpha i(t)\,. $$

From Kirchhoff's law we have $v_L(t) = v_i(t) - v_o(t)$. Replacing this in the inductor expression above we get the dynamic equation as

$$ {}_0d_t^1 i(t) = -\frac{1}{L} v_o(t) + \frac{1}{L} v_i(t) - \psi(1, i(t), a, 0, t)\,. $$

To have vector space of fractional dynamic variable it is necessary to reduce all the differential relationship to differential based on the largest common (denominator) differential fraction. In this case choice is ½. We define the following fractional order dynamic variable vector:

$$ \bar{x}(t) = \begin{bmatrix} x_1(t) \\ x_2(t) \\ x_3(t) \end{bmatrix} \equiv \begin{bmatrix} v_o(t) \\ i(t) \\ {}_0d_t^{1/2} i(t) \end{bmatrix} $$

The system input is defined as $u(t) = v_i(t)$ and system output is chosen as $y(t) = v_o(t)$. The vector representation of the system will be as follows:

$$
{}_0d_t^{1/2}\bar{x}(t) = \begin{bmatrix} 0 & \alpha & 0 \\ 0 & 0 & 1 \\ -\dfrac{1}{L} & 0 & 0 \end{bmatrix}\bar{x}(t) + \begin{bmatrix} 0 \\ 0 \\ \dfrac{1}{L} \end{bmatrix}\bar{u}(t) + \begin{bmatrix} -\psi(1/2, v_o(t), a, 0, t) \\ 0 \\ -\psi(1, i(t), a, 0, t) \end{bmatrix}
$$

$$
\bar{y}(t) = \begin{bmatrix} 1 & 0 & 0 \end{bmatrix}\bar{x}(t) + \begin{bmatrix} 0 \end{bmatrix}\bar{u}(t)
$$

Performing Laplace transformations and matrix algebraic manipulations on the above two vector space equations we obtain:

$$
\bar{X}(s) = \frac{1}{s^{3/2} + \alpha / L}\begin{bmatrix} s & \alpha s^{1/2} & \alpha \\ -\dfrac{1}{L} & s & s^{1/2} \\ -\dfrac{s^{1/2}}{L} & -\dfrac{\alpha}{L} & s \end{bmatrix}\left\{\begin{bmatrix} 0 \\ 0 \\ \dfrac{1}{L} \end{bmatrix}\bar{U}(s) - \begin{bmatrix} \psi_1(s) \\ \psi_2(s) \\ \psi_3(s) \end{bmatrix}\right\}
$$

Where $\psi_1(s) = \psi(1/2, v_o(t), a, 0, s)$, $\psi_2(s) = 0$ and $\psi_3(s) = \psi(1, i(t), a, 0, s)$, which is a constant for an inductor. The Laplace transform of the forced response $\bar{X}_F(s)$ is the first term from above relation is:

$$
\bar{X}_F(s) = \frac{(1/L)}{s^{3/2} + (\alpha / L)}\begin{bmatrix} \alpha \\ s^{1/2} \\ s \end{bmatrix}\bar{U}(s).
$$

The Laplace transform of the initialized response $\bar{X}_i(s)$ is from the second term that is:

$$
\bar{X}_i(s) = \frac{-1}{s^{3/2} + (\alpha / L)}\begin{bmatrix} s\psi_1(s) + \alpha s^{1/2}\psi_2(s) + \alpha\psi_3(s) \\ \left(\dfrac{-1}{L}\right)\psi_1(s) + s\psi_2(s) + s^{1/2}\psi_3(s) \\ \left(\dfrac{-s^{1/2}}{L}\right)\psi_1(s) + \left(\dfrac{-\alpha}{L}\right)\psi_2(s) + s\psi_3(s) \end{bmatrix}
$$

These expressions can now be evaluated for any specific inputs and initialization function.

The choice of fractional dynamic variable order basis q should imply minimal configurations. In the above example choosing $q = 1/4$ rather than ½ would yield six dynamic variables instead three. It is also important to remember that the least number of fractional dynamic variables is obtained by choosing the basis q, as the largest common (denominator) fraction of the differential order. (For $q_1 = 1/2$ and $q_2 = 1/3$ would require basis $q = 1/6$)

The input output transfer function with $L = 1$ is:

$$G(s) = \frac{V_0(s)}{V_i(s)} = \frac{\frac{1}{\sqrt{s}}}{s + \frac{1}{\sqrt{s}}} = \frac{1}{s^{3/2} + 1}$$

The time domain representation of the transfer function (without initialization) is thus:

$${}_0d_t^{3/2} v_0(t) + v_0(t) = v_i(t) .$$

Here initialization is zero $\psi(v_0, 3/2, a, 0, t) = 0$, thus, operation is ${}_0D_t^{3/2} v_0(t) = {}_0d_t^{3/2} v_0(t)$. This problem can be solved in several ways depending upon the specific input. Using $F_q[a,t]$ function as obtained for the solution of 'fundamental fractional order differential equation' the impulse response of this circuit with $q = 3/2$ as basis, and $v_i(t) = \delta(t), V_i(s) = 1$, solution is:

$$v_0(t) = \mathcal{L}^{-1}\left\{\frac{1}{s^{3/2} + 1}\right\} = F_{3/2}[-1.t] .$$

For a Heaviside step $H(t) = \begin{cases} 1 & t \geq 0 \\ 0 & t < 0 \end{cases}$ input. $V_i(s) = 1/s$, the solution is:

$$v_0(t) = \mathcal{L}^{-1}\left\{\frac{1}{s\left(s^{3/2} + 1\right)}\right\} = H(t) - E_{3/2}\left[-t^{3/2}\right]$$

For the same circuit solution basis is now chosen to be $q = 1/2$. Here w-plane 'conformal transformation' is demonstrated. This transformation was described in Chapter 7, and let $s^{1/2} = w$. Putting this we get w-plane transfer function as:

$$\frac{V_0(w)}{V_i(w)} = G(w) = \frac{1}{w^3 + 1}$$

The denominator w^3+1 has one root at $w=-1$. Dividing the polynomial w^3+1 by $(w+1)$ gives other factor as, w^2-w+1. The other two roots of this are from roots of w^2-w+1, those are $w=\frac{1}{2}\pm \mathrm{j}\frac{\sqrt{3}}{2}=e^{\pm \mathrm{j}\pi/3}$.

Therefore the factorized form of denominator is:

$$w^3+1=\left(w-\{-1\}\right)\left(w-\left\{\frac{1}{2}+\mathrm{j}\frac{\sqrt{3}}{2}\right\}\right)\left(w-\left\{\frac{1}{2}-\mathrm{j}\frac{\sqrt{3}}{2}\right\}\right)$$

This transfer function has poles in w-plane at $w=-1$, $w=e^{+\mathrm{j}\pi/3}$ and $w=e^{-\mathrm{j}\pi/3}$. Since the basis $q=1/2$, the stability wedge is at angle $\varphi=\pm q\frac{\pi}{2}=\pm\frac{\pi}{4}$. All these poles are to the left of the instability wedge in $w-$plane. The two poles $\exp(+\mathrm{j}\pi/3)$ and $\exp(-\mathrm{j}\pi/3)$ in the right half of the $w-$plane corresponds to poles at $s=e^{+\mathrm{j}2\pi/3}$ and $s=e^{-\mathrm{j}2\pi/3}$ in $s-$plane; therefore an oscillatory response is suggested. The third pole in $w-$plane $w=-1$ is in 'hyper-damped' zone. This pole adds a rapidly decaying time response, to the oscillatory response due to the other two poles.

Expanding the $G(w)$ obtained for base $q=1/2$, we get the following:

$$G(w)=\frac{1}{w^3+1}=\frac{1}{3}\left\{\frac{1}{w+1}-\frac{\frac{1}{2}+\mathrm{j}\frac{\sqrt{3}}{2}}{w-\left(\frac{1}{2}+\mathrm{j}\frac{\sqrt{3}}{2}\right)}-\frac{\frac{1}{2}-\mathrm{j}\frac{\sqrt{3}}{2}}{w-\left(\frac{1}{2}-\mathrm{j}\frac{\sqrt{3}}{2}\right)}\right\}$$

$$=\frac{0.3333}{w+1}-\frac{0.1667+\mathrm{j}0.2887}{w-(0.5+\mathrm{j}0.866)}-\frac{0.1667-\mathrm{j}0.2887}{w-(0.5-\mathrm{j}0.866)}$$

$$=\frac{0.333}{s^{1/2}+1}-\frac{0.1667+\mathrm{j}0.2887}{s^{1/2}-(0.5+\mathrm{j}0.866)}-\frac{0.1667-\mathrm{j}0.2887}{s^{1/2}-(0.5-\mathrm{j}0.866)}$$

The time response can be obtained by inverse Laplace transforming and by using $\mathcal{L}\left\{F_q\left[\pm a,t\right]\right\}=\frac{1}{s^q\mp a}=\frac{1}{w\mp a}$ pair as:

$$v_0(t)=0.3333F_{1/2}\left[-1,t\right]-(0.1667+\mathrm{j}0.2887)F_{1/2}\left[0.5+\mathrm{j}0.866,t\right]-(0.1667-\mathrm{j}0.2887)F_{1/2}\left[0.5-\mathrm{j}0.866,t\right]$$

The $v_0(t)$ thus obtained above for the circuit of semi-derivative terminal point immitance, connected to a voltage source through an inductor, is for only 'forced-response'. Similarly the solution of the 'initialization-response' may be obtained and then superimposed on this.

8.6 Realization of Fractional Order Transfer Function for $PI^{\alpha}D^{\beta}$

8.6.1 *Fractional Order PID Controller Approximation by FPP and FPZ*

This section presents an effective method for the approximation by a rational function, for a given frequency band, of the fractional-order differentiator s^m and integrator s^{-m} (m is a real positive number), and the fractional $PI^{\lambda}D^{\mu}$ controller.

First, the fractional-order integrator s^{-m} $(0 < m < 1)$ is modeled by a fractional power pole (FPP) in a given frequency band of practical interest. Next, this FPP is approximated by a rational function, using the method of singularity function approximations, discussed in earlier Chapter 7. The method in Chapter 7 was used to model the fractional-order differentiator s^m $(0 < m < 1)$ by a fractional power zero (FPZ). Then, the approximation method of the FPP is extended to the FPZ to obtain its rational function approximation. Therefore, with this method, one can achieve any desired accuracy over any frequency band, and a rational function approximation of the fractional-order differentiator and integrator. The rational function approximation of the fractional $PI^{\lambda}D^{\mu}$ controller is just an application of the above method.

8.6.2 *Fractional Order Integrator*

8.6.2.1 Rational Approximation

The transfer function of the fractional-order integrator is represented in the frequency domain by the following irrational transfer function:

$$G_I(s) = \frac{1}{s^m} \tag{8.1}$$

where $s = \mathrm{j}\omega$ is the complex frequency and m is a positive real number such that $0 < m < 1$. The modulus of $G_I(\mathrm{j}\omega) = (\mathrm{j}\omega)^{-m}$ at frequency ω_L is $\frac{1}{\sqrt[m]{\omega_L{}^2}}$. The logarithmic value of the modulus at the frequency ω_L, is therefore, $-10m\log(\omega_L)^2$.

In a given frequency band of practical interest (ω_L, ω_H), this fractional-order operator can be modeled by an FPP whose transfer function is given as follows:

$$G(s) = \frac{K_I}{\left(1 + \left(s/\omega_C\right)\right)^m} \tag{8.2}$$

Suppose that for ω is in between (ω_L, ω_H), where $\omega >> \omega_C$. Therefore

$$G(s)=\frac{K_I}{\left(s/\omega_c\right)^m}=\frac{K_I\omega_c^m}{s^m}=\frac{1}{s^m}=G_I(s)\,. \tag{8.3}$$

The modulus of

$G(\mathrm{j}\omega)=(\omega_C)^{-m}\left(1+\mathrm{j}\omega/\omega_C\right)^{-m}$, at frequency ω_L is $\dfrac{1}{[(\omega_C)^m]\sqrt[m]{1+(\omega_L)^2/(\omega_C)^2}}$.

The logarithmic value of this modulus at ω_L is thus, $-20m\log(\omega_C)-10m\log\left(\frac{\omega_L^2}{\omega_C^2}\right)$. The difference in logarithmic values of gain modulus at ω_L is denoted by ε, also called as slope error, $\varepsilon=\log\left|G_I(\mathrm{j}\omega_L)\right|-\log\left|G(\mathrm{j}\omega)\right|$. Expanding this by putting the logarithmic modulus value we have:

$$-10m\log(\omega_L)^2+20m\log\omega_C+10m\log\frac{(\omega_L)^2}{(\omega_C)^2}=\varepsilon$$

$$-10m\log(\omega_L)^2+10m\log(\omega_C)^2+10m\log\frac{(\omega_L)^2}{(\omega_C)^2}=\varepsilon$$

$$\log\left\{\frac{(\omega_C)^2}{(\omega_L)^2}\left(1+\frac{(\omega_L)^2}{(\omega_C)^2}\right)\right\}=\frac{\varepsilon}{10m}$$

$$\left(\frac{\omega_C}{\omega_L}\right)^2=10^{\left[\frac{\varepsilon}{10m}\right]}-1$$

$$\omega_C=\omega_L\sqrt{10^{\left[\frac{\varepsilon}{10m}\right]}-1}$$

With $K_I=\left(1/\omega_C{}^m\right);\omega_C$ is the $(-3m)$dB corner frequency of the FPP, which is obtained from the low frequency ω_L, as:

$$\omega_C=\omega_L\sqrt{\left(10^{(\varepsilon/10m)}-1\right)}$$

where ε is the maximum permitted error between the slopes of the fractional-order integrator of (8.1) and the FPP of (8.2) in the given frequency band of interest (ω_L, ω_H). Say the frequency of interest is from 0.1 Hz to 10 Hz, and to realize fractional integrator of order 0.6 with 10^{-5} as error in the slope of this realized transfer function with the ideal fractional integrator, the, f_C the -3m dB corner frequency is:

$$f_C = 0.1\sqrt{10^{\left[\frac{10^{-5}}{10\times 0.6}\right]} - 1} \cong 2\times 10^{-4}\ \text{Hz}.$$

In order to represent the FPP of (8.2), and consequently the fractional-order integrator, by a linear time-invariant system model, it is necessary to approximate its irrational transfer function by a rational one. The method of approximation (detailed in Chapter 7) consists of approximating the -20m dB/decade slope on the Bode magnitude plot of the FPP by a number of alternate slopes of -20 and 0 dB/decade corresponding to alternate poles and zeros on the negative real axis of the s-plane such that $p_0 < z_0 < p_1 < z_1 \ldots < z_{N-1} < p_N$. Hence, the approximation is given by:

$$G(s) = \frac{K_I}{\left(1+\left(s/\omega_c\right)\right)^m} \cong K_I \frac{\prod_{i=0}^{N-1}\left(1+\left(s/z_i\right)\right)}{\prod_{i=0}^{N}\left(1+\left(s/p_i\right)\right)} \tag{8.4}$$

Using a simple graphical method (detailed in Chapter 7) that began with a specified error y in decibels and frequency band $\omega_{\max}$ which can be $100\omega_H$ the parameters a, b, p_0, z_0 and N can be calculated as:

$$a = 10^{[y/10(1-m)]},\ b = 10^{[y/10m]},\ p_0 = \omega_C 10^{[y/20m]}$$

$$z_0 = ap_0$$

$$N = \text{Integer}\left[\frac{\log\left(\omega_{\max}/p_0\right)}{\log(ab)}\right] + 1$$

The poles (p_i) and the zeros (z_i) of (8.4) are found to be in a geometric progression form. They can then be derived from the above parameters as $p_i = (ab)^i$ for $i = 0,1,\ldots N$ and $z_i = (ab)^i$ for $i = 0,1,2\ldots(N-1)$. Hence, the fractional order integrator can be approximated by a rational function in a given frequency band of interest as:

$$G(s) = \frac{K_I}{\left(1+\left(s/\omega_C\right)\right)^m} \cong K_I \frac{\prod_{i=0}^{N-1}\left(1+\left(s/z_0(ab)^i\right)\right)}{\prod_{i=0}^{N}\left(1+\left(s/p_0(ab)^i\right)\right)} \tag{8.5}$$

8.6.3 Fractional Order Differentiator

8.6.3.1 Rational Approximation

The transfer function of the fractional-order integrator is represented in the frequency domain by the following irrational transfer function:

$$G_D(s) = s^m \tag{8.6}$$

Where $s = j\omega$ is the complex frequency and m is a positive real number such that $0 < m < 1$. In a given frequency band of practical interest (ω_L, ω_H), this fractional-order operator can be modeled by an FPP whose transfer function is given as follows:

$$G(s) = K_D\left(1+\frac{s}{\omega_C}\right)^m \tag{8.7}$$

Suppose that ω is between (ω_L, ω_H), with $\omega >> \omega_C$. Therefore

$$G(s) = K_D\left(s/\omega_C\right)^m = \frac{K_D s^m}{\omega_C{}^m} = s^m = G_D(s) \tag{8.8}$$

With $K_D = (\omega_C{}^m)$; ω_C is the -3m dB frequency corner of the FPZ, which is obtained from the low frequency ω_L, as:

$$\omega_C = \omega_L\sqrt{\left(10^{(\varepsilon/10m)} - 1\right)}$$

where ε is the maximum permitted error between the slopes of the fractional-order differentiator of (8.6) and the FPZ of (8.7) in the given frequency band of interest (ω_L, ω_H).

In order to represent the FPZ of (8.7), and consequently the fractional-order differentiator, by a linear time-invariant system model, it is necessary to approximate its irrational transfer function by a rational one. The method of approximation consists of approximating the -20m dB/decade slope on the Bode plot of the FPZ by a number of alternate slopes of -20 and 0 dB/decade corresponding to alternate poles and zeros on the negative real axis of the s-plane such that $z_0 < p_0 < z_1 < p_1 < \ldots\ldots < p_N$. Hence, the approximation is given by:

$$G(s) = K_D\left(s/\omega_C\right)^m \cong K_D\frac{\prod_{i=0}^{N}\left(1+\left(s/z_i\right)\right)}{\prod_{i=0}^{N}\left(1+\left(s/p_i\right)\right)} \tag{8.9}$$

Using a simple graphical (detailed in Chapter 7) method that began with a specified error y in decibels and frequency band $\omega_{\max}$ which can be $100\omega_H$, the parameters a, b, p_0, z_0 and N can be calculated as:

$$a = 10^{[y/10(1-m)]}, b = 10^{[y/10m]}, z_o = \omega_C 10^{[y/20m]}$$

$$p_0 = a z_0$$

$$N = \text{Integer}\left[\frac{\log(\omega_{\max}/z_0)}{\log(ab)}\right] + 1$$

The poles (p_i) and the zeros (z_i) of (8.9) are found to be in a geometric progression form. They can then be derived from the above parameters as $p_i = p_0(ab)^i$ for $i = 0,1,2...,N$ and $z_i = z_0(ab)^i$ for $i = 1,2,...(N-1)$. Hence, the fractional order differentiator can be approximated by a rational function in a given frequency band of interest as:

$$G(s) = K_D\left(s/\omega_C\right)^m \cong K_D \frac{\prod_{i=0}^{N}\left(1+\left(s/z_0(ab)^i\right)\right)}{\prod_{i=0}^{N}\left(1+\left(s/p_o(ab)^i\right)\right)} \tag{8.10}$$

8.6.4 Fractional $PI^{\lambda}D^{\mu}$ Controller

8.6.4.1 Rational Approximation

The fractional $PI^{\lambda}D^{\mu}$ controller is a generalization of the PID controller. The transfer function of this controller is given in the frequency domain by the following irrational function:

$$C(s) = K_P + \frac{T_I}{s^{\lambda}} + T_D s^{\mu} \tag{8.11}$$

where $s = j\omega$ is the complex frequency, K_P is the proportional constant, T_I is the integration constant, T_D is the differentiation constant and λ and μ are positive real numbers. In general, these real numbers are such that $\lambda > 1$ and $\mu < 2$. Hence, (8.11) can be rewritten as

$$C(s) = K_P + \left[\frac{T_I}{s}\right]\left(\frac{1}{s^{m_I}}\right) + [T_D s]\left(s^{m_D}\right) \tag{8.12}$$

where (T_I/s) is a first-order integrator, $(1/s^{m_I})$ is a fractional order integrator with $0 < m_I < 1$, $(T_D s)$ is a first-order differentiator and (s^{m_D}) is a fractional-order differentiator with $0 < m_D < 1$. In order to represent the fractional $PI^\lambda D^\mu$ controller of (8.11) by a linear time-invariant system model, it is necessary to approximate its irrational transfer function by a rational one. Hence, in a given frequency band of practical interest (ω_L, ω_H), the fractional-order integrator can be modeled by an FPP and the fractional-order differentiator by an FPZ. It has also been shown how the FPP and the FPZ can be approximated by rational functions.

Hence, (8.12) becomes

$$C(s) = K_P + \left[\frac{T_I}{s}\right]\left(K_I \frac{\prod_{i=0}^{N_I-1}\left(1+\left(s/z_{I_i}\right)\right)}{\prod_{i=0}^{N_I}\left(1+\left(s/p_{I_i}\right)\right)}\right) + \left[T_D s\right]\left(K_D \frac{\prod_{i=0}^{N_D}\left(1+\left(s/z_{D_i}\right)\right)}{\prod_{i=0}^{N_D}\left(1+\left(s/p_{D_i}\right)\right)}\right) \tag{8.13}$$

The poles (p_{I_i}), the zeros (z_{I_i}) and the parameters K_I and N_I of the rational function approximation of the fractional order integrator can be calculated as described in earlier section 8.6.2. Also, the zeros (z_{D_i}), the poles (p_{D_i}) and the parameters K_D and N_D of thc rational function approximation of the fractional-order differentiator can be calculated as described in section 8.6.3.

8.6.5 Realization of Fractional Order Element by Circuit Network

The realization of any rational function can be achieved by using lumped linear circuit components R and C. The most important consideration is the availability of the values of these passive components. Very large capacitances and very low resistances are generally not available. Here the design of simple one port and two port networks are explained.

8.6.5.1 Impedance Functions of a Single Port Network

One port network consists of series connection of $R-C$ parallel, hence problem is to find values of R and C. The partial fraction of rational transfer function, an approximate for fractional integrator will yield:

$$G(s) = \prod_{i=1}^{n} \frac{(s+z_i)}{(s+p_i)} = k_0 + \sum_{i=1}^{n} \frac{k_i}{s+p_i}$$

and the impedance function of the $R-C$ parallel combination is $z_{RC}(s) = \dfrac{1/C}{s+(1/RC)}$. Comparing this with above, for i^{th} $R-C$ parallel segment we get $C_i = 1/k_i$ and $R_i = k_i / p_i$. This means a resistance of value $R_0 = k_0$ and

n sets of $R-C$ parallel segments connected in series, will give one port network with desired impedance function $G(s)$. This implementation is possible only when $p_i < z_i$ or for only fractional integration. For fractional order differentiation the network is implemented as $1/G(s)$ and should be connected in place of $Z_i(s)$ of Figure 8.1. The problem with this implementation is that there are unique values of R_i and C_i for fixed p_i and z_i. Another limitation is that non standard value for capacitor is demanded. Further, after calculations very high capacitances are required for low values of poles and very low value of resistances are required for high frequency poles. These are difficult to obtain and impractical for operational amplifier practical circuits.

8.6.5.2 Impedance Functions of a Two Port Network

Thus another idea is implemented as to use cascaded connections of two port networks each having first order transfer function with a zero. In the circuit of Figure 8.1, $Z_i(s)$ and $Z_f(s)$ of the operational amplifier, select $R-C$ parallel for both these impedances. The indices are 1 for Z_i and 2 for Z_f, and will give the transfer function (Figure 8.1) as,

$$G(s) = \frac{C_1}{C_2}\frac{s+(1/R_1C_1)}{s+(1/R_2C_2)}.$$

Now selecting $C_{1i} = C_{2i} = C_i$, we get $R_{1i} = 1/z_iC_i$ and $R_{2i} = 1/p_iC_i$. Here easily available values of C_i's can be selected independent of p_i and z_i. The required values at the resistances can be configured by using the fixed value of available resistances with trim-pots. This circuit though realizable practically is cascaded by AC coupling, which is susceptible to noise. Hence a modified scheme is generated by this method coupled with single port network.

8.6.5.3 Improved Two Port Network

Further improvement is achieved by combining the above two ideas. The rational approximation transfer function is implemented here, by cascaded connection of the basic two port network (Figure 8.1 with Z_i and Z_f) in which one port network has first order impedance function with a zero represented as $Z'(s)$ as:

$$Z'(s) = \frac{s+z}{s+p} = k' + \frac{k}{s+p}$$

This modified impedance means resistance R' in series with $R-C$ parallel. This gives

$$Z'(s) = R'\frac{s+(1/R_pC)}{s+(1/RC)}, \text{ where } R_p = RR'/(R+R')$$

In this realization zero is always greater than pole. Hence referring to Figure 8.1, for the integration circuit chose $Z_f(s) = Z'(s)$ and $Z_i(s) = R'$ because $p_i = p$ and $z_i = z$. For differentiation circuit put in Figure 8.1, $Z_i(s) = Z'(s)$ and $Z_f(s) = R'$, because $p_i = z$ and $z_i = p$.

For implementation of ith two-port network segment select any available capacitor C_i. Then the resistance values are calculated as $R_i = 1/p_iC_i$ and $R_i' = 1/(z_i - p_i)C_i$, for fractional order integration. For fractional order differentiation we have $R_i = 1/z_iC_i$ and $R_i' = 1/(p_i - z_i)C_i$. For example for 6 pole-zero pair approximation for CPE for half order differentiation, the first zero is at $z_1 = 2.2537\text{rad/sec}$ (Chapter 7 CPE algorithm), and first pole at $p_1 = 6.0406\text{rad/sec}$. Now practically choosing $C_1 = 1\mu\text{F}$, gives $R_1 = 1/z_1C_1 = 443.7\text{K}\Omega$ and $R_1' = 1/(p_1 - z_1)C_1 = 264\text{K}\Omega$. So the first op-amp stage of two port network will comprise of impedance $Z_i(s)$ made with $R_1 = 443\text{K}\Omega$ parallel with $C_1 = 1\mu\text{F}$, series with $R_1' = 264\text{K}\Omega$. The $Z_f(s)$ comprises of $R_1' = 264\text{K}\Omega$. There will be six such circuits for different pole-zero values, to realize approximate of $s^{0.5}$, in the said frequency band (Chapter 7). The circuit is in Figure 8.18 and the values are listed in Table 8.1.

Table 8.1 Six cascaded circuits with values for realization of half order differentiator

i	z_i	p_i	C_i	$R_{zi} = 1/z_iC_i$	$R_{ii} = R_{fi}$ $= 1/(p_i - z_i)C_i$
1	2.2537	6.0406	1μF	443.7KΩ	264KΩ
2	15.955	42.764	1μF	62.67KΩ	37.3KΩ
3	112.95	302.75	680nF	18.83KΩ	11.21KΩ
4	799.65	2143.3	68nF	18.39KΩ	10.94KΩ
5	5661.1	15173	10nF	17.64KΩ	10.5KΩ
6	40078	107420	1nF	24.95KΩ	14.85KΩ

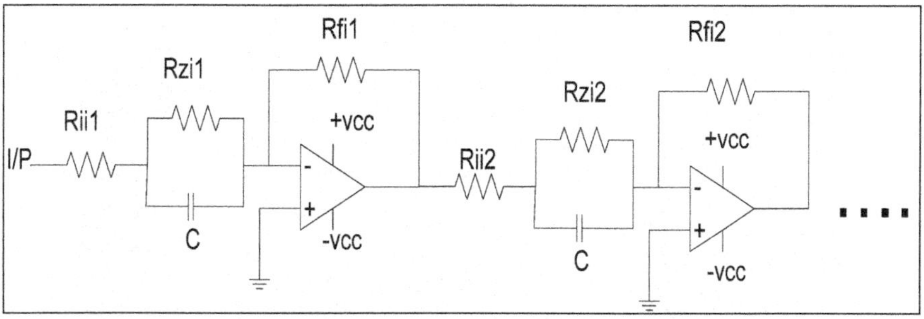

Fig. 8.18 Cascaded circuit for realization of practical fractional order differentiator

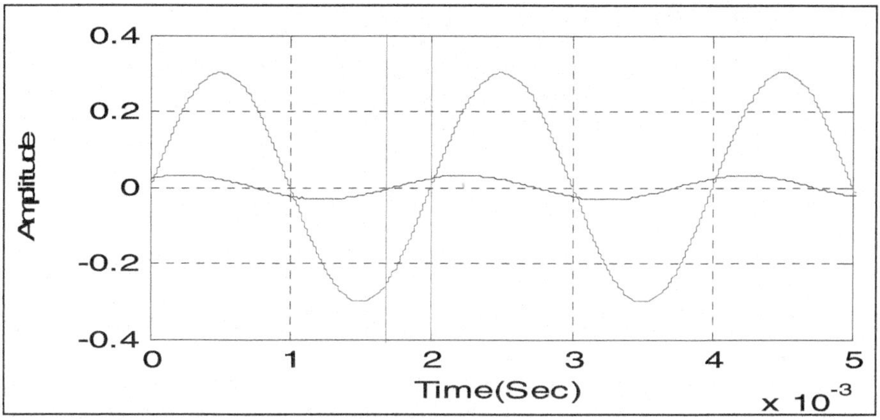

Fig. 8.19 Input and output waveform for α=+0.5

For the circuit shown in Figure 8.18, the input-output wave form at different frequency is obtained. Here it is observed that the input-output phase remains constant at 45^0 with lead, for $\alpha = +0.5$ for the desired frequency band. The waveform shown in Figure 8.19 is for frequency f=500Hz. For fractional integration realization the $Z_f(s)$, comprises of R_{zi} parallel with C_i and series with R_{ii}. The $Z_i(s)$ will comprise of R_{fi} in Figure 8.18. This way one can obtain rational approximates of the fractional order differintegrals for a desired band of frequency for any value from $\alpha \in (0,1)$. The Figure 8.20 gives picture of this technique applied to make hardwire fractional order PID. The knobs and selection switches are set so as to get fractional order differentiation or integration for order 0.2, 0.5, 0.8 and 1.0, with proportional integral and differential gain potentiometers. The experiment in Figure 8.20 shows connection with DC motor emulator circuit as devise to be controlled. This experimental platform gives hardwire environment for study of 'iso-damping' and various aspects of controls (integer order vis-à-vis fractional order PID).

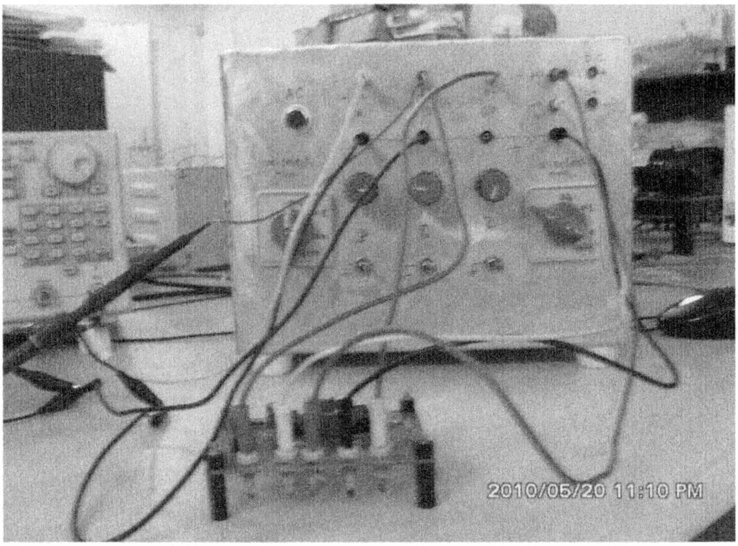

Fig. 8.20 Hardwire circuit with fractional order PID being tested in close loop with DC Motor emulator circuit

8.7 Advance Digital Algorithms Realization for Fractional Controls

As seen from the definitions of fractional differintegrations the operations are not a local or point property, rather distributed quantities. In the process therefore to obtain value of fractional differentiation and integration a bit of historical behavior are required. In spatial terms required is what behavior at the neighborhood of that function is. The GL RL definitions capture that historical behavior of the function, undoubtedly enormous amount of memory requirement be needed in spite of having short-memory principle. The historical behavior of the function is described by initialization function (instead of constant states in integer order calculus). Minimum 100 recent past history of the function is required to obtain good estimate with tolerable error and to get that initial function. There are advances in algorithms are to emulate the fractional differintegration by digital control theory of discretization lead to several other memory efficient method, as compared to GL method described herein. Few advance digital control algorithms are by power series expansion of Tustin rule Al-Alouni rule continued fraction expansion and method of interpolation through fractional order delay. These advance algorithms are to realize the fractional control digital systems with approximately one-tenth of memory requirement as compared with the short-memory principle of the GL method. Advances in the algorithm development in this direction are ongoing process, and this century will see several of them. Here in this section attempt is made to give the digital realization of the fractional differintegration operators. In this section required is reader's knowledge about z- transforms and control system basics and digital filters fundamentals.

8.7.1 Concept of Generating Function

Discretization of fractional order differentiation s^r can be expressed by so called generating function. Here the s is Laplace variable and in discrete domain is z variable. To revise the concepts of digital controls the monomial z^{-1} is the unit delay dictated by sampling or dicretization time T. The generating function is $s=\omega(z^{-1})$. For backward difference or Euler's rule the function is $G_{gf}(z^{-1})=\frac{1-z^{-1}}{T}$. Performing the power series expansion (PSE) of the $(1-z^{-1})^{\pm r}$ using short memory principle we get:

$$(s)^{\pm r}=\left(G_{gf}(z^{-1})\right)^{\pm r}=T^{\mp r}z^{-[L/T]}\sum_{j=0}^{[L/T]}(-1)^j\binom{\pm r}{j}z^{[L/T]-j}$$

T: sampling period, L: Memory length, $(-1)^j\binom{\pm r}{j}$ are binomial coefficients of the form i.e. $c_0^{(r)}=1, c_j^{(r)}=\left(1-\frac{1+(\pm r)}{j}\right)c_j^{(r)}-1$. This PSE of the backward difference rule give the digital realization of the GL Method with short memory principle. Applying this realization in digital filter realization gives the Finite Impulse Response (FIR) digital filter.

Generating function for the Trapezoidal rule (Tustin) is

$$G_T(z)=\left(G_{gf}(z^{-1})\right)^{\pm r}=\left(\frac{2}{T}\frac{1-z^{-1}}{1+z^{-1}}\right)^{\pm r}$$

Al-Alouni mixed Trapezoidal and Euler generating function is

$$G(z)=\left(G_{gf}(z^{-1})\right)^{\pm r}=\left(\frac{8}{7T}\frac{1-z^{-1}}{1+z^{-1}}\right)^{\pm r}$$

These are infinite order rational discrete time transfer function. To approximate it with a finite order rational one Continued Fraction Expansion (CFE) is an efficient way.

$$G(z)=\hat{G}(z)=a_0(z)+\cfrac{b_1(z)}{a_1(z)+\cfrac{b_2(z)}{a_2(z)+\cfrac{b_3(z)}{\cdots\cdots}}},$$

With a_i, b_i are either rational functions of the variable z or constants. With CFE method the realization obtained is Infinite Impulse Response (IIR) digital filter. In number theory the CFE method is used to represent a real number. A number can be represented as

$$x = a_0 + \cfrac{1}{a_1 + \cfrac{1}{a_2 + \dots}}.$$

For $x=\pi$, $a_0 = [\pi] = 3$, $a_1 = \left[\frac{1}{\pi - 3}\right] = 7$, $a_2 = \left[\cfrac{1}{\cfrac{1}{\pi-3} - 7}\right] = 15$.

The $[*]$ is 'FLOOR' function returns the integer part after operation.

8.7.2 Digital Filter Realization by Rational Function Approximation for Fractional Operator

Al-Alouni (1997) stated that magnitude of frequency of ideal integrator $1/s$ lies between that of Simpson and Trapezoidal digital integrators. It is reasonable to interpolate the Simpson and Trapezoidal digital integrators to compromise the high frequency response. The Simpson digital integrator is

$$H_S(z) = \frac{T(z^2+4z+1)}{3(z^2-1)},$$

The Tustin (trapezoidal) integrator is $H_T(z) = \frac{T}{2}\frac{(z+1)}{(z-1)}$.

The combined digital integrator is $H(z) = aH_S(z) + (1-a)H_T(z)$, and the tuning knob fraction $a \in [0,1]$. Putting Simpson integrator and Tustin integrator we obtain

$$H(z) = \frac{T(3-a)\left\{z^2 + [2(3+a)/(3-a)]z + 1\right\}}{6(z^2-1)} = \frac{T(3-a)(z+r_1)(z+r_2)}{6(z^2-1)}$$

With $r_1 = \frac{3+a+2\sqrt{3a}}{(3-a)}$ and $r_2 = \frac{3+a-2\sqrt{3a}}{(3-a)}$. Note that $r_1 = \frac{1}{r_2}$, and $r_1 = r_2 = 1$, when $a=0$ (Pure Tustin). For $a \neq 0$, $H(z)$ must have one Non –Minimum Phase zero.

Firstly we can obtain a family of new integer order digital differentiators from the digital integrator introduced above by mixing Simpson and Tustin. Direct inversion of $H(z)$ will give an unstable filter since $H(z)$ has non-minimum phase (NMP) zero r_1.

By reflecting the NMP r_1 to $1/r_1$ i.e. r_2 we have approximate as

$$\hat{H}(z) = K\frac{T(3-a)(z+r_2)^2}{6(z^2-1)}.$$

To determine K let the final value of the impulse response have $H(z)$ and $\hat{H}(z)$ be the same. Applying final value theorem i.e.

$\lim_{z\to 1}(z-1)H(z) = \lim_{z\to 1}(z-1)\hat{H}(z)$, which gives $K = r_1$. Therefore, the new families of the digital differentiator are given by generating function

$$G(z) = \frac{1}{\hat{H}(z)} = \frac{6(z^2-1)}{r_1 T(3-a)(z+r_2)^2} = \frac{6r_2(z^2-1)}{T(3-a)(z+r_2)^2}$$

Finally we can obtain the expression for the Digital Fractional Order Differentiator as:

$$G(z^{-1}) = \left(G_{gf}(z^{-1})\right)^r = k_0\left(\frac{1-z^{-2}}{(1+bz^{-1})^2}\right)^r \quad \text{Where } r \in [0,1],$$

$$k_0 = \left(\frac{6r_2}{T(3-a)}\right)^r \text{ and } b = r_2.$$

It is well known that compared to Power Series Expansion (PSE) the Continued Fraction Expansion (CFE) is a method of evaluation of functions with faster convergence in larger domain in complex plane. Using the CFE approximations for an irrational function $G(z^{-1})$ can be expressed as approximation. The CFE can be automated by using MATLAB symbolic toolbox $CFE\left(\frac{1-x^2}{(1+bx)^2}\right)^r$ with $x = z^{-1}$ for desired order n. The MATLAB script generates the above CFE with polynomial containing the numerator and denominator with coefficients being function of b and r. Following list is IIR transfer function for $n = 3$, $a = 0-1$ in steps of 0.25 for semi-differentiation $r = 0.5$ at sampling time of $T = 0.001s$ (1ms).

$G_{n,a}$ Means n-th order polynomial approximate at mixing value a.

$$G_{(3,0)}(z^{-1}) = \frac{357.8-178.9z^{-1}-178.9z^{-2}+44.72}{8+4z^{-1}-4z^{2}-z^{-3}}$$

$$G_{(3,0.25)}(z^{-1}) = \frac{392.9-78.04z^{-1}349.8z^{-2}+88.97z^{-3}}{11.32+4z^{-1}-5.66z^{-2}-z^{-3}}$$

$$G_{(3,0.5)}(z^{-1}) = \frac{1501-503.6z^{-1}-1289z^{-2}+446.5z^{-3}}{47.26+4z^{-1}-23.63z^{-2}-z^{-3}}$$

$$G_{(3,0.75)}(z^{-1}) = \frac{968.1 - 442z^{-1} - 820.8z^{-2}363z^{-3}}{32.47 - 4z^{-1} - 16.24z^{-2} + z^{-3}}$$

$$G_{(3,1.00)}(z^{-1}) = \frac{353.1 - 208z^{-1} - 297.4z^{-2} + 164.7z^{-3}}{12.46 - 4z^{-1} - 6.228z^{-2} + z^{-3}}$$

The fourth order approximation is for digital IIR is listed below

$$G_{(4,0)}(z^{-1}) = \frac{715.5 - 357.8z^{-1} - 536.7z^{-2} + 178.9z^{-3} + 44.72z^{-4}}{16 + 8z^{-1} - 12z^{-2} - 4z^{-3} + z^{-4}}$$

$$G_{(4,0.25)}(z^{-1}) = \frac{555.3 - 392.9z^{-1} - 477.2z^{-2} + 349.8z^{-3} - 19.56z^{-4}}{16 - 2.489z^{-1} - 12z^{-2} + 1.245z^{-3} + z^{-4}}$$

$$G_{(4,0.5)}(z^{-1}) = \frac{508.1 - 1501z^{-1} - 4.478z^{-2} + 1289z^{-3} - 382.9z^{-4}}{16 - 40.54z^{-1} - 12z^{-2} + 20.27z^{-3} + z^{-4}}$$

$$G_{(4,0.75)}(z^{-1}) = \frac{477 + 968.1z^{-1} - 919z^{-2} - 820.8z^{-3} + 422.7z^{-4}}{16 + 37.8z^{-1} - 12z^{-2} - 8.371z^{-3} + z^{-4}}$$

8.7.3 Filter Stability Consideration

For odd CFE $n = 3$, the pole-zero maps are nicely placed and behaved, that is, all pole and zeros of the transfer function lie inside the unit circle and the poles and zeros are interlaced along the segment of the real axis corresponding to $z \in (-1,1)$. However, when $n = 4$ even and if 'a' is near unity (tending towards Simpson rule) there may have one canceling pole-zero pair, which may not be desirable. Therefore suggestion is to use odd order (polynomial) expansions for CFE. When $a = 0$ for Tustin (trapezoidal) rule pole-zero (P-Z) maps always inside unit circle in an interlacing way along the segment of the real axis $z \in (-1,1)$. For special case $a = 0$ Tustin CFE for odd polynomial expansion gives

$$D^r(z) = CFE\left(\frac{1-z^{-1}}{1+z^{-1}}\right)^r = 1 + \cfrac{z^{-1}}{-\cfrac{1}{2}\cfrac{1}{r} + \cfrac{z^{-1}}{-2 + \cfrac{z^{-1}}{\cfrac{3}{2}\cfrac{r}{r^2-1} + \cfrac{z^{-1}}{2+\ldots\ldots}}}}$$

For semi-differentiation realization $r = 0.5$ for discretization time $T = 0.001s$ (1ms) the approximate odd CFE expansions are:

$$G_1(z) = 44.72\frac{z-0.5}{z+0.5}, \quad G_3(z) = 44.72\frac{z^3 - 0.5z^2 - 0.5z + 0.125}{z^3 + 0.5z^2 - 0.5z - 0.125}$$

8.8 Charge Conservation for Fractal Distribution

Let us consider a fractal distribution of charged particles. For example we assume that charged particles with constant density are distributed over the fractal. In this case, the number density of particles N enclosed in a volume with characteristic size R satisfies the scaling law $N(R) \sim R^d$, whereas for a regular n-dimensional Euclidian object we have $N(R) \sim R^n$ (n is integer). For charged particles with number density $n(r,t)$ we have that the charge density can be defined by $\rho(r,t) = \pm en(r,t)$, where $\pm e$ represents charge in coulombs. The total charge of the region W is then given as $Q(W) = \int_W \rho(r,t)dV_3$, or $Q(W) = \pm eN(W)$, where $N(W)$ is the number of particles in the region W. The fractional generalization of this equation is $Q(W) = \int_W \rho(r,t)dV_d$, where d is the fractal dimension of the distribution and dV_d is element of d-dimensional volume, such that $dV_d = K_3(r,d)dV_3$ and $K_3(r,d) = r^{d-3}\Gamma^{-3}(d/3)$ (refer Chapter 5 for details).

Let us try to define Electric current of fractal distribution. For charged particles with number density $n(r,t)$ flowing with velocity $v = v(r,t)$ (a function of position and time); the resulting current density is given by $J(r,t)$ as $J(r,t) = en(r,t)v$. Where e is the coulomb of charge of particle. For electrons it is $-e$ and for holes it is $+e$. The electric current defined as the flux of electric charge. Measuring the field $J(r,t)$ passing through a surface $S = \partial W$ gives the current (that is flux of charge) as:

$I(S) = \Phi_J(S) = \int_S (J \bullet dS_2)$, where $J = J(r,t)$ is the current vector, dS_2 is differential area. The fractional generalization of this flux equation for the fractal distribution can be written in the following form:

$I(S) = \int_S (J(r,t) \bullet dS_d)$, where $dS_d = K_2(r,d)dS_2$, and $K_2 = r^{d-2}\Gamma^{-2}(d/2)$ (Chapter 5).

The electric charge has fundamental property; the change of quantity of charge inside a region W is bounded by the surface $S = \partial W$ is always equal to the flux of charge through this surface. If we denote $J(r,t)$ as the electric current density, then the charge conservation is written as:

$\frac{dQ(W)}{dt} = -I(S)$ or in integral form as $\frac{d}{dt}\int_W \rho(r,t)dV_d = -\oint_{\partial W} (J(r,t) \bullet dS_d)$. In particular when the surface $S = \partial W$ is fixed we can write $\frac{d}{dt}\int_W \rho(r,t)dV_d = \int_W \frac{\partial \rho(r,t)}{\partial t} dV_d$. Using the fractional generalization of the Gauss's law (Chaprer-5), we have:

$$\oint_{\partial W}\left(J(r,t)\bullet dS_d\right)=\int_W\left(K_3(r,d_3)\right)^{-1}\frac{\partial}{\partial x}\left[K_2(r,d_2)J(r,t)\right]dV_d,$$

where $K_3(r,d_3)=r^{d_3-3}\Gamma^{-3}(d_3/3)$. Assuming the flux vector is flowing in x direction only. Substituting the RHS above two equations into the integral equation of charge conservation law (above); we get the following:

$$K_3(r,d_3)\frac{\partial\rho(r,t)}{\partial t}+\frac{\partial}{\partial x}\left[K_2(r,d_2)J(r,t)\right]=0$$

This equation can be considered as continuity equation for fractal distribution of particles.

8.9 Electric Field of Fractal Distribution

8.9.1 Electric Field and Coulomb's Law for Fractal Distribution

In this section the electrostatic laws will be generalized for fractal distributions, from the generalization concepts developed in Chapter 5. The Coulomb's law relates Electric field to charges, as they are the source. For a point charge Q at position r' the electric field at a point r is given as

$E=\frac{Q}{4\pi\varepsilon_0}\frac{r-r'}{|r-r'|^3}$. For a continuous stationary distribution of charges $\rho(r')$ the Electric field is: $E(r)=\frac{1}{4\pi\varepsilon_0}\int_W\frac{r-r'}{|r-r'|^3}\rho(r')dV_3'$ with $dV_3'=dx'dy'dz'$. These are well known expressions for electrodynamics. The fractional generalizations of these laws are taken in this section for fractal charge distribution, in the following sections with concept developed in Chapter 5.

The fractional generalizations of the above equations for the electric field is

$E(r)=\frac{1}{4\pi\varepsilon_0}\int_W\frac{r-r'}{|r-r'|^3}\rho(r')dV_d'$, where $2<d<3$ and $dV_d'=K_3(r,d)dV_3'$

This can be considered as Coulomb's law for a fractal stationary distribution of electric charges in a fractal dimension d enclosed in region W.

8.9.2 Gauss's Law for Fractal Distribution

Electric field passing through a surface $S=\partial W$ gives 'electric flux' as

$\Phi_E(S)=\oint_S E\bullet dS_2$, where E is the electric-vector and dS_2 is the differential area. Gauss law tells us that total flux $\Phi_E(S)$ of the electric field E through a closed surface $S = \partial W$ is proportional to the electric charge $Q(W)$ inside the surface, that is $\varepsilon_0 \Phi_E(S) = Q(W)$. The electric flux out of any closed surface is proportional to the total charge enclosed within the surface.

For the fractal distribution the Gauss's law states: $\oint_S E \bullet dS_2 = \frac{1}{\varepsilon_0} \int_W \rho(r,t) dV_d$, where $dV_d = K_3(r,d) dV_3$.

If we consider the stationary spherically symmetric fractal distribution $\rho(r,t) = \rho(r)$ and a ball region W; such that $r \le R$, then we have $Q(W) = 4\pi \int_0^R \rho(r) K_3(r,d) r^2 dr$. Putting the value of $K_3(r,d) = r^{d-3} \Gamma^{-3}(d/3)$, we have $Q(W) = 4\pi \{\Gamma^{-3}(d/3)\} \int_0^R \rho(r) r^{d-1} dr$.

Using the sphere S such that its radius is $r = R$ as surface $S = \partial W$ we get $\Phi_E(\partial W) = 4\pi R^2 E(R)$. Substituting these two in the Gauss law we get the equation of electric field. As a result of Gauss's law for fractal distribution with spherical symmetry leads to the equation for electric field as:

$$E(R) = \frac{1}{\varepsilon_0 R^2 \Gamma^3(d/3)} \int_0^R \rho(r) r^{d-1} dr$$

For example the electric field of homogeneous, uniform $\rho(r) = \rho_0$, spherically symmetric fractal distribution is defined by $E(R) = \frac{\rho_0}{\varepsilon_0 d \Gamma^3(d/3)} R^{d-2} \sim R^{d-2}$.

8.10 Magnetic Field of Fractal Distribution

8.10.1 Biot-Savart Law for Fractal Distribution

The Biot-Savart law relates magnetic fields to the currents. The sources of magnetic fields are currents. This is similar to Coulomb's law. For a continuous distribution the Biot-Savart law is $B(r) = \frac{\mu_0}{4\pi} \int_W \frac{J(r') \times (r - r')}{|r - r'|^3} dV_3'$.

Finding the magnetic field resulting from a fractal current distribution involves the vector product and is inherently a fractional calculus problem when the distance from the current to the field point is continuously changing. The fractional generalization of the above equation is: $B(r) = \frac{\mu_0}{4\pi} \int_W \frac{J(r') \times (r - r')}{|r - r'|^3} dV_d'$.

8.10.2 Ampere's Law for Fractal Distribution

The magnetic field in space around an electric current is proportional to the electric current which serves as source, just as electric field in space is proportional to the charge. In case of static electric fields the line integral of magnetic field around a close loop is proportional to the electric current flowing through the loop. The Ampere's law is equivalent to the steady state of the Maxwell integral equation in free space, and relates the spatially varying magnetic field $B(r)$ to the current density $J(r)$. Let there be a closed path around a current. The Ampere's law states that line integral of the magnetic field B along the closed path L is given by $\oint_L B \bullet dL = \mu_0 I(S)$, where dL is differential length of line element. For fractal distribution the generalization is $I(S) = \int_S (J \bullet dS_d)$, where $dS_d = K_2(r,d)dS_2$, with $1 < d < 2$ and $K_2(r,d) = r^{d-2}\Gamma^{-2}(d/2)$.

If we consider cylindrical symmetry fractal distribution we have $I(S) = 2\pi \int_0^R J(r)K_2(r,d)rdr$ and putting the value of $K_2(r,d)$ we get $I(S) = 2\pi[\Gamma^{-2}(d/2)] \int_0^R J(r)r^{d-1}dr$. Using a circle as L with radius $r = R$ we have $\oint_L (B \bullet dL) = 2\pi RB(R)$, where $B(R) = \frac{\mu_0}{R\Gamma^2(d/2)} \int_0^R J(r)r^{d-1}dr$. For example magnetic field $B(r)$, at $r = R$ of uniform and homogeneous, fractal current density $J(r) = J_0$ is $B(R) = J_0 \frac{\mu_0}{d\Gamma^2(d/2)} R^{d-1} \sim R^{d-1}$.

8.11 Maxwell Equation for Fractal Distribution

The Maxwell equations are fundamental set of equations governing electromagnetism. The equations that can be expressed in integral form expressions are:

$$\oint_S (E \bullet dS_2) = \frac{1}{\varepsilon_0} \int_W \rho dV_3$$

$$\oint_L (E \bullet dL_1) = -\frac{\partial}{\partial t} \int_S (B \bullet dS_2)$$

$$\oint_S (B \bullet dS_2) = 0$$

$$\oint_L (B \bullet dL_1) = \mu_0 \int_S (J \bullet dS_2) + \varepsilon_0 \mu_0 \frac{\partial}{\partial t} \int_S (E \bullet dS_2)$$

Suppose that the electromagnetic field can be defined on fractal as an approximation of some real case with fractal medium, and if the fields $E(r)$ and $B(r)$ can be defined on fractal and does not exist outside fractal in Euclidian space of 3D, then we can have fractional generalization of these integral Maxwell equations in the form:

$$\oint_S (E \bullet dS_d) = \frac{1}{\varepsilon_0} \int_W \rho dV_d$$

$$\oint_L (E \bullet dL_d) = -\frac{\partial}{\partial t} \int_S (B \bullet dS_d)$$

$$\oint_S (B \bullet dS_d) = 0$$

$$\oint_L (B \bullet dL_d) = \mu_0 \int_S (J \bullet dS_d) + \varepsilon_0 \mu_0 \frac{\partial}{\partial t} \int_S (E \bullet dS_d)$$

These fractional integrals are considered as an approximation of integrals on fractals.

Using the fractional generalization of the Stroke's law and Gauss's law for line-surface-volume integrations, (Chapter 5) we write the fractional Maxwell equations in the form as:

$$\int_W (K_3(r,d_3))^{-1} \text{div}\left[K_2(r,d_2)E\right] dV_d = \frac{1}{\varepsilon_0} \int_W \rho dV_d$$

$$\int_S (K_2(r,d_2))^{-1} \left(\text{curl}\left[K_1(r,d_1)\right]\right) dS_d = -\frac{\partial}{\partial t} \int_S B \bullet dS_d$$

$$\int_W (K_3(r,d_3))^{-1} \text{div}\left[K_2(r,d_2)B\right] dV_d = 0$$

$$\int_S (K_2(r,d_3))^{-1} \left(\text{curl}\left[K_1(r,d_1)\right] B.dS_d\right) = \mu_0 \int_S (J \bullet dS_d) + \varepsilon_0 \mu_0 \frac{\partial}{\partial t} \int_S (E \bullet dS_d)$$

As a result we get following Maxwell equations in differential forms:

$$\text{div}\left[K_2(r,d_2)E\right] = \frac{1}{\varepsilon_0} K_3(r,d_3)\rho$$

$$\text{curl}\left[K_1(r,d_1)E\right] = -K_2(r,d_2)\frac{\partial}{\partial t}B$$

$$\text{div}\left[K_2(r,d_2)B\right] = 0$$

$$\text{curl}\left[K_1(r,d_1)B\right] = \mu_0 K_2(r,d_2)J + \varepsilon_0 \mu_0 K_2(r,d_2)\frac{\partial E}{\partial t}$$

Note that in classical electrodynamics there is no magnetic charges or 'magnetic monopoles'. This makes $\text{div}\left[K_2(r,d_2)B\right] = 0$ .This magnetic field B is also called 'dipole' field; no matter how small the volume is, you will get equal numbers of north and south poles. This equation can be re-written as: $\mathrm{div}B = -\left(B \bullet \mathrm{grad}K_2(r,d_2)\right)$. For a general case $d \neq 2$ the vector $\mathrm{grad}\left[K_2(r,d_2)\right] \neq 0$ and the magnetic field satisfies $\mathrm{div}\left[B\right] \neq 0$, indicating presence of 'magnetic monopole' or existence of 'magnetic-charges', with magnitude as $e_m \sim B \bullet \nabla K_2(r,d_2)$.

8.12 Electric Dipole Moments for Fractal Distribution

To compute one particular case of multi-pole expansion, let $\mathbf{R}$ be a vector from a fixed point i.e., origin to the observation, $\mathbf{r}$ be the vector from the reference point (origin) to a point in the body to the observation point, and $\mathbf{d} = \mathbf{R} - \mathbf{r}$ be the vector from a point in the body to the observation point. From law of cosines we have: $d^2 = R^2 + r^2 - 2rR\cos\theta = R^2\left(1 + \frac{r^2}{R^2} - 2\frac{r}{R}\cos\theta\right)$ where $d = |\mathbf{d}|$ and $\cos\theta = (\mathbf{r}.\mathbf{R})/(rR)$ so $d = R\sqrt{1 + \frac{r^2}{R^2} - 2\frac{r}{R}\cos\theta}$. Now define $\in = r/R$ and $x = \cos\theta$, then $\frac{1}{d} = \frac{1}{R}\left(1 - 2\in x + \in^2\right)^{-1/2}$. Note that $(1 - 2\in x + \in^2)^{-1/2}$ is the generating function for the Legendre polynomial $P_n(x)$ defined as $(1 - 2\in x + \in^2)^{-1/2} = \sum_{n=0}^{\infty} \in^n P_n(x)$. Using this we have the equation $\frac{1}{d} = \frac{1}{R}\sum_{n=0}^{\infty}\left(\frac{r}{R}\right)^n P_n(\cos\theta)$.

Any potential (gravitational, electric and magnetic) which behaves as inverse square $(1/d)$ of distance can be expressed as:

$$U = \frac{1}{4\pi\varepsilon_0}\sum_{n=0}^{\infty}\frac{1}{R^{n+1}}\int_W r^n P_n(\cos\theta)\,\rho(\mathbf{r})dV_3 .$$

The $n = 0$ term in this expansion is called monopole term, can be written from above, as $P_0(x) = 1$; that is in expanded form with monopole extracted in following way:

$$U = \frac{1}{4\pi\varepsilon_0}\frac{1}{R}\int_W \rho(\mathbf{r}) + \frac{1}{4\pi\varepsilon_0}\sum_{n=1}^{\infty}\frac{1}{R^{n+1}}\int_W r^n P_n(\cos\theta)\,\rho(\mathbf{r})dV_3$$

The n-th term $U_n = \frac{1}{4\pi\varepsilon_0}\frac{1}{R^{n+1}}\int_W r^n P_n(\cos\theta)\,\rho(\mathbf{r})dV_3$ is commonly termed to the multi-pole; when $n = 0$ this is monopole, with $n = 1$ this potential is dipole, and for $n = 2$ this is quadra-pole moments.

Therefore electric multi-pole expansion is determination of the voltage U due to collection of charges obtained by performing a multi-pole expansion. This corresponds to a series expansion of charge density $\rho(\mathbf{r})$ in terms of its moments given by expression $U = \frac{1}{4\pi\varepsilon_0}\sum_{n=0}^{\infty}\frac{1}{R^{n+1}}\int_W r^n P_n(\cos\theta)\,\rho(\mathbf{r})dV_3$ where $P_n(\cos\theta)$ is the Legendre Polynomial and θ is the polar angle defined such that $\cos\theta = (\mathbf{r}.\mathbf{R})/|\mathbf{r}||\mathbf{R}|$. The first term arises from $P_0(x)=1$, while all further terms vanish as a result of $P_n(x)$ being a polynomial in x for $n \geq 1$, giving $P_n(0)=0$ for all $n \geq 1$. If we have $Q(W) = \int_W \rho(\mathbf{r})dV_3 = 0$ then the monopole term, for $n=0$ vanishes.

Set up coordinate system so that θ measures the angle from the charge-charge line with the midpoint of this line being the origin. Then $n=1$ the dipole term is given as:

$$
\begin{aligned}
U_1 &= \frac{1}{4\pi\varepsilon_0}\frac{1}{R^2}\int_W rP_1(\cos\theta)\,\rho(\mathbf{r})dV_3 \\
&= \frac{1}{4\pi\varepsilon_0}\frac{1}{R^2}\int_W r\cos\theta\rho(\mathbf{r})dV_3 \\
&= \frac{1}{4\pi\varepsilon_0}\frac{1}{R^2}\int_W \frac{(\mathbf{r}.\mathbf{R})}{R}\rho(\mathbf{r})dV_3 \\
&= \frac{1}{4\pi\varepsilon_0}\frac{1}{R^3}\int_W (\mathbf{r}.\mathbf{R})\rho(\mathbf{r})dV_3 \\
&= \frac{1}{4\pi\varepsilon_0}\frac{1}{R^3}\left(\mathbf{R}.\int_W \mathbf{r}\rho(\mathbf{r})dV_3\right) = \frac{1}{4\pi\varepsilon_0}\frac{(\mathbf{R}.\mathbf{p})}{R^3}
\end{aligned}
$$

For a continuous charge distribution the electric dipole moment is defined as $\mathbf{p} = \int_W \mathbf{r}\rho(\mathbf{r})dV_3$, where $\mathbf{r}$ points from positive to negative.

For a fractal distribution define the dipole moment as generalization of classical definition; that is: $\mathbf{p}_d = \int_W \mathbf{r}\rho(\mathbf{r})dV_d$. This gives dipole moment of fractal distribution as:

$$
U_1 = \frac{1}{4\pi\varepsilon_0}\frac{(\mathbf{R}.\mathbf{r}_d)}{R^3} = \frac{1}{4\pi\varepsilon_0}\frac{p_d\cos\alpha}{R^2}
$$

where $p_d = \sqrt{\left[p_{x(d)}\right]^2 + \left[p_{y(d)}\right]^2 + \left[p_{z(d)}\right]^2}$ and $\cos\alpha = (\mathbf{R}.\mathbf{p}_d)/(p_d R)$.

Let us consider the dipole moment for the fractal distribution, given by $\mathbf{p}_d = \int_W \mathbf{r}\rho(\mathbf{r})dV_d$ .Referring to Chapter 5, we have $K_3(x,y,z,d) = \frac{|xyz|^{\frac{d}{3}-1}}{\Gamma^3(d/3)}$ .Let us consider dipole moment for uniform homogeneous $\rho(\mathbf{r}) = \rho_0$ fractal distribution of electric charge in box region W with $0 \le x \le A$, $0 \le y \le B$ and $0 \le z \le C$. We have:

$$p_{x(d)} = \int_W x\rho_0 dV_d = \rho_0 \iiint_W x \frac{(xyz)^{(d/3)-1}}{\Gamma^3(d/3)} dxdydz$$

$$= \frac{\rho_0}{\Gamma^3(d/3)} \int_0^A dx \int_0^B dy \int_0^C dz x^{\frac{d}{3}} y^{\frac{d}{3}-1} z^{\frac{d}{3}-1} = \frac{\rho_0 (ABC)^{\frac{d}{3}}}{\Gamma^3(d/3)} \frac{A}{\left(\frac{d}{3}\right)^2 \left(\frac{d}{3}+1\right)}$$

The electric charge of the box region W is defined by

$$Q(W) = \rho_0 \int_W dV_d = \frac{\rho_0 (ABC)^{\frac{d}{3}}}{\left(\frac{d}{3}\right)^3 \Gamma^3(d/3)}$$

Using the above expression we have dipole moments for fractal distribution in box in the form

$$p_{x(d)} = \frac{\frac{d}{3}}{\frac{d}{3}+1} Q(W)A, \quad p_{y(d)} = \frac{\frac{d}{3}}{\frac{d}{3}+1} Q(W)B, \quad p_{z(d)} = \frac{\frac{d}{3}}{\frac{d}{3}+1} Q(W)C,$$

where $2 < d \le 3$.

8.13 Concluding Comments

Here variety of Electrical and Electronics circuits has given the feel of generalized fractional calculus, approaches. The intentions are, to demonstrate the broad array of uses of the fractional calculus, to clearly delineate the effects of the initialization function, to contrast generalized versus integer order differentiation and integration, to demonstrate the generalized 'zero-operation' (inverse-operation), to demonstrate the use of Laplace transform methods, to show some unusual aspects of mathematics concerning the modeling of distributed effects; broadly to give scientists and engineers a concept to model the reality of system being investigated by use of fractional calculus. The electrodynamics gets changed in fractal space giving rise to different interpretation; and even there exists possibility of finding 'isolated' magnetic charges. These ideas will generate several new laws of nature; or 'new mathematics' what Newton envisaged should nature shall follow.

Chapter 9
Application of Generalized Fractional Calculus in Other Science and Engineering Fields

9.1 Introduction

In this chapter a series of applications are described where fractional calculus is finding application. We start with diffusion model in electrochemistry, electrode electrolyte interface, capacitor theory, fractance circuits, and application in feed back control systems, viscoelasticity, and vibration damping system. This survey cannot cover complete applications like modern trends in electromagnetic theory like fractional multipole, hereditary prediction of gene behavior, fractional neural modeling in bio-sciences, communication channel traffic models, chaos theory, hence simple applications are provided for appreciation. However in the feedback control system section attempt is made to provide vector state feed back controller and observer available for multivariate control science, with explanation of fractional order feedback control and fractional phase shaper design to achieve robust iso-damped close loop performance.

9.2 Diffusion Model in Electrochemistry

One of the studies of importance in electrochemistry is the determination of concentration of analyzed electro active species near the electrode surface. The characteristic describing function is found experimentally as $m(t) = {}_0D_t^{-0.5}i(t)$, which is the fractional (half) integral of the current. Then the subject of interest is to find surface concentration $C_s(t)$ of the electro active species, can be evaluated as,

$$C_s(t) = C_0 - k({}_0D_t^{-0.5}i(t)),$$

where $k = 1/(nAF\sqrt{\mathbb{D}})$, A being electrode area, n number of electrons involved in the reaction, $\mathbb{D}$ is the diffusion coefficient and F is the Faraday constant. C_0 is the uniform concentration of the electro active species through out the electrolyte medium, at the initial equilibrium situation characterized by constant potential at which the electrochemical reaction is possible.

The relationship is derived from the classical diffusion equation:

$$\frac{\partial C(x,t)}{\partial t} = \mathbb{D}\frac{\partial^2 C(x,t)}{\partial x^2},$$

for

$$(0 < x < \infty \,\&\, t > 0) \text{ with } C(\infty,t) = C_0 \,\&\, C(x,0) = C_0;$$

$$\text{and } \left[\mathbb{D}\frac{\partial C(x,t)}{\partial x}\right]_{x=0} = \frac{i(t)}{nAF}$$

(Similar equation for lossy semi-infinite transmission line and heat flux studies)
Some interesting points are listed below.

First the $m(t)$ is characteristic intermediate between the current $i(t)$ and the charged passed $q(t)$. The charged passed is one integral as $q(t) = {}_0D_t^{-1}i(t)$. This hints at non-conservation law of charges, as $m(t)$ manifests. Secondly the kinetics of the electrode process and the surface property of electrode (alluding to heterogeneity) are not assumed. In a certain sense this approach of modeling embeds in a certain sense, contribution of particular features of the process in "non-integer order" of integration. Thirdly instead of classical diffusion equation it is possible to model with fractional-order diffusion equation as:

$${}_0D_t^{\alpha}C(x,t) = \mathbb{D}\frac{\partial^2 C(x,t)}{\partial x^2},$$

with $0 < \alpha < 1$, then the surface concentration will be related to $m_{\alpha}(t) = {}_0D_t^{\alpha/2}i(t)$.

9.3 Electrode-Electrolyte Interface Impedance

Warburg impedance in electrical battery (cell) is another motivating example of reality of fractional calculus. Limitations of electrical batteries, which always exhibit a limited current output is due to the fact that microscopic electrochemical process at the electrode electrolyte interface have "finite rate" and limit the current output. Battery manufactures use porous electrodes to circumvent this limitation, by way of increasing the surface area. It has been experimentally established that metal-electrolyte surface interface impedance does not exhibit pure capacitance behavior, instead governed as power law: $Z(\omega)=\frac{K}{(\mathrm{j}\omega)^{\alpha}}$ where $0<\alpha<1$. In Laplace domain impedance is, $Z(s)=Ks^{-\alpha}$.

This power term approaches unity (impedance tending to capacitive) as the smoothness of the interface is increased to infinity.

Warburg impedance $Z(\omega)\propto(\mathrm{j}\omega)^{-0.5}$, for any solid-state diffusion, electro chemistry, gives rise to power law in frequency. The constitutive equation is $\frac{\partial C}{\partial t}=\mathbb{D}\frac{\partial^2 C}{\partial x^2}$, which gives rise to $Z(\omega)\propto(\mathrm{j}\omega)^{-0.5}$. When diffusion takes place in a layer of thickness L driven by diffusion over voltage at $x=0$, the observed behavior is not solely Warburg impedance. In a spatially restricted situation there are at least two domains in the impedance spectra, which are separated by characteristic frequency $\omega_d=\frac{\mathbb{D}}{L^2}$. Warburg occurs at high frequency regime $\omega>\omega_d$. At low frequency $\omega<\omega_d$, the impedance behavior depends if the diffusion species are reflected or extracted at the end of region ($x=L$). The reason for this is that the frequency ω_d corresponds to the transit time for a diffusing particle injected at $x=0$, to cover a distance L. For $\omega>\omega_d$ the particles will not sense the boundary $x=L$, so that the system will behave as "semi-infinite" media. Anomalous diffusion is characterized by a mean square displacement of the diffusing particles that does not follow ordinary linear law $<r^2>\propto t$, but more generally $<r^2>\propto t^{\beta}$.

Not surprisingly many different mechanisms give rise to anomalous behavior, like complex flow, flow through porous, flow through random tubular ion exchange resin, due to structural complexity, shapes of complex geometry etc.

9.3.1 Normal Diffusion in a Finite Boundary System

The continuity equation is fundamental conservation law, $\frac{\partial C}{\partial t}=-\frac{\partial J}{\partial x}$ relating to time variation of the number density C of the diffusing species, to the macroscopic flux, J. The constitutive equation for diffusion is $J=-\mathbb{D}\frac{\partial C}{\partial x}$. To calculate diffusion impedance consider $\tilde{E}(t)=\left(\frac{dE}{dC}\right)_0\tilde{C}(t)$, where E is excitation voltage at $x=0$, the expression gives linear dependence at $x=0$. This is small amplitude oscillating condition, and the tide denotes that. The subscript (0) denotes a derivative following local equilibrium condition. The (oscillating) current at the origin (is flux J of charges q passing through area A), related to small oscillating flux as: $\tilde{i}(t)=qA\tilde{J}(t)$. Taking the Laplace as the above expressions of voltage and currents are for small changing quantities with respect to time, we obtain at $x = 0$,

$$\tilde{E}(s) = \left(\frac{dE}{dC}\right)_0 \tilde{C}(s) \text{ and } \tilde{I}(s) = qAJ(s).$$

This is first boundary condition. Thus the diffusion impedance is

$$Z(s) = \frac{\tilde{E}(s)}{\tilde{I}(s)} = R_W \frac{\mathbb{D}}{L} \frac{\tilde{C}(s)}{\tilde{J}(s)}$$

where Warburg constant is R_W given by $R_W = \frac{L}{qA\mathbb{D}}\left(\frac{dE}{dC}\right)_0$ is having unit of resistance if unit of $\mathbb{D}$ is $[\mathrm{cm}^2\mathrm{s}^{-1}]$.

Also denote diffusion capacitance as:

$$C_W = \frac{L^2}{R_W \mathbb{D}}.$$

The various diffusion equations for small amplitude sinusoidal concentration may be written as in the Laplace domain form by taking Laplace of continuity and constitutive equations:

$$s\tilde{C}(s) = -\frac{\partial \tilde{J}(s)}{\partial x} \text{ and } \tilde{J}(s) = -\mathbb{D}\frac{\partial \tilde{C}(s)}{\partial x}$$

From these we get $\frac{\partial^2 \tilde{C}}{\partial x^2} = \frac{1}{\lambda^2}\tilde{C}(s)$. Here $\omega_d = \frac{\mathbb{D}}{L^2}$, the characteristic frequency thus we have

$$\lambda(s) = \left(\frac{\mathbb{D}}{s}\right)^{1/2} = L\left(\frac{\omega_d}{s}\right)^{1/2},$$

which is function of frequency.

This Laplace (time-frequency) equation has most general solution as: $\tilde{C}(s) = B_1 \cosh\frac{x}{\lambda} + B_2 \sinh\frac{x}{\lambda}$. This equation will be used to give solution of Anomalous diffusion cases described in next sub-section by proper choice of $\lambda(s)$ and ω_d.

An absorbing boundary condition is: $C = 0$ at $x = L$, for this boundary condition, the diffusion impedance at origin is

$$Z(s) = R_W \frac{\mathbb{D}}{L} \frac{C(s)}{J(s)} = R_W \left(\frac{\omega_d}{s}\right)^{1/2} \tanh\left[\left(\frac{s}{\omega_d}\right)^{1/2}\right].$$

The solution to this is explained in Chapter 3.

At high frequency $\omega >> \omega_d$ the result is Warburg impedance ($Z(s) = R_W\left(\omega_d / s\right)^{1/2}$) and for $\omega << \omega_d$ the impedance is approximated by resistance R_W parallel with one-third of diffusion capacitance $C_W / 3$, where the diffusion capacitance is $C_W = L^2 / R_W \mathbb{D}$. This is obtained by impedance-plot, on complex plane for the normalized terminal impedance.

A reflective boundary condition is flux disappearing at $x = L$ that is $(\partial\tilde{C} / \partial x) = 0$, the diffusion impedance is for this condition is: $Z(s) = R_W (\omega_d / s)^{1/2} \coth\left[(s / \omega_d)^{1/2}\right]$. This has limiting behavior at high frequency $\omega >> \omega_d$ as Warburg impedance $Z(s) \sim (s)^{-1/2}$, as half order behavior. For low frequency $\omega << \omega_d$ one gets series combination of R_W and diffusion capacitance C_W. These approximations appear from normalized impedance-plots on complex plane method explained in Chapter 3.

The same semi-infinite conditions for high frequency approximations represented as semi-infinite lossy transmission line model too with r and c as per unit length resistance and capacitance expressed as

$$r = \frac{R_W}{L} \text{ and } c = \frac{L}{R_W \mathbb{D}}.$$

9.3.2 Anomalous Diffusion in Finite Boundary System

9.3.2.1 Diffusion with Fractional Continuity Equation

For Fick's law the Fractional Continuity Equation in Laplace Form is $s^\gamma C = -\frac{\partial J}{\partial x}$, with $\gamma < 1$. The constitutive relation of Fick's law remaining the same i.e. $J = -\mathbb{D}\frac{\partial C}{\partial x}$. This gives fractional diffusion equation as:

$$\frac{\partial^\gamma C(x,t)}{\partial t^\gamma} = \mathbb{D}\frac{\partial^2 C(x,t)}{\partial x^2},$$

with $\gamma < 1$. Gives Laplace equation as:

$$\frac{\partial^2 C(x,s)}{\partial x^2} = \frac{s^\gamma}{\mathbb{D}} C(x,s) = \frac{1}{\lambda_1(s)} C(x,s),$$

with $\tilde{C}(s) = B_1 \cosh\frac{x}{\lambda_1} + B_2 \sinh\frac{x}{\lambda_1}$ as solution; where $\lambda_1^2 = \mathbb{D}/s^\gamma$. Therefore

$$\lambda_1(s) = L\left(\frac{\omega_{d1}}{s}\right)^{\gamma/2} \text{ with } \omega_{d1} = \left(\frac{\mathbb{D}}{L^2}\right)^{1/\gamma}.$$

The reflecting boundary condition at $x = L$ where flux vanishes gives terminal impedance as: $Z(s) = R_W\,(\omega_{d1}/s)^{\gamma/2} \coth\left[(s/\omega_{d1})^{\gamma/2}\right]$. At high frequency $\omega >> \omega_{d1}$ this function is approximated by $Z(s) = R_W\,(\omega_{d1}/s)^{\gamma/2}$. At low frequency $\omega << \omega_{d1}$ this impedance can be approximated by $Z(s) = \frac{1}{3}R_W + Qs^{-\gamma}$, where $Q = R_W(\omega_d)^\gamma$. This is a series combination of resistance and a constant phase element (CPE) which gives constant lagging phase of $-\gamma(\pi/2)$ degrees.

The absorbing boundary condition at $x = L$ gives terminal impedance as $Z(s) = R_W\,(\omega_{d1}/s)^{\gamma/2} \tanh\left[(s/\omega_{d1})^{\gamma/2}\right]$, which again at high frequency returns fractional order terminal impedance $Z(s) = R_W\,(\omega_{d1}/s)^{\gamma/2}$. At low frequencies the behavior is approximated as: $\frac{1}{Z(s)} = \frac{1}{R_W} + \frac{1}{Q_1 s^{-\gamma}}$ that is, parallel combination of resistance R_W and CPE $Q_1 = 3Q$.

9.3.2.2 Diffusion with Fractional Differential Constitutive Equation

The continuity equation is integer order and its Laplace representation is: $sC = -\frac{\partial J}{\partial x}$. The Laplace of Fractional Differential Constitutive Equation is:

$$J = -\mathbb{D}s^{\gamma-1}\frac{\partial C}{\partial x} \text{ with } 0 < \gamma < 1.$$

This gives the same fractional diffusion equation as above.

Here take $\lambda_1(s) = L(\omega_{d1}/s)^{\gamma/2}$ and $\omega_{d1} = \left(\mathbb{D}/L^2\right)^{1/\gamma}$ that is same as above. The impedance with reflective boundary condition is

$$Z(s)=R_W(\omega_{d1})^{\gamma-1}(\omega_{d1}/s)^{1-\gamma/2}\coth\left[(s/\omega_{d1})^{\gamma/2}\right],$$

this function can be approximated at high frequencies by $Z(s)=R_W(\omega_{d1})^{\gamma-1}(\omega_{d1}/s)^{1-\gamma/2}$ .At low frequency the approximation is parallel combination of a new diffusion capacitance $C_X=1/\left(R_W\omega_{d1}^{\gamma}\right)$ and a CPE $Q_2=R_W/3$ that is:

$$Z(s)=\frac{1}{C_X s}+Qs^{-(1-\gamma)}.$$

The terminal impedance for absorbing boundary condition the function is: $Z(s)=R_W(\omega_{d1})^{\gamma-1}(\omega_{d1}/s)^{1-\gamma/2}\tanh\left[(s/\omega_{d1})^{\gamma/2}\right]$, the high frequency behavior is approximated as $Z(s)=R_W(\omega_{d1})^{\gamma-1}(\omega_{d1}/s)^{1-\gamma/2}$ and low frequency behavior is approximated as parallel combination of $C_Y=3R_W\omega_d$ and CPE $Q_3=R_W$ that is:

$$\frac{1}{Z(s)}=C_Y s+\frac{1}{Q_3 s^{-(1-\gamma)}}$$

Though the diffusion equations for both the above anomalous diffusion are same with the same characteristic frequency formulation ω_{d1} and the frequency dependent diffusion parameter $\lambda_1(s)$, yet the terminal impedances in both the anomalous cases are different (for same boundary condition). The reason is in one case we have generalized the continuity equation and in other we have generalized the constitutive equation, with new definition of flux.

9.3.2.3 Diffusion with Fractional Integral Constitutive Equation

The flux continuity equation is integer order and its Laplace is: $sC=-\frac{\partial J}{\partial x}$, and the Fractional Integral equation for the constitutive equation in Laplace domain is: $J=-\mathbb{D}s^{\gamma-1}\frac{\partial C}{\partial x}$, with $0<\gamma<1$. This gives fractional diffusion equation as:

$$\frac{\partial^{2-\gamma}}{\partial t^{2-\gamma}}C(x,t)=\mathbb{D}\frac{\partial^2}{\partial x^2}C(x,t).$$

Its Laplace equivalent is:

$$\frac{\partial^2 C(x,s)}{\partial x^2}=\frac{s^{2-\gamma}}{\mathbb{D}}C(x,s)=\frac{1}{(\lambda_2)^2}C(x,s)$$

The solution in Laplace domain is: $\tilde{C}(s) = B_1 \cosh\frac{x}{\lambda_2} + B_2 \sinh\frac{x}{\lambda_2}$. With

$$\lambda_2(s) = L\left(\frac{\omega_{d2}}{s}\right)^{1-\gamma/2} \text{ and } \omega_{d2} = (\mathbb{D}/L^2)^{1/(2-\gamma)}.$$

The reflective boundary condition gives the terminal impedance as $Z(s) = R_W(\omega_{d2})^{1-\gamma}(\omega_{d2}/s)^{\gamma/2}\coth\left[(s/\omega_{d2})^{1-\gamma/2}\right]$. The limiting behavior at high frequency is approximated as $Z(s) = R_W(\omega_{d2})^{1-\gamma}(\omega_{d2}/s)^{\gamma/2}$. The low frequency behavior is approximated by a series combination of capacitor $C_Z = 1/R_W(\omega_{d2})^{2-\gamma}$ and an element 'resistoductance' $H = R_w/3$ (at values $\gamma = 1$ is pure resistance and as $\gamma < 1$ the element tends toward inductor). This is also termed as Generalized Warburg Impedance

$$Z(s) = \frac{1}{C_Z s} + Hs^{1-\gamma}.$$

For the boundary condition absorbing in nature at $x = L$ the terminal impedance is: $Z(s) = R_W(\omega_{d2})^{1-\gamma}(\omega_{d2}/s)^{\gamma/2}\tanh\left[(s/\omega_{d2})^{1-\gamma/2}\right]$. In the low frequency part, $\omega \ll \omega_{d2}$ the approximating relation is by adding admittances of a capacitor which is $(1/3)C_z$ and 'resistoductance' $H_1 = 3H = R_W$ that is:

$$\frac{1}{Z(s)} = \frac{C_z}{3} + \frac{1}{3Hs^{1-\gamma}}.$$

9.4 Capacitor Theory

Capacitor is charge storage devise and is assumed that whatever charges are pumped are held between the plates (electrodes) by ideal dielectric having no loss. Therefore the impedance of non-leaky capacitor or ideal capacitor is $Z_c = \frac{1}{\mathrm{j}\omega C}$. With dielectric constant assumed to be ideal as $\varepsilon = \varepsilon' - \mathrm{j}0$. Now if we really have dielectric absorption then the real capacitor is $Z_c = \frac{1}{(\mathrm{j}\omega C)^{0.999}}$, having slight imperfection, but good for electronic circuits and sample and hold circuits etc. Now if one designs a leaky capacitor by selecting di-electric (not ideal) but having $\varepsilon = \varepsilon' - j\varepsilon''$ with $\varepsilon' = \varepsilon'' = 10^6$ for wide range of frequency and temperature then $\varepsilon = \varepsilon_r \omega^{-1/2}(1-\mathrm{j}) = \varepsilon_r \sqrt{2}(\mathrm{j}\omega)^{-1/2}$. Using this di-electric the capacitance impedance is

$$Z = \frac{1}{\mathrm{j}\omega C\varepsilon} = \frac{1}{\mathrm{j}\omega C\varepsilon_r \sqrt{2}(\mathrm{j}\omega)^{-1/2}} = \frac{1}{C\varepsilon_r \sqrt{2}(\mathrm{j}\omega)^{1/2}}.$$

The di-electric is Lithium Hydrazium Sulphate ($LiN_2H_2SO_4$). These capacitors are useful to realize the analog circuits for fractional order controller the Figure 9.1 gives operational amplifier realization of using these capacitors.

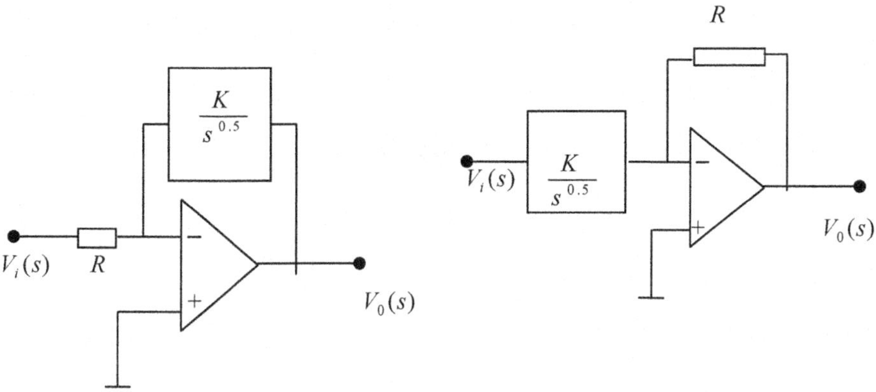

$$\frac{V_0(s)}{V_i(s)} = \frac{K}{R\sqrt{s}} \qquad \frac{V_0(s)}{V_2(s)} = K\sqrt{s}$$

Fig. 9.1 Operational amplifier circuit to have fractional integration and differentiation.

This fact was experimented in the late nineteenth century by M J Curie, who noted the current voltage relation as (power law) $i(t) = \dfrac{V}{ht^{v}}$, where $(0 < v < 1, t > 0)$. In this expression h is related to the capacitance of the capacitor and the kind of di-electric and v is related to losses of the capacitor. The transfer function of this model of capacitor is found to be $H(s) = C_{\varphi}s^{v}$ where C_{φ} is model constant close to what is called the capacitance. The capacitor impedance is described by $Z(s) = \dfrac{1}{C_{\varphi}s^{v}}$ Advance research has realized these half order capacitors by polymer-coated electrode, and has been used as constant phase element to get semi-differentigration circuits.

9.5 Fractance Circuit

Electrical circuit related to fractional calculus is fractance. An electrical circuit behaving in between capacitance and resistance. An example of fractance is tree fractance shown in Figure 9.2 a self-similar structure.

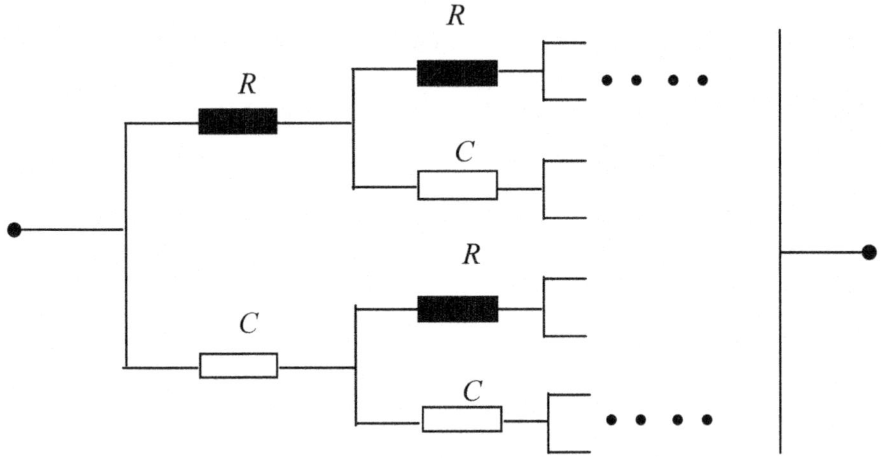

Fig. 9.2 Tree fractance circuit.

The impedance of tree fractance is

$$Z(\mathrm{j}\omega)=\sqrt{\frac{R}{C}}\frac{1}{\sqrt{\omega}}\exp\left(\frac{-\mathrm{j}\pi}{4}\right),$$

this corresponds to fractional order transfer function

$$Z(s)=\sqrt{\frac{R}{C}}\frac{1}{\sqrt{s}}.$$

A circuit exhibiting fractional order behavior is called fractance, not essentially limited to half order, as described in the self-similar circuit diagram of Figure 9.2. The order can be of any arbitrary order in general. The fractance devises have the following properties. They are constant phase elements, i.e. the phase angle is constant independent of frequency within wide range of frequency band. Second it is possible to construct a filter, which has moderate characteristics, which cannot be realized by using conventional devises. Generally speaking there are three basic fractance devises. The most popular one is domino ladder circuit network. Very often are used a binary tree structure as in Figure 9.2. Also balanced transmission line structure is also used (or symmetrical domino ladder).

Design of fractances can be easily done by any of these topological configuration as mentioned, to realize the rational approximated transfer function for the fractional order Laplace operator (Chapter 7). Truncated continued fraction expansion (CFE) does not require any further transformation, a rational approximation based on any other method as (say Newton method of Carlson, described in Chapter 7), must be transformed into the form of CFE. The values of the electrical circuit elements, which are necessary for realizing a fractance, are then determined from the obtained continued fraction. If all the coefficients of obtained CFE are positive, then the fractance can be simply made by passive elements (R, L, C). If some of the coefficients are negative then, the fractance realization requires active circuit as negative impedance converters, realized by operation amplifier circuits. In some of the methods of CFE to realize the represented transfer function negative impedances do appear. Negative impedance converters are also called current inverter, and have transfer function as $V_i / I_i = -Z$, realized by operational amplifier.

Recalling the semi-infinite lossy transmission line, where r is series resistance per unit length, and c is the shunt capacitance per unit length, used to demonstrate semi-differentigration as derived terminal impedance (in Chapter 3). The same circuit, truncated, if we mentally reconstruct the equivalent impedance by traveling from right side of the transmission line towards the left side, we will get:

$$Z_{rc}(s) = r + \cfrac{1}{cs + \cfrac{1}{r + \cfrac{1}{cs + \cfrac{1}{\cdots\cdots\cdots\cdots\cdots \cfrac{1}{cs + \cfrac{1}{r + \cfrac{1}{cs}}}}}}}$$

For the shunt element the admittance is $Y(s) = cs$ and thus impedance is $Z(s) = 1/Y(s)$. This truncated $Z_{rc}(s)$ is approximation to the driving point impedance, which is $\sqrt{r/cs}$ (fractional Laplace operator) for semi-infinite line. The generalization of this with series impedances of different values, $Z_1(s), Z_3(s).....Z_{2n-3}(s), Z_{2n-1}(s)$, and shunt admittances of different values $Y_2(s), Y_4(s)...Y_{2n-2}(s), Y_{2n}(s)$ gives the CFE form of domino ladder immitance as:

$$Z(s) = Z_1(s) + \cfrac{1}{Y_2(s) + \cfrac{1}{Z_3(s) + \cfrac{1}{\ddots \cfrac{1}{Y_{2n-2}(s) + \cfrac{1}{Z_{2n-1}(s) + \cfrac{1}{Y_{2n}(s)}}}}}}$$

The transfer function $Z(s)=\dfrac{s^2+4s+3}{s^3+6s^2+8s}$ is realized by doing CFE for admittance as:

$$Y(s)=\frac{s^3+6s^2+8s}{s^2+4s+3}=s+\cfrac{1}{\cfrac{1}{2}+\cfrac{1}{\cfrac{4}{3}s+\cfrac{1}{\cfrac{3}{2}+\cfrac{1}{\cfrac{1}{3}s}}}},$$

Meaning shunt capacitors are $1,\frac{4}{3},\frac{1}{3}$, corresponding to admittances, $Y_2(s),Y_4(s),Y_6(s)$ the series impedance in this case $Z_1(s)=0$, and the other series impedances corresponding to $Z_3(s),Z_5(s)$ are $\frac{1}{2},\frac{3}{2}$. This is standard Caurer-I form of circuit synthesis. The same can be realized as Caurer-II form. Depending on the rational polynomial the realization with CFE can have R,C,L components combination of them or even negative impedances, in the ladder form.

The fundamentals of circuit synthesis are applicable for this fractance realization, of rational transfer function synthesis for fractional order Laplace operator.

9.6 Feedback Control System

Feed back control system is one the major area where concept of fractional calculus should be applied to obtain efficient system. This concept gives overall efficiency (in terms of energy) also longevity and freedom to control engineer to compensate any shifts in the transfer function due to parametric spreads aging etc. A system is efficient if the controller were of similar order to that of a plant (system) being controlled. In reality the systems are fractional order therefore to have fractional order controller will be efficient. Even for integer order systems the fractional controls give better freedom to achieve what is "isodamping". Meaning, to achieve overall close loop behavior of overshoot independent of feed forward gain (pay-load, amplifier feed forward gain, in power systems the load current/load resistance). H W Bode envisaged this concept of having fractional integrator circuits to achieve overshoot independent of the amplifier gain in 1945. He proposed a fractional order controller, the purpose of which is to have a feed back amplifier of good linearity and stable gains even though the amplifier show non-linear characteristics and variable gain over ambient and time. Bode proposed a feed back amplifiers, whose open loop frequency characteristics $G_0(j\omega)$ is such that the gain is constant for $0<\omega<\omega_0$ and phase is constant or $-\pi(1-y)$ radians for $\omega>\omega_0$. The suggested value was $y=1/6$, which guarantees a phase margin (PM) of 30^0. The open loop transfer function is given as

$$G_0(\mathrm{j}\omega)=\frac{A_0}{\left[\sqrt{1-(\omega/\omega_0)^2}+\mathrm{j}(\omega/\omega_o)\right]^{2(1-y)}},$$

meaning $|G_0(\mathrm{j}\omega)|=A_0$ for $\omega<\omega_0$ and angle i.e. $\arg G_0(\mathrm{j}\omega)=-\pi(1-y)$ radians for $\omega>\omega_0$. This is early development of fractional order controls. Thus it was recognized that the open-loop transfer function of a good control system show a fractional order integral form with a fractional order between 1 and 2 (between totally being first order and second order). Meaning that open loop transfer function should be like $G_0(s)=1/s^k$. This gives close loop transfer function as

$$G_{CL}(s)=\frac{G_0(s)}{1+G_0(s)}=\frac{(1/s^k)}{1+(1/s^k)}=\frac{1}{s^k+1},\ \text{where } s=\mathrm{j}\omega.$$

In close loop transfer function $G_{CL}(s)=\dfrac{1}{s^k+1}$ expression, put for $s=\mathrm{j}\omega$, $\mathrm{j}=\cos\dfrac{\pi}{2}+\mathrm{j}\sin\dfrac{\pi}{2}$ then, $s=\omega\left[\cos\dfrac{\pi}{2}+\mathrm{j}\sin\dfrac{\pi}{2}\right]$ and

$$s^k=\omega^k\left[\cos\frac{k\pi}{2}+\mathrm{j}\sin\frac{k\pi}{2}\right]=\omega^k\cos\frac{k\pi}{2}+\mathrm{j}\omega^k\sin\frac{k\pi}{2},$$

Put this value of ω^k in $G_{CL}(s)$ to get:

$$G_{CL}(s)=\frac{1}{s^k+1}=\frac{1}{\omega^k\cos\frac{k\pi}{2}+\mathrm{j}\omega^k\sin\frac{k\pi}{2}+1}=\frac{1}{\left(\omega^k\cos\frac{k\pi}{2}+1\right)+\mathrm{j}\omega^k\sin\frac{k\pi}{2}}$$

$$|G_{CL}(s)|=\frac{1}{\left[\omega^{2k}\cos^2\frac{k\pi}{2}+1+2\omega^k\cos\frac{k\pi}{2}+\omega^{2k}\sin^2\frac{k\pi}{2}\right]^{0.5}}=\frac{1}{\sqrt{\left[\omega^{2k}+2\omega^k\cos\frac{k\pi}{2}+1\right]}}$$

M_r is maximum value of $|G_{CL}(s)|$ at ω_r when denominator $\omega^{2k}+2\omega^k\cos\dfrac{k\pi}{2}+1$ is minimum.

Therefore, $\frac{d}{d\omega}\left[\omega^{2k}+2\omega^{k}\cos\frac{k\pi}{2}+1\right]=0$ gives $2k\omega^{2k-1}+2k\omega^{k-1}\cos\frac{k\pi}{2}=0$, meaning at $\omega^{k}=-\cos\frac{k\pi}{2}$ the magnitude of $G_{CL}(s)$ is maximized.

$\omega_r=\left|\cos\frac{k\pi}{2}\right|^{1/k}$, and putting this value of $\omega=\omega_r$ in expression of $|G_{CL}(s)|$ we get

$$M_r=\frac{1}{\sqrt{\left[\left(-\cos\frac{k\pi}{2}\right)^2+2\left(-\cos\frac{k\pi}{2}\right)\cos\frac{k\pi}{2}+1\right]}}=\frac{1}{\sqrt{\cos^2\frac{k\pi}{2}-2\cos^2\frac{k\pi}{2}+1}}=\frac{1}{\sqrt{1-\cos^2\frac{k\pi}{2}}}$$

$$=\frac{1}{\sin\frac{k\pi}{2}}$$

For finding the damping ratio we find the poles of $G_{CL}(s)$ by transformation to w-plane and then with respect to s plane we look at the pole location.

Putting $w=s^k$ in the expression of close loop transfer function we obtain:

$$G_{CL}(w)=\frac{1}{w+1},$$

with poles at $w=1e^{\pm j\pi}$ in $w-$plane.

Therefore the s plane pole is at $w^{1/k}$ meaning poles at $s=(1)^{1/k}e^{\pm j\pi/k}$ in the $s-$plane. The line with angle $(\pm\pi/k)$ with the positive real axis of the s-plane is the locus of poles for $G_{CL}(s)$ and are called iso-damped lines for particular value of k. The damping ratio $\varsigma=\frac{\Re e(s)}{|s|}$ with respect to imaginary $(j\omega)$ axis.

The angle of the iso-damped line with respect to imaginary axis is $\left(\frac{\pi}{k}-\frac{\pi}{2}\right)$ and thus any where on this line the pole is, the damping ratio is

$$\varsigma=\frac{\Re e(s)}{|s|}=\sin\left(\frac{\pi}{k}-\frac{\pi}{2}\right)$$

This close-loop transfer function gives step response properties of controlled system output as robustness and stability measures.

$$|G_{CL}(s)|=\left|\frac{1}{\left[\omega^k\cos(k\pi/2)+1+j\omega^k\sin(k\pi/2)\right]}\right|=\frac{1}{\left[\omega^{2k}+2\omega^k\cos(k\pi/2)+1\right]^{0.5}}$$

$$M_r=\frac{1}{\sin(k\pi/2)},\ \omega_r=|\cos(k\pi/2)|^{1/k}.$$

The amplitude takes the peak value M_r at . The damping ratio can be obtained from the poles of $G_{CL}(s)$ as $\varsigma=\sin\left(\frac{\pi}{k}-\frac{\pi}{2}\right)$. The phase margin is given by $PM=\pi-k\frac{\pi}{2}=90(2-k)^0$. The overshoot can be expressed as approximate formula as, $M_P\cong(k-1)(0.8k-0.6)$ per unit. These performance specifications are also termed as robustness measures and are listed for various fractional orders:

Table 9.1 Robustness Measures for various fractional orders k

k (Order)	PM Degree	ς	M_P%	M_r
1	90	1	0	1
1.1	81	0.96	2.8	1.0125
1.2	72	0.87	7.4	1.0515
1.3	63	0.75	13.6	1.1223
1.4	54	0.62	21.1	1.2361
1.5	45	0.5	30.0	1.4142
1.6	36	0.38	40.5	1.7013
1.7	27	0.27	52.5	2.2027
1.8	18	0.17	66.3	3.2361
1.9	9	0.083	82.1	6.3925

General properties of Bode's ideal control system transfer functions are:

a) Open loop:

Type: $G_0(s)=\frac{K}{s^k}$, $(1<k<2)$

Magnitude: constant slope of $--k20\text{dB/decade}$
Cross over frequency: a function of gain K
Phase: horizontal line of $-k\pi/2$.
Nyquist: straight line at argument $-k\pi/2$

b) Closed loop:

Type: $G_{CL}(s)=\frac{K}{s^k+K}$, $(1<k<2)$

Gain margin: $A_m=\infty$ infinite.

Phase margin: $\Phi_m = \pi\left(1-\frac{k}{2}\right)$, constant

Step response: $y(t) = Kt^k E_{k,k+1}(-Kt^k)$ The solution to this is as follows:

The step response of a system with closed loop transfer function $G_{CL}(s)$ is:

$$Y(s) = \frac{1}{s} G_{CL}(s).$$

The closed loop transfer function is $G_{CL}(s) = \frac{K}{s^k + K}$.

Therefore:

$$Y(s) = \frac{1}{s} G_{CL}(s) = \frac{K}{s(s^k + K)}$$

$$Y(s) = K\frac{1/K}{s}\left(\frac{K}{s^k + K}\right) = K\frac{1/K}{s}\left[1-\frac{s^k}{s^k + K}\right] = \frac{1}{s} - \frac{s^k}{s^k + K} = \frac{1}{s} - \frac{s^{k-1}}{s^k + K}$$

Inverse Laplace transforming the above we get:

$$y(t) = \left[1 - E_k(-Kt^k)\right]$$

Using series expansion of one parameter Mittag-Leffler function we obtain:

$$y(t) = 1 - \left[1 + \frac{(-Kt^k)}{\Gamma(k+1)} + \frac{(-Kt^k)^2}{\Gamma(2k+1)} + \frac{(-Kt^k)^3}{\Gamma(3k+1)} + \ldots\right]$$

$$= -\frac{(-Kt^k)}{\Gamma(k+1)} - \frac{(-Kt^k)^2}{\Gamma(2k+1)} - \frac{(-Kt^k)^3}{\Gamma(3k+1)} - \ldots$$

$$= \frac{(Kt^k)}{\Gamma(k+1)} - \frac{(Kt^k)^2}{\Gamma(2k+1)} + \frac{(Kt^k)^3}{\Gamma(3k+1)} - \ldots$$

$$= Kt^k\left[\frac{1}{\Gamma(k+1)} + \frac{(-Kt^k)}{\Gamma(2k+1)} + \frac{(-Kt^k)^2}{\Gamma(3k+1)} + \ldots\right]$$

Now using definition of two parameter Mittag-Leffler function

$$E_{\alpha,\beta}(z) = \sum_{n=0}^{\infty} \frac{z^n}{\Gamma(\alpha n + \beta)}$$

We write

$$E_{k,k+1}(-Kt^k) = \frac{1}{\Gamma(k+1)} + \frac{(-Kt^k)}{\Gamma(2k+1)} + \frac{(-Kt^k)^2}{\Gamma(3k+1)} + ..$$

and use this to have time response expression as:

$$y(t) = Kt^k E_{k,k+1}(-Kt^k)$$

The solution with one parameter Mittag-Leffler and two parameters Mittag-Leffler are equivalent. More form of algebraic manipulations of the series can give different compact functions as solutions (say by Robotnov-Hartley and several other variants of Mittag-Leffler).

Therefore for Heaviside step input $H(t) = \begin{cases} 1 & t \geq 0 \\ 0 & t < 0 \end{cases}$, the time response of output is:

$$y(t) = Kt^k E_{k,k+1}(-Kt^k) = 1 - E_k(-Kt^k) = H(t) - E_k(-Kt^k)$$

Further if the composition of a feed back controller with fractional differentiator and α -order fractional integrator gives $k = 1 + \alpha$, while obtaining $G_{CL}(s)$, then:

$$y(t) = Kt^{1+\alpha} E_{1+\alpha,2+\alpha}(-Kt^{1+\alpha})$$

Now with especially digital and advanced analog techniques this fractional order control system is possible in having real time responses. Figure 9.3 shows the concept of fractional order PID system connection.

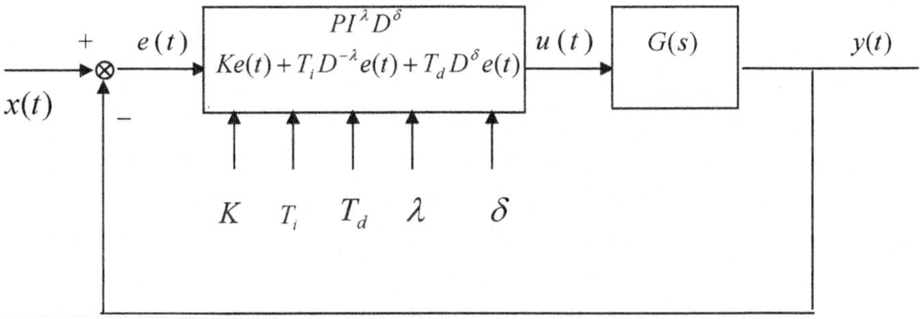

Fig. 9.3 Fractional order PID controls.

In fractional order PID control what is extra freedom to operator is in terms of two extra knobs namely in the order of differentiation and order of integration. In these fractional order controller $PI^{\lambda}D^{\delta}$ the knob values λ, δ is between 0-1. In PID type control we compensate only the dominant roots and with these extra freedoms we are able to continuously span the area shaded in Figure 9.4. The Laplace domain of this controller has the form: $H(s)=\dfrac{U(s)}{E(s)}=K_p+T_i s^{-\lambda}+T_d s^{\delta}$.

This $PI^{\lambda}D^{\delta}$ controller with complex zeros and poles located anywhere in the left-hand side $s\rightarrow w$ plane, may be rewritten as:

$$H(s)=K\frac{(s/\omega_n)^{\delta+\lambda}+(2\varsigma s^{\lambda})/\omega_n+1}{s^{\lambda}},$$

where K is a gain, ς is the dimensionless damping ratio (is chosen mostly as under-damped $\varsigma<1$, and =1 for critically damped) and ω_n is the natural frequency. This $PI^{\lambda}D^{\delta}$ is generalization to *TID* (Tilt Integral Derivative) compensator, which has a similar structure as *PID* but the proportional component is replaced with tilted component having a transfer function s, to the power of $(-1/n)$. Therefore the transfer function of *TID* is: $H(s)=\dfrac{T}{s^{1/n}}+\dfrac{I}{s}+Ds$, where T,I,D are controller constant and n is non zero real number (between 2 and 3). The above transfer function approximates Bode's ideal transfer function (US patent 5-371-670).

It can be also mentioned that the controller can be also characterized with fractional Laplace operator by band-limited lead effect. That lead lag compensator be of representation: $H(s)=K\left(\dfrac{1+\tau_1 s^q}{1+\tau_2 s^q}\right)$, $\tau_2<\tau_1$ and q being fractional number. This transfer function can be realized by rational function approximation methods by recursive distribution of poles and zeros.

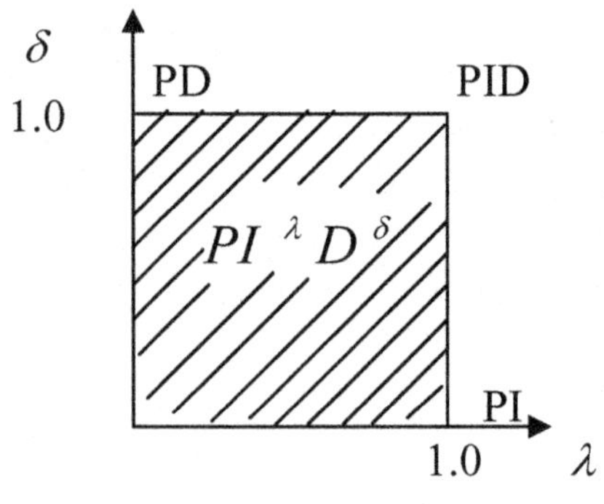

Fig. 9.4 Fractional order PID and integer order PID, PD, PI.

The example in state variable formulation gives some light about PID controls in fractional domain. Let a fractional plant transfer characteristics $G(s)$ be identified as

$$a_2 D^{\alpha} y(t) + a_1 D^{\beta} y(t) + a_0 y(t) = u(t)$$

Simply written as: $a_2 y^{(\alpha)}(t) + a_1 y^{(\beta)}(t) + a_0 y(t) = u(t)$. The fractional order PID $H(s)$ is:

$$u(t) = Ke(t) + T_i D_t^{-\lambda} e(t) + T_d D_t^{\delta} e(t)$$

Using state variable $y(t) = x(t) = x_1(t)$ and $x^{(\beta)}(t) = x_2(t)$. The G(s) is obtained as state space as:

$$\begin{bmatrix} x_1^{(\beta)}(t) \\ x_2^{(\alpha-\beta)}(t) \end{bmatrix} = \begin{bmatrix} 0 & 1 \\ -\dfrac{a_0}{a_2} & -\dfrac{a_1}{a_2} \end{bmatrix} \begin{bmatrix} x_1(t) \\ x_2(t) \end{bmatrix} + \begin{bmatrix} 0 \\ \dfrac{1}{a_2} \end{bmatrix} \begin{bmatrix} 0 \\ u(t) \end{bmatrix}, \text{ and } y(t) = x_1(t)$$

By GL definition

$${}_a D_t^{\pm r} f(t) = \lim_{T \to 0} \frac{1}{T^{\pm r}} \sum_{j=0}^{\left[\frac{t-a}{T}\right]} b_j^{\pm r} f(t - jT),$$

Where $b_0^{\pm r} = 1, b_j^{\pm r} = \left(1 - \dfrac{(1 + (\pm r))}{j}\right) b_{j-1}^{\pm r}$ are binomial coefficients.

The discretized state space expression for G(s) is:

$$\begin{aligned}
x_{1,k+1} &= -\sum_{j=1}^{k+1} b_j^{\beta} x_{1,k+1-j} + T^{\beta} x_{2,k} \\
x_{2,k+1} &= -\sum_{j=1}^{k+1} c_j^{\alpha-\beta} x_{2,k+1-j} + T^{\alpha-\beta} \left(-\frac{a_0}{a_2} x_{1,k} - \frac{a_1}{a_2} x_{2,k} + \frac{1}{a_2} u_k \right) \\
y_k &= x_{1,k}
\end{aligned}$$

The transfer characteristic equation is in the state space format is discretized using the definition of Grunwald-Letnikov GL differintegral. For the state variables fractional derivative $x_1^{(\beta)} = D_t^{\beta} x_1(t)$ and $x_2^{(\alpha-\beta)} = D_t^{(\alpha-\beta)} x_2(t)$, GL expansion is made and then written in above format.

As we see fractional order differential equations accumulate the whole information of the formula in a weighted form, this is so called "memory effect". Fractionally differentiated state space variable must be known as long as system has been operated to obtain correct response. This is known as "initialization function". For integer order systems it is constant and for fractional order systems it is time varying. In the usual integer order system theory, the set of states of the system, known at any given point in time along with the system equations are sufficient to predict the response of the system both forward and backwards in time. The fractional dynamic variables do not represent the "state" of the system at any given time in the previous sense; we need "history" of states or sufficient number by "short memory principle" for initialization function computation. Because of above mentioned memory effect from this following high memory consumption advance control algorithms with direct discretization of Tustin rule Al-Alouni rule by power series expansion (PSE) and or continued fraction expansion (CFE) is developed. In these new approaches the memory requirement is one tenth of the memory required for GL method. Several advance algorithms even to reduce this memory requirement is interesting topic of research and development.

Many generalization of integer control design are possible with freedom allowed by fractional order systems. Some of them are listed below:

a) Integral Control: Fractional integrals $H(s) = ks^{-q}$ are used as compensators. The interesting feature of fractional integrals is that they still allow closed-loop tracking of step reference signals, while allowing the freedom to tune the low frequency and high frequency behavior by tuning the value of the q, although the tracking will be slower.
b) Derivative Controls: Although the pure derivative control is seldom used, derivatives of any fractional order $H(s) = ks^{q}$ are available, and these will have less noise amplification at high frequencies than integer order derivatives.
c) PI, PD, and PID Controls: Fractional elements allow use of any value of the q, for integral and the derivative in these controllers. If a fractional PID control is implemented, the fraction in the derivative part need not be same of the integral part fraction. The different fractions are indicated in the Figure 9.3, the controller will be $H(s) = k_p + k_i s^{-q_1} + k_d s^{q_2}$. The generalizations of the PID controls give a research topic of continuum order distribution controller (Chapter 10).
d) Lead and lags: Lead compensators are often used to help stabilize marginally stable system. Lag compensators are often used to reduce the magnitude of high frequency loop gain of the system. Using fractional order components, it is possible to design fractional leads and fractional lag $H(s) = \dfrac{k(s^{q_1} + a)}{s^{q_2} + b}$. The benefit to these is that it is easier to shape the open loop and closed-loop frequency responses using them than exclusively integer-order elements, due to the extra freedom offered by the continuum values of the q.
e) Start point singularity: Design criterion unique to fractional order system deals with time domain singularity occurring at time zero. If the plant

transfer function $G(s)$ does not contain a term in the denominator with an exponent of at least 1, that is the leading term in the denominator is s^q, then by the initial value theorem $g(0)=\lim_{s\to\infty} sG(s)$, the impulse response, $g(t)=\mathcal{L}^{-1}\{G(s)\}$, will have a singularity at time zero. This may not be desirable, however using appropriate compensator in the forward path (say $H(s)=ks^{q-1}$) the singularity in the output of plant be eliminated

9.6.1 Concept of Iso-Damping

The concept of having iso-damping i.e. overshoot independent of the payload (or system gain) has remarkable usage in the field of control sciences. This is only possible by use of fractional order calculus theory. This concept was introduced as an example of DC motor controls in the Chapter 3. The plant may or may not be of fractional order, but controller with fractional order differential and or integral action is what makes the system response iso damped.

This concept when applied to Nuclear Power Plant controls elevates over all fuel efficiency robustness. The Nuclear Reactor is divided in two parts of, namely reactor core and coupled energy transfer equipments (heat exchangers, boiler, turbine generator). The reactor part error correction is carried out in terms of ratio control of the observed neutron power (feedback) to the demanded power the set point (on suitable set exponent). The effective power error is written as

$$\text{EPE} = K_1(\log\{\text{Cal}P\}-\log\{\text{Dem}P\})+K_2\left(\frac{1}{T_{\text{OBS}}}-\frac{1}{T_{\text{SET}}}\right)\propto\frac{\text{Cal}P.e^{t/T_{\text{OBS}}}}{\text{Dem}P.e^{t/T_{\text{SET}}}},$$

i.e. the ratio of observed neutronic power to the demanded power. This gives better results as compared to the existing integer order linear PID type corrections, as the governing formula is close to the reactor physics (which follows exponential and logarithmic expressions). In above expression $K_1 = 1$ and $K_2 = 0.5$ seconds (digitization time of control computer). This expression error governance gives fuel-efficient concept.

The reactor is coupled to several energy conversion devises, which governed by fractional order PID concept with iso-damping will increase overall nuclear power plant efficiency. The systems are always under dynamic corrections, and there are changes in gains (say pay-loads) and parametric shifts. This will be compensated by having iso-damped control systems where overshoot is independent of gain (pay-loads). The systems thus coupled to nuclear reactor will load the same with constant overshoot through out thus will have same energy transfer and not jittery. This process integrated over life span of nuclear power station will save nuclear fuel. This example to govern large energy transfer machines with fractional order controller with iso-damping is true for any power plant thus, to enhance overall fuel efficiency.

Start-up of nuclear plants when no inherent feedbacks are present is risky affair. More so, for experimental reactors where nuclear cores are of various experimental in nature. Although reactor trip systems will take care of any excursions in power levels at start up experiments, yet governing the start up procedure with a fractional order feed back controller where overshoot is independent of gain (fuel characteristics) will add another level of safety and confidence.

As an example of DC motor controls let the motor (plant) be expressed as $G(s)=\dfrac{K_m}{Js(s+1)}$ with J as payload inertia. Let the selected fractional open loop transfer function have robustness and stability measure as per Bode's ideal function with phase margin constant of 60^0. Then Bode's ideal open loop transfer function, which gives this phase margin, is $G_0(s)=\dfrac{1}{s^{4/3}}$. Since $G_0(s)=H(s)G(s)$, we can obtain a transfer function in the form: $H(s)=\dfrac{J}{K_m}\left(s^{2/3}+\dfrac{1}{s^{1/3}}\right)=K\left(s^{2/3}+s^{-1/3}\right)$, which is a particular case of $PI^{\lambda}D^{\delta}$ where $K=J/K_m$ is the controller constant. The phase margin of the controlled system with a forward loop controller is

$$\Phi_m=\arg\left[H(\mathrm{j}\omega_0)G(\mathrm{j}\omega_0)\right]+\pi=\arg\left[\frac{1}{(\mathrm{j}\omega)^{4/3}}\right]+\pi=\pi-\frac{4}{3}\frac{\pi}{2}=\frac{\pi}{3},$$

where ω_0 is cross over frequency.

The constant phase margin is not dependent on payload (gain) changes and the system gain K. The phase curve is a horizontal line at $-2\pi/3$.

Step-response of the closed loop can be expressed as:

$$y(t)=\mathcal{L}^{-1}\left\{\frac{1}{s(s^{1+1/3}+1)}\right\}=t^{1+1/3}E_{1+1/3,2+1/3}\left(-t^{1+1/3}\right),$$

where step-response is independent of the payload inertia at fractional order setting of 4/3. The solution is noted as follows:

Close loop transfer function from open loop transfer function $G_0(s)=s^{-4/3}$ is

$$G_{CL}(s)=\frac{G_0(s)}{1+G_0(s)}=\frac{1}{s^{4/3}+1}.$$

The step input will have output (time) response as:

$$y(t)=\mathcal{L}^{-1}\left\{\frac{1}{s\left(s^{4/3}+1\right)}\right\}=\mathcal{L}^{-1}\left\{\frac{1}{s}\left[1-\frac{s^{4/3}}{s^{4/3}+1}\right]\right\}=\mathcal{L}^{-1}\left\{\frac{1}{s}-\frac{s^{\frac{4}{3}-1}}{s^{4/3}+1}\right\}=1-E_{4/3}(-t^{4/3})$$

Expanding with definition of one-parameter Mittag-Leffler function $E_{4/3}(-t^{4/3})$ we get:

$$y(t)=1-\left[1+\frac{(-t^{4/3})}{\Gamma\left(\frac{4}{3}+1\right)}+\frac{(-t^{4/3})^2}{\Gamma\left(2.\frac{4}{3}+1\right)}+\frac{(-t^{4/3})^3}{\Gamma\left(3.\frac{4}{3}+1\right)}+\dots\right]$$

$$=-\frac{(-t^{4/3})}{\Gamma\left(\frac{4}{3}+1\right)}-\frac{(-t^{4/3})^2}{\Gamma\left(2.\frac{4}{3}+1\right)}-\frac{(-t^{4/3})^3}{\Gamma\left(3.\frac{4}{3}+1\right)}-\dots$$

$$=\frac{t^{4/3}}{\Gamma\left(\frac{4}{3}+1\right)}-\frac{(t^{4/3})^2}{\Gamma\left(2.\frac{4}{3}+1\right)}+\frac{(t^{4/3})^3}{\Gamma\left(3.\frac{4}{3}+1\right)}-\dots$$

$$=t^{4/3}\left[\frac{1}{\Gamma\left(\frac{4}{3}+1\right)}+\frac{(-t^{4/3})}{\Gamma\left(2.\frac{4}{3}+1\right)}+\frac{(-t^{4/3})^2}{\Gamma\left(3.\frac{4}{3}+1\right)}+\dots\right]$$

Using the definition of two-parameter Mittag-Leffler function the following is obtained:

$$y(t)=t^{4/3}E_{\frac{4}{3},\frac{4}{3}+1}(-t^{4/3})=t^{1+\frac{1}{3}}E_{1+\frac{1}{3},2+\frac{1}{3}}(-t^{1+\frac{1}{3}})$$

9.6.2 Frequency Domain Design for Fractional Order Plant and Fractional Order Controller Tuning

If $G(s)$ is the plant transfer function, the objective is to find a controller $H(s)$, as in Figure 9.3, so that the open loop system $G_{OL}(s)=H(s)G(s)$ would meet the following specification:

(a) Phase Margin specification:

$$\arg\left[G_{OL}(\mathrm{j}\omega_{gc})\right]=\arg\left[H(\mathrm{j}\omega_{gc})G(\mathrm{j}\omega_{gc})\right]=-\pi+\Phi_m\,.$$

Example the phase margin Φ_m is around 80^0, for a given plant ω_{gc} is of the order of 0.3rad/s.

(b) Gain crossover frequency specification: $|G_{OL}(\mathrm{j}\omega_{gc})| = |H(\mathrm{j}\omega_{gc})G(\mathrm{j}\omega_{gc})| = 1$ That is, 0dB at the cross over frequency.

(c) Robustness to Gain variation:

$$\left(\frac{d}{d\omega}\arg\left[G_{OL}(\mathrm{j}\omega_{gc})\right]\right)_{\omega=\omega_{gc}} = 0.$$

This gives the flat phase near the cross over frequency. A constant phase for a band of frequency till gain cross over frequency gives iso-damping.

(d) Complimentary Sensitivity specification:

$$|G_{CL}(\mathrm{j}\omega)| = 20\log\left|\frac{H(\mathrm{j}\omega)G(\mathrm{j}\omega)}{1+H(\mathrm{j}\omega)G(\mathrm{j}\omega)}\right| \le A\text{dB}\ \forall\omega \ge \omega_t\text{rad/s}.$$

For example selected values for tuning are around $A = -40\text{dB}$, and $\omega_t = 10\text{rad/s}$, for one application.

(e) Sensitivity Specification:

$$|S(\mathrm{j}\omega)| = 20\log\left|\frac{1}{1+H(\mathrm{j}\omega)G(\mathrm{j}\omega)}\right| \le B\text{dB}\ \forall\omega \ge \omega_s\text{rad/s}.$$

For example the value is around $B = -40\text{dB}$, and $\omega_s = 10^{-2}\text{rad/s}$, for one particular application.

Let us consider a plant which is represented as a Fractional-order plus a time delay as

$$G(\mathrm{j}\omega) = \frac{K}{T(\mathrm{j}\omega)^{\alpha}+1}e^{-\mathrm{j}L\omega} = \frac{K\left[\cos\omega L - \mathrm{j}\sin\omega L\right]}{T\omega^{\alpha}\left(\cos\frac{\alpha\pi}{2}+\mathrm{j}\sin\frac{\alpha\pi}{2}\right)+1} = \frac{K\left[\cos\omega L - \mathrm{j}\sin\omega L\right]}{\left(1+T\omega^{\alpha}\cos\frac{\alpha\pi}{2}\right)+\mathrm{j}T\omega^{\alpha}\sin\frac{\alpha\pi}{2}}$$

$$|G(\mathrm{j}\omega)| = \frac{K}{\sqrt{\left(1+T\omega^{\alpha}\cos\frac{\alpha\pi}{2}\right)^2+\left(T\omega^{\alpha}\sin\frac{\alpha\pi}{2}\right)^2}}$$

$$\arg\left[G(\mathrm{j}\omega)\right] = \tan^{-1}\left(\frac{-\sin\omega L}{\cos\omega L}\right) - \tan^{-1}\left(\frac{T\omega^{\alpha}\sin\frac{\alpha\pi}{2}}{1+T\omega^{\alpha}\cos\frac{\alpha\pi}{2}}\right) = -\omega L - \tan^{-1}\left(\frac{T\omega^{\alpha}\sin\frac{\alpha\pi}{2}}{1+T\omega^{\alpha}\cos\frac{\alpha\pi}{2}}\right)$$

The derivative of phase of the plant is:

$$\frac{d}{d\omega}\arg\left[G(\mathrm{j}\omega)\right]=\frac{d}{d\omega}\left[-\omega L-\tan^{-1}\left(\frac{T\omega^{\alpha}\sin\frac{\alpha\pi}{2}}{1+T\omega^{\alpha}\cos\frac{\alpha\pi}{2}}\right)\right]$$

$$=-L-\frac{T\alpha\omega^{\alpha-1}\sin\frac{\alpha\pi}{2}}{\left(1+T\omega^{\alpha}\cos\frac{\alpha\pi}{2}\right)^{2}+\left(T\omega^{\alpha}\sin\frac{\alpha\pi}{2}\right)^{2}}$$

Let us consider a controller with structure $H(s)=K_p+K_i s^{-\lambda}K_d s^{\mu}$, then:

$$\begin{aligned}H(\mathrm{j}\omega)&=K_p+K_i\omega^{-\lambda}\mathrm{j}^{-\lambda}+K_d\omega^{\mu}\mathrm{j}^{\mu}\\&=K_p+K_i\omega^{-\lambda}\left(\cos\frac{\lambda\pi}{2}-\mathrm{j}\sin\frac{\lambda\pi}{2}\right)+K_d\omega^{\mu}\left(\cos\frac{\mu\pi}{2}+\mathrm{j}\sin\frac{\mu\pi}{2}\right)\\&=\left(K_p+K_i\omega^{-\lambda}\cos\frac{\lambda\pi}{2}+K_d\omega^{\mu}\cos\frac{\mu\pi}{2}\right)+\mathrm{j}\left(K_d\omega^{\mu}\sin\frac{\mu\pi}{2}-K_i\omega^{-\lambda}\cos\frac{\lambda\pi}{2}\right)\end{aligned}$$

$$\left|H(\mathrm{j}\omega)\right|=\sqrt{\left(K_p+K_i\omega^{-\lambda}\cos\frac{\lambda\pi}{2}+K_d\omega^{\mu}\cos\frac{\mu\pi}{2}\right)^{2}+\left(K_d\omega^{\mu}\sin\frac{\mu\pi}{2}-K_i\omega^{-\lambda}\cos\frac{\lambda\pi}{2}\right)^{2}}$$

$$\arg\left[H(\mathrm{j}\omega)\right]=\tan^{-1}\left(\frac{K_d\omega^{\mu}\sin\frac{\mu\pi}{2}-K_i\omega^{-\lambda}\cos\frac{\lambda\pi}{2}}{K_p+K_i\omega^{-\lambda}\cos\frac{\lambda\pi}{2}+K_d\omega^{\mu}\cos\frac{\mu\pi}{2}}\right)$$

The derivative of phase of the controller is:

$$\frac{d}{d\omega}\arg\left[H(\mathrm{j}\omega)\right]=\frac{d}{d\omega}\left[\tan^{-1}\left(\frac{K_d\omega^{\mu}\sin\frac{\mu\pi}{2}-K_i\omega^{-\lambda}\cos\frac{\lambda\pi}{2}}{K_p+K_i\omega^{-\lambda}\cos\frac{\lambda\pi}{2}+K_d\omega^{\mu}\cos\frac{\mu\pi}{2}}\right)\right]$$

$$=\frac{\left(K_pK_d\mu\omega^{\mu-1}\sin\frac{\mu\pi}{2}\right)+\left(K_pK_i\lambda\omega^{-\lambda-1}\sin\frac{\lambda\pi}{2}\right)+\left(K_iK_d(\lambda+\mu)\omega^{-\lambda+\mu-1}\sin\frac{(\lambda+\mu)\pi}{2}\right)}{\left(K_p+K_i\omega^{-\lambda}\cos\frac{\lambda\pi}{2}+K_d\omega^{\mu}\cos\frac{\mu\pi}{2}\right)^{2}+\left(K_d\omega^{\mu}\sin\frac{\mu\pi}{2}-K_i\omega^{-\lambda}\cos\frac{\lambda\pi}{2}\right)^{2}}$$

For this controller five parameters namely $K_p, K_i, K_d, \lambda, \mu$ and need to be found out for tuning purpose, the five above criteria is therefore required. The problem is that an analytical explicit solution is not easy to derive. For the above coupled transcendental non-linear equations MATLAB Optimization tool box is employed, where Powell's Trust-Region-Dogleg algorithm is implemented.

9.6.3 Family of Fractional Order Controllers

The classical PID structure can be extended to a family of Fractional Order Controller, structures, for controller transfer function $H(s)$ (Figure 9.3) as follows:

a) Integer Order PID or IOPID controller: $H_1(s) = K_p + K_i s^{-1} + K_d s$

b) Fractional Order PI or FOPI: $H_2(s) = K_p + K_i s^{-\lambda}$

c) Fractional Order (PI) or FO(PI) controller: $H_3(s) = \left(K_p + K_i s^{-1}\right)^{\lambda}$

d) Fractional Order (PID) or FO(PID) controller:

$$H_4(s) = \left(K_p + K_i s^{-1} + K_d s\right)^{\lambda}$$

e) Fractional Order PID or FOPID controller:

$$H_5(s) = K_p + K_i s^{-\lambda} + K_d s^{\mu}$$

Clearly integer order PID is special case of the generalized controller structures described above. Generally the PD controllers cannot remove the steady state error due to the absence of the integral element and also suffers from initial derivative kick. That is why it is used for processes mostly having integrator present in the process plant model.

9.6.4 Fractional Vector Feedback Controller

This section considers the use of fractional vector feed back for control system design, especially for multivariate fractional state space. The modeling of the system in fractional vector state-space has been covered in Chapter 7, and example given in Chapter 8. The vector representation is:

$${}_0d_t^q \overline{x}(t) + \overline{\psi}(\overline{x}, q, a, 0, t) = \mathrm{A}\overline{x}(t) + \mathrm{B}\overline{u}(t) \text{ and } \overline{y}(t) = \mathrm{C}\overline{x}(t) + \mathrm{D}\overline{u}(t).$$

Typically, vector feedback is implemented as $\overline{u}(t) = -K\overline{x}(t) + \overline{r}(t)$, where $\overline{r}$ is vector set point and K is the feedback gain matrix to be determined. The closed loop system then becomes:

$$\begin{aligned} {}_0d_t^q \overline{x}(t) &= [\mathrm{A} - \mathrm{B}K]\overline{x}(t) - \overline{\psi}(\overline{x}, q, a, 0, t) + \mathrm{B}\overline{r}(t) \\ \overline{y}(t) &= [\mathrm{C} - \mathrm{D}K]\overline{x}(t) + \mathrm{D}\overline{r}(t) \end{aligned}$$

By choosing K appropriately and using standard pole-placement methods it is possible to place the eigenvalues anywhere in the w-plane. The $\det[w\mathrm{I} - \mathrm{A} + \mathrm{B}K]$ will determine the pole placement. While doing this, the eigenvalues should be placed to the left of the instability wedge (Chapter 7). In this $s^q = w$, is the transformation from $s \to w$ plane. Presently without worrying about the instability due to initialization vector $\overline{\psi}$, the pole placement can be carried to give the stable closed loop performance as desired. At present it is unclear how linear quadratic regulator (LQR) and other optimal feedback regulator rules be redrawn for fractional vector state problems.

Consider

$$\mathrm{A}=\begin{bmatrix}0 & 1 & 0\\ 0 & 0 & 1\\ -2 & -4 & -6\end{bmatrix}, \mathrm{B}=\begin{bmatrix}0\\0\\1\end{bmatrix}, \mathrm{C}=\begin{bmatrix}1 & 2 & 3\end{bmatrix}, \mathrm{D}=\begin{bmatrix}0\end{bmatrix},$$

the system is ${}_0D_t^q\overline{x}(t)=\mathrm{A}\overline{x}+\mathrm{B}\overline{u}$ and $\overline{y}(t)=\mathrm{C}\overline{x}(t)$ with controller as $\overline{u}(t)=-K\overline{x}(t)$. The control parameter (gains) will be chosen such that the poles in the w-plane are $w_{1,2}=-2\pm \mathrm{j}4$ and $w_3=-6$. The Laplace transformation of the vector system equation with feedback regulator is $\overline{X}(s)=-[s^q\mathrm{I}-\mathrm{A}+\mathrm{B}K]^{-1}\overline{\psi}(s)$. The close loop system matrix is $\left[\mathrm{I}s^q-\mathrm{A}+\mathrm{B}K\right]=\left[w\mathrm{I}-\mathrm{A}+\mathrm{B}K\right]$, from here the value of K is evaluated as:

$$\det\left[w\mathrm{I}-\mathrm{A}+\mathrm{B}K\right]=\det\left\{\begin{bmatrix}w & 0 & 0\\ 0 & w & 0\\ 0 & 0 & w\end{bmatrix}-\begin{bmatrix}0 & 1 & 0\\ 0 & 0 & 1\\ -2 & -4 & -6\end{bmatrix}+\begin{bmatrix}0\\0\\1\end{bmatrix}\begin{bmatrix}k_1 & k_2 & k_3\end{bmatrix}\right\}$$

$$\det\begin{bmatrix}w & -1 & 0\\ 0 & w & -1\\ 2+k_1 & 4+k_2 & w+6+k_3\end{bmatrix}=w^3+(6+k_3)w^2+(4+k_2)w+(2+k_1)$$

From the poles given in the w-plane the characteristic equation is $\alpha(w)=(w+2-\mathrm{j}4)(w+2+\mathrm{j}4)(w+6)=w^3+10w^2+44w+120$, Comparing this with coefficients of the determinant we get $K=\begin{bmatrix}k_1 & k_2 & k_3\end{bmatrix}=\begin{bmatrix}4 & 40 & 118\end{bmatrix}$.

Here we have obtained the controller matrix and the gains because the system is controllable as rank of matrix $\begin{bmatrix}\mathrm{B} & \mathrm{AB} & \mathrm{A}^2\mathrm{B}\end{bmatrix}$ is 3 full ranks. These are usual theory of multivariate control science. Refer Figure 9.5 for fractional vector state feed back block diagram.

9.6.5 Observer in Fractional Vector System

Just as in integer order theory, it is important to create observers or vector estimators for fractional order system. This section will present the theory necessary for designing fractional order observer. The fractional order vector estimator is:

$${}_0d_t^q\hat{\bar{x}}(t)+\hat{\bar{\psi}}(\bar{x},q,a,0,t)=\mathrm{A}\hat{\bar{x}}(t)+\mathrm{B}\bar{u}(t)-\mathrm{L}\left[\bar{y}(t)-\hat{\bar{y}}(t)\right]$$

$\hat{\bar{y}}(t)=\mathrm{C}\hat{\bar{x}}(t)+\mathrm{D}\bar{u}(t)$, where a non-zero initialization function $\hat{\bar{\psi}}$ has been assumed for the observer. The vector error $\bar{e}(t)$ is defined as difference between the real system outputs $\bar{x}(t)$ and estimated observer outputs $\hat{\bar{x}}(t)$: $\bar{e}(t)=\bar{x}(t)-\hat{\bar{x}}(t)$. The observer gain L is determined as to force the error between the plant vectors to go to zero. The dynamics of the error are obtained by fractionally differentiating the error equation as: ${}_0d_t^q\bar{e}(t)={}_0d_t^q\bar{x}(t)-{}_0d_t^q\hat{\bar{x}}(t)$.

In this substitute the system equations to get:

$${}_0d_t^q\bar{e}(t)=\left[\mathrm{A}\bar{x}(t)+\mathrm{B}\bar{u}(t)-\bar{\psi}(\bar{x},q,a,0,t)\right]-\left[\mathrm{A}\hat{\bar{x}}(t)+\mathrm{B}\bar{u}(t)-\hat{\bar{\psi}}(\hat{\bar{x}},q,a,0,t)\right]-\mathrm{L}\left[\bar{y}(t)-\hat{\bar{y}}(t)\right]$$

Now replacing the sensed system outputs $\bar{y}(t)$ and $\hat{\bar{y}}(t)$ with the vector variable $\mathrm{C}x(t)+\mathrm{D}u(t)$ we obtain:

$${}_0d_t^q\bar{e}(t)=\left[\mathrm{A}\bar{x}(t)+\mathrm{B}\bar{u}(t)-\bar{\psi}(\bar{x},q,a,0,t)\right]-\left[\mathrm{A}\hat{\bar{x}}(t)+\mathrm{B}\bar{u}(t)-\hat{\bar{\psi}}(\hat{\bar{x}},q,a,0,t)\right]$$

$$-\mathrm{L}\left[\mathrm{C}\bar{x}(t)+\mathrm{D}\bar{u}(t)-\mathrm{C}\hat{\bar{x}}(t)-\mathrm{D}\hat{\bar{u}}(t)\right]$$

Eliminating and doing simple algebra and substituting $\bar{x}(t)-\hat{\bar{x}}(t)=\bar{e}(t)$ we get:

$${}_0d_t^q\bar{e}(t)=\left[\mathrm{A}-\mathrm{LC}\right]\bar{e}(t)-\left\{\bar{\psi}(\bar{x},q,a,0,t)-\hat{\bar{\psi}}(\hat{\bar{x}},q,a,0,t)\right\}$$

The matrix L is determined to force the observer error to go to zero by placing the eigenvalues of $\left[\mathrm{A}-\mathrm{LC}\right]$ in the stable region of *w*-plane (Chapter 7). The initialization function response eventually decays to zero for any system with $0<q<1$, and only has transient effect on the observer error. However proper choice of $\hat{\bar{\psi}}$ will help drive the error to zero sooner than if $\hat{\bar{\psi}}$ was simply zero.

Consider ${}_0D_t^q\bar{x}(t)=\mathrm{A}\bar{x}(t)+\mathrm{B}\bar{u}(t)$ and $\bar{y}(t)=\mathrm{C}\bar{x}(t)$, observer gain has to set such that the closed loop poles follow the *w*-plane as $w_{1,2}=-2\pm \mathrm{j}\sqrt{12}, w_3=-5$.

$$\mathrm{A}=\begin{bmatrix}0 & 1 & 0\\ 0 & 0 & 1\\ -6 & -11 & -6\end{bmatrix},\mathrm{B}=\begin{bmatrix}0\\0\\1\end{bmatrix},\mathrm{C}=\begin{bmatrix}1 & 0 & 0\end{bmatrix},\mathrm{D}=\begin{bmatrix}0\end{bmatrix}$$

The normal observability matrix is

$$\begin{bmatrix} C \\ CA \\ CA^2 \end{bmatrix} = \begin{bmatrix} 1 & 0 & 0 \\ 0 & 1 & 0 \\ 0 & 0 & 1 \end{bmatrix}$$

has full rank. Therefore the system is observable. The characteristic polynomial is:

$$\alpha(w) = (w+2-j\sqrt{12})(w+2+j\sqrt{12})(w+5) = w^3+9w^2+36w+80 .$$

$$\det[wI - A + LC] = \det\left\{\begin{bmatrix} w & 0 & 0 \\ 0 & w & 0 \\ 0 & 0 & w \end{bmatrix} - \begin{bmatrix} 0 & 1 & 0 \\ 0 & 0 & 1 \\ -6 & -11 & -6 \end{bmatrix} + \begin{bmatrix} l_1 \\ l_2 \\ l_3 \end{bmatrix}[1 \quad 0 \quad 0]\right\}$$

$$\det\begin{bmatrix} w+l_1 & -1 & 0 \\ l_2 & w & -1 \\ l_3+6 & 11 & w+6 \end{bmatrix} = w^3+(l_1+6)w^2+(6l_1+l_2+11)w+(11l_1+6l_2+l_3+6)$$

Equating the determinant to the characteristic polynomial we obtain $(l_1)+6=9$, $6(l_1)+(l_2)+11=36$ $(11)l_1+(6)l_2+l_3+6=80$ gives the value for observer gain as $L=[l_1 \quad l_2 \quad l_3]=[3 \quad 7 \quad -1]$.

Refer Figure 9.6 for fractional vector observer block diagram.

9.6.6 Modern Aspects of Fractional Control

The multivariate control aspects are discussed in the previous sections with the examples. The controllable and observable issues are discussed and disregarding the time varying initialization vector the gains of fractional state feed back and gains of observer were calculated. In the example coefficient equalization scheme is chosen to equalize the determinants of controller and observer matrices. Here the Bass-Gura and Ackerman formula from the rank determination approach is also suitable, to place the poles as desired in the w-plane stability wedge.

Question of controllability and observability and minimality arise when the systems are expressed in the vector fractional dynamic variable form. There are two directions to go in this regard. One is to completely redefine and derive all of the related vector system properties. This is not done at this time. Though effort is on in this regard. It is important to note that the fractional dynamic variable vector alone does not contain all the information about the state of the system, but requires vector $\bar{\psi}$ the initialization vector. The implication of this is totally not clear at this stage at present; however, one important consideration in observability and controllability issue is the inclusion of the time dependent initialization function. Rather than re-deriving all the system theory results on controllability and observability, useful results can still be obtained by simply using the observabilty and controllability matrices directly for particular fractional vector system-while neglecting the initialization vector. This can be done on usual *ABCD* matrices completely without regard to their deeper theoretic implementation.

Finally it should be noted that the vector space of fractional dynamic variables allows direct use of standard state variable feedback and observer theory, with the understanding that the closed loop poles are being placed in stability zone of *w*-plane. It is not clear at this point how to interpret any optimal control theory rules. Although we could use Lyapunov and Riccati equations for design, their interpretation is not clear with regard to optimality. One would expect that the resulting controllers which always guarantee to have closed loop poles in left half plane (LHP) of *s*-domain, would now place all closed loop poles in the LHP of *w*-domain, which would guarantee some form of stable 'hyper-damped' response.

Kalman' decomposition laws are still valid with initialization function taken as one of the plant disturbance input. Also for state estimator 'fractional Kalman filter' and extended fractional Kalman filter' for linear and non-linear systems are recent developmental field of multivariate control science.

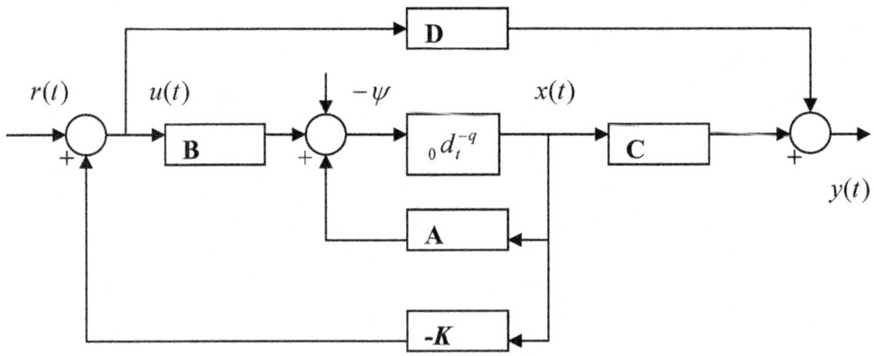

Fig. 9.5 State variable fractional vector controller

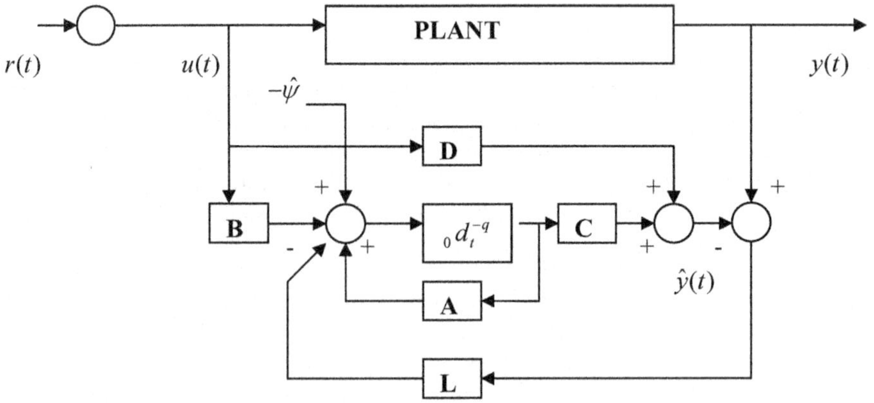

Fig. 9.6 Fractional vector dynamic variable observer.

9.7 Fractional Compensator

9.7.1 Generalized Compensator

Generalization of the traditional lead compensator is discussed in this section. This is obtained by introducing as a new parameter the fractional order, α, of the structure. By doing so the proposed fractional order lead compensator (FOLC) has the form

$$C(s)=k_c\left(\frac{s+\frac{1}{\lambda}}{s+\frac{1}{x\lambda}}\right)^{\alpha}=k_c\left(\frac{\lambda s+1}{x\lambda s+1}\right)^{\alpha} \tag{9.1}$$

In this generalization, the above controller (9.1) is used just as a lead compensator that is, its purpose is not to ensure a constant phase ($\alpha\ \pi/2$) in a frequency interval, but to ensure the fulfilling of the design specifications traditionally used for a lead compensator. Providing the designer with greater flexibility in shaping the frequency plot of the compensator, by the slope modification factor α, the fractional lead compensator enjoys a distinctive edge over its integer order counterpart.

9.7.2 Frequency Characteristics of the Lead Compensator

For the conventional lead compensator

$$C^*(s)=k_c\left(\frac{s+\frac{1}{\lambda}}{s+\frac{1}{x\lambda}}\right)=k_c\left(\frac{\lambda s+1}{x\lambda s+1}\right) \tag{9.2}$$

The zero frequency is given by $\omega_{zero}=(1/\lambda)$ and the pole frequency by $\omega_{pole}=(1/x\lambda)$. The key idea in the design of a lead compensator is to increase the phase margin of the open loop system, by adding phase in the neighborhood of the gain crossover frequency, ω_c. The bode plots of (9.2) are shown in Figure 9.7, where it can be observed that the maximum phase Φ_m is given at frequency ω_m, which is the geometric mean of the corner frequencies ω_{zero} and ω_{pole}.

Therefore,

$$\arg[C^*(s)]_{\omega=\omega_m}=\Phi_m=\tan^{-1}(\lambda\omega_m)-\tan^{-1}(x\lambda\omega_m)=\sin^{-1}\left(\frac{1-x}{1+x}\right) \tag{9.3}$$

$$\omega_m = \frac{1}{\lambda\sqrt{x}} \tag{9.4}$$

$$\left|\frac{C^*(\mathrm{j}\omega_m)}{k_c x}\right| = \left|\frac{\mathrm{j}\omega_m\lambda+1}{\mathrm{j}\omega_m x\lambda+1}\right| = \frac{1}{\sqrt{x}} \tag{9.5}$$

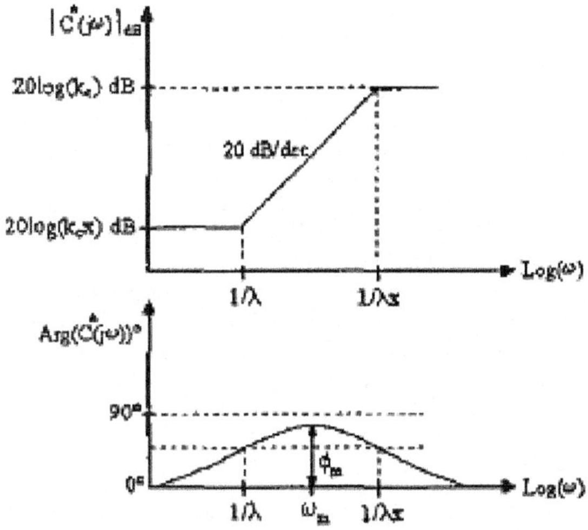

Fig. 9.7 Bode plots of the transfer function C*(s)

It must be taken into account that this compensator modifies the magnitude curve, moving the frequency ω_c to the right, of the Bode-plot, after compensation and producing a phase boost at the uncompensated ω_c point (Figures 9.9 and 9.10). In order to maintain the specification of the phase margin, the Φ_m of the compensator, this phase lag must be compensated by increasing this by few degrees the maximum phase Φ_m that the compensator, must give (over phase). This over phase is estimated by trial and error method.

However, analytical methods can be used for the design of the integer compensator in order to guarantee the desired Φ_m at the desired ω_c without a trial and error process, though the over phase is always present.

Now, considering the fractional lead compensator of (9.1), as can be seen from Figure 9.8, the parameter x sets the distance between the fractional zero and pole, and the parameter λ sets their position on the frequency axis. The choice of these two parameters depends on the value of α. Higher the value of α, the higher the slope of the magnitude of $C(s)$ and the higher the maximum phase Φ_m that the compensator can give.

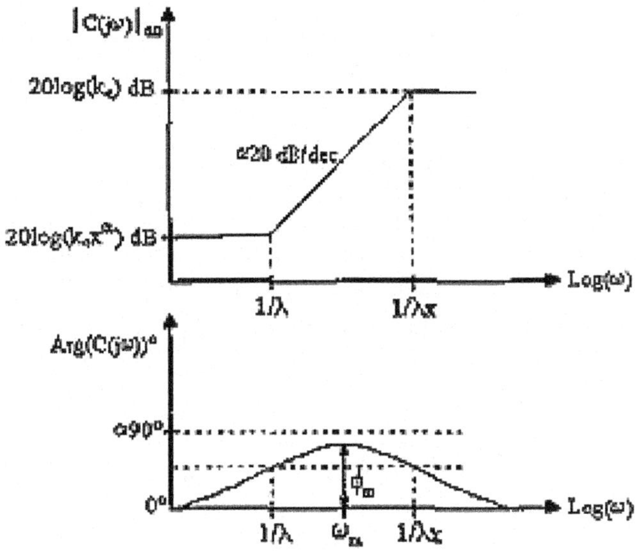

Fig. 9.8 Bode plots of the transfer function C(s) fractional order compensator.

At the frequency ω_m

$$\left|\frac{C(\mathrm{j}\omega_m)}{k_c x^{\alpha}}\right| = \left|C'(s)\right|_{\omega=\omega_m} = \left(\sqrt{\frac{(\lambda\omega_m)^2+1}{(x\lambda\omega_m)^2+1}}\right)^{\alpha} = \left(\frac{1}{\sqrt{x}}\right)^{\alpha} \tag{9.6}$$

$$\arg(C'(s))_{\omega=\omega_m} = \Phi_m = \alpha\left[\tan^{-1}(\lambda\omega_m) - \tan^{-1}(x\lambda\omega_m)\right] = \alpha\sin^{-1}\left(\frac{1-x}{1+x}\right) \tag{9.7}$$

9.7.3 Compensation Using a Fractional Lead Compensator

In the most general case, first of all, the value of compensator gains $k' = k_c x^{\alpha}$ can be set in order to fulfill a static error constant specification for the compensated system. For a general plant model of a form (system type n)

$$G(s) = \frac{k\prod_i(\tau_i s+1)}{s^n\prod_j(\tau_j s+1)} \tag{9.8}$$

the static error constant k_{ss} , has the expression:

$$k_{ss} = \lim_{s\to 0}\left[s^n C(s)G(s)\right] = \lim_{s\to 0}\left[s^n k'\left(\frac{\lambda s+1}{x\lambda s+1}\right)^{\alpha}\frac{k\prod_i(\tau_i s+1)}{s^n\prod_j(\tau_j s+1)}\right] = \lim_{s\to 0}\left[\frac{s^n k' k}{s^n}\right] = k'k \tag{9.9}$$

that is, $k' = k_c x^{\alpha} = k_{ss}/k$, setting the relation between parameters k_c, x and α. Knowing the value of k', the bode plots of the system $G'(s) = k' G(s)$ are obtained, in which the static error constant specification is already fulfilled. Now, specifications of gain crossover frequency and phase margin must be achieved. Through the Bode plots of the plant $G'(s)$ the maximum phase (Φ_m) and the magnitude that the compensator $C'(s)$ must give to fulfill these two frequency specifications is observed. Then, the relations for the parameters of the fractional structure are given by the equations:

$$\lambda = \frac{1}{\omega_c\sqrt{x}} \tag{9.10}$$

$$\left|C'(s)\right|_{\omega_m=\omega_c} = \left(\frac{1}{\sqrt{x}}\right)^{\alpha} = \left|\frac{1}{k'G(s)}\right|_{\omega=\omega_c} \tag{9.11}$$

$$\arg(C'(s))_{\omega_m=\omega_c} = \Phi_m = \alpha\sin^{-1}\left(\frac{1-x}{1+x}\right) = -\pi + \Phi_m - \arg(k'G(s))_{\omega=\omega_c} \tag{9.12}$$

The above expressions come from the fact that open loop transfer function after compensation is $G'(s)C'(s)$ must have at gain cross over point (ω_c) the phase as $-\pi + \Phi_m$ for stability. Thus

$$\arg[G'(s)C'(s)]_{@\,\omega=\omega_c} = \arg[C'(s)] + \arg[k'G(s)] = -\pi + \Phi_m$$

So, a set of three nonlinear equations (9.10) to (9.12)) and three unknown parameters (x, λ, α) are obtained. Solving these equations, the value of parameter k_c can be easily obtained, since

$$k_c = \frac{k_{ss}}{x^{\alpha}k} \tag{9.13}$$

Therefore we have a total of four non linear equations and four unknown parameters to fulfill three design specifications, independently, ensuring the maximum phase, Φ_m at the frequency ω_c.

One of the major issues in obtaining an analytical design structure for the fractional lead compensator is the solution of the above mentioned set of non linear transcendental equations, whose solution and convergence are strongly governed by factors such as selection of starting point (initial solution guess) and the number of iterations. The only issue which has to be dealt carefully is the issue of convergence of solutions of the non linear equations. However, design issues such as accuracy of fractional order devices and hardware requirements for integer order approximations for implementing the fractional order design, need to be thoroughly analyzed.

The introduction of parameter α, fractional order of the structure, allows flexibility on the fulfillment of specifications of phase margin, Φ_m, at the gain crossover frequency, ω_c and the static error constant k_{ss}. The proposed method of design is based on the condition of null over phase, forcing the compensator to give it maximum phase at the gain crossover frequency.

As mentioned at the beginning there is distinct advantages of the fractional order design over the integer order structure, by providing greater Gain and Phase margins throughout the frequency range of interest.

Analogously, lag and lead-lag fractional order compensators can be designed based on similar non linear relations and compared with their integer order counterparts. The Figures 9.9 and 9.10 compares the Bode gain and phase plots for integer order compensator.

To fulfill the design specifications for the given plant $G(s)$, using both integer and fractional order designs, and then compare the performances, we consider plant as:

$$G(s) = \frac{2}{s(0.5s+1)}$$

With, velocity error constant, $k_v = 20$. Gain crossover frequency, $\omega_c = 10$ rad/sec., Phase margin, $\Phi_m = = 0.27\pi = 50°$.

For both the compensators, from the velocity error constant specifications, it is obtained as $k' = k_v / k = 10$. Now for the system $G'(s) = k'G(s)$, it is observed that at the gain crossover frequency, $\omega_c = 10$ rad/sec, the magnitude of $G'(s)$ is -8.16 dB and the phase is -168.7°. Thus design of the requisite compensators, to fulfill the conditions is given by:

$$\left[\left| C'(s) \right|_{\omega=\omega_c} \right]_{dB} = 8.16\text{dB} \ \text{ and}$$

$$\arg[C'(s)]_{\omega=\omega_c} = (-180^0 + 50^0 - \{-168.7^0\}) = 38.7^0$$

Integer order Compensator:

$$C_I(s) = \frac{0.2082s+1}{0.048s+1}$$

Fractional Order Compensator:

$$C_F(s) = \left(\frac{0.5266s+1}{0.0178s+1} \right)^{0.55}$$

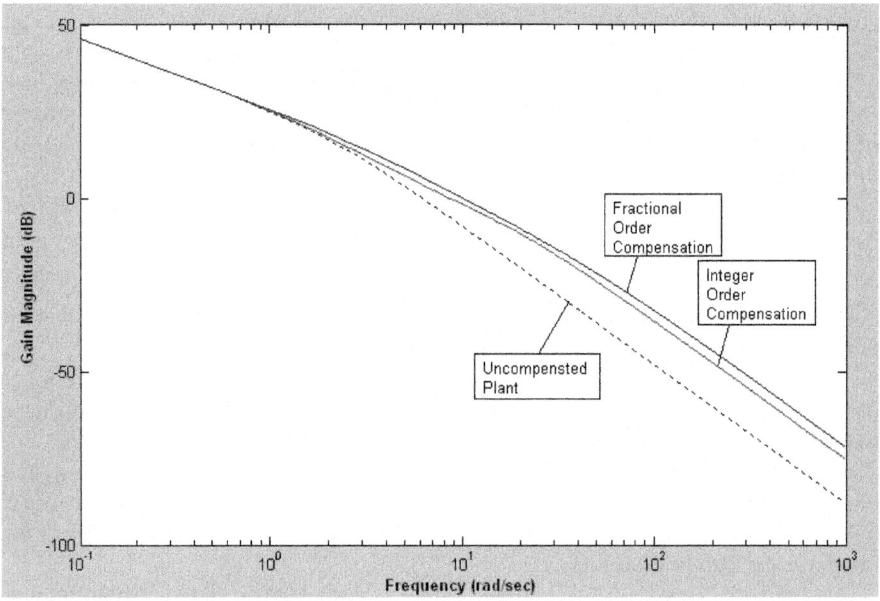

Fig. 9.9 Compensated Gain plots

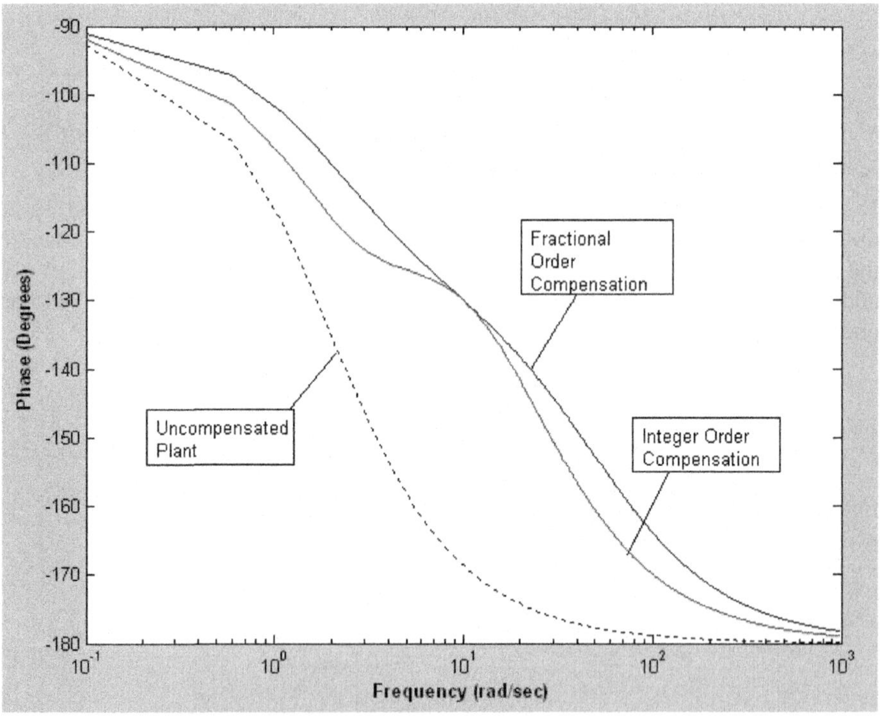

Fig. 9.10 Compensated Phase plots

9.8 Phase Shaping with Fractional Order Differ-Integrator

In this section, the methodology of designing a phase shaper with a fractional order differ-integrator is presented for a given system $G(s)$ comprising the plant $G_{pl}(s)$ and its PID controller $G_c(s)$, so that open loop transfer function is $G(s) = G_c(s) \times G_{pl}(s)$, which is tuned by standard techniques. On this tuned plant with controller if one has to achieve iso-damping, the extra fractional order differ-integrator block is required, and is demonstrated.

9.8.1 Application of Bode's Phase Integral

Bode's phase integral formula is used to approximate the derivative of the phase of transfer-function $G(s)$ around its gain crossover frequency ω_{gc}. Using Bode's phase integral, it can be shown, that for a stable, minimal phase system in the neighborhood of any ω, (Chapter 3).

$$\omega \frac{d\angle G(j\omega)}{d\omega} = \angle G(j\omega) + \frac{2}{\pi}[\ln|k_g| - \ln|G(j\omega)|] \tag{9.14}$$

where, k_g is the static gain of $G(s)$. It is established that equation (9.14) is valid for both minimal and non-minimal phase systems alike. Substituting $\omega = \omega_{gc}$ in equation (9.14), where $G(j\omega_{gc}) = 1$, and $\angle G(j\omega_{gc}) = -\pi + \Phi_m$ yields

$$\left.\frac{d\angle G(j\omega)}{d\omega}\right|_{\omega=\omega_{gc}} = \frac{\Phi_m - \pi}{\omega_{gc}} + \frac{2}{\pi\omega_{gc}} \ln|k_g| \tag{9.15}$$

where, Φ_m is the phase margin of $G(s)$.

Now, if it is attempted to flatten the phase around $\omega = \omega_{gc}$ using a phase shaper $G_{ph}(s)$, then the relationship

$$\frac{d}{d\omega}\angle G(j\omega) + \frac{d}{d\omega}\angle G_{ph}(j\omega) = 0 \tag{9.16}$$

(9.16) must be satisfied over a frequency band $\Delta\omega$, around ω_{gc}. For a FOPTD (First Order Plant with Time Delay) or a SOPTD (Second order Plant with Time Delay) system, $\left.\frac{d\angle G(j\omega)}{d\omega}\right|_{\omega=\omega_{gc}}$ is negative and therefore, for equation (9.16) to be valid in the neighborhood of ω_{gc}, $\left.\frac{d}{d\omega}\angle G_{ph}(j\omega)\right|_{\omega=\omega_{gc}}$ should be positive. This can be achieved by using a phase shaper of the (general) form as:

$$G_{ph}(s) = (1 + as^q) \ , \text{with} \ \frac{1}{a} \le \omega_{gc}^q \ \text{and} \ 0 \le q \le 1 \tag{9.17}$$

If it be assumed that the phase shaper does not introduce any change in static gain or phase of the original plant, then the phase shaper described by equation (9.17), which is general, must be modified as, a lagging phase shaper (integrator)

$$G_{ph}(s) = \frac{(1 + as^q)}{s^q} \tag{9.18}$$

With, $G_{ph}(\mathrm{j}\omega) = a + (\mathrm{j}\omega)^{-q} = \left(a + \omega^{-q} \cos\left(\frac{\pi q}{2} \right) \right) - \mathrm{j}\left(\omega^{-q} \sin \frac{\pi q}{2} \right)$

$$\angle G_{ph}(s) = -\frac{q\pi}{2} + \tan^{-1}\left(\frac{a\omega^q \sin \frac{q\pi}{2}}{1 + a\omega^q \cos \frac{q\pi}{2}} \right), \text{ and}$$

$$\frac{d\angle G(s)}{d\omega} = \left(\frac{aq\omega^q \sin \frac{q\pi}{2}}{\omega(1 + 2a\omega^q \cos \frac{q\pi}{2} + a^2 \omega^{2q}} \right)$$

The derivative of the phase curve of plant along with PID controller is (9.15) that is

$$\left. \frac{d\angle G(\mathrm{j}\omega)}{d\omega} \right|_{\omega=\omega_{gc}} = \frac{\Phi_m - \pi}{\omega_{gc}} + \frac{2}{\pi \omega_{gc}} \ln|k_g| .$$

By putting the derivative of phase shaper (9.18) that is

$$\angle G_{ph}(s) = -\frac{q\pi}{2} + \tan^{-1}\left(\frac{a\omega^q \sin \frac{q\pi}{2}}{1 + a\omega^q \cos \frac{q\pi}{2}} \right),$$

and using (9.16) that is,

$$\frac{d}{d\omega} \angle G(\mathrm{j}\omega) + \frac{d}{d\omega} \angle G_{ph}(\mathrm{j}\omega) = 0 \quad ,$$

we get the following:

$$\frac{\Phi_m - \pi}{\omega_{gc}} + \frac{2}{\pi \omega_{gc}} \ln\left|k_g\right| + \frac{aq\omega^q \sin\frac{q\pi}{2}}{\omega(1+2a\omega^q \cos\frac{q\pi}{2} + a^2\omega^{2q})} = 0 \tag{9.19}$$

Equation (9.19) assumes that the slope of the phase curve of $G(s)$ remains constant around ω_{gc}. The advantage of a Fractional Order differ-integrator as a phase shaper arises from the fact that it allows a flexibility in making the phase curve of $G_{OL}(s)$ flat over a varying frequency spread by suitable selection of a and q.

However, the addition of $G_{ph}(s)$ alters the phase of $G(s)$ and the net phase of $G_{OL}(s)$ at ω_{gc} can be expressed, by putting the phase of phase shaper (9.18) with known minimum desired phase margin Φ_{md}, the phase angle of the overall system at gain cross-over frequency ω_{gc} is by following expressions

$$\Phi_{\omega_{gc}} = \angle G(\mathrm{j}\omega) + \angle G_{ph}(\mathrm{j}\omega)$$

$$\Phi\Big|_{\omega=\omega_{gc}} = (\Phi_m - \pi) + \left[-\frac{q\pi}{2} + \tan^{-1}\left(\frac{a\omega^q \sin\frac{q\pi}{2}}{1 + a\omega^q \cos\frac{q\pi}{2}} \right) \right] \tag{9.20}$$

From equation (20) it follows that the phase of $G_{OL}(s)$ at ω_{gc} is less than the phase of $G(s)$ at the same frequency. Since the phase shaper flattens the phase curve of $G_{OL}(s)$ around ω_{gc}, it follows that the phase margin may reduce with the introduction of the phase shaper. Thus if the minimum desired phase margin with the phase shaper be Φ_{md}, (that is total phase of plant plus PID plus phase shaper, added to -180^0 at ω_{gc} is LHS of (20)); then it follows that

$$\Phi_{md} - \Phi_m + \frac{q\pi}{2} - \tan^{-1}\left(\frac{a\omega^q \sin\frac{q\pi}{2}}{1 + a\omega^q \cos\frac{q\pi}{2}} \right) \le 0 \tag{21}$$

Thus, the problem is of finding a phase shaper of the form represented by equation (9.18) that produces maximum flatness in terms of frequency spread can be viewed as a constrained optimization that finds (q, a) maximizing the value of $\left|\omega - \omega_{gc}\right|$ and satisfying the constraints represented by equations (9.16), (9.17), (9.19) and (9.21) using MATLAB's Optimization Toolbox function *fmincon(.)*

9.8.2 *Plant with Tuned with Integer Order PID Made Iso-Damped with Additional Fractional Differ-Integrator*

Consider a plant identified by reduced order model as a 'first order with time delay' (FOPTD) represented by Transfer function $G_{pl}(s)=\dfrac{1}{Ts+1}e^{-Ls}$ is controlled by, normal integer order PID that is:

$$G_c(s)=k_p+\frac{k_i}{s}+k_d s\,.$$

The Fractional Order phase shaper $G_{ph}(s)$ is so designed that the resultant closed-loop system exhibits iso-damped response to step changes in input over a range of gain variations. The design methodology for the phase shaper maximizes the width in terms of frequency of the flat-phase region in the asymptotic phase curve around the gain crossover frequency of $G(s)$. These ensure a constant phase margin and hence gain independent overshoot for the time response of the system. The open loop transfer function is thus: $G_{OL}(s)=G_{ph}(s)\times G_c(s)\times G_{pl}(s)$.

A flat phase curve around the gain cross-over frequency also ensures enhanced parametric robustness. The methodology presented in this section assumes the following:

a). The plant transfer function $G_{pl}(s)$ is can be approximated by a First Order plus Time Delay (FOPTD).

b). The PID controller may be tuned by any standard method and the closed loop system comprising the plant and the controller is a stable system.

c). The phase shaper does not add any additional gain, nor does it add a net phase change to $G(s)$.

Specified gain-cross over frequency ω_{gc}, and at that frequency Phase margin specified is Φ_{md}; thus

$$\Phi_{md}=\arg[G_{OL}(\mathrm{j}\omega_{gc})]=\arg[G_{ph}\times G_c(\mathrm{j}\omega_{gc})\times G_{pl}(\mathrm{j}\omega_{gc})]=-\pi+\Phi_m\,.$$

For robustness, we have

$$\left.\frac{d}{d\omega}\arg(G_{OL}(\mathrm{j}\omega))\right|_{\omega=\omega_{gc}}=0$$

With these conditions that the phase derivative with respect to the frequency is zero i.e. the phase from bode plot is flat, and at the gain crossover frequency and the system became robust for gain variations.

For amplitude we have the condition

$$\left|G_{OL}(\mathrm{j}\omega_{gc})\right| = \left|G_{ph}(\mathrm{j}\omega_{gc})\times G_c(\mathrm{j}\omega_{gc})\times G_{pl}(\mathrm{j}\omega_{gc})\right| = 1$$

around the gain crossover point magnitude of open loop transfer function is equal to one. Steps to fulfill the above criteria are as follows:

From the transfer function of the PID controller, we have:

$$G_c(\mathrm{j}\omega) = k_p + \mathrm{j}\left(k_d\omega - \frac{k_i}{\omega}\right).$$

This gives argument as:

$$\arg[G_c(\mathrm{j}\omega)] = \tan^{-1}\left(\frac{k_d\omega - \frac{k_i}{\omega}}{k_p}\right)$$

and gain modulus of controller is

$$\left|G_c(\mathrm{j}\omega)\right| = \sqrt{k_p{}^2 + \left(k_d\omega - \frac{k_i}{\omega}\right)^2}$$

From the transfer function of the plant as identified as a first order with time delay $\arg[G_{pl}(\mathrm{j}\omega)] = -\tan^{-1}(\omega T) - L\omega$ and the modulus is,

$$\left|G_{pl}(\mathrm{j}\omega)\right| = \frac{1}{\sqrt{1+(\omega T)^2}}$$

Transfer function of the fractional order phase shaper is:

$$G_{ph}(s) = \frac{1+as^q}{s^q} = a + s^{-q}$$

$$G_{ph}(\mathrm{j}\omega) = a + (\mathrm{j}\omega)^{-q} = \left(a + \omega^{-q}\cos\left(\frac{\pi q}{2}\right)\right) - \mathrm{j}\left(\omega^{-q}\sin\frac{\pi q}{2}\right),$$

which gives modulus and argument as follows:

$$\arg\left[G_{ph}(\mathrm{j}\omega)\right] = -\tan^{-1}\left(\frac{\omega^{-q}\sin\frac{q\pi}{2}}{a+\omega^{-q}\cos\frac{q\pi}{2}}\right),$$

$$\left|G_{ph}(\mathrm{j}\omega)\right| = \sqrt{\left(a+\omega^{-q}\cos\frac{\pi q}{2}\right)^2 + \left(\omega^{-q}\sin\frac{\pi q}{2}\right)^2}$$

The open loop transfer function of a plant controlled by PID controller with fractional order phase shaper is $G_{OL}(s) = G_{ph}(s) \times G_c(s) \times G_{pl}(s)$. The net gain and phase of the system is as follows:

$$|G_{OL}(\mathrm{j}\omega)| = \frac{\sqrt{k_p^{\,2} + \left(k_d\omega - \frac{k_i}{\omega}\right)^2} \times \sqrt{\left(a + \omega^{-q}\cos\frac{\pi q}{2}\right)^2 + \left(\omega^{-q}\sin\frac{\pi q}{2}\right)^2}}{\sqrt{1 + (\omega T)^2}}$$

$$\arg\left[G_{OL}(\mathrm{j}\omega)\right] = \tan^{-1}\left(\frac{k_d\omega - \frac{k_i}{\omega}}{k_p}\right) - \tan^{-1}\left(\frac{\omega^{-q}\sin\frac{q\pi}{2}}{a + \omega^{-q}\cos\frac{q\pi}{2}}\right) - \tan^{-1}(\omega T) - L\omega$$

From robustness criterion,

$$\left.\frac{d}{d\omega}\arg G_{OL}(\mathrm{j}\omega)\right|_{\omega=\omega_{gc}} = 0,$$

applying this on above expression, we get:

$$\frac{k_p(k_d\omega^2 + k_i)}{(k_p\omega)^2 + (k_d\omega^2 - k_i)^2} + \frac{aq\omega^{q-1}\sin\frac{q\pi}{2}}{1 + 2a\omega^q\cos\frac{q\pi}{2} + a^2\omega^{2q}} = \frac{T}{1 + (\omega T)^2} + L$$

$$B = \frac{T}{1 + (\omega T)^2} + L - \frac{k_p(k_d\omega^2 + k_i)}{(k_p\omega)^2 + (k_d\omega^2 - k_i)^2},$$

where

$$B = \frac{aq\omega^{q-1}\sin\frac{q\pi}{2}}{1 + 2a\omega^q\cos\frac{q\pi}{2} + a^2\omega^{2q}}$$

From gain criterion, $|G_{OL}(\mathrm{j}\omega_{gc})| = |G_{ph}(\mathrm{j}\omega_{gc}) \times G_c(\mathrm{j}\omega_{gc}) \times G_{pl}(\mathrm{j}\omega_{gc})| = 1$ we get

$$\frac{\sqrt{k_p^{\,2} + \left[k_d\omega - \frac{k_i}{\omega}\right]^2}\sqrt{\left(a + \omega^{-q}\cos\frac{q\pi}{2}\right)^2 + \left(\omega^{-q}\sin\frac{q\pi}{2}\right)^2}}{\sqrt{1 + (\omega T)^2}} = 1$$

From above after simplification we obtain:

$$a^2 + 2a\omega^{-q}\cos\frac{q\pi}{2} + \omega^{-2q} = A,$$

where

$$A = \frac{1+(\omega T)^2}{k_p{}^2 + \left(k_d\omega - \frac{k_i}{\omega}\right)^2}$$

From $a^2 + 2a\omega^{-q}\cos\frac{q\pi}{2} + \omega^{-2q} = A$ and $\dfrac{aq\omega^{q-1}\sin\frac{q\pi}{2}}{1+2a\omega^q\cos\frac{q\pi}{2} + a^2\omega^{2q}} = B$,

we get

$$q\sin\frac{q\pi}{2} = \frac{AB}{a}\omega^{q+1}$$

Phase margin specification gives:

$$\arg\left[G_{OL}(\mathrm{j}\omega)\right] = \tan^{-1}\left(\frac{k_d\omega - \frac{k_i}{\omega}}{k_p}\right) - \tan^{-1}\left(\frac{\omega^{-q}\sin\frac{q\pi}{2}}{a+\omega^{-q}\cos\frac{q\pi}{2}}\right) - \tan^{-1}(\omega T) - L\omega = -\pi + \Phi_m$$

Write $\left(\dfrac{k_d\omega - \frac{k_i}{\omega}}{k_p}\right) = C$ and $-\pi + \Phi_m + L\omega = D$, and substitute in above to get simplification for the following terms as:

$$\left(\frac{\omega^{-q}\sin\frac{q\pi}{2}}{a+\omega^{-q}\cos\frac{q\pi}{2}}\right) = \frac{C\cos D - \sin D - \omega T\cos D + \omega TC\sin D}{\cos D - C\sin D + \omega T\cos D - \omega T\sin D} = E$$

$$\left(\frac{\omega^{-q}\sin\frac{q\pi}{2}}{a+\omega^{-q}\cos\frac{q\pi}{2}}\right)_{\omega=\omega_{gc}} = E$$

From the all above steps we get three conditions:

1. $a^2 + 2a\omega^{-q}\cos\frac{q\pi}{2} + \omega^{-2q} = A$
2. $q\sin\frac{q\pi}{2} = \frac{AB}{a}\omega^{q+1}$
3. $\left(\frac{\omega^{-q}\sin\frac{q\pi}{2}}{a+\omega^{-q}\cos\frac{q\pi}{2}}\right)_{\omega=\omega_{gc}} = E$

The problem is to find (a,q) that is, fractional order phase shaper, $G_{ph}(s)$ for given Φ_{md}, for a tuned plant $G_{pl}(s)$ already working with tuned PID controller $G_c(s)$. This optimization problem is solved using MATLAB's Optimization Toolbox function *fsolve (*)*. The rationale behind use of *fsolve (*)* arises from the fact that this is an optimization problem with non-linear constraints and it has been shown that for such optimization problems *fsolve (*)* can be used effectively.

Take a plant, identified as $G_{pl}(s) = \frac{1}{1.11s+1}e^{-0.105s}$ which is controlled by a *PI* controller, which is tuned with parameters and represented as $G_c(s) = 7.73 + \frac{10.5}{s}$.

Transfer function of the phase shaper obtained by above methods, for desired phase margin of $\Phi_{md} > 35^0$, gives the following phase shaper. This limit is to ensure that not too much oscillatory behavior is obtained after iso-damping. Phase moving toward zero degrees is less oscillatory.

$$G_{ph}(s) = k\frac{1+1.3104s^{0.7895}}{s^{0.7895}}$$

Figures 9.11 and 9.12 represent the iso-damped step response under varying k, (the scalar gain introduced in order to evaluate the gain spread studies) the frequency response with phase flattening due to the phase shaper, shows the peak overshoot is fixed for 200% variation in parameter k, whereas without this additional phase shaper the PID only has large change in overshoot with same parametric spreads. The gain and phase curves show the reduction in phase margin and increase in gain margin and lowering of gain cross over frequency; while phase shaper of fractional order is introduced.

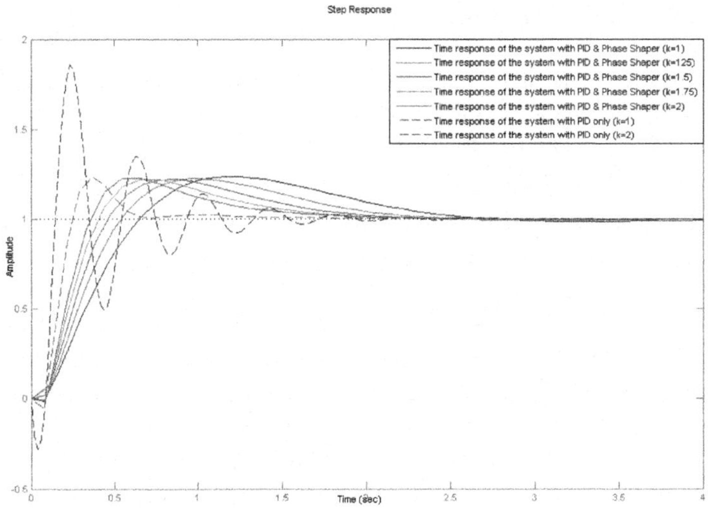

Fig. 9.11 Iso-damped step responses with the variation of loop gain, dashed line represents original system controlled by PID controller and continuous lines represent responses along with PID and phase shaper for varying gains.

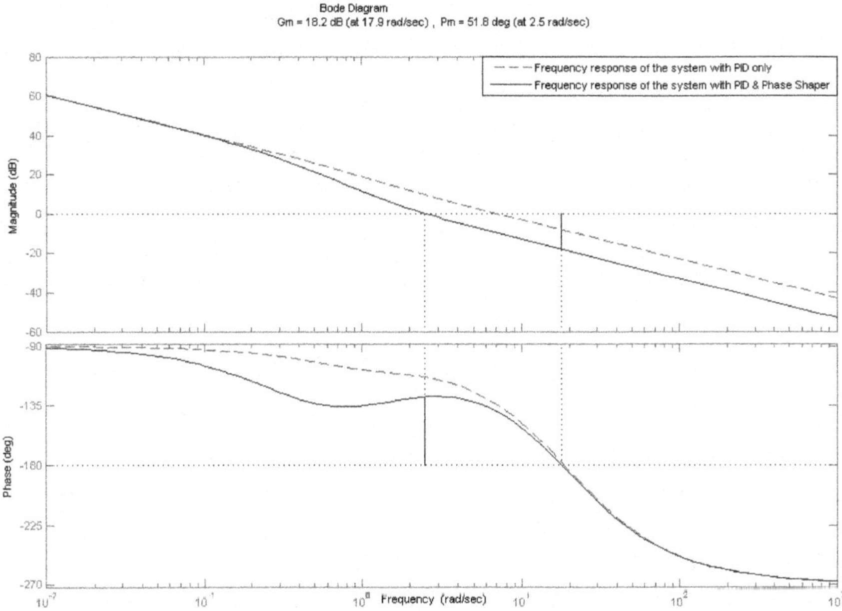

Fig. 9.12 Open Loop Frequency responses (Gain and Phase) of the plant controlled by PID with (continuous line) and without (dashed line) phase shaper.

9.9 Viscoelasticity (Stress-Strain)

Macromolecular systems show in many cases viscous-elastic behavior, which combines feature of solids (elasticity) and of liquids (viscosity). The expression for solids is by Hook's law i.e. $\sigma(t) = E\varepsilon(t)$, and liquid behavior, of viscosity by $\sigma(t) = \eta\varepsilon'(t)$, i.e. Newton's law. $\varepsilon(t)$, is the strain (displacement), $\sigma(t)$ is stress, E modulus of elasticity, η is coefficient of viscosity and $\varepsilon'(t)$ is strain rate. In general the real objects show a behavior which combines characteristic features of solids and liquids. Let us model a visco-elastic behavior by series connected pure elastic Hook's element with pure viscous element of Newton's law. Coupling $\sigma_1(t) = E\varepsilon_1(t)$ with series with $\sigma_2(t) = \eta\varepsilon_2'(t)$. In this configuration the elongations (displacements) gets added to give total displacement (across variable). The total displacement (strain) is $\varepsilon = \varepsilon_1 + \varepsilon_2$. The stress being the through variable will be common to these series element, i.e. $\sigma_1 = \sigma_2 = \sigma$. Taking $\tau = \eta / E$, the differential equation representation is $\tau \frac{d}{dt}\sigma(t) + \sigma(t) = \tau E \frac{d}{dt}\varepsilon(t)$. The relaxation modulus for 'shear-jump', i.e. $\varepsilon(t) = H(t)$, unit step at $t = 0$, gives $\delta(t)$ as source term. Then, $G(t) = \sigma(t) = E\exp\left(-\frac{t}{\tau}\right)$ is the Green's function, and solution to homogeneous differential equation, and also the relaxation modulus for the shear-jump excitation. This model leads to exponential stress decay. Real materials however show a more general decay profile as power law $G(t) \propto t^{-\alpha}$. This is due to linear superposition of normal modes leads to non-exponential power-law decay, in a 'fractal' chain network of connected monomers of a long polymer chain (Rouse dynamics Chapter 3). The decay coefficient α is related to the spectral dimension d_s of the fractal network. In general late time asymptote of relaxation modulus gives indication of 'spectral' dimension of the fractal network.

Let us try to generalize the integer order model, with fractional derivatives as:

$$\tau^\alpha \frac{d^\alpha}{dt^\alpha}\sigma(t) + \sigma(t) = \tau^\beta E \frac{d^\beta}{dt^\beta}\varepsilon(t), \text{ with } 0 < \alpha, \beta < 1$$

For short time $t << \tau$, the relaxation modulus will be as $G(t) \propto t^{\alpha-\beta}$. In this generalization for $\alpha > \beta$, the $G(t)$ grows with time, which is not physically correct. Therefore, for this method of generalization for modeling to be correct $\alpha \le \beta$. This way direct replacement of integer order derivative with fractional derivative operator is meaningful only if physically correct picture is obtained.

Assuming, the relaxation response function as power law to shear-jump in strain is $G(t) = \frac{E}{\Gamma(1-\gamma)}\left(\frac{t}{\tau}\right)^{-\gamma}$, with $0 < \gamma < 1$, then from above reasoning the visco-elastic relationship in fractional order model is:

$$\sigma(t) \equiv E\tau^{\gamma}\frac{d^{\gamma}}{dt^{\gamma}}\varepsilon(t)$$

that is, making $\alpha = 0$ and $\beta = \gamma$.

In Chapter 3 we have seen the relation between Force F_0 and average displacement $\langle x_0(t)\rangle$, for tagged bead in chained network as half order integral equation, for semi-infinite case. The case was discussed there for a linear chain in network. That model is thought of as monomers of beads connected to form a polymer chain. Now instead of a linear chain consider the network be of fractal nature, meaning that the connectivity instead of linear is random and heavily connected from one monomer (bead) to all others. Thus making a mesh of monomers to form a fractal lattice. In case of linear chain the average-displacement of tagged bead follows $\sqrt{t}$,form the number or group $g(t) \propto (t)^{1/2}$ moving collectively after application of step-force at $n = 0$ monomer. This slows down the tagged monomer-bead (due to collective motion). It is evident that pulling a monomer of a macro network which has more connectivity than the linear chain involves even a slower drift. Refer the Chapter 3 for linear chain of network, for which the Rouse's dynamics gave one dimensional equation of diffusion type. It was discussed that its Green's function is Gaussian whose width increases with time as $\sqrt{t}$; stated differently the number of monomers (group of beads) $g(t)$ which are involved in a collective motion obey $g(t) \propto \sqrt{t}$. Now if this were to happen in fractal network, would lead to different equation on the corresponding fractal lattice. This fractal lattice is having scales of fractal dimensions different from Euclidean lattice; for linear chain the dimension $d_s = 1$.

Thus for fractal lattice we expect, the group of beads obey $g_f(t) \propto (t)^{d_s/2}$, with $d_s < 2$ as 'spectral' dimension of the network. It is well known that in a random walk model of a fractal lattice, the number of different sites $S(t)$ visited by random walker during time t grows as $S(t) \propto t^{d_s/2}$ for $d_s < 2$ and $S(t) \propto t$ for $d_s \geq 2$, where d_s is spectral dimension of the network. In case of regular lattice d_s is Euclidean dimension. Now the dynamical process has single parameter combination with dimension of time namely $\zeta b^2 / T$ (refer Chapter 3 for these constants as discussed) then:

$$g_f(t) = \begin{cases} \dfrac{T^{d_s/2}}{\zeta^{d_s/2} b^{d_s}} t^{d_s/2} & ; d_s < 2 \\ \dfrac{T}{\zeta b^2} t & ; d_s \geq 2 \end{cases}$$

At later times that is, $t >> \tau_R$; $g_f(t) = N$ holds. The cross over time τ_G follows by setting $g_f(\tau_G) = N$,and for $d_s < 2$ this leads to $\tau_G \cong \zeta b^2 N^{2/d_s} / T$. For linear chain $d_s = 1$, one gets Rouse time τ_R (Chapter 3). The domain which moves collectively has the, average velocity $\langle v_x(t) \rangle = (\zeta g(t))^{-1} F_0$. For $d_s < 2$ this leads to:

$$\langle x_0(t) \rangle \cong \frac{F_0 b^{d_s}}{\zeta^{1-d_s/2} T^{d_s/2}} t^{1-d_s/2} .$$

Using the argument of fractional integration of a constant as for linear chain, we get similar expression for fractal network as:

$$\langle x_0(t) \rangle \cong \frac{b^{d_s}}{\zeta^{1-d_s/2} T^{d_s/2}} \frac{d^{(d_s/2)-1}}{dt^{(d_s/2)-1}} F(t) .$$

With $\alpha = 1-(d_s/2)$, we get fractional integral equation as $\overline{x}(t) = \dfrac{b^{2-2\alpha}}{\zeta^{\alpha} T^{1-\alpha}} \dfrac{d^{-\alpha}}{dt^{-\alpha}} F(t)$, where $0 < \alpha \leq (1/2)$. This expression is similar to the generalization of stress strain relationship as $\varepsilon(t) = E^{-1} \tau^{-1} \dfrac{d^{-\gamma}}{dt^{-\gamma}} \sigma(t)$ expressed above. Thus from microscopic chain dynamics of elastic connected monomer beads in a fractal network one can get macroscopic stress strain fractional differential equations for viscoelasticity.

Stress relaxation and creep behavior in stress strain relationship is well described by fractional order models. A stress strain law for viscoelastic materials is described as $\varepsilon(t) = \dfrac{1}{K} {}_c D_t^{-v} \sigma(t)$ or a new specimen where no initialization is required and is thus memory less be represented as (un-initialized derivative) $\varepsilon(t) = \dfrac{1}{K} {}_a d_t^{-v} \sigma(t)$. Here ε is strain and σ is stress. For $v = 0$ the material is elastic solid and for $v = 1$ the material is viscous liquid. K is constant depends on material. In the above stress strain relationship, both effects i.e. instantaneous elastic and long term viscous flow are neglected.

Consider a unit step load is applied from $t = 0$ and $t = d$ on a new specimen (un-initialized). $\sigma(t) = \kappa[H(t) - H(t-d)]$, H is Heaviside step unity function and κ is the magnitude. Then:

$$\varepsilon(t) = \frac{\kappa}{K\Gamma(v)} \int_0^t (t-\tau)^{v-1} [H(\tau) - H(\tau - d)] d\tau$$

$$= \frac{\kappa}{K\Gamma(v)} \left[\int_0^t (t-\tau)^{v-1} d\tau H(t) - \int_0^t (t-\tau)^{v-1} d\tau H(t-d) \right]$$

$$= \frac{\kappa}{K\Gamma(v+1)} \left[t^v H(t) - (t-d)^v H(t-d) \right]$$

defines the strain (creep) response, for the prescribed loading. The Figure 9.7 gives the strain curve for this prescribed loading. For $t < d = 1$ the curve is essentially a creep function and for $t > d = 1$ is relaxation period.

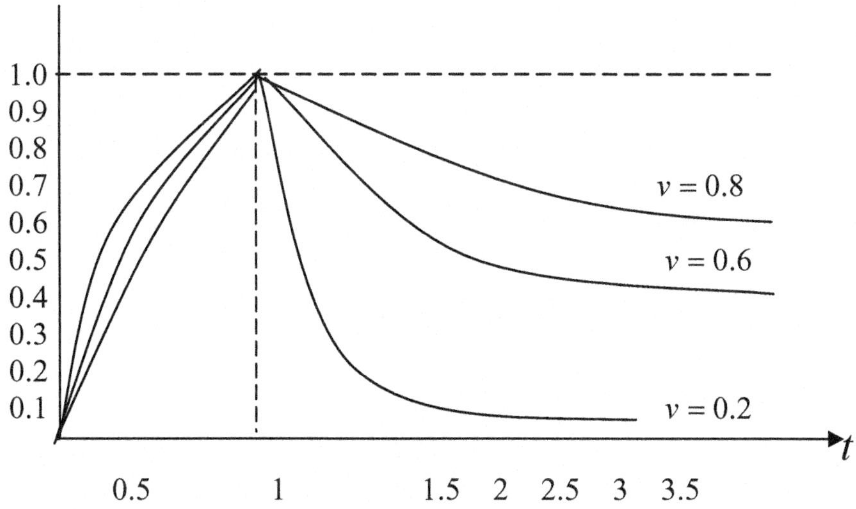

Fig. 9.13 strain versus time (creep function & relaxation function)

The above example material was new. For creep initialization function associated with constant past loading may be readily inferred from

$$\varepsilon(t) = \frac{\kappa}{K\Gamma(v+1)} \left[t^v H(t) - (t-d)^v H(t-d) \right]$$

as follows:

Consider this problem initialized at point c, then ${}_cD_t^{-q}f(t) = {}_aD_t^{-q}$, for $t > c$, therefore:

$${}_cd_t^{-q}f(t) + \psi(f, -q, a, c, t) = {}_aD_t^{-q}f(t) = {}_ad_t^{-q}f(t),$$

for $t > c$. Thus ψ may be expressed as:

$$\psi(f, -q, a, c, t) = {}_ad_t^{-q} - {}_cd_t^{-q}, \text{ for } t > c.$$

For obtaining the initialization function of creep problem initialized at $t = c = d$, where d is as used in above equation i.e.

$$\varepsilon(t) = \frac{\kappa}{K\Gamma(v+1)}\left[t^v H(t) - (t-d)^v H(t-d)\right].$$

Applying this result to equation $\psi(f, -q, a, c, t) = {}_ad_t^{-q} - {}_cd_t^{-q}$, to the creep problem, gives:

$${}_cD_t^{-v}\sigma(t) = {}_cd_t^{-v}\sigma(t) + \psi(\sigma, -v, a, c, t), \text{ for } t > c$$

Then taking $c = d$ and since $\sigma(t) = 0$ for $t > d$, we get:

$${}_cd_t^{-v}\sigma(t) = {}_dd_t^{-v}\sigma(t) = 0, \text{ for } t > c = d$$

and K times the response equation of

$$\varepsilon(t) = \frac{\kappa}{K\Gamma(v+1)}\left[t^v H(t) - (t-d)^v H(t-d)\right],$$

is the initialization function $\psi(t)$, therefore:

$$\psi(\sigma, -v, a, c, t) = \frac{\kappa}{\Gamma(v+1)}\left[t^v H(t) - (t-c)^v H(t-c)\right], \text{ for } t > c = d.$$

This initialization function with proper time shifts may thus be employed for the material, which had “creep history”.

The stress relaxation (or creep) function $\chi(t)$ is defined as the stress required to produce a strain $H(t)$, excluding terms that are initially infinite or do not tend to zero a time grows. For a new material this stress is:

$$\chi(t)=\sigma(t)=K\,{}_a d_t^{v}\varepsilon(t)=K\,{}_a d_t^{1}\,{}_a d_t^{-(1-v)}\varepsilon(t)=K\frac{d}{dt}\frac{K}{\Gamma(1-v)}\int_0^t (t-\tau)^{-v}H(\tau)d\tau$$

$$\chi(t)=\frac{K}{\Gamma(1-v)}t^{-v}H(t)$$

yielding as a result the stress (creep) relaxation function for this formulation.

This stress-strain with memory and history definitions with fractional calculus is a suitable method of describing the fundamentals of 'shape-memory' alloys.

9.10 Vibration Damping System

The addition of damping is important in many applications for stability enhancement. One such is requirement to have vibration-damping augmentation in gas turbines blades (especially for aerospace). Figure 9.8 gives spring-mass-viscodamped dynamic system diagram and we derive the transfer function of the same.

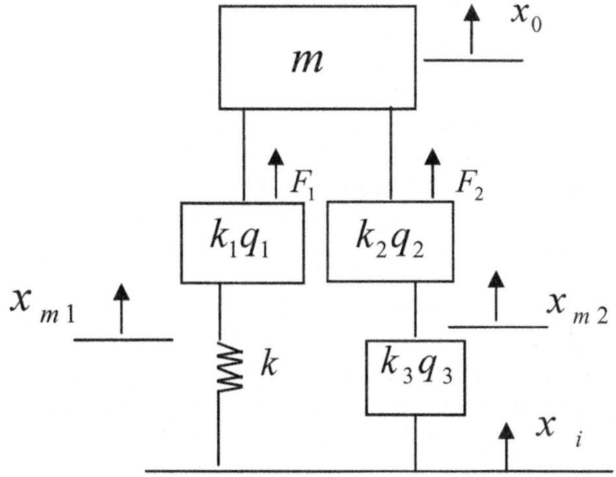

Fig. 9.14 Spring mass viscodamped dynamics.

Since the transfer function requires the initialization to be zero, the equations describing the system uses un-initialized operators as follows:

$$F_1=-k_1\,{}_0d_t^{q1}(x_0-x_{m1})$$
$$F_1=k(x_{m1}-x_i)$$

$$F_2 = -k_2\,{}_0d_t^{q2}(x_0 - x_{m2})$$
$$F_2 = k_3\,{}_0d_t^{q3}(x_{m2} - x_i)\text{, and}$$

$$F_1 + F_2 = m\,{}_0d_t^2 x_0$$

Where k's are damping coefficients, and the spring constants, and m, F, and x are the mass force and displacement respectively.

After taking Laplace transforms and considerable algebra and elimination process the generalized transfer function is:

$$\frac{X_0(s)}{X_i(s)} = \frac{\frac{1}{k_2}s^{q1+q3} + \frac{1}{k_3}s^{q1+q2} + \frac{1}{k_1}s^{q2+q3} + \frac{1}{k}s^{q1+q2+q3}}{\frac{m}{k_1k_2}s^{2+q3} + \frac{m}{kk_2}s^{2+q1+q3} + \frac{m}{k_1k_3}s^{2+q2} + \frac{m}{kk_3}s^{2+q1+q2} + \frac{1}{k_2}s^{q1+q3} + \frac{1}{k_3}s^{q1+q2} + \frac{1}{k_1}s^{q2+q3} + \frac{1}{k}s^{q1+q2+q3}}$$

This general transfer function can be studied for special cases, by appropriate choices of $k, k_i, and\ q_i$. For example to allow damper 3 to represent a conventional dashpot-damper lets $q3 = 1$ and select appropriate k_3. Now making the $q1 = q2 = 0$, means eliminating the dampers and spring of infinite stiffness i.e. $k_1 = k_2 = \infty$ gives simple transfer function of fractional order $q3$ with viscoelastic element 3. The transfer function is:

$$\frac{X_0(s)}{X_i(s)} = \frac{\frac{k_3}{m}s^{q3} + \frac{k}{m}}{s^2 + \frac{k_3}{m}s^{q3} + \frac{k}{m}}$$

The frequency response may now be evaluated by putting $s \to j\omega$, and determining the magnitude and phase angle using the derived transfer function, for various values of $k/m, k_3/m$, and letting the fractional order $q3$ vary from 0-2 in steps of 0.2, to arrive at designated design measures.

In nature we always get mix of mass spring and damper systems and thus to have reality modeling this fractional calculus plays important role.

9.11 The Non-newtonian Fluid Anamolous Behavior with Memory

The non-Newtonian fluid behavior is depicted in Figure 9.15. A non-Newtonian fluid reveals anomalous visco-elastic properties as compared to Newtonian fluid, the 'beads-on- a- string' structure gets formed in visco-elastic fluid of non-Newtonian in nature. The spreading experiment is conducted where arrowroot solution is kept between two glass plates and steady load is applied. A camera is kept below to capture snap shots regularly to record the spreading pattern. The area is calculated graphically later, and its plot with respect to time for various loads is shown in Figure 9.16. Interestingly the observation reveals oscillatory nature of the spreading. We relate area to strain and following analysis shows that the fractional differential equation gives suitable explanation of this anomalous behavior, of non-Newtonian relaxation (with memory). Our experiment shows that the fractional order corresponding to this oscillatory relaxation behavior is of fractional order $q \cong 1.5$.

We write for Newtonian fluids a lumped spring and a lumped dashpot model as

$$\beta \frac{d}{dt}\varepsilon(t) + E\varepsilon(t) = \sigma(t)$$

The above equation is generalized to representation of the stress-strain in distributed spring and dashpot system for a non-Newtonian fluid as a fractional differential equation:

$$\frac{d^q}{dt^q}\varepsilon(t) + B\varepsilon(t) = \frac{1}{\beta}\sigma(t)$$

where, $B = E/\beta$. The constant of this expression β is generalized viscous coefficient the units of which are having the non-integer order q imbibed into it, and E, the modulus of elasticity. When the order $q = 1$, then normal constant of viscosity is returned. The unit of B for order $q = 1$ is per seconds i.e. $[s^{-1}]$, but for any other order $q \neq 1$; the unit modifies as $[s^{-q}]$.

Mathematically one has to see the Green's function for general relaxation in equation given above by fractional differential equation, so we write the homogeneous equation with RHS equal to zero. To that, we give delta function stress excitation. The strain built up for any relaxation process may be treated as convolution integral of a strain variable with integral kernel $K_q(t)$ as:

$$\frac{d}{dt}\varepsilon(t) = -\int_0^t K_q(t-\tau)\varepsilon(\tau)d\tau$$

Well if the memory kernel is $K_0(t) = B_0\delta(t)$, we have the above system without memory and the Green's function will be $\varepsilon(t) = \varepsilon_0 e^{-B_0 t}$ that is, the impulse response quickly decays to zero. Here ε_0 is initial strain of the system at $t = 0$. Derivation is following:

$$K(t) - B_0\delta(t)$$

$$\frac{d}{dt}\varepsilon(t) = -\int_0^t B_0\delta(t-\tau)\varepsilon(\tau)d\tau = -B_0\varepsilon(t)$$

$$\varepsilon(t) = \varepsilon_0 \exp\{-B_0 t\}$$

The homogeneous strain relaxation equation for no-memory case is first order Ordinary Differential Equation i.e. the Newtonian case, with $B_0 = E / \beta$

$$\frac{d}{dt}\varepsilon(t) + B_0\varepsilon(t) = 0$$

If the memory kernel is a constant say $K_1 = B_1$, then we will have oscillatory Green's function, which never decays to zero.

$$K_1(t) = B_1$$
$$\frac{d^2}{dt^2}\varepsilon(t) = -B_1\varepsilon(t)$$
$$\varepsilon(t) = \varepsilon_0 \cos(\sqrt{B_1}t)$$

Above is representation of a constant memory system.

The generalized memory integral is as follows and the case is for non-Newtonian fluids.

$$K(t) = B_q t^{q-2}; \qquad 0 < q \le 2$$
$$\frac{d}{dt}\varepsilon(t) = -\frac{1}{\tau^q}\left[\frac{d^{1-q}}{dt^{1-q}}\varepsilon(t)\right]; \qquad \tau^q = \left[B_q\Gamma(q-1)\right]^{-1}$$

Its corresponding generalized differential equation, obtained from above derivation, is the system with memory index coming as fractional order of the Fractional Differential Equation with, $0 < q \le 2$,

$$\frac{d^q}{dt^q}\varepsilon(t) - \varepsilon_0\frac{t^{-q}}{\Gamma(1-q)} = -\tau^{-q}\varepsilon(t).$$

These concepts were earlier discussed in Chapter 2.

In our experiment, the oscillatory response to a step input we say that the order is between $1 < q < 2$ and thus system has "long lingering and decaying" memory.

The above derived general equation we say $\varepsilon_0 = 0$, at initial time and the stress be Heaviside's step input then it modifies to our original fractional differential equation (FDE), what we had assumed, with $\tau^{-q} \equiv B$.

The fractional order of the FDE corresponds to system with memory. The non-Newtonian fluids without oscillatory behavior will have fractional order $0 < q < 1$, and the step-response, for input $\sigma(t) = \sigma_0 H(t), H(t)$ is Heaviside's step; will have monotonically increasing strain response, given by one argument Mittag-Leffler function $\varepsilon(t) = \beta(B_q)^{-1}\sigma_0[1 - E_q(-B_q t^q)]$. Its impulse response will be having long tailed decay, that is, the response will have long-range temporal correlation. The Newtonian fluid will have integer order in with $q = 1$, the system without memory, and the step-response will have the monotonically increasing strain as $\varepsilon(t) \approx 1 - \exp(-B_0 t)$; where as its impulse response will decay quickly as $\varepsilon(t) \approx \exp(-B_0 t)$. This Newtonian system can be modeled with a discrete ideal spring and a ideal dashpot. Whereas, the non-Newtonian cases require a different representation; like fractal chain of the ideal spring and ideal dashpot. We have observed oscillatory case of strain and thus our fractional order is at $q \cong 1.5$.

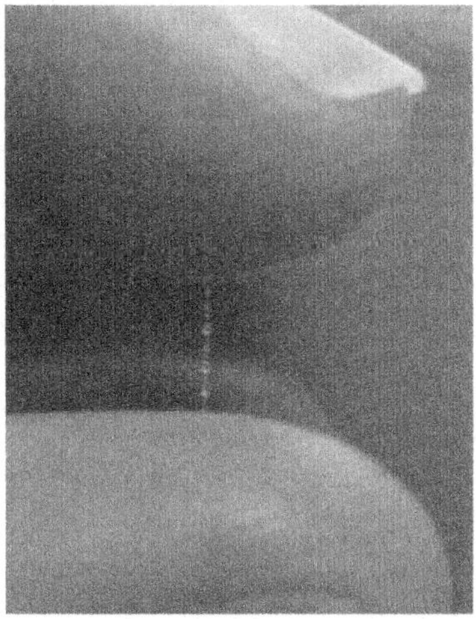

Fig. 9.15 Spreading of Non-Newtonian fluid

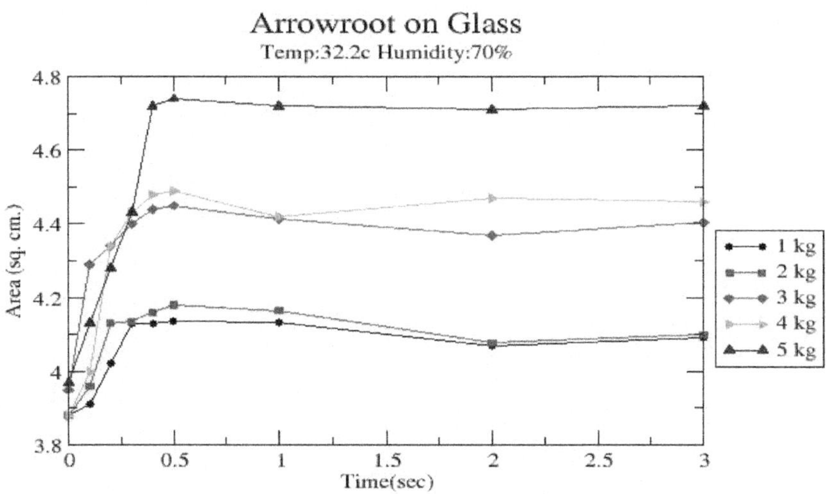

Fig. 9.16 Plot of area with respect to time showing oscillation in relaxation.

A small comment may be made regarding percolation of strain. Well in normal integer order system the percolation phenomena is by Brownian motion with regular irregularity, giving a standard stochastic process, here the Brownian motion has no memory, that its past walk is independent of present step. Here in non-Newtonian case we have a case where the Fractional Brownian motion is prevalent with memory that is, a walker remembers its past steps to move the present. Well its memory may be persistent or anti-persistent, but the walker certainly has a memory, and system relaxation through FDE is with memory.

9.12 Concluding Comments

As far the imagination goes, the reality systems have the effect of past. The present behavior is someway definitely related to past history. The actual systems therefore are non-Markovian systems with past history, heredity and memory. The fractional calculus thus is a language what nature understands the best. This way system modeling and control becomes efficient. The application here included simple aspects of system modeling and control and further research is required to have optimal descriptor of dynamic vector space controllers. Not only for science and engineering this tool is efficient descriptor, but also for finance and stock market analysis the tool is being recently explored and applied. Therefore this 300 years old (new) subject will revive in this century to speak, what nature understand the best.

Chapter 10
System Order Identification and Control

10.1 Introduction

For unknown systems, 'system identification' has become the standard tool of the control engineer and scientists. Identifying a given system from data becomes more difficult, however when fractional orders are allowed. Here identification process is demonstrated using assumption that system order distribution is a continuous one. Frequency domain system identification can thus be performed using numerical methods demonstrated in this chapter. Here one concept of *r*-Laplace transforms is discussed (Laplace transform in log domain), to discuss the system order distribution. Also mentioned is variable order identification as further development where the system order also varies with ambient and time is highlighted. Here in this chapter, an identification method based on continuous order distribution, is discussed. This technique is suitable for both the standard integer order and fractional order systems. This is topic for further advance research as to qualify the procedure of system order identification and to have technique of tackling variable order. Extending this continuous order distribution discussion, the advance research of having a continuum order feedback and generalized PID control is elucidated. Also in this chapter some peculiarities of the pole property of fractional order system as ultra-damping, hyper-damping and fractional resonance is explained. Elaborate research in this direction is ongoing process; to crisply define the system identification, crisply define the variable order structure, along with generalized controller for future applications. The system identification in presence of disorder is what is challenging and some unification of disordered time-response that is relaxation is too discussed. This is general process of returning to equilibrium for say any stable system or properties of condense matter physics. The introduction to complex order calculus in system identification is too touched upon, in this chapter, along with identification of main parameters of 'irregular' stochastically behaving systems.

10.2 Fractional Order Systems

As the concept of order is central to the understanding of fractional (or integer) order systems, some discussion of this concept now follows. In this discussion, single-input-output systems are considered. The examples in the earlier chapters for heat flow and transmission line (lossy, and lossless) and several examples, gave the stage for half order system or zero order system, integer order systems and fractional order systems. Recalling the characteristic equations or transfer function definition we call a system first order, second order third order etc. similarly the system can be of fractional order too i.e. the characteristic equation having powers of s-variable of non integers numbers (real). We also consider that system representation is generally of 'minimal phase' and they are linear. For non-minimal phase the system behavior is for a positive step demand the output initially goes in reverse and then changes direction to follow the demand. This is peculiar of gas turbines where for non-minimal phase system the velocity first will marginally decrease and then eventually increase to follow increment in the demanded velocity.

Mathematical order conventionally is defined as highest derivative occurring in a given differential equation. The concept of mathematical order is applicable to both ordinary and fractional differential equations. Normally, when the order is used without qualifier, it implies the meaning of mathematical order.

For linear dynamic systems that are described by ordinary differential equations the system mathematical order implies or is equivalent to the following:

(1). The highest derivative in ordinary differential equation. (2). The highest power of Laplace variable-s, in the characteristic equation. (3). The number of initializing constants required for the differential equation. (4). The length of the state vector. (5) The number of singularities in the characteristic equation. (6). The number of energy storage elements. (7). The number of independent spatial directions in which a trajectory can move. (8). The number of devices that can add 90^0 sinusoidal steady state phase lag and (9). The number of devises that retain some memory of the past.

The utility of the definition of mathematical order is that infers all the system characteristics for system, with integer order components.

Thus the benefit of having a definition for order for linear ordinary differential equations is that it allows a direct understanding of the behavior of given dynamic system. Unfortunately, for fractional differential equations, the order of the highest derivative does not infer all of the previously mentioned properties. Indeed, the most important characteristics of order in integer order ordinary differential equation are probably item (3) i.e. it indicates number of initializing constants, which together with the differential equations allow prediction of the future behavior. In system terminology this information provides initial 'states', of the system being analyzed. Clearly the order of highest derivative in a fractional differential equations does not have this property nor does it predict the associated number of energy/memory elements associated with fractional differential equation, nor does it infer the number of integrations (even fractional), required to solve simulate the given fractional differential equation. Thus the issue of order and the information required together with the fractional differential equation to predict the future is fundamental and should be treated differently. This is explained as seeming looking first order characteristic polynomial with fractional order components may go into resonance.

For integer order systems once the maximum order of the system to be identified is chosen, the parameters of the model can be optimized directly. For fractional order systems, the identification requires a) the choice of the number of fractional operators, b) the fractional powers of the operator, c) the coefficients of the operators. Specifically in electrode-electrolyte interface experiments for determining the interface impedances the frequency domain techniques are followed, for chosen transfer function. But to identify the form of transfer function itself, with order and coefficient an approach from experimental data of frequency domain analysis should be the starting point, to identify unknown systems.

10.3 Continuous Order Distribution

A very basic of mass spring damper system of force balance is taken here to study the concept of continuous order distribution. The familiar system is represented (with un-initialized derivative) as:

$$m\,{}_0d_t^2 x(t) + b\,{}_0d_t^1 x(t) + kx(t) = f(t) \tag{10.1}$$

where $x(t)$ is position of the mass m, $f(t)$ is the forcing function on the mass, b is the damping and k is the spring (restoring) force. In the Laplace domain this takes following form.

$$(ms^2 + bs + k)X(s) = F(s) \tag{10.2}$$

It is well known that some element intermediate between spring and dashpot behaves and balances the forces called viscoelastic element. Such element is described as:

$$k_q\,{}_0d_t^q x(t) = f(t) \qquad 0 \le q \le 1 \tag{10.3}$$

The Laplace representation is:

$$k_q s^q X(s) = F(s) \tag{10.4}$$

Adding this viscoelastic element to the original assumed (lumped) system (10.2) we get:

$$\left(ms^2 + bs + k_q s^q + k\right) X(s) = F(s) \tag{10.5}$$

It is known that viscoelastic elements will posses any order q between 0 and 1, so we can add another viscoelastic element and then keep on adding several others too, like the following:

$$\left(ms^2 + bs + k_{q2} s^{q2} + k_{q1} s^{q1} + k\right) X(s) = F(s) \tag{10.6}$$

Similar concepts were first introduced by Prof Michelle Caputo as Distributed Order Differential equations. We keep on adding fractional order to (10.6) and obtain (10.7).

$$\left(ms^2 + k_{q4}s^{q4} + k_{q3}s^{q3} + bs + k_{q2}s^{q2} + k_{q1}s^{q1} + k\right)X(s) = F(s) \qquad (10.7)$$

This process could be continued so that the system can therefore be expressed as power series with $0 \le q_n \le 2$, with N as integer as following:

$$\left(\sum_{n=0}^{N} k_n s^{q_n}\right)X(s) = F(s) \qquad (10.8)$$

Now in reality the order q_n is temperature dependent, and the entire system will be considered as layered of such material. Therefore if the material is subjected to temperature distribution, then the material will exhibit order distribution too. In the limit of infinitesimally small elements the above (10.8) will tend to continuum, and the summation be replaced then by integral. This gives fundamental motivating procedure for concept with continuous order distribution. This continuous order is expressed as:

$$\left(\int_0^2 k(q)s^q dq\right)X(s) = F(s) \qquad (10.9)$$

This is very general representation of a dynamic system of any type taken for system identification studies. For demonstration the familiar integer order dynamic spring damper mass element equation (10.2) can be re-written with form expressed in (10.9) as:

$$\left(\int_0^2 [m\delta(q-2) + b\delta(q-1) + k\delta(q)]s^q dq\right)X(s) = F(s) \qquad (10.10)$$

Figure 10.1 show the plot of $k(q)$ and q, for the classical (10.2) mass –spring – damper, i.e. order distribution. Here the order is discrete, dirac-delta functions at 0,1,2.

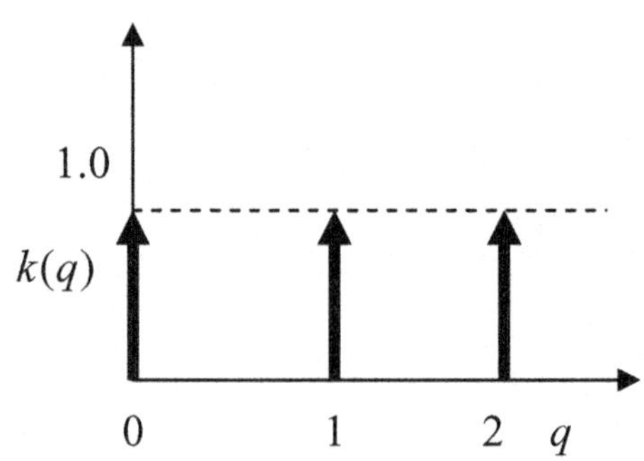

Fig. 10.1 Order distribution of mass-spring- damper integer order system.

The Figure 10.1 demonstrates for ideal (classical) systems the orders are concentrated at a single number, in this case at 0, 1, and 2. This is accepted if the mass spring and damper are (really) ideal elements. In the classical calculus terminology these orders are point property, the question is if at all these are point quantities? If so then we should have mass assigned to a point? But in reality the mass is assigned a distributed volume and thus the idealism vanishes. Therefore the order indeed need not be appoint property. Well even for integer order systems these dirac-deltas concentrated at 0, 1, 2 should spread out may be like Gaussian distribution with peaks at 0, 1, 2 and spread from $q=0$ to $+\infty$, in continuous spectrum. Though the Figure 10.1 show the height of these dirac-delta functions to be same, but actually the height is proportional to the coefficients. As per the equation (10.10) the height will depend on the relative values of the multiplier coefficients m,b,k. To demonstrate the concept all dirac-delta at the discrete order points are taken as of unity height, actually the heights should be different.

Allowed the restriction on the maximum possible order (i.e. second order in the present case) the equation (10.9) can be still generalized as continuum power series as following.

$$\left(\int_{0}^{\infty} k(q)s^{q}\,dq\right)X(s)=F(s) \tag{10.11}$$

The time domain representation of the (10.11) is

$$\int_{0}^{\infty} k(q)\,{}_0d_t^q x(t)dq = f(t) \tag{10.12}$$

Mathematically system can be also described by continuum asymptotic series instead of power series (10.11) as:

$$\left(\int_{0}^{\infty} k(q)s^{-q}\,dq\right)X(s)=F(s) \tag{10.13}$$

Combining (10.11) and (10.13) we obtain a very general system descriptive method as:

$$\left(\int_{-\infty}^{+\infty} k(q)s^{q}\,dq\right)X(s)=F(s) \tag{10.14}$$

The time domain symbolic representation of (10.14) is

$$\int_{-\infty}^{+\infty} [k(q)\,{}_0d_t^q x(t)]dq = f(t)$$

Let us try and get system response from the equation (10.14). In inverted form we can write the equation (10.14) in Laplace domain for obtaining $X(s)$ as

$$X(s)=\frac{F(s)}{\int_{-\infty}^{+\infty}k(q)s^q dq}=\frac{F(s)}{P(s)},$$

With

$$P(s)=\int_{-\infty}^{+\infty}k(q)s^q dq$$

Consider a simple case of a second order system with uniformly distributed order of value $k(q)=\mathrm{K}$ for the order interval $0\le q\le 2$. Then calculation of $P(s)$ is:

$$P(s)=\int_{-\infty}^{\infty}k(q)s^q dq=\int_0^2 \mathrm{K}s^q dq=\int_0^2\left(\mathrm{K}e^{q\ln(s)}\right)dq=\mathrm{K}\left[\frac{s^2-1}{\ln(s)}\right]$$

Forcing function as delta-function $F(s)=1$ gives the response function in Laplace as:

$$X(s)=\frac{\ln(s)}{\mathrm{K}\left[s^2-1\right]}$$

Consider another simple case that order is uniformly distributed equally at $q=\overline{q}$ with spread of $\pm\Delta\overline{q}$, and strength as $k(q)=\mathrm{K}$, a constant. The response for $f(t)=\delta(t)$, the unit delta will consider $F(s)=1$, and the integral expression for the response is thus, as demonstrated above:

$$P(s)=\int_{-\infty}^{+\infty}k(q)s^q dq=\int_{\overline{q}-\Delta\overline{q}}^{\overline{q}+\Delta\overline{q}}\mathrm{K}s^q dq\,.$$

With change of variable as $p=q-\overline{q}$ we have with new limits:

$$P(s)=\int_{-\Delta\overline{q}}^{+\Delta\overline{q}}\mathrm{K}s^{(p+\overline{q})}dp=\mathrm{K}s^{\overline{q}}\int_{-\Delta\overline{q}}^{+\Delta\overline{q}}e^{(p)\ln s}dp=\mathrm{K}s^{\overline{q}}\left(\frac{e^{(p)\ln s}}{\ln s}\right)_{p=-\Delta\overline{q}}^{p=+\Delta\overline{q}}=2\mathrm{K}s^{\overline{q}}\frac{\sinh\left([\Delta\overline{q}]\ln s\right)}{\ln s}$$

Therefore the response to the impulse input is (in Laplace variables):

$$X(s) = s^{-(\Delta\overline{q})} \frac{\ln(s)}{2\mathrm{K}\sinh(\Delta\overline{q}\ln s)}$$

This is basic response from which other responses are formed by convolution.

10.4 Determination of Order Distribution From Frequency Domain Experimental Data

Take the general power series representation of a system (10.11)

$$\left(\int_0^\infty k(q)s^q\,dq\right)X(s) = F(s)$$

It is desirable to get the transfer function:

$$\frac{X(s)}{F(s)} = \frac{1}{\int_0^\infty k(q)s^q\,dq} = \frac{1}{P(s)} = G(s) \tag{10.15}$$

Inverting (10.15) we get:

$$P(s) = \int_0^\infty k(q)s^q\,dq = \frac{1}{G(s)} \tag{10.16}$$

For an unknown system the measured frequency response is available experimentally, as $G(\mathrm{j}\omega)$. In equation (10.16), we replace $s \to \mathrm{j}\omega$, and assume the representation of the system is by form (10.16). Then

$$\int_0^\infty k(q)(\mathrm{j}\omega)^q\,dq = \frac{1}{G(\mathrm{j}\omega)} \tag{10.17}$$

System identification problem thus boils down to finding the order distribution $k(q)$ given $G(\mathrm{j}\omega)$.

Obviously analytical approach is difficult and we resort to numerical approach. In reality the order distribution decays as $q \to \infty$. So to assume $k(q) \to 0$ as q grows will not be an offset from reality. In that event the integral expressed in (10.17) converges and Euler's formula to numerically solve (10.17) is used. The integral of (10.17) will take the summation form as:

$$\sum_{n=0}^{N} k_n (\mathrm{j}\omega)^{nQ} Q = \frac{1}{G(\mathrm{j}\omega)} \tag{10.18}$$

where Q is the constant sample width, in the variable q and k_n is the height of the sampled order distribution. Remembering that we usually have sampled frequency response data the equation (10.18) should be satisfied at each data point or for any frequency (ω_i). Thus we write the (10.18) as:

$$\sum_{n=0}^{N} k_n (\mathrm{j}\omega_i)^{nQ} Q = \frac{1}{G(\mathrm{j}\omega_i)}, \text{ For any } i \tag{10.19}$$

The equation (10.19) can be written for 1 to M, frequency points as:

$$\begin{aligned}
&Qk_0 + Qk_1(\mathrm{j}\omega_1)^{Q} + Qk_2(\mathrm{j}\omega_1)^{2Q} + Qk_3(\mathrm{j}\omega_1)^{3Q} + \ldots + Qk_N(\mathrm{j}\omega_1)^{NQ} = \frac{1}{G(\mathrm{j}\omega_1)}\\
&Qk_0 + Qk_1(\mathrm{j}\omega_2)^{Q} + Qk_2(\mathrm{j}\omega_2)^{2Q} + Qk_3(\mathrm{j}\omega_2)^{3Q} + \ldots + Qk_N(\mathrm{j}\omega_2)^{NQ} = \frac{1}{G(\mathrm{j}\omega_2)}\\
&\ldots\ldots\ldots\ldots\ldots\ldots\ldots\ldots\ldots\ldots\ldots\ldots\ldots\ldots\ldots\ldots\ldots\ldots\\
&\ldots\ldots\ldots\ldots\ldots\ldots\ldots\ldots\ldots\ldots\ldots\ldots\ldots\ldots\ldots\ldots\ldots\ldots\\
&Qk_0 + Qk_1(\mathrm{j}\omega_M)^{Q} + Qk_2(\mathrm{j}\omega_M)^{2Q} + Qk(\mathrm{j}\omega_M)^{3Q} + \ldots + Qk_N(\mathrm{j}\omega_M)^{NQ} = \frac{1}{G(\mathrm{j}\omega_M)}
\end{aligned} \tag{10.20}$$

Rearranging the (10.20) in Matrix form we obtain:

$$Q\begin{bmatrix} 1 & (\mathrm{j}\omega_1)^{Q} & (\mathrm{j}\omega_1)^{2Q} & . & (\mathrm{j}\omega_1)^{NQ} \\ 1 & (\mathrm{j}\omega_2)^{Q} & (\mathrm{j}\omega_2)^{2Q} & . & (\mathrm{j}\omega_2)^{NQ} \\ . & . & . & . & . \\ . & . & . & . & . \\ 1 & (\mathrm{j}\omega_M)^{Q} & (\mathrm{j}\omega_M)^{2Q} & . & (\mathrm{j}\omega_M)^{NQ} \end{bmatrix} \begin{bmatrix} k_0 \\ k_1 \\ . \\ . \\ k_N \end{bmatrix} = \begin{bmatrix} 1/G(\mathrm{j}\omega_1) \\ 1/G(\mathrm{j}\omega_2) \\ . \\ . \\ 1/G(\mathrm{j}\omega_M) \end{bmatrix} \tag{10.21}$$

The compact form representation is shown below:

$$[W].[k] = [g] \tag{10.22}$$

The $[W]$ matrix includes Q and the vectors are $[k]$ and $[g]$. Clearly if $M > N$, then least square solution or matrix inverse solution may be applied, and can be solved to get sampled order-distribution $k(q)$. The problem with the solution of (10.22) is the matrix $[W]$ tends towards singularity if N (the number of order samples) grows large, or the order distribution sample size Q is made smaller.

Therefore numerically the order distribution $k(q)$ is $[k]=[W]^{-1}[g]$ or can be written as pseudo inverse expression as $[k]=\left[[W]^T[W]\right]^{-1}[W]^T[g]$. The proven concepts of numerical robustness be maintained in these calculations too.

10.5 Analysis of Continuous Order Distribution

By rewriting the integral in system equation (10.11) i.e. $\int_0^\infty k(q)s^q dq$, the exponent $s^q=\exp[q\ln(s)]$, we obtain:

$$\left(\int_0^\infty k(q)e^{q\ln(s)}dq\right)X(s)=P(s)X(s)=F(s) \tag{10.23}$$

The expression (10.23) is effectively a Laplace transform of the function $k(q)$ with the new Laplace variable $r=-\ln(s)$. As long as the order distribution $k(q)$ is of exponential order then the resulting $P(s)$, is easy to evaluate using this r-Laplace transform.

Table 10.1 presents the transfer function $P(s)$, for second order systems with continuous order distribution.

The integral in equation (10.23) resembles Laplace transform in log frequency and could be used for frequency domain analysis by replacing $s \to \mathrm{j}\omega_m$. Then,

$$\int_0^\infty dq\{k(nQ)e^{nQ\ln(\mathrm{j}\omega_m)}\}=P(\mathrm{j}\omega_m);$$

rather difficult to solve, but solvable to obtain $k(nQ)$.

The order distribution $k(q)$ is taken for all $q \geq 0$. The system descriptions with the characteristic equations are expressed in differential equations with differentiation order greater than zero. So the integration terms are also converted to differentiation and the characteristic equations are in polynomial of powers of s^q with $q \geq 0$. In the following derivations thus, the q is always taken as greater than zero, and $k(q)=0$ for all $q<0$.

To evaluate r-Laplace transform from given order distribution (continuous spiked or mixed), the Laplace identities are used. So $P(r)=\int_0^\infty k(q)e^{-rq}dq$, obtained is r-Laplace transform, and then in the obtained expression of $P(r)$ substitution for $r=-\ln(s)$ is carried out to get $P(s)$. Meaning obtain r-Laplace as $\mathcal{L}_{\mathcal{R}}\{k(q)\}=P(r)$, and then get $P(s)$.

Following examples demonstrates derivation of r-Laplace $P(r)$ and then Laplace transform $P(s)$ of continuous order, spiked ordered and mixed order distributions $k(q)$.

For exponential order distribution $k(q) = e^{-q}$ for all $q \geq 0$.

r-Laplace transform is

$$P(r) = \int_0^\infty e^{-q} e^{-rq} dq = \int_0^\infty e^{-q(r+1)} dq = \frac{1}{r+1},$$

in this expression putting $r = -\ln(s)$, gives

$$P(s) = \frac{1}{1 - \ln(s)}$$

For $k(q) = \dfrac{1}{q+a}$ for all $q \geq 0$,

$$P(r) = \int_0^\infty \frac{1}{q+a} e^{-qr} dq .$$

This integral we solve by using definition of Exponential Integral as:

$$Ei(x) \overset{\text{def}}{=} -\int_{-x}^\infty \frac{e^{-t}}{t} dt$$

Put $u = q + a$, then

$$P(r) = \int_a^\infty \frac{1}{u} e^{-r(u-a)} du = e^{ar} \int_a^\infty \frac{1}{u} e^{-ru} du .$$

In this take $ru = v$, then

$$P(r) = e^{ar} \int_{ar}^\infty \frac{r}{v} e^{-v} \frac{dv}{r} = e^{ar} \int_{-(-ar)}^\infty \frac{1}{v} e^{-v} dv = -e^{ar} \left\{ -\int_{-(-ar)}^\infty \frac{e^{-v}}{v} dv \right\} = -e^{ar} Ei(-ar)$$

Substitute $r = -\ln(s)$ to get

$$P(s) = -e^{a[-\ln(s)]} Ei\left(-a[-\ln(s)]\right) = -\frac{1}{s^a} Ei\left[\ln(s^a)\right]$$

For a train of spikes alternating at $q = 1, 2, 3, 4...$, we have

$$k(q) = \delta(q-1) - \delta(q-2) + \delta(q-3) - \delta(q-4) + ...$$

By using the shifted Dirac delta and its transform as $\delta(t - t_0) \leftrightarrow e^{-st_0}$, we get:

$$P(r) = e^{-r} - e^{-2r} + e^{-3r} - e^{-4r} + ...$$

$$= (e^{-r} + e^{-3r} + e^{-5r} + ...) - (e^{-2r} + e^{-4r} + e^{-6r} + ...)$$

$$= \frac{e^{-r}}{1-e^{-2r}} - \frac{e^{-2r}}{1-e^{-2r}} = \frac{1-e^{-r}}{e^{r} - e^{-r}} = \frac{1}{2\sinh(r)}(1-e^{-r})$$

Substituting $r = -\ln(s)$ we obtain

$$P(s) = \frac{1}{2\sinh[-\ln(s)]}\left(1 - e^{-[-\ln(s)]}\right) = \frac{1}{\sinh[\ln(s)]}\left(\frac{s-1}{2}\right)$$

For

$$k(q) = \begin{cases} 1 & q \geq 0 \\ 0 & q > 1 \end{cases},$$

call this as 'window order-one' $\mathrm{WIN}(0,1)$.
Then

$$P(r) = \int_0^{\infty} 1.e^{-rq} dq \text{ is } P(r) = \int_0^{1} e^{-rq} dq = \frac{1-e^{-r}}{r}$$

and substituting $r = -\ln(s)$ we have:

$$P(s) = \frac{1-e^{-r[-\ln(s)]}}{-\ln(s)} = \frac{s-1}{\ln(s)}$$

For

$$k(q) = \begin{cases} 1 & q \geq 0 \\ 0 & q > 2 \end{cases},$$

call this as 'window order-two' $\mathrm{WIN}(0,2)$,
Then

$$P(r) = \frac{1-e^{-2r}}{r}$$

and substituting $r = -\ln(s)$ we have

$$P(s) = \frac{s^2 - 1}{\ln(s)}$$

For order distribution with continuous from 0-1 as $\mathrm{WIN}(0,1)$ and spike at $q = 2$ is

$$k(q) = \mathrm{WIN}(0,1) + \delta(q-2),$$

then

$$P(r) = \mathcal{L}_{\mathcal{R}}\{\mathrm{WIN}(0,1)\} + \mathcal{L}_{\mathcal{R}}\{\delta(q-2)\}$$

gives:

$$P(r) = \frac{1-e^{-r}}{r} + e^{-2r} = \frac{1-e^{-r} + re^{-2r}}{r}$$

and substituting $r = -\ln(s)$ we have

$$P(s) = \frac{1-e^{-[-\ln(s)]} + [-\ln(s)]e^{-2[-\ln(s)]}}{-\ln(s)} = \frac{s^2\ln(s) + s - 1}{\ln(s)}$$

For

$$k(q) = \begin{cases} q & q \geq 0 \\ 0 & q > 2 \end{cases},$$

call this as $+\mathrm{RAMP}(0,2)$.

In this derivation we use Laplace identity:

$$\mathcal{L}\{t^n f(t)\} = (-1)^n \frac{d^n F(s)}{ds^n}.$$

This observation makes construction of $+\mathrm{RAMP}(0,2)$ from $\mathrm{WIN}(0,2)$ as:

$$+\mathrm{RAMP}(0,2) = q\mathrm{WIN}(0,2)$$

So Laplace will be $\mathcal{L}_{\mathcal{R}}\{q\mathrm{WIN}(0,2)\} = (-1)^1 \dfrac{d}{dr}\mathcal{L}_{\mathcal{R}}\{\mathrm{WIN}(0,2)\}$

$$P(r) = (-1)\frac{d}{dr}\frac{1-e^{-2r}}{r} = \frac{1-2re^{-2r}-e^{-2r}}{r^2} = \frac{1-(2r+1)e^{-2r}}{r^2}$$

and substituting $r = -\ln(s)$ we have

$$P(s) = \frac{1-(2[-\ln(s)]+1)e^{-2[-\ln(s)]}}{[\ln(s)]^2} = \frac{1+[2\ln(s)-1]s^2}{[\ln(s)]^2}$$

For

$$k(q) = \begin{cases} q & q \ge 0 \\ 0 & q > 1 \end{cases},$$

call this as $+\mathrm{RAMP}(0,1)$, using similar procedure as above we get:

$$P(r) = \frac{1-(1+r)e^{-r}}{r^2}$$

and substituting $r = -\ln(s)$ we get

$$P(s) = \frac{1+[\ln(s)-1]s}{[\ln(s)]^2}$$

For

$$k(q) = \begin{cases} -q+2 & q \ge 0 \\ 0 & q > 2 \end{cases},$$

call this as $-\mathrm{RAMP}(0,2)$, can be composed as:

$$k(q) = -[+\mathrm{RAMP}(0,2)] + 2[\mathrm{WIN}(0,2)],$$

By using derived Laplace of these constituents we get:

$$P(r) = -\left(\frac{1-e^{-2r}-2re^{-2r}}{r^2}\right) + 2\left(\frac{1-e^{-2r}}{r}\right) = \frac{2r-1+e^{-2r}}{r^2},$$

and substituting $r = -\ln(s)$ we get:

$$P(s)=\frac{s^2-1-2\ln(s)}{[\ln(s)]^2}$$

For

$$k(q)=\begin{cases}-q+1 & q\ge 0\\ 0 & q>1\end{cases},$$

call this as $-\text{RAMP}(0,1)$ and by the above procedure we compose this and write:

$$k(q)=-[+\text{RAMP}(0,1)]+1[\text{WIN}(0,1)],$$

then taking Laplace of the constituents we get:

$$P(r)=-\left(\frac{1-(1+r)e^{-r}}{r^2}\right)+1\left(\frac{1-e^{-r}}{r}\right)=\frac{r-1+e^{-r}}{r^2}$$

and substituting $r=-\ln(s)$ we get: $P(s)=\dfrac{s-1-\ln(s)}{[\ln(s)]^2}$

For

$$k(q)=\begin{cases}0 & q\ge 0\\ -q+2 & q\ge 1\\ 0 & q>2\end{cases},$$

is $-\text{RAMP}(0,1)$ shifted from $q=0$ to $q=1$.

Here Laplace shift identity $\mathcal{L}\{f(t-t_0)\}=e^{-st_0}F(s)$ is used to get

$$P(r)=e^{-r}\left(\frac{r-1+e^{-r}}{r^2}\right)=\frac{re^{-r}-e^{-r}+e^{-2r}}{r^2}=\frac{(r-1)e^{-r}+e^{-2r}}{r^2}$$

and substituting $r=-\ln(s)$ we get:

$$P(s)=\frac{([-\ln(s)]-1)e^{-[-\ln(s)]}+e^{-2[-\ln(s)]}}{[\ln(s)]^2}=\frac{s^2+s[\ln(s)+1]}{[\ln(s)]^2}$$

For

$$k(q) = \begin{cases} q & q \geq 0 \\ -q+2 & q \geq 1 \\ 0 & q > 2 \end{cases}$$

can be composed by $+\text{RAMP}(0,1) + [-\text{RAMP}(0,1)]_{@\, q=1}$.

From above obtained results for shifted $-\text{RAMP}(0,1)$ and $+\text{RAMP}(0,1)$ we get:

$$P(r) = \frac{1 - e^{-r} - re^{-r}}{r^2} + \frac{re^{-r} - e^{-r} - e^{-2r}}{r^2} = \frac{1 - 2e^{-r} + e^{-2r}}{r^2}$$

and substituting $r = -\ln(s)$ we get:

$$P(s) = \frac{1 - 2s + s^2}{[\ln(s)]^2}$$

For

$$k(q) = \frac{1}{2} - \frac{1}{2}\cos(2\pi q)$$

we use standard Laplace transform for Heaviside step and cosine expressions to obtain

$$P(r) = \frac{1}{2} \times \frac{1}{r} - \frac{1}{2}\left(\frac{r}{r^2 + 4\pi^2}\right) = \frac{2\pi^2}{r(r^2 + 4\pi^2)}$$

and substituting $r = -\ln(s)$ we get:

$$P(s) = \frac{-2\pi^2}{\ln(s)\left\{[\ln(s)]^2 + 4\pi^2\right\}}.$$

We now use this derived expression to get truncated

$$k(q) = \begin{cases} 0.5 - 0.5\cos(2\pi q) & q \geq 0 \\ 0 & q > 2 \end{cases},$$

this can be composed by continuous function $\frac{1}{2} - \frac{1}{2}\cos(2\pi q)$ and from this subtracting shifted function at $q = 2$, that is $\frac{1}{2} - \frac{1}{2}\cos[2\pi(q-2)]$. Using shift identity of Laplace operation we get:

$$P(r)=\frac{2\pi^2}{r(r^2+4\pi^2)}-e^{-2r}\left(\frac{2\pi^2}{r(r^2+4\pi^2)}\right)=\frac{2\pi^2(1-e^{-2r})}{r(r^2+4\pi^2)}$$

and substituting $r=-\ln(s)$ we get:

$$P(s)=\frac{2\pi^2(s^2-1)}{\ln(s)\left\{[\ln(s)]^2+4\pi^2\right\}}$$

If $k(q)$ is composed of one continuous function $f(q)$ multiplied by $a\mathrm{WIN}(0,q_n)$ then to get r-Laplace transform, convolution identity is used.

Meaning

$$P(r)=\mathcal{L}_{\mathcal{R}}\{f(q)\}*\mathcal{L}_{\mathcal{R}}\{a\mathrm{WIN}(0,q_n)\}$$

From the system order distribution, continuous spiked or mixed $k(q)$, it thus possible to obtain the Laplace transform $P(s)$ of the same (by going through intermediate r-Laplace i.e. $P(r)$). The reciprocal of the Laplace transform of the system order distribution gives the frequency response or the system transfer function, $G(s)=1/P(s)$. When Laplace transform of the input excitation is multiplied by reciprocal of Laplace transform of order distribution we get the output i.e. $X(s)=F(s)/P(s)$, $F(s)$ is excitation. Conversely by controlling or manipulating the shape/form of $P(s)$, the systems transfer function or the output shape/form can be controlled, with infinite freedom. The manipulation of the order distribution $k(q)$ thus gives a thought for 'continuum order feed back controller'.

Table 10.1

Order-distribution	r-Laplace-Transform	Laplace transfer function
$k(q)$ vs. q	$P(r)$	$P(s)$ $r=-\ln(s)$
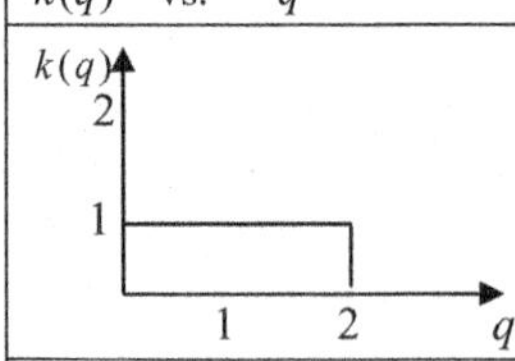	$\frac{1-e^{-2r}}{r}$	$\frac{s^2-1}{\ln(s)}$
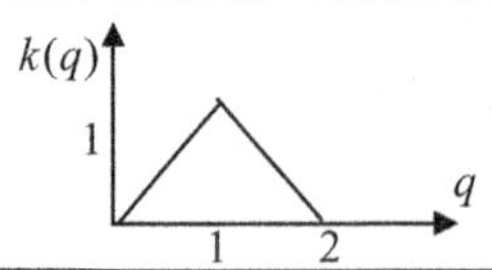	$\frac{1-2e^{-r}+e^{-2r}}{r^2}$	$\frac{1-2s+s^2}{[\ln(s)]^2}$

Table 10.1 (*continued*)

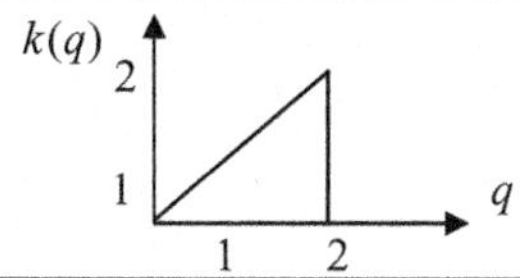	$\dfrac{1-(1+2r)e^{-2r}}{r^2}$	$\dfrac{1-[1-2\ln(s)]s^2}{[\ln(s)]^2}$
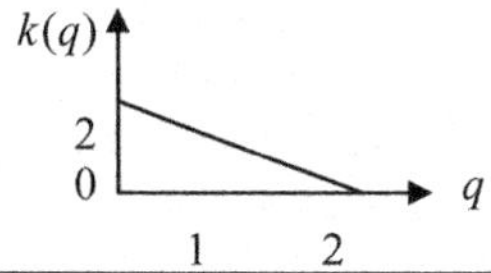	$\dfrac{2r-1+e^{-2r}}{r^2}$	$\dfrac{s^2-1-2\ln(s)}{[\ln(s)]^2}$
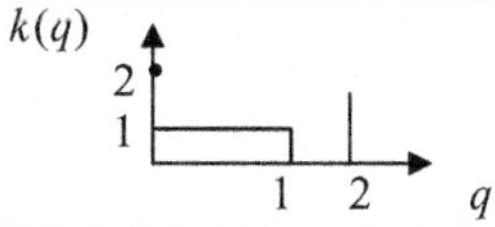	$\dfrac{re^{-2r}+1-e^{-r}}{r}$	$\dfrac{s-1+s^2\ln(s)}{\ln(s)}$
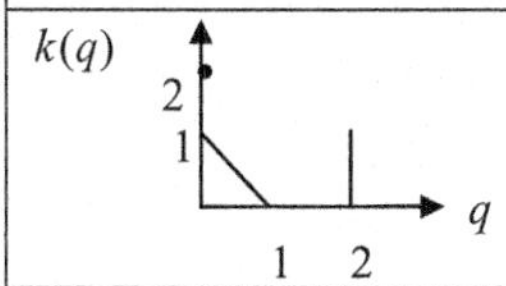	$\dfrac{r-1+e^{-r}+r^2e^{-2r}}{r^2}$	$\dfrac{s^2[\ln(s)]^2-\ln(s)+s-1}{[\ln(s)]^2}$
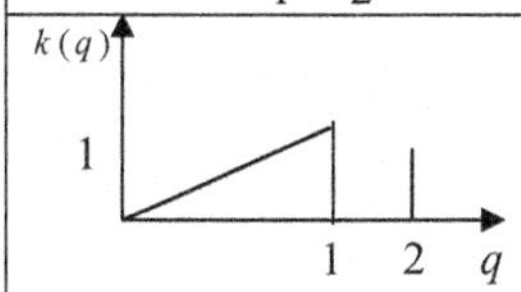	$\dfrac{1-(1+r)e^{-r}+r^2e^{-2r}}{r^2}$	$\dfrac{1-s+s\ln(s)+s^2[\ln(s)]^2}{[\ln(s)]^2}$
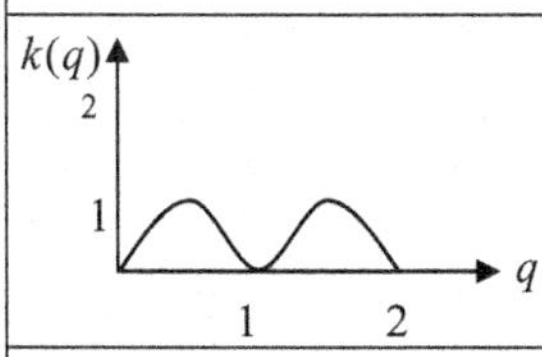	$\dfrac{4\pi^2(1-e^{-2r})}{2r(r^2+4\pi^2)}$	$\dfrac{4\pi^2(s^2-1)}{2\ln(s)\left[\{\ln(s)\}^2+4\pi^2\right]}$
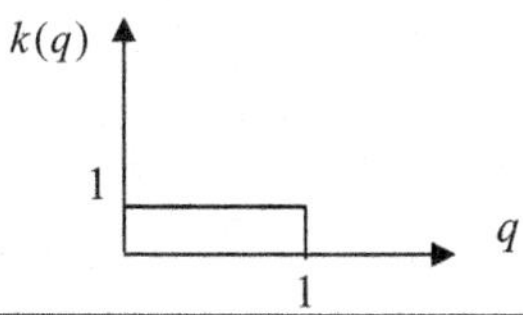	$\dfrac{1-e^{-r}}{r}$	$\dfrac{s-1}{\ln(s)}$
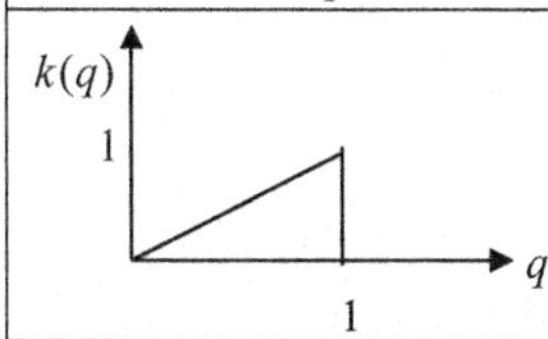	$\dfrac{1-(1+r)e^{-r}}{r^2}$	$\dfrac{1-s[1-\ln(s)]}{[\ln(s)]^2}$

The Figure 10.2 represents the system identification method applied for the transfer function $G(s)=\dfrac{1}{s^2+1.5s^{1.5}+s+1.5s^{0.5}+1}$. The equation (10.20) with

$Q = 0.1, N = 25$ with ω spaced logarithmically between $10^{-2} - 10^{2}$, gives the discretized plots for $k(q)$, the order distribution. Similarly Figure 10.3 represents the order distribution obtained for $G(s) = \frac{1}{s^2 + 0.5s + 1}$. The observation in figure 10.2 and Figure 10.3 is that the order spikes that one would have obtained are somewhat smeared and are blunt. The techniques to concentrate the $k(q)$ distribution peaks into specific discrete q values is topic of advance research and development. The concentration of the spikes should ideally as in Figure 10.1, for integer as well as fractional order system to be identified. Given order distribution in discrete form the transfer function is reconstructed by

$$G(s) \equiv \frac{1}{\sum_{n=0}^{N} k_n s^{nQ} Q}$$

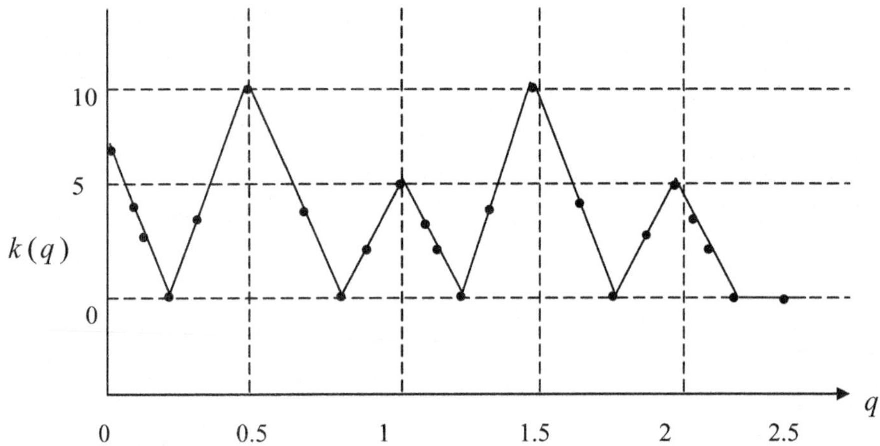

Fig. 10.2 Order distribution for transfer function $G(s) = \frac{1}{s^2 + 1.5s^{1.5} + s + 1.5s^{0.5} + 1}$

Sampling issues are described here for different types of order distribution, (spiked, continuous and mixed). In evaluating the order-distribution integral, it will be important to distinguish between types, i.e. spiked (impulsive) and continuous. If the composition of the order distribution is assumed to be purely

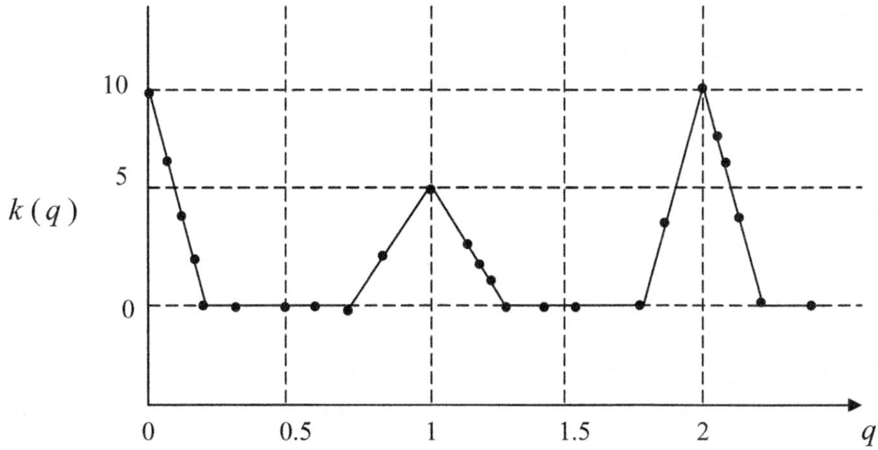

Fig. 10.3 Order distribution of transfer function $G(s) = \dfrac{1}{s^2 + 0.5s + 1}$

impulsive entirely, or more generally

$$k(q) = \sum_{n=0}^{N} k_{\delta}(nQ)\delta(q - nQ)\,, q_{\max} = NQ\,,$$

then the integral can be evaluated by Euler expansion. The approximation then for the integral is

$$\int_0^{q_{\max}} k(q)s^q dq \cong \sum_{n=0}^{N} k_{\delta}(nQ)s^{nQ}Q\,.$$

Here in this discussion a note may be taken as it is observed that Q appears explicitly in the summation. The unit impulses are spikes of unit area, their heights when identified by the above method, are amplified by $1/Q$. This height gets reduced by Q in the summation to give correct results.

For continuous order distribution for type as say $k(q) = \exp(-q)$ and others in the Table 10.1, then the integral can be evaluated by inter-sample reconstruction technique. This is analogous to sampled data reconstruction problem in time domain. Assuming the order-distribution to be piecewise constant, then the approximate integral is expressed as

$$\int_0^{q_{\max}} k(q)s^q dq \cong \sum_{n=0}^{N} k_c(nQ) \int_{nQ}^{(n+1)Q} s^q dq = \frac{s^Q - 1}{\ln(s)} \sum_{n=0}^{N} k_c(nQ)s^{nQ}\,.$$

Evaluation of

$$\int_{nQ}^{(n+1)Q} s^q dq$$

is obtained by writing $s^q = e^{q\ln(s)}$.
So

$$\int_{nQ}^{(n+1)Q} s^q dq = \int_{nQ}^{(n+1)Q} e^{q\ln(s)} dq = \left[\frac{e^{q\ln(s)}}{\ln(s)}\right]_{nQ}^{(n+1)Q} = \left[\frac{s^q}{\ln(s)}\right]_{nQ}^{(n+1)Q} = \frac{1}{\ln(s)}\left[s^{(n+1)Q} - s^{nQ}\right] = \frac{s^{nQ}\left[s^Q - 1\right]}{\ln(s)}$$

Notice that, Q does not explicitly multiply the summation, though implicitly being present in numerator. Using this discussion the expression-10.22 is modified and presented as:

$$\begin{bmatrix} \frac{(j\omega_1)^Q - 1}{\ln(j\omega_1)}\left\{1 \quad (j\omega_1)^Q \quad (j\omega_1)^{2Q} \quad * \quad (j\omega_1)^{NQ}\right\} \\ \frac{(j\omega_2)^Q - 1}{\ln(j\omega_2)}\left\{1 \quad (j\omega_2)^Q \quad (j\omega_2)^{2Q} \quad * \quad (j\omega_2)^{NQ}\right\} \\ * \\ \frac{(j\omega_M)^Q - 1}{\ln(j\omega_M)}\left\{1 \quad (j\omega_M)^Q \quad (j\omega_M)^{2Q} \quad * \quad (j\omega_M)^{NQ}\right\} \end{bmatrix} \begin{bmatrix} k_{c0} \\ k_{c1} \\ * \\ k_{cN} \end{bmatrix} = \begin{bmatrix} 1/G(j\omega_1) \\ 1/G(j\omega_2) \\ * \\ 1/G(j\omega_M) \end{bmatrix}$$

If the order distribution is assumed to be with mixed type that is with both continuous order distribution and the impulsive type then the reconstruction procedure will be mixed of the above discussed procedure, for the system identification experiments. The expression-16 gets modified as:

$$\left[\sum_{n=0}^{N} k_\delta(nQ)\, s^{nQ} Q + \frac{s^Q - 1}{\ln(s)} \sum_{n=0}^{N} k_c(nQ)\, s^{nQ}\right]_{s=j\omega_1} = \frac{1}{G(j\omega_1)}.$$

The expression-10.22 thus for the system identification with mixed order distribution is:

$$\begin{bmatrix} Q\left\{1 \quad (j\omega_1)^Q \quad * \quad (j\omega_1)^{NQ}\right\} & \frac{(j\omega_1)^Q - 1}{\ln(j\omega_1)}\left\{1 \quad (j\omega_1)^Q \quad * \quad (j\omega_1)^{NQ}\right\} \\ Q\left\{1 \quad (j\omega_2)^Q \quad * \quad (j\omega_2)^{NQ}\right\} & \frac{(j\omega_2)^Q - 1}{\ln(j\omega_2)}\left\{1 \quad (j\omega_2)^Q \quad * \quad (j\omega_2)^{NQ}\right\} \\ * & * \\ Q\left\{1 \quad (j\omega_M)^Q \quad * \quad (j\omega_M)^{NQ}\right\} & \frac{(j\omega_M)^Q - 1}{\ln(j\omega_M)}\left\{1 \quad (j\omega_M)^Q \quad * \quad (j\omega_M)^{NQ}\right\} \end{bmatrix} \begin{bmatrix} k_{\delta 0} \\ k_{\delta 1} \\ * \\ k_{\delta N} \\ k_{c0} \\ k_{c1} \\ * \\ k_{cN} \end{bmatrix} = \begin{bmatrix} 1/G(j\omega_1) \\ 1/G(j\omega_2) \\ * \\ 1/G(j\omega_M) \end{bmatrix}$$

10.6 Variable Order System

While developing the concept of continuous order distribution, some thought was put in if the system is subjected to ambient variations say temperature variation the fixed order does show distribution. For instance taking the transfer impedance of a half order element at fixed ambient is $Z(\mathrm{j}\omega) \propto \dfrac{1}{(\mathrm{j}\omega)^{0.5}}$ may well show a change in order 0.5 with the variable ambient. Therefore the differintegrals of the fractional or integer order may well have the order q, which varies with time i.e. becomes $q(t)$. Consider the fractional differential equation ${}_cD_t^q y(t) = f(t)$ and inferred integral equation is ${}_cD_t^{-q} f(t) = y(t)$.

Since q can take any real value, the development of calculus for varying q with t and y is essential field or research. The variable order system will have ${}_cD_t^{-q(t,y)} f(t) = y(t)$.

In experiments of condense matter physics, relaxation process where the relaxation is intermittent and thus fractional order differential equations govern the process, the ambient temperature changes give rise to changes in relaxation curves. The expression governing fractional relaxation process is:

$$\phi(t) - \phi_0 = -(\tau)^{-q}\, {}_0D_t^{-q}\phi(t).$$

With, $0 < q \le 1$; stating degree of intermittency in relaxation process. Taking initial conditions as $\phi_0 = 1$, the relaxation expression (for delta input excitation) is $\phi(t) = E_q\left(-\{t/\tau\}^q\right)$. The power law exponent here $q(T) = CT$, is temperature dependent, with C as constant and T defining absolute temperature. Several power-law curves give different degree of intermittency of relaxation curves. This is the motivation to develop variable order differintegration systems.

10.6.1 RL Definition for Variable Order

Consider only time variation of the order i.e. $q \to q(t)$ the RL definition with zero initial condition $\psi(f, -q(t), a, c, t) = 0$ yields:

$$ {}_0D_t^{-q(t)} f(t) \equiv \int_0^t \frac{(t-\tau)^{q_e(t,\tau)-1}}{\Gamma(q_g(t,\tau))} f(\tau)\,d\tau$$

Here the arguments of exponent are q_e and the Gamma function argument is q_g may be different. This is basic difference from fixed order system. The above formulation of the definition can have $q(t,\tau) \to q(t), q(t,\tau) \to q(\tau), q(t,\tau) \to q(t-\tau)$. Substitution of these three definitions into above formulation of $q_e(t,\tau)$ and $q_g(t,\tau)$ yields nine expressions of the variable order fractional RL integration. These nine definitions are subjected to the criteria and desirable properties of fractional integration, to rule out the undesirable formulations. The most important ones are linearity and index law (composition), leaving aside backward compatibility and zero property.

Consider the linearity property for the definition of variable order integration as mentioned above:

$$
\begin{aligned}
{}_0D_t^{-q(t)}\left(af(t)+bg(t)\right) &= \int_0^t \frac{(t-\tau)^{q_e(t,\tau)-1}}{\Gamma(q_g(t,\tau))}\{af(\tau)+bg(\tau)\}d\tau \\
&= a\int_0^t \frac{(t-\tau)^{q_e(t,\tau)-1}}{\Gamma(q_g(t,\tau))} f(\tau)d\tau + b\int_0^t \frac{(t-\tau)^{q_e(t,\tau)-1}}{\Gamma(q_g(t,\tau))} g(\tau)d\tau \\
&= a\,{}_0D_t^{-q(t)} f(t) + b\,{}_0D_t^{-q(t)} g(t)
\end{aligned}
$$

The above derivation shows that linearity is satisfied for all arguments of q in numerator as well as denominator for the definition of fractional integral.

Out of all these substitutions the definition of the variable order fractional RL integration when $q_e(t,\tau) = q_g(t,\tau)$, gives interesting observation, for the substitution case with, $q(t,\tau) \to q(t-\tau)$; which provided adherence to the index law, but failed to satisfy composition law under following definition:

$$
{}_0D_t^{-q(t)} f(t) \equiv \int_0^t \frac{(t-\tau)^{q(t-\tau)-1}}{\Gamma(q(t-\tau))} f(\tau)d\tau
$$

Here it was inferred that

$$
{}_0D_t^{-q(t)}\,{}_0D_t^{-v(t)} f(t) = {}_0D_t^{-v(t)}\,{}_0D_t^{-q(t)} f(t) \neq {}_0D_t^{-q(t)-v(t)} f(t).
$$

The detailed proof uses convolution theory and convolution nature of this particular definition, which appears to be the most satisfactory one for usage.

It should be mentioned that different physical processes might effectively use different definitions and all the three forms namely

$$
{}_0D_t^{-q(t)} f(t) \equiv \int_0^t \frac{(t-\tau)^{q(t-\tau)-1}}{\Gamma(q(t-\tau))} f(\tau)d\tau, \quad {}_0D_t^{-q(t)} f(t) \equiv \int_0^t \frac{(t-\tau)^{q(t)-1}}{\Gamma(q(t))} f(\tau)d\tau
$$

and

$$ {}_0D_t^{-q(t)} f(t) \equiv \int_0^t \frac{(t-\tau)^{q(\tau)-1}}{\Gamma(q(\tau))} f(\tau)d\tau $$

may prove useful. This is because this formulations adherence to the index law and because of the convolution forms, which makes available all of the results of the associated theory, the most compelling definition is

$$ {}_0D_t^{-q(t)} f(t) \equiv \int_0^t \frac{(t-\tau)^{q(t-\tau)-1}}{\Gamma(q(t-\tau))} f(\tau)d\tau \qquad q(t) > 0 $$

Figure 10.4 describes a variable order structure.

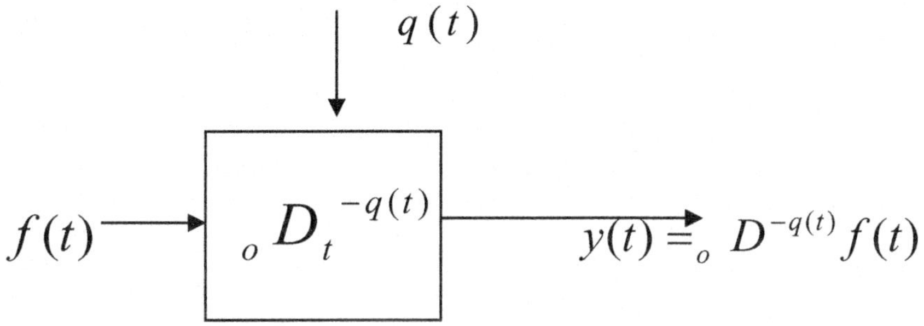

Fig. 10.4 Variable order structure

If the above form of equations is assumed, then parallel definition for the variable structure derivative might be considered as:

$$ {}_0D_t^{q(t)} f(t) \equiv {}_0D_t^m \, {}_0D_t^{-p(t)} f(t), \; q(t) > 0 $$

where $q(t) = m - p(t)$, and m is positive integer. If it is assumed that $q(t)$ be always positive, the m could be taken as the least integer greater than $q(t)$. However since composition does not hold it is not clear that if this could be reasonable definition. Matters are further complicated by the fact that it may be desirable to allow $q(t)$ to range over both positive and negative values. This places a 'seam' at $q = 0$, which may make any approach based on RL definition implausible, and perhaps requires approach based on GL definition.

10.6.2 Laplace Transforms and Transfer Function of Variable Order System

The derivation of the Laplace transforms of the variable order structure integral follows that for the fixed order case exactly since the convolution theorem can be applied. Then, considering the un-initialized case of fractional integration of variable structure, we get:

$$\mathcal{L}\left\{{}_0D_t^{-q(t)} f(t)\right\} = \int_0^\infty e^{-st}\left(\int_0^t \frac{(t-\tau)^{q(t-\tau)-1}}{\Gamma(q(t-\tau))} f(\tau)d\tau\right)dt\,,\ q(t) > 0, t > 0$$

With definition of convolution

$$\mathcal{L}\{h(t) * g(t)\} = H(s)G(s) = \mathcal{L}\left(\int_0^t h(\tau)g(t-\tau)d\tau\right),$$

and taking $h(t) = f(t)$ and $g(t) = t^{q(t)-1} / \Gamma(q(t))$, the convolution theorem yields:

$$\mathcal{L}\left\{{}_0D_t^{-q(t)} f(t)\right\} = F(s)G(s) = \mathcal{L}\{f(t)\}\mathcal{L}\left\{\frac{t^{q(t)-1}}{\Gamma(q(t))}\right\}$$

The variable order structure differintegration allows the introduction of a new transfer function concept. Refer Figure 10.4; the conventional transfer function relates the Laplace transform of the output to the transform of the input (ratio of the two) by:

$$TF_1 \equiv \frac{\mathcal{L}\{y(t)\}}{\mathcal{L}\{f(t)\}} = \frac{\mathcal{L}\left\{{}_0D_t^{-q(t)} f(t)\right\}}{\mathcal{L}\{f(t)\}} = \frac{\mathcal{L}\{f(t)\}\mathcal{L}\left\{\frac{t^{q(t)-1}}{\Gamma(q(t))}\right\}}{\mathcal{L}\{f(t)\}} = \mathcal{L}\left\{\frac{t^{q(t)-1}}{\Gamma(q(t))}\right\}$$

Since the $q(t)$ is now a variable, it may be thought of as an input (Figure 10.4), and for this new transfer function may be defined as: $TF_2 \equiv \frac{\mathcal{L}\{y(t)\}}{\mathcal{L}\{q(t)\}}$. For the considered definition, the process from $q(t)$ to $y(t)$ has not been shown to be linear.

That is ${}_0D_t^{-q_1(t)-q_2(t)} f(t) = {}_0D_t^{-q_1(t)} f(t) + {}_0D_t^{-q_2(t)} f(t)$ has not been shown. Thus for meaning and utility of TF_2, for these definitions requires further considerations.

The relationship of two transfer functions may be determined as follows. Consider $f(t)$ and $q(t)$ to be related by $g(t)$, where:

$$q(t) = \int_0^t f(t)g(t-\tau)d\tau\,,\ \text{then}\ \mathcal{L}\{q(t)\} = \mathcal{L}\{f(t)\}\mathcal{L}\{g(t)\},$$

by convolution theorem. Then:

$$TF_2 = \frac{\mathcal{L}\{y(t)\}}{\mathcal{L}\{q(t)\}} = \frac{\mathcal{L}\{f(t)\}TF_1}{\mathcal{L}\{f(t)\}\mathcal{L}\{g(t)\}} \text{ gives } \frac{TF_1}{TF_2} = \mathcal{L}\{g(t)\}$$

10.6.3 GL Definition for Variable Order

A variable structure differintegral is formed by GL definition too. Let $\Delta T = (t-a)/N$ and limit consideration to $q \to q(t)$ or $q \to q(t - j\Delta T)$. Then expressing q generally as $q \to q(t, j\Delta T)$, a generalized GL form may be written as:

$${}_aD_t^{q(t)} f(t) = \lim_{\substack{N\to\infty \\ \Delta T \to 0}} \sum_{j=0}^{N-1} \Delta T^{q_E(t, j\Delta T)} \frac{\Gamma(j - q_N(t, j\Delta T))}{\Gamma(-q_D(t, j\Delta T))\Gamma(j+1)} f(t - j\Delta T)$$

It is observed that $q(t, j\Delta T)$ occurs three times in this expression, in the exponent as q_E, in the numerator q_N, and in the denominator q_D. Thus the number of formula combination (permutation) is eight, as compared to nine in RL type definition previously discussed.

For the combinations $q_E(t, j\Delta T) \to q_E(t - j\Delta T)$, $q_N(t, j\Delta T) \to q_N(t - j\Delta T)$, $q_D(t, j\Delta T) \to q_D(t - j\Delta T)$ the GL formulation is following:

$${}_aD_t^{q(t)} f(t) = \lim_{\substack{N\to\infty \\ \Delta T \to 0}} \sum_{j=0}^{N-1} \Delta T^{q_E(t - j\Delta T)} \frac{\Gamma(j - q_N(t - j\Delta T))}{\Gamma(-q_D(t - j\Delta T))\Gamma(j+1)} f(t - j\Delta T)$$

This is the most promising definition out of the eight permutations.

Consider two time interval evaluations of GL differeintegration of ${}_aD_t^{q(t)} f(t)$, for common $q(t)$ with one evaluation for $a < t \le t_1$, and another in interval $a < t \le t_2$ where $t_2 > t_1$ (Figure 10.5). The function did not exist before the time $t = a$, so $f(t) = 0$ for $t < a$. In particular consider the order $q(t)$ steps from constant value $q(t_1) \to q(t_2)$ a new constant value at time $t = t_1$. The evaluation of ${}_aD_t^{q(t)} f(t)$, to $t = t_2$ can be viewed as an evaluation from $t = a \to t_1$, summed to an evaluation from $t = t_1 \to t_2$. In view of the hereditary nature of the fractional differintegration, the evaluation to $t = t_1$ is part of history in the evaluation a to t_2; thus it is apparent that for the evaluation of ${}_aD_t^{q(t)} f(t)$ to $t = t_2$ based on the value $q_E(t, j\Delta T) \to q_E(t - j\Delta T)$, $q_N(t, j\Delta T) \to q_N(t - j\Delta T)$, $q_D(t, j\Delta T) \to q_D(t - j\Delta T)$, will yield desirable result all others will give undesirable result. This is also because, over the period $t < t_1$, part of evaluation the value of $q(t, j\Delta T) = q(t) = q(t_2)$ will be used in the summation, essentially changing history. Thus it appears that the only satisfactory definition is as

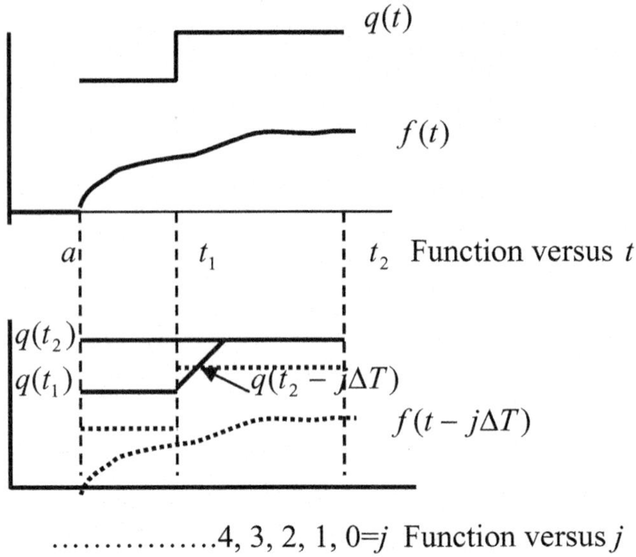

Fig. 10.5 Variable $q(t)$ differintegration by GL method

mentioned above. This definition parallels the RL promising formulation, and adds credibility of convolution related variable structure form. The intuitive nature of GL form, in terms of available conceptualization allows the hope that a form of it may evolve which would satisfy composition property. The detailed study in this direction is still matter of research.

10.7 Generalized PID-Controls

Here in this section generalization of the PI-PID controller is presented. The generalization is possible only because the availability of the fractional order elements. The standard integer order PI-controller has a transfer function of the form:

$$H(s) = k_p + \frac{k_i}{s} \tag{10.24}$$

This can be written as:

$$H(s) = k_p + k_i s^{-1} \tag{10.25}$$

The controller can now be generalized using fractional order integrals. Assuming the base fraction $q = 1/N$, for now, a generalized PI-controller can be written as:

$$H(s) = k_0 + k_1 s^{-q} + k_2 s^{-2q} + \ldots + k_{N-1} s^{-(N-1)q} + k_N s^{-1}, \quad 0 < q < 1 \tag{10.26}$$

$$H(s) = \sum_{n=0}^{N} k_n s^{-nq}, Nq = 1 \tag{10.27}$$

Performing manipulation on above expressions we obtain:

$$H(s) = \frac{k_0 s + k_1 s^{(N-1)q} + \ldots + k_{N-1} s^q + k_N}{s}, Nq = 1 \tag{10.28}$$

$$H(s) = \frac{\sum_{n=0}^{N} k_n s^{(N-n)q}}{s}, \quad Nq = 1 \tag{10.29}$$

Clearly, this controller allows a much degree and much variety of compensation results. Inserting this into the standard closed-loop control configuration with plant transfer function $G(s)$, gives closed loop transfer function $T(s) = G(s)H(s) / (1 + G(s)H(s))$ as:

$$T(s) = \frac{\left[\sum_{n=0}^{N} k_n s^{(N-n)q}\right] G(s)}{s + \left[\sum_{n=0}^{N} k_n s^{(N-n)q}\right] G(s)}, \quad Nq = 1 \tag{10.30}$$

With $G(s) = \mathrm{N}(s) / \mathrm{D}(s)$ the close loop transfer function becomes:

$$T(s) = \frac{\left[\sum_{n=0}^{N} k_n s^{(N-n)q}\right] \mathrm{N}(s)}{s\mathrm{D}(s) + \left[\sum_{n=0}^{N} k_n s^{(N-n)q}\right] \mathrm{N}(s)} \tag{10.31}$$

This generalization of a PI-control process gives considerable more design capability and freedom. Both close-loop poles and close-loop zeros can be placed by proper selection of the gains. The compensator can be further generalized by considering the powers of q to be unrelated, in this case the controller is:

$$H(s) = k_0 + k_1 s^{-q_1} + k_2 s^{-q_2} + \ldots + k_{N-1} s^{-q_{N-1}} + k_N s^{-1} = \frac{\sum_{n=0}^{N} k_n s^{q_{N-n}}}{s} \tag{10.32}$$

The summation assumes $q_0 = 0$, and $q_N = 1$, and $0 \le q_i \le 1$ and $0 \le i \le N$.

The above-discussed approach can be extended to PID controllers where derivative action controller is allowed. However the derivative control is seldom used because its high frequency attenuation of noise issues. Still to generalize the concept the PID controller, can thus be generalized as:

$$H(s) = k_N' s^{+1} + k_{N-1}' s^{+(N-1)q} + \ldots + k_2' s^{+2q} + k_1' s^{+q} + k_0 + k_1 s^{-q} + k_2 s^{-2q} + \ldots + k_{N-1} s^{-(N-1)q} + k_N s^{-1} \tag{10.33}$$

or equivalently with $q = 1/N$, the generalized PID is $H(s) = \sum_{n=-N}^{N} h_n s^{nq}$. Based on the previous discussions the same compensator can be decomposed into numerator and denominator and be expressed as

$$H(s) = \frac{\sum_{n=0}^{N} b_n s^{n/N}}{\sum_{n=0}^{N} a_n s^{n/N}} = \frac{\sum_{n=0}^{N} b_n s^{p_n}}{\sum_{n=0}^{N} a_n s^{q_n}}.$$

In a closed loop feed back configuration the plant $G(s) = \mathrm{N}(s)/\mathrm{D}(s)$ we get the generalized transfer function:

$$T(s) = \frac{\left[\sum_{n=0}^{N} b_n s^{p_n}\right] \mathrm{N}(s)}{\left[\sum_{n=0}^{N} a_n s^{q_n}\right] \mathrm{D}(s) + \left[\sum_{n=0}^{N} b_n s^{p_n}\right] \mathrm{N}(s)} \tag{10.34}$$

To determine the effectiveness of the generalized PID controller the analysis must be done on Nyquist plane, the quality of which depends on the approximation limited by approximation size.

10.8 Continuum Order Feed Back Control System

As discussed in system identification section taking the summation to the limit we get the transfer function of (10.33) in the form:

$$H(s)=\frac{\int_a^b K_N(q)s^q dq}{\int_a^b K_D(q)s^q dq} \tag{10.35}$$

This is continuum order compensator, where the order distribution $K(q)$ must be selected for numerator and denominator so that $H(s)$ remains causal and so that the integrand must converge. The general PID expression (10.33) can be expressed as power series as

$$H(s)=\frac{\sum_{n=0}^{N} b_n s^{p_n}}{\sum_{n=0}^{N} a_n s^{q_n}}=\sum_{n=0}^{\infty} c_n s^{r_n}$$

or asymptotic series as $H(s)=\sum_{n=0}^{\infty} c_n s^{-w_n}$ representations. Thus the power series integral representation for the continuum compensator would look as:

$$H_{PS}(s)=\int_0^a K_{PS}(q)s^q dq$$

And the asymptotic series would be $H_{AS}(s)=\int_0^{\infty} K_{AS}(q)s^q dq$. The combined form will be having a form as $H(s)=\int_{-\infty}^{a} K(q)s^q dq$, where the order distribution function $K(q)$ is chosen to make the integral convergent. For an order distribution for generalizing the PID controller of equation (10.33) will be:

$$H(s)=\int_{-1}^{+1} K(q)s^q dq \tag{10.36}$$

Given the $K(q)$ function it is easy task to perform the analysis of closed loop system in Nyquist plane. The selection of suitable $K(q)$ is field of research; following example of an oscillating plant will give some insight of the open issues.

Consider a plant transfer function $G(s)=\frac{1}{s^2+1}$, (an oscillator un-damped). A possible compensator using the order distribution would be $H(s)=\int_0^2 K(q)s^q dq$, and if this controller were placed before the plant the resulting close-loop system will be having a transfer function as

$$T(s)=\frac{\int_0^2 K(q)s^q dq}{s^2+\int_0^2 K(q)s^q dq+1},$$

this compensator allows an infinite number of frequencies in the closed-loop system and thus allows considerable freedom to design the appropriate $K(q)$. The question of selection of this function is to minimize some desired error, for a given input, or more appropriately a cost function, but much research is required in this direction.

10.9 Time Domain Response of Sinusoidal Inputs for Fractional Order Operator

Replacing $s^q \to (\mathrm{j}\omega)^q$ gives frequency response of the transfer function. Frequency domain approaches assumes that the time responses are sinusoidal steady state. Whenever, an input is applied to a system, the response will consist of transient part and steady state part. The frequency response approach assumes that the transient has decayed away, and that the response is in sinusoidal steady state. For fractional order systems sinusoid steady state also implies that the initialized response due to initialization function (ψ) has decayed to near zero.

A periodic function with period T is represented as Fourier series:

$$f(t)=\sum_{k=1}^{\infty}\left(c_k e^{\mathrm{j}2\pi kt/T}+\bar{c}_k e^{-\mathrm{j}2\pi kt/T}\right).$$

Fourier integral obtains the coefficient and the coefficients are complex conjugate. The coefficient is $c_k=\frac{1}{T}\int_0^T f(t)e^{-\mathrm{j}2\pi kt/T}dt$. Oldham Spanier gives method for evaluating fractional differentigration of repeated periodic function by using lower-incomplete Gamma function as:

$${}_0d_t^q\left(\exp(\pm\mathrm{j}2\pi k/T)\right)=\left(\frac{\pm\mathrm{j}2\pi k}{T}\right)^q \gamma(-q,\pm\mathrm{j}2\pi kt/T).$$

This term contains both the transient and the steady state fractional derivative of the periodic function. An asymptotic expansion for larger values of t for incomplete Gamma function γ term gives:

$$ {}_0d_t^q f(t) = \sum_{k=1}^{\infty} \left(\frac{2\pi k}{T}\right)^q \left(c_k \exp\left[\mathrm{j}2\pi\{(kt/T)+(q/4)\}\right] + \bar{c}_k \exp\left[-\mathrm{j}2\pi\{(kt/T)+(q/4)\}\right]\right) $$

Defining the radian frequency $\omega_0 = 2\pi/T$ gives equivalent response as:

$$ {}_0d_t^q f(t) = \sum_{k=1}^{\infty} (k\omega_0)^q \left(c_k \exp \mathrm{j}\left[k\omega_0 t + (\pi q/2)\right] + \bar{c}_k \exp\left\{-\mathrm{j}\left[k\omega_0 t + (\pi q/2)\right]\right\}\right) $$

For any given input frequency $k\omega_0$, it can be seen that the magnitude of the magnitude of the corresponding fractionally differintegrated steady state output sinusoid has its magnitude scaled by $(k\omega_0)^q$, and is phase shifted by $q\pi/2$. For example after the decay of the transient

$$ {}_0d_t^q (\sin \omega t) \to \lim_{t\to\infty} (\omega)^q \sin\left(\omega t + [\pi q/2]\right) $$

This result generalizes the response obtained for integer order systems. Here it is important to note that frequency response results require sinusoidal steady state, and initialization function plays less important role.

10.10 Frequency Domain Response of Sinusoidal Inputs for Fractional Order Operator

In earlier section it was mentioned that replacement of $s^q \to (\mathrm{j}\omega)^q$ gives the steady-state response of the transfer function. However replacement $s^{nq} \to (\mathrm{j}\omega)^{nq}$, where q is fraction and nq, not necessary an integer, be carried on, question of multiple solution exists. The question here is actually which root of $(\mathrm{j})^{nq}$ to use. The primary root is considered to be one with smallest angle from the positive real axis; with remaining roots being secondary roots. The answer to this question is given by time domain result of the previous section. That is, using the primary roots given by the frequency domain substitution $s^{nq} \to (\mathrm{j}\omega)^{nq}$ will give the frequency response corresponding to the correct time domain response. For example, $s^{0.5}$, substituted by $s = \mathrm{j}\omega$, gives $(\mathrm{j}\omega)^{0.5}$ which is $\mathrm{j}^{0.5}\omega^{0.5}$. Recognizing, $\sqrt{\mathrm{j}}$ has roots $\exp(\mathrm{j}\pi/4)$ and $\exp(\mathrm{j}5\pi/4)$, the primary root is always chosen for the frequency response i.e. the root $e^{\mathrm{j}\pi/4}$. This observation allows using the standard tools for control system analysis as Bode plot, Nyquist plot and others.

The frequency response of fractional order differintegral gives insight into the use of the control system tools. The un-initialized Laplace relation is $\mathcal{L}\left\{{}_0d_t^q f(t)\right\} = s^q F(s)$. Thus the transfer function of the fractional operator is

$$H(s) = \frac{\mathcal{L}\left\{{}_0d_t^q f(t)\right\}}{F(s)} = s^q .$$

To obtain frequency response replaces $s^q \to (\mathrm{j}\omega)^q$.

The magnitude response is simply $\left|H(\mathrm{j}\omega)\right| = \omega^q$, which rolls off at $20q$ dB/decade on Bode plot, and the phase shift is given by the angle of the primary root of $(\mathrm{j})^q$ which is $\arg H(\mathrm{j}\omega) = q\pi/2$. The derivative operation is for $q > 0$, and integration operation is for $q < 0$.

10.11 Ultra-Damped System Response

The properties of temporal behavior of systems were discussed in Chapter 7, with respect to pole-locations in w-plane. In Chapter 7 it was discussed that poles lying to the left of the wedge, given by lines with angle $q\pi$ in w-plane, are on the secondary Riemann sheet of the s-plane. These pole properties were termed as hyper-damped, as they were damped more than usual integer-order over-damped poles. Now with respect to the w-plane it is with some necessity that distinction is made, between, the poles that are on the negative real axis of w-plane, (which are called ultra-damped) and complex conjugate poles of w-plane (which are still hyper-damped).

An ultra-damped system will consist of parallel combination of the form $H(s) = \frac{Y(s)}{U(s)} = \frac{k}{s^q + a}, a > 0$ and real. Here k is system gain, a is ultra-damped pole in w-plane and $q > 0$. For simplicity take $a = 1, k = 1$ and frequency response transfer function is

$$H(\mathrm{j}\omega) = \frac{1}{(\mathrm{j}\omega)^q + 1} .$$

For small values of frequency $H(\mathrm{j}\omega) = 1$, and magnitude is thus unity and phase angle is zero. For large values of frequency, the transfer function becomes $H(\mathrm{j}\omega) = (\mathrm{j}\omega)^{-q}$, and frequency response reverts back to that of a simple fractional order operator discussed in previous section.

10.12 Hyper-Damped System Response

Hyper-damped system have a pair of complex conjugate poles off the negative real axis of *w*-plane, but are farther to the left than the under-damped region (Chapter 7). These poles are at $|\varphi| > \pm q\pi$ lines. Here there is several type of behavior depending upon the specific location of the poles in *w*-plane.

First we address system that can be realized with passive energy storage element. A passive energy storage element is one that cannot return more energy to the system than placed into the element by the system in the past. An active element is one that can return more energy to a system than placed into it in past. Typically an active element will have associated with it, either a large gain or negative gain. Necessary (but not sufficient) condition for fractional order system be passive are that minimal transfer function denominator have all positive coefficients, and that all poles lie to the left of stability wedge in *w*-plane. Another concept traditionally associated with passivity is positive real concept. To be positively real system, the frequency response of transfer function must always lie in the right half of Nyquist plot. Meaning that the $\arg H(\mathrm{j}\omega) \le \pm\pi/2$ and $\Re e H(\mathrm{j}\omega) > 0$. This means that maximum phase-shift of positive-real transfer function is bounded by $\pm 90^0$. A minimum phase system has smallest possible phase shift for given magnitude response. An implication of this in integer order system is that all the system poles and zeros must lie in left half plane of the *s*-plane. For fractional order system the implication of this being minimum phase all pole-zero lie left of instability wedge of *w*-plane, $|\varphi| > \pm q\pi/2$. Consider as an example all pole system, passive and positive real, minimum phase, transfer function $H(s) = \dfrac{1}{s + as^{0.5} + 1}$. (A fractional order system can be of minimum phase without being positive-real). The properties of this transfer function depends on pole location as decided by value of *a*, is summarized in Table 10.3.

The under-damped active region for $0 > a > -\sqrt{2}$ has further consideration. The negative value of '*a*' indicative of an active system, while *w*-plane poles remain in 'under-damped' region. The fact that there exist under-damped poles implies, that this system has a resonance and a resonance peak should appear in the frequency response Bode diagram too. Thus even though the high frequency asymptotes as well as Laplace transformed transfer function, of this example, indicate that this system is only a first order; it can still go into resonance. Clearly adding more fractional order terms with smaller value of *q*, would allow even more resonance.

Consequently, it appears that the highest power of Laplace variable in transfer function is no longer an indicator of the order effective order of fractional order system, or of number of resonance to expect in its frequency response. We call the resonance as fractional resonance.

Table 10.3

Value of a	$w=\rho \exp \mathrm{j}\phi$ w-plane	$s = r\exp \mathrm{j}\theta$ s-plane	Property
$a<-\sqrt{2}$	$\Re e(w)>0$ $\lvert\phi\rvert<\frac{\pi}{4}$	$\Re e(s)<0$ $\lvert\theta\rvert<\frac{\pi}{2}$ Right Half Plane	UNSTABLE
$0>a>-\sqrt{2}$	$\Re e(w)>0$ $\frac{\pi}{4}<\lvert\phi\rvert<\frac{\pi}{2}$	$\Re e(s)<0$ $\frac{\pi}{2}<\lvert\theta\rvert<\pi$ Left Half Plane	STABLE UNDERDAMPED MIN-PHASE
$2>a>0$	$\Re e(w)<0$ $\frac{\pi}{2}<\lvert\phi\rvert<\pi$	$\pi<\lvert\theta\rvert<2\pi$ Secondary Riemann Sheet	STABLE HYPERDAMPED
$a>2$	$\lvert\phi\rvert>\pi$	$\lvert\theta\rvert>2\pi$ Secondary Riemann Sheet	STABLE ULTRADAMPED

10.13 Complex Order Differintegrations

Let us consider a function $f(t)$ differentiated with order q, where $q=u+\mathrm{j}v$ is a complex quantity. Represent this operation by a function $g(t)$ as:

$$g(t) = {}_0 d_t^q f(t) = \frac{d^{u+\mathrm{j}v}}{dt^{u+\mathrm{j}v}} f(t),$$

considering un-initialized differintegration for simplicity. The Laplace transform of the function $g(t)$ enables us to express the above operation as:

$$\mathcal{L}\{g(t)\} = G(s) = s^{u+\mathrm{j}v}F(s) = s^u s^{\mathrm{j}v}F(s) = s^u e^{\mathrm{j}v\ln(s)}F(s)$$

We have used in above the identity $x^{\mathrm{j}(y)} = \exp\left[\ln(x^{\mathrm{j}(y)})\right] = \exp\left[\mathrm{j}(y)\ln(x)\right]$; and expressing $\exp(\mathrm{j}\theta) = \cos\theta + \mathrm{j}\sin\theta$ we get:

$$G(s) = s^u[\cos(v\ln s) + \mathrm{j}\sin(v\ln s)]F(s)$$

The invert function of $g(t)$ is 'complex-integration' expressed as: $f(t) = {}_0 d_t^{-(u+\mathrm{j}v)} g(t)$ and obtaining Laplace expression we obtain:

$$F(s) = s^{-(u+\mathrm{j}v)} G(s) .$$

Using

$$\mathcal{L}^{-1}\{s^{-q}\} = (t^{q-1})[\Gamma(q)]^{-1} ,$$

we write, $\mathcal{L}\{s^{-(u+\mathrm{j}v)}\} = (t^{u+\mathrm{j}v-1})[\Gamma(u+\mathrm{j}v)]^{-1}$. Consider $g(t)$ to be unit impulse so $G(s) = 1$. To this unit impulse we obtain the $f(t)$ as:

$$f(t) = [\Gamma(u+\mathrm{j}v)]^{-1} t^{u-1} t^{\mathrm{j}v} = [\Gamma(u+\mathrm{j}v)]^{-1} t^{u-1} e^{\mathrm{j}v \ln t}$$

that is, when simplified gives real as well as imaginary response in time! Well, does imaginary response constitute response in dream!

$$f(t) = \frac{t^{u-1}}{\Gamma(u+\mathrm{j}v)} [\cos(v \ln t) + \mathrm{j} \sin(v \ln t)]$$

However, this imaginary time response is what is difficult to explain physically at present, but we can consider from above derivation a formation of 'conjugated differintegrations' operation with $q = u + \mathrm{j}v$ and $\bar{q} = u - \mathrm{j}v$. The operation is described as follows:

$$g(t) = {}_0 d_t^q f(t) + {}_0 d_t^{\bar{q}} f(t) = {}_0 d_t^{u+\mathrm{j}v} f(t) + {}_0 d_t^{u-\mathrm{j}v} f(t) .$$

Doing Laplace operation on this defined operation with conjugated differintegration we obtain:

$$\mathcal{L}\{g(t)\} = (s^{u+\mathrm{j}v} + s^{u-\mathrm{j}v}) F(s) = (s^u s^{\mathrm{j}v} + s^u s^{-\mathrm{j}v}) F(s) .$$

Simplifying this we obtain:

$$G(s) = s^u [e^{\mathrm{j}v \ln s} + e^{-\mathrm{j}v \ln s}] F(s) = 2 s^u \cos(v \ln s) F(s)$$

Similarly, we construct operation as: $\bar{g}(t) = {}_0 d_t^{u+\mathrm{j}v} f(t) - {}_0 d_t^{u-\mathrm{j}v} f(t)$, on repeating this in the above procedure we get:

$$\bar{G}(s) = 2\mathrm{j} s^u \sin(v \ln s) F(s)$$

Note that $(s^{u+jv} \times s^{u-jv})F(s) = s^{2u}F(s)$ and $(s^{u+jv} \div s^{u-jv})F(s) = s^{2jv}F(s)$. We can express a corollary from this and above derivation that, 'A real order differentiation process can be decomposed or broken into product of two complex conjugated derivatives.'

We express the complex conjugated order integration process as:

$$g(t) = {}_0D_t^{-q} f(t) \equiv {}_0D_t^{-(u+jv)} f(t) + {}_0D_t^{-(u-jv)} f(t).$$

Here doing the Laplace operation we obtain $\mathcal{L}\{g(t)\} = s^{-u}(s^{-jv} + s^{jv})F(s)$. Since $f(t)$ is unit impulse $F(s) = 1$, so we get:

$$g(t) = \mathcal{L}^{-1}\{s^{-(u+jv)} + s^{-(u-jv)}\} = \mathcal{L}^{-1}\left\{2s^{-u}\cos(v\ln s)\right\} = \frac{t^{u+jv-1}}{\Gamma(u+jv)} + \frac{t^{u-jv-1}}{\Gamma(u-jv)}$$

Note the following properties of reciprocal gamma function with complex arguments:

$$\frac{1}{\Gamma(u+jv)} = \mathrm{Re}\left[\frac{1}{\Gamma(u+jv)}\right] + j\,\mathrm{Im}\left[\frac{1}{\Gamma(u+jv)}\right] \text{ and}$$
$$\frac{1}{\Gamma(u-jv)} = \mathrm{Re}\left[\frac{1}{\Gamma(u+jv)}\right] - j\,\mathrm{Im}\left[\frac{1}{\Gamma(u+jv)}\right].$$

Using this property we obtain:

$$g(t) = t^{u-1}\left\{\mathrm{Re}\left[\frac{1}{\Gamma(u+jv)}\right](t^{-jv} + t^{jv}) + j\,\mathrm{Im}\left[\frac{1}{\Gamma(u+jv)}\right](t^{-jv} - t^{jv})\right\}$$

$$g(t) = 2t^{u-1}\left\{\mathrm{Re}\left[\frac{1}{\Gamma(u+jv)}\right]\cos(v\ln t) + \mathrm{Im}\left[\frac{1}{\Gamma(u+jv)}\right]\sin(v\ln t)\right\}$$

This response $g(t)$ to a unit impulse excitation $f(t) = \delta(t)$ has purely real response. Observing the time response one finds a power function in time with 'log periodic oscillations' superimposed. If one plots the Bode frequency (log-log) graph with $G(s) = 2s^{-u}\cos(v\ln s)$, with $s = j\omega$, and $u, v > 0$ we get a graph depicted in Figure 10.6.

The figure 10.6 is of normal fractional integration process with a downward decreasing slope of $-20u$ dB per decade, but impressed on this is an oscillatory ripple in log frequency $\approx \cos(v\log\omega)$. The normal power function in time is due to the real part of the differintergal and the log-periodic oscillation is due to presence of imaginary part of the 'complex' differintegral process. So one may

$$G(s) = 2s^{-u}\cos(v \ln s)$$

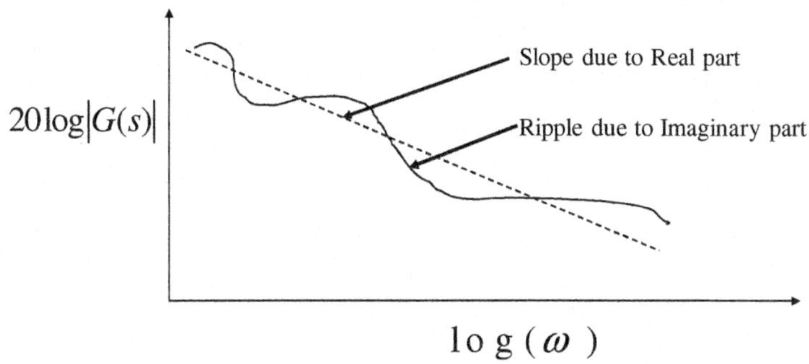

In a frequency response plot of a system, if the roll off, contains ripple of log of frequency then this is indicator of presence of complex differentials/integrals in the system characteristics. Also frequency is same in all scales of log a concept of 'fractal ripple'.

Fig. 10.6 Appearance of Complex Order in System

generalize this observation that if one observes in a frequency response graph of a system a 'fractal ripple' impressed that is, the frequency of the ripple is same in all decades of log scale; the system may be having possibility of complex order in the differential equation system description!

In the section three, equation (10.9) describes the continuous order system as:

$$\left(\int_{-\infty}^{+\infty} k(q)s^q dq\right) X(s) = F(s)$$

We have generalized the limits of integration from the example in (10.9).

For a continuous order system we can therefore extract the $X(s)$ for a suitable forcing function $F(s)$ in following possible way

Let

$$P(s) = \int_{-\infty}^{+\infty} k(q)s^q dq .$$

The $P(s)$ is defined in section three. Now taking the order $q = u + jv$ makes the extraction of $P(s)$ as:

$$P(s) = \int_{-\infty}^{+\infty}\int_{-\infty}^{+\infty} k(u,v)s^{u+jv} dudv$$

That is, carrying out integration in (u,v) plane where the order distribution is a surface given by function $k(u,v)$.

Let us consider a surface $k(u,v)=\mathrm{K}$, where K is constant. This surface is centered at $u=\bar{u}$ with width of $\pm\Delta u$, and $v=0$ with width $\pm\Delta v$. Consider forcing function to be unit delta function so $F(s)=1$. This enables us to determine the response function for this 'continuous complex order' distributed system as $X(s)=F(s)/P(s)=1/P(s)$

$$P(s)=\int_{-\Delta v}^{+\Delta v}\int_{\bar{u}-\Delta u}^{\bar{u}+\Delta u} ks^{u+\mathrm{j}v}\,du\,dv\,.$$

In this make change of variables as $w=u-\bar{u}$ so $du=dw$, with new limits of integration we obtain the response as:

$$P(s)=\int_{-\Delta v}^{+\Delta v}\int_{-\Delta u}^{+\Delta u} \mathrm{K}s^{w+\bar{u}+\mathrm{j}v}\,dw\,dv$$

$$=\mathrm{K}s^{\bar{u}}\int_{-\Delta v}^{+\Delta v}\int_{-\Delta u}^{+\Delta u} s^{w}s^{\mathrm{j}v}\,dw\,dv$$

$$=\mathrm{K}s^{\bar{u}}\left(\int_{-\Delta u}^{+\Delta u}\exp\left[w\ln(s)\right]dw\right)\left(\int_{-\Delta v}^{+\Delta v}\exp\left[\mathrm{j}v\ln(s)\right]dv\right)$$

$$=\mathrm{K}s^{\bar{u}}\left[\left(\frac{\exp(w\ln s)}{\ln s}\right)\right]_{w=-\Delta u}^{w=+\Delta u}\left[\left(\frac{\exp(\mathrm{j}v\ln s)}{\mathrm{j}v\ln s}\right)\right]_{v=-\Delta v}^{v=+\Delta v}$$

$$=4\mathrm{K}s^{\bar{u}}\left[\frac{\sinh(\Delta u\ln s)}{\ln s}\right]\left[\frac{\sin(\Delta v\ln s)}{\ln s}\right]$$

Thus

$$X(s)=s^{-\bar{u}}\frac{(\ln s)^2}{4\mathrm{K}\left[\sinh(\Delta u\ln s)\right]\left[\sin(\Delta v\ln s)\right]}\,.$$

is the response to delta function input.

The Laplace invert of this will give time domain response $x(t)$ for the above. In all above examples we demonstrated obtaining impulse response for a forcing function as delta function that is, Greens function for solution of complex order system. Any other response to any other function can be obtained in usual rule, as to convolute the Greens function obtained with the forcing function (in time domain), or multiplication of Laplace of forcing function with the Laplace of Greens function, in s (frequency) domain.

10.14 Ordering the Disorder of System

The system with disorder can have anomalous temporal or spatial response compared to system without disorder. Thus, the differential equation governing the system in background of disorder has to be identified; here the Fractional Order Differential Equation plays important role in system identification. Therefore suitable techniques need to evolve to order the disorder in system and identify the same. In classical sense the system having no memory can be modeled by linear differential equations with constant coefficients. However, system with memory has inbuilt fractional order differential operator which arise due to convolution in memory integrals. Chapter 2 has briefed about memory integrals and kernels-for non exponential relaxations. Rubber molecules presumably cannot remember past, here linear differential equations with constant coefficients suffice, and such systems have exponential decay in time (with one time constant) to an impulse excitation. Damping behavior of systems if expressed using linear constant coefficients show that linear differential equations cannot include long memory; that fractional derivative has. However, sufficiently high micro structural disorder can lead statistically to macroscopic behavior well approximated by fractional derivatives.

10.14.1 Disordered Relaxation with Multiple States and Relaxation Constants

Consider a partial differential equation (PDE)

$$\frac{\partial}{\partial t}u(\lambda,t)+(\lambda)^{1/\alpha}u(\lambda,t)=\delta(t),$$

with $u(\lambda,t)=0$, for $t<0$, and $\alpha>0$.

The above PDE is having free parameter λ. Now if the free parameter $\alpha=1$, then we have single time constant system ($\lambda=\tau^{-1}$) with solution as $u(\lambda,t)=\exp(-\lambda t)$, with initial condition $u(\lambda,0)=1$. This is similar to capacitor discharge into a resistance as $v(t)=V_0\exp(-t/RC)$, with $\lambda=1/RC$, as unique time constant (discharge rate), remaining the same throughout the discharge period. The situation can be seen as non-exponential decay; where the time constant (discharge rate) is changing and can have infinite number of values. The situation may have capacitor and resistance distributed in space, the capacitance and resistance may be related with voltage and several imperfections (disorder) and may lead to several time constants of discharge during the discharge time. In other terms we may call these disorders as non-linearity resulting in several time constants, and thus 'weak' discharge. The situation may be like $v(t) = V_0 \exp\{-\lambda(t)t\}$ where, $\lambda(t) = 1/R(t)C(t)$. One can express this 'non-exponential' discharge with several time constants during the discharge period, by method of weighted residue and variation principle, expressed as:

$$\begin{bmatrix} \psi_1(t) \\ \psi_2(t) \\ * \\ \psi_N(t) \end{bmatrix} = \begin{bmatrix} a_{11} & a_{12} & * & a_{1N} \\ a_{21} & a_{22} & * & a_{2N} \\ * & * & * & * \\ a_{N1} & a_{N2} & * & a_{NN} \end{bmatrix} \begin{bmatrix} e^{-\lambda_1 t} \\ e^{-\lambda_2 t} \\ * \\ e^{-\lambda_N t} \end{bmatrix}$$

$$[\Psi] = [\psi_1(t) \quad \psi_2(t) \quad \psi_3(t) \quad * \quad * \quad \psi_N(t)], \; v(t) = \|[\Psi]\| = \sqrt{[\Psi]^T [\Psi]}$$

The discharging voltage may be expressed as several states given by matrix $\Psi(t)$ and combination of several decay parameters $\lambda_1, \lambda_2, \ldots \lambda_N$ and the decay voltage by the norm of state matrix. This several state representation points to a fact that decay discharge or mixing taking place in disordered way leads to several internal states, each without memory. The representation

$$\|[\Psi]\| = \sqrt{[\Psi]^T [\Psi]}$$

is a way matrix representation of the autocorrelation function.

10.14.2 Appearance of Fractional Derivative in Disordered Relaxation

The several time constants (discharge rate) have taken power law distribution as $(\lambda)^q$ and $q = 1/\alpha$. The strong-discharge or exponential discharge with one time constant follow a normal distribution with well defined average that represents average time constant or discharge rate, and that normal distribution has well defined standard deviation. Unlike the normal distribution the 'power-law' distribution has no defined average or moments (standard deviation); and is representation of system which has a variety. The heterogeneity or the disordered system thus has varieties of ways by which dissipation mechanism takes place; as described in above section. Taking the PDE of section 14.1 we have solution to that PDE with free variable λ that can take infinite values as $u(\lambda,t) = h(\lambda,t) = \exp\left(-\lambda^{1/\alpha} t\right)$, the $h(\lambda,t)$ denotes 'impulse response function'. On integrating this 'impulse response function' for free variable λ from 0 to ∞, we get the function of time and that is called 'impulse response' $g(t)$. The impulse response is:

$$g(t)=\int_0^\infty h(\lambda,t)d\lambda=\int_0^\infty \exp\left(-\lambda^{1/\alpha}t\right)d\lambda=\frac{\Gamma(1+\alpha)}{t^\alpha}$$

The above is obtained by using definition of Gamma function. This is a power-law. The $g(t)$ 'impulse-response' of linear constant coefficient system starting from rest is also called Green's function.

Let us replace the PDE excitation function from delta function to any other forcing function $\dot{f}(t)=df(t)/dt$ and re-write the PDE as:

$$\frac{\partial}{\partial t}u(\lambda,t)+(\lambda)^{1/\alpha}u(\lambda,t)=\dot{f}(t)$$

Then the response to this new excitation is convolution of Green's function $g(t)$ obtained above with the forcing function that is:

$$r(t)=g(t)*\dot{f}(t)=\int_0^t g(t')\dot{f}(t-t')dt'=\Gamma(1+\alpha)\int_0^t\frac{\dot{f}(t-t')}{t'^\alpha}dt'$$

Multiplying and dividing the above expression with $\Gamma(1-\alpha)$ and assuming $0<\alpha<1$,with $f(t)=0$ for $t\le 0$, and using Riemann-Liouvelli definition of fractional derivative we obtain response as:

$$r(t)=\Gamma(1+\alpha)\Gamma(1-\alpha)\,{}_0D_t^{-(1-\alpha)}\left[\dot{f}(t)\right]=\Gamma(1+\alpha)\Gamma(1-\alpha)\,{}_0D_t^{\alpha}f(t)$$

Implying the appearance of fractional derivative for cases where several time-constants define a relaxation process. Therefore a disordered relaxation (response) may well be formulated by fractional differential equation, the order giving the 'intermittency' of relaxation disordered process!

Many systems materials with complex microscopic dissipative mechanism may macroscopically show fractional order behavior. Damping (relaxation, mixing) models that use fractional order terms may involve relatively few fitted parameters for explaining the dissipative mechanism. The fractional order behavior observed may be artifact of many complex dissipative mechanisms (each without memory).

10.14.3 Generalization of Disordered Relaxation

Fractional Calculus methods have been invoked (recently) to model relaxation processes in complex systems. This has lead to interesting discussions into nature of transport coefficients appropriately changed in equations to describe these complex materials and systems. This observation is leading to a thought to have

'Universality' of Disordered Material (System) Relaxation", and classify them accordingly. The system identification needs to take these observations for disorders in the system.

Relaxation integral I_t^{ϕ} where ϕ characterizes 'degree of intermittency' in the relaxation process-and exact solutions of these integrals describes relaxation in condense matter (system)-'intermittency in relaxation'.

Extension in this above model of relaxation, to include β 'dynamic heterogeneity' arising out from particle clustering (too may be included) which is ubiquitous in condense matter, with $I_t^{\phi,\beta}$ intermittency plus heterogeneity as disorder. The solution to this gives hybrid between stretched exponential relaxation and a relaxation described by Mittag-Leffler function. Both these disorders are 'ordered' and modeled by fractional differential equation; helps in identifying the system.

10.14.3.1 Intermittency Disorder

Here we discuss a general model of relaxation that is developed to address the issue of disorder in the system. First we discuss an intermittent relaxation and generalize the concept. We assume that system is in equilibrium and observable property of the system A (stress strain rate, voltage current charge rate velocity position) is represented by an autocorrelation function (normalized):

$$\Psi(t) \equiv \frac{\langle A(t)A(0)\rangle}{\langle A^2(0)\rangle},$$

obeying relaxation law as:

$$\frac{d}{dt}\Psi(t) = -\int_0^t dt' K(t-t')\Psi(t'),$$

where

$$\Psi(t) = 0 \text{ for } t < 0 \text{ and } \Psi(0) = 1 .$$

The autocorrelation function is

$$R_{AA}(t) = \langle A(t)A(0)\rangle = \lim_{T\to\infty}\frac{1}{T}\int_{-T/2}^{+T/2} A(t)A(t+t')dt' .$$

By Wiener-Khintchine theorem of signal processing, the Fourier Transform of this autocorrelation function gives the power spectral density;

$$S_{AA}(f) = 2\int_{-\infty}^{+\infty} R_{AA}(t)e^{-i2\pi ft'}dt' .$$

The relation between the power law exponent of power spectral density, to Hurst exponent, and to the fractional order governing equation, is dealt with in Chapter 4 and Chapter 5.

This is convolution of evolution of the relaxation process represented by memory integral, (Chapter 2). We now integrate the above described memory integral and re-write the same as:

$$\Psi(t)-\Psi(0)=-\int_0^t dt'\int_0^t dt' K(t-t')\Psi(t'), \quad \text{write} \quad R(t,t')=R(t-t')=\int_0^t dt' K(t-t')$$

to get:

$$\Psi(t)=1-\int_0^t dt' R(t-t')\Psi(t')=1-\int_0^t dt' R(t,t')\Psi(t')$$

The new parameter is rate term of transition of state that is: $R(t)=\int_0^t dt' K(t')$ is integral of memory kernel of the basic memory (convolution) integral of the process.

The $R(t)$ can be considered as 'renewal-rate' which members drop out through death (or otherwise) and new members are added in order to keep the relative number of policy holders a constant. The parameter $\Psi(t)$ is relative number of charter members of an insurance policy. Here we consider $\Psi(t)$ as probability that initial state of dynamic system property $A(0)$ persists that is, survives up to time t $(t>0)$. This renewal rate can be obtained by differential equation which generalizes Poisson's process.

The system begins at state-0 at initial time $t=0$ and will change to state-1 at time T, where T is randomly drawn from $f(x)=\lambda\exp(-\lambda x)$ the Poisson's process. The system will be in state-1 at some time t_1

$$p_1(t_1)=\int_0^{t_1}\lambda\exp(-\lambda t')dt'=1-\exp(-\lambda t_1).$$

The probability of system still being at state-0 at time t_1 is complement of above that is:

$$p_0(t_1)=\exp(-\lambda t_1).$$

Also at any time t the absolute rate of change of probability of being in state-1 is $dp_1/dt=\lambda p_0$, off course, $p_0+p_1=1$, and thus $(\lambda)^{-1}(dp_1/dt)+p_1=1$.

For the first transition from state-0 to state-1 we can write the equation as:

$\frac{dp_0}{dt} = -\lambda p_0$, with λ independent of time, a constant. Now if we assume $R(t) = \lambda$ which is decreasing with time we generalize the first transition decay equation as:

$$\frac{d}{dt}p(t) = -R(t)p(t) + \frac{d}{dt}R(t) \text{ or } R(t) = p(t) + p(t) * R(t) = p(t) + \int_0^t d\tau p(t-\tau)R(\tau)$$

This is 'Feller Fluctuating Theory' where $p(t)$ is the probability density describing the 'first passage' time, between relaxation increment events. The solution to this equation

$$R(t) = p(t) + \int_0^t dt' p(t-t')R(t')$$

gives $R(t)$ of the random process governing the large scale relaxation process.

The occurrence of universality and classification in $\Psi(t)$ and $R(t)$ in this model of relaxation is from the observations, of 'moments' of $p(t)$ that is if they are finite or infinite (diverging).

$$\langle t^n \rangle = \int_0^\infty dt'(t')^n p(t')$$

We will discuss the cases where average and variances are finite and infinite, that is if $\langle t^2 \rangle, \langle t \rangle < \infty$ and $\langle t^2 \rangle, \langle t \rangle \to \infty$. This corresponds to different degree of intensity in the fluctuations governing relaxation process.

10.14.3.2 Strong Intense Relaxation

With finite mean and variance that is $\langle t^2 \rangle < \infty$ relaxation event occurs with well defined average period $\langle t \rangle < \infty$ leading to rapid and strong mixing (relaxation) that is, $\Psi(t) \to 0$ as $t \to \infty$, very fast. This is 'strong relaxation' and is without memory. Say if we take relaxation time as Poisson's process

$$p(t') = (\tau_0)^{-1} \exp(-t'/\tau_0),$$

and then using this in the Rate equation

$$R(t) = p(t) + \int_0^t dt' p(t-t') R(t'),$$

we obtain $R(t) = 1/\tau_0$, which in turn implies from

$$\Psi(t) = 1 - \int_0^t dt' R(t-t') \Psi(t') = 1 - \int_0^t dt' R(t,t') \Psi(t')$$

that $\Psi(t) = \exp(-t/\tau_0)$. More generally the rate $R(t)$ for any $p(t)$ with finite variance and mean has asymptotic relation $R(t) \approx (1/\tau_0) + C/t$. For large times approach a constant rate of relaxation, for constant rate $R(t) = (1/\tau_0)$, the Kernel of Memory-Integral is:

$$K(t) = \frac{d}{dt} R(t) = \frac{1}{\tau_0} \delta(t).$$

Giving relaxation without memory, (discussed in Chapter 2) .Exponential decay (strong mixing) is commonly found in idealized systems. The integral equation and corresponding differential equation without memory in case of strong relaxation is listed as obtained from convolution form of memory integral (Chapter 2).

$$\Psi(t) = 1 - (\tau_0)^{-1} I_t^1 \Psi(t) \text{ and } D_t^1 \Psi(t) = -(\tau_0)^{-1} \Psi(t).$$

10.14.3.3 Weak Intermittent Relaxation

In rate equation if we take power law decaying rate as $R(t) \approx C t^{\phi-1}$, with $0 < \phi < 1$, it indicates that intensity of relaxation process goes down with time. Here the relaxation process is non-exponential in nature, with a power law kernel of memory integral (as described in Chapter 2). The exponent ϕ indicates time points at which relaxation occurs as a fractal dimension and can also be termed as degree of intermittency. As this degree of intermittency approaches unity we get stronger relaxation with constant rate.

10.14.3.4 Oscillating Relaxation

One case of oscillatory relaxation was discussed in non-Newtonian fluid compression experiment in Chapter 9. In the case where rate is given as $R(t) \approx t^{\phi-1}$, with $1 < \phi \le 2$ that is rate increases with time, the relaxation is oscillatory. For the case $\phi \to 2$ the rate linearly grows with time as $R(t) \approx (\tau_0)^{-1} t$

and the kernel of the memory integral approaches constant value, as $K(t) \approx (\tau_0)^{-1}$ giving relaxation (mixing) as oscillatory function (refer Chapter 2); that is: $\Psi(t) \approx \cos(t/\tau_0)$. For the oscillatory case

$$\int_0^\infty \Psi(t')dt' = 0,$$

that is the associated transport properties either vanish or diverge. This situation is significant for systems driven far from equilibrium condition. This relaxation of oscillatory nature is observed in very small time scales. For example in case of constant rate where asymptotically constant value is reached; the initial part of the relaxation will show oscillations and then the asymptotic fast exponential decay.

The integral equation and corresponding differential equation with memory in case of weak intermittent relaxation is listed as obtained from convolution form of memory integral (Chapter 2).

The memory kernel is:

$$K(t) = K_0 t^{\phi-2}; 0 < \phi \le 2,$$

which gives following from memory integral as:

$$\frac{d}{dt}\Psi(t) = -\frac{1}{\tau^\phi}\, {}_0D_t^{1-\phi}\Psi(t),$$

with

$$\tau^\phi = \left[K_0\Gamma(\phi-1)\right]^{-1}.$$

Apply integration on both sides to get the integral equation:

$$\Psi(t) - \Psi_0 = -\tau^{-\phi}\, {}_0D_t^{-\phi}\Psi(t) = -\tau^{-\phi} I_t^\phi \Psi(t)$$

Apply fractional differentiation to both sides to get:

$${}_0D_t^\phi\Psi(t) - \Psi_0 \frac{t^{-\phi}}{\Gamma(1-\phi)} = -\tau^{-\phi}\Psi(t)$$

10.14.3.5 Generalized Dynamic Critical Index of Relaxation with Intermittency

All the above discussed relaxations can be generalized as:

$R_\phi(t) \approx \frac{\Omega_0}{\Gamma(\phi)} t^{\phi-1}$, for $0 < \phi \le 1$ with Ω_0 as coupling constant governing relaxation rate intensity and the gamma function $\Gamma(\phi)$ is used as normalizing factor. We can verify that: $\lim_{\phi\to 1} R_\phi(t) = \Omega_0$, giving constant rate and thus strong relaxation.

With this definition we have relaxation equation:

$$\Psi(t) = 1 - \int_0^t dt' R(t,t')\Psi(t') = 1 - \int_0^t dt' \frac{\Omega_0}{\Gamma(\phi)}(t-t')^{\phi-1}\Psi(t'),$$

inserting Riemann-Liouvelli definition we obtain: $\Psi(t) = 1 - \Omega_0 I_t^\phi \Psi(t)$, where ϕ denoting intermittency as described earlier is termed as ' dynamic critical index' characterizing the relaxation process. The solution to this generalized relaxation equation is Mittag-Leffler function $\Psi(t) = E_\phi(-\Delta_\tau)$ with $\Delta_\tau = (t/\tau_0)^\phi$ and $\tau_0 = (\Omega_0)^{-1/\phi}$. The asymptotic expansion of this Mittag-Leffler function gives asymptotic decay as power-law with scale-invariance; that is: $\Psi(t,\phi) \approx (t/\tau_0)^{-\phi}$; for late times. This, property of Mittag-Leffler function that is monotonic decay is characteristic of large scale relaxation decay functions, which is often taken for granted by experimentalists. Also this Mittag-Leffler function approximates, at early times a stretched exponential,

$$E_\alpha\left(-[t/\tau]^\alpha\right) \sim \exp\left(-[t/\tau]^\alpha / \Gamma(1+\alpha)\right)$$

and at late times approximates as

$$E_\alpha(-[t/\tau]^\alpha) \sim \left([t/\tau]^\alpha \Gamma(1-\alpha)\right)^{-1}.$$

The Figure 10.7 gives relaxation (mixing) curves for different dynamic critical index values. As the value tends to unity the Mittag-Leffler curve approaches exponential decay. The curve in Figure 10.7 assumes homogeneous system without clustering disorder $\beta = 0$.

In experiments it is also observed that by changing ambient (temperature) changing the magnitude of coupling constant Ω_0, the functional form of the relaxation function remains invariant provided the dynamic critical index ϕ does not change in the temperature interval. This time-temperature invariant property is due to scale invariant memory kernel of the relaxation integral. The rate and its corresponding memory kernel, and its scale invariance are represented as follows:

$$K(t;\phi) = \frac{dR(t)}{dt} \approx \frac{\Omega_0}{\Gamma(\phi-1)} t^{\phi-2}, \quad 0 < \phi \le 1$$

$$K(\chi t;\phi) = \chi^{\phi-2} K(t;\phi)$$

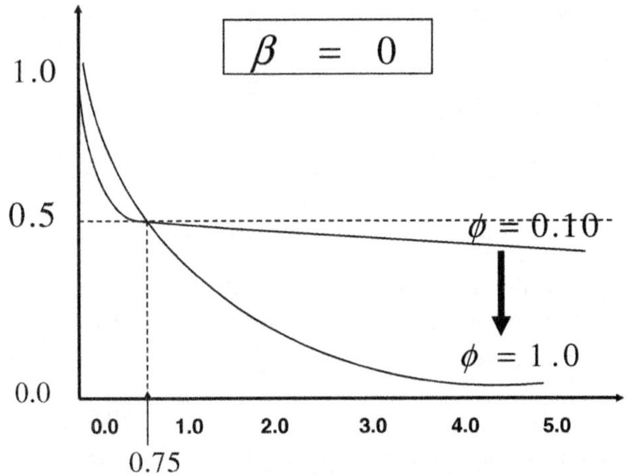

Fig. 10.7 Relaxation curves for different dynamic critical index.

10.14.3.6 Spatial Disorder

Previous section discussed the issue of disordered relaxation (mixing) for system without heterogeneity in this system (materials), which develops transiently through inter-particle interactions. As an example, cooled liquids develop large scale heterogeneity, which has significant influence on their properties. Material disorder can lead to, 'stretched exponential' relaxation' that is:

$$\Psi(t) = \exp\left[-\Omega_0 t^{1-\beta} / (1-\beta)\right]$$

This can be viewed as 'locally strong relaxation (mixing)' with β as spatial dimension of the 'clustering' heterogeneity involved. The distribution of the cluster size is assumed to be governed by Boltzmann's distribution. The experiments on stress relaxation in condense matter suggest the values of β as 2/3 for linear disorder, ½ for sheet type disorder and 2/5 for clumps type disorder. With $\beta = 0$, the relaxation is 'strong relaxation' with expression as $\Psi(t) = \exp(-\Omega_0 t)$, with memory integral kernel is:

$K(t) = \Omega_0 \delta(t)$. For the stretched exponential relaxation process the memory integral kernel is $K(t-\tau) = \Omega_0 \tau^{-\beta} \delta(t-\tau)$. To get the differential equation for this stretched exponential law, write the basic memory integral in convolution form:

$$\frac{d\Psi(t)}{dt} = -\int_0^t dt' K(t-t')\Psi(t'),$$

with $\Psi(0) = 1$, and put the memory integral to get:

$$\frac{d\Psi(t)}{dt} = -\int_0^t dt'\Omega_0 (t')^{-\beta} \delta(t-t')\Psi(t') = -\Omega_0 \int_0^t dt'\delta(t-t')\left\{\Psi(t')(t')^{-\beta}\right\}.$$

Delta function integration gives

$$\frac{d}{dt}\Psi(t) = -\Omega_0 t^{-\beta}\Psi(t)$$

The graph of stretched exponential is depicted in Figure 10.8. In this figure the dynamic critical index is assumed to be unity.

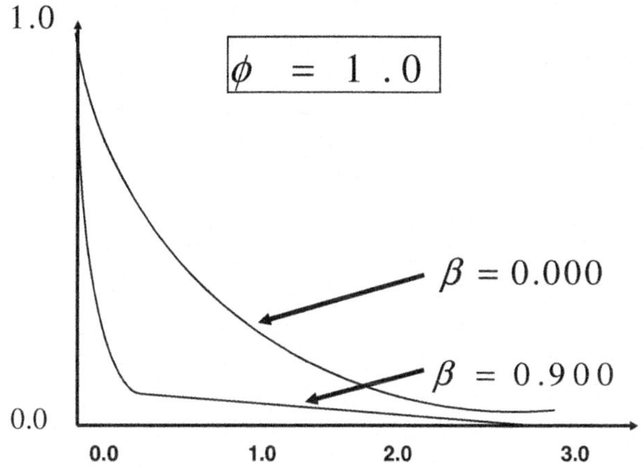

Fig. 10.8 Stretched exponential relaxation process

Kohrash-William's Watts' law is widely used to correlate relaxation data in complex liquids.

10.14.3.7 Hybrid Disorder with Intermittency and Spatial Heterogeneity

With both the types of disorder present in the system we compose a hybrid of Mittag-Leffler with stretched exponential relaxation function. The rate term is expressed as:

$$R_{\phi,\beta}(t,\tau) = \Omega_0 \tau^{-\beta}(t-\tau)^{\phi-1}/\Gamma(\phi), \ \phi > 0; \beta \le 1$$

The parameter β signifies measure of spatial geometrical clustering disorder-heterogeneity whereas ϕ is measure of intermittency of relaxation intensity the temporal heterogeneity. With these two parameters the relaxation equation is:

$$\Psi(t;\phi,\beta) = 1 - \int_0^t d\tau R_{\phi,\beta}(t,\tau)\Psi(\tau;\phi,\beta)$$

$$\Psi(t;\phi,\beta) = 1 - \Omega_0 \int_0^t d\tau \left[(t-\tau)^{\phi-1}\tau^{-\beta} / \Gamma(\phi)\right]\Psi(\tau;\phi,\beta)$$

The integral expression is Erdelyl-Kober fractional operator and is expressed as:

$$I_t^{p,q} f(t) = t^{-(p+q)} \int_0^t d\tau \left[(t-\tau)^{q-1}\tau^{p} / \Gamma(q)\right] f(\tau).$$

This is related to Riemann-Liouvelli fractional integral operator as:

$$I_t^{p,q} f(t) = t^{-(p+q)} I_t^q \left[t^p f(t)\right].$$

The solution is listed as follows:

$$\Psi(t;\phi,\beta) = \sum_{k=0}^{\infty} a_k(\phi,\beta)\left[z_\Omega(\phi,\beta)\right]^k, \; a_0(\phi,\beta) = 1$$

$$a_k(\phi,\beta) = \prod_{m=1}^{k} \frac{\Gamma(1+m\hat{\phi}-\phi)}{\Gamma(1+m\hat{\phi})}, \; k>0, \; \hat{\phi} = \phi - \beta, \; 0 < \hat{\phi},\phi;\beta \le 1$$

$$z_\Omega(\phi,\beta) = \Omega_0 \tau^{\hat{\phi}} = \left(t/\tau^*\right)^{\hat{\phi}}, \; \tau^*(\phi,\beta) = \Omega_0^{-1/\hat{\phi}}$$

The function $\Psi(t;\phi,\beta)$ is Mittag-Leffler for $\beta = 0$ and is stretched exponential for $\phi = 1$. Figure 10.9 demonstrates hybrid relaxation with clustering heterogeneity as well as intermittency.

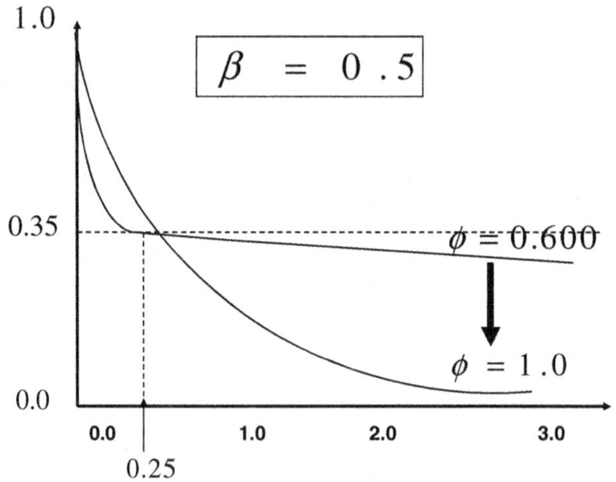

Fig. 10.9 Hybrid of relaxation with Mittag-Leffler and Stretched Exponential

10.15 Identification of Fractional Stochastic Processes

The systems of stochastic nature like in biophysics, physiology, economics and many others give identification with long term probability densities that extend far beyond typical tail of region of Gaussian distribution. One way in terms of Fractional Order Signal Processing method, these processes have been classified as $1/f$ phenomena, since their time series have spectrum that are inverse square law of frequency or their probability are inverse power law. In either case the underlying structure is fractal in nature. For example the differential return plot of stock market price index looks similar at several time scales (two hourly, weekly or monthly). Modern economics relies heavily on these statistics. Most of the statistics are based on Gaussian distribution, these kinds of data occurs in financial data. For example the one day return on stock does closely conform to a Gaussian curve. In this case the return of yesterday has nothing to do with return of today. However, as the return period moves to say weekly period or fortnight period returns the distribution changes to long memory with lingering tail. Here 'tails' of the curve follow a power-law. These longer 'time-series' of data have some amount of autocorrelation and a non-random Hurst exponent ($H \neq 0.5$). The process has some memory of past event, which is forgotten as time moves forward. For example, larger trades in market will move the market price up or down. A large purchase order will increase the price and a large sellout will bring the stock price down; these are called market impact. When an order has measurable market impact, the prices does not rebound to previous price after the order is filled. The market acts as if it has some memory of what took place and the effect of this impact takes time to decay; well market with memory! Similar long ranged distribution with memory is observed in network delays, human gait time pattern, mortality rate for say epidemic spreads and noise in reactor systems and several others. The idea is to identify and characterize these stochastic systems.

10.15.1 Fitting Stochastic Data into Parameters of Levy Stable Distribution

As discussed in various sections of the book, Levy stable distributions are generalization of Gaussian, and the index of the Levy distribution gives the idea of long and lifted tail and it's Long Range Dependence. It was pointed out earlier that Levy stable family of statistical distribution, with Gaussian being one of them. It's explicit pdf generally does not exist, instead is given by a characteristic function as:

$$\varphi(\theta) = \exp\left\{-C^{\alpha}\left|\theta\right|^{\alpha}(1-\mathrm{i}B\operatorname{sign}(\theta)\tan[\alpha\pi/2]+\mathrm{i}\mu\theta\right\},$$

with α as Levy index, B as skewness, C is scale parameter where, $0 < C < \infty$; corresponds to standard deviation of Gaussian distribution, μ is shift parameter.

The function $\text{sign}(\theta)$ takes value -1, for $\theta < 0$, zero for $\theta = 0$ and +1 for $\theta > 0$. Parameter C is related to Hurst exponent (H) as $C = C_0 l^H$, where l is sampling interval of the data field.

A standard symmetric distribution with zero mean is obtained when $B = 0$, the characteristic function becomes, $\varphi(\theta) = \exp(-C^\alpha |\theta|^\alpha)$. The pdf of this is given as,

$$f(x) = \frac{1}{\pi} \int_0^\infty dk \exp(-|Ck|^\alpha \cos(kx),$$

where k is frequency or 'wave-number'.

The sample-quantile based estimation technique of Famma & Roll (1972) helps to identify the parameters is as follows:

Choose a constant interval of length l, the sampling interval or its multiple, to calculate the 'increments' of the physical quantity; and then plot the cumulative distribution function (CDF).

Calculate the quantile range $Q_f - Q_{1-f}$ from the obtained CDF, where f (fractile) is truncated part of the CDF and f can be as small as 0.72. Quantile Q_f can be interpolated from the CDF. The scale parameter C is calculated as:

$$C = \frac{Q_{0.72} - Q_{0.28}}{1.654}$$

Then define

$$A = \frac{Q_{0.95} - Q_{0.05}}{Q_{0.75} - Q_{0.25}},$$

with this calculate Levy index as:

$$\alpha = 0.02A^4 - 0.36A^3 + 2.49A^2 - 7.77A + 10.52.$$

This equation is valid for $1 < \alpha < 2$ and $2.327 < A < 6.426$.

Estimate Hurst index H by $C = C_0 |l|^H$, or $\log C = \log C_0 + H \log|l|$. The Hurst exponent is the slope of $\log C$ versus $\log|l|$. Given α and H the data can be thus described by fractional Levy motion. For $(1/\alpha) < H < 1$, the data is persistent. For $0 < H < (1/\alpha)$ the data is anti-persistent; and for, $H = 1/\alpha$, the data is pure 'white-noise'.

Well quantile are points taken at regular interval for the CDF. Dividing ordered data into q essential equal sized data subsets. The $k-\text{th}-q$ quantile for a random variable is value x-such that x is at most k/q and the probability the random variable will be more than x is at most $(q-k)/q$. $Q_{0.75}$, of a random variable X is such a value x such that $\Pr(X \le x) = 0.75$.

10.15.2 Estimation of Hurst Index by Rescaled Range (R/S method) for Stochastic Data

Estimation of the Hurst index is demonstrated in earlier section, along with the important Levy distribution parameter. The relation of Hurst exponent to average local Holder exponents, the decay fractional order of the power spectral density, and the relation with fractional differencing parameters and the fractal dimensions were already discussed earlier in this book. Let us recall the Hurst exponent it not only identifies the LRD degree, but also gives the degree of roughness. A smaller Hurst exponent gives higher fractal dimension, and indicating rougher surface. The classical estimation of Hurst parameter is explained here in this section.

From the average of all the values of random time series we calculate first for a period τ. Say τ represents yearly period. Call it average mean of the time series for a period τ as $\langle x \rangle_\tau \equiv \frac{1}{\tau}\sum_{i=1}^{\tau} x(t_i)$. Say for any period (year) u, $x(u)$ represents the total value obtained for $x(t_i)$; then deviation from mean for that period is $x(u) - \langle x \rangle_\tau$. Note that $\langle x \rangle_\tau$ is mean calculated over same multi-τ (yearly) period.

Let $X(t,\tau) = \sum_{u=1}^{\tau}\left\{x(u) - \langle x \rangle_\tau\right\}$ represent running sum of accumulated deviation from mean for period (year) $u = 1$ to $u = \tau$. The following algorithm will clarify the meaning of this $X(t,\tau)$.

$$\text{sum} = 0;$$
$$\text{for}(t = 0; t < \tau; t++)\{$$
$$\text{sum} = \text{sum} + x(t);$$
$$\}$$
$$\text{mean} = \frac{\text{sum}}{\tau}$$
$$X(0) = 0;$$
$$for(t = 1; t \le \tau; t++)\{$$
$$X(t) = X(t-1) + \left(x(t-1) - \text{mean}\right);$$
$$\}$$

Then range $\mathbf{R}(\tau)$ is defined as $\mathbf{R}(\tau) = \max\left[X(t,\tau)\right] - \min\left[X(t,\tau)\right]$ for $1 < t < \tau$. Dividing this by the standard deviation 'Rescaled Range' is obtained as

$$\mathbf{R}/\mathbf{S} = \mathbf{R}(\tau) \div \mathbf{S}(\tau) \text{ where, } \mathbf{S}(\tau) = \sqrt{\frac{1}{\tau}\sum_{t=1}^{\tau}\left\{x(t) - \langle x \rangle_{\tau}\right\}^2} .$$

The Hurst exponent is calculated then by average Rescaled Range over multiple regions in data, that is $\langle \mathbf{R}/\mathbf{S} \rangle = Cn^H$, as $n \to \infty$, where n is region size.

The Hurst exponent applies to data that are statistically self similar, meaning that the properties of the entire data set are same for the subsection of the data set. For example the two halves of the data set have the same statistical properties as the entire data set. Calculate $\mathbf{R}/\mathbf{S}$ for the entire data set. Call it $\mathbf{R}/\mathbf{S}_0$, and then make this first calculation as $\mathbf{R}/\mathbf{S}_0 \to \mathbf{R}/\mathbf{S}_{\text{SAVE-0}}$. Then, $\mathbf{R}/\mathbf{S}$ is calculated for the two halves $\mathbf{R}/\mathbf{S}_0$ and $\mathbf{R}/\mathbf{S}_1$. These two $\mathbf{R}/\mathbf{S}_0$ and $\mathbf{R}/\mathbf{S}_1$ are averaged and saved as $\mathbf{R}/\mathbf{S}_{\text{SAVE-1}}$. In this case the process continues by dividing each of the previous sections by half and calculating Rescaled Range for each new section. The Rescaled Range values for each new section are averaged. At some point this subdivision stops; since the region is getting smaller and smaller. Usually the smallest region will have eight data points. Let us take the example of 1024 data points of a time series of fractional stochastic process. The division of these data points would give region-size $\{n_i\}$ as $\{1024, 512, 256, 128, 64, 32, 16, 8\}$ for these let the calculated $\mathbf{R}/\mathbf{S}_{\text{Average}}$, for individual n be as a set:

$$\{\mathbf{R}/\mathbf{S}_{\text{Average-}i}\} = \{96.4451, 55.7367, 30.2581, 20.9820, 12.6513, 7.2883, 4.4608, 2.7399\}$$

Let $\mathrm{X}_i = \log_2 n_i$ thus we have $\{\mathrm{X}_i\} = \{10, 9, 8, 7, 6, 5, 4, 3\}$ and

$$\mathrm{Y}_i = \log_2 \mathbf{R}/\mathbf{S}_{\text{Average}-i}, \text{ to get}$$

$$\{\mathrm{Y}_i\} = \{6.5916, 5.8006, 4.9193, 4.3911, 3.6612, 2.8656, 2.1573, 1.4541\} .$$

A linear regression on $\{\mathrm{X}_i\}$ and $\{\mathrm{Y}_i\}$ fits a straight line as $\mathrm{Y} = m\mathrm{X} + c$, with $m = 0.7270$ as slope, and $c = 0.7455$ as Y-intercept. The estimated slope gives $H \cong 0.73$.

The above mentioned method uses non-overlapping data regions, where size of data set is in power of two. This is simplification, the overlapping region with data set size as other power is too possible.

10.16 The Concept of System Order and Disadvantage of Fractional Order System

As the concept of order is central to the understanding of fractional (or integer) order systems, some discussion of this concept now follows. In the discussion, mostly single-input-output systems are considered. The examples in preceding chapters for heat flow and transmission line (lossy and lossless), dynamics of chain network, normal Fickian diffusion and many other processes, gave the stage for half order system or zero order system. Thereby we had generalized the observed half-order of differintegrations to arbitrary 'fractional' order, relating sometimes to 'fractal-lattice', non-Euclidian geometry, weakly-intermittent relaxation, or disordered background. Recalling the characteristic equations or transfer function we call a system first order, second order third order etc. similarly the system can be of fractional order too. We also consider that system representation is minimal and they are linear.

Mathematical order is defined as highest derivative occurring in a given differential equation. The concept of mathematical order is applicable to both ordinary and fractional differential equations. Normally, when the order is used without qualifier, it implies the meaning of mathematical order.

For linear dynamic systems that are described by ordinary differential equations the system mathematical order implies or is equivalent to the following:

(1). The highest derivative in ordinary differential equation.
(2). The highest power of Laplace variable s, in the characteristic equation.
(3). The number of initializing constants required for the differential equation.
(4). The length of the state vector.
(5) The number of singularities in the characteristic equation.
(6). The number of energy storage elements.
(7). The number of independent spatial directions in which a trajectory can move.
(8). The number of devices that can add 90^0 sinusoidal steady state phase lag.
(9). The number of devises that retain some memory of the past.

The utility of the definition of mathematical order is that infers all the system characteristics for system, with integer order components.

Thus the benefit of having a definition for order for linear ordinary differential equations is that it allows a direct understanding of the behavior of given dynamic system. Unfortunately, for fractional differential equations, the order of the highest derivative does not infer all of the previously mentioned properties. Indeed, the most important characteristics of order in integer order ordinary differential equation is probably item (3) i.e. it indicates number of initializing constants, which together with the differential equations allow prediction of the future behavior. In system terminology this information provides initial states, of the system being analyzed. Clearly the order of highest derivative in a fractional differential equations does not have this property nor does it predict the associated number of energy/memory elements associated with fractional differential equation, nor does it infer the number of integrations (even fractional), required to solve simulate the given fractional differential equation. Thus the issue of order and the information required together with the fractional differential equation to predict the future is fundamental and should be treated differently.

In the examples of preceding chapters it has been demonstrated that when specific differintegral operators (q=1/2 & 0) are considered as semi-infinite systems, a time-dependent term resulting from the initial spatial condition should be added to the forced response. This is an important observation for solution of fractional differential equations that can have rather arbitrary initial conditions.

Here we points out to some of the disadvantage specially occurs in computational efforts, to realize fractional order systems. This area of science is evolving field presently disadvantages are pointed in order to make distinction between classical integer order calculus. The points are summarized below:

a) Fractional order differential equations accumulate the entire information of the function in weighted form.
b) Fractionally differentiated state variables must be known as long as system has been operated. This is known as initialization function.
c) For integer order systems, the initialization function is constant and for fractional order systems it is time varying.
d) Integer order system set of state along with system equations is sufficient to predict the response.
e) The fractional dynamic variables do not represent the state of the system.
f) Fractional dynamics require history of states or sufficient number of points by short memory principle, for initialization function computation.
g) The above memory effect requires large memory. The evolving developments to reduce this requirement in form of power series expansion and continued fraction expansion of the Generating function in digital domain is on going process.
h) Diverging definitions of fractional derivatives make a confusing start as what to do with fractional initial states, or any conversion needs to be done to have fractional differential equation solution irrespective of diverging definitions?

10.17 Concluding Comments

A very advanced topic is touched upon as on today, about the reality of an order of a system. Well, can there be total certainty about the fixed integer order definition of a system, or experimental determination of characteristic system equation, forces one to think, that order of a system be distributed in the neighborhood of some integer or real numbers or even complex numbers. If the physical quantities, were absolutely a point property then having fixed integer order be of reality, but having seen that all physical quantities are distributed property well, and then having a distributed order should therefore be a reality. Well physical processes with disordered background too have anomalous behavior which requires non-integer order differential equations, has been focused along with identification of order of irregular stochastic process and its memory relation is deliberated. When order of a system has distribution, then having controller with distributed order may be wise to have for efficient governance. Also ambient can change the order of the system and future development in variable fractional order system mathematics is enrich area of research.

Chapter 11
Solution of Generalized Differential Equation Systems

11.1 Introduction

Mathematical modeling of many engineering and physics problems leads to extraordinary differential equations (Non-linear, Delayed and Fractional Order). We call them Generalized Dynamic System. An effective method is required to analyze the mathematical model which provides solutions conforming to physical reality. For instance a Fractional Differential Equation (FDE), where the leading differential operator is Reiman-Liouvelli (RL) type requires fractional order initial states which are sometimes hard to physically relate. Therefore, we must be able to solve these dynamic systems, in space, time, frequency, area, and volume, with physical reality conserved. The usual procedures, like Runga-Kutta, Grunwarld-Letnikov Discretization with short memory principle etc, necessarily change the actual problems in essential ways in order to make it mathematically tractable by conventional methods. Unfortunately, these changes necessarily change the solution; therefore, they can deviate, sometimes seriously, from the actual physical behavior. The avoidance of these limitations so that physically correct solutions can be obtained would add in an important way to our insight into natural behavior of physical systems and would offer a potential for advances in science and technology. Adomian Decomposition Method (ADM) is applied here in this by physical process description; where a process reacts to external forcing function. This reactions-chain generates internal modes from zero mode reaction to first mode, second mode and to infinite modes; instantaneously in parallel time or space-scales; at the origin and the sum of all these modes gives entire system reaction. By this approach formulation of Fractional Differential Equation (FDE) by RL method it is found that there is no need to worry about the fractional initial states; instead one can use integer order initial states (the conventional ones) to arrive at solution of FDE. ADM method was first explored by mathematicians Prof Rasajit Bera and Prof S Saharay, for obtaining solutions to Fractional Order Differential equations.

11.2 Generalized Dynamic System and Evolution of Its Solution by Principle of Action Reaction

General physics law states that a system will react to external stimulus and will have opposition to the changes; and the process is described by system dynamic equations.

Let there be general differential equation system described as (11.1).

$$\begin{aligned} &D_x^m u(x) + a_1 D_x^{m-1} u(x) + a_2 D_x^{m-2} u(x) + ... + a_{m-1} D_x^1 u(x) + +a_m u(x) + \\ &+b_k [u(x)]^k + b_{k-1} [u(x)]^{k-1} + ...b_0 u(x) = G(x) \end{aligned} \tag{11.1}$$

We can decompose this as linear part (11.2).

$$\mathrm{L_O} u(x) = G(x) - R(u) - N(u) \tag{11.2}$$

The operator $\mathrm{L_O}$ represents a linear operator representing the highest orders of change in the process parameter. $\mathrm{L_O}(u) = d^m u(x) / dx^m = D_x^m u(x)$. This is easily invertible. This order of change (m) could be one, two or any positive integer or even fractional (say half, one-fourth, one and one-fourth). This order of change could be with respect to time, space, space square (area) frequency or time-square-depending on the process description where,

$$R(u) = a_1 D_x^{m-1} u(x) + a_2 D_x^{m-2} u(x) + ... + a_{m-1} D_x^1 u(x) \tag{11.3}$$

is the remainder differential operator of order less than m. This R could be of integer or fractional order. The rest of the terms are put as (11.4) contains nonlinear as well as the linear terms; and assume this (11.4) as analytic function N.

$$N(u) = a_m u(x) + b_k [u(x)]^k + b_{k-1} [u(x)]^{k-1} + ..b_0 u(x) \tag{11.4}$$

The $G(x)$ is sum of all external stimulus source/sink.

General physics law states that a system will react to external stimulus and will have opposition to the changes; by the system reaction terms defined by R and N defined in the system description equation (11.1). These R and N generates internal stimulus when excited by external source/sink as to oppose the cause. The reactions are causal in nature. If the external stimulus and the internal reactions to the stimulus get balanced then the process parameter remains static without any growth (accumulation) or decay (loss). Else, the process parameter will have a solution as infinite (or finite) decomposed modes; generated by system itself to oppose stimulus generated internally by previous modes.

Adomian Decomposition Method (ADM) Computational method yields analytical solution; has certain advantages over standard numerical techniques. However, the ADM was discovered in mid-late eighties and utilized to tackle non-linear problems of physics ADM is free from rounding off errors as it does not involve discretization and does not require large computer memory. ADM is qualitative rather than quantitative, analytic, requiring neither linearization nor perturbation and continues with no resort to discretization. ADM splitting gives equation into linear (L_0) that is, containing highest order of change, the remainder part (*R*) that has change rates less than (L_0) and then the non-linear (*N*) part. Thereafter, inverting the "highest order" derivative (L_0) in the linear operator on both sides of the differential equation is the first step. Second is to identify the initial/boundary conditions and terms involving the independent variables alone; as initial approximation. Decomposing the unknown functions (*N*) into series whose components are to be determined by ADM. The decomposed parts of ADM method are related physically to system reactions of various modes from zeroth mode to infinity mode. The sum of all these modes is the solution of Differential Equation (Non-Linear, Linear Integer Order or Fractional Order). Physically the zeroth mode reaction comes from external stimulus plus the initial integer order states; which instantly generates the internal stimuli of infinite modes-to oppose this first action (change) in opposite way, in time or space (at the origin). Exactly the ADM in mathematics generates these infinite modes reactions; and therefore ADM is close to physical reality. The ADM helps to physically visualize the reaction of system by decomposing the total gross reaction into all these infinite modes. If the differential equation system with (L_0) is of Riemann-Liouvelli type fractional operator, then classically one needs the initial states as fractional order like $u^{\alpha}(0), u^{\alpha-1}(0)$ etc; where α is not an integer. These states are hard to visualize physically. With this ADM the RL formulation does not need these fractional initial states instead requires $u(0), \dot{u}(0)$, the integer order states give the solution and thus physically realizable easily. This new finding too is highlighted in this chapter along with several other problems to give physical insight to the solution of extraordinary differential equation systems. This way one gets insight to Physics of General Differential Equation Systems and its solution by Physical Principle and equivalent mathematical decomposition method. This facilitates ease in modeling systems close to physical reality. The system transfer function for analysis is delta function excitation (forcing function). This gives solution to homogeneous set of system of differential equations, called Green's function. Response to any other type of forcing function (say Heaviside's Step, Ramp or Sinusoidal function) is obtained by convoluting this delta function's response with (other) excitation function. Therefore most of the examples are discussed with delta function as forcing function; however the solution to any other type of forcing function is similar. One example is thus solved for mass spring damper with (half) fractional order element with Heaviside's step as input to demonstrate this methods utility.

11.3 Physical Reasoning to Solve First Order System and Its Mode Decomposition

The application of ADM to simple ordinary differential equation will give insight into the action reaction theory of physics. Thus the ADM will try to explain the physical behavior too. It will be therefore demonstrated that ADM is actually translating the physics of the process where any change is opposed by the system itself. Consider the first order differential equation, $\dot{x}(t)+ax(t)=f(t)$ or in terms of (11.1) as:

$$D_t^1 x(t)+ax(t)=f(t) \tag{11.5}$$

With initial condition and forcing function (source) as $x(0)=0, f(t)=K\delta(t)$.

Practically, let $x(t)$ be instantaneous current of RC circuit connected to battery, by a switch. The voltage excitation is a step function $V_{BB}H(t)$.

Where $H(t)=0; t<0$ and $H(t)=1; t\geq 0$ a Heaviside's step function. The circuit equation is

$$\frac{1}{C}\int_{0^+}^{t} i(t)dt+Ri(t)=V_{BB}H(t) \text{ or } \frac{1}{RC}\int_{0^+}^{t} i(t)dt+i(t)=\frac{V_{BB}H(t)}{R} \tag{11.6}$$

Differentiating (11.6), we get:

$$\frac{d}{dt}i(t)+\frac{1}{RC}i(t)=\frac{V_{BB}}{R}\delta(t) \tag{11.7}$$

The (11.6) is a voltage equation and (11.7) is current equation, re-written with compliance with (11.5). This basic equation like (11.5) and (11.7) gives rate of change of current (function) as related to external stimulus. The current excitation is impulse excitation in (11.7). Have the initial current in the system be zero $i(0)=0$. This system has characteristic time constant RC seconds; meaning that current in system changes e times in RC seconds. We shall consider response at larger time scales than RC. In (11.5) this characteristic time constant is $(1/a)$. Here time scales and concept of time-constant is mentioned. This could be length scale, frequency scale time-square scale area volume or any other scales depending on the units of a in (11.5).

At initial time zero, the switch closer instance gives impulse excitation of current, and assuming if the capacitance of the circuit were absent at this initial instance (capacitance comes into action at later time), then the current in the resistance is,

$$i(t)=\frac{V_{BB}}{R}\int_{0}^{t}\delta(t)dt=\frac{V_{BB}}{R}D_t^{-1}\delta(t)={}^{V_{BB}}\!/_{R} \text{ a constant.}$$

Here a point is mentioned that integration of the forcing function comes because the inertial element capacitance is present in the circuit and equation (11.7). If the capacitor is completely absent then the current reaction will be simple $(V_{BB}/R)\delta(t)$;

meaning that the output current would have vanished instantaneously with the impulse voltage input and the current would be then zero i.e. $i(t) = 0$ at $t \geq 0$. This initial moment current in resistor is V_{BB}/R at $t = 0$; since initially uncharged capacitor acts as short circuit impedance. The capacitor presence is making the current linger for time greater than zero. The circuit as natural reaction to any force will oppose this flow of current that is the change in current from zero to V_{BB}/R -is the capacitor action.

Therefore, the first (and the foremost) reaction comes from the resistive element (without lag or lead) that is, $i_0 = i(0) + D_t^{-1}[V_{BB}\delta(t)/R] = V_{BB}/R$, in the absence of the capacitive element (initially short-circuited); and this is the first reaction due to external force (and initial current if at all be present in the circuit). This sudden change in charges (Coulombs) cannot flow into capacitor, as the voltage across it cannot change instantaneously. This constant action of current gives rise to a rate of change of current in the system (per unit time constant) and is (i_0/RC) A/s. The first reactionary constant current thus is opposed by internal generated current as $i_1 = -D_t^{-1}[i_0/RC]$, which is in opposition to this first reactionary current initial reaction, therefore negative. This action reaction summed up to give $i(t) = i_0 + i_i$, the total current. The internally generated reactionary current gives a rate of change as (i_1/RC) A/s; which will generate opposition current to the cause i_1; as $i_2 = -D_t^{-1}[i_1/RC]$, which again is added to give total reaction as $i(t) = i_0 + i_1 + i_2$, as the total current. This way infinite set of stimulus currents are generated as chain reaction giving the total current as $i(t) = i_0 + \sum_{n=1}^{\infty} i_n$; where i_0 is the reaction due to external stimulus (and initial current if present in the circuit) and rest are internally generated modes; acting in opposition to the rate of change in current.

The reaction $i(t)$, for (11.7) can therefore be written as:

$$i(t) = [i(0) + D_t^{-1}(V_{BB}\delta(t)/R)] + \frac{1}{RC}(-1)^n D_t^{-1}[\sum_{n=1}^{\infty} i_{n-1}] \tag{11.8}$$

This (11.8) is appearing as physical reasoning as infinite series as:

$$i(t) = i_0 - \frac{1}{RC}D_t^{-1}[i_0] + \frac{1}{RC}D_t^{-1}[i_1] - \frac{1}{RC}D_t^{-1}[i_2] + \frac{1}{RC}D_t^{-1}[i_3] + \ldots.. \tag{11.9}$$

In recursion, we obtain:

$$\begin{aligned} i_0 &= i(0) + D_t^{-1}[V_{BB}\delta(t)/R] \\ i_n &= -\frac{1}{RC}D_t^{-1}[i_{n-1}]; n \geq 1 \end{aligned} \tag{11.10}$$

Applying (11.10), we obtain:

$$
\begin{aligned}
i_0 &= V_{BB} / R \\
i_1 &= -\frac{1}{RC} D_t^{-1}[V_{BB} / R] = -\frac{V_{BB}}{R}\frac{t}{RC} \\
i_2 &= -\frac{1}{RC} D_t^{-1}\left[-\frac{V_{BB}}{R}\frac{t}{RC}\right] = \frac{V_{BB}}{R}\frac{1}{2!}\left(\frac{t}{RC}\right)^2 \\
i_3 &= -\frac{1}{RC} D_t^{-1}\left[\frac{V_{BB}}{R}\frac{1}{2!}\left(\frac{t}{RC}\right)^3\right] = -\frac{V_{BB}}{R}\frac{1}{3!}\left(\frac{t}{RC}\right)^3
\end{aligned}
\tag{11.11}
$$

Giving the total reaction of the system (11.7) as:

$$
i(t) = \frac{V_{BB}}{R}\left(1 - \frac{t}{RC} + \frac{1}{2!}\left(\frac{t}{RC}\right)^2 - \frac{1}{3!}\left(\frac{t}{RC}\right)^3 + \ldots\ldots\ldots\right) = \frac{V_{BB}}{R} e^{-\left(\frac{t}{RC}\right)}
\tag{11.12}
$$

The physical reasoning logic "opposite reaction to action" gives Mode-Decomposition and addition of all these infinite modes gives the entire system response. The observation is that zeroth mode reaction is formed by the external source/sink stimulus plus due to any initial condition. To oppose that rate of change an opposite internal reaction integral action takes place. This internal action is the first mode-reaction which causes a rate of change; and again integral action to this first mode, in opposition makes the second modal reaction. So on and so forth to make sum of "converging" analytical solution to the system's differential equation. The above (11.12) can be decomposed as follows with change in symbol.

$$
u(t) = \frac{V_{BB}}{R}\left(1 - \frac{t}{RC} + \frac{1}{2!}\left(\frac{t}{RC}\right)^2 - \frac{1}{3!}\left(\frac{t}{RC}\right)^3 + \ldots\ldots\ldots\right) = \frac{V_{BB}}{R} e^{-\left(\frac{t}{RC}\right)}
$$

$$
\begin{aligned}
u_0 &= u(0) = \frac{V_{BB}}{R} \\
u_1 &= u'(0) = -\frac{V_{BB}}{R^2 C} \\
u_2 &= u''(0) = \frac{V_{BB}}{R^3 C^2} \\
u_3 &= u'''(0) = -\frac{V_{BB}}{R^4 C^3}
\end{aligned}
$$

....

$$
u(t) = u_0 + u_1 t + u_2 t^2 + u_3 t^3 + \ldots\ldots\ldots\ldots\ldots\ldots + u_n t^n + \ldots..
$$

$$
u(\lambda) = \sum_{n=0}^{\infty} \lambda^n u_n
$$

$$
u_n = \frac{1}{n!}\left[\frac{d^n u(\lambda)}{d\lambda^n}\right]_{\lambda=0}
$$

Meaning that the series solution can be decomposed as Maclurain series at origin with all derivative components. This is fundamental of decomposition.

11.4 Physical Reasoning to Solve Second Order System and Its Mode-Decomposition

Consider a classical oscillator of integer-second order, mass spring system represented as:

$$mD_t^2 x(t) + kx(t) = f(t) \tag{11.13}$$

With initial conditions and forcing function defined as

$$x(0) = 0, \dot{x}(0) = 0, f(t) = \delta(t)$$

We can re-write the equation as:

$$D_t^2 x(t) = \ddot{x}(t) = \frac{1}{m} f(t) - \frac{k}{m} x(t) \tag{11.14}$$

This above equation gives insight into physical aspect of the process. The RHS states the opposing action to a forcing function, which is manifested as motion is given by LHS of (11.14). At the initial time the displacement being zero with the velocity implies that the displacement at just time $t = 0^+$ is due to the forcing function; alone. This displacement action is without any opposition.

This (zeroth mode) displacement call it

$$x_0 = D_t^{-2} \frac{1}{m} f(t)$$

Due to nature of this forcing function as an impulse, the displacement (zeroth mode) takes the form as

$$x_0 = \frac{1}{m} D_t^{-2} \delta(t) = \frac{1}{m} t$$

This displacement action would be true, in the absence of any retarding or opposing element say spring or friction. (In case of (11.14) it is spring action). In absence of any opposition, the constitutive equation will be $m\ddot{x}(t) = f(t)$, and for impulse force, the displacement will be linear function of time $x(t) = t / m$, with constant velocity

$$\dot{x}(t) = 1 / m .$$

The presence of spring makes the equation of motion as:

$$\ddot{x}(t) = \frac{1}{m} f(t) - \frac{k}{m} x(t) \,,$$

the external excitation being opposed by the spring action by opposite spring force and is internally generated

$$f_i = -\frac{k}{m} x_{i-1} \,.$$

The primary and the zeroth mode of displacement is due to external force on the mass that is:

$$x_0 = \frac{1}{m} D_t^{-2} f(t) = \frac{1}{m} D_t^{-2} \delta(t) = \frac{1}{m} t \,.$$

This zeroth mode of displacement is solely due to external excitation; since the initial conditions are at rest. This displacement is now opposed by spring. Due to this opposing element, the displacement caused by external force, the spring generates an opposing force (first mode, from zeroth order mode displacement), as:

$$f_1 = -\frac{k}{m} x_0 \,,$$

and to this, new (internal force) the displacement would be

$$x_1 = D_t^{-2} f_1 = -\frac{k}{m} D_t^{-2} x_0 = -\frac{k}{m} D_t^{-2} \left(\frac{1}{m} t \right) = -\frac{k}{m^2} \frac{t^3}{3!}$$

(This is the first order mode reaction-displacement), this displacement, again generates an internal force; inside the spring as

$$f_2 = -\frac{k}{m} x_1 = +\frac{k^2}{m^3} \frac{t^3}{3!} \,,$$

and to this force the displacement is second order mode.

$$x_2 = D_t^{-2} f_2 = -\frac{k}{m} D_t^{-2} x_1 = -\frac{k}{m} D_t^{-2} \left(-\frac{k}{m} \frac{1}{m} \frac{t^3}{3!} \right) = \frac{k^2}{m^3} \frac{t^5}{5!} \,,$$

and so on.

In the absence of the spring, the opposing forces will be zero. We can call this as displacement as sum of all the modal displacements from zero to infinity modes, with zero modes being the only reaction to the bare excitation (and if any initial displacement and velocity be present) and all other modes are opposing reactions taking place in the spring. The modes can be tabulated as in Table 11.1.

Table 11.1 Decomposing the action reaction of second order mass spring system

MODE	FORCE	DISPLACEMENT
0	$f_0 = \frac{1}{m} f(t) = \frac{1}{m}\delta(t)$	$x_0 = D_t^{-2} f_0 = \frac{1}{m} t$
Higher Modes	Higher Modal Internal Forces	Higher Modal Internal displacements
1	$f_1 = -\frac{k}{m} x_0 = -\frac{k}{m^2} t$	$x_1 = D_t^{-2} f_1 = -\frac{k}{m^2}\frac{t^3}{3!}$
2	$f_2 = -\frac{k}{m} x_1 = +\frac{k^2}{m^3}\frac{t^3}{3!}$	$x_2 = D_t^{-2} f_2 = +\frac{k^2}{m^3}\frac{t^5}{5!}$
3	$f_3 = -\frac{k}{m} x_2 = -\frac{k^3}{m^3}\frac{t^5}{5!}$	$x_3 = D_t^{-2} f_3 = -\frac{k^3}{m^4}\frac{t^7}{7!}$
....	....	....

The process block diagram is represented in Figure 11.1, with $k = 1$ and $m = 1$.
Adding up all the (modal displacements reactions), the solution to (11.13) is obtained as infinite series:

$$x(t) = x_0 + x_1 + x_2 + x_3 + ... = \frac{1}{m} t - \frac{k}{m^2}\frac{t^3}{3!} + \frac{k^2}{m^3}\frac{t^5}{5!} - \frac{k^3}{m^4}\frac{t^7}{7!} = \frac{1}{m}\left[t - \frac{k}{m}\frac{t^3}{3!} + \frac{k^2}{m^2}\frac{t^5}{5!} - ....\right] \tag{11.15}$$

Multiplying the above series by $\sqrt{k/m}$ and dividing by same we get:

$$x(t) = \frac{1}{\sqrt{km}}\left[\sqrt{\frac{k}{m}} t - \left(\frac{k}{m}\right)^{3/2}\frac{t^3}{3!} + \left(\frac{k}{m}\right)^{5/2}\frac{t^5}{5!} - ...\right] = \frac{1}{\sqrt{km}}\sin\left(\sqrt{\frac{k}{m}} t\right) \tag{11.16}$$

This is oscillator with natural frequency $\omega = \sqrt{k/m}$ radians per second. In (11.15) we decompose with symbol change to show that the series solution is Maclurain series at origin, and expressed as follows:

$$u(t)=\frac{1}{m}t-\frac{k}{m^2}\frac{t^3}{3!}+\frac{k^2}{m^3}\frac{t^5}{5!}-\frac{k^3}{m^4}\frac{t^7}{7!}=\frac{1}{m}\left[t-\frac{k}{m}\frac{t^3}{3!}+\frac{k^2}{m^2}\frac{t^5}{5!}-....\right]$$

$$u_0=u(0)=0$$

$$u_1=u'(0)=\frac{1}{m}$$

$$u_2=u''(0)=0$$

$$u_3=u'''(0)=\frac{k}{m^2}$$

....

$$u(t)=u_0+u_1t+u_2t^2+u_3t^3+......u_nt^n+........$$

$$u(\lambda)=\sum_{n=0}^{\infty}\lambda^n u_n$$

$$u_n=\frac{1}{n!}\left[\frac{du(\lambda)}{d\lambda^n}\right]_{\lambda=0}$$

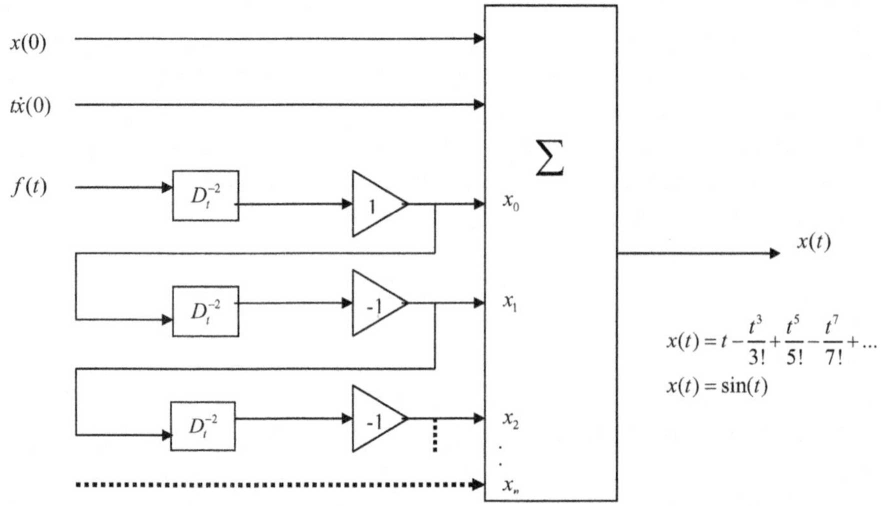

Fig. 11.1 Block diagram showing decomposition and solution of second order differential equation

11.5 Adomian Decomposition Fundamentals and Adomian Polynomials

We symbolize the general differential equation as:

$$Fu=G \tag{11.17}$$

F : General non-linear Ordinary Differential Operator, this can also be Fractional Differential Operator also of Riemann-Liouvelli (RL) or Caputo type. This operator can be decomposed as:

$$Fu = \mathrm{L_O}u + Ru + Nu = G \tag{11.18}$$

$\mathrm{L_O}$: Highest Order Derivative (Integer or Fractional Order) which is invertible.

R : Linear differential (remainder) operator of order less than that of $\mathrm{L_O}$; this can also be fractional differential linear operator.

N : Is the Non-Linear Part which will be decomposed into infinite sum of Adomian Polynomial (This term can too be of linear or constant still decomposition is valid). For decomposition this needs to be analytic.

G : Is the source term.

The decomposed equation can be re-written as:

$$\mathrm{L_O}u = G - Ru - Nu \tag{11.19}$$

Applying invert operator on both sides we get:

$$u = \Phi + \mathrm{L_O}^{-1}G - \mathrm{L_O}^{-1}[R(u)] - \mathrm{L_O}^{-1}[N(u)] \tag{11.20}$$

where Φ is solution of the homogeneous equation $\mathrm{L_O}u = 0$; so that $\mathrm{L_O}\Phi = 0$, this comes from initial/boundary conditions. The LHS of (11.17) physically is, reaction of each components of physical system; with RHS of (11.17) representing source/sink or forcing term. For example a mass spring and damper system has the constituent equation as $m\frac{d^2x}{dt^2} + c\frac{dx}{dt} + kx = f(t)$, the LHS of this is reaction of each elements, the sum of which balances the RHS and the external force. In terms of (11.17) (11.18) in this physical system

$$\mathrm{L_O} = D_t^2, R = D_t^1 \text{ and } N(x) = x.$$

The solution to this is

$$x(t) = \Phi + \frac{1}{m}\mathrm{L_O}^{-1}f - \frac{c}{m}\mathrm{L_O}^{-1}R - \frac{k}{m}\mathrm{L_O}^{-1}N$$

In this example the order of $\mathrm{L_O}$ is two then $\Phi = u(0) + t\dot{u}(0)$. Assuming the time dependent differential equation system; and the invert operator in this case is,

$$\mathrm{L_O}^{-1}f(t) \equiv D_t^{-2}f(t) = \int_0^t\int_0^t f(t)dtdt.$$

If the order of $\mathrm{L_O}$ is of order one then,

$$\Phi = u(0) \text{ and } L_O^{-1} f(t) \equiv D_t^{-1} f(t) = \int_0^t f(t)dt .$$

For decomposition of the $N(u)$ part in the (11.4) define a "grouping" parameter, close to one as λ. The function u can be expressed as:

$$u(\lambda) = \sum_{n=0}^{\infty} \lambda^n u_n = u_0 + \lambda u_1 + \lambda^2 u_2 + \ldots\ldots\ldots \tag{11.21}$$

This (11.21) is Maclurain series with respect to λ; with u_n's as coefficients of the Maclurain series around $\lambda = 0$ that is $u_n = u^{(n)}(0)/n!$. Then $N(u)$ in Maclurain, series with respect to λ we obtain:

$$N(u) = \sum_{n=0}^{\infty} \lambda^n A_n \tag{11.22}$$

where

$$A_n = \frac{1}{n!}\left[\frac{d^n}{d\lambda^n} N\left(\sum_{k=0}^{\infty} \lambda^k u_k\right)\right]_{\lambda=0} \tag{11.23}$$

The parameter λ is just an identifier for collecting terms in suitable way such that u_n depends on $u_0, u_1, u_2, \ldots u_{n-1}$, and later on, we will set $\lambda = 1$ .Paremetrizing the equation (11.20) we get:

$$u = \Phi + L_O^{-1} G - \lambda L_O^{-1}[R(u)] - \lambda L_O^{-1}[N(u)] \tag{11.24}$$

Expanding with decomposition the (11.24) we obtain:

$$u = \sum_{n=0}^{\infty} \lambda^n u_n = \Phi + L_O^{-1} G - \lambda L_O^{-1} R\left(\sum_{n=0}^{\infty} \lambda^n u_n\right) - \lambda L_O^{-1} \sum_{n=0}^{\infty} \lambda^n A_n \tag{11.25}$$

Equating the coefficients of equal powers of λ in the expression for $n = 0$, to get u_0, then $n = 1$, to get u_1 and so on, in (11.25), we get:

$$\begin{aligned}
u_0 &= \Phi + L_O^{-1} G \\
u_1 &= -L_O^{-1}\left[R(u_0)\right] - L_O^{-1}(A_0) \\
u_2 &= -L_O^{-1}\left[R(u_1)\right] - L_O^{-1}(A_1) \\
&\ldots\ldots\ldots\ldots \\
u_n &= -L_O^{-1}\left[R(u_{n-1})\right] - L_O^{-1}(A_{n-1}), n \ge 1
\end{aligned} \tag{11.26}$$

Finally

$$\varphi_N(t) = \sum_{n=0}^{N-1} u_n(t)$$

with $N \geq 1$; and exact solution of (11.17) is

$$u(t) = \lim_{N \to \infty} \varphi_N(t)$$

This method is applied in various problems of physics. The convergence of this method is very well proved too. In the ADM method described in the expression (11.26) contains Adomian polynomials A_n as recurring formulations where the invert operator is operational. Finding these A_n from (11.23) is demonstrated in this section. Suppose that the non-linear part of (11.18) that is $N(u)$ is represented as

$$\sum_{n=0}^{\infty} \lambda^n A_n;$$

that is $N(u(\lambda))$ is assumed to be analytic in λ. So we write

$$Nu = N(u(\lambda)) = \sum_{n=0}^{\infty} \lambda^n A_n.$$

The A_n's are polynomial defined in such a way that each A_n depends only on $u_0, u_1, u_2, \ldots u_n$.

Thus,

$$A_0 = A_0(u_0), A_1 = A_1(u_0, u_1), A_2 = A_2(u_0, u_1, u_2) \ldots\ldots \text{etc.}$$

Therefore, one possible formulation is listed below (11.27):

$$A_0 = A_0(u_0) = N(u(\lambda))_{\lambda=0} = N(u_0 + \lambda u_1 + \lambda^2 u_2 + \ldots)_{\lambda=0} = N(u_0)$$
$$A_1 = A_1(u_0, u_1) = \left(\frac{\partial N}{\partial u}\right)\left(\frac{\partial u}{\partial \lambda}\right)_{\lambda=0}$$
$$A_2 = A_2(u_0, u_1, u_2) = \frac{1}{2}\left[\left(\frac{\partial^2 N}{\partial u^2}\right)\left(\frac{\partial u}{\partial \lambda}\right)^2 + \left(\frac{\partial N}{\partial u}\right)\left(\frac{\partial^2 u}{\partial \lambda^2}\right)\right]_{\lambda=0}$$
$$A_3 = A_3(u_0, u_1, u_2, u_3) = \frac{1}{6}\left[\left(\frac{\partial^3 N}{\partial u^3}\right)\left(\frac{\partial u}{\partial \lambda}\right)^3 + 3\left(\frac{\partial^2 N}{\partial u^2}\right)\left(\frac{\partial u}{\partial \lambda}\right)\left(\frac{\partial^2 u}{\partial \lambda^2}\right) + \left(\frac{\partial N}{\partial u}\right)\left(\frac{\partial^3 u}{\partial \lambda^3}\right)\right]_{\lambda=0} \quad (11.27)$$

The A_n's can be re-formatted, from (11.27) in the following form as (11.28):

$$
\begin{aligned}
A_0 &= N(u_0)\\
A_1 &= u_1\left(\frac{d}{du_0}\right)N(u_0) = u_1 N'(u_0)\\
A_2 &= u_2\left(\frac{d}{du_0}\right)N(u_0) + \frac{u_1^2}{2!}\left(\frac{d^2}{du_0^2}\right)N(u_0) = u_2 N'(u_0) + \frac{1}{2!}u_1^2 N''(u_0)\\
A_3 &= u_3\left(\frac{d}{du_0}\right)N(u_0) + u_1 u_2\left(\frac{d^2}{du_0^2}\right)N(u_0) + \frac{u_1^3}{3!}\left(\frac{d^3}{du_0^3}\right)N(u_0)\\
A_3 &= u_3 N'(u_0) + u_1 u_2 N''(u_0) + \frac{u_1^3}{3!}N'''(u_0)
\end{aligned}
\tag{11.28}
$$

For the case where non-linear term is linear, that is say $N(u) = u$; then $A_n = u_n$, else $A_n = A_n(u_0, u_1, u_2 ... u_n)$ for all $n = 0,1,2,3......$.

For examples if $N(u) = u^3$, then Adomian Polynomials for this non-linearity are:

$$A_0 = u_0^3;\ A_1 = 3u_0^2 u_1;\ A_2 = 3u_0^2 u_2 + 3u_1^2 u_0;\ A_3 = u_1^3 + 3u_0^2 u_3 + 6u_0 u_1 u_2$$

and so on.

The derivation of obtaining Adomian Polynomials comes from Generalized Taylor's series (Maclurain series) of several variables from linear analysis. This is described as follows:

$$
\begin{aligned}
N(u) &= \sum_{n=0}^{\infty} \lambda^n A_n\\
N(u_0 + \lambda u_1 + \lambda^2 u_2 + ...) &= A_0 + \lambda A_1 + \lambda^2 A_2 + \lambda^3 A_3 + ...
\end{aligned}
\tag{11.29}
$$

Put $\lambda = 0$, to get $N(u_0) = A_0$.

Differentiate once (11.29), with respect to λ to get:

$$\frac{d}{d\lambda}[N(u_0 + \lambda u_1 + \lambda^2 u_2 + ...)] = \frac{d}{d\lambda}(A_0 + \lambda A_1 + \lambda^2 A_2 + ...) \tag{11.30}$$

Using partial derivative expansion on the LHS of (11.30) and differentiating RHS of (11.30), we obtain the following:

$$\frac{\partial N(u_0 + \lambda u_1 + \lambda^2 u_2 + ..)}{\partial(u_0 + \lambda u_1 + \lambda^2 u_2 + ...)}\frac{\partial}{\partial\lambda}(u_0 + \lambda u_1 + \lambda^2 u_2 + ..) = A_1 + 2\lambda A_2 + 3\lambda^2 A_3 + ... \tag{11.31}$$

Put $\lambda = 0$ in above (11.31) to get:

$$A_1 = u_1 \frac{dN(u_0)}{du_0} = u_1 N'(u_0) \tag{11.32}$$

Differentiating once more (11.32) with respect to λ we get:

$$\frac{d}{d\lambda}\left[\frac{\partial N(u_0+\lambda u_1+..)}{\partial(u_0+\lambda u_1+...)}\frac{\partial}{\partial\lambda}(u_0+\lambda u_1+..)\right] = \frac{d}{d\lambda}(A_1+2\lambda A_2+3\lambda^2 A_3+...) \tag{11.33}$$

Using partial derivative expansion on the LHS of (11.33) and differentiating RHS of (11.33), we obtain the following:

$$\frac{\partial}{\partial\lambda}(u_0+\lambda u_1+...)\frac{d}{d\lambda}\left[\frac{\partial N(u_0+\lambda u_1+..)}{\partial(u_0+\lambda u_1+...)}\right]+\frac{\partial N(u_0+\lambda u_1+..)}{\partial(u_0+\lambda u_1+..)}\frac{d}{d\lambda}\frac{\partial}{\partial\lambda}(u_0+\lambda u_1+..) = 2A_2+3!\lambda A_3+... \tag{11.34}$$

Using partial derivative expansion on the LHS of (11.34) and differentiating RHS of (11.34), we obtain the following:

$$(u_1+2\lambda u_2+..)\frac{\partial^2 N(u_0+\lambda u_1+..)}{\partial(u_0+\lambda u_1+..)^2}\frac{\partial}{\partial\lambda}(u_0+\lambda u_1+..)+\frac{\partial N(u_0+\lambda u_1+..)}{\partial(u_0+\lambda u_1+..)}(2u_2+3\lambda u_3+..) = 2A_2+.. \tag{11.35}$$

Putting $\lambda = 0$ in above expression (11.35), we obtain:

$$u_1{}^2\frac{\partial^2 N(u_0)}{\partial u_0{}^2}+2u_2\frac{\partial N(u_0)}{\partial u_0} = 2A_2 \tag{11.36}$$

Implying,

$$A_2 = u_2\frac{dN(u_0)}{du_0}+\frac{u_1{}^2}{2}\frac{d^2N(u_0)}{du_0{}^2} = u_2 N'(u_0)+\frac{1}{2}u_1{}^2 N''(u_0) \tag{11.37}$$

Continuing like this we get set of the Adomian Polynomials for the function $N(u)$.

If the non-linearity part is

$$N_1(x) = x^2 = \sum_{n=0}^{\infty}\lambda^n A_n$$

then the Adomian Polynomials are

$$A_0 = x_0^2; A_1 = 2x_0x_1; A_2 = x_1^2 + 2x_0x_2; A_3 = 2x_0x_3 + 2x_0x_2 .....$$

If the non-linearity part is,

$$N_2(x) = x^3 = \sum_{n=0}^{\infty} \lambda^n A_n$$

the Adomian Polynomials are

$$A_0 = x_0^3; A_1 = 3x_0^2x_1; A_2 = 3x_0^2x_2 + 3x_1^2x_0; A_3 = 3x_0^2x_3 + 6x_0x_1x_2 + x_1^3 ....$$

For the linear term

$N_0(x) = x = \sum_{n=0}^{\infty} \lambda^n A_n$, $A_0 = x_0$ .and $A_n = x_n$. For constant $N(u)$, the Adomian polynomials are:

$$N_0(x) = K , A_0 = N_0(x_0) = K ; A_1 = A_2 = ... = A_\infty = 0 .$$

The series solution $\sum_{n=0}^{\infty} u_n$ thus may have finite terms with higher modes as zero; depending on $N(u)$.

11.6 Generalization of Physical Law of Nature Vis-À-Vis ADM

The physical description and then obtaining decomposed solution matches well with the ADM. From the section, 2, 3 and 4 we generalize the system of General Differential Equations and give action-reaction laws to it so that one can obtain the solution by decomposition into infinite (or finite) modes. Let there be general differential equation system as:

$$\begin{aligned} &D_x^m u(x) + a_1 D_x^{m-1} u(x) + a_2 D_x^{m-2} u(x) + ... + a_m u(x) \\ &+ b_k [u(x)]^k + b_{k-1} [u(x)]^{k-1} + ...b_0 u(x) = G(x) \end{aligned} \tag{11.38}$$

We can write (11.38) as:

$$\mathrm{L}_0 u(x) = G(x) - R(u) - N(u) \tag{11.39}$$

where

$$R(u) = a_1 D_x^{m-1} + a_2 D_x^{m-2} + ... + a_{m-1} D_x^1$$

is the remainder differential operator of order less than m. This $N(u) = a_m u(x) + b_k [u(x)]^k + b_{k-1} [u(x)]^{k-1} + ...b_0 u(x)$ contains nonlinear as well as the linear terms. The $G(x)$ is sum of all external stimulus source/sink.

The L_O represents a linear operator representing the highest orders of change in the process parameter. $L_O(u) = d^m u(x)/dx^m = D_x^m u(x)$; which is easily invertible. This order of change (m) could be one, two or any positive integer or even fractional (say half, one-fourth). Then with ADM we have:

$$\begin{aligned}
&u(\lambda) = u_0 + \lambda u_1 + \lambda^2 u_2 + \lambda^3 u_3 + ...\\
&u_0 = \Phi + L_O^{-1} G\\
&N[u(\lambda)] = N(u_0 + \lambda u_1 + \lambda^2 u_2 + ...) = A_0 + \lambda A_1 + \lambda^2 A_2 + \lambda^3 A_3 + ...\\
&A_0 = N(u_0)\\
&u_1 = -L_O^{-1} R(u_0) - L_O^{-1}(A_1)\\
&A_1 = u_1 N'(u_0)\\
&u_2 = -L_O^{-1} R(u_1) - L_O^{-1}(A_1)\\
&A_2 = u_2 N'(u_0) + \frac{1}{2} u_1^{\,2} N''(u_0)\\
&u_3 = -L_O^{-1} R(u_2) - L_O^{-1}(A_2)\\
&A_{n-1} = \frac{1}{(n-1)!}\left[\frac{d^{n-1} N(\sum_{k=0}^{\infty} \lambda^k u_k)}{d\lambda^{n-1}}\right]_{\lambda=0}\\
&u_n = -L_O^{-1} R(u_{n-1}) - L_O^{-1}(A_{n-1})\\
&u(x) = \sum_{n=0}^{\infty} u_n
\end{aligned} \tag{11.40}$$

The (11.40) is exactly what was described in section 2, 3 with physical principles of action reaction generating all internal modes.

11.7 ADM Applied to First Order Linear Differential Equation and Mode-Decomposition Solution

Comparing with ADM described by (11.26) and (11.40); and with solution obtained by physical reasoning, to arrive at solution of (11.5), (11.7) gives a similarity. Therefore, the ADM is close to physical system behavior where the system reacts naturally in opposite way to resist any change this is physical law, which is described by ADM, for solving system of differential equations (11.40).

The initial reaction to the external disturbances and the complete set of opposing reaction due to self opposed elements to the change, gets summed up to get the overall reaction yielding solution for (11.5) as set in (11.41) obtained vis-à-vis (11.40):

$$x(t) = x(0) + x_0 + x_1 + x_2 + x_3 + ...$$

$$x(t) = x(0) + D_t^{-1} f(t) + aD_t^{-1} f_1(t) + aD_t^{-1} f_2(t) + ...$$

$$x(t) = x(0) + D_t^{-1} f(t) + a\sum_{n=1}^{\infty}(-1)^n (D_t^{-1})^n x_{n-1}$$

$$x_0 = x(0) + D_t^{-1} f(t)$$

$$x(t) = 1 - at + \frac{(at)^2}{2!} - \frac{(at)^3}{3!} + ... = \exp(-at) \qquad (11.41)$$

The x_n's are internal reaction to internal generated forces for $n > 0$.

The action reaction process described is represented in Figure 11.2, with Parameter $a = K = 1$, the D_t^{-1} is anti-derivative operator of unity order is

$$D_t^{-1} f(t) \equiv \int_0^t f(t)dt$$

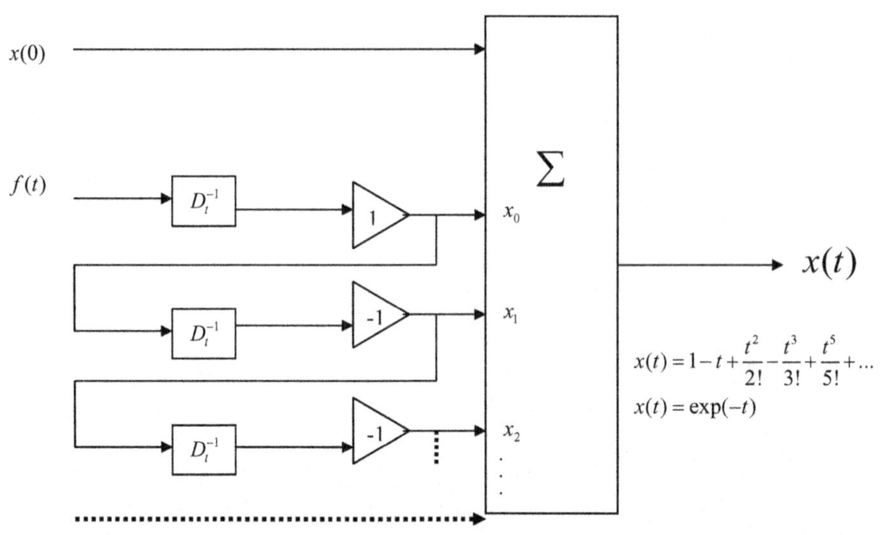

Fig. 11.2 Block showing solution of first order differential equation by decomposition

In the ADM, as described in (11.26) we can write the set of modes (reactions) as:

$$
\begin{aligned}
x_0 &= \Phi + \mathrm{L_O}^{-1} G \\
x_1 &= -\mathrm{L_O}^{-1}[R(x_0)] - \mathrm{L_O}^{-1}(A_0) \\
x_2 &= -\mathrm{L_O}^{-1}[R(x_1)] - \mathrm{L_O}^{-1}(A_1)
\end{aligned}
\tag{11.42}
$$

In the case of (11.5) $\mathrm{L_O}^{-1} = D_t^{-1}, G = f(t) = K\delta(t), N(x) = ax$, is linear and $x_0 = A_0$, with Adomian Polynomials (11.26) as: $A_n = ax_n$ with no remainder term as $R = 0$. With this decomposition we get (11.43)

$$
\begin{aligned}
x_0 &= \Phi + \mathrm{L_O}^{-1} G = D_t^{-1} K\delta(t) = K \\
A_0 &= ax_0 = aK \\
x_1 &= -\mathrm{L_O}^{-1}[R(x_0)] - \mathrm{L_O}^{-1}(A_0) = -D_t^{-1}[aK] = -Kat \\
A_1 &= ax_1 = -Ka^2 t \\
x_2 &= -\mathrm{L_O}^{-1}[R(x_1)] - \mathrm{L_O}^{-1}(A_1) = -D_t^{-1}[-Ka^2 t] = K\frac{(at)^2}{2!} \\
A_2 &= ax_2 = Ka^3 \frac{t^2}{2!} \\
x_3 &= -\mathrm{L_O}^{-1}[R(x_2)] - \mathrm{L_O}^{-1}(A_2) = -D_t^{-1}\left[Ka^3 \frac{t^2}{2}\right] = -K\frac{(at)^3}{3!} \\
x(t) &= K\left(1 - at + \frac{(at)^2}{2!} - \frac{(at)^3}{3!} + \ldots\right) = Ke^{-at}
\end{aligned}
\tag{11.43}
$$

The infinite set of reactionary currents at instance ($t \to 0$) is formed. The first (or zeroth) reaction current is due to initial state of the circuit and solely due to external force represented by x_0. Then recurring opposite reactions occur as set of internal forces due to opposing the changes, giving rise immediately the first mode and second mode (and to infinity-modes) of currents; adding up giving the total current reaction as, $x(t) = x_0 + \sum_{n=1}^{\infty} x_n$. Therefore the ADM method is related to physical process of physics as to any "action" there is equal and opposite "reaction" maybe external or internal to the system. In other words, all system reacts in opposite way to any change (external or internal).

11.8 ADM Applied to Second Order Linear Differential Equation System and Mode-Decomposition

The ADM method for (11.13) has

$$
\mathrm{L_O}^{-1} = D_t^{-2}, \Phi = x(0) + t\dot{x}(0) = 0, G = \frac{1}{m} f(t) = \frac{1}{m}\delta(t),
$$

$N(x) = \frac{k}{m}x$ which generates Adomian Polynomials (11.26) as

$$A_0 = \frac{k}{m}x_0 \text{, and}$$

$$A_n = \frac{k}{m}x_n .$$

Here in (11.13) the remainder part is $R(x) = 0$. Using the ADM for (11.13) one gets the modal displacements as described by physical reasoning also as:

$$\begin{aligned}
x_0 &= \Phi + \mathrm{L_O}^{-1}G = D_t^{-2}\frac{1}{m}\delta(t) = \frac{1}{m}t \\
A_0 &= \frac{k}{m}x_0 = \frac{k}{m^2}t \\
x_1 &= -\mathrm{L_O}^{-1}R(x) - \mathrm{L_O}^{-1}(A_0) = -D_t^{-2}\left(\frac{k}{m^2}\right) = -\frac{k}{m^2}\frac{t^3}{3!} \\
A_1 &= \frac{k}{m}x_1 = -\frac{k^2}{m^3}\frac{t^3}{3!} \\
x_2 &= -\mathrm{L_O}^{-1}R(x) - \mathrm{L_O}^{-1}(A_1) = -D_t^{-2}(-t^3/3!) = \frac{k^2}{m^3}\frac{t^5}{5!}
\end{aligned} \tag{11.44}$$

Giving the solution to (11.13) as

$$\begin{aligned}
x(t) &= x_0 + x_1 + x_2 + x_3 + \ldots.. = \frac{1}{m}t - \frac{k}{m^2}\frac{t^3}{3!} + \frac{k^2}{m^3}\frac{t^5}{5!} - \frac{k^3}{m^4}\frac{t^7}{7!} + \ldots + .. \\
x(t) &= \frac{1}{m}\left[t - \frac{k}{m}\frac{t^3}{3!} + \frac{k^2}{m^2}\frac{t^5}{5!} - \ldots.\right]
\end{aligned} \tag{11.45}$$

This too demonstrates the decomposition by ADM gives the physical modes of reaction process, generated as infinite series.

Multiplying the above series by $\sqrt{k/m}$ and dividing by same we get:

$$x(t) = \frac{1}{\sqrt{km}}\left[\sqrt{\frac{k}{m}}t - \left(\frac{k}{m}\right)^{3/2}\frac{t^3}{3!} + \left(\frac{k}{m}\right)^{5/2}\frac{t^5}{5!} - \ldots\right] = \frac{1}{\sqrt{km}}\sin\left(\sqrt{\frac{k}{m}}t\right) \tag{11.46}$$

This is oscillator with natural frequency $\omega = \sqrt{k/m}$ radians per second; obtained earlier by physical law of action-reaction process.

11.9 ADM for First Order Linear Differential Equation System with Half Order Element and Mode-Decomposition

Consider a first order differential equation with presence of a fractional (half) order element as (11.47)

$$\dot{x}(t) + D_t^{1/2} x(t) + x(t) = f(t) \tag{11.47}$$

With initial conditions as $x(0) = 0, f(t) = \delta(t)$. With the physical explanation as done in section 2 and 3 gives following solution (11.48):

$$\begin{aligned} x(t) &= x(0) + D_t^{-1} f(t) + \sum_{n=1}^{\infty} (-1)^n (D_t^{-1} + D_t^{-1/2})^n f_{n-1}(t) \\ f_0 &= x(0) + D_t^{-1} f(t) \end{aligned} \tag{11.48}$$

For this system the $R = D_t^{1/2}, \mathrm{L_O} = D_t^1$ and nonlinear part is $N(x) = x$; (is linear in nature) thus as per (11.26) $A_n = x_n$. This demonstration also shows the fact because the fractional order component is of lesser order than the main component, which is of, in this case is integer order the initial condition does not depend on the fractional derivative definition. Here the initial states are always of integer order in nature.

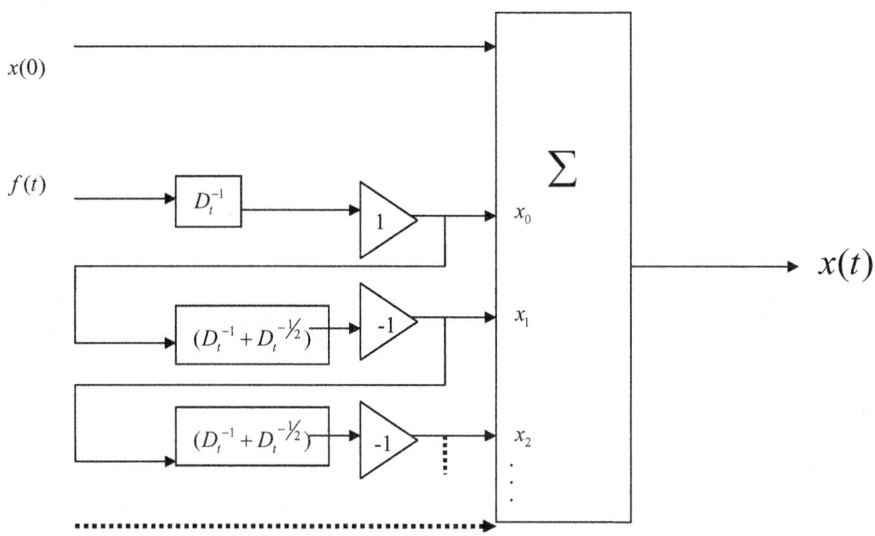

Fig. 11.3 Block showing solution of first order differential equation by decomposition in presence of fractional half order term

The application of ADM method gives the components as, represented in Fig: 11.3.

$$x_0 = \Phi + \mathrm{L_O}^{-1}G = x(0) + D_t^{-1} f(t) = D_t^{-1}\delta(t) = 1$$

$$A_0 = x_0 = 1$$

$$x_1 = -\mathrm{L_O}^{-1}[R(x_0)] - \mathrm{L_O}^{-1}[A_0] = -D_t^{-1}[D_t^{1/2}(1)] - D_t^{-1}[1] = -\frac{t^{1/2}}{\Gamma(1.5)} - t$$

$$A_1 = x_1$$

$$x_2 = -\mathrm{L_O}^{-1}[R(x_1)] - \mathrm{L_O}^{-1}[A_1] = -D_t^{-1}[D_t^{1/2}(x_1)] - D_t^{-1}[x_1] = t + \frac{t^{3/2}}{\Gamma(2.5)} + \frac{2}{3}\frac{t^{3/2}}{\Gamma(1.5)} + \frac{t^2}{2} \tag{11.49}$$

The physics of this process may be viewed as, *RC* circuit reacting to an impulse current reaction in the presence of a semi-infinite *RC* cable (CRO Probe) connected to a shunt to measure the current. The semi-infinite lossy cable acts as half order element and the first order circuit reaction thus will be modified by presence of this half order element. Refer Figure 11.4.

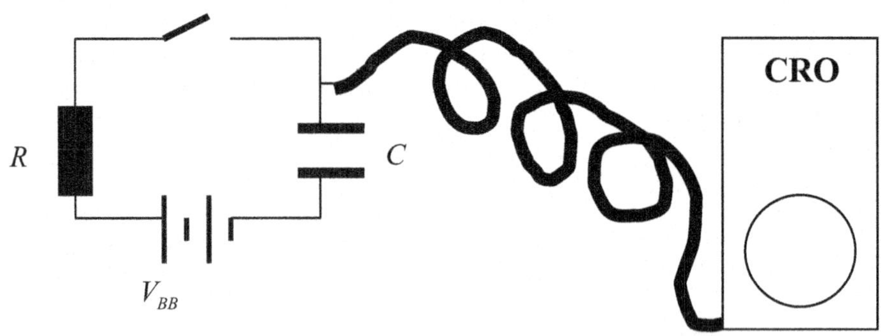

Fig. 11.4 The *RC* circuit (a first order differential equation), with semi-infinite cable as fractional half order element

11.10 ADM for Second Order System, with Half Order Element and It's Physics

11.10.1 Forcing Function as Delta Function

Solution of second order differential equation with presence of half order element is considered in (11.50)

$$\ddot{x}(t) + {}_0D_t^{1/2}x(t) + x(t) = f(t) \tag{11.50}$$

With the initial condition as $x(0) = 0, \dot{x}(0) = 0, f(t) = \delta(t)$. Rearranging the above equation (11.50), we rewrite (double integrating both sides) as

$$x(t) = x(0) + t\dot{x}(0) + D^{-2} f(t) - D^{-2}x(t) - D^{-3/2}x(t) \tag{11.51}$$

The physical reasoning generates (11.52)

$$\begin{aligned}
x_0 &= D_t^{-2}\delta(t) = t \\
x_1 &= -(D_t^{-3/2} + D_t^{-2})x_0 = -\left(\frac{t^{5/2}}{\Gamma(7/2)} + \frac{t^3}{\Gamma(4)}\right) \\
x_2 &= -(D_t^{-3/2} + D_t^{-2})x_1 = \left(\frac{t^4}{\Gamma(5)} + 2\frac{t^{9/2}}{\Gamma(11/2)} + \frac{t^5}{\Gamma(6)}\right)
\end{aligned} \tag{11.52}$$

The modal displacements are generated after the application of external forcing function is depicted in the Table 11.2. The block diagram of the process is shown in Figure 11.5.

The ADM method generates the modes as follows in (11.53).

$$\begin{aligned}
x_0 &= \Phi + \mathrm{L_O}^{-1}G = D_t^{-2}\delta(t) \\
A_0 &= x_0 \\
x_1 &= -\mathrm{L_O}^{-1}R(x_0) - \mathrm{L_O}^{-1}A_0 = -(D_t^{-3/2} + D_t^{-2})x_0 = -\left(\frac{t^{5/2}}{\Gamma(7/2)} + \frac{t^3}{\Gamma(4)}\right) \\
A_i &= x_1 = -\left(\frac{t^{5/2}}{\Gamma(7/2)} + \frac{t^3}{\Gamma(4)}\right) \\
x_2 &= -\mathrm{L_O}^{-1}R(x_1) - \mathrm{L_O}^{-1}A_1 = -(D_t^{-3/2} + D_t^{-2})x_1 = \left(\frac{t^4}{\Gamma(5)} + 2\frac{t^{9/2}}{\Gamma(11/2)} + \frac{t^5}{\Gamma(6)}\right)
\end{aligned} \tag{11.53}$$

Table 11.2 Modal force and displacements for second order system with fractional order damping

MODE	FORCE	DISPLACEMENT
0	$\delta(t)$	t
1	$-t$	$-\frac{t^{5/2}}{\Gamma(3.5)} - \frac{t^3}{\Gamma(4)}$
2	$\frac{t^{5/2}}{\Gamma(3.5)} + \frac{t^3}{\Gamma(4)}$	$\frac{t^4}{\Gamma(5)} + 2\frac{t^{9/2}}{\Gamma(5.5)} + \frac{t^5}{\Gamma(6)}$
3	$-\frac{t^4}{\Gamma(5)} - 2\frac{t^{9/2}}{\Gamma(5.5)} - \frac{t^5}{\Gamma(6)}$	$-\frac{t^{11/2}}{\Gamma(6.5)} - \frac{t^6}{\Gamma(7)} - \frac{t^{13/2}}{\Gamma(7.5)} - \frac{t^7}{\Gamma(8)}$
.....	.....	.....

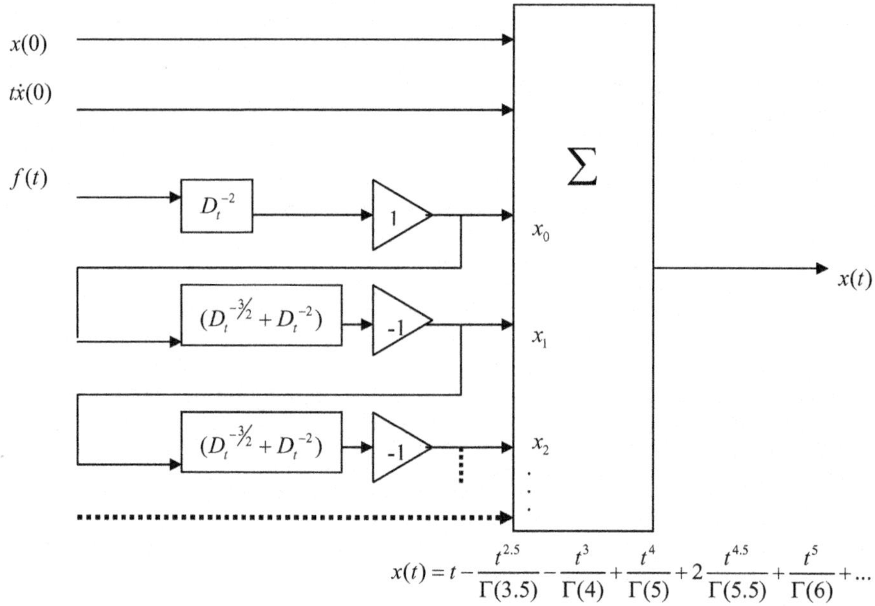

Fig. 11.5 Block diagram showing solution of by decomposition of a second order differential equation in presence of fractional order term.

11.10.2 Forcing Function as Step Function

Consider a fractional oscillator described by (11.54)

$$D_t^2 x(t)+\frac{c}{m}D_t^{1/2}x(t)+\frac{k}{m}x(t)=\frac{1}{m}f(t) \tag{11.54}$$

The system is initially stationary (i.e. $x(0)=0$ and $D_t^1\left[x(t)\right]_{t=0}=\dot{x}(0)=0$ are initial conditions). Let this stationary fractional oscillator is subjected to and excitation function $f(t)=FH(t)$, where $H(t)$ is Heaviside's unit step function and F is a constant. By application of ADM (11.26), (11.27) and with the help of Figure 11.3 we obtain the reactionary displacement modes as in set (11.55) below:

$$x_0(t)=\frac{1}{m}D_t^{-2}f(t)=\frac{Ft^2}{2m}$$

$$x_1(t)=-\frac{c}{m}D_t^{-3/2}x_0(t)-\frac{k}{m}D_t^{-2}x_0(t)=-\frac{cF}{m^2}\frac{t^{7/2}}{\Gamma\left(\frac{9}{2}\right)}-\frac{kF}{m^2}\frac{t^4}{\Gamma(5)}$$

$$x_2(t) = -\frac{c}{m} D_t^{-3/2} x_1(t) - \frac{k}{m} D_t^{-2} x_1(t) = \frac{c^2 F}{m^3} \frac{t^5}{\Gamma(6)} + \frac{2kcF}{m^3} \frac{t^{11/2}}{\Gamma\left(\frac{13}{2}\right)} + \frac{k^2 F}{m^3} \frac{t^6}{\Gamma(7)}$$

$$x_3(t) = -\frac{c}{m} D_t^{-3/2} x_2(t) - \frac{k}{m} D_t^{-2} x_2(t)$$

$$x_3(t) = -\frac{c^3 F}{m^4} \frac{t^{13/2}}{\Gamma\left(\frac{15}{2}\right)} - \frac{3kc^2 F}{m^4} \frac{t^7}{\Gamma(8)} - \frac{3k^2 cF}{m^4} \frac{t^{15/2}}{\Gamma\left(\frac{17}{2}\right)} - \frac{k^3 F}{m^4} \frac{t^8}{\Gamma(9)} \tag{11.55}$$

The total solution is sum of all these component modes to infinity as series in (11.56).

$$x(t) = x_0(t) + x_1(t) + x_2(t) + x_3(t) + \ldots + \ldots$$

$$x(t) = \frac{F}{m} \sum_{r=0}^{\infty} \frac{(-1)^r}{r!} \left(\frac{k}{m}\right)^r t^{2(r+1)} \sum_{j=0}^{\infty} \left(\frac{-c}{m}\right)^j \frac{(j+r)!\, t^{3j/2}}{j!\,\Gamma\left(\frac{3j}{2} + 2r + 3\right)} \tag{11.56}$$

$$x(t) = \frac{F}{m} \sum_{r=0}^{\infty} \frac{(-1)^r}{r!} \left(\frac{k}{m}\right)^r t^{2(r+1)} E^{(r)}_{\frac{3}{2},\frac{r}{2}+3} \left(\frac{-c}{m} t^{3/2}\right)$$

where $E_{\alpha,\beta}(x)$ is two parameter Mittag-Leffler function, and $E^{(r)}_{\alpha,\beta}(x)$ is the derivative of Mittag-Leffler function, defined as follows (11.57):

$$E^{(r)}_{\alpha,\beta}(x) = \frac{d^r}{dx^r} E_{\alpha,\beta}(x) = \sum_{j=0}^{\infty} \frac{(j+r)!\, x^j}{j!\,\Gamma(\alpha j + \alpha r + \beta)}, \quad (r = 0,1,2,3\ldots) \tag{11.57}$$

11.10.3 Explanation Physical Action Reaction Process Vis-À-Vis ADM

The fundamentals of mode decomposition as explained above in case of second order differential equation, in the presence of fractional order component may be explained in slightly elaborated way as follows with a mass spring fractional viscous system as

$$\frac{d^2}{dt^2} x(t) + a \frac{d^{1/2}}{dt^{1/2}} x(t) + bx(t) = f(t) \tag{11.58}$$

where a is the constant of half order property and b is the spring stiffness constant for spring. The above equation of motion is for unit mass attached to an ideal spring with half order visco-elastic element.

The initial conditions are: $x(0)=0, \dot{x}(0)=0, f(t)=\delta(t)$, for delta function input case.

The above equation (11.58) can be re-written in terms of external force and opposing internal forces as (11.59):

$$\frac{d^2}{dt^2}x(t)=f(t)-a\frac{d^{1/2}}{dt^{1/2}}x(t)-bx(t) \tag{11.59}$$

Decomposing the above by modal decomposition, we get the zero order mode as first reaction (immediate) that is: $x_0=D_t^{-2}f(t)$. This zero order displacement is the reaction without presence of the spring or any other opposing elements. Due to this zero order, displacement there will be opposing forces which appear as

$$f_{11}=-aD_t^{1/2}x_0 \text{ and } f_{12}=-bx_0$$

giving rise to first order displacements as

$$x_{11}=-D_t^{-2}f_{11}=-aD_t^{-3/2}x_0 \text{ and } x_{12}=-D_t^{-2}f_{12}=-bD_t^{-2}x_0$$

The overall first modal displacement is: $x_1=x_{11}+x_{12}$. From this the reaction force for second modes are generated as:

$$f_{21}=-aD_t^{1/2}x_1 \text{ and } f_{22}=-bx_1$$

giving rise to second modal displacement as:

$$x_{21}=-D_t^{-2}f_{11}=-aD_t^{-3/2}x_1 \text{ and } x_{22}=-D_t^{-2}f_{22}=-bD_t^{-2}x_1$$

Similarly, we can carry on for infinity as this self-similar pattern of reactions generated within the system to external stimulus. The observation is that the half order element adds second force to the ideal spring restoring force as obtained in case of pure second order pure oscillator (11.13).

The practical way of explaining the fractional order behavior is by considering a LC-oscillator and then trying to measure the oscillation by CRO probe, which is semi-infinite cable (with losses) acting as half order element. The constant a of the half order element (11.58) is depending on the distributed loss parameter that is per unit series resistance and per unit shunt capacitance.

The constitutive equation for the circuit in (Figure 11.6) is described in (11.60)

$$\begin{aligned}
&\frac{1}{C}\int_0^t i(t)dt+L\frac{d}{dt}i(t)+a\frac{d^{-1/2}}{dt^{-1/2}}i(t)=V_{BB}H(t)\\
&\frac{i(t)}{C}+L\frac{d^2}{dt^2}i(t)+a\frac{d^{1/2}}{dt^{1/2}}i(t)=V_{BB}\delta(t)
\end{aligned} \tag{11.60}$$

This is demonstration of the oscillator with fractional order element, shown in Figure 11.6. Practically in circuit experiments it is observed that a purely oscillating circuit when connected by shunt to a long CRO-probe, goes to damped oscillations removal of probe that again gives the oscillations. Use of a very short probe to CRO gives oscillation. This is due to fact the long CRO probe may act as lossy transmission line, behaving as half-order damping element.

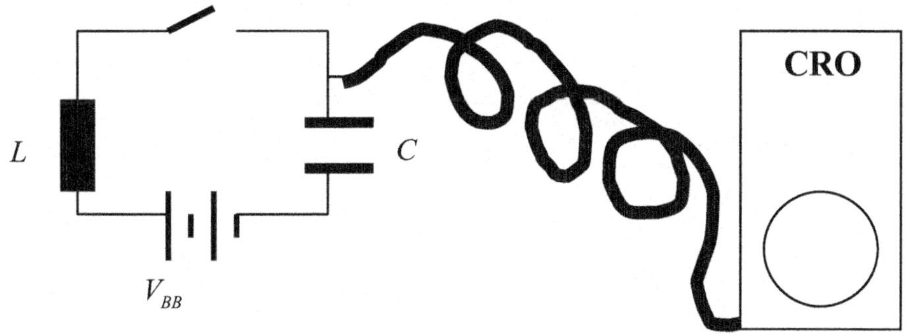

Fig. 11.6 The oscillator circuit (a second order differential equation), with semi-infinite cable CRO-probe acting as half order element.

11.11 Application of Decomposition Method in RL-Formulated Partial Fractional Differential Equations Linear Diffusion Wave Equation and Solution to Impulse Forcing Function

Here attempt is made to obtain series solution with ADM and the physical explanation of several modes generating as reaction. This problem example elaborates that the time evolution of process parameter takes place without the forcing function present, $G = 0$; only the effect is due to initial value (in this case is Dirac's delta function as process parameter present at space-origin).

The fractional time rate of change of the process variable is related to spatial double derivative of the same as expressed in (11.61). In (11.61) the formulation of fractional derivative is of Riemann-Liouvelli (RL) scheme.

Let us consider the problem of fractional time diffusion as:

$$\left[\frac{\partial^\alpha}{\partial t^\alpha}u(x,t)\right]_{RL} = \frac{\partial^2}{\partial x^2}u(x,t) \tag{11.61}$$

with $1 < \alpha < 2$ and have the initial condition as $u(x,0^+) = \delta(x)$ and $u_t(x,0^+) = 0$. The integer order highest to the fractional order in case of (11.61) is $m = 2$.

Converting the (11.61) into Caputo derivative formulation, we obtain:

$$ {}^{C}\left[\frac{\partial^\alpha}{\partial t^\alpha}u(x,t)\right]+\frac{t^{-\alpha}}{\Gamma(1-\alpha)}u(x,0)+\frac{t^{1-\alpha}}{\Gamma(2-\alpha)}\dot{u}(x,0)=\frac{\partial^2}{\partial x^2}u(x,t) \tag{11.62} $$

Observation here states that the (11.61) RL derivative of fractional order when changed to Caputo formulation (11.62) gives rise to extra source/sink terms of the inverse power function of the independent variables in the constituent equation. In (11.62) after applying the definition of Caputo derivative in the fractional operator and rearranging, we get:

$$ D_t^{-(2-\alpha)}\left[\frac{\partial^2}{\partial t^2}u(x,t)\right]=\frac{\partial^2}{\partial x^2}u(x,t)-\frac{t^{-\alpha}}{\Gamma(1-\alpha)}u(x,0)-\frac{t^{1-\alpha}}{\Gamma(2-\alpha)}\dot{u}(x,0) \tag{11.63} $$

Inverting (11.63), we get:

$$ \left[\frac{\partial^2}{\partial t^2}u(x,t)\right]=D_t^{(2-\alpha)}\left[\frac{\partial^2}{\partial x^2}u(x,t)\right]-\frac{u(x,0)}{\Gamma(1-\alpha)}D_t^{(2-\alpha)}[t^{-\alpha}]-\frac{\dot{u}(x,0)}{\Gamma(2-\alpha)}D_t^{(2-\alpha)}[t^{1-\alpha}] \tag{11.64} $$

Applying fractional derivative of the power functions in RHS of (11.64) we get:

$$ \left[\frac{\partial^2}{\partial t^2}u(x,t)\right]=D_t^{(2-\alpha)}\left[\frac{\partial^2}{\partial x^2}u(x,t)\right]-\frac{u(x,0)}{\Gamma(1-\alpha)}\left[\frac{\Gamma(-\alpha+1)t^{-\alpha-2+\alpha}}{\Gamma(-\alpha+1-2+\alpha)}\right]-\frac{\dot{u}(x,0)}{\Gamma(2-\alpha)}\left[\frac{\Gamma(1-\alpha+1)t^{1-\alpha-2+\alpha}}{\Gamma(1-\alpha+1-2+\alpha)}\right] \tag{11.65} $$

Simplifying (11.65)

$$ \left[\frac{\partial^2}{\partial t^2}u(x,t)\right]=D_t^{(2-\alpha)}\left[\frac{\partial^2}{\partial x^2}u(x,t)\right]-u(x,0)\left[\frac{t^{-2\alpha}}{\Gamma(-1)}\right]-\dot{u}(x,0)\left[\frac{t^{-1}}{\Gamma(0)}\right] \tag{11.66} $$

The last two terms of (11.66) gives zero; since reciprocal Gamma function is zero at values zero and negative integer points giving the modified diffusion equation as:

$$ \left[\frac{\partial^2}{\partial t^2}u(x,t)\right]=D_t^{(2-\alpha)}\left[\frac{\partial^2}{\partial x^2}u(x,t)\right] \tag{11.67} $$

Taking the (space) Fourier Transform of (11.67), we get:

$$ \frac{\partial^2}{\partial t^2}\overline{u}(k,t)=-k^2D_t^{(2-\alpha)}\overline{u}(k,t) \tag{11.68} $$

With transformed initial condition as $\overline{u}(k,0)=\frac{1}{\sqrt{2\pi}}$; and $\overline{u}_t(k,0)=0$.

where

$$\overline{u}(k,t)=\frac{1}{\sqrt{2\pi}}\int_{-\infty}^{+\infty}e^{ikx}u(x,t)dx\,,\; k\in\Re$$

is the spatial Fourier Transform definition. The parameter k is "wave-vector". The system of equations (11.67) has been transformed to (11.68) so we solve for $\overline{u}(k,t)$ and write with ADM (11.26) the solution as:

$$\begin{aligned}
&\overline{u}(k,t)=\Phi+\mathrm{L_O}^{-1}G-\mathrm{L_O}^{-1}R(u)-\mathrm{L_O}^{-1}N(u)\\
&\overline{u}(k,t)=\Phi+\mathrm{L_O}^{-1}[D_t^{2-\alpha}\overline{u}(k,t)]\\
&\overline{u}(k,t)=\overline{u}(k,0)+t\overline{u}_t(k,0)-k^2D_t^{-2}[D_t^{2-\alpha}\overline{u}(k,t)]=\overline{u}(k,0)-k^2D_t^{-2}[D_t^{2-\alpha}\overline{u}(k,t)]
\end{aligned}\tag{11.69}$$

In (11.69) the $\mathrm{L_O}^{-1}=D_t^{-2}$, $G=R=0$, $N(u)=\overline{u}(k,t)$ is the linear.

Therefore, $\overline{u}_0=\Phi=\overline{u}_0(k,0)$ and for $n\geq 1$; $\overline{u}_n=-\mathrm{L_O}^{-1}D_t^{2-\alpha}\overline{u}_{n-1}$ Following the ADM, (11.26) we get

$$\overline{u}(k,t)=\sum_{n=0}^{\infty}\overline{u}_n(k,t)$$

where the components are listed in (11.70)

$$\begin{aligned}
&\overline{u}_0=\frac{1}{\sqrt{2\pi}}\\
&A_0=\overline{u}_0\\
&\overline{u}_1=-\mathrm{L_O}^{-1}D_t^{2-\alpha}[A_0]-k^2D_t^{-2}[D_t^{2-\alpha}\overline{u}_0]=-\frac{k^2}{\sqrt{2\pi}}\left(\frac{t^\alpha}{\Gamma(\alpha+1)}\right)\\
&A_1=\overline{u}_1\\
&\overline{u}_2=-k^2D_t^{-2}[D_t^{2-\alpha}\overline{u}_1]=\frac{k^4}{\sqrt{2\pi}}\left(\frac{t^{2\alpha}}{\Gamma(2\alpha+1)}\right)\\
&A_2=\overline{u}_2\\
&\overline{u}_3=-k^2D_t^{-2}[D_t^{2-\alpha}\overline{u}_2]=-\frac{k^6}{\sqrt{2\pi}}\left(\frac{t^{3\alpha}}{\Gamma(3\alpha+1)}\right)\\
&\ldots\ldots\ldots\ldots\ldots\ldots
\end{aligned}\tag{11.70}$$

Therefore,

$$\begin{aligned}
&\overline{u}(k,t)=\frac{1}{\sqrt{2\pi}}\left[1-\frac{k^2t^\alpha}{\Gamma(\alpha+1)}+\frac{k^4t^{2\alpha}}{\Gamma(2\alpha+1)}-\frac{k^6t^{3\alpha}}{\Gamma(3\alpha+1)}+\ldots\right]\\
&\overline{u}(k,t)=\frac{1}{\sqrt{2\pi}}\sum_{n=0}^{\infty}\frac{(k^2)^nt^{n\alpha}}{\Gamma(n\alpha+1)}=\frac{1}{\sqrt{2\pi}}E_\alpha(-k^2t^\alpha)
\end{aligned}\tag{11.71}$$

The (11.71) is series solution of (11.68) in space-Fourier Transformed system, in terms of Mittag-Leffler function, here, is one parameter Mittag-Leffler function $E_{\alpha}(-k^2t^{\alpha})$. Taking the Inverse Fourier Transform of (11.71), we get solution to (11.61) with impulse excitation as:

$$u(x,t) = \frac{1}{2}t^{-\alpha/2}M_{\alpha/2}(|x|/t^{\alpha/2})\text{; for } -\infty < x < +\infty \text{ and } t \geq 0 \tag{11.72}$$

where Inverse Fourier transform is:

$$u(x,t) = \frac{1}{\sqrt{2\pi}}\int_{-\infty}^{+\infty} e^{-ikx}\overline{u}(k,t)dk \text{ , } x \in \Re \text{ .}$$

The $M_{\alpha/2}$ is special case of Wright function defined as in (11.73)

$$M_{\gamma}(|z|) = \sum_{n=0}^{\infty} \frac{(-|z|)^n}{n!\Gamma(-n\gamma + \{1-\gamma\})}\text{;}$$

where $0 < \gamma < 1$

$$M_{\alpha/2}\left(|x|/t^{\alpha/2}\right) = \sum_{n=0}^{\infty} \frac{\left(-|x|/t^{\alpha/2}\right)^n}{n!\Gamma\left[-\frac{n\alpha}{2} + \left(1-\frac{\alpha}{2}\right)\right]}\text{; where } 0 < \frac{\alpha}{2} < 1 \tag{11.73}$$

11.12 Generalization of Fractional Order Leading Terms in Differential Equations Formulated with Riemann-Liouvelli and Caputo Definitions–and Use of Integer Order Initial/Boundary Conditions–with Decomposition Method

In this section merger of two classical definitions of Fractional Derivatives with decomposition, technique is demonstrated; where only integer order initial/boundary conditions will be employed to get to the modal solutions-in decomposition method. This generalization and unification is important as to eliminate the need of much difficult fractional order initial states required classically by RL formulation of FDE.

11.12.1 Decomposition of Caputo Derivative in Fractional Differential Equations

Let the linear part of the equation $\mathrm{L_O}u + Ru + Nu = G$ be of Caputo Fractional Derivative represented as $\mathrm{L_O} = {}^{C}D_t^{\alpha}$, which is composed of integer order derivative of function followed by fractional integration. That is if an integer m

is just greater than fraction $\alpha > 0$; that is $(m-1) < \alpha < m$, then; Caputo operator is: ${}^{C}D_t^{\alpha} \equiv D_t^{-(m-\alpha)} D_t^{m}$. This gives the differential equation system as:

$$D_t^{-(m-\alpha)} D_t^{m}(u) = G - R(u) - N(u) \tag{11.74}$$

Inverting this we get:

$$D_t^{m}(u) = D_t^{(m-\alpha)} G - D_t^{(m-\alpha)} R(u) - D_t^{(m-\alpha)} N(u) \tag{11.75}$$

The solution is:

$$u(t) = \Phi + \mathrm{L_O}^{-1} D_t^{m-\alpha} G - \mathrm{L_O}^{-1} D_t^{m-\alpha} R(u) - \mathrm{L_O}^{-1} D_t^{m-\alpha} N(u) \tag{11.76}$$

where $\mathrm{L_O}^{-1} = D_t^{-m}$; we have used complementation property that is $D^m D^{-m} = I$.

The Φ is solution to integer order homogeneous condition of $D_t^{m} u(t) = 0$; and is same as in the case of integer order general differential equation solution, described above. Due to this fact, researchers like to formulate with Caputo derivative.

11.12.2 Riemann-Liouvelli (RL) Derivative and Its Decomposition for Solving Fractional Differential Equation with Integer Order Initial Condition

The RL derivative operator is $D_t^{\alpha} \equiv D_t^{m} D_t^{-(m-\alpha)}$ i.e., the function is first fractionally integrated and then differentiated by integer order, which is just greater than the fractional order. In this solution, the homogeneous equation formed by RL operator requires fractional initial states; though sometimes difficult to interpret physically. However, one can relate these fractional initial states to physical quantities provided the laws of physics are known (As Ohm's Law, Stress-Strain relations and flow-pressure relations etc.) Here in this chapter, it is demonstrated that decomposing fractional derivative with RL definition by transforming to Caputo expression first then applying decomposition rules, one can solve the fractional differential equations with RL formulations and with integer order initial states.

Generally, the Caputo and RL definitions of fractional derivatives are not equal, but are equated by initial conditions as:

$$\left[D_t^{\alpha} f(t)\right]_{RL} = {}^{C}D_t^{\alpha} f(t) + \sum_{k=0}^{m-1} \frac{t^{k-\alpha}}{\Gamma(k-\alpha+1)} f^{(k)}(0^{+}),$$

Where $(m-1) < \alpha < m$; $m \in \mathrm{N}$ $t > 0$

(11.77)

Let the linear part L_O be of RL derivative type. Then the formulation with definition of RL derivative gives the system as:

$$D_t^m D_t^{-(m-\alpha)}(u) = G - R(u) - N(u)$$

In the expression with RL we change to Caputo and relate with RL-Caputo relation and get:

$$ {}^C D_t^{\alpha}(u) = G - R(u) - N(u) - \sum_{k=0}^{m-1} \frac{t^{k-\alpha}}{\Gamma(k-\alpha+1)} u^{(k)}(0^+) \qquad (11.78)$$

Here the RL differential equation is changed to Caputo formulation. This is equivalent to original differential equation, but with extra power series term with integer order initial conditions appearing as extra source/sink term. Let us follow the decomposition method, as obtained for Caputo formulation in section 11.0.1, thereby giving the solution as:

$$u(t) = \Phi + L_O^{-1} D_t^{m-\alpha} G - L_O^{-1} D_t^{m-\alpha} R(u) - L_O^{-1} D_t^{m-\alpha} N(u) - L_O^{-1} \left\{ D_t^{m-\alpha} \sum_{k=0}^{m-1} \frac{t^{k-\alpha}}{\Gamma(k-\alpha+1)} u^{(k)}(0^+) \right\} \qquad (11.79)$$

Let us examine the bracketed term of the RHS of, the source sink term.

$$\begin{aligned} D_t^{m-\alpha} \sum_{k=0}^{m-1} \frac{t^{k-\alpha}}{\Gamma(k-\alpha+1)} u^{(k)}(0^+) &= u^{(k)}(0^+) \sum_{k=0}^{m-1} \frac{\Gamma(k-\alpha+1) t^{k-\alpha-m+\alpha}}{\Gamma(k-\alpha+1)\Gamma(k-\alpha+1-m+\alpha)} \\ &= u^{(k)}(0^+) \sum_{k=0}^{m-1} \frac{t^{k-m}}{\Gamma(k-m+1)} \end{aligned} \qquad (11.80)$$

The expression above contains reciprocal of Gamma function at negative integer points and zero point, the values of which are zero. The reciprocal of Gamma functions $\Gamma(-(m-1)), \Gamma(-(m-2)) \ldots \Gamma(0)$ are zeros. This reciprocal Gamma function is multiplied by powers of t at $t \to 0^+ = \varepsilon$; as $\varepsilon^{-m}, \varepsilon^{1-m}, \varepsilon^{2-m}, \ldots . \varepsilon^{-1}$. Therefore, at $t \to 0^+$, the inverted source/sink extra term be collapsed to zero, i.e.,

$$D_t^{m-\alpha} \sum_{k=0}^{m-1} \frac{t^{k-\alpha}}{\Gamma(k-\alpha+1)} u^{(k)}(0^+) = 0 \cdot$$

The fractional differentiation of a power function is given by $D^{\mu} t^{\lambda} = \Gamma(\lambda+1) t^{\lambda-\mu} / \Gamma(\lambda+1-\mu)$, Euler's rule of generalized differ-integration; where $\mu \in \Re^+$, with $\lambda > -1$. Let us take monomial of type say x^n, with n as integer. We differentiate this with integer order m such that $(m-n) = 1, 2, 3, \ldots; m > n$. Then in the integer order calculus $D_x^m x^n = 0$. Say a square function x^2 differentiated thrice, four times, and so on will give zero.

Same is the observation for $D_t^{(m-\alpha)} t^{k-\alpha}$, returns zero since differential order (fractional) minus the power order that is $(m-\alpha)-(k-\alpha)=m-k$, is $m,(m-1),(m-2)......1$, for $k=0,1,2,3,.....(m-1)$. This is new observation not used elsewhere earlier in RL fractional calculus context. This new observation and its application is now useful for solving FDE with RL formulation by decomposition technique where the extra source sink term appearing in FDE (changing from RL to Caputo) collapses to zero thus giving ease and uniformity in the two definitions of fractional calculus.

The above argument suggests that with RL derivative formulations too one can have solution in ADM approach to solve fractional differential equation, with the help of integer order initial/boundary condition.

Therefore the solution of General Fractional Order Differential Equation where the leading terms are of Fractional Derivative of Caputo or RL type is:

$$u(t)=\Phi+\mathrm{L}_{\mathrm{O}}^{-1} D_t^{m-\alpha} G-\mathrm{L}_{\mathrm{O}}^{-1} D_t^{m-\alpha} R(u)-\mathrm{L}_{\mathrm{O}}^{-1} D_t^{m-\alpha} N(u) \tag{11.81}$$

where Φ comes from integer order initial/boundary conditions. This unifies the two definitions of Caputo/RL to solve FDE with only integer order initial states.

11.13 Application of Decomposition Method in RL Formulated Fractional Differential Equations (Non-Linear) and Its Solution

So far, we considered linear systems and reasoned out physically the decomposition and the action-reaction concepts to solve the differential equation systems, by ADM. The non-linear part $N(u)$ in the earlier cases were linear in nature and thus the Adomian Polynomials for each mode were same ($A_n=u_n$) for $n\geq 1$; for obtaining the subsequent parallel modes and thereby the solution. The non-linear part which is described gives different Adomian Polynomials for the different modes to get solution of non-linear systems.

Consider RL formulated Fractional differential equation with non-linearity as in (11.82)

$$\frac{d^\alpha y}{dt^\alpha}=(1-y)^4;\ \text{With}\ 0<\alpha<1;\ \text{and}\ y(0^+)=0 \tag{11.82}$$

The nearest integer in this case is one; for the fractional order α. The invert operator is $\mathrm{L}_{\mathrm{O}}^{-1}=D_t^{-1}$ and $\Phi=C=0$, $N(u)=(1-y)^4=\sum_{n=0}^{\infty}\lambda^n A_n$, $\lambda\approx 1$ the solution is thus; as in (11.61)–(11.67) is:

$$y(t)=\sum_{n=0}^{\infty} y_n=\Phi+\mathrm{L}_{\mathrm{O}}^{-1} D_t^{1-\alpha}\sum_{n=0}^{\infty} A_n=D_t^{-1}\left[D^{1-\alpha}\sum_{k=0}^{\infty} A_n\right],\ \lambda\approx 1 \tag{11.83}$$

The source/sink term that appears in the RL to Caputo change in (11.82) is $-\frac{t^{-\alpha}}{\Gamma(1-\alpha)}y(0^+)$. Taking $D_t^{m-\alpha}$ of this source/sink term gives;

$$-\frac{y(0^+)\Gamma(-\alpha+1)t^{-\alpha-m+\alpha}}{\Gamma(1-\alpha)\Gamma(-\alpha+1-m+\alpha)}=-\frac{y(0^+)t^{-1}}{\Gamma(0)}=0 \tag{11.84}$$

Therefore, the decomposed solution of (11.82) is (11.84).

The non-linearity $N(y)=(1-y)^4=\sum_{n=0}^{\infty}\lambda^n A_n$, $\lambda\approx 1$ and to find the Adomian Polynomials following are the steps (11.26) and (11.27)

The iterations are listed in (11.85)

$$y_0=0,\ A_0=(1-y_0)^4=1$$

$$y_1(t)=D_t^{-1}D_t^{1-\alpha}\left[A_0\right]=D_t^{-1}D_t^{1-\alpha}(1)=\frac{t^{\alpha}}{\Gamma(\alpha+1)}$$

$$\begin{aligned}
A_1 &= y_1 N'(y_0)\\
N'(y_0) &= \frac{d}{dy_0}(1-y_0)^4=-4(1-y_0)^3=-4\\
A_1 &= y_1(-4)=-\frac{4t^{\alpha}}{\Gamma(\alpha+1)}
\end{aligned} \tag{11.85}$$

From this, we obtain the next term of solution as (11.86)

$$y_2(t)=D_t^{-1}D_t^{1-\alpha}\left[A_1\right]=D_t^{-1}D_t^{1-\alpha}\left[\frac{-4t^{\alpha}}{\Gamma(\alpha+1)}\right]=\frac{-4t^{2\alpha}}{\Gamma(2\alpha+1)} \tag{11.86}$$

Next step is to obtain A_2 as (11.87)

$$\begin{aligned}
A_2 &= y_2 N'(y_0)+\frac{y_1^2}{2!}N''(y_0)\\
N''(y_0) &= \frac{d^2}{dy_0^2}(1-y_0)^4=12(1-y_0)^2=12\\
N'(y_0) &= -4\\
A_2 &= \left(\frac{-4t^{2\alpha}}{\Gamma(2\alpha+1)}\right)(-4)+\frac{12}{2}\left(\frac{t^{\alpha}}{\Gamma(\alpha+1)}\right)^2=\frac{6\Gamma(2\alpha+1)-16\left[\Gamma(\alpha+1)\right]^2}{\Gamma(2\alpha+1)\left[\Gamma(\alpha+1)\right]^2}t^{2\alpha}
\end{aligned} \tag{11.87}$$

From above (11.87) we obtain (11.88):

$$y_3(t) = D_t^{-1} D_t^{1-\alpha}[A_2] = \frac{6\Gamma(2\alpha+1) - 16[\Gamma(\alpha+1)]^2}{\Gamma(3\alpha+1)[\Gamma(\alpha+1)]^2} t^{3\alpha} \quad (11.88)$$

The series form solution to (11.82) is (11.89)

$$y(t) = \frac{1}{\Gamma(\alpha+1)} t^{\alpha} - \frac{4}{\Gamma(2\alpha+1)} t^{2\alpha} + \frac{6\Gamma(2\alpha+1) - 16[\Gamma(\alpha+1)]^2}{\Gamma(3\alpha+1)[\Gamma(\alpha+1)]^2} t^{3\alpha} + \ldots\ldots\ldots\ldots \quad (11.89)$$

11.14 Application of Decomposition Method in RL-Formulated Partial Fractional Differential Equations Non-linear Diffusion-Wave Equation and Solution

Consider the Non-Linear equation (11.90), formulated by RL fractional differential operator

$$\left[\frac{\partial^{\alpha}}{\partial t^{\alpha}} u(x,t)\right]_{RL} = \gamma^2 \frac{\partial^2}{\partial x^2} u(x,t) - c^2 u(x,t) + \sigma[u(x,t)]^3 \quad (11.90)$$

with $1 < \alpha \le 2$ and $u(x,0) = \varepsilon \cos kx$, and the first time derivative $u_t(x,0) = 0$ as initial condition.

The argument about vanishing extra source/sink term while RL is changed to Caputo formulation in this case is same as in the previous problem (11.61)– (11.67).

In this case, $m = 2$, that is, integer number just greater than α the order of fractional derivative of (11.90). The non-linear part is

$N(u) = c^2 u - \sigma u^3 = \sum_{n=0}^{\infty} \lambda^n A_n$, $\lambda \approx 1$.

We simplify the (11.90) in easier notation as

$$D_t^{\alpha} u = \gamma^2 u_{xx} - c^2 u + \sigma u^3;$$

Where

$$u_{xx} \equiv \frac{\partial^2}{\partial x^2} u(x,t) .$$

From (11.61)–(11.67) we write the solution of (11.90) as (11.91):

$$\begin{aligned} u(x,t) &= \Phi + \gamma^2 D_t^{-2} D_t^{2-\alpha}[u_{xx}] - D_t^{-2} D_t^{2-\alpha} N(u) \\ u(x,t) &= u(x,0) + t u_t(x,0) + \gamma^2 D_t^{-2} D_t^{2-\alpha}[u_{xx}] - D_t^{-2} D_t^{2-\alpha} \sum_{n=0}^{\infty} A_n \end{aligned} \quad (11.91)$$

The iteration follows as shown in steps in (11.92)

$$u_0 = \Phi = u(x,0) = \varepsilon \cos kx$$
$$A_0 = c^2 u_0 - \sigma u_0^{\,3} = c^2 \varepsilon \cos kx - \sigma \varepsilon^3 \cos^3 kx$$
$$u_1 = \gamma^2 D_t^{-2} D_t^{2-\alpha} u_{0_{(xx)}} - D_t^{-2} D_t^{2-\alpha} A_0$$
$$u_{0_{(xx)}} = \frac{\partial^2 u_0}{\partial x^2} = -\varepsilon k^2 \cos kx$$
$$D_t^{-2} u_{0_{(xx)}} = \int_0^t \int_0^t u_{0_{(xx)}} dtdt = -\frac{\varepsilon k^2 t^2}{2} \cos kx$$
$$D_t^{-2} A_0 = \int_0^t \int_0^t A_0 dtdt = \frac{t^2 c^2}{2} \varepsilon \cos kx - \frac{\sigma \varepsilon^3 t^2}{2} \cos^3 kx$$
$$u_1 = -\frac{\gamma^2 \varepsilon k^2 \cos kx}{\Gamma(\alpha+1)} t^\alpha - \frac{(c^2 \varepsilon \cos kx - \sigma \varepsilon^3 \cos^3 kx)}{\Gamma(\alpha+1)} t^\alpha$$
$$A_1 = u_1 (c^2 - 3\sigma u_0^{\,2})$$
$$u_2 = \gamma^2 D_t^{-2} D_t^{2-\alpha} u_{1_{(xx)}} - D_t^{-2} D_t^{2-\alpha} A_1$$
$$u_{1_{(xx)}} = \frac{\partial^2 u_1}{\partial x^2}$$
$$\begin{aligned} u_2 = {} & \frac{\varepsilon \gamma^4 k^4 \cos kx}{\Gamma(2\alpha+1)} t^{2\alpha} - \left[\frac{\gamma^2}{\Gamma(2\alpha+1)}\right]\left[-c^2 \varepsilon k^2 \cos kx + \frac{3k^2 \sigma \varepsilon^3}{4} \cos kx + \frac{9k^2 \sigma \varepsilon^3}{4} \cos^3 kx\right] t^{2\alpha} \\ & + \frac{c^2 \varepsilon \gamma^2 k^2 \cos kx}{\Gamma(2\alpha+1)} t^{2\alpha} + \frac{c^2 (c^2 \varepsilon \cos kx - \sigma \varepsilon^3 \cos^3 kx)}{\Gamma(2\alpha+1)} t^{2\alpha} - \frac{3\sigma \varepsilon^3 \gamma^2 k^2 \cos^3 kx}{\Gamma(2\alpha+1)} t^{2\alpha} \\ & - \frac{3\sigma \varepsilon^2 \cos^2 kx (c^2 \varepsilon \cos kx - \sigma \varepsilon^3 \cos^3 kx)}{\Gamma(2\alpha+1)} t^{2\alpha} \end{aligned} \tag{11.92}$$

Therefore the approximate (three term) solution to (11.90) is

$$u(x,t) \approx u_0 + u_1 + u_2 = \sum_{n=0}^{2} u_n$$

11.15 Decomposition Method for Generalized Equation of Motion

Consider the equation of motion (11.93) a non-linear one; where two bodies of mass m_1 and m_2 collide and the variable $x(t)$ denotes indentation with respect to time.

The contact force is Hertz force given by $f = Kx^n$. For spherical particles this index of non-linearity is $n = 3/2$. The parameter s is depending on the contact properties of material.

$$\frac{m_1 m_2}{2(m_1+m_2)}\frac{d^2x}{dt^2} + K(n+1)x^n = 0 \tag{11.93}$$

We consider the (11.93) in simple way as

$$\frac{d^2x}{dt^2} = -ax^{3/2}\text{, where, } n = 3/2 \text{ and } a = \frac{5(m_1+m_2)K}{m_1 m_2} \tag{11.94}$$

with $x(0)=0$ and $D_t^1\left[x(t)\right]_{t=0} = \dot{x}(0) = u$ as the initial conditions. The (11.94) has

$$G = 0,\ R = 0\ ,\ \mathrm{L_O}^{-1} = D_t^{-2},\ N(x) = ax^{3/2}.$$

The ADM for (11.94) gives set as (11.95).

$$\Phi = x(0) + t\dot{x}(0) = ut$$

$$x_0 = \Phi + \mathrm{L_O}^{-1}G = ut$$

$$A_0 = N(x_0) = ax_0^{3/2} = au^{3/2}t^{3/2}$$

$$x_1 = -\mathrm{L_O}^{-1}R(x_0) - \mathrm{L_O}^{-1}(A_0) = -D_t^{-2}\left[au^{3/2}t^{3/2}\right] = -\frac{4}{35}au^{3/2}t^{7/2}$$

$$A_1 = x_1 N'(x_0) = \frac{3}{2}ax_0^{1/2}x_1 = -\frac{12}{70}a^2u^2t^4$$

$$x_2 = -\mathrm{L_O}^{-1}R(x_1) - \mathrm{L_O}^{-1}(A_1) = -D_t^{-2}\left[-\frac{12}{70}a^2u^2t^4\right] = \frac{12}{210}a^2u^2t^6 \tag{11.95}$$

and so on. The solution, with decomposed modes of displacements as series, to (11.93) and (11.94) is

$$x(t) = ut - \frac{4au^{3/2}}{35}t^{7/2} + \frac{12a^2u^2}{210}t^6 + ... \tag{11.96}$$

For completion sake let $n = 0$, in, (11.93) and (11.94). This makes $N(x_0) = a = A_0$, and $N'(x) = 0$, therefore $A_1 = 0 = A_2 = ... = A_\infty$.

Here, $x_0 = \Phi + \mathrm{L_O}^{-1}G = ut$ and $x_1 = -\mathrm{L_O}^{-1}A_0 = -D_t^{-2}(a) = -\frac{1}{2}at^2$, with higher modes all zero that is $x_2 = x_3 = .. = x_\infty = 0$. The solution is just sum of zeroth and first mode of reaction that is $x(t) = ut - \frac{1}{2}at^2$; this is familiar to Newton's laws of motion.

For, $n = 1$ the equation (11.93) becomes oscillatory system (mass spring) described earlier.

11.16 Decomposition Method for Delay Differential Equation System

In this concluding section, application of mode decomposition in delay differential equation is demonstrated. A delayed differential equation system is system like (11.97).

$$\mathrm{L_O}(u) + R(u) + N(u) = G + f(x, u(x), u(g(x))) \tag{11.97}$$

Consider a non-linear delayed differential equation system as (11.98).

$$\frac{dy}{dx} = 1 - 2y^2(x/2) \tag{11.98}$$

with $y(0) = 0$, as initial condition.

Here in (11.98) $G = 1, R = 0, \mathrm{L_O}^{-1} = D_x^{-1}, N(y) = 2y^2(x/2)$. Applying ADM we get the set as mentioned in (11.99)

$$y_0(x) = \Phi + \mathrm{L_O}^{-1}G = D_x^{-1}[1] = x$$

$$y_0(x/2) = \frac{x}{2}$$

$$A_0 = N(y_0) = 2y_0^2(x/2) = 2\left(\frac{x}{2}\right)^2 = \frac{1}{2}x^2$$

$$y_1(x) = -\mathrm{L_O}^{-1}(A_0) = -D_x^{-1}\left(\frac{x^2}{2}\right) = -\frac{x^3}{6}$$

$$y_1(x/2) = -\frac{(x/2)^3}{6}$$

$$N'(y_0) = 4y_0(x/2); N''(y_0) = 4$$

$$A_1 = y_1 N'(y_0) = y_1(x/2)4y_0(x/2) = -\frac{(x/2)^3}{6} \times 4\left(\frac{x}{2}\right) = -\frac{x^4}{24}$$

$$y_2(x) = -\mathrm{L}_{\mathrm{O}}^{-1}(A_1) = -D_x^{-1}\left(-\frac{x^4}{24}\right) = \frac{x^5}{120}$$

$$y_2(x/2) = \frac{(x/2)^5}{120}$$

$$A_2 = y_2 N'(y_0) + \frac{1}{2!} y_1^2 N''(y_0) = y_2(x/2) \times 4y_0(x/2) + \frac{1}{2} y_1^2(x/2) \times 4 = \frac{x^6}{720}$$

$$y_3(x) = -\mathrm{L}_{\mathrm{O}}^{-1}(A_2) = -D_x^{-1}(A_2) = \frac{x^7}{5040} \tag{11.99}$$

The solution to (11.98) is

$$y(x) = x - \frac{x^3}{6} + \frac{x^5}{120} - \frac{x^7}{5040} + ... = \sin x\,.$$

This is exact solution to (11.98) as the RHS of (11.98) is

$$1 - 2y^2(x/2) = 1 - 2\sin^2(x/2) = \cos x = d(\sin x)/dx$$

11.17 Proposition

11.17.1 Fractional Initial States Classical Solution to FDE

Classical theory of fractional differential equation solution gives the idea of fractional initial states. Consider the fractional differential equation system (Homogeneous) as:

$$[D + aD^{1/2} + bD^0]y(t) = 0 \tag{11.100}$$

The Laplace Transforming the (11.100) gives:

$$[sY(s) - y(0)] + a[s^{1/2}Y(s) - D^{-1/2}y(0)] + bY(s) = 0\,,$$

gives the following arranged relation, as

$$[s + as^{1/2} + b]Y(s) - y(0) - aD^{-1/2}y(0) = 0 \tag{11.101}$$

giving

$$Y(s) = C / P(\sqrt{s}) \text{, with } C = y(0) + aD^{-1/2} y(0) \text{,}$$

and indicial polynomial as: $P(x) = x^2 + ax + b = (x-\alpha)(x-\beta)$, with α, β as roots of indicial polynomial. Now, we question, how do we, know that $C = y(0) + aD^{-1/2} y(0)$ is finite? How do we physically find the meaning and value of $D^{-1/2} y(0)$ that is fractional initial state? If $C = y(0) + aD^{-1/2} y(0)$ is not finite then problem is serious and this approach is meaningless. With partial fraction approach, we get partial fraction of inverse indicial polynomial and with assumption that C is finite, and nonzero constant, we proceed to find solution to (11.100). If $C = 0$, then the only solution is trivial solution that is $y(t) = 0$. The partial fraction of indicial polynomial is

$$\frac{1}{P(x)} = \frac{1}{\alpha - \beta}\left[\frac{1}{x-\alpha} - \frac{1}{x-\beta}\right] \tag{11.102}$$

Putting $x = (s)^{1/2}$, we get

$$\frac{1}{P(\sqrt{s})} = \frac{1}{\alpha - \beta}\left[\frac{1}{\sqrt{s}-\alpha} - \frac{1}{\sqrt{s}-\beta}\right] \tag{11.103}$$

The inverse Laplace is solution to (11.100), and in the form of Robotnov-Harley function, as one possibility we obtain, solution to (11.100) as

$$y(t) = \frac{C}{\alpha - \beta}\left[F_{1/2}(\alpha, t) - F_{1/2}(\beta, t)\right],$$

for $\alpha \neq \beta$, and $y(t) = C[F_{1/2}(\alpha, t) * F_{1/2}(\alpha, t)]$, this is for $\alpha = \beta$. The Robotnov-Hartley function, defined as

$$F_q(a, t) = \sum_{n=o}^{\infty} \frac{a^n t^{(n+1)q-1}}{\Gamma(\{n+1\}q)},$$

with its Laplace-Transform as

$$F_q(a, s) = \frac{1}{s^q - a}.$$

In this approach, the fractional initial state is arising from Generalization of Laplace Transform, and for this example is explained as:

$$D^{1/2}y(t) = D^1[D^{-1/2}y(t)],$$

is Riemann-Liouvelli definition of fractional derivative.

Let $D^{-1/2}y(t) = y_1(t)$. Then Laplace pair is

$$Y_1(s) = s^{-1/2}Y(s).$$

Then Laplace of RL

$$D^{1/2}y(t) = D^1 y_1(t) \text{ is } [sY_1(s) - y_1(0)],$$

putting values of

$$y_1(0) = D^{-1/2}y(0), Y_1(s),$$

the Laplace of RL

$$D^{1/2}y(t) = s^{1/2} - D^{-1/2}y(0),$$

requiring fractional initial state.

If the half-derivative operator in (11.100) had been of Caputo type, then

$${}^{C}D^{1/2}y(t) = D^{-1/2}[D^1 y(t)],$$

With $y_1(t) = D^1 y(t)$, has $Y_1(s) = sY(s) - y(0)$. The Laplace of Caputo half derivative is

$$s^{-1/2}[Y_1(s)] = s^{1/2}Y(s) - s^{-1/2}y(0)$$

and requires integer order initial state.

Therefore the dichotomy persists in (11.100) if formulated by RL scheme then requires $y(0)$ and $y^{-1/2}(0)$, and if formulated by Caputo requires $y(0)$. This is one reason for Caputo derivative being popular. In RL, definition $y^{-1/2}(0)$ is hard to visualize.

Generally, the Caputo and RL definitions of fractional derivatives are not equal, but are equated by initial conditions as

$$\left[D_t^{\alpha} f(t)\right]_{RL} = {}^{C}D_t^{\alpha} f(t) + \sum_{k=0}^{m-1} \frac{t^{k-\alpha}}{\Gamma(k-\alpha+1)} f^{(k)}(0^+) \qquad (11.104)$$

These two definitions are equal only when, the initial states are zero. The (11.100) is of first-order differential equation with remaining lesser orders, classically, should have required only one initial state yet the treatment is asking for two initial states. However, the definition of order in FDE is not as simple as Integer Order Calculus as the first order system with half order element may behave as classical second order system of Integer Order Calculus, yet the application scientists and engineers will be comfortable if the RL formulations of FDE requires only integer order states to get practical realizable responses to FDE system.

11.17.2 Basic Fractional Order Differential Equation System and its Classical Solution

A fractional differential equation system is described as with RL definition of derivative as

$$[D^{nv} + a_1 D^{(n-1)v} + a_2 D^{(n-2)v} + \dots\dots.. a_n D^0]y(t) = G(t) \tag{11.105}$$

If of order (n,q), where n,q are positive integer and v is positive fractional number, $v = 1/q$. If we have integer $N \geq nv$, where homogeneous initial /boundary conditions are given as:

$$D^j y(0) = 0,\ j = 0,1,2,\dots..N-1.$$

For (11.105) the indicial polynomial is $P(x)$, and the Green's function (the solution of homogeneous system of (11.105) is $K(t)$ which is Laplace Inverse of $[P(s^v)]^{-1}$, then solution of (11.105) is

$$y(t) = \int_0^t K(t-\xi)G(t)d\xi .$$

Here, point is to be mentioned that the homogeneous integer order initial states are given for this simple convolution of Green's function and the source (forcing) function. In the homogeneous differential equation system of (11.105) the source term is zero, requires the idea of fractional initial states, as explained in example (11.100). The homogeneous system of (11.105) has fundamental solution $y_1(t)$ as Laplace Inverse of $[P(s^v)]^{-1}$, with $y_1, y_2, y_3 \dots y_N$ as linearly independent solutions with $y_{j+1} = D^j y_1(t)$, $j = 0,1,2,\dots N-1$.This is classical theorem of Fractional Calculus to have solution to FDE.

11.17.3 Classical Solution to Fractional Fokker-Plank Kolmogorov Equation (FFPK) by Fourier-Laplace Technique

Consider generalized FFPKE given as:

$$\frac{\partial^{\beta} P(x,t)}{\partial t^{\beta}} = \frac{\partial^{\alpha} P(x,t)}{\partial |x|^{\alpha}} + \frac{t^{-\beta}}{\Gamma(1-\beta)} \delta(x),$$

we would like the solution and asymptotic solution, where the expression of the source is taken to satisfy the normalized condition

$$\int_{-\infty}^{+\infty} P(x,t)dx = 1.$$

Let us introduce Laplace-Fourier Transformation as

$$P(k,s) = \int_{0}^{\infty} dt \int_{-\infty}^{+\infty} dx e^{-st+ikx} P(x,t).$$

This concept and the FFPKE were discussed in earlier chapters. Using this Laplace-Fourier transform to the FFPKE we obtain the following:

$$s^{\beta} P(k,s) + |k|^{\alpha} P(k,s) = s^{\beta-1}.$$

On applying only Fourier transform we get:

$$\frac{\partial^{\beta}}{\partial t^{\alpha}} P(k,t) + |k|^{\alpha} P(k,t) = \frac{t^{-\beta}}{\Gamma(1-\beta)}, \text{for } t > 0.$$

The solution of this is

$$P(k,t) = \sum_{m=0}^{\infty} \frac{(-1)^m}{\Gamma(m\beta+1)} \left(|k|^{\alpha} t^{\beta}\right)^m,$$

it is easy to prove this as solution by Euler's formula that is

$$\frac{\partial^{\beta} t^{\gamma}}{\partial t^{\beta}} = \frac{\Gamma(\gamma+1)}{\Gamma(\gamma+1-\beta)} t^{\gamma-\beta}.$$

In compact form the solution of the FDE in Fourier space is

$$P(k,t) = E_{\beta}\left(-|k|^{\alpha} t^{\beta}\right);$$

that is individual (Fourier) modes decay with Mittag-Leffler pattern with time. If we do term by term Fourier invert that is writing $\Im^{-1}E_\beta\left(-|k|^\alpha t^\beta\right)$, we will get $P(x,t)$, that is as follows

$$P(x,t)=\frac{1}{\pi|y|}\frac{1}{t^{\mu/2}}\sum_{m=1}^{\infty}\left\{\frac{(-1)^m}{|y|^{m\alpha}}\frac{\Gamma(m\alpha+1)}{\Gamma(m\beta+1)}\cos\left[\frac{\pi}{2}(m\alpha+1)\right]\right\},$$

where $y=x/t^{\mu/2}$, $\mu=2\beta/\alpha$.

It follows the asymptotic form as

$$P(x,t)\sim\frac{1}{\pi}\frac{t^\beta}{|x|^{\alpha+1}}\frac{\Gamma(1+\alpha)}{\Gamma(1+\beta)}\sin\left(\frac{\pi\alpha}{2}\right),$$

where, $|x|>>t^{\mu/2}=t^{\beta/\alpha}$. From this we also can conclude that Fractional Order moments that is $|x|^\delta$, $\delta>0$ converge if $0<\delta<\alpha<2$. Similar solution was obtained in previous section earlier (11.71).

11.17.4 Decomposition of Fractional Differential Equation Principle and Equivalence of RL and Caputo Definitions to Solve FDE with Integer Order Initial States

In this section a system of FDE with RL formulation is decomposed by changing to Caputo and step by step convergence is shown. Convert the (11.105) to Caputo System and write as:

$${}^{C}D^{nv}y(t)+\sum_{k=0}^{N-1}\frac{t^{k-nv}y^{(k)}(0)}{\Gamma(k-nv+1)}+a_1\,{}^{C}D^{(n-1)v}y(t)+a_1\sum_{k=0}^{N-2}\frac{t^{k-(n-1)v}y^{(k)}(0)}{\Gamma(k-(n-1)v+1)}+..a_nD^0y(t)=G(t)$$

(11.106)

The nearest integer for first term of LHS of (11.106) is N, for second term $N-1$, and so on.

Decomposing the (11.106) with Caputo definition, we obtain:

$$[D^{-(N-nv)}D^Ny(t)+a_1D^{-(N-1-(n-1)v)}D^{N-1}y(t)+...a_nD^0y(t)]+[\sum_{k=0}^{N-1}\frac{t^{k-nv}y^{(k)}(0)}{\Gamma(k-nv+1)}+$$
$$+a_1\sum_{k=0}^{N-2}\frac{t^{k-(n-1)v}y^{(k)}(0)}{\Gamma(k-(n-1)v+1)}+.................]=G(t)$$

(11.107)

Differentiating (11.107) $(N-nv)$ times and applying identity expression for differential operator $D^m D^{-m}=I$, we obtain

$$[D^N y(t)+a_1 D^{-(1-v)} D^{N-1} y(t)+a_2 D^{2(1-v)} D^{N-2} ... a_n D^{N-nv} D^0 y(t)]+[D^{N-nv}\sum_{k=0}^{N-1}\frac{t^{k-nv}y^{(k)}(0)}{\Gamma(k-nv+1)}+$$

$$+a_1 D^{(N-nv)}\sum_{k=0}^{N-2}\frac{t^{k-(n-1)v}y^{(k)}(0)}{\Gamma(k-(n-1)v+1)}+................]=D^{(N-nv)}G(t)$$

(11.108)

Let us examine the extra power series terms in (11.108) and apply Euler's formula

$$D^{\mu}t^{\lambda}=\Gamma(\lambda+1)t^{\lambda-\mu}/\Gamma(\lambda+1-\mu),$$

where $\mu\in\Re^+$, with $\lambda>-1$, then

$$D^{N-nv}t^{k-nv}=\frac{\Gamma(k-nv+1)}{\Gamma(k-nv+1-N+nv)}t^{k-nv-N+nv}=\frac{\Gamma(k-nv+1)}{\Gamma(k+1-N)}t^{k-N}$$

$$D^{N-nv}\sum_{k=0}^{N-1}\frac{t^{k-nv}y^{(k)}(0)}{\Gamma(k-nv+1)}=y^{(k)}(0^+)\sum_{k=0}^{N-1}\frac{\Gamma(k-nv+1)t^{k-nv-N+nv}}{\Gamma(k-nv+1)\Gamma(k-nv+1-N+nv)}$$
$$=y^{(k)}(0^+)\sum_{k=0}^{m-1}\frac{t^{k-N}}{\Gamma(k-N+1)} \qquad (11.109)$$

The expression (11.109) contains reciprocal of Gamma function at negative integer points, the values of which are zero. The reciprocal of Gamma functions $\Gamma(-(N-1)),\Gamma(-(N-2))...\Gamma(0)$ are zeros. This reciprocal Gamma function is multiplied by powers of t at $t\to 0^+=\varepsilon$; as $\varepsilon^{-N},\varepsilon^{1-N},\varepsilon^{2-N},....\varepsilon^{-1}$. Therefore, at $t\to 0^+$, the (11.109) be collapsed to zero, i.e.,

$$\frac{\varepsilon^{-N}}{\Gamma(1-N)}+\frac{\varepsilon^{1-N}}{\Gamma(2-N)}+\frac{\varepsilon^{2-N}}{\Gamma(3-N)}+...+\frac{\varepsilon^{-1}}{\Gamma(0)}=0;$$

giving

$$D_t^{N-nv}\sum_{k=0}^{m-1}\frac{t^{k-nv}}{\Gamma(k-nv+1)}y^{(k)}(0^+)=0 \qquad (11.110)$$

The 0^+ is put in above argument is because the differentiation is valid for half-open interval $(0,t)$. The Euler expression has $\lambda>-1$, for the functions for ease in differintegrable class, has been put for convenience of existence. This (11.110) is new observation not used elsewhere earlier in RL fractional calculus context. This new observation and its application is now useful for solving FDE with RL formulation by decomposition technique where the extra source sink term appearing in FDE (changing from RL to Caputo) collapses to zero thus giving ease and uniformity in the two definitions of fractional calculus. This reduces (11.108) to

$$[D^{N}y(t)+a_1D^{-(1-v)}D^{N-1}y(t)+a_2D^{2(1-v)}D^{N-2}...a_nD^{N-nv}D^{0}y(t)]+$$
$$[a_1D^{(N-nv)}\sum_{k=0}^{N-2}\frac{t^{k-(n-1)v}y^{(k)}(0)}{\Gamma(k-(n-1)v+1)}+................]=D^{(N-nv)}G(t) \tag{11.111}$$

Look at the second differential term in (11.109) that is $D^{1-v}D^{N-1}$. The fractional derivative is Caputo type and the nearest integer is "one". Therefore, $D^{1-v}=D^{-v}D^{1}$, is the Caputo definition. Differentiating v times

$$[D^{v}D^{N}y(t)+a_1D^{1}D^{N-1}y(t)+a_2D^{v}D^{2(1-v)}D^{N-2}...a_nD^{v}D^{N-nv}D^{0}y(t)]+$$
$$[a_1D^{v}D^{(N-nv)}\sum_{k=0}^{N-2}\frac{t^{k-(n-1)v}y^{(k)}(0)}{\Gamma(k-(n-1)v+1)}+................]=D^{v}D^{(N-nv)}G(t) \tag{11.112}$$

Observing, power series term of the term in LHS of (11.112) the term is:

$$D^{v}D^{(N-nv)}\sum_{k=0}^{N-2}\frac{t^{k-(n-1)v}y^{(k)}(0)}{\Gamma(k-(n-1)v+1)}=D^{N-nv+v}\sum_{k=0}^{N-2}\frac{t^{k-(n-1)v}y^{(k)}(0)}{\Gamma(k-(n-1)v+1)},$$

has derivative order $N-nv+v$, and power function exponent as $k-nv+v$; the expanded term will be the following:

$$D^{N-nv+v}\sum_{k=0}^{N-2}\frac{t^{k-(n-1)v}y^{(k)}(0)}{\Gamma(k-(n-1)v+1)}=\sum_{k=0}^{N-2}\frac{t^{k-N}}{\Gamma(k+1-N)}=\lim_{t\to 0^{+}}\left[\frac{\varepsilon^{-N}}{\Gamma(1-N)}+\frac{\varepsilon^{1-N}}{\Gamma(2-N)}+...\frac{\varepsilon^{-2}}{\Gamma(-1)}\right]=0$$

The (11.112) becomes:

$$[D^{v}D^{N}y(t)+a_1D^{1}D^{N-1}y(t)+a_2D^{v}D^{2(1-v)}D^{N-2}y(t)...a_nD^{v}D^{N-nv}D^{0}y(t)]+$$
$$[a_2D^{v}D^{N-nv}\sum_{k=0}^{N-3}\frac{t^{k-(n-2)v}y^{(k)}(0)}{\Gamma(k-(n-2)v+1)}+................]=D^{v}D^{(N-nv)}G(t) \tag{11.113}$$

Reducing the fractional integral operator D^{v}, gives the (11.113) as:

$$[D^N y(t)+a_1D^{1-v}D^{N-1}y(t)+a_2D^{2(1-v)}D^{N-2}y(t)...a_nD^{N-nv}D^0y(t)]+$$
$$[a_2D^{N-nv}\sum_{k=0}^{N-3}\frac{t^{k-(n-2)v}y^{(k)}(0)}{\Gamma(k-(n-2)v+1)}+.................]=D^{(N-nv)}G(t) \quad (11.114)$$

The (11.113) and (11.109) are of similar the power term after this second iteration has become zero, and in (11.113) remaining power terms are present. It appears that the power terms for the coefficients $a_1, a_2 + ...$, will keep on reducing to zero if we follow this stepwise decomposition. Let us repeat the steps for the remaining power functions of (11.113). The differential Caputo is

$$D^{2(1-v)}D^{N-2} = D^{-2v}D^2D^{N-2}.$$

We differentiate (11.113) by $2v$ to get

$$[D^{2v}D^N y(t)+a_1D^{2v}D^{1-v}D^{N-1}y(t)+a_2D^{2v}D^{2(1-v)}D^{N-2}y(t)...a_nD^{2v}D^{N-nv}D^0y(t)]+$$
$$[a_2D^{2v}D^{N-nv}\sum_{k=0}^{N-3}\frac{t^{k-(n-2)v}y^{(k)}(0)}{\Gamma(k-(n-2)v+1)}+.................]=D^{2v}D^{(N-nv)}G(t)$$
$$(11.115)$$

The differentiation of power term, of exponent $(k-(n-2)v)$ by $(N-nv+2v)$, returns zero as the discussed for the power exponents for other previous coefficients. That is

$$D^{2v}D^{N-nv}\sum_{k=0}^{N-3}\frac{t^{k-(n-2)v}}{\Gamma(k-(n-2)v+1)}=\sum_{k=0}^{N-3}\frac{t^{k-N}}{\Gamma(k+1-N)}=\lim_{t\to 0^+}\left[\frac{\varepsilon^{-N}}{\Gamma(1-N)}+\frac{\varepsilon^{1-N}}{\Gamma(2-N)}+...+\frac{\varepsilon^{-3}}{\Gamma(-2)}\right]=0$$

Thus, power term of (11.114) goes to zero and again integrating on both sides of (11.114) by $2v$, we get

$$[D^N y(t)+a_1D^{1-v}D^{N-1}y(t)+a_2D^{2(1-v)}D^{N-2}y(t)...a_nD^{N-nv}D^0y(t)]+$$
$$[a_3D^{N-nv}\sum_{k=0}^{N-4}\ +.................]=D^{(N-nv)}G(t) \quad (11.116)$$

This way by induction we arrive at decomposed equation, above formulated by RL definition as:

$$[D^N+a_1D^{1-v}D^{N-1}+a_2D^{2(1-v)}D^{N-2}+......+a_nD^{N-nv}]y(t)=D^{N-nv}G(t) \quad (11.117)$$

In one example of $(3,2)$ order

$$[D^{3/2}-2D-D^{1/2}+2D^0]y(t)=0,$$

formulated by RL derivative, this method gives equivalent one as

$$[D^2 - 2D^{3/2} - 2D^1 + D^{1/2}]y(t) = 0\,,$$

is matching with above criteria. For a system where $N = nv$, the reduction need not be carried out. This procedure states that with RL formulated system of differential equations one can decompose the same, with above described method, where the leading order is integer order N, and thus should require N integer order initial states-not fractional to get the fundamental solution. If the formulation were of Caputo then this decomposition method also holds.

The example of (11.100) formulated is having highest order as one, should have required one initial states, but classically we saw that it had asked for fractional order states too when formulated by RL definition. Thus, decomposition method obtained here for RL and Caputo gives uniformity in solving Fractional Differential Equations, with decomposition.

11.17.5 Application to Fractional Diffusion-Wave Equation with Input Sine Excitation with RL-Formulation

Here demonstration is made to have ADM for RL formulation giving the same result as with Caputo and obtained analytical result. For simplicity and demonstration of this method, the diffusion-wave equation constant is assumed unity.

Consider

$$\left[\frac{\partial^\alpha}{\partial t^\alpha}u(x,t)\right]_{RL} = \frac{\partial^2}{\partial x^2}u(x,t)\,,\ 1 < \alpha \le 2 \tag{11.118}$$

with

$$u(x,0^+) = \sin x\,,\quad \frac{\partial u(x,0^+)}{\partial t} = u_t(x,0^+) = 0\,.$$

The $m = 2$ in the case of above diffusion-wave equation and while changing RL to Caputo formulation gives the extra source/sink term as

$$-\frac{t^{-\alpha}}{\Gamma(1-\alpha)}u(x,0^+) - \frac{t^{1-\alpha}}{\Gamma(2-\alpha)}u_t(x,0^+)\,.$$

Operating fractional derivative $D_t^{m-\alpha} \equiv D_t^{2-\alpha}$ on this source/sink term gives

$$-u(x,0^+)\frac{t^{-2}}{\Gamma(-1)} - u_t(x,0^+)\frac{t^{-1}}{\Gamma(0)} = 0$$

Referring to earlier discussion we can re-write the above as:

$$u(x,t) = \Phi + L_O^{-1}\left(D_t^{m-\alpha}\left[\frac{\partial^2}{\partial x^2} u(x,t) \right] \right),$$

for this system $m = 2$, $L_O^{-1} = D_t^{-2}$ and thus

$$\Phi = u(x,0) + t\frac{\partial}{\partial t}u(x,0) = \sin x .$$

By the decomposition technique, we write the following:

$$u_0 = \Phi = \sin x$$

$$u_1 = L_O^{-1}\left(D_t^{2-\alpha}\left[\frac{\partial^2}{\partial x^2} u_0 \right] \right) = D_t^{-2}\left(D_t^{2-\alpha}\left[\frac{\partial^2}{\partial x^2} \sin x \right] \right) = -\frac{t^\alpha}{\Gamma(\alpha+1)}\sin x$$

$$u_2 = L_O^{-1}\left(D_t^{2-\alpha}\left[\frac{\partial^2}{\partial x^2} u_1 \right] \right) = D_t^{-2}\left(D_t^{2-\alpha}\left[\frac{\partial^2}{\partial x^2} \left\{ -\frac{t^\alpha}{\Gamma(\alpha+1)} \right\}\sin x \right] \right) = \frac{t^{2\alpha}}{\Gamma(2\alpha+1)}\sin x$$

$$u_3 = L_O^{-1}\left(D_t^{2-\alpha}\left[\frac{\partial^2}{\partial x^2} u_2 \right] \right) = D_t^{-2}\left(D_t^{2-\alpha}\left[\frac{\partial^2}{\partial x^2} \left\{ \frac{t^{2\alpha}}{\Gamma(2\alpha+1)}\sin x \right\} \right] \right) = -\frac{t^{3\alpha}}{\Gamma(3\alpha+1)}\sin x$$

....

(11.119)

Therefore the solution to is:

$$u(x,t) = u_1 + u_2 + u_3 + ... = \left(\sum_{n=0}^{\infty} \frac{(-t^\alpha)^n}{\Gamma(n\alpha+1)} \right)\sin x$$

The expression above can be exactly put in close form by use of Mittag-Leffler function as $u(x,t) = E_\alpha(-t^\alpha)\sin x$.

11.18 Observations

From the above arguments, the following is stated: Assume that $G(t)$ and $y(t)$ are "well-behaved" in half open interval $(0,t]$, and are fractionally differ-integrable.

A Fractional Differential Equation (11.120)

$$[D^{nv} + a_1 D^{(n-1)v} + a_2 D^{(n-2)v} + ...a_n D^0]y(t) = G(t) \tag{11.120}$$

with RL Derivative or Caputo Derivative be decomposed as

$$[D^N + a_1 D^{1-v} D^{N-1} + a_2 D^{2(1-v)} D^{N-2} + ... + a_n D^{N-nv}] y(t) = D^{N-nv} G(t) \qquad (11.121)$$

With $N \geq nv$, and solved with initial states

$$y^j(t),\ j = 0,1,2........., (N-1).$$

The fundamental solution is

$$y(t) = y_0 - a_1 D^{-v} \sum_{k=1}^{\infty} y_{k-1} - ..... - a_n D^{-nv} \sum_{k=1}^{\infty} y_{k-1} \qquad (11.122)$$

With Φ a solution of $D^N y(t) = 0$, and $y_0 = \Phi + D^{-nv} G(t)$.

Requiring only integer order initial states, for Fractional Differential Equation with leading terms as fractional order. This we can call Lemma ,derived from all previous discussions and examples.

11.19 Concluding Comments

The Decomposition Method of Mathematics of Linear Analysis, as demonstrated is generalization of the physical law of nature that is, the process reacts in opposite way to thwart any changes in the process variable. The Decomposition Method as explained for physical systems gives insight into micro-scale reactions, to the external or internal stimulus so as to oppose the changes; thereby generating infinite (or finite) modes of reactions, the sum of which gives the total system behavior. Fractional Differential Equation (FDE) by RL method it is found that there is no need to worry about the fractional initial states; instead one can use integer order initial states (the conventional ones) to arrive at solution of FDE. This new finding too is highlighted in this chapter-along with several other problems to give physical insight to the solution of extraordinary differential equation systems. This way one gets insight to Physics of General Differential Equation Systems–and its solution–by Physical Principle and equivalent mathematical decomposition method. This facilitates ease in modeling to get approximate analytic behavior of General Dynamic System-involving Extra Ordinary Differential Equations. Well to conclude the fractional differential equations be it of Riemann-Liouvelli, or be it Caputo formulated, can be solved in the practical way with integer order initial states, and with physical principle of decomposition and action reaction. These fractional order system of differential equations are as rigorous as their counterparts in integer order systems, well if the system are solvable then there is close relation to physics and series reaction and series solution.

References

1. Abbaoui, K., Cherruault, Y.: The Decomposition Methods Applied to the Cauchy Problems. Kybernetes 28, 103–108 (1999)
2. Abbaoui, K., Cherruault, Y.: Convergence of Adomian Method Applied to Differential Equations. Computers Maths Applic. 28(5), 103–109 (1994)
3. Abbaoui, K., Cherruault, Y.: New Ideas for Proving Convergence of Decomposition Methods. Computers Math. Applic. 29, 103–108 (1995)
4. Abramowitz, M., Stegun, I.A. (eds.): Handbook of Mathematical Functions. Nat. Bur. Stand. Appl. Math. Ser. 55. U.S. Govt. Printing Office Washington D C (1964)
5. Adomian, G.: Nonlinear Stochastic System theory and Application to Physics. Kluwer Academic Publishers, Netherlands (1989)
6. Adomian, G.: An Analytical Solution of Stochastic Navier-Stokes Systems. Foundations of Physics 21(7), 831–843 (1991)
7. Adomian, G.: Solution of Physical Problems by Decomposition. Computers Math. Appl. 27(9/10), 145–154 (1994)
8. Adomian, G.: Solving Frontier Problems of Physics: The decomposition method. Kluwer Academic, Boston (1994)
9. Adomian, G.: Solutions of Nonlinear PD.E. Appl. Math. Lett. 11(3), 121–123 (1998)
10. Adomian, G., Rach, R.: Linear and Nonlinear Schrodinger Equations. Foundations of Physics 21, 983–991 (1991)
11. Agarwal, O.P.: Solution for a Fractional Diffusion-Wave Equation Defined in a Bounded Domain. Nonlinear Dynamics 29, 145–155 (2002)
12. Agarwal, O.P.: A General Formulation and Solution for Fractional Optimal Control Problem. In: Nonlinear Dynamics, vol. 38, pp. 323–337. Kluwer Academic Publishers, Dordrecht (2004)
13. Akoi, Y., Sen, M., et al.: Approximation of transient temperature in complex geometries using fractional derivatives. Department of Aerospace and Mechanical Engineering University of Notre Damme. Technical Note (December 19, 2005)
14. Al-Alouni, M.A.: A class of second order integrators and low pass differentiators. IEEE Trans. on circuit and systems 1: Fundamental Theory and Applications 42(4), 220–223 (1995)
15. Al-Alouni, M.A.: Filling the gap between the bilinear and backward difference transforms. an interactive design approach. Int. J of Electrical Engineering Education 34(4), 331–337 (1997)
16. Alexander, S., Orbach, R.: Density of states on fractals (fractons). J-Physique-Letters 43, L-625–L-631 (1982)
17. Almenas, K., Lee, R.: Nuclear Engineering: An introduction. Springer, Berlin (1992)

18. Anh, V.V., et al.: Fractional differential equation driven by Levy noise. J. Appl. Math and stochastic analysis 16(2), 97–119 (2003)
19. Arora, H.L., Abdelwahid, F.I.: Solution of Non-Linear Order Differential Equations via Adomian Decomposition Method. Appl. Math. Lett. 6(1), 21–23 (1993)
20. Aware, M.V., Dhabale, A., Bhandarkar, S., Das, S.: Analogue Emulator for DC Motor with Integral & Fractional Order PID. Patent Reference Application 334/MUM/2010 Intellectual Property Office of India Mumbai (2010)
21. Aware, M.V., Dhabale, A., Bhandarkar, S., Das, S.: Modular Circuit for realizing any order of Fractional Differ Integrals. Patent Reference Application 344/MUM/2010 Intellectual Property Office of India Mumbai (2010)
22. Aware., M.V., Dhabale., A.S., Bhandarkar, S.G.: Design algorithm for two port fractional order differintegrals-Patent Document at Indian Patent Office(Mumbai-India) (2010)
23. Axtell, M., Bise, M.E.: Fractional calculus applications in control systems. In: Proc. of IEEE Nat. Aerospace and Electronics Conf., New York, pp. 563–566 (1990)
24. Bachelier, L.: Theorie de la speculation. Ann. Sci. Ecole. Norm. Suppli. 3(17), 21–86 (1900); Reprinted Cootner. P. H. (Ed.). The Random Character of Stock Market Price, Cambridge MA: MIT: Press (1964)
25. Bagley, R.L.: Power law and fractional calculus model of viscoelasticity. AIAA Journal 27(10), 1412–1417 (1989)
26. Bagley, R.L.: The thermorheological complex materials. Int. J. Engg. Sci. 29(7), 797–806 (1991)
27. Bagley, R.L., Calico, R.A.: Fractional order state equations for control of viscoelastic structures. J. Guid. Cont. and Dyn. 14(2), 304–311 (1991)
28. Bagley, R.L., Torvik, P.J.: Fractional calculus in transient analysis of viscoelastically damped structures. AIAA Journal 23(6), 918–925 (1985)
29. Balakrishnan, V.: Anomalous diffusion in one-dimension. Physica. A 132, 569–580 (1985)
30. Barcena, R., et al.: Auto tuning of fractional order hold circuit. In: IEEE. Int. Conf on Control applications, Mexico City, September 5-7, pp. 7–12 (2001)
31. Barcena, et al.: Discrete control for computer hard disc by fractional order hold devise, pp. 1–34. University of Pais Vasc Spain (2000)
32. Batman, H.: Higher transcendental function, vol. 3. Mc Graw Hill, New-York (1954)
33. Benson: The fractional order governing levy motion. Water resources research 36(6), 1413–1423 (2000)
34. Bhambhani, V., Chen, Y., Xue, D.: Optimal Fractional Order Proportional Integral Controller for Varying Time-Delay Systems. In: IFAC World Congress Seoul Korea (2008)
35. Bhambhani, V., Han, Y., Mukhoopadhyay, S., Luo, Y., Chen, Y.: Random Delay Effect Minimization on Hardware-in-loop Networked Control System Using Optimal Fractional Order PI Controllers. In: 3rd IFAC FDA, November 5-7 (2008)
36. Bhattacharya, R., Gupta, V.K.: Application of central limit theorems to solute transport in saturated porous media, from kinetic to field scales: Dynamics of fluids in hierarchical porous media. In: Cushman, J.H. (ed.), pp. 61–96. Academic Press, San Diego (1990)
37. Bisquert, J., Compte, A.: Theory of electrochemical impedance of anomalous diffusion. J. of Electro analytical Chemistry 499, 112–120 (2001)
38. Biswas, K., Sen, S., Dutta, P.K.: Modeling of a capacitive probe in a polarizable medium. Sensors actuators: Phys. 120, 115–122 (2005)

39. Biswas, K., Sen, S., Dutta, P.K.: A constant phase element sensor for monitoring microbial growth. Sensors and Actuators SNB9063 9063, 1–6 (2005)
40. Black, F., Scholes, M.: Pricing Options and Corporate Liability. J. Political Econ. 81, 637 (1973)
41. Bode, H.W.: Network analysis and feed back amplifier design. Van Nostrand, New-York (1949)
42. Bohannan, G.: Analog Realization of fractional order control element revisited. Department of Physics Montonna State Univ. Technical Note (October 27, 2002)
43. Bologna, M.: Short Introduction to fractional calculus. Lecture Notes, pp.41–54. Univ. Tarpeca Chile (2005)
44. Bush, V.: Operational Circuit Analysis. Wiley, New-York (1929)
45. Buslowicz, M.: Stability of linear continuous-time fractional order system with delays of retarded type. Bulletin of the Polish Academy of Sciences Technical sciences 56(4), 319–324 (2008)
46. Caputo, M.: Distributed Order Differential Equation Modeling Dielectric Induction and Diffusion. Fractional Calculus and Applied Analysis 4(4), 421–442 (2001)
47. Caputo, M.: Linear Models of Dissipation whose Q is almost Frequency Independent. Geophysics Journal of Royal Astronomical Society 13, 529–539 (1967)
48. Caputo, M.: Mean Fractional Order-Derivative Differential Equations and Filters. Ann. Univ. Ferrara-Ser. VII-Sc. Mat. 1, 73–84 (1995)
49. Caputo, M.: The rheology of an anelastic medium studied by means of the observation of splitting of its eigenfrequencies. J. Acoust. Soc. Am. 86(5), 1984–1987 (1989)
50. Caputo, M.: The Riemann Sheet Solutions of Anelasticity. Annali di Matematica pura Ed applicata (IV) 1, 335–342 (1994)
51. Caputo, M., Mainardi, F.: A new dissipation model based on memory mechanism. Pure and applied Geophysics 91(8), 134–147 (1971)
52. Carlsaw, H.S., Jaeger, J.C.: The Conduction of Heat in Solids. Oxford Univ. Press, London (1947)
53. Carlson, et al.: Approximation of fractional capacitors$(1/s)^{1/n}$ by regular Newton process. IEEE Trans. on Circuit Theory 11(2), 210–213 (1964)
54. Carlson, Halijak: Simulation of the fractional derivative operator operator $\sqrt{s}$ *and fractional integral operator* $1/\sqrt{s}$. Kansas State University Bulletin ~45, 1-22 (1961)
55. Carpinteri, A., Mainardi, F.: Fractals and Fractional Calculus in Continuum Mechanics. Springer, New York (1997)
56. Chandra, S., Nag, S., Tarafdar, S., Das, S.: Analysis of the Spreading of non-Newtonian fluid using Fractional Calculus. In: National Symposium on Mathematical Modeling of Natural Phenomena NMMNP-2010, October 29, Bose Institute, Calcutta-India (2010)
57. Charef, A., Sun, H.H., Tsao, Y.Y., Onaral, B.: Fractal systems as represented by singularity function. IEEE Trans. on Automatic Control 37(9), 1465–1470 (1992)
58. Chaves, A.S.: A fractional diffusion equation to describe Levy flights. Phys. Letts. A 239, 13–16 (1998)
59. Chen, Y., Vinagre, B.M., Podlubny, I., et al.: On fractional order disturbance observer. In: DETC03-VIB48371 ASME Design Engineering Technical Conference Chicago USA, September 2-6, pp. 1–8 (2003)

60. Chen, Y.Q.: Analytical stability bound for class of delayed fractional order dynamic system. In: IEEE Conf. On Decision and control Florida WeA05-1, pp. 1421–1424 (2001)
61. Chen, V.: A new type of IIR digital fractional order differentiator. Signal Processing 83, 2359–2365 (2003)
62. Chen, Y., Vinagre, B.M., Podlubny, I.: Fractional order disturbance observer for robust vibration suspension. Non Linear Dynamics 38, 355–367 (2004)
63. Cherruault, Y.: Convergence of Adomian's Method. Kybernetes 18, 31–38 (1989)
64. Choudhury, M.D., Chandra, S., Nag, S., Tarafdar, S., Das, S.: Oscillatory Spreading and Surface Instability of a Non-Newtonian Fluid under Compression, December 10 (2010); Condense Matter Archives Ref: ARXIV: 1012.0882v2
65. Churchill Ruel, V., Brown James, W., Verhey Roger, F.: Complex Variables & Applications, 3rd edn. Mc Graw Hill Kogakusha Ltd (1974)
66. Churchill, R.V.: Operational Mathematics, 2nd edn. McGraw-Hills, New-York (1948)
67. Compte, A.: Continuous time random walks on moving fluid. Phys. Rev. E 55(6), 6821–6831 (1997)
68. Coronel-Brizo, Hernandez-Montoya: On fitting the Patreo-Levy distribution to stock market index data. Selecting a suitable cut-off value. Physica A 354, 437–449 (2005)
69. Crank, J.: The Mathematics of Diffusion. Oxford Univ. Press, London (1956)
70. Daoud, M.: Spectral dimension and conductivity exponent of the percolating cluster. J-Physique Letters 44, L925–L928 (1983)
71. Das, S., Biswas, B.B.: Fuel Efficient Nuclear Reactor Controls. In: Int. Conf. on Nuclear Engineering, ICONE-13 (50843) Beijing (May 2005)
72. Das, S., Biswas, B.B.: Total Energy Utilization from Nuclear Source. In: PORT-2006, Nuclear Energy for New Europe Slovenia (2006)
73. Das, S., Majumder, B., Pan, I., Gupta, A., Das, S.: A New Fractional Fourier Transform Based design of a Band Pass FIR filter for Power Feed Back in Nuclear Reactor under noisy environment. In: ICETECT (IEEE Conference Kanyakumari-India) (2011)
74. Das, S., Mukherjee, S., Pan, I., Gupta, A., Das, S.: Identification of Core Temperature in Fractional Order Noisy Environment for Thermal Feed Back in Nuclear Reactor. In: IEEE Student's Technology Symposium IIT, Kharagpur India, January 14-16 (2011)
75. Das, S.: Efficient control of Nuclear Plants IAEA-TM- Control & Instrumentation (2007)
76. Das, S.: Mathematico-Physics of Generalized Calculus University Course work book for PhD (Applied Mathematics and Physics) for Department of Applied Mathematics, University of Calcutta and Department of Physics, University of Jadavpur; in limited prints at University of Calcutta and University of Jadavpur-Calcutta
77. Das, S.: Solution of Extraordinary Differential Equation with Physical Reasoning by Obtaining Modal Reaction Series. In: Modeling and Simulation in Engineering, pp. 1–19. Hindawi Publishing Corp. (2010); ID-739675
78. Das, S.: Convergence of Riemann-Liouvelli and Caputo Derivative for Practical Solution of Fractional Order Differential Equations. International Journal of Applied Mathematics & Statistics 23(D11), 64–73 (2011)

79. Das, S., et al.: Ratio control with logarithmic logics in new P&P control algorithm for a true fuel-efficient reactor. Int. J. Nuclear Energy Science & Technology 3(1), 1–18 (2007)
80. Das, S.: Fractional Stochastic Modeling for Random Dynamic Delays in Computer Control System. International Journal of Applied Mathematics & Statistics 21(J11), 131–140 (2011)
81. Das, S.: Generalized Dynamic Systems Solution by Decomposed Physical Reactions. International Journal of Applied Mathematics & Statistics J10(special Issue), 44–75 (2010)
82. Das, S., Bhattacharya, S., Keswani, R.K., Sunder Rajan, S.: Circuit Theory Approach with Fractional Calculus to describe half space Geophysical Analysis for Transient Electromagnetic Method. Geophysical Journal T-31(2), 147–159 (2009)
83. Das, S., Biswas, B.B.: Fractional divergence for Neutron Flux profile in Nuclear Reactor. Europe Int. J. of Nuclear Energy Science & Technology 3(2), 139–159 (2007)
84. Das, S., Biswas, B.B.: Embedding Nuclear Physics Rules & Fuel Chemistry Limits in Control Algorithm for Fuel Efficient Reactor. Int. J. of Nuclear Governance Economy and Ecology 1(3), 244–264 (2007)
85. Das, S., Biswas, B.B.: Shaped Normalized Reactor Period Function, definitions properties & application. Int. J. Nuclear Energy Science & Technology 2(4), 309–327 (2006)
86. Das, S., Biswas, B.B.: Controlling Nuclear Plants with Fuel Efficiency. Int. J. of Nuclear Power-atw-Gmbh 2, 107–116 (2007)
87. Davidson, B.: Neutron Transport Theory. Oxford Press, London (1957)
88. Diethelm, K.: An algorithm for the numerical solution of differential equations of fractional order. In: Electronics transactions on Numerical Analysis, March 1997, vol. 5, pp. 1–6 (1997); ISSN 1068-9613
89. Diethelm, K.: An Algorithm for Numerical Solutions of Differential Equations of Fractional Order. Elec. Transact. Numer. Anal. 5, 1–6 (1997)
90. Diethelm, K., Ford, N.J., Freed, A.D.: A predictor-Corrector Approach for the Numerical Solution of Fractional Differential Equations. Nonlinear Dyn. 29, 3–22 (2002)
91. Dive, R., Aware, M.V., Dhabale, A.S., Das, S.: New Method of Fractional Order Differintegral Representation with Fractional Approximation. In: National Symposium of Mathematical Modeling of Natural Phenomena NMMNP-2010, Calcutta-India, October 29 (2010)
92. Dive, R., Dhabale, A.S., Aware, M.V., Das, S.: Calculation of Pole Zero Position for Realization of Fractional Order Impedance by Interlacing Poles and Zeros. In: International Conference of System Dynamics & Controls ICSDC, Manipal-India, August 19–22 (2010)
93. Dominik, S.: Fractional Kalman Filter algorithm for states, parameters and order of fractional system estimation. XV Krajowa Konferencja Automatyki, Warszawa, Polska, pp. 1–14 (2005)
94. Dorcak, L.: Numerical Models for simulation of fractional order control system. In: UER04-1994, pp.1–15. Slovak Academy of Science (1994)
95. Dorcak, Petras, I., et al.: Fractional order state space models. In: Int. Carpathian Control Conference ICCC Malenovice Czech Rep., May 27–30, pp. 193–198 (2002)
96. Dostal, J.: Operational Amplifiers. Butterworth-Heinemann, Boston (1993)

97. Dubois, F., Mengue, S.: Mixed Collocation for fractional differential equations. Numerical Algorithms 34(2), 303–311 (2005)
98. Dutta Roy, S.C.: On realization of constant-argument immitances of fractional operator. IEEE Trans. on Circuit Theory 14(3), 264–374 (1967)
99. Einstein, A.E.: Investigations on theory of the Brownian movement (Translation). Dover, Mineola N.Y (1956)
100. El Sayeed, A.M.A.: Fractional order diffusion wave equation. Int. J. of Theor. Phy. 35, 311–322 (1996)
101. Engheta: The fractional curl operator in electromagnetic. Microwave Opt. Tech. Letter 17, 86–91 (1998)
102. Engheta, N., Werner, D.H., Mitra, R. (eds.): Frontiers in Electromagnetism, pp. 523–552. IEEE, New-York (2000)
103. Erdelyi, A.: On some functional transformation. Univ. Potitec. Torino (1950)
104. Erdelyi, A.: Asymptotic expansions. Dover, New York (1954)
105. Erdelyi, A. (ed.): Tables of Integral Transforms, vol. 1. McGraw-Hill, New York (1954)
106. Falconer, K.: Fractal Geometry. John Wiley, New-York (1990)
107. Fama, E.: Test of the Multi-period Two Parameter Model. J. Business 35, 420 (1963)
108. Famma, E., Roll, R.: Parameter Estimate for Symmetric Stable Distribution. J. Am. Stat. Assoc. 63, 817–836 (1972)
109. Feller, W.: An introduction to probability theory and its applications, 2nd edn., vol. II. John Wiley, New York (1971)
110. Gary, B., et al.: Electrical components with fractional order impedances. Patent US 20060267595 (November 2006)
111. Gel'fand, I.M., Shilov, G.E.: Generalized functions, vol. 1. Academic Press, New-York (1964)
112. George, A.J., Chakrabarti, A.: The Adomian Method Applied to Some Extra Ordinary Differential Equations. Appl. Math. Letts. 8(3), 91–97 (1995)
113. Glockle, W.G., Nonnenmacher, T.F.: Fractional Integral Operators and Fox Functions in the Theory of Viscoelasticity. Macromolecules 24, 6426–6434 (1991)
114. Glockle, W.G., Nonnenmacher, T.F.: Fractional integrators and Fox functions in the theory of viscoelelasticity. Macromolecules 24, 6426–6436 (1991)
115. Gopal, M.: Modern Control System Theory, 2nd edn. Wiley Eastern Limited (April 1993)
116. Gopikrishnan, P., Plerou, V., Nunes Amaral, L.A., Meyer, M., Stanley, H.E.: Scaling of the distribution of fluctuations of financial market indices. Phys. Rev. E 60, 5305 (1999)
117. Gorenflo, R., Luchko, Y., Rogosin, S.: Mittag-Leffler Type Functions: Notes on Growth Properties and Distribution of Zeros, no. A-97-04. Department of Mathematics and Informatics, Free University of Berlin (1997)
118. Gorenflo, R., Mainardi, F.: Fractional oscillators and Mittag-Leffler function. Technical Report Preprint A-14/96, University of Berlin (1996)
119. Gorenflo, R.: Fractional Calculus: Some numerical methods. In: Fractals and Fractional Calculus in Continuum Mechanics, Springer, New York (1997)
120. Gorenflo, R., Mainardi, F.: Fractional calculus: integral and differential equations of fractional order. In: Fractals and Fractional Calculus in Continuum Mechanics, Springer, New York (1997)
121. Gradshteyn, I.S., Ryzhik, I.M.: Tables of Integral Series & Products. Academic Press, San Diago (1992)

122. Halijak, C.A.: An RC impedance approximation to $(1/s)^{1/2}$. IEEE Circuit Theory CT-11(4), 494–495 (1964)
123. Hartley, T., Lorenzo, C.: Fractional order system identification by continuous order distribution. In: Elsevier Signal Processing, pp. 2287–2300 (2003)
124. Hartley, T., Lorenzo, C.F.: Chaos in fractional Chua's circuit. IEEE Trans. On Circuits and Systems-I: Fundamental Theory and Applications 42(8), 485–490 (1995)
125. Heaviside, O.: Electromagnetic Theory, Chelsea Edition New-York, vol. II (1922/1971)
126. Heymans, Podlubny, et al.: Physical interpretation of initial conditions for fractional differential equation with Reimann-Liouvelli fractional derivative. Rheological Acta, November 29 (2005)
127. Hifler, R.: Applications of fractional calculus in Physics. World Scientific, Singapore (2000)
128. Hifler, R.: Fractional Diffusion based on Riemann-Liouvelli Fractional Derivative. J. Phys. Chem. B 104, 3914 (2000)
129. Hifler, R., Anton, L.: Fractional master equation and fractal random walk. Phys. Rev. A 51, 818–851 (1995)
130. Hiroshi, W.: Spectral dimension of a wire network. J. Phys A: Math. Gen. 18, 2807–2823 (1985)
131. Hwang, C., Cheng: A numerical algorithm for stability testing of fractional delay systems. Elsevier Automatca 42, 825–831 (2006)
132. Ichise, M., Nagayanagi, Y., Kojima, T.: An analog simulation of non-integer order transfer function analysis for electrode process. J. Electroanal. Chem. 33(1), 253–265 (1971)
133. Jens-Uwe, S., Alexader, B.: On the statistics of generalized Gaussian structures: collapse and random external fields. J. Phys. A: Math. Gen. 28, 6669–6674 (1995)
134. Jifeng, W., Yunkai, et al.: Frequency domain analysis and application for fractional order control system. J. of Phy. Conference Series 13, 268–273 (2005)
135. Jones, H.E., Shenoi, B.A.: Maximally flat lumped element approximation to fractional operator immitance function. IEEE Trans. on Circuits & Systems 17, 125–128 (1970)
136. Kapoor, V., Gelhar: Transport in three dimensional aquifer: 1. Dynamics of concentration fluctuations. 2. Predictions and observations of concentration fluctuations. Water Resources. Res. 30(6), 1775–1801 (1994)
137. Karimi, A., Garcia, D., Longchamp, R.: PID controller design using Bode's integral. In: Proceedings of the American Control Conference, vol. 6, pp. 5007–5012 (2002)
138. Kaya, D.: An Explicit and Numerical Solutions of Some Fifth Order KdV Equation by Decomposition Method. Applied Math. Comp. 144(2-3), 353–363 (2003)
139. Kaya, D.: A Numerical Simulation of Solitary-Wave Solution of the Generalized Regularized Long-Wave Equations. Appl. Math. Comp. 149(3), 833–841 (2004)
140. Kaya, D., El-Sayeed, S.M.: An Application of the decomposition method for the Generalized KdV and RLW Equations. Chaos, Solitons and Fractals 17(5), 869–877 (2003)
141. Kaya, D., El-Sayeed, S.M.: On a Generalized Fifth Order KdV Equations. Phys. Lett. A 310(1), 44–51 (2003)
142. Kaya, D., Yokus, A.: A Numerical Comparison of Partial Solutions in the Decomposition Method for Linear and Nonlinear Partial Differential Equations. Math. Comp. Simul. 60(6), 507–512 (2002)

143. Kirayakova, V.: Generalized Fractional Calculus and Applications. In: Pitman Research Notes in Maths. No. 301, Longman, Harlow (1994)
144. Klafter, J., Blumen, A., Shlesinger, M.F.: Stochastic pathways to anomalous diffusion. Phys. Rev. A 35 (1987)
145. Kolwankar, K.M., Gangal, A.D.: Local Fractional Derivatives and fractal functions of several variables. J. of Physics 9801010 1 (1998)
146. Kolwankar, Gangal: Fractional differentiability of nowhere differentiable functions and dimensions. Chaos 6(4), 505–513 (1996)
147. Kolwankar, K.M., Gangal, A.D.: Local Fractional Fokker-Plank Equation. Physical Review Letters 80(2), 214–217 (1998)
148. Koponen, I.: Analytical approach to problem of convergence of truncated Levy flights towards the Gaussian Stochastic Process. Phys. Rev. E 52, 1197 (1995)
149. Kundert, K.: Modeling di-electric absorption in capacitors. The Designer Guide Community Technical Note (2004)
150. Kuo, D., Ford: Analysis of fractional differential equations. In: Numerical Analysis Report No 377, pp. 1–18. University of Manchester (2003)
151. Kuo, F.F.: Network Analysis and Synthesis. Willey, Chichester (1966)
152. Lankarani, H.M., Nikravesh, P.E.: A contact Force Model with Hysteresis Damping for Impact Analysis of Multibody System. J. of Mech. Design 112, 369–376 (1990)
153. Launsse, P., Oustaloup, et al: A restricted complexity controller with CRONE control system design and tuning. University Bordeaux-ENSEIRB, Technical Note (2003)
154. Lazarevie, M.P.: Finitc Time Stability Analysis PD$^{\alpha}$ of fractional control of robotic time delays systems. Mechanics Research Communications 33, 269–279 (2006)
155. LePage, W.R.: Complex variables and the Laplace transform for Engineers. Dover, New-York (1961)
156. Leszcynski, J., et al.: A numerical method solution for ordinary differential equation of fractional order. Technical university of Czestochowa Poland Lecture Notes (2005)
157. Li, X., et al.: On concept of local fractional differentiation. Department of applied Mathematics University of Western Ontario London. Lecture Notes (2004)
158. Loverro, A.: Fractional Calculus history definition and application. Department of Aerospace and Mechanical Engg. University of Notre Damme. Technical Note (May 8, 2004)
159. Dorcak, L., Petraset, I., et al.: Comparison of the methods for discrete approximations of fractional order operator. Acta Montanisca Slovak Rocnik 8, cislo 4, 236–239 (2003)
160. Lurie, B.J.: Three parameter tunable Tilt-Integral-Derivative TID Controller. US Patent, 5-371-670 (1994)
161. Lynch, V.E., et al.: Numerical methods of solution of partial differential equations of fractional order. J. of Comput. Phy. 192, 406–421 (2003)
162. Lévy, P.: Théorie de l'Addition des Variables Aléatoires. Gauthier-Villars, Paris (1937)
163. Mainardi, F.: Fractional relaxation-oscillation and fractional diffusion wave phenomena. Chaos Solitons and Fractals 7, 1467–1477 (1996)
164. Manabe, S.: Early development of fractional order control. ASME DETC2003/VIB-48370 pp. 1-8
165. Manabe, S.: The Non-Integer Integral and its application to Control Systems. ETJ. of Japan 6(3-4), 83–87 (1961)

166. Mandelbrot, B.B.: The variation of certain speculative prices. J. Business 36, 394–419 (1963)
167. Mandelbrot, B.B.: The Fractal Geometry of Nature. Freeman, New-York (1977)
168. Mandelbrot, B.B.: Self-Affine and Fractal Dimension. Physica Scripta 32, 257–260 (1985)
169. Mandelbrot, B.B.: The Variation of Certain Speculative Prices. J. Business 36, 294 (1963)
170. Mandelbrot, B.B., van Ness, J.W.: Fractional Brownian motion, fractional noise and applications. SIAM Rev. 10(4), 422 (1968)
171. Mantega, R.N., Stanley, H.E.: Scaling behavior in the dynamics of an economic index. Nature 376, 46 (1995)
172. Mantega, R.N., Stanley, H.E.: Stochastic Process with Ultraslow Convergence to a Gaussian: The Truncated Levy Flight. Phys. Rev. Lett. 73, 2946 (1994)
173. Matignon, D.: Generalized Fractional Differential and Difference Equation: Stability Properties and Modeling Issues. In: Proc. of Maths. Theory of Networks and Systems Symposium, Padova, Italy (1998)
174. Metzler, R., Klafter, J.: The Random Walk's Guide to Anomalous Diffusion: A fractional Dynamics Approach. Physics Report 339, 1–77 (2000)
175. Middleton, D.: An introduction to Statistical Communication Theory. McGraw-Hill, New York (1988)
176. Miller, K.S., Ross, B.: An Introduction to Fractional Calculus and Fractional Differential Equations. Willey, New York (1993)
177. Moshrefi-Torbati, M., Hammond, J.K.: Physical and geometrical interpretation of fractional operators. Journal of Franklin Institute 335(6), 1077–1086 (1998)
178. Muller, U.A., Dacorogna, M.M., Pictet, O.V.: A Practical Guide to Heavy Tails. In: Adler, R.J., Feldman, R.E., Taqqu, M.S. (eds.) Statistical Techniques and Applications, p. 283. Birkhäuser, Basel (1998)
179. Munkhammar, J.D.: Fractional calculus and Taylor-Riemann series. UUDM Thesis Report (2005)
180. Nakagava, M., Sorimachi, K.: Basic characteristics of fractance device. IEICE Trans. fundamentals E75-A(12), 1814–1818 (1992)
181. Naqvi, Q.A., Abbas, M.: Fractional duality solutions and corresponding sources. PIER 25, 290–294 (2000)
182. Oberhettinger, F., Badii, L.: Tables of Laplace transforms. Springer, Berlin (1973)
183. Ogata, A., Banks, R.B.: A solution of differential equation of longitudinal dispersion in porous medium. U.S Geological Survey Professional Papers 411-A (1961)
184. Oldham, K.B.: Interrelation of current and concentration at electrodes. J. Appl. Electrochem. 21, 1068–1072 (1991)
185. Oldham, Spanier: Fractional calculus. Academic Press, San Diego (1974)
186. Oldham, Spanier: The replacement of Fick's law by formulation involving semi differentiation. J. Electroanal. Chem. 26, 331–341 (1970)
187. Ortigueira, M.D.: An Introduction to the Fractional Continuous Time Linear Systems. The 21st Century Systems, IEEE Circuits and Systems Magazine, Third Quarter, 19–26 (2008)
188. Oustaloup, A.: From fractality to non-integer order derivation through recursivity, a property common to these two concepts: a fundamental idea for a new process control strategy. In: Proceedings of the 12th IMACS World Congress Paris, July 18-22, vol. 3, pp. 203–208 (1988)

189. Oustaloup, A., et al.: Frequency band complex non-integer differentiator characterization and synthesis. IEEE Trans. on circuits and systems-I: Fundamental Theory and Application 47(1), 25–39 (2000)
190. Parvate, A., Gangal, A.D.: Calculus on Fractal Subset on Real line-I: Formulation. In: Fractals, vol. 17(1), pp. 53–81. World Scientific, Singapore (2009)
191. Paul, M., Stanley Eugene, H.: Spectral Dimension for the Diffusion-Limited Aggregation Model of Colloid Growth. Physical Review Letters 51(16), 1457–1460 (1983)
192. Petras, I., Vinagre, B.M.: Practical Applications of digital fractional order controller to Temperature Control. In: Acta Montanistica, Slovaca Rocnik 7, vol. 2, pp. 131–137 (2002)
193. Petras, I., Vinagre,BM.: Practical application of digital fractional order controller to temperature control. In: Acta Montansitica Slovak (2002)
194. Petras, I.: The fractional order controller method for their synthesis and application. BERG University Koissce Slovak Republic Technical Note (2004)
195. Podlubny, I., Dorcak, Kostial, J.: On fractional derivatives, fractional order dynamic systems and PI^{λ}-D^{μ} controllers. In: Procc. of the 36th IEEE CDC, San Diego, December 2004, pp. 4985–4990 (2004)
196. Podlubny, I., Petras, I., et al.: Realization of fractional order controllers. In: Podlubny, I., Petras, I., et al. (eds.) Acta Montanistica Slovak, Rocnik, vol. 8, pp. 233–235(2003)
197. Podlubny, I., et al.: Using continued fraction expansion to discretize fractional order derivative. FDTA Non Linear Dynamic, For submission to special issue on Fractional Derivative and their application (2003)
198. Podlubny, I.: Fractional derivatives: A new stage in process modeling and control. In: 4th International DAAAM Symposium, Brno, Czech Republic, September 16-18, pp. 263–264 (1993)
199. Podlubny, I.: Matrix approach to discrete fractional calculus. Fractional Calculus and Applied Analysis 3(4) (2000)
200. Podlubny, I.: Presentation on fractional calculus. IEEE chapter control, San Diego, September 29 (2005)
201. Podlubny, I.: The Laplace Transform Method for Linear Differential Equations of the Fractional Order. Inst. Exp. Phys. Slovak Acad. Sci., No. UEF-02-94, Kocice (1994)
202. Podlubny, I.: Fractional differential equations. Academic Press, San Diego (1999)
203. Podlubny, I.: Geometric and Physical meaning of fractional derivative. In: Fractional Calculus and Applied Analysis, vol. 5(4), pp. 367–386. Technical University, Kosice Slovak (2002); ISSN 1311-045
204. Podlubny, I.: Fractional order systems and PI^{λ} D^{μ} controllers. IEEE Trans. Auto Cont. 44(1), 208–214 (1999)
205. Poularikas, A.D.: The handbook of formulas and tables for signal processing. In: The Electrical Engineering Handbook Series. CRC Press and IEEE Press, New York (1999)
206. Quan, Y., Sun, R., Zhou, A.: An Overview of Fractional Order Signal Processing (FOSP) Techniques. In: Proceedings of DET 2007. ASME Design Engineering Technical Conference, Las Vegas, Nevada, USA, September 4–7, pp. 1–16 (2007)
207. Raberto, M., Scalas, E., Mainardi, F.: Waiting times and returns in high frequency financial data: an empirical study, pp. 1–18. Elsevier, Amsterdam (2006); ARXIX:COND-MAT/0203596VI 28

208. Ramiro, S., Barbosa, et al.: Time domain design of fractional differintegrators using least square. Signal Processing 86, 2567–2581 (2006)
209. Rammal, R., Toulouse, G.: Random walks on fractal structures and percolation clusters. J. Physique-Letters 44, L13–L22 (1983)
210. Rayes-Melo, M.E., Martinez-Vega, J.J., Guerrero-Salazar, C.A., Ortiz-Mendez, U.: Modeling of relaxation phenomena in organic dielectric materials, applications of differential and integral operator of fractional order. Journal of Optoelectronics and Advanced materials 6(3), 1037–1043 (2004)
211. Reis, C., et al.: Synthesis of logic circuit using fractional order dynamic fitness function. Transaction on engineering computing and Technology, 77–80 (2004); ISSN ENFORMATICA VI 2004 1305-5313
212. Riew, F.: Mechanics with fractional derivatives. Phys. Rev. E 55(3), 3581–3592 (1997)
213. Robotnov, Y.N.: Elements of Hereditary Solid Mechanics. English MIR Publishers, Moscow (1980)
214. Roceo, A., West, B.J.: Fractional Calculus and evolution of fractal phenomena. Elsevier Physica A. 265, 535–546 (1999)
215. Roman, H.E., Alemany, P.A.: Continuous time random walks and fractional diffusion equation. J. Phys A: Math. Gen. 27, 3407–3410 (1994)
216. Rouse Prince Jr., E.: A theory of Linear Viscoelastic Properties of Dilute Solutions of Coiling Polymer. The Journal of Chemical Physics 21(7), 1272–1280 (1953)
217. Sabatier, J., Agrawal, O.P., Machado, J.A. (eds.): Advances in Fractional Calculus. Springer, Heidelberg (2007)
218. Saha Ray, S., Bera, R.K.: An approximate solution of non-linear fractional differential equation by Adomian's Decomposition. Analytical Applied Math. Computer 167, 561–571
219. Saha Ray, S., Bera, R.K.: Analytical solution of Bagley Torvik equation by Adomian's decomposition method. Appl. Math. Comp. 168, 389–410
220. Saha Ray, S., Bera, R.K.: Analytical solution of dynamic system containing fractional derivative of order one-half by Admonian decomposition method. ASME J. of Applied Mechanic 72(1) (July 2005)
221. Saha Ray, S.: A new approach for the application of Adomian decomposition method for solution of fractional space diffusion equation with insulated ends. Applied Mathematics and Computations 202(2), 544–549
222. Saha Ray, S.: Exact solution for time fractional diffusion wave equation by decomposition method. Physics Scripta, 53–61 (2007)
223. Saha Ray, S.: Exact solution for time-fractional diffusion-wave equations by decomposition method. Physics Scripta 75, 53–61
224. Saha, S., Das, S., Ghosh, R., Goswami, B., Gupta, A., Balasubramanium, R., Chandra, A.K., Das, S.: Fractional Order Phase-Shaper Design With Routh's Criterion for Iso-Damped Control System. IEEE-India Council Gujarat-Section (Paper ID 740) INDICON (December 2009)
225. Saha, S., Das, S., Ghosh, R., Goswami, B., Gupta, A., Balasubramanium, R., Chandra, A.K., Das, S.: Design of Fractional Order Phase Shaper for Iso-Damped Control of PHWR Under Set-Back. IEEE Transactions on Nuclear Science 57(3), 1602–1612 (2010)
226. Saha, S., Das, S., Ghosh, R., Goswami, B., Gupta, A., Balasubramanium, R., Chandra, A.K., Das, S.: Fractional Order Phase Shaper Design With Bode's Integral for Iso-Damped Control System. ISA-Transactions 49, 196–206 (2010)

227. Samko, S.G., Kilbas, A.A., Maritcheva, O.I.: On Fractional Integrals and derivatives: Theory and Applications. Gordon and Breach, New York (1993)
228. Samorodnitsky, G., Taqqu, M.S.: Stable Non-Gaussian Random Process. Chapman and Hall, New York (1994)
229. Sapagovas, M., Vileiniskis, V.: The solution of two-dimensional neutron diffusion equations with delayed neutrons. INFORMATICA 12(2), 337–342
230. Sardar, T., Saha Ray, S., Bera, R.K., Biswas, B.B.: The solution of the multi-term fractionally damped Van-der-Pol Equation. Phys. Scr. 80, 1–5 (2009)
231. Sardar, T., Saha Ray, S., Bera, R.K., Biswas, B.B., Das, S.: The Solution of Coupled Fractional Neutron Diffusion Equations with Delayed Neutron. Int. J. of Nuclear Energy Science & Technology 5(2), 105–113 (2010)
232. Saxena, R.K., Mathai, A.M., Haubold, H.J.: On Generalized Fractional Kinetic Equations. Int. J. of Math. Phys. 0406046 v1 (2004)
233. Saxena, R.K.: Fractional calculus and fractional differential equations. In: Saxena, R.K. (ed.) Lecture Note (ch.3), pp. 1–39. Jai Narayan Vyas University, Rajasthan (2006)
234. Scafetta, N., Griffin, L., West, B.J.: Holder Exponent Spectrum for Human Gait. Physica. A 328, 561–583 (2003)
235. Scalas, E.: The application of continuous time random walks in finance and economics. Physica-A 362, 225–239 (2006)
236. Scalas, E., Gorenflo, R., Mainardi, F., Raberto, M.: Revisiting the derivation of the fractional diffusion equation. Elsevier, Amsterdam (2002); ARXIV:COND-MAT/0210166 V2 11.02.2002
237. Schneider, W.R., Wyss, W.: Fractional Diffusion and Wave Equations. Math. Phys. 30, 134–144 (1989)
238. Schumer, R., Benson, D.A., Merscaert, M.M., Wheatcraft, S.W.: Eulerian derivation of fractional advection dispersion equation. J. of Contaminant Hydrology 48, 69–88 (2001)
239. Schummer, R., Benson, D.A.: Multiscaling fractional advection dispersion equation and their solution. Water Resource Research 39(1), 2–11 (2003)
240. Scott Blair, G.W.: Measurement of Mind and Matter. Dennis Dobson, London (1950)
241. Scott Blair, G.W.: Psychoreology: links between the past and the present. Journal of Texture Studies 5, 3–12 (1974)
242. Shawagfeh, N.T.: The Decomposition Method for Fractional Differential Equations. J. Frac. Calc. 16, 27–33 (1999)
243. Simons, G.F., Robertson, J.S.: Differential Equations with Applications & Historical Notes, 2nd edn. McGraw-Hills Inc., New York (1991)
244. Sinha, S., Tarafdar, S.: Viscous Fingering Patterns and Evolution of Their Fractal Dimensions. Industrial Engineering & Chemical Research (2009)
245. Sinha, S., Dutta, T., Tarafdar, S.: Adhesion and fingering in the lifting Hele-Shaw cell: Role of the substrate. The European Physical Journal 25, 267–275 (2008)
246. Srivastava, H.M.: A certain family of sub-exponential series. Int. J. Math. Educ. Sci. Tech. 25(2), 211–216 (1994)
247. Srivastava, H.M.: On extension of Mittag-Leffler function. Yokohama Math. J. 16(2), 77–88 (1968)
248. Srivastava, H.M., Owa, S.: Some characterization and distortion theorems involving fractional calculus, generalized hyper geometric functions, Hadamard products, Linear Operators and certain subclasses of Analytical functions. Nagoya Math. J. 106, 1–128 (1987)

249. Stacey, W.M.: Nuclear Reactor Physics. Wiley-Interscience Publications, New-York (2001)
250. Stauffer, D.: Hausdroff Dimension and Fluctuations for the Largest Cluster at the Two-Dimensional Percolation Threshold. In: Z-Physik B, vol. 37, pp. 89–91. Springer, Heidelberg (1980)
251. Sugi, M., et al.: $\sqrt{\omega}$-Variations of AC admittance in the inhomogeneous distributed RC lines. Jpn. J. Appl. Phys. 39(9A), 5367–5368 (2000)
252. Sugi, M., Hirano, Y., Miura, Y.F., Saito, K.: Simulation of fractal immitance by analog circuits: an approach to optimized circuits. IEICE Trans. Fundamentals 82(8), 1627–1635 (1999)
253. Sun, H.H., Abelwahab, A.A., Onaral, B.: Linear Approximation of Transfer functions with a pole of fractional power. IEEE Trans. on Automatic Control 29(5), 441–444 (1984)
254. Tarasov, V.E.: Electromagnetic field of fractal distribution of charged particles. Physics of Plasmas 12, 1–9 (2005)
255. Trenceveski, K., et al.: On fractional derivative of some function of exponential type. In: Ser. Mat., vol. 13, pp. 77–84. Ser Mat. University Beograd PUBL, ELEKTROTEHN FAK (2002)
256. Tsao, Y.Y., Onaral, B., Sun, H.H.: An algorithm for determining global parameters of minimum phase systems with fractional power spectra. IEEE Trans. Inst. And Meas. 38(3), 723–729 (1989)
257. Tseng, C.C.: Improved design of digital fractional order differentiator using fractional sample delay. TCAS, 1-1815 (2004)
258. Tseng, C.-C., Pei, S.-C., Hsia, S.-C.: Computation of fractional derivatives using Fourier transforms and digital FIR differentiator. Elsevier Signal Processing 80, 151–159 (2000)
259. Tu, S.T., Srivastava, H.M., et al.: A certain family of fractional differential equations. Taiwanese Journal of Mathematics 4(3), 417–426 (2000)
260. Tustin, A., et al.: The design of systems for automatic control of the position of massive objects. The proceedings of Institute of Electrical Engineers 105C(1) (1958)
261. Valimaki, et al.: Principles of fractional delay filters. In: IEEE Conf on Acoustic Speech Processing (ICASSP), Istanbul, Turkey, June 5-9 (2000)
262. Vinagre, Petras, I.: Two direct Tustin discretization methods for fractional order differentiation / integration. J. Franklin Institute 340, 349–362 (2003)
263. Vingare, Petras, I., Podlubny, et al.: Using fractional order adjustment rules and fractional order reference model in model reference adaptive control. Non linear Dynamics 29, 269–279 (2002)
264. Wallis, M.: Computer experiments with fractional Gausian noises PART 3: Mathematical Appendix. Water Resource research 5, 328–338 (1969)
265. Wang, J.C.: Realization of generalized Wartburg impedance with RC ladder networks and transmission lines. J. of Electrochem. Soc. 134(8), 1915–1920 (1987)
266. Wazwaz, A.: A reliable Modification of Adomian Decomposition Method. Appl. Math. Comp. 102(1), 77–86 (1999)
267. West, B.J., Sheshadri, V.: Linear Systems with Levy Fluctuations. Physica 113A, 205–216 (1982)
268. Westerlund, et al.: Capacitor theory. IEEE Trans. on Dielectric and Electrical Insulation 1(5), 826–839 (1994)
269. Westerlund: Dead Matter has memory. Physics Scripta 43, 174–179 (1991)
270. Wyss: The fractional diffusion equation. J. Math. Phys. 27(11), 2782–2785 (1986)

271. Yang, C., et al.: A conceptually effective predictor corrector method for simulating fractional order dynamic control system. ANZIAM. J. 47 (EMAC 2005) 47, 168–184 (2005)
272. Yang, C.-Y., Hsu, K.-C., Chen, K.-C.: The use of the Levy stable distribution for geophysical data. Hydrology Journal 17, 1265–1273 (2009)
273. Yu, K.W., Hui, P.M., Stroud, D.: Anomalous Transport in random resistor-capacitor networks: Application to composite materials. Physics Letters-A 118(6), 305–308 (1986)
274. Zhang, S.: Positive solutions for boundary value problems on non-linear fractional differential equations. Electronic J. of differential equations 36, 1–12 (2006)

“This Is Not the End”

CPSIA information can be obtained at www.ICGtesting.com
Printed in the USA
BVOW06*1045201215

430671BV00006B/12/P

9 783642 205446